Desalination Engineering

About the Author

Nikolay Voutchkov, PE, BCEE, is an internationally recognized expert in the field of membrane water and wastewater treatment with extensive experience in the development of seawater desalination projects. He is the president of Water Globe Consulting, LLC, which provides independent engineering advisory services to the desalination and water reuse industry. Mr. Voutchkov is an active member of the WateReuse Association's Desalination Committee, the American Water Works Association's Desalination and Membrane Processes Committees, the American Membrane Technology Association, and the International Desalination Association.

Desalination Engineering

Planning and Design

Nikolay Voutchkov, PE, BCEE
Water Globe Consulting, LLC

McGraw Hill

New York Chicago San Francisco
Lisbon London Madrid Mexico City
Milan New Delhi San Juan
Seoul Singapore Sydney Toronto

The McGraw-Hill Companies

1 2 3 4 5 6 7 8 9 0 DOC/DOC 1 8 7 6 5 4 3 2

ISBN 978-0-07-177715-5
MHID 0-07-177715-6

Sponsoring Editor
Larry S. Hager

Editing Supervisor
Stephen M. Smith

Production Supervisor
Richard C. Ruzycka

Acquisitions Coordinator
Bridget L. Thoreson

Project Manager
Rohini Deb, Cenveo Publisher Services

Copy Editors
Jeff Anderson and James Dryden

Proofreader
Eina Malik

Art Director, Cover
Jeff Weeks

Composition
Cenveo Publisher Services

Contents

Preface

Over the past decade, water scarcity, changes in global climate patterns, and urban growth have led to a great shift in the paradigm of municipal water resource management. Low-cost surface and/or groundwater sources are practically depleted in many highly urbanized regions of the world. Therefore, the water supply planning paradigm is evolving from almost exclusive reliance on traditional freshwater resources toward building an environmentally sustainable diversified water portfolio in which low-cost conventional water sources are balanced with more costly but also more reliable and sustainable water supply alternatives such as desalination.

While only 0.5 percent of the world's available water resources are brackish in nature, brackish water desalination has found widespread application because it allows the production of freshwater at reasonably low cost and energy expenditures. At present, over 77 percent of the existing desalination plants in the United States are brackish water desalination facilities. Approximately 220 brackish water desalination plants produce freshwater for municipal water supplies in states such as Florida, Texas, California, New Mexico, and Virginia. Worldwide, brackish water desalination also contributes to municipal and agricultural water supplies in many arid regions, such as southern Spain, the Middle East, Australia, South America, and southern Israel.

Seawater desalination, while more costly at present, allows access to the world's ultimate water resource—the ocean. This water supply alternative has experienced a continuous exponential growth over the last 20 years, a pattern that is projected to continue well into the next decade.

This book provides detailed background information on the planning and engineering of brackish and seawater desalination projects for municipal water supply. While it includes a brief overview of key widely used desalination technologies, it focuses on reverse osmosis (RO) desalination, which at present is the most widely used technology for the production of freshwater from saline water sources.

The book's chapters address practically all aspects of brackish and seawater desalination, from basic principles to planning and environmental review of projects to the design of key desalination plant components such as intake, pretreatment facilities, the reverse osmosis system, post-treatment of desalinated water, and concentrate management. The book also provides guidance and examples for sizing and cost estimation of desalination plant facilities.

It is important to note that the facility and equipment sizing procedures presented in this book are not intended to serve as standard all-inclusive design procedures; their main purpose is to illustrate typical methodologies and approaches used by desalination professionals. References to particular technologies, equipment, and membrane manufacturers should not be construed as endorsement by the author or a recommendation for preferential use or consideration. Cost graphs included in the book are recommended for use in preparing initial order-of-magnitude estimates of a project's construction costs. Site-specific project conditions may result in significant differences from the values determined based on the cost curves.

The book includes a total of 17 chapters, which follow a typical process of project planning, environmental review, and selection and sizing of key desalination plant components. Chapter 1 ("Desalination Engineering: An Overview") provides a brief review of the most commonly used desalination technologies at present, including thermal desalination, electrodialysis, and reverse osmosis separation. This chapter mainly emphasizes the basic principles of and differences between these technologies. All other chapters are focused only on reverse osmosis desalination processes, equipment, and technologies.

Chapter 2 ("Source Water Quality Characterization") provides guidance for the characterization of saline source water quality. The chapter describes the main source water constituents that can impact the performance of RO system membranes and identifies commonly used techniques for their measurement. It also provides practical direction on how the source water quality data can be applied in selecting desalination treatment processes and technologies.

Chapter 3 ("Fundamentals of Reverse Osmosis Desalination") features an overview of RO membrane structures, materials, and configurations that have found practical application for desalination. It presents basic theoretical principles and models for water and salt transport through membranes, with an emphasis on the nonporous solution-diffusion transport model, which is the most widely used model in practice at present. The chapter introduces desalination system performance parameters applied in planning and designing RO systems and describes phenomena that influence desalination efficiency, including concentration polarization, membrane fouling, and flux distribution in membrane vessels. It discusses the effects of salinity, recovery, feed pressure, and permeate back pressure on RO membrane performance.

Chapter 4 ("Planning Considerations") discusses the process of planning for a new desalination project. It describes factors, issues, and alternatives to consider in project planning, such as plant's service area and site; intake type and location; source water type and quality; product water quality; plant discharge; selection of key plant treatment processes, configuration, and layout; project implementation schedule and phasing; project economics; contractor procurement alternatives; and project funding considerations.

Chapter 5 ("Environmental Review and Permitting") provides an overview of the next step in the process of implementing a desalination project—assessing its potential impacts on the surrounding environment and developing measures to mitigate such impacts if they are found to be significant. The chapter describes the main environmental challenges associated with the impact of desalination plant intakes and discharges and identifies proven practical solutions to quantify and address these challenges. In addition, it presents the typical set of permits and permitting conditions for source water intakes, discharges, and product water quality.

Chapter 6 ("Intakes for Source Water Collection") focuses on the type and configuration of intakes used for brackish and seawater desalination, as well as design considerations for open and subsurface intakes. This chapter contains construction cost curves for onshore and offshore open intakes as well as for vertical well intakes.

Chapter 7 ("Intake Pump Stations") presents key advantages and disadvantages of alternative intake pump station configurations, including wet-well, dry-well, and canned pump systems, and provides guidance for pump station sizing and cost estimation.

Chapter 8 ("Source Water Screening") discusses alternative types of intake screens (including bar, band, and drum screens), microscreens, and cartridge filters. It also describes design and cost considerations for the selection of desalination plant screening facilities.

Chapter 9 ("Source Water Conditioning") is dedicated to systems for chemical conditioning of saline source water prior to its further pretreatment or direct application of membrane separation. The chapter addresses commonly used chemicals such as coagulants, flocculants, scale inhibitors, biocides, and pH adjustment compounds.

Chapter 10 ("Sand Removal, Sedimentation, and Dissolved Air Flotation") presents alternative pretreatment technologies that are commonly used for removing relatively large particulate solids from the source water. The chapter includes construction cost curves for lamella settlers and dissolved air flotation clarifiers.

Chapter 11 ("Pretreatment by Granular Media Filtration") is dedicated to the most commonly used type of technology for removing fine solid particles from the source water—granular media filtration. It discusses alternative filter configurations and their performance and applicability. The chapter includes cost curves for dual-media gravity and pressure filters.

Chapter 12 ("Pretreatment by Membrane Filtration") discusses the use of microfiltration and ultrafiltration membrane systems for pretreatment of saline source water. The chapter presents key design and planning considerations for most commonly used commercially available membranes, from Norit, Hydranautics, Filmtec, GE Zenon, and Memcor/Siemens. It contains design examples for submerged and pressure-driven ultrafiltration systems and cost curves for membrane pretreatment.

Chapter 13 ("Comparison of Granular Media and Membrane Pretreatment") provides a comparative evaluation of granular media and membrane pretreatment systems in terms of effect of source water quality on their performance, surface area requirements, generated residuals, chemical and power use, and overall water production costs. The chapter also includes guidelines for selecting a pretreatment system.

Chapter 14 ("Reverse Osmosis Separation") features a detailed overview of key RO system components—high-pressure pumps, RO racks, energy recovery system, RO membrane cleaning system, and instrumentation and controls. The chapter discusses the performance and configuration of state-of-the-art RO membrane elements used for nanofiltration, brackish water desalination, and seawater desalination. This chapter also addresses commonly applied RO system configurations and discusses full-scale applications of such systems. It includes design examples and provides guidance information for cost estimation.

Chapter 15 ("Post-Treatment of Desalinated Water") describes commonly applied technologies for remineralization of permeate produced by desalination systems, including lime–carbon dioxide feed systems and limestone (calcite) contactors. It includes guidelines for the application, configuration, and design of such systems, as well as example cost estimates for lime–carbon dioxide and calcite–carbon dioxide conditioning systems. The chapter also features an overview of alternative disinfection systems and guidance for their use in desalination applications.

Chapter 16 ("Desalination Plant Discharge Management") includes a comprehensive overview of commonly applied technologies and systems for disposal of concentrate and other waste streams generated in desalination plants. The featured concentrate disposal alternatives are near-shore and offshore concentrate discharge, disposal to a sanitary sewer, deep injection wells, land application, evaporation ponds, and zero liquid discharge systems. Construction cost curves are presented for each of these systems.

Chapter 17 ("Desalination Project Cost Estimates") describes the main components of the capital and operation and maintenance costs of brackish and seawater desalination plants and provides guidelines for their assessment based on site-specific project conditions and components. This chapter also includes an example cost estimate of a 40,000 m^3/day (10.6 mgd) seawater desalination plant.

This book is intended for water utility engineers, managers, and planners; consulting engineers and designers; students and teachers in the desalination field; and staff members of federal and state regulatory agencies involved in the permitting of desalination projects.

Nikolay Voutchkov, PE, BCEE

Acknowledgments

I would like to express my appreciation to the fellow professionals, colleagues, and associates who have reviewed various portions of this book and provided constructive comments and suggestions. In particular, I am very grateful to Dr. Boris Liberman and Dr. Mark Wilf for the stimulating discussions and shared experience, and for their views on key book topics such as source water quality characterization, intakes, seawater pretreatment, alternative reverse osmosis system configurations, and energy use minimization.

Special appreciation is extended to Dr. Steven La Page of M-REP Consulting and Dr. Jeffrey Graham and Dr. Scott Jenkins of the Scripps Institution of Oceanography (University of California, San Diego) for their work on the development of the methodology for assessing the salinity tolerance of marine microorganisms.

I gratefully appreciate the assistance of Mr. Daniel Marler, manager of the Carlsbad seawater desalination demonstration plant in Southern California, and of Dr. Sunny Jiang, professor at the University of California, Irvine, and their contributions toward the advancement of the knowledge and understanding of seawater pretreatment phenomena addressed in this book.

I would like to offer a special note of gratitude to the many equipment and membrane manufacturers and public utilities in the United States, Australia, Japan, and other parts of the world who have contributed valuable information, details, specifications, and photos used in this book.

Finally, my deep and sincere gratitude goes to Mr. Wade Miller, executive director of the WateReuse Association, for his inspiration, encouragement, and impetus to write this book.

Abbreviations

A	Water permeability coefficient
Å	Angstrom
ac	Acre
AOC	Assimilable organic carbon
AWA	Australian Water Association
AWWA	American Water Works Association
B	Beta (concentration polarization factor)
B	Boron
Ba	Barium
BOD	Biological oxygen demand
BOOT	Build-own-operate-transfer
Br	Bromide
BWRO	Brackish water reverse osmosis
°C	Degree Celsius (unit of temperature)
ΔC	Concentration gradient
C_f	Contingency factor
Ca	Calcium
CA	Cellulose acetate
$CaCO_3$	Calcium carbonate
CCPP	Calcium carbonate precipitation potential
CF	Concentration factor
CFD	Computational fluid dynamics
CIP	Clean-in-place
cm	Centimeter
CMF	Conventional media filtration
CO_2	Carbon dioxide
CWA	Clean Water Act

DAF	Dissolved air flotation
DBB	Design-bid-build
DBO	Design-build-operate
DBP	Disinfection byproduct
DC	Direct current
DO	Dissolved oxygen
DOC	Dissolved organic carbon
DW	Dry weight
EC	Electrical conductivity
ED	Electrodialysis
EDR	Electrodialysis reversal
EPA	Environmental Protection Agency
EPC	Engineering, procurement, and construction
°F	Degree Fahrenheit (unit of temperature)
F_p	Feed pressure
FRP	Fiberglass-reinforced plastic
ft	Foot
ft^2	Square foot
ft^3	Cubic foot
gal/min	Gallons per minute
gfd	Gallons per square foot per day
GOR	Gained output ratio
GRP	Glass-reinforced plastic
GWA	Government of Western Australia
GWI	Global Water Intelligence
h	Hour
ha	Hectare
HBMP	Hydrobiological monitoring program
HDPE	High-density polyethylene
hp	Horsepower (unit of power)
H_2S	Hydrogen sulfide
IDA	International Desalination Association
IPP	Industrial pretreatment program
IX	Ion exchange
J	Membrane permeate flux
kg	Kilogram

km	Kilometer
kW	Kilowatt (unit of power)
kWh	Kilowatt-hour (unit of energy)
kWh/m^3	Kilowatt-hours per cubic meter (unit of energy)
L	Liter
lb	Pound
lb/in^2	Pounds per square inch (unit of pressure)
lmh	Liters per square meter per hour
LSI	Langelier Saturation Index
LT2ESWTR	Long Term 2 Enhanced Surface Water Treatment Rule
m	Meter
m^2	Square meter
m^3	Cubic meter
m^3/day	Cubic meters per day
MED	Multi-effect distillation
MED-TC	Multi-effect distillation with thermal compression
MENA	Middle East and North Africa
meq	Milliequivalent
MF	Microfiltration
Mg	Magnesium
mg/L	Milligrams per liter
mgd	Million gallons per day
mi	Mile
min	Minute
mL	Milliliter
mm	Millimeter
Mn	Manganese
MSF	Multistage flash distillation
MTBE	Methyl tertiary butyl ether (gasoline additive)
mV	Millivolt
MVC	Mechanical vacuum compression
μg/L	Micrograms per liter
μm	Micrometer
μS/cm	Microsiemens per centimeter (unit of specific conductance)
Na	Sodium
NDMA	*N*-nitrosodimethylamine

NDP	Net driving pressure
NF	Nanofiltration
NOM	Natural organic matter
NP	Nanoparticle
NPDES	National Pollutant Discharge Elimination System
NTU	Nephelometric turbidity unit
O&M	Operations and maintenance
O_{pfc}	Osmotic pressure on the concentrate side of the membrane
OCSD	Orange County Sanitation District
ORP	Oxidation-reduction potential
P	Phosphorus
P_r	Permeate recovery rate
PA	Polyamide
PDFB	Percent difference from balance
pH	Indication of the acidity or basicity of solution
PP	Polypropylene
ppb	Parts per billion (1 ppb = 1 μg/L)
ppm	Parts per million (1 ppm = 1 mg/L)
ppt	Parts per thousand (1 ppt = 1000 mg/L)
PREN	Pitting resistance equivalent number
psu	Practical salinity unit (1 psu = 1000 mg/L)
PVC	Polyvinyl chloride
Q_f	Fresh water flow
R	Permeate recovery rate
RCRA	Resource Conservation and Recovery Act
RIB	Rapid infiltration basin
RO	Reverse osmosis
RWQCB	Regional water quality control board
s	Second
S	Membrane area of an element
S_p	Salt passage
S_r	Salt rejection
S. AZ	Southern Arizona
S. CA	Southern California
SAR	Sodium adsorption ratio
SARI	Santa Ana River Interceptor

SCADA	Supervisory Control and Data Acquisition
SDI	Silt Density Index
SDSI	Stiff–Davis Saturation Index
SER	Standard evaporation rate
STE	Salinity tolerance evaluation
SUVA	Specific UV absorbance
SWRO	Seawater reverse osmosis
TCLP	Toxicity characteristic leaching procedure
TDS	Total dissolved solids
TDS_f	TDS concentration of feed water
TDS_p	TDS concentration of permeate
TFN	Thin-film nanocomposite
THC	Total hydrocarbons
TMP	Transmembrane pressure
TN	Total nitrogen
TOC	Total organic carbon
TP	Total phosphorus
TSS	Total suspended solids
u	Atomic mass unit
UAE	United Arab Emirates
UF	Ultrafiltration
UIC	Underground Injection Control
US	United States (of America)
USBR	U.S. Bureau of Reclamation
USDW	Underground source of drinking water
UV	Ultraviolet radiation
VC	Vapor Compression
W. TX	Western Texas
WET	Whole effluent toxicity
WHO	World Health Organization
WPA	Water purchase agreement
WTP	Water treatment plant
WWTP	Wastewater treatment plant
Y	Plant recovery
ZID	Zone of initial dilution
ZLD	Zero liquid discharge

CHAPTER 1

Desalination Engineering: An Overview

1.1 Introduction

Approximately 97.5 percent of the water on our planet is located in the oceans and therefore is classified as seawater. Of the 2.5 percent of the planet's freshwater, approximately 70 percent is in the form of polar ice and snow and 30 percent is groundwater, river and lake water, and air moisture. So even though the volume of the earth's water is vast, less than 35 million km^3 of the 1386 million km^3 (8.4 million mi^3 of the 333 million mi^3) of water on the planet is of low salinity and is suitable for use after applying conventional water treatment only (Black and King, 2009). Desalination provides a means for tapping the world's main water resource—the ocean.

Over the past 30 years, desalination has made great strides in many arid regions of the world, such as the Middle East and the Mediterranean. Technological advances and the associated decrease in water production costs over the past decade have expanded its use in areas traditionally supplied with freshwater resources.

At present, desalination plants operate in more than 120 countries worldwide; some desert states, such as Saudi Arabia and the United Arab Emirates, rely on desalinated water for over 70 percent of their water supply. According to the 2011–2012 IDA Desalination Yearbook [Global Water Intelligence (GWI) and International Desalination Association (IDA), 2012], by the end of 2011 worldwide there were approximately 16,000 desalination plants, with a total installed production capacity of 71.9 million m^3/day [19,000 million gal/day (mgd)].

While currently desalination provides only 1.5 percent of the water supply worldwide, it is expected that in the next decade the construction of new desalination plants will grow exponentially due to the ever-changing climate patterns triggered by global warming combined with population growth pressures, limited availability of new and inexpensive terrestrial water sources, and dramatic advances in membrane technology, which are projected to further reduce the cost and energy use of desalination.

The brackish water quantity on the planet is fairly limited (0.5 percent), and most of the large and easily accessible brackish water aquifers worldwide are already in use. A significant portion of the new capacity growth is expected to come from the development of seawater desalination plants. While brackish water sources, especially brackish aquifers, are finite in terms of capacity and rate of recharging, the ocean has two unique and distinctive features as a water supply source—it is droughtproof and practically limitless.

Over 50 percent of the world's population lives in urban centers bordering the ocean. In many arid parts of the world, such as the Middle East, Australia, North Africa, and Southern California, the population concentration along the coast exceeds 75 percent. Usually coastal zones are also the highest population growth hot spots. Therefore, seawater desalination provides the logical solution for a sustainable, long-term management of the growing water demand pressures in coastal areas. Brackish desalination is also expected to increase in capacity, especially in inland areas with still untapped brackish water aquifers.

A clear recent trend in seawater desalination is the construction of larger-capacity plants, which deliver an increasingly greater portion of the freshwater supply of coastal cities around the globe. While most of the large desalination plants built between 2000 and 2005 were typically designed to supply only 5 to 10 percent of the drinking water of large coastal urban centers, today most regional or national desalination project programs in countries such as Spain, Australia, Israel, Algeria, and Singapore aim to fill 20 to 25 percent of their long-term drinking water needs with desalinated seawater. Increased reliance on seawater desalination is often paralleled with ongoing programs for enhanced water reuse and conservation, with a long-term target of achieving near-even contributions of conventional water supply sources, seawater desalination, water reuse, and conservation to the total water portfolio of large coastal communities.

1.2 Terminology

The mineral or salt content of water is usually measured by the water quality parameter called *total dissolved solids* (TDS), the concentration of which is expressed in milligrams per liter (mg/L) or parts per thousand (ppt). The World Health Organization, as well as the United Sates Environmental Protection Agency (US EPA) under the Safe Drinking Water Act, have established a maximum TDS concentration of 500 mg/L as a potable water standard. This TDS level can be used as a classification limit to define potable (fresh) water.

Typically, water with a TDS concentration higher than 500 mg/L and not higher than 15,000 mg/L (15 ppt) is classified as brackish. Natural water sources such as sea, bay, and ocean waters that have TDS concentrations higher than 15,000 mg/L are generally classified as seawater. For example, Pacific Ocean seawater along the West Coast of the United States has an average TDS concentration of 35,000 mg/L. This concentration can actually range from 33,000 to 36,000 mg/L at various locations and depths along the coast.

1.3 Overview of Desalination Technologies

Sea and brackish water are typically desalinated using two general types of water treatment technologies: thermal evaporation (distillation) and reverse osmosis (RO) membrane separation.

In *thermal distillation*, freshwater is separated from the saline source by evaporation. In *reverse osmosis* desalination, freshwater is produced from saline source water by pressure-driven transport through semipermeable membranes. The main driving force in RO desalination is pressure, which is needed to overcome the naturally occurring osmotic pressure that in turn is proportional to the source water's salinity.

Besides thermal distillation and RO membrane separation, two other mainstream desalination technologies widely applied at present are *electrodialysis* (ED) and *ion*

Separation Process	Range of Source Water TDS Concentration for Cost-Effective Application, mg/L
Distillation	20,000–100,000
Reverse osmosis separation	50–46,000
Electrodialysis	200–3000
Ion exchange	1–800

Table 1.1 Desalination Process Applicability

exchange (IX). Electrodialysis is electrically driven desalination in which salt ions are removed out of the source water through exposure to direct electric current. The main driving force for ED separation is electric current, which is proportional to the salinity of the source water.

IX is the selective removal of salt ions from water by adsorption onto ion-selective resin media. The driving force in this desalination process is the ion charge of the IX resin, which can selectively attract and retain ions of the opposite charge contained in the saline source water.

Table 1.1 provides a general indication of the range of source water salinity for which distillation, RO separation, ED, and IX can be applied cost effectively for desalination. For processes with overlapping salinity ranges, a life-cycle cost analysis for the site-specific conditions of a given desalination project is typically applied to determine the most suitable desalination technology for the project.

Currently, approximately 60 percent of the world's desalination systems are RO membrane separation plants and 34 percent are thermal desalination facilities (GWI and IDA, 2012). The percentage of RO desalination installations has been increasing steadily over the past 10 years due to the remarkable advances in membrane separation and energy recovery technologies, as well as the associated reductions of overall water production costs. At present, ED- and IX-based technologies contribute less than 6 percent of the total installed desalination plant capacity worldwide.

1.4 Thermal Desalination

1.4.1 Overview

All thermal desalination technologies apply distillation (i.e., are based on heating the source water) to produce water vapor, which is then condensed into a low-salinity water. Since the energy for water evaporation is practically not dependent on the source water salinity concentration, thermal evaporation is very suitable for desalination of high-salinity waters and brine. This is one of the reasons that thermal desalination has been widely adopted by Middle Eastern countries such as Saudi Arabia, Oman, Qatar, the United Arab Emirates, Bahrain, and Kuwait, which use some of the most saline water bodies on the planet for water supply (namely, the Red Sea, Persian Gulf, Gulf of Oman, and Indian Ocean). At present, approximately 75 percent of the world's thermal desalination plants are located in the Arabian Peninsula—half of those in Saudi Arabia.

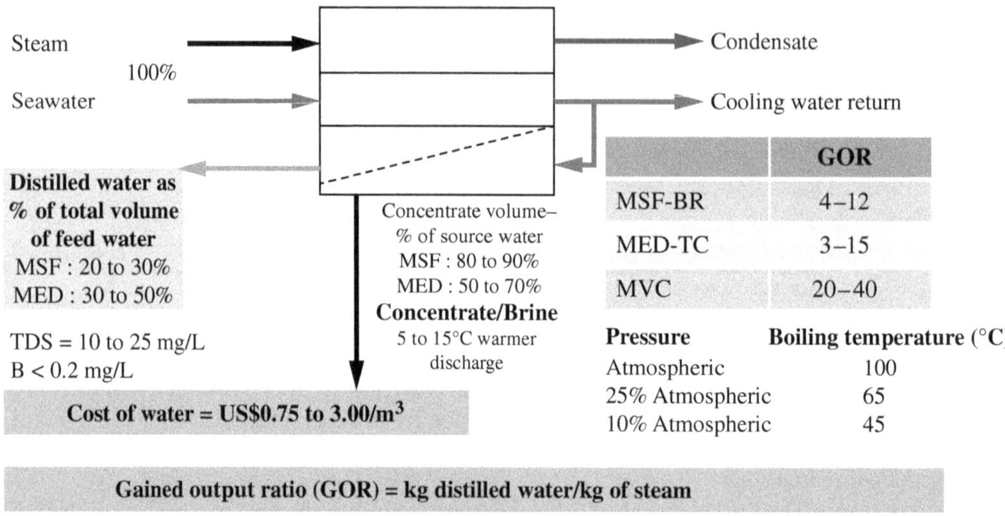

FIGURE 1.1 General schematic of thermal evaporation technologies.

All thermal desalination plants have five key streams: source water (seawater, brackish water, or brine) used for desalination; steam needed for evaporation of the source water; cooling water to condense the freshwater vapor generated from the source water's evaporation; low salinity distilled water (distillate); and concentrate (brine), which contains the salts and other impurities separated from the source water (Fig. 1.1).

The three most commonly used types of thermal desalination technologies are *multistage flash distillation* (MSF), *multieffect distillation* (MED), and *vapor compression* (VC). Each of these classes of technology has evolved over the past 40 to 60 years toward improvements in efficiency and productivity. For example, *MSF-BR* (see Fig. 1.1) is the abbreviation for a multistage flash distillation process with brine recycle, which reduces the source water volume and the steam needed for evaporation. Similarly, *MED-TC* stands for *multieffect distillation with thermal compression*, a state-of-the art MED technology; and *MVC* is an acronym for *mechanical vapor compression*, a VC technology that can run without the need for an outside source of steam.

The three types of thermal technologies mainly differ by the temperature and pressure at which the source water is boiled to generate freshwater vapor. The oldest thermal evaporation process—MSF—boils water at near-atmospheric pressure and a temperature close to 100°C (212°F). This type of process requires a large quantity of high-temperature steam.

MED and VC are newer thermal desalination technologies, whose improved efficiency stems from the fact that water can be boiled at a lower temperature if the boiling process occurs at a pressure lower than the atmospheric pressure. Boiling water at a lower temperature allows the use of less and lower-quality steam for the production of the same volume of water.

As shown in Fig. 1.1, in MED vessels the boiling process typically occurs at lower temperatures and pressures than in MSF systems. VC thermal desalination systems operate at lower pressures than either MSF or MED, which allows these systems to evaporate water at even lower temperatures and to generate their own steam rather than depend on outside steam sources.

The ratio of the mass of low-salinity water (distillate) produced to the mass of heating steam used to produce this water is commonly referred to as the *gained output ratio* (GOR) or performance ratio. Depending on the thermal desalination technology used, the site-specific conditions, and the source water quality, GOR typically varies between 4 and 40—i.e., thermal desalination technologies produce 4 to 40 kg of freshwater using 1 kg of steam. The higher the GOR, the more efficient the technology, because it produces more freshwater from the same amount of steam.

As seen in Fig. 1.1, all thermal desalination technologies generate very low-salinity water (TDS in a range of 5 to 25 mg/L). This freshwater also has a very low content of pathogens and other contaminants of concern, such as boron, bromides, and organics (Cotruvo et al., 2010).

Thermal desalination is most popular in the Middle East, where seawater desalination is typically combined with power generation that provides low-cost steam for the distillation process. Thermal desalination requires large quantities of steam.

Most power plants outside the Middle East are not designed to yield significant amounts of waste steam as a side product of power generation. This is one of the key reasons why thermal desalination has not found wider application outside of the region.

1.4.2 Multistage Flash Distillation

In the multistage flash distillation (MSF) evaporator vessels (also referred to as *flash stages* or *effects*), the high-salinity source water is heated to a temperature of 90 to 115°C (194 to 239°F) in a vessel (the heating section in Fig. 1.2) to create water vapor. The pressure in the first stage is maintained slightly below the saturation vapor pressure of the water. So when the high-pressure vapor created in the heating section enters into the first stage, its pressure is reduced to a level at which the vapor "flashes" into steam.

Steam (waste heat) for the heating section is provided by the power plant co-located with the desalination plant. Each flash stage (effect) has a condenser to turn the steam into distillate. The condensers are equipped with heat exchanger tubes, which are cooled by the source water that is fed to the condensers.

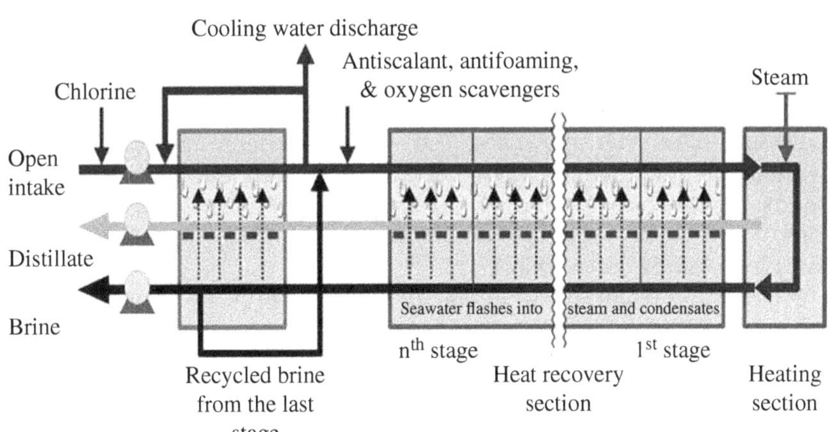

FIGURE 1.2 Schematic of an MSF distillation system.

Entrainment separators (mist eliminators or demister pads) remove the high-salinity mist from the low-salinity rising steam. This steam condenses into pure water (distillate) on the heat exchanger tubes and is collected in distillate trays, from where it is conveyed to a product water tank. Distillate flows from stage to stage and is collected at the last stage.

The concentrate (brine) is generated in each stage and after collection at the last stage some of it typically is recycled to the source water stream in order to reduce the total volume of source water that must be collected by the intake for desalination. The recirculated brine flowing through the interior of the condenser tubes also removes the latent heat of condensation. As a result, the recirculated brine is also preheated close to maximum operating temperature, thereby recovering the energy of the condensing vapor and reducing the overall heating needs of the source water. This "brine recycle" feature has been adopted in practically all of the most recent MSF facility designs and allows significant improvement of the overall cost competitiveness of MSF installations.

Each flash stage typically produces approximately 1 percent of the total volume of the desalination plant's condensate. Since a typical MSF unit has 19 to 28 effects, the total MSF plant recovery (i.e., the volume of distillate expressed as a percentage of the total volume of processed source water) is typically 19 to 28 percent. For comparison, RO seawater desalination plants have a recovery of 40 to 45 percent. The latest MSF technology has 45-stage units—i.e., can operate at 45 percent recovery. This feature allows it to compete with RO systems in terms of recovery.

Historically, MSF was the first commercially available thermal desalination technology applied to production of potable water on a large-scale, which explains its popularity. Over 80 percent of thermally desalinated water today is produced in MSF plants. The GOR for MSF systems is typically between 2 and 8; the latest MSF technology has a GOR of 7 to 9. The pumping power required for the operation of the MSF systems is 2.0 to 3.5 kWh/m^3 (7.6 to 13.3 kWh/1000 gal) of product water.

1.4.3 Multiple-Effect Distillation

In multiple-effect distillation (MED) systems, saline source water is typically not heated; cold source water is sprayed via nozzles or perforated plates over bundles of heat exchanger tubes. This feed water sprayed on the tube bundles boils, and the generated vapor passes through mist eliminators, which collect brine droplets from the vapor. The feed water that turned into vapor in the first stage (effect) is introduced into the heat exchanger tubes of the next effect. Because the next effect is maintained at slightly lower pressure, although the vapor is slightly cooler, it still condenses into freshwater at this lower temperature. This process of reducing the ambient pressure in each successive stage allows the feed water to undergo multiple successive boilings without the introduction of new heat. Steam flowing through the exchanger tubes is condensed into pure water (Fig. 1.3) and collected from each effect. Heating steam (or vapor) introduced in the heat exchanger tubes of the first effect is provided from an outside source by a steam ejector.

The MED system shown in Fig. 1.3 is also equipped with a brine recycle system, which allows the introduction of warmer-than-ambient water in the first effects of the system, thereby reducing both the volume of feed water that must be collected by the plant intake system and the overall energy needs of the system.

The main difference between the MED and MSF processes is that while vapor is created in an MSF system through flashing, evaporation of feed water in MED is achieved through heat transfer from the steam in the condenser tubes into the source

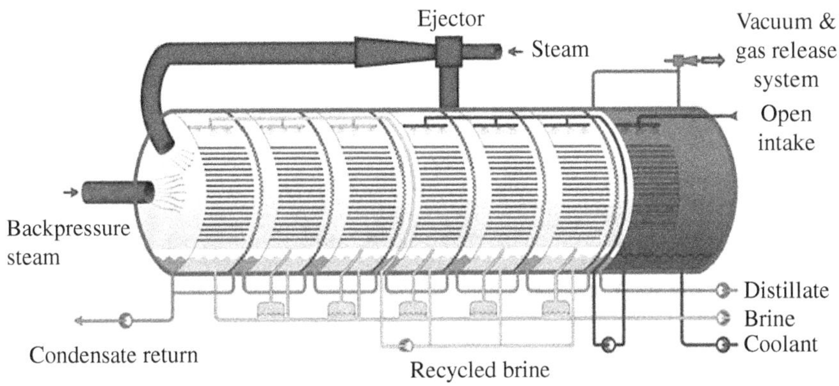

FIGURE 1.3 Schematic of an MED system.

water sprayed onto these tubes. This heat transfer at the same time results in condensation of the vapor to freshwater.

MED desalination systems typically operate at lower temperatures than MSF plants (maximum brine concentrate temperature of 62 to 75°C versus 115°C) and yield higher GORs. The newest MED technologies, which include vertically positioned effects (vertical tube evaporators), may yield a GOR of up to 24 kg of potable water per kilogram of steam. The pumping power required for the operation of MED systems is also lower than that typically needed for MSF plants (0.8 to 1.4 kWh/m^3/3.0 to 5.3 kWh/1000 gal of product water). Therefore, MED is now increasingly gaining ground over MSF desalination, especially in the Middle East, where thermal desalination is still the predominant method for producing potable water from seawater.

1.4.4 Vapor Compression

The heat source for vapor compression (VC) systems is compressed vapor produced by a mechanical compressor or a steam jet ejector rather than a direct exchange of heat from steam (Fig. 1.4).

In VC systems the source water is evaporated and the vapor is conveyed to a compressor. The vapor is then compressed to increase its temperature to a point adequate to evaporate the source water sprayed over tube bundles through which the vapor is conveyed. As the compressed vapor exchanges its heat with the new source water being sprayed on the evaporation tubes, it is condensed into pure water. A feed water preheater (plate-type heat exchanger) is used to start the process and reach evaporation temperature.

VC and MED work based on similar principles. However, while in MED the steam produced by source water evaporation is introduced and condensed in a separate condenser located in the downstream effect, in VC the steam generated from evaporation of new source water sprayed on the outside surface of the heat exchanger tubes is recirculated by the vapor compressor and introduced into the inner side of the of the same heat exchanger tubes in which it condenses to form distillate.

VC desalination has found applications mostly in small municipal and resort water supply systems, as well as industrial applications. The total amount of power required for the operation of mechanical VC systems is typically 8 to 12 kWh/m^3 (30 to 45 kWh/1000 gal) of product water.

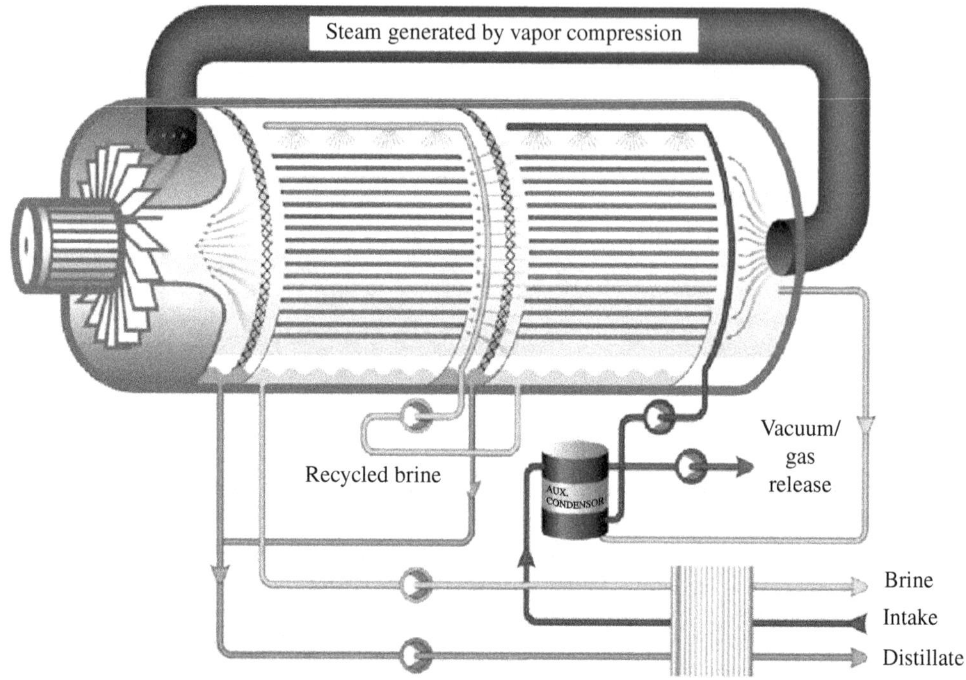

FIGURE 1.4 Schematic of a VC system.

1.5 Membrane Desalination

1.5.1 Overview

Membrane desalination is the process of separating minerals from the source water using semipermeable membranes. Two general types of technologies currently used for membrane desalination are electrodialysis (ED) and RO. In ED systems, salts are separated from the source water through the application of direct current. RO is a process in which the product water (permeate) is separated from the salts contained in the source water by pressure-driven transport through a semipermeable membrane.

1.5.2 Electrodialysis

In electrodialysis (ED)–based desalination systems, the separation of minerals and product water is achieved through the application of direct electric current to the source water. This current drives the mineral ions and other ions with strong electric charge that are contained in the source water through ion-selective membranes to a pair of electrodes of opposite charges (Fig. 1.5).

As ions accumulate on the surface of the electrodes, they cause fouling over time and have to be cleaned frequently in order to maintain a steady-state ED process. A practical solution to this challenge is to reverse the polarity of the oppositely charged electrodes periodically (typically two to four times per hour) in order to avoid frequent electrode cleaning. An ED process that includes periodic change of the polarity of the

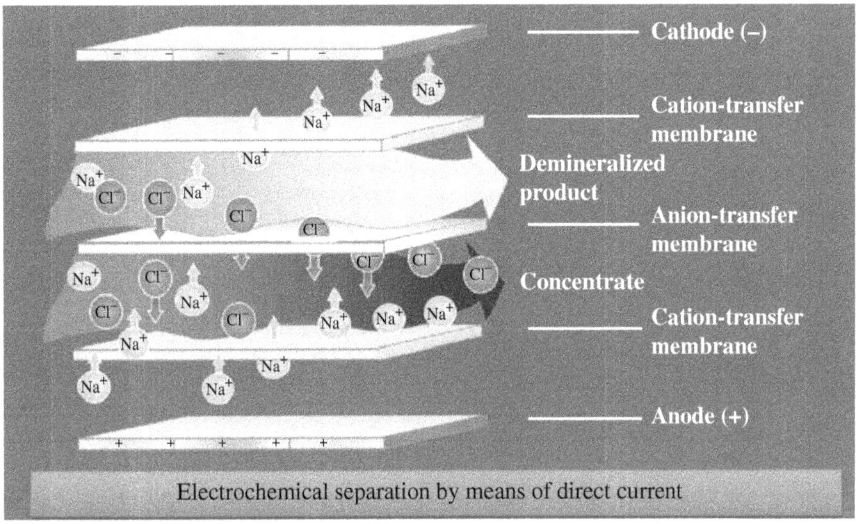

FIGURE 1.5 Schematic of the electrodialysis process.

system's electrodes is referred to as an *electrodialysis reversal* (EDR) process. At present, practically all commercially available ED systems are of the EDR type.

ED systems consist of a large number (300 to 600 pairs) of cation and anion exchange membranes separated by dilute flow dividers (spacers) to keep them from sticking together and to convey the desalinated flow through and out of the membranes. Each pair of membranes is separated from the adjacent pairs above and below it by concentrate spacers which collect, convey, and evacuate the salt ions retained between the adjacent membranes.

The membranes used for ED are different from those applied for RO desalination—they have a porous structure similar to that of microfiltration and ultrafiltration membranes. RO membranes do not have physical pores. ED membranes are more resistant to chlorine and fouling and are significantly thicker than RO membranes.

It is important to note that a single set of EDR stacks can only remove approximately 50 percent of salts. As a result, multiple EDR stacks connected in series are often used to meet more stringent product water TDS targets. It should be pointed out that compared to brackish water RO membranes, which typically yield only up to 85 to 90 percent recovery, EDR systems can reach freshwater recovery of 95 percent or more.

The energy needed for ED desalination is proportional to the amount of salt removed from the source water. TDS concentration and source water quality determine to a great extent which of the two membrane separation technologies (RO or ED) is more suitable and cost effective for a given application. Typically, ED membrane separation is found to be cost competitive for source waters with TDS concentrations lower than 3000 mg/L. This applicability threshold, however, is a function of the unit cost of electricity and may vary from project to project.

The TDS removal efficiency of ED desalination systems is not affected by nonionized compounds or objects with a weak ion charge (i.e., solids particles, organics, and microorganisms). Therefore, ED membrane desalination processes can treat source waters of higher turbidity and biofouling and scaling potential than can RO systems.

Contaminant	Distillation (%)	ED/EDR (%)	RO (%)
TDS	>99.9	50–90	90–99.5
Pesticides, Organics/VOCs	50–90	<5	5–50
Pathogens	>99	<5	>99.99
TOC	>95	<20	95–98
Radiological	>99	50–90	90–99
Nitrate	>99	60–69	90–94
Calcium	>99	45–50	95–97
Magnesium	>99	55–62	95–97
Bicarbonate	>99	45–47	95–97
Potassium	>99	55–58	90–92

Table 1.2 Contaminant Removal by Alternative Desalination Technologies

However, the TDS removal efficiency of ED systems is typically lower than that of RO systems (15.0 to 90.0 percent versus 99.0 to 99.8 percent), which is one of the key reasons why they have found practical use mainly for brackish water desalination.

In general, EDR systems can only effectively remove particles that have a strong electric charge, such as mono- and bivalent salt ions, silica, nitrates, and radium. EDR systems have a very low removal efficiency with regard to low-charged compounds and particles—i.e., organics and pathogens. Table 1.2 provides a comparison of the removal efficiencies of distillation, ED, and RO systems for key source water quality compounds. One important observation from this table is that, as compared to distillation and RO separation, ED desalination only partially removes nutrients from the source water. This fact explains why EDR is often considered more attractive than RO or thermal desalination (which remove practically all minerals from the source water) if the planned use of the desalinated water is for agricultural purposes—i.e., generating fresh or reclaimed water for irrigation of agricultural crops.

Construction and equipment costs for *brackish water reverse osmosis* (BWRO) and EDR systems of the same freshwater production capacity are usually comparable, or EDR is less costly, depending on the RO membrane fouling capacity of the source water. However, since the amount of electricity consumed by EDR systems is directly proportional to the source water's salinity, at salinities of 2000 to 3000 mg/L the energy use of EDR systems usually exceeds that of BWRO or nanofiltration systems for source waters. Therefore, EDR systems are not as commonly used as RO systems for BWRO desalination and are never applied for *seawater reverse osmosis* (SWRO) desalination.

It should be pointed out, however, that salinity is not the only criterion for evaluating the cost competitiveness of EDR and BWRO systems. Often, other compounds such as silica play a key role in the decision making process. For example, at the largest operational EDR plant worldwide at present—the 200,000 m^3/day Barcelona desalination facility in Spain—this technology was preferred to BWRO desalination because the brackish surface water source for this plant—the Llobregat River—contains very high level of silica, which would limit recovery from a BWRO plant to only 65 percent; the EDR system can achieve 90 percent recovery. In addition, the Llobregat River was found to have very high organic content, which was projected to cause heavy fouling and operational constraints on a BWRO plant of similar size.

1.5.3 Reverse Osmosis

Reverse osmosis (RO) is a process where water containing inorganic salts (minerals), suspended solids, soluble and insoluble organics, aquatic microorganisms, and dissolved gases (collectively called *source water constituents* or *contaminants*) is forced under pressure through a semipermeable membrane. *Semipermeable* refers to a membrane that selectively allows water to pass through it at much higher rate than the transfer rate of any constituents contained in the water.

Depending on their size and electric charge, most water constituents are retained (rejected) on the feed side of the RO membrane while the purified water (permeate) passes through the membrane. Figure 1.6 illustrates the sizes and types of solids removed by RO membranes as compared to other commonly used filtration technologies.

RO membranes can reject particulate and dissolved solids of practically any size. However, they do not reject well gases, because of their small molecular size. Usually RO membranes remove over 90 percent of compounds of 200 daltons (Da) or more. One Da is equal to 1.666054×10^{-24} g. In terms of physical size, RO membranes can reject well solids larger than 1 (Angstrom) Å. This means that they can remove practically all suspended solids, protozoa (i.e., *Giardia* and *Cryptosporidium*), bacteria, viruses, and other human pathogens contained in the source water.

While RO membranes can retain both particulate and dissolved solids, they are designed to primarily reject soluble compounds (mineral ions). The structure and

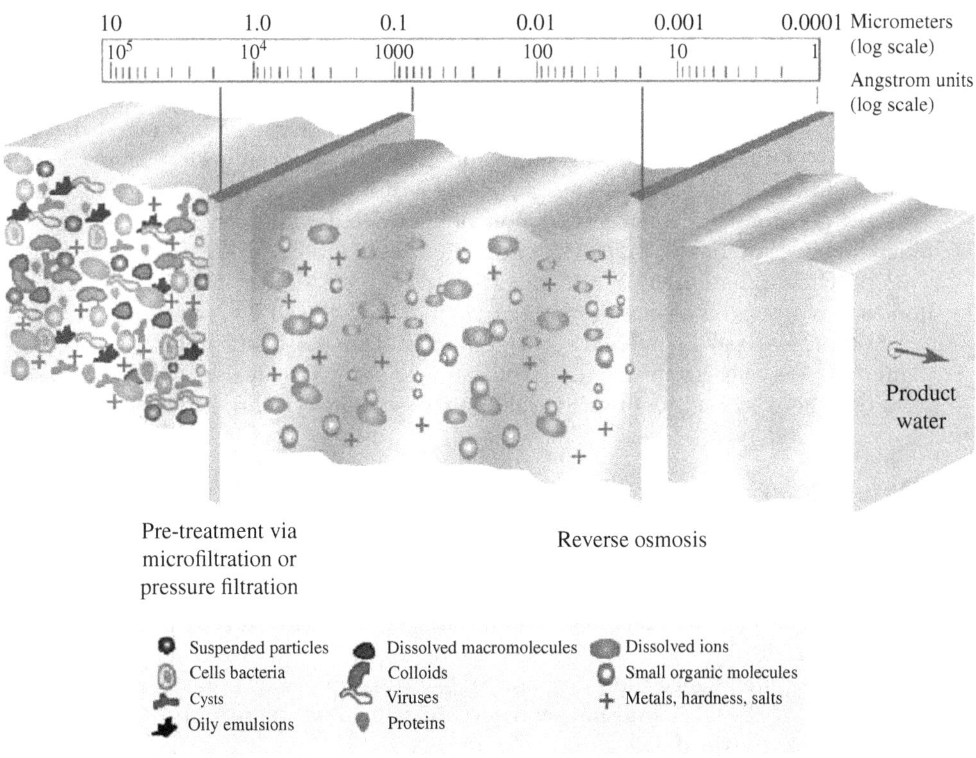

FIGURE 1.6 Contaminant removal by RO membranes.

Energy Type	MED	MSF	VC	BWRO	SWRO
Steam pressure, ata	0.2–0.4	2.5–3.5	Not needed	Not needed	Not needed
Electric energy equivalent, kWh/m^3 (kWh/1000 gal)	4.5–6.0 (17.0–22.7)	9.5–11.0 (35.9–41.6)	NA	NA	NA
Electricity consumption, kWh/m^3 (kWh/1000 gal)	1.2–1.8 (4.5–6.8)	3.2–4.0 (12.1–15.1)	8.0–12.0 (30.3–45.4)	0.3–2.8 (1.1–10.6)	2.5–4.0 (9.5–15.1)
Total energy use, kWh/m^3 (kWh/1000 gal)	5.7–7.8 (21.5–29.5)	12.7–15.0 (48.0–56.7)	8.0–12.0 (30.3–45.4)	0.3–2.8 (1.1–10.6)	2.5–4.0 (9.5–15.1)
Water production costs, US$ per cubic meter (US$ per 1000 gal)	0.7–3.5 (2.6–13.2)	0.9–4.0 (3.4–15.1)	1.0–3.5 (3.8–13.2)	0.2–1.8 (0.8–6.8)	0.5–3.0 (1.9–11.3)

Note: NA = Not applicable.

Table 1.3 Energy and Water Production Costs for Alternative Desalination Technologies

configuration of RO membranes is such that they cannot store and remove from their surface large amounts of suspended solids. If left in the source water, the solid particulates would accumulate and quickly plug (foul) the surface of the RO membranes, not allowing the membranes to maintain a continuous steady-state desalination process. Therefore, the suspended solids (particulates) contained in source water used for desalination have to be removed before they reach the RO membranes.

The following chapters of this book focus exclusively on the planning and engineering of RO membrane desalination systems. Over the past 20 years, RO membrane separation has evolved more rapidly than any other desalination technology, mainly because of its competitive energy consumption and water production costs (Table 1.3). The energy and cost analysis presented in Table 1.3 indicates that the all-inclusive energy consumption for freshwater production of thermal desalination plants is typically much higher than that for brackish or seawater desalination.

BWRO desalination yields the lowest overall production costs of all the desalination technologies. It is also important to note that the latest MED projects built over the last 5 years have been completed at costs comparable to those for similarly sized SWRO plants. For the majority of medium and large projects, however, SWRO desalination usually is more cost competitive than thermal desalination technologies.

1.6 References

Black, M., and J. King, *The Atlas of Water: Mapping the World's Most Critical Resource*, 2d ed., Earthscan Ltd, London, 2009.

Cotruvo, J., N. Voutchkov, J. Fawell, P. Payment, D. Cunliffe, and S. Lattemann, *Desalination Technology, Health and Environmental Impacts*, CRC Press and IWA Publishing, Boca Raton, FL, 2010.

Global Water Intelligence and International Desalination Association, *IDA Desalination Yearbook 2011–2012*, International Desalination Association and Global Water Intelligence, Oxford, UK.

CHAPTER 2

Source Water Quality Characterization

2.1 Introduction

A desalination plant's source water quality has an impact on the treatment needed and ultimately on the quality of the produced drinking water. Desalination plants can use either subsurface intakes, for collection of source water from brackish aquifers or coastal seawater aquifers, or open intakes, for collection of saline surface source water. Subsurface intakes collect saline water which is prefiltered through the surrounding geological formations; often this water is not influenced by anthropogenic contamination and does not contain pathogens. However, when the aquifer from which saline source water is taken is of alluvial origin or is influenced by an adjacent alluvial aquifer, the water may contain elevated concentrations of iron and manganese, arsenic, cyanide, ammonia, sodium bisulfite, and other undesirable compounds of natural origin.

Open intakes receive raw water directly from the water column of the surface saline source (ocean, river, lake, etc.). Therefore, this type of intake is almost always under the influence of potential natural and anthropogenic sources of contamination, such as surface runoff, commercial and recreational ship traffic, and storm water and wastewater discharges. Therefore, it may contain pathogens.

Source water characterization and assessment is a key component of the planning and design of desalination projects. The type and detection level of the measured source water pollutants are mainly determined by the applicable product water quality regulations and regulatory requirements governing the desalination plant's discharge. In addition, saline source water quality is typically analyzed for parameters that may not be regulated but do have an impact on the final use of the desalinated water and on the desalination plant's performance, such as silt density index, algal content, silica, total and biologically assimilable organic carbon, minerals that can foul the *reverse osmosis* (RO) membranes.

2.2 Watershed Sanitary Survey

In accordance with the requirements of the federal Safe Drinking Water Act in the United States, the source water quality of desalination plants with open intakes has to be determined by completing a watershed sanitary survey. This survey quantifies pollutants regulated by the Safe Drinking Water Act and state regulatory requirements, and identifies potential sources of water contamination in terms of pathogens and other anthropogenic pollutants located within a 1.64-km (1.0-mi) radius of the location of the intake.

A typical watershed sanitary survey contains descriptions and evaluation of the following:

- Source water intake—physical components, configuration, and condition
- Source water treatment approach that addresses the site-specific water quality challenges of the project
- Potential impacts of the new source of water supply on the receiving water distribution system
- Location and configuration of finished water storage facilities
- Location and configuration of product water pump facilities and controls
- Source water quality monitoring data collected at the intake area over a period of 6 to 12 months
- Proposed water distribution system modifications to accommodate the new water source
- Plan for water treatment plant operator compliance with all applicable regulatory requirements for drinking water production and distribution from the new source

Development (and subsequent five-year updates) of the watershed sanitary survey for a given desalination project with an open intake is a costly effort; completion typically takes over one year. However, the survey is required by the current regulations to document potential changes in source water quality and new sources of contamination in the source water intake area.

For desalination plants using subsurface intakes that are not under the influence of surface water contamination (i.e., confined brackish water aquifers), current Safe Drinking Water Act regulations allow the desalination project's proponent to complete a more simplified characterization of the source water quality, referred to as a *source water assessment*. However, a practical challenge that many desalination projects with subsurface intakes face is how to determine that the collected source water is not under the influence of surface water contamination. Current regulations do not specify a clear path for such a determination, nor the types of studies and water quality collection efforts needed to adequately complete such a determination. If deep confined brackish water aquifers are used to collect source water for a given desalination project, the determination is relatively easier. However, if the desalination plant is equipped with subsurface intake such as vertical beach wells, infiltration galleries, and horizontal and slant wells that collect water from an unconfined shallow aquifer, at present there is no clearly defined methodology or criteria for determining whether the subsurface intake is under the influence of surface water contamination. This issue is further complicated by the fact that beach erosion and storm impacts can result in a change of the depth of the filtration layer separating the subsurface intake well screens from the surface water over the useful life of the desalination project.

2.3 Assessment of the Pathogen Content of Source Water

Key pathogens of concern for human health that may be contained in the source water used for desalination are viruses, *Giardia* cysts, and *Cryptosporidium*. Removal and inactivation requirements for viruses and *Giardia* in the saline source water collected by open intakes or subsurface intakes collecting groundwater under the direct influence of surface water are established in the US Environmental Protection Agency's Surface Water Treatment Rule, while the *Cryptosporidium* log removal requirements are addressed in the agency's Long Term 2 Enhanced Surface Water Treatment Rule.

Most human pathogens are of freshwater origin; they are less likely to survive in seawater or in highly saline brackish water. However, at present there are no comprehensive studies that allow the establishment of a clear correlation between the survival rates of viruses, *Escherichia coli*, *Giardia* cysts, and *Cryptosporidium* oocysts and source water salinity. Therefore, source water characterization in terms of pathogens has to be completed for each individual desalination project for a period of 12 to 24 months. During this testing period, grab samples have to be collected at least once per month at the same intake location and time of day. These samples, at a minimum, have to be analyzed for *Cryptosporidium*, *Giardia*, total and fecal coliform count, total heterotrophic plant count, and turbidity.

2.4 Overview of Source Water Constituents

The constituents contained in source water used for desalination can be classified in four main groups: (1) dissolved minerals and gases, (2) colloids and suspended solids, (3) organics, and (4) microorganisms. During membrane separation, all of these constituents either are removed by the membranes, pass through the membranes into the low-salinity water stream, or are retained and accumulate on the surface and in the molecular matrix of the membranes.

The source water constituents retained and accumulated on the membrane surface or in the molecular membrane structure over time change the membranes' ability to reject new constituents and to produce freshwater of desired quantity and quality. Such constituents are referred as *foulants* and are of particular importance in the design and engineering of desalination systems.

Depending on their nature, membrane foulants can be classified as follows:

- Particulate foulants (mainly suspended solids and silt)
- Colloidal foulants—compounds of relatively small size (0.2 to 1.0 μm) that are not in fully dissolved form, which when concentrated during the membrane separation process may coalesce and precipitate on the membrane surface (mainly claylike substances)
- Mineral scaling foulants—inorganic compounds (i.e., calcium, magnesium, barium, and strontium salts) which during the salt separation process may precipitate and form a scale on the membrane surface (such as calcium carbonate and sulfate or magnesium hydroxide) or may block the membrane diffusion layer (such as iron and manganese)
- Organic foulants—organic matter of natural or anthropogenic origin that can attach to and foul membranes
- Microbial foulants—aquatic organisms and soluble organic compounds that can serve as food to the microorganisms which inhabit the source water and can form a fouling biofilm that reduces membrane transport

2.5 Minerals

The primary purpose of desalination is to remove dissolved minerals (salts) contained in the saline source water. A commonly used measure of the content of dissolved minerals is the concentration of total dissolved solids (salinity). This parameter encompasses all ions in the source water, including sodium, potassium, bromide, boron, calcium, magnesium, chloride, sulfate, bicarbonate, nitrate, metals, etc.

2.5.1 Mineral Content of Seawater

Table 2.1 shows key ion content and *total dissolved solids* (TDS) concentrations of typical Pacific Ocean water and of permeate produced from this water by *seawater reverse osmosis* (SWRO) membrane separation. Concentrations in this table are expressed both in milligrams per liter (mg/L) and in milliequivalents per liter (meq/L).

The milligrams-per-liter parameter indicates the ratio of ion weight to solution volume. The milliequivalent-per-liter designation reflects the capacity of ions to react with one another. The atomic or formula weight of an ion divided by its valence (number of positive or negative charges) is called the *equivalent weight* (eq) of the ion. One-thousandth of this weight is termed a *milliequivalent* (meq). For example the milliequivalent-per-liter concentration of calcium in Table 2.1 can be calculated as follows:

1. Calcium has a molecular weight of 40.08 g/mol.
2. Calcium has a valence of +2.

	TDS Concentration, mg/L		TDS Concentration, meq/L	
Parameter	**Raw Water**	**Permeate**	**Raw Water**	**Permeate**
Cations				
Calcium	403	0.6	20.1	0.03
Magnesium	1298	1.3	106.2	0.11
Sodium	10,693	88.0	464.9	3.82
Potassium	387	4.3	12.9	0.14
Boron	4.6	0.8	1.5	0.26
Bromide	74	0.7	0.9	0.01
Total Cations	**12,859.6**	**95.7**	**606.5**	**4.37**
Anions				
Bicarbonate	142	2.2	2.24	0.03
Sulfate	2710	7.1	56.6	0.16
Chloride	19,287	145.0	542.6	4.24
Fluoride	1.4	0.0	0.06	0.00
Nitrate	0.00	0.0	0.0	0.00
Total Anions	**22,140.4**	**154.3**	**601.5**	**4.43**
TDS	**35,000.0**	**250.0**	**1208.0**	**8.80**

Table 2.1 Typical Pacific Ocean Water Quality

3. The equivalent weight of calcium is (40.08 g/mol)/(2 eq/mol) = 20.04 g/eq (or mg/meq).
4. Since the seawater sample in Table 2.1 contains 403 mg/L of calcium, then the concentration of calcium in milliequivalents per liter is (403 mg/L of calcium)/(20.04 mg/meq) = 20.1 meq/L of calcium.

The main reason why the TDS concentration is often measured in milliequivalents per liter instead of in milligrams per liter is to check the accuracy of the measurement for the water for which analysis is completed. When added together, the milliequivalent-per-liter concentrations of cations (positively charged ions) contained in the water should approximately equal the total milliequivalent-per-liter concentrations of anions (negatively charged ions) in the solution. These two values are usually not exactly equal, since other ions beside those listed in Table 2.1 are present in the water. If the difference between total cation and anion content exceeds 5 to 10 percent, then the accuracy of the laboratory analysis is inadequate or other ions are present in the water which may not have been reported or which are not typically contained in saline water of the particular type of source.

Analysis of Table 2.1 indicates that sodium chloride contributes over 85 percent of the TDS concentration of Pacific Ocean water. The three other large contributors to TDS are sulfate (8 percent), magnesium (4 percent), and calcium (1 percent). All other ions in seawater combined contribute only 2 percent of the TDS in the water.

The ion makeup of seawater at the same location may vary seasonally, typically within a range of 10 percent. However, seawater salinity varies in a much wider range in different parts of the world. Table 2.2 shows the typical salinity and temperature of seawater in various arid areas throughout the world; the highest-salinity seawater occurs in the Middle East (the Persian Gulf and Red Sea). The table also contains the typical values of another key source water quality factor for RO desalination—temperature. Warmer water has lower viscosity (i.e., is less dense), which in turn increases the production rate of the RO membranes, and vice versa.

2.5.2 Mineral Content of Brackish Water

Table 2.3 presents the TDS content of several brackish water sources located in the United States. Analysis of this table indicates that brackish water TDS content may vary significantly from one location to another and that sodium and chloride may not always

Seawater Source	Typical TDS Concentration, mg/L	Temperature, °C
Pacific and Atlantic Oceans	35,000	9–26 (avg 18)
Caribbean Sea	36,000	16–35 (avg 26)
Mediterranean Sea	38,000	16–35 (avg 26)
Gulf of Oman and Indian Ocean	40,000	22–35 (avg 30)
Red Sea	41,000	24–32 (avg 28)
Persian Gulf	45,000	16–35 (avg 26)

Note: Seawater TDS and temperature may be outside the table ranges for any specific location.

Table 2.2 Seawater TDS and Temperature of Various Ocean Water Sources

Parameter	Orange County, California	Rio Grande, Texas	Tularosa, New Mexico	Cape Hatteras, North Carolina
Cations, mg/L				
Calcium	140.0	163.0	420.0	545.0
Magnesium	10.0	51.0	163.0	1398.0
Sodium	300.0	292.0	114.0	4961.0
Potassium	35.0	0.0	2.30	99.0
Boron	0.8	0.0	0.14	1.2
Bromide	7.4	4.5	0.70	12.5
Total Cations	**493.2**	**510.5**	**700.14**	**7016.7**
Anions, mg/L				
Bicarbonate	275.0	275.0	270.0	223.0
Sulfate	350.0	336.0	1370.0	173.0
Chloride	350.0	492.0	170.0	6523.0
Fluoride	0.8	0.08	0.0	1.3
Silica	10.0	35.0	22.0	22.0
Nitrate	1.0	1.5	10.0	1.0
Total Anions	**986.8**	**1139.58**	**1842.0**	**6943.3**
TDS mg/L	**1480.0**	**1650.0**	**2542**	**13,960.0**

Table 2.3 Brackish Water Quality of Several Sources

be the main contributors to the TDS content of the water. Usually, brackish water has a higher content of silica and nitrates than does seawater, which often necessitates additional pretreatment.

Because the existing standard method for measurement of TDS concentration of water involves collection and evaporation of a discrete water sample at 105°C followed by weighing of the solids remaining after evaporation, this parameter can only be measured discretely. On the other hand, the continuous monitoring of source water and product water TDS concentrations is essential for the cost-effective and efficient operation of RO systems.

Therefore, in practice, TDS concentration is often monitored continuously by measurement of the *electrical conductivity* (EC) of the water. Electrical conductivity (also known as *specific conductance*) is a measure of a solution's ability to conduct electricity. Conductivity is expressed in microsiemens per meter (μS/m).

The ratio between TDS and EC in source water is site specific and usually varies in a range between 0.67 and 0.70. For example, seawater with a TDS concentration of 35,000 mg/L would typically have a conductivity of 50,000 to 52,000 μS/m. The ratio between TDS and EC depends on the content of sodium chloride in the water and on the temperature. If the TDS is made of 100 percent sodium chloride, the TDS/EC ratio is typically 0.5. This ratio increases as the content of sodium chloride decreases (i.e., ions other than sodium and chloride have a measurable contribution to salinity).

Figure 2.1 Crystalline scale on the surface of an RO membrane.

Since product water TDS consists of over 91 percent sodium chloride (as compared to source seawater, where sodium chloride contributes only 86 percent of the salinity), the typical ratio of TDS to EC in permeate is 0.5.

The TDS of the source water is the most important water quality parameter in RO desalination for two main reasons. This parameter is a main factor in determining the feed pressure and therefore, the energy needed to produce freshwater from a given saline water source. Every 100 mg/L of TDS in the source water creates approximately 0.07 bar (1 lb/in^2) of osmotic pressure, which will need to be overcome by the pressure applied to the saline water fed to the RO membranes. For example, seawater that contains 35,000 mg/L of TDS will create approximately 24.5 bar (355 lb/in^2) of osmotic pressure. In addition, TDS concentration of the source water is a key factor in determining the expected product water quality, since RO membranes reject a given percentage of the feed water's TDS.

Besides sodium and chloride, which need to be removed in order to produce freshwater, other key inorganic constituents of TDS are various minerals (mainly salts of calcium and magnesium). These bivalent salts can precipitate on the surface of the membrane and form a thin layer of crystalline scale (Fig. 2.1), which in turn can plug the membrane surface and significantly reduce membrane productivity.

2.6 Dissolved Gases

Both seawater and brackish water often contain various dissolved gases. The most common gases are oxygen, carbon dioxide, hydrogen sulfide, and ammonia. All of these gases pass through RO membranes, and thus desalination is typically not a suitable technology for degasification of brackish and seawater for drinking water production.

Usually, ocean water and desalinated water originating from it are supersaturated with dissolved oxygen. The dissolved oxygen concentration of these waters typically varies between 5 and 8 mg/L. For comparison, most brackish waters originating from groundwater aquifers have a very low oxygen content (0.5 mg/L or less). Because of its high oxygen content, seawater does not contain hydrogen sulfide. This gas is, however, frequently encountered in source waters from brackish aquifers.

2.7 Particulate Membrane Foulants

2.7.1 Description

Particulate foulants are organic and inorganic particles contained in the source water, such as fine debris, plankton, detritus, and silt. These solids cannot pass through RO membranes. All suspended solids which naturally occur in insoluble form, if not removed by pretreatment, would be retained on the feed side of the RO membranes; depending on the hydrodynamic conditions on the membrane surface and the size and charge of these particles, they would either migrate along the membrane leafs and ultimately exit with the concentrate or will be trapped on the membrane surface and begin to accumulate there, causing loss of membrane productivity over time. This type of foulant can be effectively removed by prefiltering of the source water prior to RO membrane separation.

Particulate foulants in raw source seawater vary in size. However, most of them, including picophytoplankton, are larger than 0.1 μm. Usually over 90 percent of particulate foulants are larger than 1 μm. A well-designed and properly operating pretreatment system will produce permeate that does not contain particles larger than 20 μm. Typically, suspended solids larger than 100 μm contained in seawater and surface brackish water are settleable and can be removed by clarification of the source water prior to filtration.

2.7.2 Parameters and Measurement Methods

Turbidity

Turbidity is a parameter which measures the content of particulate foulants in the source seawater. The turbidity level in the source water is indicative of the content of clay, silt, suspended organic matter, and microscopic aquatic life, such as phyto- and zooplankton. It is expressed in *nephelometric turbidity units* (NTU).

The turbidity of open ocean and surface brackish waters can vary between 0.1 and several hundred NTU, although under normal dry weather conditions, it is typically between 0.5 and 2.0 NTU. Rain events, algal blooms, storms, snowmelt, river discharges, and human activity (such as wastewater discharges, ship traffic, etc.) can cause significant turbidity increases and variations. Usually, saline water with a turbidity below 0.05 NTU causes very low particulate fouling. Most RO membrane manufacturers have a maximum feed water turbidity of 1.0 NTU, although this level is relatively high in practical terms. Usually, filtered water turbidity below 0.1 NTU is desirable.

Although turbidity is a good measure of the overall content of particulates in the source water, on its own it is not an adequate parameter to characterize water's potential for particulate or other fouling. Turbidity measurement does not provide information regarding the type and size of particles in the source water and does not measure the content of dissolved organic and inorganic foulants. The size of particles contained in the source water matters because RO membrane feed and concentrate spacers, through which the saline source water is distributed inside the membranes, are of limited width (typically 0.7 to 0.9 mm).

Silt Density Index

Silt density index (SDI) is a parameter that provides an indication of the particulate fouling potential of source water. If RO system is operated at a constant transmembrane pressure, particulate membrane fouling will result in a decline of system productivity (membrane flux) over time. SDI gives an indication of the rate of flux decline through a filter of standard size and diameter operated at constant pressure for a given period of time.

A standard SDI_{15} test procedure is described in ASTM Standard D4189-07 (American Water Works Association, 2007) and is based on the measurement of the time in seconds it takes to collect a 500-mL sample through a paper filter of size 0.45 μm and diameter 45 mm both at the start of the test (t_0 = 0 min) and after the source water has flowed through the filter under a driving filtration pressure of 2.1 bar (30 lb/in^2) for 15 min (t_{15} = 15 min). The two sample durations (t_0 and t_{15}) are applied to a formula (Eq. 2.1), and the resulting SDI_{15} value indicates the particulate fouling potential of the source water:

$$SDI_{15} = \frac{1-(t_0/t_n)}{n}100 \tag{2.1}$$

where t_0 and t_n = the respective times in seconds it takes to collect 500 mL of filtered water at the beginning of the test and after running water through the filter for the duration of the selected test run time.

It should be pointed out that while the standard SDI test requires a test run time of 15 min between the first and second measurements, the test can also be run for 5 or 10 min, depending on the solids concentration. Based on this formula, the maximum value of SDI_{15} is 6.7; this condition would occur if the time to collect 500 mL after 15 min of filtration were infinite.

Typically, source water with an SDI_{15} lower than 4 is considered to have adequately low RO membrane particulate fouling potential, and its use in membrane desalination is expected to result in a reasonably slow flux decline over time. Source water with SDI_{15} lower than 2 is considered to have a very low fouling potential and to be of good quality.

In order to maintain their performance warranties, membrane manufacturers usually require that the SDI of the source water fed to the RO membranes be less than 5. If a source water's SDI_{15} is higher than 5, this typically means that this source water has a very high content of particulate foulants and therefore is not directly suitable for desalination, because it would likely cause accelerated fouling of RO membranes. For source water with such a high SDI_{15} value, it is often useful to complete the SDI test at shorter (5- or 10-min) intervals, which will usually provide more meaningful results (Mosset et al., 2008).

Figure 2.2 shows filtration pads from SDI_{15} tests along with the SDI_{15} values of pretreated seawater. The SDI test pads marked "D2" are from iron-salt coagulated seawater pretreated via two-stage upflow sand media filters, while these marked "Zenon" are from noncoagulated *ultrafiltration* (UF) filter effluent.

The top two D2 SDI filter pads (5.3 SDI and 5.2 SDI) have a brown discoloration caused by an overdose of ferric coagulant. These SDI tests indicate that the pretreated influent is not suitable for SWRO treatment, due to high SDI levels. The other SDI pads are indicative of pretreated filter effluent of relatively low particulate fouling potential.

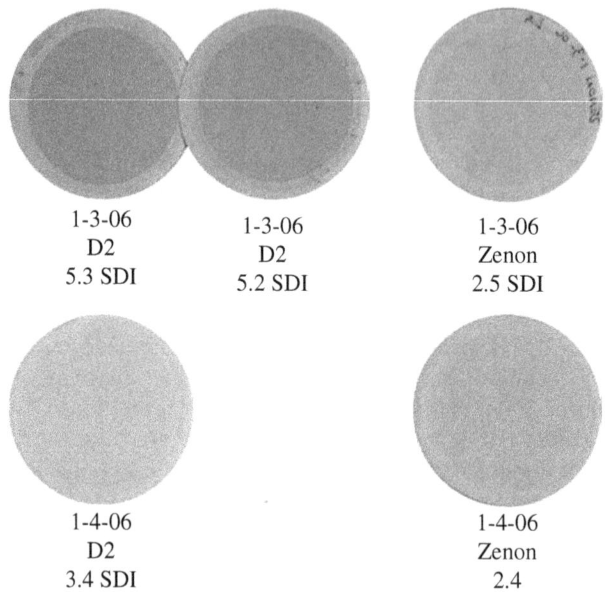

Figure 2.2 SDI pads from tests of pretreated seawater of various quality.

It should be pointed out that although widely used in operation practice today, the SDI_{15} test has only a limited ability to measure the RO membrane fouling potential of saline water. The test is based on filtration of a saline water sample through a *microfiltration* (MF) membrane pad with 0.45-μm pore openings. Therefore, the SDI measurement is mainly indicative of the content of large-size particulates and colloids in the source water. The mechanisms of fouling for permeable MF and UF membranes and semipermeable RO membranes may differ significantly, depending on the types of foulant contained in the source water.

Fouling of MF and UF membranes usually occurs due to a combination of micropore plugging and cake formation on the membrane surface. In contrast, RO membranes are typically fouled by the formation of cake of deposits on the membrane surface (without pore plugging). Since smaller particles create significantly higher resistance in the filter cake than big particulates, their effect on RO membrane fouling can be much more pronounced than that of the large-size particles captured by the SDI test.

In order to address the effect of smaller particles on SWRO membrane fouling, a number of alternative filtration indices have been developed over the past 10 years. Advantages and disadvantages of these modified fouling indices are discussed elsewhere (Khirani et al., 2006; Boerlage, 2007; Mosset et al., 2008).

Total Suspended Solids

Total suspended solids (TSS) concentration is a measure of the total weight of solid residuals contained in the source water; and it is customarily presented in milligrams per liter. TSS is measured by filtering a known volume of water (typically 1 L) through a preweighed glass-fiber filter, drying the filter with the solids retained on it at 103 to

105°C, and then weighing the filter again after drying. The difference between the weight of the dried filter and of the clean filter, divided by the volume of the filtered sample, reflects the total amount of particulate (suspended) solids in the source water.

It should be pointed out that because saline water contains dissolved solids which will crystallize and convert into particulate solids when the sample is heated at 103 to 105°C, often TSS analysis of saline water completed in accordance with the standard methods for water and wastewater analysis yields an erroneously high TSS content in the water. The higher the source water's salinity and the lower its particulate content, the more inaccurate this measurement is. In order to address the challenge associated with the standard method of TSS measurement, it is recommended to wash the solids retained on the filter by spraying the filter with deionized water before drying. Unless the source water solids sample is washed before drying, the results of this sample are meaningless.

If the laboratory TSS test is completed properly and the filtered sample is well washed, this parameter usually provides a much more accurate measure of the actual content of particulate solids in the source water than does turbidity, because it accounts for the actual weight of the particulate material present in the sample. For comparison, turbidity measurement is dependent on particle size, shape, and color, and typically is not reflective of particles of very small size (i.e., particles of 0.5 μm or less), such as fine silt and picoalgae.

In fact, a change in the ratio of TSS to turbidity is a good indicator of a shift in the size of particles contained in the source water, which may be triggered by algal blooms, storms, strong winds, and other similar events which can result in resuspension of solids from the bottom sediments into the water column.

Typically, an increase in the TSS/turbidity ratio is indicative of a shift of particulate solids toward smaller-size particles. For example, during non-algal-bloom conditions, the TSS/turbidity ratio of an appropriately processed sample is typically in the range of 1.5 to 2.5—i.e., water with a turbidity of 2 NTU would have a TSS concentration of 3 to 5 mg/L. During heavy algal blooms dominated by small-size (pico- and micro-) algae, the TSS of the source water could increase over 10 times (e.g., to 40 mg/L), while the source water turbidity could be multiplied by 2 to 3 times only—for this example it would be in a range of 4 to 6 NTU, with a corresponding increase in the TSS/NTU ratio from 2/1 to between 6/1 and 10/1.

Chlorophyll *a*

The chlorophyll *a* concentration of a source water is an indicator of the content of algae with green pigmentation in the water. This parameter is measured using a fluorometer or spectrophotometer. The content of chlorophyll *a* is proportional to the light transmission through the water sample at a given wavelength, which is detected by the instrument and converted into concentration units, typically either micrograms per liter or milligrams per liter.

As a rule of thumb, source waters with chlorophyll *a* content below 0.5 μg/L have low fouling potential, and algae levels are indicative of non-algal-bloom conditions. During severe algal blooms, the chlorophyll *a* level could exceed 10 μg/L.

It should be pointed out that content of algae and chlorophyll *a* naturally varies diurnally and seasonally, and also changes with depth. In general, algal content is proportional to the intensity of solar irradiation and typically increases significantly in the summer, as compared to the average annual algal content.

Tracking chlorophyll *a* on a daily basis and trending the collected data allows an operator to determine the occurrence of algal bloom events, because during algal blooms, the chlorophyll *a* content in the source water typically increases several times in a matter of only several days.

Algal Count

Algal count is a measure of the number of algal particles per unit volume of source water. The algal count is expressed in total number of algal cells per milliliter of water. Under normal non-algal-bloom conditions, the algal count in saline source water is usually below 1000 cells per milliliter. Algal blooms are considered to be of concern when the algal count exceeds 2000 cells per milliliter. Algal blooms are often referred as mild if the algal content is between 2000 and 20,000 cells per milliliter, of medium intensity when the algal count varies between 20,000 and 60,000 cells per milliliter, and severe when algal content exceeds 60,000 cells per milliliter.

Total algal count can be measured by online instrumentation. The algal count of individual species in the water is completed by laboratory analysis and is a very useful tool to verify the presence of algal bloom and determine the size and type of algae triggering the bloom. Typically, algal bloom is defined as an event in which over 75 percent of the algae are from the same species. Knowing the type and size of the dominant algal species is critical to optimizing the source water chemical conditioning and pretreatment approach.

Particle Distribution Profile

The particle distribution profile presents the number of solid particles in the source water for different particle size ranges. Usually particles are classified in the following ranges: 1 μm or less; > 1 and ≤ 2 μm; > 2 and ≤ 5 μm; > 5 and ≤ 10 μm; > 10 and ≤ 20 μm; > 20 and ≤ 50 μm; and > 50 μm. Particles of sizes smaller than 1 μm are poorly removed by conventional granular media filtration and dissolved air flotation clarification; key pathogens such as *Giardia* and *Cryptosporidium* are in the range of 2 to 10 μm and are typically removed at 2 log by conventional granular media pretreatment filtration. Particles which have a size of 20 μm or more are well removed by both membrane and granular media pretreatment filters. Typically, a pretreatment system is considered to perform well if the pretreated water contains less than 100 particles per milliliter equal to or smaller than 2 μm and does not contain larger-size particles.

2.7.3 Threshold Levels of Particulate Foulants

Table 2.4 presents a list of source water quality parameters used for characterizing particulate content that operators are recommended to measure when deciding upon the type of pretreatment needed for a given source water. Table 13.3 provides general guidelines for selecting the type of pretreatment needed for the site specific conditions of a desalination project.

2.8 Colloidal Membrane Foulants

2.8.1 Description

Colloidal foulants are inorganic and organic compounds that naturally exist in suspension and may be concentrated by the RO separation process and precipitate on the membrane surface, thereby causing membrane flux to decline over time. Colloidal solids

Source Water Quality Parameter	Pretreatment Issues and Considerations
Turbidity, NTU	Levels above 0.1 mg/L are indicative of a high potential for fouling. Spikes above 50 NTU for more than 1 h would require sedimentation or dissolved air flotation treatment prior to filtration.
Silt density index (SDI_{15})	Source seawater levels consistently below 2 all year round indicate that no pretreatment is needed. An SDI greater than 4 indicates that pretreatment is necessary.
Total suspended solids, mg/L	Needed to assess the amount of residuals generated during pretreatment. It does not correlate well with turbidity beyond 5 NTU.
Chlorophyll *a*	Indicative of algal bloom occurrence. If water contains more 0.5 μg/L, the source water may be in an algal bloom condition.
Algal count, cells per milliliter	Indicative of algal bloom occurrence. If water contains more than 2000 cells per milliliter, the source water is in an algal bloom condition.

Table 2.4 Water Quality Parameters for Characterization of Particulate Foulants

have a particle size of 0.001 to 1 μm. For prevention of colloidal fouling, RO membrane manufacturers usually recommend a feed turbidity of less than 0.1 NTU and an SDI_{15} of less than 4.

Metal oxide and hydroxide foulants most frequently encountered during brackish and seawater desalination are iron, manganese, copper, zinc, and aluminum. Typically, open ocean seawater contains very low levels of these metal foulants, and therefore, if such fouling is encountered on the membrane elements, the usual sources are overdosing of coagulant (iron salt) or corrosion of pipes, fittings, tanks, and other metal equipment located upstream of the SWRO system.

Iron and Manganese

Iron and manganese fouling may occur if source water is collected via subsurface intake from a brackish aquifer with a naturally high content of these metals or from a coastal aquifer which is under the influence of fresh groundwater that contains high levels of these compounds in reduced form (iron of more than 2 mg/L as ferrous or manganese of more than 0.5 mg/L).

If iron and manganese are in reduced form and they are below 1.0 and 0.1 mg/L, respectively, they can be removed by RO membranes without causing accelerated fouling. However, if iron and manganese are in oxidized form, their levels should be reduced below 0.05 and 0.02 mg/L, respectively, to prevent mineral fouling of the membranes.

In addition to naturally occurring colloidal matter, colloidal iron fouling on the surface of RO membranes may be caused by corrosion of upstream piping and equipment or by overdosing or poor mixing of iron-based coagulant used for conditioning the source water. If the source water contains chlorine, this colloidal iron tends to catalyze the oxidation process caused by chlorine, which in turn enhances the damage to the RO membranes even when residual chlorine in the saline source water is in very small doses.

Silica

Another mineral fouling compound frequently encountered in brackish aquifers is silica. Total silica (silicon dioxide) in the source water consists of reactive silica, which is in soluble form, and unreactive silica, which is in colloidal form. While reactive silica is not a challenge for RO membranes, colloidal silica in the saline source water can cause significant membrane fouling. It should be pointed out, however, that elevated content of silica in colloidal form is mainly found in brackish water sources; unreactive silica is present in very low levels in seawater and is fouling challenge only when its level in the concentrate exceeds 100 mg/L.

The stability of colloids is reduced with an increase in source water salinity, and therefore typical seawater with a TDS concentration in a range of 30,000 to 45,000 mg/L would contain silica in dissolved and precipitated forms rather than in colloidal form. Open ocean seawater typically contains silica of less than 20 mg/L, and therefore this compound does not cause mineral fouling of SWRO membranes.

However, if the source seawater is collected via a subsurface well intake which is under the influence of a brackish coastal aquifer with a high content of colloidal silica, or it is collected near an area where a silt-laden river enters into the ocean, then colloidal fouling may become a challenge. Colloidal foulants can be removed by coagulation, flocculation, sedimentation and filtration, similar to particulate foulants.

Polymers

Another common source of colloidal fouling is overdosing or poor mixing of polymers used for conditioning of the saline source water prior to filtration. Such a problem usually occurs when the source water contains fine particulates of relatively low quantity and electric charge (i.e., the water is of very low turbidity, typically below 0.5 NTU) and a polymer is added to enhance the flocculation of such particles.

Hydrocarbons

The most common organic colloidal foulants are oil-product-based hydrocarbons. Such compounds are not contained naturally in most brackish waters or in open ocean seawater, and their occurrence indicates that the saline water intake area or aquifer is under influence of man-made sources of contamination—typically discharge from a wastewater treatment plant, from storm drains collecting surface runoff from urban areas (parking lots, industrial sites, etc.), or from waste discharges or oil leaks released by ships, boats, or near-shore oil storage tanks in port areas.

Even in very small quantities (0.02 mg/L or more), oil and grease can cause accelerated fouling of RO membranes. Therefore, it is desirable that oil and grease content in the source water be maintained below 0.02 mg/L at all times.

2.8.2 Parameters and Measurement Methods

Colloidal Iron, Manganese, and Silica

Colloidal iron, manganese, and silica are measured by standard laboratory tests. Usually colloidal (also referred to as *unreactive*) silica is determined by measuring total and reactive silica in the source water.

Total Hydrocarbons

Total hydrocarbon concentration can be measured by both laboratory analytical methods and online analyzers. Most desalination plants using surface water sources (i.e., a saline river estuary or ocean water) are equipped with online total hydrocarbon analyzers.

Source Water Quality Parameter	Pretreatment Issues and Considerations
Iron, mg/L	If iron is in reduced form, RO membranes can tolerate up to 2 mg/L. If iron is in oxidized form, a concentration of more than 0.05 mg/L will cause accelerated fouling.
Manganese, mg/L	If manganese is in reduced form, RO membranes can tolerate up to 0.1 mg/L. If manganese is in oxidized form, a concentration of more than 0.02 mg/L will cause accelerated fouling.
Silica, mg/L	Concentrations higher than 100 mg/L in concentrate may cause accelerated fouling.
Total hydrocarbons, mg/L	Concentrations higher than 0.02 mg/L will cause accelerated fouling.

TABLE 2.5 Water Quality Parameters for Characterization of Colloidal Foulants

2.8.3 Threshold Levels of Colloidal Foulants

Table 2.5 presents key source water quality parameters which operators are recommended to measure in order to characterize potential colloidal foulants in the source water.

2.9 Mineral Membrane-Scaling Foulants

2.9.1 Description

All minerals contained in the source water are concentrated during the process of membrane salt separation. As their concentration increases during the desalination process, ions of calcium, magnesium, barium, strontium, sulfate, and carbonate can form sparingly insoluble salts, which could precipitate on the RO membrane surface. The mineral scales that typically form during desalination are those of calcium carbonate, calcium and magnesium sulfate, and barium and strontium sulfate.

Formation of mineral scales on the membrane surface is balanced by the high salinity of the source water, which tends to increase the solubility of all salts. This means that the higher the salinity of the source seawater, the less likely a mineral scale is to form on the membrane surface at typical seawater pH of 7.6 to 8.3 and desalination system recovery of 45 to 50 percent. In brackish seawater desalination systems, which typically operate at much higher recoveries (75 to 85 percent) and use source waters that have relatively lower ionic strength, mineral scaling is a frequent problem.

In typical seawater desalination systems, mineral-scale fouling is usually not a challenge, unless the source seawater pH has to be increased to 8.8 or more in order to enhance boron removal by the SWRO membranes. Calcium carbonate is the most commonly encountered mineral foulant in brackish water desalination plants. Calcium sulfate and magnesium hydroxide are the most frequent causes of SWRO membrane scaling. Scale formation can be prevented by the addition of an antiscalant or dispersant to the source water.

It is important to note that although the source seawater's temperature usually has a limited influence on scale formation, when this temperature exceeds 35°C, calcium carbonate scale will form at an accelerated rate.

2.9.2 Parameters and Measurement Methods

Commonly used parameters which can be used to predict a source water's potential to form mineral scale of calcium carbonate are the *Langelier Saturation Index* (LSI) and the *Stiff–Davis Saturation Index* (SDSI). These indices are functions of the source seawater's pH, calcium concentration, alkalinity, temperature, and TDS concentration or ionic strength.

The Langelier Saturation Index can be calculated using the following equation:

$$\mathrm{LSI} = \mathrm{pH} - \mathrm{pH_s} \tag{2.2}$$

where pH = the actual pH of the saline source water

$$\mathrm{pH_s} = (9.30 + A + B) - (C + D) \tag{2.3}$$

$A = [\log(\mathrm{TDS} - 1]/10$, TDS in mg/L
$B = -13.2 \times \log(\text{temperature} + 273) + 34.55$, temperature in °C
$C = \log[\mathrm{Ca^{2+}}] - 0.4$, $[\mathrm{Ca^{2+}}]$ in mg/L as $CaCO_3$
$D = \log(\text{alkalinity})$, alkalinity in mg/L as $CaCO_3$

The LSI value is indicative of the source water's ability to form calcium carbonate scale only, and is not reflective of the formation of other scalants. If LSI is higher than 0.2, the source water is likely to cause slight scaling; if it is above 1, the source water will cause severe scaling on membranes. If LSI is negative, the water has a tendency to dissolve scaling.

The LSI value predicts the scaling potential of the source water only for a TDS lower than 4000 mg/L. For saline source waters of higher TDS, the Stiff–Davis Saturation Index is applied. This index can be calculated as follows:

$$\mathrm{SDSI} = \mathrm{pH} - \mathrm{pCa} - p_{\mathrm{alk}} - K \tag{2.4}$$

where $\mathrm{pCa} = -\log[\mathrm{Ca^{2+}}]$, $[\mathrm{Ca^{2+}}]$ in mg/L as $CaCO_3$
$p_{\mathrm{alk}} = -\log(\text{alkalinity})$, alkalinity in mg/L as $CaCO_3$
K = a constant which is a function of total ionic strength and temperature

Values of K, as well as nomographs and examples for calculating LSI and SDSI, are presented elsewhere (American Water Works Association, 2007).

2.9.3 Threshold Levels of Mineral Foulants

Table 2.6 presents the concentration threshold levels of common mineral scalants above which the compounds will begin to accumulate on the membrane surface and form

Scalant	Maximum Threshold Measured in the Concentrate for RO Membrane Scaling
Calcium carbonate in source water without scale inhibitor, LSI	0.2
Calcium sulfate, %	230
Strontium sulfate, %	800
Barium sulfate, %	6000
Calcium fluoride, %	12,000

TABLE 2.6 Threshold Levels of Common Scalants

mineral scales (*Chemical Pretreatment for RO and NF*, 2008). The mineral scaling potential of the saline source water can be determined using proprietary software available from the manufacturers of antiscalants. The topic of scale formation and prevention is discussed in greater detail in Chap. 9.

2.10 Natural Organic Foulants

2.10.1 Description

Depending on their origin, all saline waters can contain naturally occurring or man-made organic compounds and aquatic microorganisms. Since all microorganisms and most organic molecules are relatively large in size, they are well rejected by RO membranes. However, some organic compounds and aquatic species may accumulate on the membrane surface and form a cake layer that can significantly hinder the membrane's main function—rejection of dissolved solids. Depending on the main source of the fouling cake layer, these RO membrane foulants are typically divided into two separate groups: natural organic foulants and microbiological foulants (or biofoulants). The phenomenon of accumulation of aquatic organisms and their metabolic products on the membrane surface is known as *biological fouling* or *biofouling*.

Natural organic matter (NOM) is typically contained in surface saline waters (brackish or open ocean seawater) and includes compounds which are produced by naturally decaying algae and other aquatic vegetation and fauna: proteins, carbohydrates, oils, pigments (i.e., tannins), and humic and fulvic substances (acids). High NOM content in the source water used for production of drinking water is undesirable because it causes discoloration of the water, forms carcinogenic disinfection byproducts when disinfected with chlorine, and results in complexation with heavy metals, which in turn causes accelerated membrane fouling.

Under normal non-algal-bloom conditions, typical seawater and brackish waters collected using open intakes do not contain quantities of NOM great enough to present a significant challenge to desalination plant operations. High NOM content is usually observed during algal blooms and when the desalination plant intake is located near a confluence with a river or other freshwater source or near wastewater treatment plant discharge. The easily biodegradable organic matter released by algae during their growth and respiration is referred to as extracellular organic matter (Edzwald and Haarhoff, 2011). When algae die off during the end phase of an algal bloom or their cells are broken by treatment or pumping processes, they also release intracellular organic matter in the source water. The combination of extracellular and intracellular organic matter from algae is referred to as *algogenic organic matter* (AOM). This NOM is easily biodegradable and provides a food source for biogrowth of bacteria on the RO membrane surface.

Humic acids are polymeric (polyhydroxyl aromatic) substances which have the ability to form chelates with metal ions in the water, such as iron. This feature of humic acids is very important for seawater or surface brackish water pretreatment systems using iron coagulants, because the humic acids can form a gellike layer of chelates on the surface of the membranes, which would cause fouling. Typically, such a fouling layer can be dissolved at a pH of 9 or more, at which condition both the membranes and the humic substances carry a negative charge. This feature is used for membrane cleaning.

Humic substances are hydrophobic, and therefore hydrophilic membranes are less prone to fouling by humic acids.

Most NOM in seawater consists of compounds of relatively large molecular weights (500 to 3000 daltons). A typical SWRO desalination membrane would reject over 90 percent of compounds that have molecular weight higher than 200 Da.

Humic and fulvic substances mainly differ in their ability to dissolve in strong acids. While humic substances are easily precipitated upon acidification of the source water, fulvic substances remain in solution at low pH.

Humic acids in their natural state are not a food source for most aquatic organisms. However, when oxidized with chlorine or other oxidants, they can become easily biodegradable and serve as a food source for aquatic bacteria growing on the RO membrane surface. Therefore, continuous chlorination of source water containing large amounts of humic acids often causes more membrane biofouling problems than it solves.

Negatively charged NOM, which dominates in surface brackish water and seawater collected by open intakes, has a tendency to adhere to the surface of thin-film composite RO membranes. Once adsorption occurs, the NOM begins to form a cake or gel on the membrane surface and to affect membrane performance. It should be noted that NOM, depending on its properties and origin, may also adhere to the surface of UF and MF pretreatment membranes and cause significant productivity loss by plugging membrane pores, adsorbing to the internal matrix of the membranes, and forming a cake of organic matter on the membrane surface. Usually, saline water from surface water sources contains NOM that causes moderate fouling of UF and MF membranes which can be removed by routine chemically enhanced backwash and periodic cleaning of the pretreatment membranes.

This NOM can be removed from UF and MF membranes with very little (typically less than 2 mg/L) or no coagulant addition to the saline source water. However, if the source water is influenced by surface runoff or a large amount of alluvial organics, or if an alluvial brackish aquifer is used for water supply, the NOM's properties and ability to cause significant membrane productivity loss may increase dramatically. Under these circumstances, the efficient removal of NOM may require very high dosages of coagulant (usually over 20 mg/L).

A NOM fractionation study on surface waters completed by the U.S. Bureau of Reclamation in 2002 indicates that this type of fouling depends on the type of membrane material and characteristics, the polarity and molecular weight of the NOM, and the chemistry of the source water.

The largest natural organic foulants are the polysaccharides, organic colloids, and proteins, followed in size by humic substances, organic acids, and low-molecular-weight organics of neutral charge. These compounds have different potentials to cause membrane fouling. Polysaccharides excreted by living bacteria on the membrane surface have the highest potential to cause RO membrane biofouling, and therefore they are classified as a separate group of foulants—microbial foulants.

2.10.2 Parameters and Measurement Methods

The most frequently used parameters for quantification of organic fouling are total organic carbon, UV_{254}, and biofilm formation rate.

Total Organic Carbon

Total organic carbon (TOC) is one of the most widely used measures for the organic content of saline source water. TOC concentration measures the content of both NOM and easily biodegradable organics, such as polysaccharides, released during algal blooms. This water quality parameter is widely used because it is relatively easy to measure and it is indicative of the tendency of the source water to cause organic fouling and biofouling of an RO membrane. TOC is measured by converting organic carbon to carbon dioxide in a high-temperature furnace in the presence of a catalyst.

Typically, open ocean seawater which is not influenced by surface freshwater influx (a nearby river confluence), human activities (i.e., wastewater or storm water discharges or ship traffic), or algal bloom events (i.e., red tide) has a very low TOC content (≤ 0.2 mg/L). When an algal bloom occurs, however, the TOC concentration of ocean water can increase by an order of magnitude or more (2 to 12 mg/L). A similar magnitude of TOC increase could be triggered by a storm water or river discharge during a high-intensity rain event, such as those that occur during rainy seasons in tropical and equatorial parts of the world. Usually, an increase of TOC content in the source water above a certain threshold (2.0 to 2.5 mg/L) triggers accelerated biofouling of RO membranes.

Observations at the Carlsbad seawater desalination demonstration plant in California, USA (which is supplied by seawater collected using near-shore open ocean intake), indicate that when TOC concentration in the source water at that location exceeds 2.0 mg/L during algal bloom events, within a one- to two-week period the SWRO system experiences measurable biofouling and an associated increase in operating pressure.

Similar TOC-level observations at the Tampa seawater desalination plant in Florida, USA (where the typical background TOC level of the seawater is less than 4 mg/L) indicate that accelerated biofouling occurs when the TOC concentration exceeds 6 to 8 mg/L. Usually, accelerated biofouling at the Tampa facility is triggered by one of two key events—rain events, which increase the content of alluvial organics in the source seawater, or algal blooms, which cause elevated organic levels due to massive dying off of algae. The increase in alluvial organics during rain events is caused by the elevated flow and alluvial content of the Alafia River, which discharges into Tampa Bay several kilometers upstream of the desalination plant's intake. During high-intensity rains in the summer months, the TOC level in the river water discharging in the bay may exceed 20 mg/L.

Analysis of various sources of seawater (Leparc et al., 2007) indicates that TOC in seawater may contain various fractions of organics, depending on the origin of the water and the type of seawater intake (Table 2.7). These fractions may also change depending on the season. Review of Table 2.7 leads to the conclusion that low-molecular-weight organic compounds are typically the greatest fraction of the TOC in seawater (at least 40 percent). Most of these compounds, however, have limited fouling potential; therefore, the higher the percentage of low-molecular-weight compounds in the source water, the lower the water's fouling potential is. A combination in a source water of a high percentage of compounds from the "other low-molecular-weight" category, low TOC, and low polysaccharide content is a clear indication of low fouling potential. Based on this rule of thumb, the well seawater at the Gibraltar SWRO facility would have the lowest fouling potential among all sources listed in Table 2.7.

Seawater Source	Total Organic Carbon (TOC), mg/L	Polysaccharides, % of Total TOC	Humic Substances and Building Blocks, % of Total TOC	Low-Molecular-Weight Acids and Neutrals, % of Total TOC	Other Low-Molecular-Weight Compounds, % of Total TOC
Surface raw seawater—Perth, Australia	0.9	3	30.1	25	41
Surface raw seawater—Ashkelon, Israel (May 2005)	1.2	14	39	25	22
Surface raw seawater—Ashkelon, Israel (Nov. 2005)	1.0	7	52	22	19
Surface raw seawater—Carboneras, Spain	0.9	8	31.1	18	42
Well seawater—Gibraltar, Spain	0.6	1	26	22	51
Surface seawater—Gibraltar, UK	0.8	5	28	25	42

Table 2.7 TOC Content and Fractions of Various Seawater Sources

Comparison of the data from the Ashkelon seawater desalination plant in Israel indicates that the most easily biodegradable organics (polysaccharides) change seasonally, increasing during the summer season along with the content of algal biomass in the Mediterranean Sea. These data also show that the TOC concentration of seawater may not always correlate with the content of polysaccharides in the water. In this case, humic substances are the main contributor of fouling of the SWRO membranes at the Ashkelon plant.

UV_{254} Absorbance

The *ultraviolet* (UV) absorbance of a seawater sample at 254 nm is an indirect measure of NOM. The UV_{254} absorbance of a saline water sample is determined by filtering the sample through a 0.45-μm filter and measuring the filtrate's absorbance of UV with a spectrophotometer. This measurement is based on the fact that specific molecular structures (chromophores) within the NOM molecules absorb UV light.

Because the NOM composition may vary from one water source to another, UV_{254} absorbance is not always easy to use for comparing the fouling potentials of different water sources. In addition, this parameter may not be reflective of the content of microbial foulants if the NOM contained in the source water is not easily biodegradable.

Specific UV absorbance (SUVA) is defined as the UV absorbance divided by the concentration of dissolved organic carbon in the source water. Edzwald and Haarhoff (2011) indicate that SUVA can be used as an indirect indicator of the occurrence of algal blooms in the source water. If the SUVA is higher than 4, then NOM in the source water consists predominantly of aquatic humic matter and it is not exhibiting algal bloom conditions. If the SUVA is between 2 and 4, the source water's NOM is a mix of assimilable organic matter (AOM) and aquatic humic matter, and the source water body from which the water originates is in the early stages of an algal bloom. When the SUVA is less than 2, the NOM in the source water consists predominantly of AOM and the source water body is experiencing an algal bloom.

2.10.3 Threshold Levels of Organic Foulants

Table 2.8 presents threshold levels of key water quality parameters used for assessing the organic fouling potential of saline source water.

Source Water Quality Parameter	Pretreatment Issues and Considerations
Total organic carbon, mg/L	If this parameter is below 0.5 mg/L, biofouling is unlikely. Above 2 mg/L, biofouling is very likely.
UV_{254}, cm^{-1}	If this value is below 0.5 cm^{-1}, the saline source water has low potential for organic fouling and biofouling.
Specific UV absorbance	If this is greater than 4, the source water is dominated by aquatic humic matter and biofouling is unlikely. If it is less than 2, the source water is experiencing algal bloom and biofouling is likely.

TABLE 2.8 Water Quality Parameters for Characterization of Organic Foulants

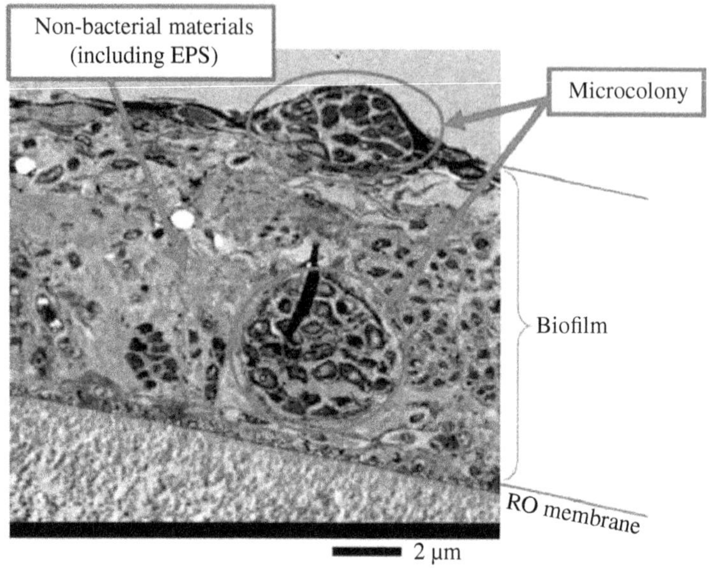

FIGURE 2.3 Biofilm on the surface of an RO membrane. (*Source: Nitto Denko.*)

2.11 Microbial Foulants

2.11.1 Description

Microbial foulants are aquatic microorganisms and organic compounds excreted by them (i.e., extracellular polymeric substances, proteins, and lipids) which are deposited on the surface of RO membranes. The biofilm formed on the membrane surface (Fig. 2.3) contributes additional resistance (pressure head losses) to the osmotic pressure that must be overcome in order to maintain steady production of freshwater by the membrane elements (Konishi et al., 2011).

Recent research indicates that a biofilm accumulated on the surface of RO membranes can cause performance decline by increasing the hydraulic resistance of the membranes and by a "cake enhanced osmotic pressure" effect (Herzberg and Elimelech, 2007). Therefore, if a microbial cake layer is formed on the surface of the membranes, membrane productivity (flux) declines and membrane salt passage increases over time. In order to compensate for the loss of productivity due to biofouling, the feed pressure of the RO membrane system would need to be increased, which in turn would result in elevated energy use to produce the same volume of freshwater. Feed pressure increase beyond a certain level may cause irreversible damage to the membrane structure and ultimately may result in the need to replace all RO membrane elements.

Although bacteria constitute the majority of the membrane biofilm, other microorganisms such as fungi, algae, and protozoa can also attach to the membrane surface and contribute to biofilm formation. Usually the most predominant bacteria causing biofouling are *Pseudomonas, Bacillus, Arthrobacter, Corynebacterium, Flavobacterium,* and *Aeromonas.* Other microorganisms such as fungi (e.g., *Penicillium, Trichoderma, Mucor,*

Fusarium, and *Aspergillus*) are typically present in the membrane biofilm in significantly lower levels than bacteria.

Biofouling is usually a significant operational challenge for saline waters of naturally elevated organic content and temperature (such as the seawater in the Middle East). Biofouling is also a challenge during intense algal blooms (i.e., red tides) or periods when surface runoff from rain precipitation or nearby river water of high organic content enters the plant's open intake. The biofouling potential of a given source water depends on many factors, including: (1) the concentration and speciation of microorganisms contained in the source water, (2) the content of easily biodegradable compounds in the water, (3) the concentration of nutrients and the balance (ratio) between organic compounds and the biologically available nitrogen and phosphorus in the source water, and (4) the source water temperature.

Bacteria contained in brackish water and seawater typically exist in two states—metabolically active and inactive. The active state of bacterial cells is characterized by fast growth and the formation of extracellular material and bacterial colonies that can accumulate on an RO membrane surface. The inactive state of existence of aquatic bacteria is characterized by low metabolic and growth rates; the cells appear in the form of single cells or small cell clusters that behave as microparticles and have a protective cellular cover which allows them to survive unfavorable environmental conditions such as a low content of food and oxygen or a high content of harmful substances such as chlorine and other biocides. At any given time, some of the aquatic bacteria naturally occurring in surface water bodies are in an active state and others are in an inactive state.

The predominant state of aquatic bacteria (active or inactive) depends on how favorable the ambient environment is for bacterial survival and growth. Most aquatic bacteria will transfer from an inactive to an active state under favorable environmental conditions, such as algal bloom events, when high concentrations of easily biodegradable organics released from the decaying algal biomass (which serve as food to these bacteria) are readily available in the source water. Since bacteria can attach on a membrane surface and grow colonies there at a very high rate, red tide or other intense algal bloom events are usually the most frequent cause of RO membrane biofouling, especially in seawater desalination plants.

The membrane biofouling process (i.e., the formation of a microbial cake layer on the surface of an RO membrane) usually follows several key steps: (1) formation of a primary organic conditioning film, (2) attachment of colonizing bacteria, (3) formation of a biopolymer matrix, (4) establishment of a mature secondary biofilm, and (5) biofilm equilibrium and die-off. The primary organic conditioning film is a microthin layer on the surface of the membrane that is rich in nutrients and easily biodegradable organics, and that creates suitable conditions for bacteria to convert from an inactive (particulatelike) state into an active state. In this state they are capable of producing extracellular polysaccharides, which are adhesive substances that allow bacteria to attach to the membrane surface and to each other.

During the first step of the biofilm formation process, active bacteria adsorb to only 10 to 15 percent of the membrane surface, but they multiply at an exponential rate, and within 5 to 15 days they colonize the entire membrane surface and form a biopolymer matrix layer that is several micrometers thick. The mucoid biopolymer matrix formed on the membrane surface entraps organic molecules, colloidal particles, suspended solids, and cells of other microorganisms (fungi, microalgae, etc.) over time to form a thicker cake with a higher resistance to permeate flow.

In order for biological fouling to occur, aquatic bacteria need to have suitable low-velocity conditions to attach to the RO membrane surface or to the surface of facilities upstream of the membrane system, such as cartridge filters; or they need low-velocity cavities or vessels along the feed water route to the RO membrane system, such as dead-end valves, fittings, or oversized or hydraulically flawed cartridge filter housings.

The formation of a permanent cake layer occurs when membrane flux exceeds a certain level (critical flux) at which aquatic microorganisms can attach to the RO membrane surface (Winters et al., 2007). When critical flux through the membrane is reached, the velocity of the feed water/concentrate flow along the surface of the membrane (cross-flow velocity) drops low enough to allow colonizing bacteria to attach to the membrane surface.

The critical flux for aquatic bacteria is dependent upon the cross-flow velocity and increases with the increase of this velocity. The most widely used operational approach to increase cross-flow velocity is to reduce RO system recovery. Operating at lower recovery leaves more flow on the concentrate/feed side of the membranes, which in turn creates a higher scouring velocity on the membrane surface that deters microorganisms from attaching to the surface.

The critical flux is also a function of the concentration of active bacteria in the source water; and it decreases as the concentration of bacteria rises. The concentration of active aquatic bacteria in turn mainly depends on the type of bacteria, the availability of easily biodegradable organic matter in the source water, and the water temperature. For a given SWRO system, decreasing recovery from 50 to 35 percent would result in approximately two times lower fouling potential for system operation in a typical flux range of 13.5 to 18.0 $L/(m^2{\cdot}h)$ [8.0 to 10.5 $gal/(ft^2{\cdot}d)$ (gfd)].

Although operation at low recovery may be attractive from the point of view of minimizing membrane biofouling, designing RO plants for low recovery is usually not cost effective because of the associated increased size of the desalination plant intake, pretreatment, and RO systems, and the 30 to 40 percent higher capital costs. Therefore, other approaches for biofouling reduction—such as control of the organic content in the source water and inactivation of aquatic bacteria by disinfection or UV irradiation—have found wider practical application.

The source of biofouling may not only be a natural event (such as an algal bloom) that triggers an increase in the content of easily biodegradable organics in the source water; it may also be the type and operation of the pretreatment processes and systems used upstream of the RO facility. One reason for accelerated biofouling could be continuous chlorination of the source water, which often is applied to inactivate aquatic microorganisms and reduce biofouling. Since chlorine is a strong oxidant, it can destroy the cells of active aquatic bacteria and algae which naturally occur in the source water at any given time.

The destroyed algal and bacterial cells release easily biodegradable organic compounds (such as polysaccharides) in the ambient water, which become food for the remaining aquatic bacteria that have survived chlorination by being in an inactive state. If the concentration of these organics reaches a certain threshold, it could trigger the conversion of these surviving bacteria from an inactive to an active state, followed by their attachment and excessive growth on the RO membrane surface, which in turn would manifest as membrane biofouling. Therefore, continuous chlorination often creates more membrane biofouling problems than it solves. On the other hand, intermittent chlorination has been found to provide effective control of microbial growth without generating a steady influx of easily biodegradable organics that can trigger a large-scale transfer of aquatic bacteria from an inactive to an active state of existence.

Another pretreatment technology that could potentially cause biofouling, especially during periods of severe algal bloom events, is the use of pressure-driven granular media filters, UF, or MF membrane filters for pretreatment. Although pressure filters provide effective removal of particulate and colloidal foulants, the high filtration driving pressure applied by these systems could break some of the algal cells in the source water and cause the release of easily biodegradable organics, which in turn could result in accelerated RO membrane biofouling. Examples of seawater algal species susceptible to cell breakage as a result of relatively low pressure (0.3 to 0.6 bar; 4 to 8 lb/in^2) are shown in Fig. 2.4.

From the point of view of minimizing biofouling associated with algal cell breakage, the most suitable pretreatment technologies are those that provide a gentle removal of the algal cells in the source water, such as downflow gravity granular media filtration and dissolved air flotation.

Another potential source of biofouling is the use of impure source water conditioning chemicals, such as antiscalants, polymers, or acids (Vrouwenvelder and Van der Krooij, 2008). Therefore, it is important to analyze these chemicals for easily biodegradable organic content.

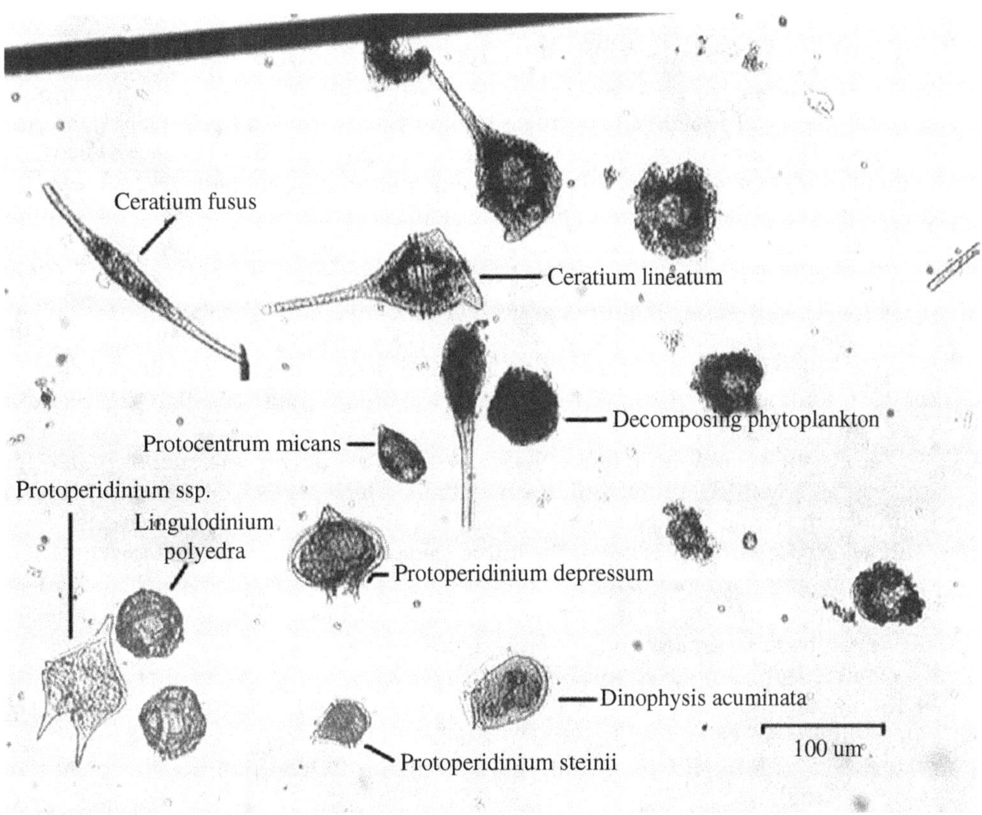

FIGURE 2.4 Seawater algal species susceptible to cell breakage.

2.11.2 Parameters and Measurement Methods

Biofilm Formation Rate

The *biofilm formation rate* (BFR) is an online monitoring tool that allows operators to measure the accumulation of biomass on the surface of a glass ring as a function of time; it is measured in picograms of adenosine triphosphate (ATP) per square centimeter (pg-ATP/cm^2; Vrouwenvelder and Van der Krooij, 2008). If BFR exceeds 120 pg-ATP/cm^2 and AOC is higher than 80 μg/L, severe biofouling is expected to occur. For a source water with BFR lower than 1 pg-ATP/cm^2, membrane biofouling is not expected to be a challenge. Veza and colleagues (2008) have found a correlation between the concentration of ATP biomass and the heterotrophic plate count of seawater.

2.11.3 Threshold Levels of Microbial Foulants

Table 2.8 includes a summary of measures commonly used to assess biofouling caused by microbial foulants. The most commonly measured source water quality parameters for characterization of biofouling caused by bacterial foulants are TOC and UV_{254}. While other parameters such as BFR have been developed, they are not widely used because of the complexity of the source water quality analysis and the monitoring equipment needed to complete the tests.

2.12 Combined Impacts of Various Types of Foulants

Membrane fouling is a complex process which often results from the additive impacts of several types of foulants. Particulate fouling usually causes a relatively quick and definitive deterioration of membrane permeate flux and RO system productivity, without significant impact on salt rejection. For comparison, colloidal fouling typically causes a marked deterioration in the RO system's salt rejection over time. In addition to deterioration of salt rejection, colloidal fouling also results in permeate flux decline over time, which is caused not only by the accumulation of a flow-resistant cake layer of colloidal particles on the surface of the membrane but also by backfusion of salt ions within the colloidal cake, which results in elevated salt concentration and osmotic pressure at the membrane surface; this in turn decreases the net driving pressure. In contrast, fouling caused by NOM is accompanied by almost constant salt rejection over time.

Because of its complex nature, NOM often creates diverse interactions with other foulants and with the surface of the membrane elements. As indicated previously, hydrophobic humic substances are typically a major foulant in source seawater, especially when the seawater is under the influence of river discharge. The high content of calcium in seawater reduces the solubility of humic acids and increases their aggregation, which in turn accelerates the accumulation of these NOM compounds on the surface of the RO membranes.

The calcium complexation of NOM in this case often forms a gel on the surface of the membranes, which is very difficult to remove. Therefore, source seawater originating from an area of river confluence into the ocean—especially if the river water were high in NOM—would be very difficult to treat and use for membrane desalination.

Biofouling of the feed channels and spacers of spiral-wound membranes (Fig. 2.5) typically results in a significant increase in the membrane differential pressure (i.e., the

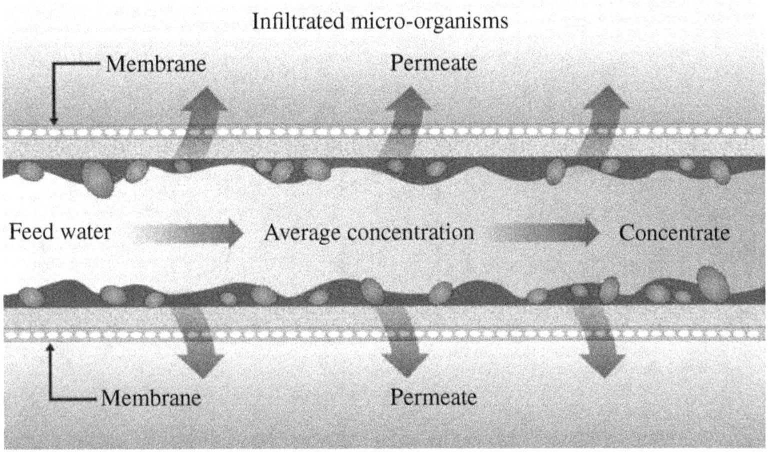

Figure 2.5 Biofilm accumulation in the spacer of an RO membrane.

pressure difference between the feed and concentrate sides of the membranes) over a very short period of time (one to several weeks).

As biofilm-forming bacteria colonize the RO membrane surface, they often block sections of the feed channels and spacers between the membrane leaves over time, so that the flow pattern within the membrane elements changes and the feed flow is completely blocked in some portions of the feed channels and increased in others. Flow channeling caused by random blockages of the feed channels and spacers results in a sharp increase in salt concentration in the affected areas, which in turn triggers precipitation of sparingly soluble salts such as calcium carbonate and sulfate in the feed channels and further exacerbates and accelerates the membrane fouling problem.

2.13 Membrane Fouling Diagnostics

2.13.1 Laboratory Autopsy of Biofouled Membranes

A laboratory autopsy is a method of characterizing the fouling nature of source water constituents that is applied when the fouling process has already taken place; it is typically used when the source water contains a number of different types of foulants and it is difficult or practically impossible to determine the source of fouling by analyses of source water quality and RO system performance only. Standard laboratory autopsy investigation involves the dissection of a fouled membrane element and physical observation of the condition of the membrane leaves and spacer (Fig. 2.6).

A sample is collected from the front end of the membrane for microbial and chemical analysis. Results are expressed in colony-forming units per square centimeter (cfu/cm^2) and by foulant composition, respectively.

The moisture content and the chemical composition of the dried deposit of the collected biofilm sample are expressed as a percentage of the foulant. Often, samples

FIGURE 2.6 Autopsy of a fouled RO membrane.

are inspected using scanning electron microscopy and surface analysis such as x-ray photoelectron spectroscopy, especially if the predominant fouling is of a complex multicause nature.

2.14 Water Quality Analysis for RO Desalination

Based on the discussion of the factors influencing RO system performance presented in the previous sections, it is recommended that, at a minimum, the source water for a given RO desalination project be analyzed for the parameters listed in Table 2.9.

Besides key minerals and parameters associated with various types of membrane fouling, Table 2.9 also contains water quality characteristics that may have an impact on the performance, integrity, and longevity of the RO membranes, such as pH, free chlorine, ammonia, and oxidation reduction potential; metals which if present above certain thresholds could induce membrane oxidation and integrity loss, such as copper, and iron. The list presented in Table 2.9 is not all-inclusive. Additional source water quality parameters may need to be measured, depending on the project and site-specific regulatory requirements governing source water intake, product water quality, and concentrate discharge.

Parameter		Unit
Key Minerals		**mg/L**
Cations	**Anions**	
Sodium	Chloride	
Magnesium	Carbonate	
Calcium	Bicarbonate	
Potassium	Sulfate	
Barium	Nitrate	
Strontium	Fluoride	
Boron	Phosphate	
Bromide	Sulfur	
Other Key RO System Design-Related Parameters		
Salinity		mg/L
Conductivity		µS/cm
Temperature		°C
pH		units
Total alkalinity		mg/L as $CaCO_3$
Total hardness		mg/L as $CaCO_3$
Iron in reduced form (Fe^{2+}), mg/L		Foulant if > 0.05 mg/L
Iron in oxidized form, mg/L		Foulant if > 2.0 mg/L Membrane damage by chlorine if > 0.5 mg/L
Manganese, mg/L		Foulant if > 0.02 mg/L
Aluminum, mg/L		Foulant if > 0.1 mg/L
Copper, µg/L		Potential membrane damage if > 50 µg/L
Arsenic		Potential toxicant if > 10 µm/L in permeate
Turbidity		Accelerated fouling if > 0.1 NTU
Total suspended solids (TSS), mg/L		Accelerated fouling if > 1 mg/L
Silt density index (SDI)		Accelerated fouling if > 5
Total hydrocarbons, mg/L		Foulant if > 0.02 mg/L
Silica (colloidal), mg/L		Foulant if > 100 mg/L in concentrate
Total organic carbon (TOC), mg/L		Potential for accelerated fouling if > 2 mg/L
UV_{254}, cm^{-1}		Potential for accelerated fouling if > 0.5 cm^{-1}
Total algal count, algal cells per milliliter		Algal bloom if > 2000 algal cells per milliliter
Hydrogen sulfide		Odor and membrane fouling if > 0.1 mg/L.
Ammonia, mg/L		Membrane damage if bromide > 0.4 mg/L
Free chlorine, mg/L		Membrane damage if > 0.01 mg/L
Oxidation reduction potential, mV		Membrane damage if > 250 mV
Total coliform count, Most Probable Number (MPN) per 100 mL		Potential pathogen contamination if > 10^4

Table 2.9 Source Water Quality Analysis for Reverse Osmosis Desalination

2.15 References

American Water Works Association, *Manual of Water Supply Practices, M46, Reverse Osmosis and Nanofiltration*, 2d ed., American Water Works Association, Washington, D.C., 2007.

Boerlage, S. F. E., "Understanding SDI and Modified Fouling Indices ($MFI_{0.45}$ and MFL_{UF})," *Proceedings of IDA World Congress*, Masspalomas, Gran Canaria, Spain, October 21–26, 2007. Ref: IDAWC/MP07-143.

Chemical Pretreatment for RO and NF, Technical Application Bulletin No. 111, Revision C, Hydranautics, Oceanside, California, 2008.

Edzwald, J. K., and J. Haarhoff, "Seawater Pretreatment for Reverse Osmosis: Chemistry, Contaminants and Coagulation," *Water Research*, 45 (17): 5428–5440, 2011.

Herzberg, M., and M. Elimelech, "Biofouling of Reverse Osmosis Membranes: Role of Biofilm-Enhanced Osmotic Pressure," *Journal of Membrane Science*, 295: 11–20, 2007.

Khirani, S., B. R. Aim, and M. H. Manero, "Improving the Measurement of the Modified Fouling Index Using Nanofiltration Membranes (NF-MFI)," *Desalination*, 191: 1–7, 2006.

Konishi, T., K. Maruyama, M. Namba, M. Kobuke, M. Nozoe, E. Liu, T. Akiyama, et al., "LD (Low-Differential-Pressure) Technology for Biofouling Control—Design Concept Supported by Microbiological Analysis," *Proceedings of IDA World Congress*, Perth, Western Australia, September 4–9, 2011. Ref: IDAWC/PER-081, 1–9.

Leparc, J., S. Rapenne, C. Courties, P. Lebaron, J. P. Croute, V. Jacquemet, and G. Tyrner, "Water Quality Performance Evaluation at Seawater Reverse Osmosis Plants Through The Use of Advanced Analytical Tools" *Desalination*, 203: 243–255, 2007.

Mosset, A., V. Bonnelyle, M. Petry, and M. A. Sanz, "The Sensitivity of SDI Analysis: From RO Feed Water to Raw Water," *Desalination*, 222: 17–23, 2008.

US Bureau of Reclamation, *Membrane Fouling: Influence of Natural Organic Matter Properties and Membrane Surface Treatment on Nanofiltration Treatment*, Desalination and Research Development Report No. 83, US Bureau of Reclamation Washington, D.C., 2002.

Veza, J. M., M. Ortiz, J. J. Sadhwani, J. E. Gonzalez, and F. J. Santana, "Measurement of Biofouling in Seawater: Some Practical Tests," *Desalination*, 220: 326–334, 2008.

Vrouwenvelder, J. S., and D. Van der Krooij, "Diagnosis, Prediction, and Prevention of Biofouling on NF and RO Membranes," *Desalination*, 139: 65–71, 2008.

Winters, H. S., A. Fane, and B. Nicolaisen, "Microbial Membrane Fouling; Microbial Cake Enhanced Osmotic Pressure (CEOP) or Thermodynamic Restriction?," *Proceedings of IDA World Congress*, Masspalomas, Gran Canaria, Spain, October 21–26, 2007. Ref: IDAWC/MP07-267.

CHAPTER 3

Fundamentals of Reverse Osmosis Desalination

3.1 Introduction

If water of high salinity is separated from water of low salinity via a semipermeable membrane, a natural process of transfer of water will occur from the low-salinity side to the high-salinity side of the membrane until the salinity on both sides reaches the same concentration. This natural process of water transfer through a membrane driven by the salinity gradient occurs in every living cell; it is known as *osmosis*.

The hydraulic pressure applied on the membrane by the water during its transfer from the low-salinity side of the membrane to the high-salinity side is termed *osmotic pressure*. Osmotic pressure is a natural force similar to gravity and is proportional to the difference in concentration of *total dissolved solids* (TDS) on both sides of the membrane, the source water temperature, and the types of ions that form the TDS content of the source water. This pressure is independent of the type of membrane itself.

In order to remove fresh (low-salinity) water from a high-salinity source water using membrane separation, the natural osmosis-driven movement of water must be reversed, i.e., the freshwater has to be transferred from the high-salinity side of the membrane to the low-salinity side. For this reversal of the natural direction of freshwater flow to occur, the high-salinity source water must be pressurized at a level higher than the naturally occurring osmotic pressure (Fig. 3.1). If the high-salinity source water is continuously pressurized at a level higher than the osmotic pressure and the pressure losses for water transfer through the membrane, a steady-state flow of freshwater from the high-salinity side of the membrane to the low-salinity side will occur, resulting in a process of salt rejection and accumulation on one side of the membrane and freshwater production on the other. This process of forced movement of water through a membrane in the opposite direction to the osmotic force driven by the salinity gradient is known as *reverse osmosis* (RO).

The rate of water transport through the membrane is several orders of magnitude higher than the rate of passage of salts. This significant difference between water and salt passage rates allows membrane systems to produce freshwater of very low mineral content.

The applied feed water pressure counters the osmotic pressure and overcomes the pressure losses that occur when the water travels through the membrane, thereby keeping the freshwater on the low-salinity (permeate) side of the membrane until this water exits the membrane vessel. The salts contained on the source water (influent) side of the

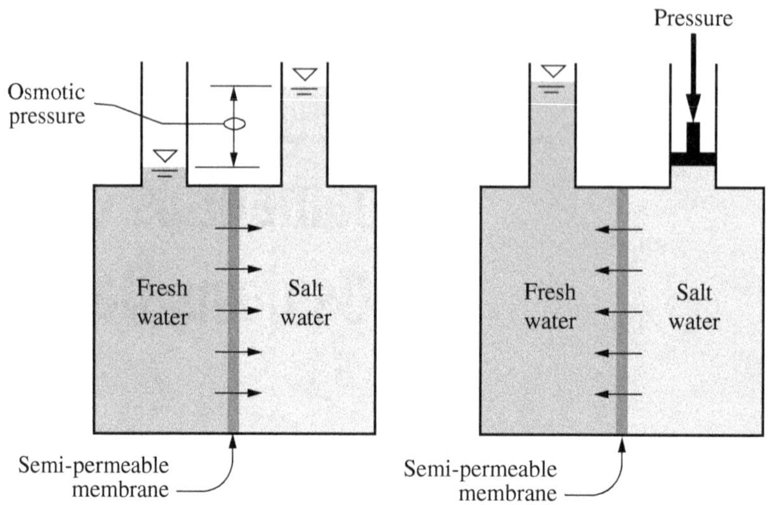

FIGURE 3.1 Osmosis and reverse osmosis.

membrane are retained and concentrated; they are ultimately evacuated from the membrane vessel for disposal. As a result, the RO process results in two streams—one of freshwater of low salinity (permeate) and one of feed source water of elevated salinity (concentrate, brine or retentate), as shown in Fig. 3.2.

While semipermeable RO membranes reject all suspended solids, they are not an absolute barrier to dissolved solids (minerals and organics alike). Some passage of dissolved solids will accompany the passage of freshwater through the membrane. The rates of water and salt passage are the two key performance characteristics of RO membranes.

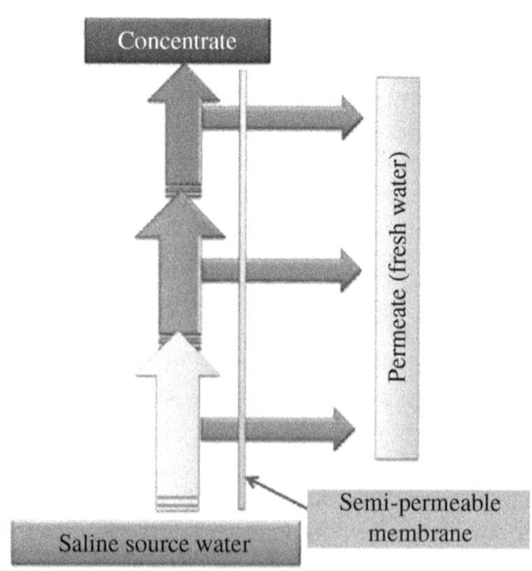

FIGURE 3.2 Reverse osmosis process.

3.2 Reverse Osmosis Membrane Structures, and Materials

Reverse osmosis membranes differ by the material of the membrane polymer and by structure and configuration. Based on their structure, membranes can be divided into two groups: conventional thin-film composite and thin-film nanocomposite. Based on the thin-film material, conventional membranes at present are classified into two main groups: polyamide and cellulose acetate. Depending on the configuration of the membranes within the actual membrane elements (modules), RO membranes are divided into three main groups: spiral-wound, hollow-fiber, and flat-sheet (plate-and-frame).

3.2.1 Conventional Thin-Film Composite Membrane Structure

The reverse osmosis membranes most widely used for desalination at present are composed of a semipermeable thin film (0.2 μm), made of either *aromatic polyamide* (PA) or *cellulose acetate* (CA), which is supported by a 0.025- to 0.050-mm microporous layer that in turn is cast on a layer of reinforcing fabric (Fig. 3.3 for a membrane with an ultrathin PA film). The 0.2-μm ultrathin polymeric film is the feature that gives the RO membrane its salt rejection abilities and characteristics. The main functions of the two support layers underneath the thin film are to reinforce the membrane structure and to maintain membrane integrity and durability.

The dense semipermeable polymer film is of a random molecular structure (matrix) that does not have pores. Water molecules are transported through the membrane film by diffusion and travel on a multidimensional curvilinear path within the randomly structured molecular polymer film matrix.

While the thin-film RO membrane with conventional random matrix-based structure shown in Fig. 3.3 is the type of membrane that dominates the desalination industry,

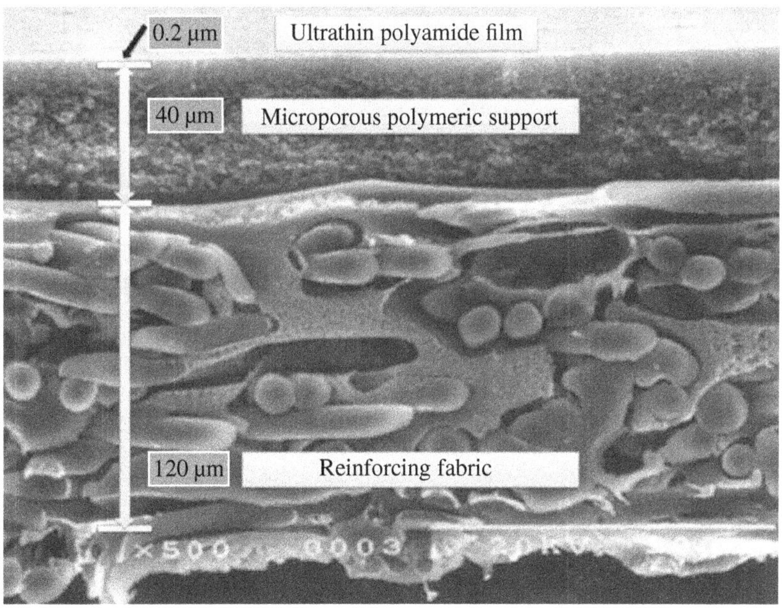

Figure 3.3 Structure of a typical RO membrane.

new thin-film membranes of more permeable structure are currently under development in research centers worldwide.

3.2.2 Thin-Film Nanocomposite Membrane Structure

Nanocomposite membranes either incorporate inorganic nanoparticles within the traditional membrane polymeric film structure (Fig. 3.4) or are made of highly structured porous film consisting of a densely packed array of nanotubes (Fig. 3.5). In Fig. 3.4, part A shows the thin film of a conventional PA membrane, supported by the polysulfone

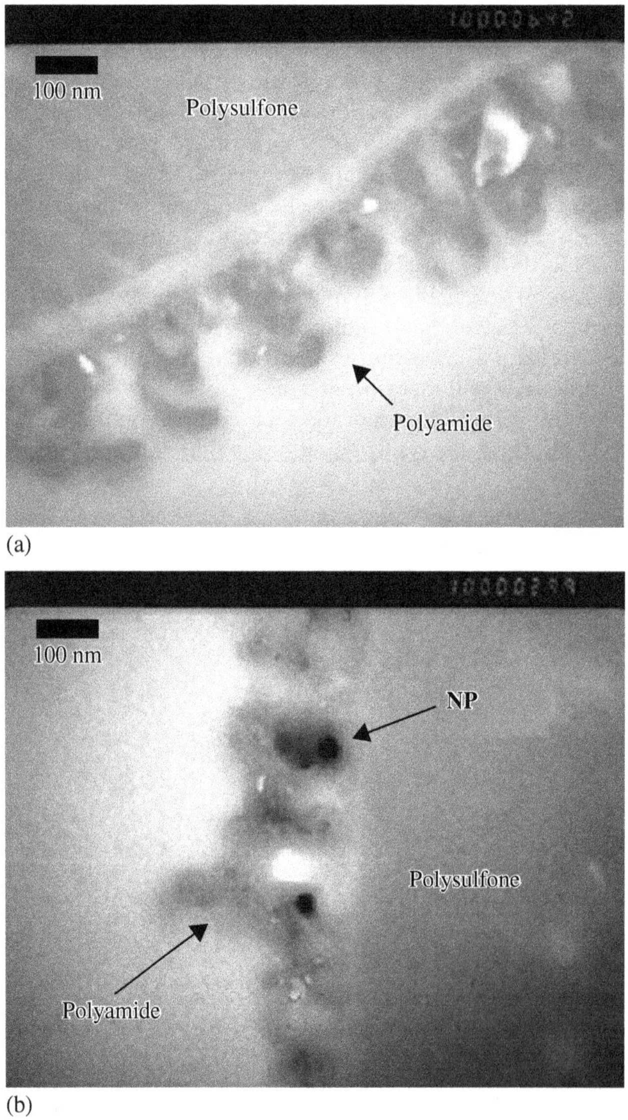

FIGURE 3.4 Polyamide RO membrane with nanoparticles.

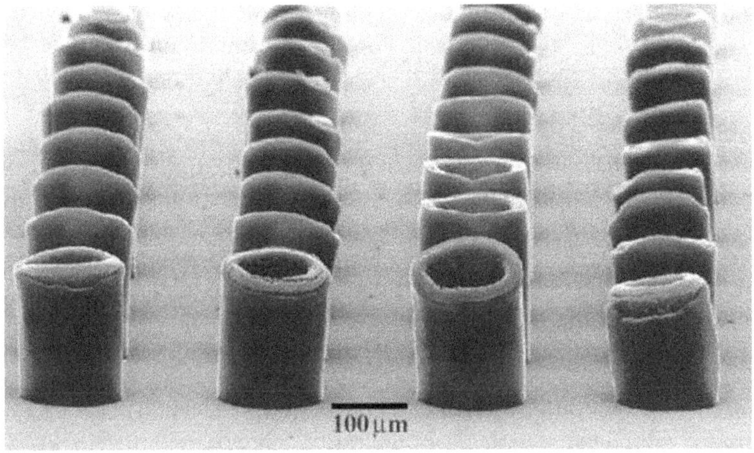

Figure 3.5 Membrane with carbon nanotubes.

support layer. Part B shows the same type of membrane with embedded nanoparticles (labeled "NP").

Nanocomposite membranes reportedly have a higher specific permeability (i.e., ability to transport more water through the same surface area at the same applied pressure) than conventional RO membranes at comparable salt rejection. In addition, thin-film nanocomposite membranes have comparable or lower fouling rates in comparison to conventional thin-film composite RO membranes operating at the same conditions, and they can be designed for enhanced rejection selectivity of specific ions.

If membrane material science evolved to a point where the membrane structure could be made of tubes of completely uniform size, theoretically the membrane could produce up to 20 times more water per unit surface area than the RO membranes commercially available on the market today. As membrane material science evolves toward the development of membranes with more uniform structure, the further development of RO desalination membrane technology has the potential to yield measurable savings in terms of water production costs.

3.2.3 Cellulose Acetate Membranes

The thin semipermeable film of the first RO membranes—developed in the late 1950s at the University of California, Los Angeles—was made of cellulose acetate (CA) polymer. While CA membranes have a three-layer structure similar to that of PA membranes, the main structural difference is that the top two layers (the ultrathin film and the microporous polymeric support) are made of different forms of the same CA polymer. In PA membranes these two layers are made of completely different polymers—the thin semipermeable film is made of polyamide, while the microporous support is made of polysulfone (see Fig. 3.3). Similar to PA membranes, CA membranes have a film layer that is typically about 0.2 μm thick; but the thickness of the entire membrane (about 100 μm) is less than that of a PA membrane (about 160 μm).

One important benefit of CA membranes is that the surface has very little charge and is considered practically uncharged, as compared to PA membranes, which have negative charge and can be more easily fouled with cationic polymers if such polymers are used for source water pretreatment. In addition, CA membranes have a smoother surface than PA membranes, which also renders them less susceptible to fouling.

CA membranes have a number of limitations, including the ability to perform only within a narrow pH range of 4 to 6 and at temperatures below 35°C (95°F). Operation outside of this pH range results in accelerated membrane hydrolysis, while exposure to temperatures above 40°C (104°F) causes membrane compaction and failure. In order to maintain the RO concentrate pH below 6, the pH of the feed water to the CA membranes has to be reduced to between 5 and 5.5, which results in significant use of acid for normal plant operation and requires RO permeate adjustment by addition of a base (typically sodium hydroxide) to achieve adequate boron rejection.

CA membranes experience accelerated deterioration in the presence of microorganisms capable of producing cellulose enzymes and bioassimilating the membrane material. However, they can tolerate exposure to free chlorine concentration of up to 1.0 mg/L, which helps to decrease the rate of membrane integrity loss due to destruction by microbial activity.

Since CA membranes have a higher density than PA membranes, they create a higher headloss when the water flows through the membranes; therefore they have to be operated at higher feed pressures, which results in elevated energy expenditures.

Despite their disadvantages, and mainly because of their high tolerance to oxidants (chlorine, peroxide, etc.) as compared to PA membranes, CA membranes are used in municipal applications for saline waters with very high fouling potential (mainly in the Middle East and Japan) and for ultrapure water production in pharmaceutical and semiconductor industries.

3.2.4 Aromatic Polyamide Membranes

Aromatic polyamide (PA) membranes are the most widely used type of RO membranes at present. They have found numerous applications in both potable and industrial water production. The thin polyamide film of this type of semipermeable membrane is formed on the surface of the microporous polysulfone support layer (Fig. 3.3) by interfacial polymerization of monomers containing polyamine and immersed in solvent containing a reactant to form a highly cross-linked thin film.

PA membranes operate at lower pressures and have higher productivity (specific flux) and lower salt passage than CA membranes, which are the main reasons they have found a wider application at present. While CA membranes have a neutral charge, PA membranes have a negative charge when the pH is greater than 5, which amplifies co-ion repulsion and results in higher overall salt rejection. However, it should be noted that when the pH is lower than 4, the charge of a PA membrane changes to positive and rejection is reduced significantly, to lower than that of a CA membrane.

Another key advantage of PA membranes is that they can operate effectively in a much wider pH range (2 to 12), which allows easier maintenance and cleaning. In addition, PA membranes are not biodegradable and usually have a longer useful life—5 to 7 years versus 3 to 5 years. Aromatic polyamide membranes are used to produce membrane elements for brackish water and seawater desalination, and nanofiltration.

It should be noted that PA membranes are highly susceptible to degradation by oxidation of chlorine and other strong oxidants. For example, exposure to chlorine longer than 1000 mg/L-hour can cause permanent damage of the thin-film structure and can significantly and irreversibly reduce membrane performance in terms of salt

Parameter	Polyamide Membranes	Cellulose Acetate Membranes
Salt rejection	High (> 99.5%)	Lower (up to 95%)
Feed pressure	Lower (by 30 to 50%)	High
Surface charge	Negative (limits use of cationic pretreatment coagulants)	Neutral (no limitations on pretreatment coagulants)
Chlorine tolerance	Poor (up to 1000 mg/L-hours); feed dechlorination needed	Good; continuous feed of 1 to 2 mg/L of chlorine is acceptable
Maximum temperature of source water	High (40 to 45°C; 104 to 113°F)	Relatively low (30 to 35°C; 86 to 95°F)
Cleaning frequency	High (weeks to months)	Lower (months to years)
Pretreatment requirements	High (SDI < 4)	Lower (SDI < 5)
Salt, silica, and organics removal	High	Relatively low
Biogrowth on membrane surface	May cause performance problems	Limited; not a cause of performance problems
pH tolerance	High (2 to 12)	Limited (4 to 6)

Table 3.1 Comparison of Polyamide and Cellulose Acetate Membranes

rejection. Oxidants are widely used for biofouling control with RO and nanofiltration membranes; therefore, the feed water to PA membranes has to be dechlorinated prior to separation. A comparison of key parameters of polyamide and cellulose acetate RO membranes in terms of their sensitivity to feed water quality is presented in Table 3.1.

Mainly because of their higher membrane rejection and lower operating pressures, polyamide membranes are the choice for most RO membrane installations today. Exceptions are applications in the Middle East, where the source water is rich in organics and thus cellulose acetate membranes offer benefits in terms of limited membrane biofouling and reduced cleaning and pretreatment needs.

Because of the relatively lower unit power costs in the Middle East, cellulose acetate membranes provide an acceptable tradeoff between lower fouling rates and chemical cleaning costs on one hand and higher operating pressures and power demand on the other. However, as newer generations of lower-fouling PA membranes are being introduced on the market, the use of CA elements is likely to diminish in the future.

3.3 Spiral-Wound, Hollow-Fiber, and Flat-Sheet RO Membrane Elements

The CA and PA membranes described in the previous section are configured into commercially available membrane elements that pack a large surface area and have standard sizes and performance. The two most widely used configurations of membrane elements at present are spiral-wound and hollow-fiber.

Until the mid-1990s, hollow-fiber elements were the most prevalent technology used for desalination, but at present the marketplace is dominated by spiral-wound

RO membrane elements. Other configurations of membrane elements, such as tubular and plate-and-frame, have found application mainly in the food and dairy industries—they are practically never used in conventional municipal brackish or seawater desalination plants, because of their higher costs and equipment space requirements. Tubular membrane elements have very limited application at present and are not discussed in detail in this book.

3.3.1 Spiral-Wound RO Membrane Elements

Spiral-wound membrane elements (modules) are made of individual flat membrane sheets that have the three-layer structure described in the previous section (i.e., ultrathin CA or PA film; microporous polymeric support; and reinforcing fabric—see Fig. 3.3). A typical 8-in.-diameter spiral-wound RO membrane element has 40 to 42 flat membrane sheets.

The flat sheets are assembled into 20 to 21 membrane envelopes (leafs), each of which consists of two sheets separated by a thin plastic net (referred to as a *permeate spacer*) to form a channel that allows evacuation of the permeate separated from the saline source water by the flat sheets (permeate carrier). Three of the four sides of the two-membrane flat-sheet envelope are sealed with glue and the fourth side is left open (Fig. 3.6). The membrane leafs are separated by a feed spacer approximately 0.7 or 0.9 mm (28 or 34 mils) thick, which forms feed channels and facilitates the mixing and conveyance of the feed-concentrate stream along the length of the membrane element (Fig. 3.7). Membranes with the wider 34-mil spacers have been introduced relatively recently and are more suitable for highly fouling waters. In order to accommodate the wider spacers, fewer membrane leafs are installed within the same RO membrane module, which results in a tradeoff between reduced membrane fouling and lower membrane element productivity.

Pressurized saline feed water is applied on the outside surface of the envelope; permeate is collected in the space inside the envelope between the two sheets and

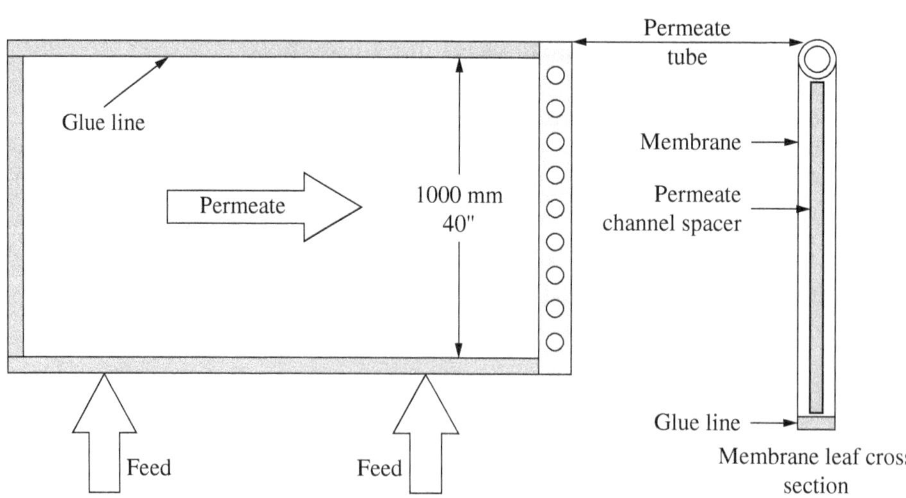

FIGURE 3.6 Flat-sheet membrane envelope. (*Source: Hydranautics.*)

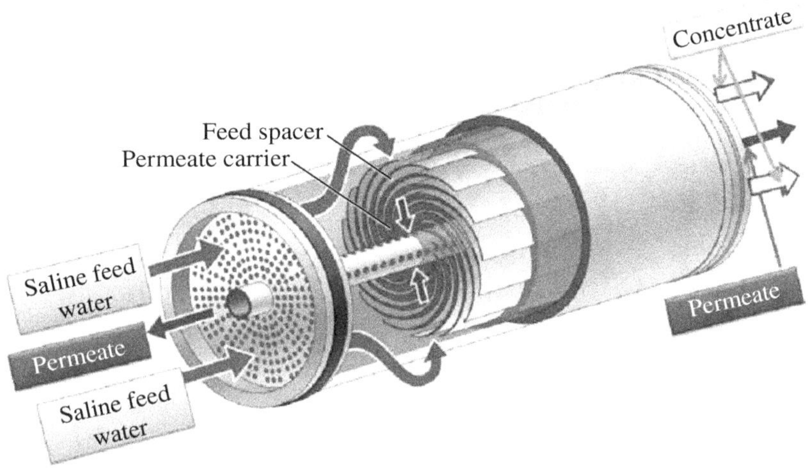

FIGURE 3.7 Spiral-wound membrane element.

directed toward the fourth, open edge of the envelope, which is connected to a central permeate collector tube. This collector tube receives desalinated water (permeate) from all flat-sheet leaves (envelopes) contained in the membrane element and evacuates it out of the element.

The assembly of flat-sheet membrane leafs and separating spacers is wrapped (rolled) around the perforated permeate collector tube. The membrane leafs are kept in the spiral-wound assembly with a tape wrapped around them and contained by an outer fiberglass shell. The two ends of each RO element are finished with plastic caps referred to as *end caps, antitelescoping devices,* or *seal carriers*. The plastic caps are perforated in a pattern that allows even distribution of the saline feed flow among all membrane leafs in the element (Fig. 3.8). The plastic caps' flow distribution pattern varies between membrane manufacturers.

The reason the plastic caps are often also referred to as *seal carriers* is that one of their functions is to carry a chevron-type U-cup-style rubber brine seal that closes the space between the membrane and the pressure vessel in which the membrane is installed. This seal prevents the feed water from bypassing the RO element (Fig. 3.9).

The source water flow is introduced from one end of the element and travels in a straight tangential path on the surface of the membrane envelopes and along the length of the membrane element (see Figs. 3.6 and 3.7). A portion of the feed flow permeates through the membrane and is collected on the other side of the membrane as freshwater. The separated salts remain on the feed side of the membrane and are mixed with the remaining feed water. As a result, the salinity of the feed water increases as this water travels from one end of the membrane element to the other. The rejected mix of feed water and salts exits at the back end of the membrane element as concentrate (brine).

As shown in Figs. 3.8 and 3.9, the permeate collector tubes of the individual RO membrane elements installed in the pressure vessel are connected to each other and to the permeate line evacuating the fresh water from the pressure vessel via interconnectors (adaptors) with integral O-rings that seal the connection points and prevent

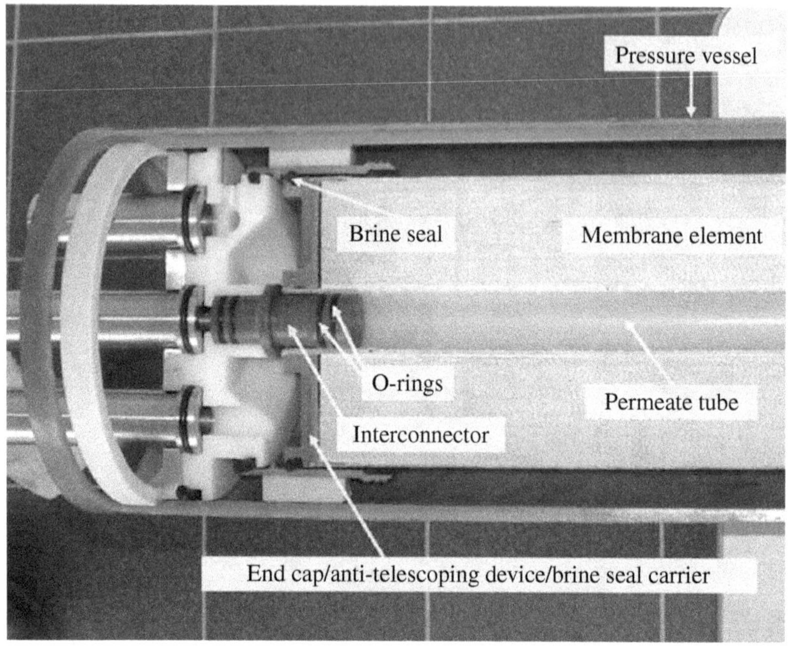

FIGURE 3.8 Cross-section of an RO membrane element installed in a pressure vessel.

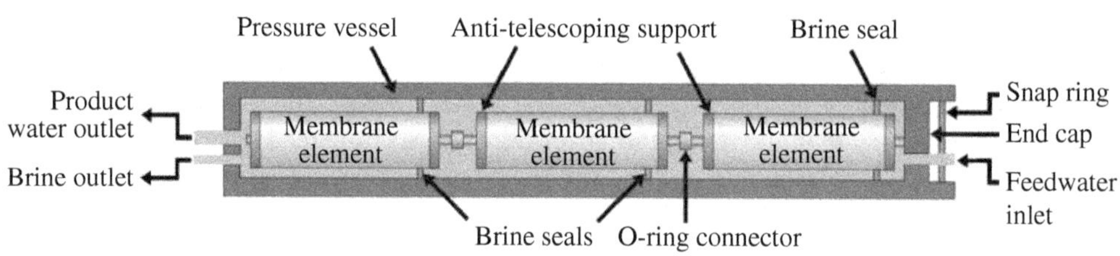

FIGURE 3.9 Membrane elements installed in a pressure vessel.

concentrate from entering the permeate collector tubes. The interconnectors with O-rings provide flexible connections between the elements, which allow for their limited movement within the vessel, for some level of flexibility in loading membranes and also facilitate handling transient pressure surges created in the vessels as a result of abrupt shutdown and start-up of the RO system. While Fig. 3.8 shows an interconnector to the permeate line, Fig. 3.10 depicts an interconnector between two RO elements.

Since broken O-rings and interconnectors are one of the most common operations challenges in RO systems, the Dow Chemical Company has introduced a different interconnection configuration (iLEC) between RO elements that requires the elements to have special interlocking end caps and allows them to be connected directly to each other rather than through conventional interconnectors (Fig. 3.11). The end caps of the

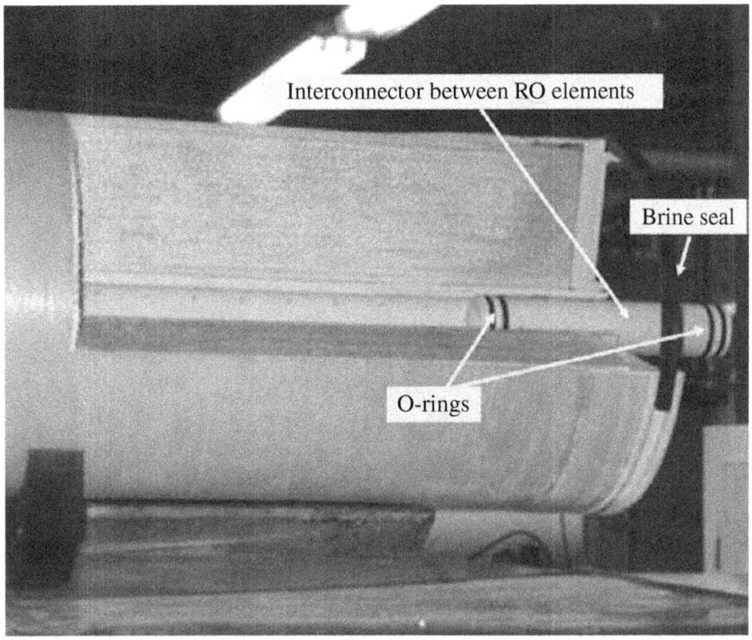

FIGURE 3.10 Interconnector between RO elements.

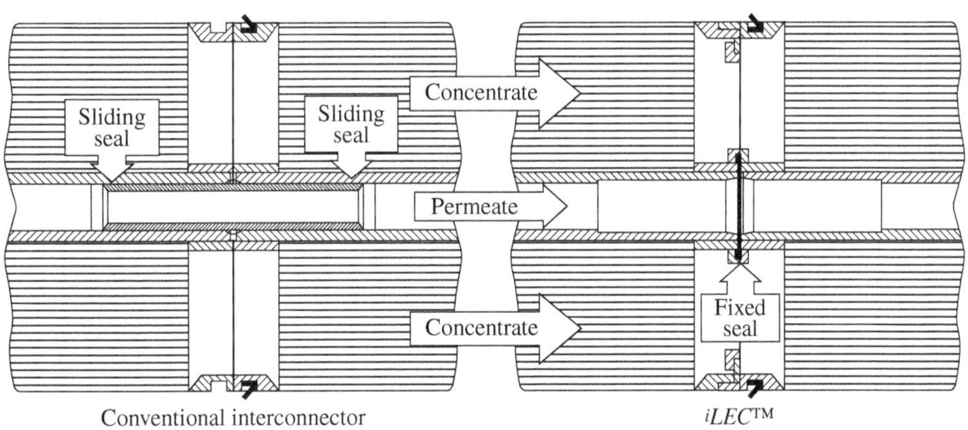

FIGURE 3.11 Comparison of conventional and iLEC membrane interconnectors. (*Source: DOW FilmTec.*)

iLEC RO membrane elements are configured so that they can be interlocked by twisting the installed RO element until its end cap locks with the end cap of a previously installed element. The end caps of the two elements are connected by one O-ring only, which is integrated into the end cap and cannot be rolled or pinched during installation. This minimizes the wear and tear on the O-rings from hydraulic surges and reduces the pressure drop caused by conventional interconnectors.

FIGURE 3.12 Typical 8-in. membrane element.

Commercially available RO membrane elements are standardized in terms of diameter and length and usually are classified by diameter. Spiral-wound RO membranes are available in 2.5-in., 4-in., 6-in., 8-in., 16-in., 18-in., and 19-in. sizes. A typical 8-in. RO membrane element is shown in Fig. 3.12.

At present, the most widely used and commercially available RO elements have a diameter of 20 cm (8 in.), length of 100 cm (40 in.) and brine spacer thickness of 28 mils (0.7 mm). Standard 8-in. seawater and brackish water elements in a typical configuration of seven elements per vessel can produce between 13 and 25 m^3/day (3500 and 6500 gal/day) and 26 and 38 m^3/day (7000 and 10,000 gal/day) of freshwater (permeate), respectively.

Larger 16-in., 18-in., and 19-in. RO brackish and seawater membrane elements are also commercially available. However, to date these large elements have received limited full-scale application. While 8-in. elements and smaller can be handled manually by a single person (Fig. 3.13), larger RO elements can only be loaded and unloaded by special equipment because of their significant weight.

Standard and large-size spiral-wound thin-film composite PA membrane elements have limitations with respect to a number of performance parameters: feed water temperature (45°C), pH (2 to 10), silt density index (less than 4), chlorine content (no measurable amounts), and feed water operating pressure (maximum of 41 or 83 bar/600 to 1,200 lb/in^2 for brackish and seawater RO membranes, respectively). A more detailed description of commercially available brackish and seawater membrane elements from key manufacturers is presented in Chap. 14.

FIGURE 3.13 Unloading an RO membrane element from a pressure vessel.

3.3.2 Hollow-Fiber RO Membrane Elements

In hollow-fiber membrane elements, the 0.1- to 1.0-μm semipermeable film is applied as a coating to the surface of hollow fibers of diameter comparable to that of human hair (42 μm internal diameter, 85 μm external diameter). The hollow fibers are assembled in bundles and folded in a half to a length of approximately 48 in. (1200 mm).

The hollow-fiber bundle is 4 to 8 in. (101.6 to 203.2 mm) in diameter and is located inside a cylindrical housing that is 6 to 12 in. (152.4 to 307.2 mm) across and 54 in. (1370 mm) long. Both ends of the bundle are epoxy-sealed to encapsulate the water introduced in the tube in a way that allows all of the concentrate generated in the tube to exit from only one location—the back end of the membrane (Fig. 3.14).

The feed water is introduced in the bundle (membrane element) through a plastic perforated tube (feed water distributor) that extends over the entire length of the

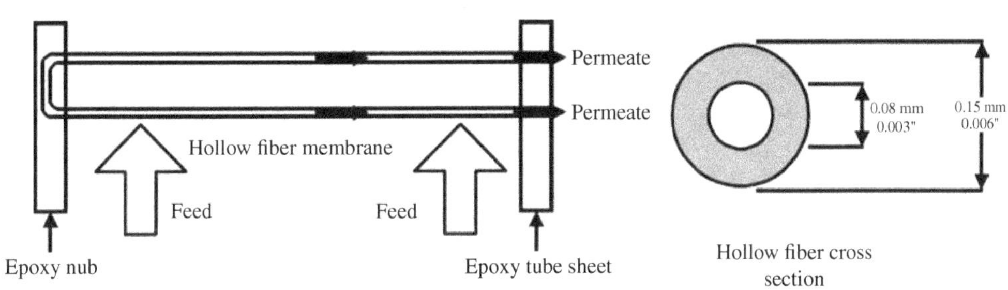

FIGURE 3.14 Hollow-fiber RO vessel with two membrane elements. (*Source:* Toyobo.)

membrane and is located in the center of the bundle. The feed water flows radially and permeates through the thin membrane film of the hollow fibers; the salts and impurities contained in this water are collected on the outer side of the fibers and evacuated through the concentrate pipe at the back end of the membrane element. Permeate is collected in the inner tubes of the hollow fibers and conveyed to the product water connection, which is located on the back or feed end of the membrane element.

As compared to spiral-wound membrane configuration, hollow-fiber membrane configuration allows approximately 4 times more membrane surface per cubic foot of membrane volume. This higher surface area results in a proportionally lower permeate flux for the same volume of processed water, which in turns reduces concentration polarization and associated scaling potential when the source seawater is of high mineral content.

As a result, a typical hollow-fiber vessel contains only two membrane elements but produces approximately the same volume of water as a conventional RO vessel that contains seven or eight elements. These features make hollow-fiber membrane elements very suitable for high-salinity waters with elevated scaling potential, such as those of the Persian Gulf, the Gulf of Oman (Indian Ocean), and the Red Sea. Therefore, this type of membrane element configuration has found a wider application in the Middle East than in other parts of the world.

Because of the lower permeate flux and higher membrane surface area, the feed water flow regime in a hollow-fiber membrane element is laminar (as compared to nearly turbulent flow that occurs in the spiral-wound elements). This low-energy laminar flow results in little to no "scrubbing effect" of the feed flow on the surface of the membranes. This low velocity along the membrane surface allows solids and biofilm to attach to and accumulate more easily on the membranes, which in turn makes hollow-fiber membranes more susceptible to particulate fouling and biofouling and more difficult to clean. As a result, this type of element requires more enhanced source water pretreatment to remove particulate foulants from the water and it operates better on waters of low turbidity and SDI, such as those obtained from well intakes. For comparison, the turbulent flow on the surface of a spiral-wound membrane element makes that membrane configuration more resistant to particulate fouling and biofouling, but because of the higher permeate flux and concentration polarization, it is more prone to mineral scaling. Currently, the only large company that makes hollow-fiber membrane elements is Toyobo Company, Japan. Their membranes are made of cellulose triacetate.

3.3.3 Flat-Sheet RO Membrane Elements

Flat-sheet membrane elements are used in plate-and-frame RO systems (Fig. 3.15). In this case, the elements consist of flat membrane sheets similar to those that are rolled to create spiral-wound elements. Typically, two flat-sheet membranes are placed in filtration plates with the membrane film site outward so that they form an envelope. The filtration plates are integral parts of the RO system stacked within its frame structure. Permeate spacers are installed between each pair of membrane sheets, forming an envelope to facilitate permeate collection and prevent the membrane sheets from sticking to each other. Feed water/brine spacers are installed between the membrane envelopes to allow feed water to flow through.

Because of its low membrane packing density—which is approximately half that of a spiral-wound system—this type of RO system is significantly larger and more

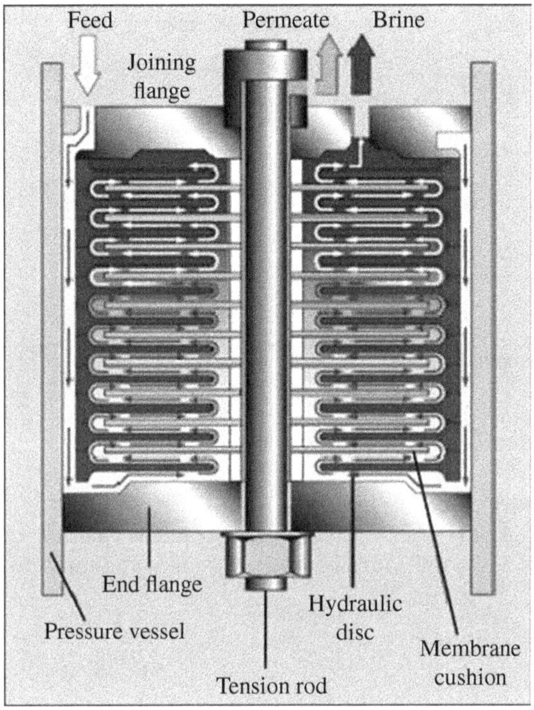

FIGURE 3.15 Plate-and-frame RO unit.

costly than a conventional spiral-wound RO system. Therefore, plate-and-frame systems have not found application for municipal water RO desalination. However, under the plate-and-frame configuration, the flat membrane sheets can easily be removed from the module and can individually be hand-cleaned. This allows for better cleaning and facilitates the use of this type of system for high-solids applications such as food processing.

3.4 Reverse Osmosis System—General Description

This section describes basic configuration and performance parameters of RO systems using spiral-wound membrane elements.

3.4.1 Configuration

As indicated in the previous sections, RO membranes in full-scale installations are assembled in membrane elements (modules) installed in vessels in series of six to eight elements per vessel, and the feed water is introduced to the front membrane elements and applied tangentially on the surface of the membranes in a cross-flow direction at pressure adequate to overcome the osmotic pressure of the saline water and the energy losses associated with the separation process. A general schematic of an RO system is shown in Fig. 3.16. Key parameters associated with the performance of reverse osmosis systems are discussed in the following subsections.

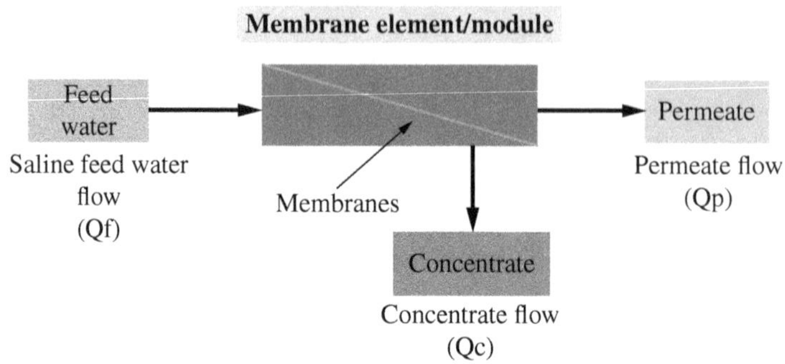

FIGURE 3.16 General schematic of an RO system.

3.4.2 Reverse Osmosis Process Parameters

Osmotic Pressure

The osmotic pressure P_o of a given saline water is calculated by measuring the molar concentrations of the individual dissolved salts in the solution and applying the following equation:

$$O_p = R \times (T + 273) \times \sum m_i \tag{3.1}$$

where O_p = the osmotic pressure of the saline water (in bars—1 bar = 14.5 lb/in^2)
R = the universal gas constant [0.082 (L·atm)/(mol·K) = 0.0809 (L·bar)/(mol·K)]
T = the water temperature in degrees Celsius, and Σm_i is the sum of the molar concentrations of all constituents in the saline water.

This formula is derived from Van't Hoff's thermodynamic law, which is applied to pressure caused by dissociation of ions in solution (Fritzmann et al., 2007).
As an example, let us calculate the osmotic pressure of Pacific Ocean seawater with a TDS concentration of 35,000 mg/L (see Table 2.1). Table 3.1 shows the estimate of the molar concentration of all salts in the source Pacific Ocean seawater (Σm_i). Based on Eq. 3.1, the osmotic pressure of the Pacific Ocean seawater at 25°C is calculated as:

$$P_o = 0.0809 \times (25 + 273) \times 1.1135 = 26.8 \text{ bar } (388.6 \text{ lb/in}^2)$$

The relative osmotic pressure per 1000 mg/L of TDS of Pacific Ocean water is 26.8/(35,000/1000) = 0.77 bar (11 lb/in^2). This ratio is often used as a rule-of-thumb relationship between source water salinity and osmotic pressure, i.e., every 1000 mg/L of salinity results in an osmotic pressure of 0.77 bar (11 lb/in^2).

Depending on the source water quality and temperature, the osmotic pressure may vary significantly from one saline source water to another. For example, in the case of the brackish water from the source in Tularosa, New Mexico, presented in Table 2.3—where calcium sulfate contributes over 70% of the TDS concentration and sodium chloride is only 11%—the actual osmotic pressure of the source water will be significantly lower than the one estimated using the rule of thumb. If the calculations shown in Table 3.1 are completed for the brackish water from Tularosa, New Mexico, shown in Table 2.3, the osmotic pressure of this water will be only 1.12 bar (16 lb/in^2), while the rule of thumb would result in an osmotic pressure estimate of 0.77 × (2542/1000) = 1.96 bar

Seawater Constituents	Concentration, mg/L	Number of milligrams per mole	Molar Concentration (m_i), mol/L
Cations			
Calcium	403	40,000	0.0101
Magnesium	1298	24,300	0.0534
Sodium	10,693	23,000	0.4649
Potassium	387	39,100	0.0099
Boron	4.6	10,800	0.0004
Bromide	74	79,900	0.0009
Total Cations	**12,859.6**	—	**0.5396**
Anions			
Bicarbonate	142	61,000	0.0023
Sulfate	2,710	96,100	0.0282
Chloride	19,287	35,500	0.5433
Fluoride	1.4	19,000	0.0001
Nitrate	0.00	62,000	0.0000
Total Anions	**22,140.4**	—	**0.5739**
Total	**TDS = 35,000 mg/L**		**Σm_i = 1.1135 mol/L**

TABLE 3.1 Molar Concentrations of Pacific Ocean Water Salts

(28 lb/in^2), which is 75% higher than the actual value. This calculation underlines the facts that osmotic pressure is a parameter that should be calculated individually for the specific source water quality and that rules of thumb for this parameter may often over- or underestimate its actual value.

Permeate Recovery

Due to mineral scaling, concentration polarization, and standard equipment and facility constraints, only a portion of the saline source water flow fed to the RO membrane system can be converted into freshwater (permeate). The percentage of the feed source water flow Q_f that is converted into freshwater flow Q_p is defined as the permeate recovery rate P_r (Fig. 3.17):

$$P_r = (Q_p/Q_f) \times 100\% \tag{3.2}$$

As indicated in Fig. 3.17, for typical *seawater reverse osmosis* (SWRO) systems the recovery rate is 40 to 65%. Brackish water desalination plants are designed and operated at higher recoveries (typically 65 to 85%).

The TDS of the concentrate TDS_c can be calculated based on the RO system permeate recovery rate P_r, the actual TDS concentration of the permeate TDS_p, and the feed water TDS (TDS_f) using the following formula:

$$TDS_c = \frac{TDS_f - TDS_p \dfrac{P_r}{100}}{1 - \dfrac{P_r}{100}} \tag{3.3}$$

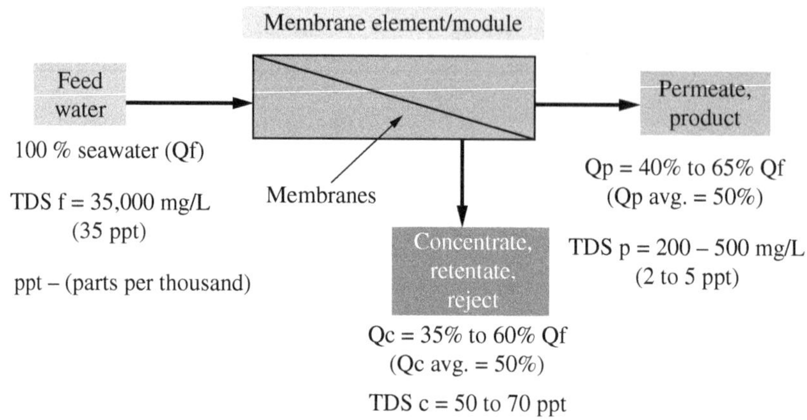

$$\% \text{ Recovery} = \frac{\text{Permeate Flow (Qp)} \times 100}{\text{Feed Flow (Qf)}} = \frac{50\% \text{ Qf}}{\text{Qf}} = 50\%$$

FIGURE 3.17 Recovery of a typical SWRO system.

For the example in Fig. 3.17, the TDS of the concentrate—assuming a recovery rate of 50% and a permeate salinity of 200 mg/L—is calculated as follows:

$$TDS_c = \frac{35{,}000 - 200\dfrac{50\%}{100}}{1 - \dfrac{50\%}{100}} = 69{,}800 \text{ mg/L}$$

As can be seen from Eq. 3.3, the higher the RO system permeate recovery rate, the more freshwater is produced from the same volume of saline source water. For example, a brackish water system designed at 75% recovery will produce 75 m^3 of low-salinity water (permeate) and 25 m^3 of concentrate out of every 100 m^3 of brackish water. Since practically all dissolved solids contained in the source water will be retained in one-fourth the volume (100/25 = 4), the RO system concentrates the source water by a factor of 4 (400%). This factor is of critical importance in terms of the ability of various salts and other compounds, such as silica, to form crystal scale on the surface of the membranes. As indicated in Table 2.6, at a concentration factor of 4 (400%), calcium sulfate scale is likely to form (as are other scales). Knowing the concentration factor of the RO system is also important for making decisions regarding the most suitable type of concentrate management system. It should be pointed out that plant recovery rate (and the associated concentration factor) also has an impact on product water quality. The more water passes through the membranes, the more salts pass as well; therefore, the overall permeate water quality decreases with an increase in system recovery. The passing of salts through the membrane can be controlled to some extent by the membrane structure, i.e., membranes with tighter structure will pass fewer salt ions when the RO system is operated at higher recovery.

Membrane Salt Passage

Salt passage S_p of a membrane is defined as the ratio between the concentration of salt in the permeate TDS_p and in the saline feed water TDS_f (see Fig. 3.17); it is indicative of the amount of salts that remain in the RO permeate after desalination.

$$S_p = (TDS_p / TDS_f) \times 100\% \tag{3.4}$$

Membrane Salt Rejection

Salt rejection S_r is a relative measure of how much of the salt that was initially in the source water is retained and rejected by the RO membrane:

$$S_r = 100\% - S_p = [1 - (TDS_p / TDS_f)] \times 100\% \tag{3.5}$$

For the seawater desalination example depicted in Fig. 3.17, the salt passage of the SWRO membrane for the high end of performance (TDS_p = 200 mg/L) is S_p = (200/35,000) × 100% = 0.57%. The total salt rejection of this membrane is S_r = 100% − 0.57% = 99.43%.

It should be pointed out that salt passage and rejection can be applied not only to the total dissolved solids contained in the source water as a whole but also to individual ions contained in the water. Not all ions are rejected equally by an RO membrane. Usually, the larger the ions and the higher their electrical charge, the better rejected they are. This means, for example, that bivalent ions such as calcium and magnesium will be rejected better than monovalent ions such as sodium and chloride.

It is also important to note that RO membranes do not reject gases. So if the source water contains ammonia and hydrogen sulfide, which are gases commonly encountered in brackish water, these gases will remain in the permeate and will be at elevated concentration as compared to their content in the source water. RO membranes also do not reject free chlorine gas, carbon dioxide, or oxygen. Ammonia and carbon dioxide gases can be rejected if they are converted into ammonium ion and bicarbonate ion by pH adjustment.

Also, low-charge monovalent ions such as boron will be rejected at a lower rate than higher-charge monovalent ions such as chloride and sodium. This is a very important feature of RO membranes, because—often for practical purposes—RO membrane structure can be modified to selectively reject specific ions better.

For example, the RO membrane material can be modified to have a "looser" structure and remove mainly bivalent ions when the key goal of water treatment is water softening (i.e., removal of calcium and magnesium). Such reverse osmosis membranes are often referred to as nanofiltration membranes. Nanofiltration membranes usually reject less than 30% of TDS (as compared to RO membranes, which reject over 90% of TDS), but they reject over 99% of calcium and magnesium and they do that at higher productivity and lower feed pressure.

Because the weight and valence of the rejected ions are very important factors influencing RO membrane salt passage and rejection, and since the source water's ion makeup can vary significantly from one location to another, sodium chloride and sodium sulfate are typically used as "standard" salts against which the rejection of different commercial RO membranes is measured.

Membrane rejection is tested against a standard salt feed solution of predetermined salinity; testing is performed at standard test feed pressure and feed flow rate. For example, for most commercial SWRO membranes, salt rejection is determined using a standard test solution of sodium chloride at a salinity concentration of 32,000 mg/L, test feed pressure of 55.2 bar (800 lb/in^2), and test recovery rate of 10%. Salt rejection for most commercially available SWRO membranes at these test conditions is 99.60 to 99.85%.

Brackish water elements are typically tested using a standard sodium chloride solution with a TDS of 500 to 2000 mg/L, test pressure of 6.7 to 15.5 bar (100 to 225 lb/in^2), and recovery rate of 15%. Standard salt rejection of most brackish RO elements varies between 99.0 and 99.6%. It should be noted that when used to desalinate actual saline water, commercially available RO membrane elements usually have a lower overall TDS rejection (higher salt passage) than their standard level reported in manufacturer specifications—mainly because not all ions contained in the saline source water are rejected as well as sodium and chloride. In addition, the actual feed rate and recovery of the RO system impact membrane rejection. Performance parameters of commercially available membranes commonly used for brackish and seawater desalination are presented in greater detail in Chap. 14.

Net Driving Pressure (Transmembrane Pressure)

Net driving pressure (NDP), also known as *transmembrane pressure*, is the actual pressure that drives the transport of freshwater from the feed side to the freshwater side of the membrane. The average NDP of a membrane system is defined as the difference between the applied feed pressure F_p of the saline water to the membrane and all other forces that counter the movement of permeate through the membrane, including the average osmotic pressure O_p which occurs on the permeate side of the RO membrane, the permeate pressure P_p existing the RO pressure vessel, and the pressure drop P_d across the feed/concentrate side of the RO membrane. The NDP can be calculated as follows:

$$\text{NDP} = F_p - (O_p + P_p + 0.5P_d) \tag{3.6}$$

The applied feed pressure F_p is controlled by the RO system operator and delivered through high-pressure feed pumps. The average osmotic pressure O_p of the membrane is determined by the salinity and the temperature of the source water and the concentrate.

The permeate pressure P_p (also known as *product water back pressure*) is a variable that is controlled by the RO plant operator and is mainly dependent on the energy needed to convey permeate to the downstream treatment and/or delivery facilities. Typically, permeate pressure is set at 1 to 2 bar (15 to 30 lb/in^2). The osmotic pressure of permeate is usually very small, because the salinity of this stream is low. Therefore, for practical purposes it is typically omitted from the calculations of the NDP.

The pressure drop P_d across the feed/concentrate side of the RO membrane depends mainly on the membrane fouling and the RO membrane and system configuration. This pressure drop is usually between 1.0 and 3.5 bar (14.5 and 50.7 lb/in^2).

For the example in Fig. 3.17, assuming an SWRO system recovery of 50%, saline water feed pressure of $F_p = 56$ bar (812 lb/in^2), permeate pressure $P_p = 1.4$ bar (20.3 lb/in^2), and pressure drop across the RO system $P_d = 3.2$ bar (46.4 lb/in^2), the NDP at which the system operates is determined as follows:

1. Calculate the average salinity on the feed/concentrate side of the RO membrane.

$$\text{TDS}_{\text{fc}} = (\text{TDS}_{\text{f}} + \text{TDS}_{\text{c}})/2 = (35{,}000 + 69{,}800)/2 = 52{,}400 \text{ mg/L}$$

2. Calculate the average osmotic pressure of the saline water on the feed/concentrate side.

$$O_p = (0.77/1000) \times 52{,}400 = 40.3 \text{ bar } (584 \text{ lb/in}^2)$$

3. Calculate the NDP.

$$\text{NDP} = F_p - (O_p + P_p + 0.5P_d) = 56 - (40.3 + 1.4 + 0.5 \times 3.2) = 12.7 \text{ bar } (184 \text{ lb/in}^2)$$

Membrane Permeate Flux

Membrane permeate flux (J), also referred to as *membrane flux*, is defined as the permeate flow a membrane produces per unit membrane area. It is calculated by dividing the flow rate Q_p of permeate produced by a RO membrane element [usually expressed in *gallons per day* (gpd) or *liters per hour* (lph)] by the total membrane area S of the element (in square feet or square meters). The flux unit is therefore gal/(ft^2·day), also referred to as *gfd*, or L/(m^2·h), also known as *lmh*.

$$J = Q_p/S \tag{3.7}$$

Since a full-scale RO system consists of a number of membrane elements, the average permeate flux of the system is calculated by dividing the total flow of permeate produced by all membranes by the total surface area of the membranes.

For design purposes, flux is selected as a function of the source water quality and the type of RO membrane used for desalination. The higher the quality of the source water applied to the membranes, the higher the acceptable design flux. This is the reason why the use of well water or water pretreatment with ultrafiltration or microfiltration membranes that produce water with SDI below 3 can often be designed and operated at higher fluxes.

Conversely, if the source water has a higher solids content (SDI > 4 most of the time), then operating the RO system at a higher flux would result in an unreasonably high frequency of cleaning, because of the rapid rate of accumulation of solids on the membranes and reduction of their permeability.

The higher the permeability of a given membrane (i.e., its ability to transport water), the higher the maximum flux the membrane can be designed for at the same source water quality. For example, brackish water RO membranes have a looser (i.e., more permeable) molecular structure; therefore, when they process high-quality well water (SDI < 2) or RO permeate, they can be operated at approximately two times higher fluxes than SWRO membranes.

Specific Membrane Permeability (Specific Flux)

Specific membrane permeability (SMP), also known as *specific membrane flux*, is a parameter that characterizes the resistance of the membrane to water flow. It is calculated as the membrane permeate flux (J) divided by the net driving pressure (*NDP*):

$$\text{SMP} = J/\text{NDP} \tag{3.8}$$

The standard specific permeability of a given membrane is typically determined for a feed temperature of 25°C and expressed in lmh/bar or gfd/(lb/in^2). For example, most commercially available seawater desalination RO membranes at present have an

SMP of 1.0 to 1.4 lmh/bar [0.04 to 0.06 gfd/(lb/in^2)]. For comparison, brackish water RO membranes have a significantly higher specific permeability of 4.9 to 8.3 lmh/bar [0.2 to 0.35 gfd/(lb/in^2)].

Nanofiltration membranes, which have a "looser" membrane structure, have an even higher specific permeability than brackish and seawater RO membranes: 7.4 to 15.8 lmh/bar [0.3 to 0.6 gfd/(lb/in^2)]. Usually, membranes of lower specific permeability also have higher salt rejection, so there is a tradeoff between lower production and higher water quality. The specific membrane permeability is determined by the chemical and physical nature of the membrane.

3.5 Models for Water and Salt Transport through Membranes

3.5.1 Overview

As indicated previously, during the process of reverse osmosis water separation, two key processes occur at the same time—transport of water (solvent) and transport of salt (solute) from the high-salinity side of the membrane to the low-salinity side. At present, there are a number of models that describe the mechanism and rate of transport of water and salts through membrane. Detailed overviews of such models and their key features are presented elsewhere (Malaeb and Ayoub, 2011). The three types of transport models that have found widest acceptance are the nonporous (homogeneous) solution-diffusion transport model, pore model, and irreversible thermodynamics model.

The most commonly used type of model for water transfer through a membrane at present is the nonporous solution-diffusion transport model (American Water Works Association, 2007). This model has found wider application than the others because of its simplicity and relatively easy confirmation by empirical tests. This model assumes that the membranes are nonporous and that water passes through them as a result of the applied net driving pressure. On the other hand, solutes (i.e., salts) pass through the membrane driven by the concentration gradient (difference) formed between the source water and freshwater sides of the membranes.

The pore model is more complex; assumes either that membranes are built of one-dimensional long and narrow pores or that membranes have pores which are generated as imperfections during the production process, and as a result, some solute leaks through the membrane into the permeate side. This more complex model was created to explain why the actual water quality is somewhat lower than that projected by the nonporous solution-diffusion model.

The irreversible thermodynamics model describes the membrane performance via a dissipation function that reflects the division of the flow field into small systems that are in local thermodynamic equilibrium. The model is based on the use of differential equations to describe the flow field of these systems and to calculate the transport of multiple solutes through the membranes (Malaeb and Ayoub, 2011). Due to its complexity, this type of model have found limited application for desalination plant planning and engineering.

3.5.2 Nonporous Solution-Diffusion Transport Model

The nonporous solution-diffusion transport model assumes that RO membranes do not have actual pores and that water travels from the high-salinity side of a membrane to the low-salinity side by convection and diffusion in randomly shaped curvilinear intermolecular channels formed by the membrane polymer chains. The transport of water through the membranes is a three-step process: (1) adsorption of water molecules on the membrane surface, (2) convection and diffusion through the membrane, and (3) desorption from the permeate side of the membrane surface into the bulk permeate water. The rates of convection and diffusion are controlled by the net driving pressure and are also a function of the membrane permeability.

The process of salt transport is driven by the concentrate gradient between the two sides of the membrane and by the membrane's ability for solute retention through size exclusion and charge (dielectric) exclusion. Uncharged molecules (solutes or salts) are rejected due to the membrane's ability to act as a sieve for molecules larger than the molecular weight cutoff of the membranes. The molecular weight cutoff is representative of the average size of molecules that can be retained by a given membrane. For example, SWRO membranes have a molecular weight cutoff of 120 to 200 Daltons (Da). This means that most of the molecules of compounds contained in the source water that do not have charge or have only a very weak charge and are larger than 200 Da (such as most algal toxins) will be rejected by SWRO membranes.

Charged compounds (solutes) such as the ions of strong acids and bases are rejected by charged exclusion—repulsion of the salt ions by the fixed electric charges attached to the membrane surface (Hassan et al., 2007). This observation leads to the conclusion that the larger the molecules and the higher the electric charge they have, the more likely they are to be rejected by the membranes.

Water and Salt Transport Rates

The reverse osmosis membrane separation process is closely related to two key features of the membranes: their abilities to transport water and salts. Desalination is possible because the rate of water transport of RO membranes is significantly higher than the rate of salt transport. In accordance with the solution-diffusion model, the transports of water and salts through the membrane are actually two independent processes driven by different forces.

Water Transport Rate The water (permeate) transport rate Qp of an RO membrane is proportional to its water transport (water permeability) coefficient A—which is a unique constant for each membrane material—as well as the total membrane area S, and the net driving pressure (NDP):

$$Q_p = A \times S \times \text{NDP} \tag{3.9}$$

As indicated in Eq. 3.7, the ratio between the water transport rate and the surface area through which water is conveyed is referred to as the membrane permeate flux J. Therefore, membrane permeate flux (also often referred to as the membrane flux) can be represented as follows:

$$J = A \times \text{NDP} \tag{3.10}$$

This formula indicates that membrane flux is controlled by two parameters—water permeability coefficient A, which is unique for each type of commercial membrane, and net driving pressure (NDP), which can be controlled by adjusting feed and permeate pressures.

Salt Transport Rate The salt transport rate Q_s is proportional to the salt transfer coefficient B—which, as the water transfer coefficient, is unique for each membrane type—the surface area S of the membrane, and the salt concentration gradient ΔC, which collectively for all salts is measured as the difference between the TDS levels of the concentrate and the permeate:

$$Q_s = B \times S \times \Delta C \tag{3.11}$$

where

$$\Delta C = C_b - C_p \tag{3.12}$$

Here, C_b is the concentration of the solute (salt) at the boundary layer/bulk feed flow and C_p is the concentration of solute (salt) in the permeate. The boundary layer is a layer of laminar feed water flow and elevated salinity that forms near the surface of the membranes as a result of the tangential source water feed flow in the spacers and of permeate flow in a perpendicular direction through the membranes on the two sides of the spacer (Fig. 3.18).

In Fig. 3.18, C_b is the concentration of the solute (i.e., salt) in the feed water, C_s is the concentration at the inner membrane surface (which typically is higher than that in the feed flow), and C_p is the concentration of the solvent (i.e., freshwater salinity) on the low-salinity (permeate) side of the membrane.

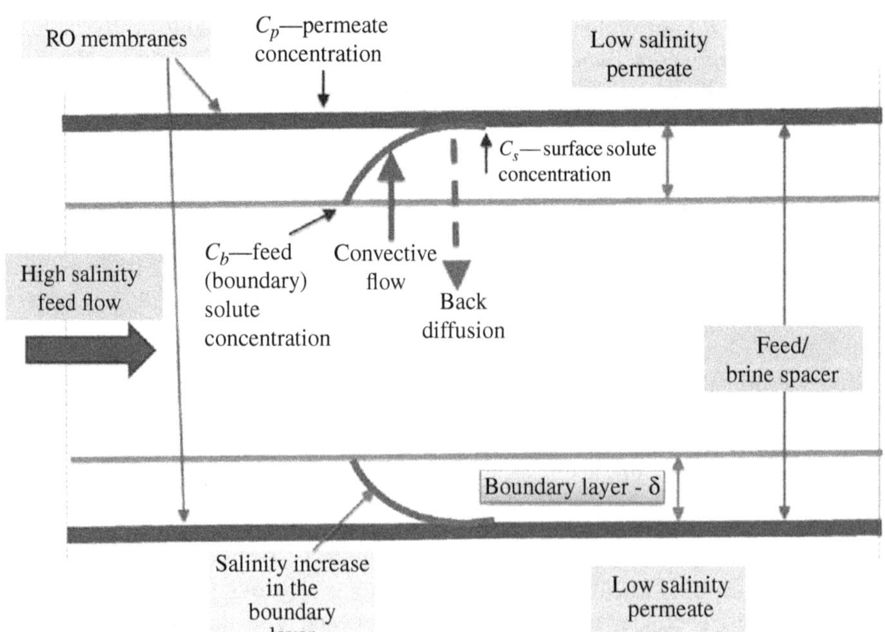

FIGURE 3.18 Boundary layers in a membrane feed spacer.

3.6 Membrane Performance Factors and Considerations

3.6.1 Concentration Polarization

A very important factor that may have a significant impact on membrane performance is concentration polarization, i.e., an increase in salinity in the boundary layer to levels, which are significantly higher than the salinity in the feed water. This phenomenon entails the formation of a boundary layer along the membrane feed surface due to the laminar flow in the membrane feed spacers and concentration of solute (salt) in this layer. Two different types of flow occur in the boundary layer—a convective flow of freshwater from the bulk of the feed water though the membranes and a diffusion flow of rejected solutes (salts) from the membrane surface back into the feed flow (see Fig. 3.18). Since the rate of convective flow of water is typically higher than that of the diffusion flow of salts, the salts rejected by the membrane tend to accumulate in the boundary layer, with the highest salt concentration (shown as C_s in Fig. 3.18) occurring at the inner surface of the membrane. Besides salts, boundary layer also accumulates particulate solids, for the same reasons.

This phenomenon of concentration of salts and solids in the boundary layer has four significant negative impacts on membrane performance: it (1) increases osmotic pressure at the membrane surface, (2) increases salt passage through the membranes, (3) creates hydraulic resistance of water flow through the membrane, and (4) creates the potential for accelerated scale formation and particulate fouling on the membrane surface because of the concentration of salt and solids in the boundary layer. The ratio between the solute (salt) content at the surface of the membrane C_s and in the bulk feed water C_b is referred to as the concentration polarization factor β:

$$\beta = C_s / C_b \tag{3.13}$$

The higher the value of β, the higher the concentration difference and the worse the impacts of that difference on membrane performance. Taking into consideration the impact of concentration polarization, Eq. 3.10 for membrane flux could be rewritten as follows:

$$J = A \times [F_p - (\beta \times O_p + P_p + 0.5P_d)] \tag{3.14}$$

Similarly, Eq. 3.11 for salt transport rate could be modified as follows:

$$Q_s = B \times S \times (\beta \times C_b - C_p) \tag{3.15}$$

Because NDP is reduced for the same feed pressure when osmotic pressure is elevated (see Eq. 3.6) and the elevated salinity in the boundary layer increases the osmotic pressure at the membrane surface, the actual permeate flow produced by the RO system is reduced. In other words, an increase in the concentration polarization factor β results in an increase of the osmotic pressure, which in turn causes a reduction in flux (i.e., lower production of freshwater).

Salt passage through the membranes is increased (i.e., salt rejection is decreased) because, as per Eqs. 3.11 and 3.12, the salt transport through the membrane is proportional to the difference in salinity from one side of the membrane to the other. Since the salinity at the feed side of the membrane is higher than the salinity in the feed solution, the salt transport is proportionally higher as well. This means that as β increases, salt rejection is reduced and the salinity of the produced freshwater is increased.

Due to the salinity gradient and accumulation of particulate solids within the boundary layer, the hydraulic resistance of that layer is higher than that of the feed water. As a result, the available NDP is decreased and the membrane flux is reduced. The hydraulic resistance compounds with the elevated osmotic pressure to reduce flux.

If the salt concentration in the boundary layer exceeds the solubility of sparingly soluble salts (such as calcium carbonate and sulfate) contained in the source water, these salts will begin precipitating on the membrane surface and form mineral scale. As indicated in Chap. 2, membrane scaling will result in reduced permeability and flux.

The magnitude of the concentration polarization factor β is driven by three key factors: (1) permeate flux, (2) feed flow, and (3) the configuration and dimensions of the feed channels and feed spacer. An increase in permeate flux (i.e., freshwater production) increases exponentially the quantity of salt ions and particulate solids conveyed to the boundary layer and therefore exacerbates concentration polarization and particulate fouling of the membrane. An increase in feed flow, however, intensifies turbulence in the boundary layer and, as a result, decreases the thickness and concentration of the layer. Depending on configuration and geometry, the RO membrane feed/concentrate spacer and feed concentrate channel may cause more or less turbulence in the concentrated boundary layer and therefore, may reduce or increase concentration polarization.

Since feed spacer configuration and channel size are constant for a given standard RO membrane element, permeate flux and feed flow are the two key factors that determine the magnitude of concentration polarization. As indicated previously, the ratio between the permeate flow and the feed flow of a given RO membrane element is the permeate recovery rate of the element. Similarly, the ratio between permeate flow and the feed flow of an entire RO system is termed the RO system recovery rate.

As the recovery rate increases, the magnitude of concentration polarization increases as well. For example, for seawater reverse osmosis systems using standard membrane elements, operation at a recovery rate of 50% would typically result in approximately 1.2 to 1.5 times greater salinity concentration at the membrane surface than that in the source seawater. Beyond 75% recovery, the concentration polarization factor would exceed 2, which would have a significant negative impact on the efficiency of the membrane separation process.

In addition, at a recovery rate above 75% and ambient salinity pH, many of the salts in seawater would begin precipitating on the membrane surface, which would require the addition of large amounts of antiscalant (scale inhibitor) and would make SWRO desalination impractical. Since scaling is pH dependent, an increase in pH to 8.8 or more (which often is practiced for enhanced boron removal) may result in scale formation at significantly lower SWRO recovery rates (50 to 55%). While the example above refers to SWRO systems, concentration polarization is a phenomenon that occurs in BWRO systems as well. However, in such systems, similar concentration polarization impacts are observed at higher recovery.

The concentration polarization phenomenon and its effect on the decline of membrane productivity (flux) is not unique to RO membranes, but also occurs on the surface of ultrafiltration and microfiltration membranes used for saline water pretreatment. In this case, concentration polarization is the accumulation of rejected particles (rather than salts) near the membrane surface, causing particle concentration in the boundary layer that is greater than that in the raw seawater fed to the pretreatment system (which in turn results in ultrafiltration or microfiltration flux decline).

In order to keep concentration polarization within reasonable limits, membrane manufacturers recommend maintaining the maximum recovery rate per membrane element in a vessel within 10 to 20%. As a result, with a typical configuration of six to eight elements per vessel and taking into consideration the actual flux of the individual elements in the vessel, a single SWRO system is practically limited to 50 to 70% recovery. Depending on the ion makeup of the saline water, the practical limit of BWRO system recovery is 85 to 95 %.

3.6.2 Membrane Fouling

Ideally, saline source water would mainly contain dissolved minerals. As long as the desalination system is operated in a manner that prevent these minerals from precipitating on the membrane surface, the RO membranes could produce freshwater of consistent quality at a high rate for a very long time without needing cleaning. Practical experience shows that for desalination plants with high source water quality and/or a well-designed pretreatment system, the RO membranes may not need to be cleaned for one or more years, and their useful life could extend beyond 10 years.

In actuality, however, pretreatment systems remove most but not all of the insoluble solids contained in the source water and they may not always effectively prevent some of the soluble solids from precipitating on the membrane surface. The suspended solids, silt, and natural organic matter that remain in the source water after pretreatment may accumulate on the surface of the RO membranes and cause loss of membrane productivity over time. In addition, because saline source water—especially if it originates from surface water source—may naturally contain microorganisms as well as dissolved organics that could serve as food for these microorganisms, a biofilm could form and grow on the RO membrane surface, causing loss of membrane productivity as well. The types of foulants contained in saline source water are described in detail in Chap. 2.

The process of reduction or loss of active membrane surface area, membrane permeability, and subsequently productivity of RO membranes due to accumulation of suspended solids and organics, precipitation of dissolved solids, and/or formation of biofilm on the RO membrane surface is called *membrane fouling*. Excessive membrane fouling is undesirable; besides having a negative impact on RO membrane productivity, it could also result in an increased use of energy for salt separation and in a deterioration of product water quality.

Most RO systems are operated to produce a constant flow of fresh (desalinated) water at a target TDS content. For a given source water salinity and temperature, and target freshwater TDS level, continuous production of a constant volume of desalinated water will require the source water to be fed to the desalination system at a constant pressure, typically in a range of 5 to 20 bar (75 to 290 lb/in^2) for *brackish water reverse osmosis* (BWRO) systems and 55 to 70 bar (800 to 1000 lb/in^2) for SWRO systems. If RO membrane fouling occurs, in order to maintain constant membrane productivity (flux) and water quality, the desalination system would need to be operated at increasingly higher NDP, which in turn means that the energy needed to produce the same volume and quality of freshwater would need to be increased. The increase in RO system's NDP over time is an evidence of accumulation and/or adsorption of fouling materials on the surface of the membranes (i.e., membrane fouling).

It should be pointed out that membrane fouling is dependent not only upon the source water quality and the performance of the pretreatment system but also upon membrane properties such as charge, roughness, and hydrophilicity (Hoek et al., 2003; Hoek and Agarwal, 2006), as well as upon the flow regime on the membrane surface (Wilf et al., 2007). Membranes with higher surface roughness and lower hydrophobicity usually are prone to higher fouling.

Typically, compounds causing RO membrane fouling can be removed by periodic cleaning of the membranes using a combination of chemicals (biocides, commercial detergents, acids, and bases). In some cases, however, membrane fouling can be irreversible, and cleaning may not recover membrane productivity, which in turn may require the replacement of some or all of the RO membranes of the desalination plant.

All RO membranes foul over time. However, the rate and reversibility of fouling are the two key factors that have the most profound effect on the performance and efficiency of the seawater reverse osmosis separation process. These factors in turn are closely related to the source seawater quality and the performance of the desalination plant's pretreatment system.

External and Internal Fouling

Depending on the location of the accumulation of insoluble rejected matter that is causing the decline of membrane performance, fouling can be classified as either external (surface) fouling or internal fouling. It should be pointed out that membranes can experience both types of fouling at the same time and therefore, membrane performance challenges are often caused by combined fouling.

External Fouling External fouling results from accumulation of deposits on the surface of the membrane by three distinct mechanisms: (1) accumulation of mineral deposits (scale); (2) formation of cake of rejected solids, particulates, colloids, and other organic and/or inorganic matter; and (3) biofilm formation, i.e., the growth and accumulation of colonies of microorganisms on the surface of the membranes that attach themselves by the excretion of extracellular materials. Although the three membrane fouling mechanisms can occur in any combination at any given time, external fouling of BWRO and SWRO membranes is most frequently caused by mineral scale formation (scaling) and biofilm accumulation (biofouling), respectively.

Internal Fouling Internal fouling is a gradual decline of membrane performance caused by changes in the chemical structure of the membrane polymers and triggered by physical compaction or chemical degradation. Physical compaction of the membrane structure may result from long-term application of feed water at pressures above what the membranes are designed to handle [typically 41 bar (600 lb/in^2) for BWRO membranes and 83 bar (1200 lb/in^2) for SWRO membranes] and/or from prolonged operation at feed water temperatures above the limit of safe membrane operation (typically 40 to 45°C).

Chemical degradation is a decline in membrane performance resulting from continuous exposure to chemicals that alter the membrane's structure, such as strong oxidants (chlorine, bromamine, ozone, peroxide, etc.) or very strong acids and alkalies (typically with a pH below 2 and above 12). While external fouling can usually be reversed by chemical cleaning of the membranes, internal fouling most often causes permanent damage of the microvoids and polymeric structure of the membranes and is therefore largely irreversible.

3.6.3 Flux Distribution within Membrane Vessels

Membrane RO elements of a typical desalination system are installed in vessels, often referred to as *membrane pressure vessels*. Usually, six to eight RO membrane elements are housed in a single membrane vessel (Fig. 3.19). Under this membrane element configuration, all of the saline feed water is introduced at the front of the membrane vessel and all permeate and concentrate are collected at the back end. As a result, the first (front) membrane element is exposed to the entire vessel feed flow and operates at flux significantly higher than that of the subsequent membrane elements.

Theoretically, if even flow distribution between all membrane elements in the vessel were possible, each element in a typical configuration of seven membrane elements per vessel would produce one-seventh (14.3%) of the total permeate flow of the vessel.

However, in actual RO systems the flow distribution in a vessel is uneven; the first membrane element usually produces about 25% of the total vessel permeate flow, while the last element yields only 6 to 8% of the total vessel permeate (see Fig. 3.19). The decline of permeate production along the length of the membrane vessel is mainly due to the increase in feed salinity and associated osmotic pressure as the permeate is removed from the vessel. The concentrate rejected from all elements remains in the vessel and is fed to the subsequent downstream elements until it exits the last element.

Since the first element processes the largest portion of the feed flow, it also receives and retains the largest quantity of the particulate and organic foulants contained in the source water, and is therefore most impacted by particulate and organic fouling and by biofouling. The remainder of the feed water that does not pass through the first RO element combines with the concentrate from this element and enters the feed channels of

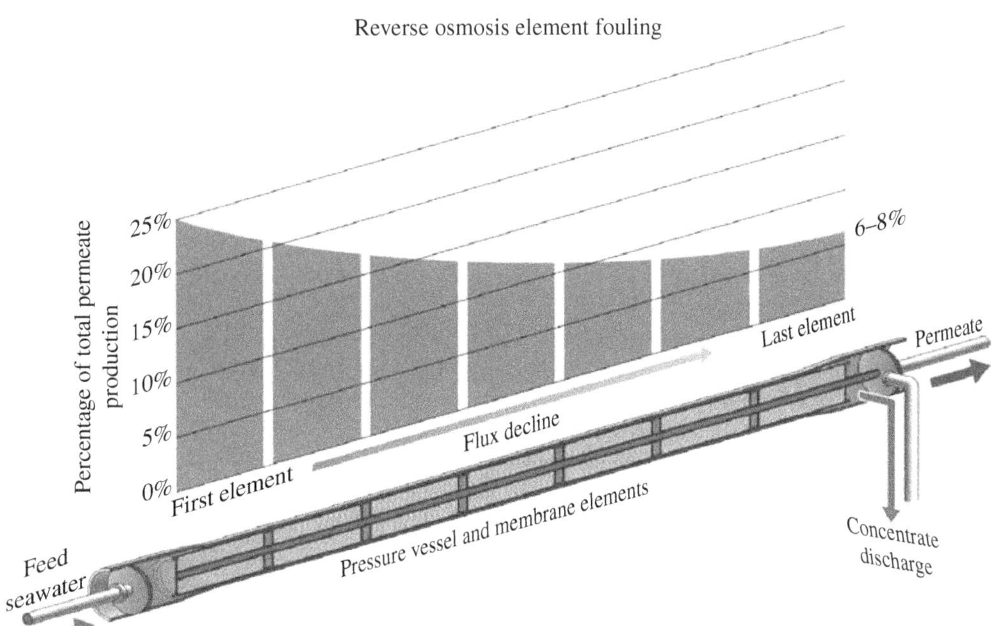

FIGURE 3.19 Membrane fouling and flux distribution in a membrane vessel.

the second RO element of the vessel. This element, therefore, is exposed to higher-salinity feed water and lower feed pressure (energy), because some of the initially applied pressure (energy) has already been used in the first RO element of the vessel to produce permeate. As a result, the permeate flow rate (flux) of the second element is lower, and the concentration polarization on the surface of the element higher, than that of the first RO element.

The subsequent membrane elements are exposed to increasingly higher feed salinity and elevated concentration polarization, which results in progressive reduction of their productivity (flux). As flux through the subsequent elements is decreased, accumulation of particulate and organic foulants on these elements diminishes and biofilm formation is reduced. However, the possibility of mineral scale formation increases, because the concentration of salts in the boundary layer near the membrane surface increases due to the increasingly higher feed salinity. Therefore, in RO systems fouling caused by accumulation of particulates, organic matter, and biofilm formation is usually most pronounced on the first and second membrane elements of the pressure vessels, whereas the last two RO elements are typically more prone to mineral scaling than other types of fouling.

The flux distribution pattern in an RO vessel can be altered significantly by the membrane fouling process itself. If the source water contains a large amount of foulants of persistent occurrence, then as the first element is completely fouled, its flux over time will be reduced below its typical level (about 25%) and the flux of the second element will be increased instead. After the fouling of the second RO element reaches its maximum, a larger portion of the feed flow will be redistributed down to the third RO element, until all elements in the vessel begin to operate at a comparable lower flux.

Flux redistribution caused by particulate and colloidal fouling, deposition of natural organic matter, and/or biofouling can trigger scale formation on the last RO element. The main reason for this phenomenon is that the concentration polarization at the surface of the last RO element typically more than doubles as a result of this flux redistribution.

As indicated previously, in a typical seven-element-per-vessel configuration under nonfouling conditions, the last element will operate at a flux of only 6 to 8% of the average vessel flux. Under fouling-driven flux redistribution in the membrane vessel, the flux of the last element will increase to 12 to 14% (i.e., approximately 2 times as high as usual). Since membrane polarization is proportional to flux, if the RO system is operated at the same recovery, then the likelihood for scale formation on the last one or two RO elements increases.

In addition to increasing the potential for mineral fouling (scaling) on the last one or two membrane elements, another reason that long-term operation of a fouled RO system is not advisable is the higher feed pressure (energy) needed to overcome the decreased membrane permeability if the system is operated to produce the same permeate flow. As the RO system feed pressure reaches a certain level—typically 50 bar (7 lb/in^2) for BWRO membranes and 83 bar (1200 lb/in^2) for SWRO membranes—the external membrane fouling will be compounded by internal fouling due to the physical compaction of the membrane structure, which could cause irreversible damage to the membranes. Therefore, understanding the causes and mechanisms of RO membrane fouling is of critical importance for the successful design and operation of RO desalination plants. This is also the reason why membrane suppliers recommend limiting the maximum operating feed pressure of RO membranes to 41 bar (600 lb/in^2) for BWRO elements and 83 bar (1200 lb/in^2) for SWRO elements.

3.6.4 Effect of Salinity on Membrane Performance

Figure 3.20 illustrates the effect of source water salinity (TDS concentration) fed to an RO system on the system's productivity of freshwater (permeate flux) and product water quality (salt rejection). Higher feed water salinity reduces the net driving pressure (assuming that the system is operating at the same feed pressure and recovery) because of the increased osmotic pressure of the feed water, which in turn decreases permeate flux (freshwater production).

In terms of salt transport, an increase in feed water salinity increases the salt concentration gradient (ΔC in Eq. 3.11), which results in accelerated salt transport through the membranes and therefore, in lower salt rejection (deteriorating product water quality).

3.6.5 Effect of Recovery on Membrane Performance

As indicated in Fig. 3.21, an increase in recovery results in a slow decrease in permeate flux until it reaches the point at which osmotic pressure exceeds the applied pressure

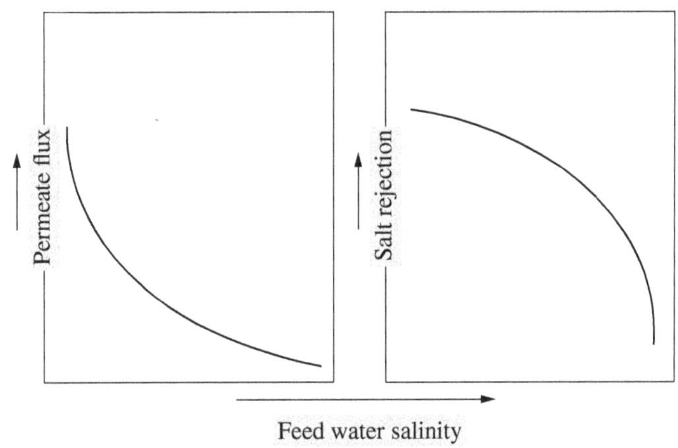

FIGURE 3.20 Effect of salinity on RO system performance.

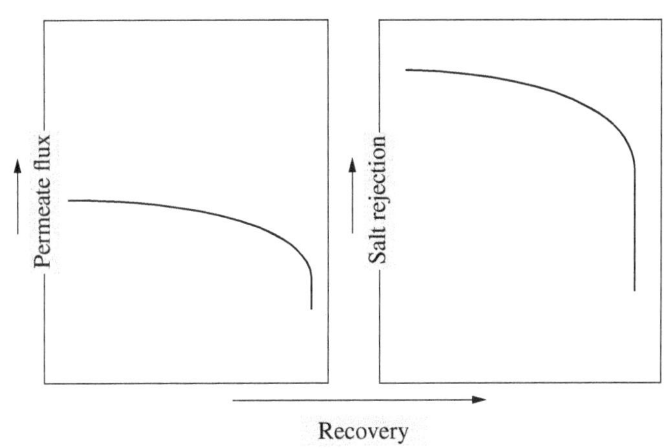

FIGURE 3.21 Effect of recovery on RO system performance.

and NDP is inadequate to drive flow through the membrane; at that point, freshwater flow production is discontinued.

3.6.6 Effect of Temperature on Membrane Performance

The use of warmer water reduces saline water viscosity, which in turn increases membrane permeability. Some of this beneficial impact is reduced by the increase of osmotic pressure with temperature (see Eq. 3.1). However, the overall impact of temperature for most membranes is typically beneficial (Fig. 3.22). As a rule of thumb, the permeate flux increases by 3% for every 1°C of temperature increase. Because most RO membranes are made of plastic materials (polymers), warmer temperatures result in a loosening up of the membrane structure, which in turn increases salt passage (i.e., deteriorates permeate water quality).

It should be pointed out that, as seen in Fig. 3.22, the rate of permeate flux gain is typically much higher than the rate of deterioration of product water quality. For source water temperatures up to 30°C (86°F), using warmer water allows reduction of the feed pressure and energy used for desalination (Greenlee et al., 2009). Because of the negative impact of temperature on osmotic pressure, operation at higher temperatures may or may not be beneficial. In addition, the use of warmer water accelerates biological fouling, which in turn also reduces membrane permeability.

As discussed previously, operating conventional spiral-wound RO membranes at temperatures above 40°C (104°F) accelerates the compaction of the membrane support layer and is undesirable because it results in a premature and irreversible loss of membrane permeability. Most membrane suppliers recommend that the temperature of the source water processed by RO membranes should be maintained below 45°C (113°F) at all times to avoid permanent membrane damage.

3.6.7 Effect of Feed Pressure on Membrane Performance

As can be seen from analysis of Eq. 3.14, membrane flux (productivity) increases along with operating feed pressure at the same source water salinity and temperature. This occurs because the increase of feed pressure results in a proportional increase of the net driving pressure through the membrane (Fig. 3.23).

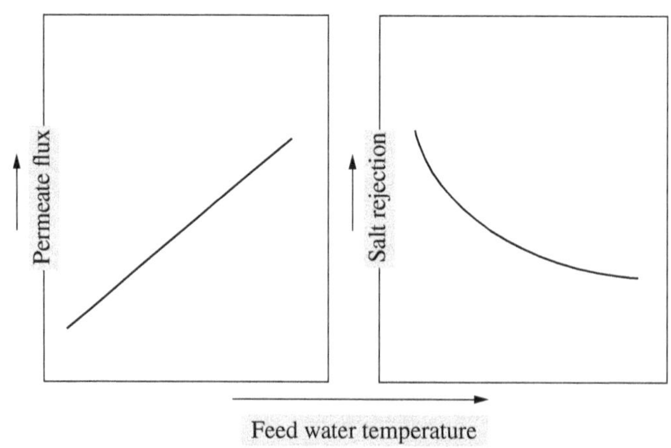

FIGURE 3.22 Effect of temperature on RO system performance.

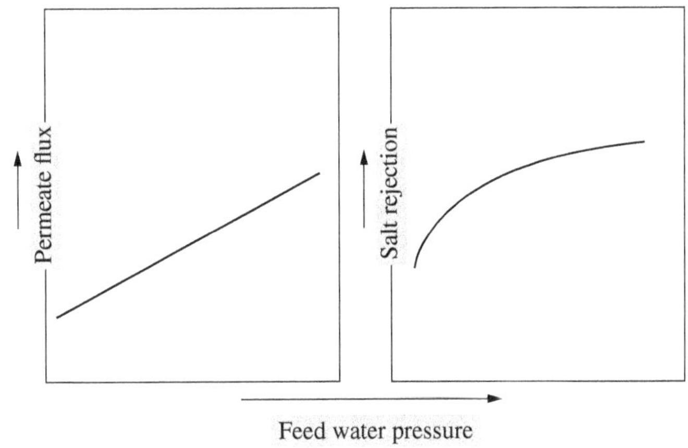

Figure 3.23 Effect of feed pressure on RO system performance.

As per Eq. 3.15, salt transport is unaffected by pressure. However, because more water is produced at higher pressure and the same amount of salt is contained in this water, the permeate salinity concentration decreases (i.e., it appears that salt rejection increases) with pressure.

3.6.8 Effect of Permeate Back Pressure on Membrane Performance

An analysis of Eq. 3.14 indicates that permeate pressure (often also referred to as *permeate back pressure*) has a direct negative impact on flux. However, if the pressure on the permeate side of the membrane is increased, this increase will result in a reduction of the diffusion rate through the membrane and therefore of β. Since practical experience shows that reduction of β and the osmotic pressure is higher than the direct decrease of flux the overall effect of permeate back pressure is positive, i.e., membrane flux could be increased by operating at elevated back pressure up to a point where β approaches 1.1. Creating additional back pressure beyond this point would have a negative impact on plant performance. The amount of permeate back pressure is limited by the impact this pressure can have on the thin film—if the pressure is higher than 0.3 bar (4.3 lb/in^2), then it may cause delamination of the thin-film membrane layer.

3.7 Key Membrane Desalination Plant Components

3.7.1 General Overview

As with any other natural water source, seawater contains solids in two forms: suspended and dissolved. Suspended solids occur in the form of insoluble particles (particulates, debris, marine organisms, silt, colloids, etc.). Dissolved solids are present in soluble form (ions of minerals such as chloride, sodium, calcium, magnesium, etc.).

At present, practically all RO desalination plants incorporate two key treatment steps designed to sequentially remove suspended and dissolved solids from the source water. The purpose of the first step—source water pretreatment—is to remove the suspended solids and prevent some of the naturally occurring soluble solids from turning into solid form and precipitating on the RO membranes during the salt separation process.

The second step—the RO system—separates the dissolved solids from the pretreated source water, thereby producing fresh low-salinity water suitable for human consumption, agricultural uses, and for industrial and other applications.

Once the desalination process is complete, the freshwater produced by the RO system is further treated for corrosion and health protection and disinfected prior to distribution for final use. This third step of the desalination plant treatment process is referred to as post-treatment.

Figure 3.24 presents a general schematic of a seawater desalination plant. In general, brackish water desalination plants incorporate similar source water treatment steps and technologies. The differences between the two types of plants are discussed in detail in the next chapters of this book.

The plant shown in Fig. 3.24 collects water using open ocean intake, which is conditioned by coagulation and flocculation and filtered by granular media pretreatment filters to remove most particulate and colloidal solids, and some organic and microbiological foulants. The filtered water is conveyed via transfer pumps through micron-size filters (referenced on the figure as cartridge filters) into the suction headers of high-pressure pumps. These pumps deliver the filtered water into the RO membrane vessels at a net driving pressure adequate to produce the target desalinated water flow and quality.

The reverse osmosis vessels are assembled in individual sets of independently operating units referred to as *RO trains* or *racks*. All RO trains collectively are termed the *reverse osmosis system*. The RO system usually has energy recovery equipment that allows it to reuse the energy contained in the concentrate for pumping of new source water into the membrane system.

The permeate generated by the RO trains is stabilized by addition of lime or contact with calcite and by the addition of carbon dioxide to provide an adequate level of alkalinity and hardness for protection of the product water distribution system against corrosion. The conditioned water is stored and disinfected prior to delivery to the final users.

The particulate solids removed from the source water by the pretreatment filters are collected in the filter backwash and further concentrated by thickening and dewatering for ultimate off-site disposal to a sanitary landfill. While this solids handling approach is adopted by many of the most recently built desalination plants, in some older facilities the concentrate and backwash water are mixed and disposed to the water body used for source water collection.

3.7.2 Plant Intake

Plant intake is designed to collect source water at a quality and quantity adequate to produce the target volume and quality of desalinated water. Chapters 6, 7, and 8 discuss various types of intakes, pump stations, and screening facilities used in seawater and brackish water desalination plants and provide guidelines for their selection and design.

3.7.3 Pretreatment

The fine microstructure of thin-film composite membranes presently used for desalination by reverse osmosis does not permit passage of particulates contained in the source

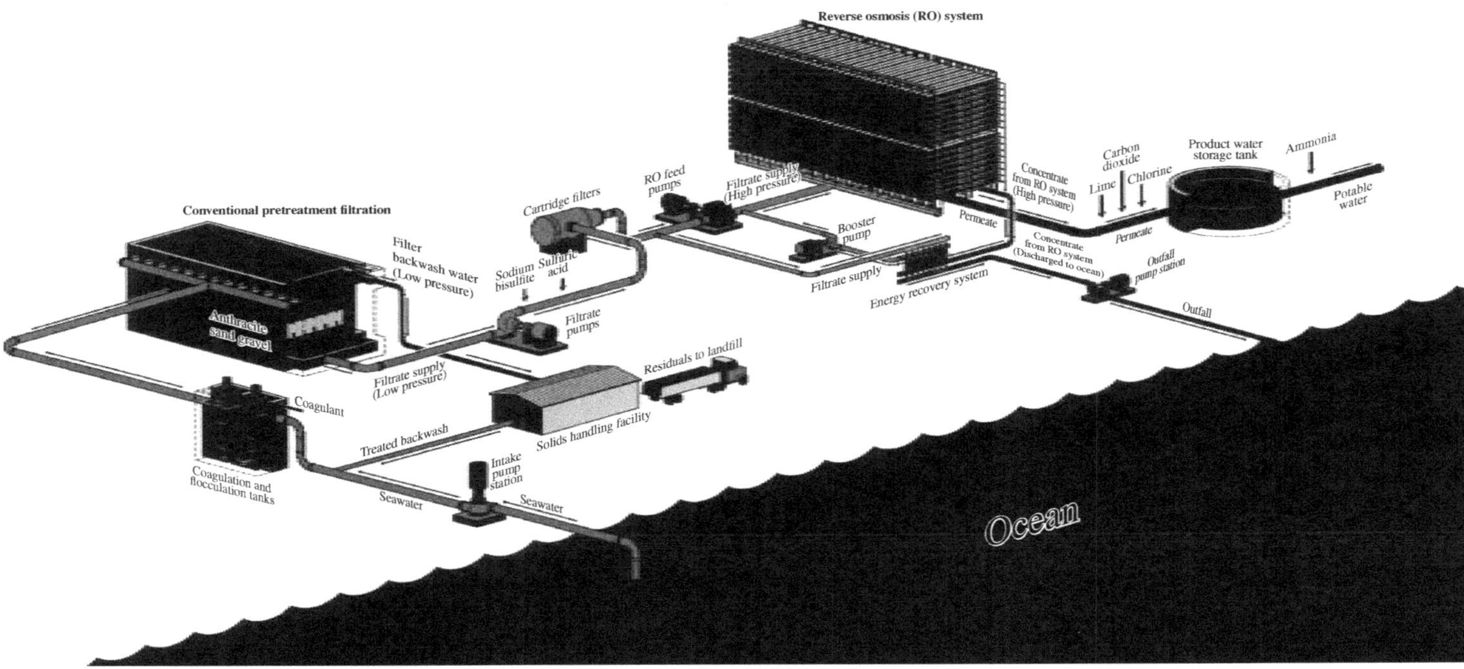

FIGURE 3.24 Schematic of a typical seawater desalination plant.

water or formed during the desalination process. Therefore, if they are present in the source water in significant amounts, these particulates may cause membrane fouling, which in turn may rapidly decrease membrane productivity and result in performance failure of the desalination plant. Membrane foulants are typically organic and inorganic colloids and particulates, naturally occurring in the source water or generated on the surface of the membranes by marine microorganisms or physical-chemical processes that occur during reverse osmosis salt separation and concentration.

The purpose of the pretreatment system is to adequately and effectively remove foulants from the source water and to secure consistent and efficient performance of the downstream reverse osmosis membranes. The pretreatment system is typically located downstream of the desalination plant's intake facilities and upstream of the reverse osmosis membrane system.

Depending on the source water quality, this system may consist of one or more water treatment processes, including screening, chemical conditioning, dissolved air flotation or gravity clarification, granular media filtration, membrane microfiltration or ultrafiltration, and cartridge filtration. Chapter 9 of this book addresses commonly used source water conditioning processes. Chapter 10 discusses sand removal, sedimentation, and dissolved air flotation clarification; Chaps. 11, 12, and 13 provide detailed overview of alternative pretreatment filtration technologies.

3.7.4 Reverse Osmosis Separation System

The key components of the RO separation system include filter effluent transfer pumps, high-pressure pumps, reverse osmosis trains, energy recovery equipment, and the membrane cleaning system. These facility components are discussed in detail in Chap. 14.

3.7.5 Post-Treatment

Post-treatment facilities include equipment for remineralization and disinfection of RO permeate. Some brackish water plants have additional post-treatment facilities for removal of odorous gases naturally contained in the source water, such as hydrogen sulfide. Alternative post-treatment technologies are addressed in detail in Chap. 15.

3.7.6 Desalination Plant Discharge Management

Desalination plants typically generate source water pretreatment and concentrate waste streams, which have to be handled in an environmentally safe and cost-effective manner. Chapter 16 describes the most commonly used desalination discharge management alternatives and provides guidance for their implementation.

3.8 References

American Water Works Association, *Manual of Water Supply Practices, M46, Reverse Osmosis and Nanofiltration*, 2nd ed., American Water Works Association, Denver, Colorado, 2007.

Fritzmann, C., J. Lowenberg, T. Wintgens, and T. Melin, "State-of-the-Art of Reverse Osmosis Desalination," *Desalination*, 216 (X), 1–76, 2007.

Greenlee, L. F., D. F. Lawler, B. D. Freeman, and B. Marrot, "Reverse Osmosis Desalination: Water Sources, Technology, and Today's Challenges," *Water Research*, 43 (X), 2317–2348, 2009.

Hassan, A., N. Ali, N. Abdull, and A. Ismail, "A Theoretical Approach on Membrane Characterization: The Deduction of Fine Structural Details of Asymmetric Nanofiltration Membranes," *Desalination*, 206 (X), 107–126, 2007.

Hoek, E. M. V., S. Bhattacharjee, and M. Elimelech, "Effect of Surface Roughness on Colloid Membrane DLVO Interactions," *Langmuir*, 19 (X), 4836–4847, 2003.

Hoek, E. M. V., and G. Agarwal, "Extended DLVO Interactions between Spherical Particles and Rough Surfaces," *Journal of Colloid and Interface Science*, 298 (1), 50–58, 2006.

Malaeb, L., and G. M. Ayoub, "Reverse Osmosis Technology for Water Treatment: State of the Art Review," *Desalination*, 267 (X), 1–8, 2011.

Wilf, M., L. Awerbuch, C. Bartels, M. Mickley, G. Pearce, and N. Voutchkov, *The Guidebook to Membrane Desalination Technology*, Balaban Desalination Publications, L'Aquila, Italy, 2007.

CHAPTER 4

Planning Considerations

4.1 Introduction

The main purpose of project planning are to define the size, location, and scope of the desalination project and to chart a roadmap for project implementation. The first step of project planning is to determine the service area of the desalination facility, identify the types of users of desalinated water in the area, and assess the water demand and water quality requirements of each water customer over the useful life of the desalination project (25 to 30 years).

Once the size and service area of the desalination project are determined, the next step of the planning process is to define the project; this encompasses identification of the most viable plant site location and intake and discharge types and configurations; characterization of plant source water quality; and selection of the treatment process configuration that can produce the target desalinated water quality and quantity at the lowest life-cycle cost and with the least impact on the surrounding terrestrial and aquatic environments.

Usually, the selection of the most cost effective and environmentally sound location and configuration for a desalination project is based on a thorough evaluation of a number of alternatives for the key desalination project components, including source water intake, concentrate discharge, pretreatment facilities, reverse osmosis (RO) system, post-treatment facilities, and product water delivery system. This project scoping effort also involves the preliminary evaluation of desalination plant's energy and chemical consumption, development of project site layout and hydraulic profile, and preparation of project implementation schedule.

After the project scope and schedule are defined, the next steps of the planning process are to prepare project's environmental impact assessment and obtain project entitlements such as legal rights to use the land for the plant site, water rights for source water collection, easements for project-related infrastructure, rights-of-way, an electric power supply agreement, and environmental permits, licenses, and other regulatory, legal, and contractual documents that are needed for project implementation.

In parallel with these activities, project planning also includes development of budgetary estimates for construction, operation and maintenance (O&M), and water production costs, as well as identification of funding sources and contractors needed for project implementation.

4.2 Plant Service Area, Capacity, and Site

4.2.1 Service Area

The service area supplied with fresh water from the desalination plant is typically determined based on jurisdictional boundaries, the demand and location of the main water users in the area, the configuration and size of the existing water distribution system servicing this area, and the distance between the key water distribution system infrastructure (i.e., water storage reservoirs, aqueducts, etc.) and the potential site of the desalination plant.

The boundaries of desalination plant's service area are also often influenced by the ability of the water purveyor to supply lower-cost water to the same area from another source or to increase the level of water conservation and/or water reuse in the area in order to sustainably balance water demand and supply.

Other important factors associated with the size of the service area of the desalination plant are the cost of water production and of water delivery. Usually, a larger service area will result in a larger plant, which in turn will yield cost savings from economies of scale. On the other hand, delivering water to a larger service area may require the construction of additional costly freshwater conveyance and distribution infrastructure, which could negate the savings associated with the construction of a larger plant.

Because the configuration of the existing water distribution system, the distance between main water users and the plant site, and the costs for construction and conveyance are very site specific, the optimum size and boundaries of the service area would have to be established based on a comprehensive life-cycle cost benefit analysis that balances the cost savings stemming from a larger plant with the cost penalties for freshwater delivery over a greater distance.

4.2.2 Plant Capacity

Typically, project's freshwater production capacity is determined based on a comprehensive comparative evaluation of the balance between the water demand in the plant's service area and the cumulative capacity of all available traditional sources of water supply and alternative freshwater resources, (i.e., desalination, water reuse, rainwater harvesting, conservation, water importation, etc.) that can be used to cover the water demand over the entire planning period or useful life of the desalination project.

Usually, desalination is one of the most costly sources of water supply for a given service area. Therefore, desalination plant's capacity is often determined based on the freshwater flow that this water supply alternative can provide during periods of prolonged drought as compared to other traditional and alternative water supply resources, and on the incremental costs of new water supplies. For example, countries exposed to long drought cycles—such as Australia, Israel, Spain, South Africa, countries in the Middle East and North Africa, etc.—have established an internal target to provide at least 25 to 50 percent of the total drinking water supply for their large coastal urban centers from seawater desalination. In many of these countries, brackish water desalination projects already supply close to the maximum capacity they can produce from saline inland aquifers and brackish surface water sources.

Another important factor for the selection of optimum plant size is the economy-of-scale benefit of building one or more large desalination plants supplying freshwater for the entire service area compared to installing a number of smaller facilities located closer to the main water users within the service area. Additional discussion of this factor is provided in Sec. 4.9 of this chapter.

Capacity analysis for a desalination project usually considers annual and diurnal water supply patterns, hydraulic limitations of the water distribution system, water quality and quantity requirements of key users in the service area, and projections of future water demand. This analysis also includes requirements and costs for conveying the desalinated water to the distribution system, potential connection points and associated capacity limits, hydraulic system requirements (i.e., size of piping and equipment, as well as operating pressure of the water distribution system at the point of delivery of desalinated water), and limitations of system conveyance capacity and potential solutions.

4.2.3 Plant Site

Site selection for a desalination plant is most often based on land availability near the main users of desalinated water and on the location of the delivery points of this water to the distribution system. The land requirements for a typical desalination plant are summarized in Table 4.1. These requirements apply for both seawater and brackish water desalination plants.

The plant site footprints in Table 4.1 are based on a comparative review of over 40 desalination projects worldwide. However, sometimes environmental and zoning regulations, physical constraints, and/or soil conditions associated with a particular site may require a smaller or larger site.

In some cases, when the available site is located in a densely populated area or land costs are very high, the desalination plant can be located at a site a fraction of the footprint shown in Table 4.1.

The development of a more compact plant layout often requires that some of the main treatment equipment and systems—such as the plant pretreatment filters, RO racks, energy recovery devices and pumps, and/or chemical feed and solids handling facilities—be installed in multistory buildings. While possible, this is usually more costly than housing all plant treatment facilities in single-story buildings with slab on grade.

	Typical Plant Site Land Requirement	
Plant Capacity, m³/day	**m²**	**acres**
1000	800–1600	0.2–0.4
5000	2500–3200	0.6–0.8
10,000	4500–6100	1.1–1.5
20,000	10,100–14,200	2.5–3.5
40,000	18,200–24,300	4.5–6.0
100,000	26,300–34,400	6.5–8.5
200,000	36,400–48,600	9.0–12.0
300,000	58,700–83,000	14.5–20.5

*Land requirements are based on a conventional plant layout. Compact plants may require less land.

Table 4.1 Desalination Plant Land Requirements*

Typically, the most viable site location for a given project is determined through a cost-benefit analysis of several alternative sites within the plant service area. Potential alternative sites are selected that, at a minimum, meet the following requirements:

- Adequate land size and footprint available to construct a desalination plant of the selected capacity (see Table 4.1)
- Accessibility from existing main roads, highways, etc.
- Proximity (usually less than 8 km or 5 mi) to the points of delivery of the desalinated water to the local water distribution system and to the points of electrical grid interconnection for plant power supply
- Relatively short distance (within 1.0 km or 0.6 mi) from the source of saline water (ocean, waterway, brackish aquifer, etc.) and the points of concentrate discharge
- Compatibility with local land planning and zoning requirements
- Limited or no soil and groundwater contamination, vegetation, debris, and existing surface and underground structures and utilities
- Location outside of environmentally sensitive areas such as wildlife reserves, migratory bird stopover sites, and natural habitats of endangered species
- Reasonable costs for obtaining entitlements associated with the use of the site (i.e., site lease, purchase, etc.)— preferably less than 0.5 percent of the total plant construction costs
- Adequate distance (at least 30 m or 100 ft) from residential dwellings, hotels, hospitals, and other developments whose inhabitants could be sensitive to increased levels of noise and traffic during plant construction and operation

After potential sites are identified, the following key engineering and environmental review activities are typically completed for each site:

- Geological reconnaissance survey
- Traffic and access survey
- Survey of existing above- and underground utilities and structures
- Biological and archeological surveys
- Evaluation of near- and offshore marine resources with a focus on the type, environmental sensitivity, and location of aquatic species inhabiting the desalination plant's intake and discharge areas
- Review of near- and offshore bathymetry, hydrology, and geology
- Assessment of the site risks associated with potential impacts on the plant intake, outfall, and facilities from beach erosion or siltation, flooding, severe storms, hurricanes, and tornadoes, and earthquakes and tsunamis
- Preliminary analysis of the saline source water quality in terms of mineral and organic content
- Schedule of conceptual plant design, environmental review, and implementation
- Identification of alternative routes for delivery of the desalinated water to the distribution system

The engineering information collected from these site studies and investigations is typically compiled into project site alternatives, and the potential project sites are then ranked based on their merits and potential disadvantages.

4.3 Intake Type and Location

Intakes are a key component of every desalination plant—their type and location have a measurable impact on source and product water quality, cost of water production, and potential environmental impacts of plant operations.

The purpose of the desalination plant intake is to collect saline source water of adequate quantity and quality in a reliable and sustainable fashion so as to produce desalinated water cost effectively and with minimal impact on the environment. Currently, there are two categories of source water collection facilities that are widely used in desalination plants: open intakes and subsurface intakes (wells and infiltration galleries).

Open intakes collect source water directly from a surface water body (brackish river or lake, ocean, etc.) via an onshore or offshore inlet structure and pipeline interconnecting this structure to the desalination plant. Subsurface intakes, such as vertical wells, horizontal wells, slant wells and infiltration galleries, are typically used to collect saline water from brackish aquifers for brackish water reverse osmosis (BWRO) desalination and from near- or offshore coastal aquifers for seawater reverse osmosis (SWRO) desalination.

Both open and subsurface intakes have functional and capacity constraints, and their operation can potentially have environmental impacts. Chapter 5 provides detailed information on potential environmental impacts of desalination plant intakes and measures for their minimization and mitigation. Chapter 6 describes key design criteria and considerations for the installation and operation of various intake types and configurations.

4.3.1 Brackish Water Intake Planning Considerations

Subsurface Intakes

Most brackish water sources worldwide are located in groundwater aquifers. Therefore, usually the prime choice and focus in the initial planning phases of brackish water desalination projects is to find one or more aquifers of adequate size and water quality that can sustainably provide source water over the useful life of the project. Since in many locations groundwater ownership is attached to the ownership of the land, securing the rights for groundwater extraction, use, and ownership is of critical importance for the viability of the project.

Once the location of an adequate source water aquifer is identified, the aquifer must be characterized in terms of transmissivity, thickness, water quality, and potential interconnection with other aquifers in the area that could be impacted by the operation of the desalination plant's well intake or could have a negative impact on plant's source water quality. Usually, the productivity of the target source water aquifer and the projected capacity of the individual extraction wells are determined based on: (1) a preliminary geological survey, which includes the collection of aquifer formation deposits for visual classification and analysis of grain size distribution; (2) installation and operation of test and observation (monitoring) wells; (3) collection of samples for groundwater quality and contamination analysis; and (4) hydrogeological modeling of well yield, radius of influence, and water quality changes over time.

Quality assessment of brackish groundwater is a key component of the project planning process. If the target brackish source water aquifer is already in use, water quality information from existing wells typically can be applied in projecting the source water quality of the new wells. However, this water quality information alone may not be adequate to predict changes of aquifer water quality as a result of the increased rate of extraction from the aquifer.

Extracting an additional volume of water from a given aquifer may result in the modification of the natural groundwater movement regime in this aquifer. Some brackish groundwater aquifers are density-stratified, and as lower-salinity water is extracted from the top portion of the aquifer, higher-salinity groundwater propagates upwards and increases the salinity of the desalination plant's feed water over time. In addition, semi-confined aquifers could allow groundwater from an adjacent aquifer to move to the main water extraction aquifer for a given project and impact its quality.

If the water quality of the adjacent aquifer is very different from that of the main aquifer, the overall quality of the plant source water, including salinity, content of key minerals and gases, color, temperature, and odor, may change over time. Therefore, it is of critical importance to complete predictive modeling of the plant's source water quality as a part of the project planning process.

The prime criteria for selecting the most suitable location for a BWRO project's source water aquifer are safe yield capacity and proximity to the desalination plant site. Another key selection factor is the presence of potential sources of subsurface or surface contamination that can propagate and contaminate the plant source water (e.g., proximity of the intake well to unlined sanitary or hazardous waste landfills, leaking fuel oil storage tanks, cemeteries, industrial or military sites known to have groundwater or surface water contamination, etc.).

Another issue of key importance is the proximity of the intake wells to existing freshwater supply wells and the potential for the operation of the desalination plant wells to result in a decrease in production capacity of the freshwater wells.

Surface Water Intakes

At present, less than 10 percent of the brackish water desalination plants worldwide have surface water intakes. Such intakes are typically located in the confluence of a river and an ocean or sea. One of the largest desalination plants at present with such intake is the Beckton desalination plant near London, in the United Kingdom (Fig. 4.1); the plant has a capacity of 150,000 m^3/day (40 mgd) and is operated by Thames Water. The criteria for selection of the location and configuration of brackish surface water intakes are similar to those of open seawater intakes, discussed in the next section.

4.3.2 Seawater Intake Planning Considerations

Subsurface Intakes

Subsurface intakes, and more specifically vertical beach wells, are the most commonly used type of intake for small seawater desalination plants. The individual production capacity of such wells can range between several hundred and 10,000 m^3/day [1.0 and 2.5 mgd or 690 to 1730 gallons per minute (gpm)]. Shallow vertical wells are also the lowest-cost type of intake. Because such intakes filter the source water slowly through the aquifer soils, they usually have minimal environmental impact and produce better-quality water than do open ocean intakes.

Figure 4.1 Surface water intake of the Beckton desalination plant, London, United Kingdom.

Therefore, if a seawater coastal aquifer of adequate hydrogeological characteristics and yield (source water production capacity) is available within 10 km (6 mi) of the desalination plant site, such intakes are often the preferred choice. Typically, permeable sand and limestone- or dolomite-type geological formations with a transmissivity of 1000 m^3/day per meter (0.088 mgd/ft) or higher are the most suitable types of strata for the construction of seawater well intakes.

Productivity of the Coastal Aquifer The capacity of the source water coastal aquifer and the quality of the water that this aquifer can yield are the two most important factors that define the size of the seawater desalination plant, and often its location. Therefore, the completion of a hydrogeological study that allows the determination of the aquifer's water production capacity and quality, and the safe yield of the individual intake wells is a critical component of the planning process for a SWRO desalination project.

Beaches and shallow bays that have low transmissivity, contain a large quantity of silted beach deposits, and are poorly flushed are typically unsuitable for the installation of beach well intakes. It should be pointed out that both beach wells and near-shore open intakes use the same seawater as a source. In desalination plants with open intakes, the solids contained in the source seawater are removed in the pretreatment filtration system. In plants with beach well intakes, the same amount of solids is retained on the ocean floor in the area where the well source water is collected, while the filtered water is slowly conveyed through the ocean floor and the beach subterranean formation until it reaches the well collectors.

The wave action near the ocean floor is the force that allows the solids separated from the beach well source water to be dissipated in the ocean. If the bay area is not well flushed and the naturally occurring wave movement is inadequate to transport the solids away from the beach well collection area at a rate higher than the rate of solids deposition, these solids will begin to accumulate on the ocean floor and ultimately reduce the well capacity and source water quality.

Useful Life of Beach Well Intakes Depending on the specific site conditions, beach wells often have a shorter useful life than do open ocean intakes. The useful life of open ocean intakes is typically between 30 and 100 years, depending on their configuration and on the quality and type of their materials of construction. Without major refurbishment, beach wells typically operate at design capacity for a period 10 to 20 years. Over time, the beach well yield may diminish due to naturally occurring scaling of the well collectors, caused by chemical precipitates and/or bacterial growth. Prediction of the rate of decline of well yield over time is difficult and requires specialized expertise and detailed studies. Therefore, beach well intakes are usually designed with 20 to 25 percent reserve or standby well capacity, which adds to their capital costs and the size of the impacted shore area.

Beach erosion is an additional factor that can significantly impact the useful life of the intake wells. As seen in Fig. 4.2, if the well intake area is exposed to a high rate of beach erosion, within several years of operation the wells may lose soil support, productivity, and structural integrity. Therefore, beach erosion may shorten the useful life of the beach wells significantly and may increase the overall life-cycle cost of water production.

Due to its significant potential impact on the operation and costs of the intake system, beach erosion in the vicinity of the targeted intake location has to be thoroughly investigated during selection of the type and location of desalination plant intake. If the selected beach site has a high potential for erosion, the beach wells have to be provided with antierosion measures. However, the preferred approach in such conditions is to use deep open intake or to install the intake wells inland in an area that is outside of the zone of active beach erosion.

Figure 4.2 Seawater intake well exposed to beach erosion.

Beach erosion may also impact the useful life and integrity of open intakes. Therefore, in coastal zones exposed to active beach erosion, the first several hundred to one thousand meters of the intake pipeline closest to the shore are typically installed under the ocean floor, at a depth outside of the zone of active beach erosion.

The useful life of a well-designed and well-operated seawater desalination plant is 25 to 30 years. Because the beach wells may often have a shorter useful life than that of the desalination plant, in the worst-case scenario two sets of beach wells may need to be constructed over the useful life of the SWRO plant. The need for replacement of some or all of the original beach wells after the first 10 to 20 years of operation of a desalination plant would magnify the shoreline impacts of the beach wells and increase the overall cost of water production. Therefore, the potential difference between the useful life of beach wells and open intakes has to be reflected in the life-cycle cost comparison associated with the selection of the most viable type of desalination plant intake.

Source Water Pretreatment Requirements As mentioned previously, seawater beach wells typically yield better intake water quality than do open intakes, in terms of turbidity, algal content, and silt density index—which are key parameters associated with the selection, sizing, complexity, and costs of a desalination plant's pretreatment system. Therefore, it is often assumed that the use of beach wells will eliminate the need for seawater pretreatment prior to reverse osmosis desalination. This assumption, however, holds true only for very specific favorable hydrogeological conditions (i.e., the wells are located in a well-flushed ocean bottom or shore, are sited away from the influence of surface freshwater, and are collecting seawater from a coastal aquifer of uniformly porous structure, such as limestone or dolomite). Long-term operational experience at numerous small seawater desalination plants in the Caribbean and several medium-size plants in Malta which have well intakes located in limestone and other favorable rock formations indicates that such plants can successfully operate with minimal pretreatment (typically bag or cartridge filters and/or sand strainers) ahead of the SWRO system. However, most seawater desalination plants using subsurface intakes have to include an additional granular or membrane filtration step prior to RO membrane salt separation in order to be able to process source water collected by subsurface intakes.

Experience with the use of beach wells for seawater desalination in California and at the largest beach well seawater desalination plant on the Pacific coast of North America (in Salina Cruz, Mexico) indicates that some desalination plants using beach wells may face a costly problem—high concentrations of manganese and/or iron in the intake water. Unless removed ahead of the RO membrane system, particulate/colloidal iron and manganese in oxidized form may quickly foul the cartridge filters and SWRO membranes, rendering the desalination plant inoperable.

The treatment of beach well water which naturally contains high concentrations of iron and/or manganese requires chemical conditioning and installation of conservatively designed "greensand" pretreatment filters ahead of the SWRO system. This costly pretreatment requirement may significantly reduce the benefits of the use of beach wells rather than open intakes. Open seawater intakes typically do not have problems with source water quality related to iron and manganese, because open ocean water does not contain these compounds in quantities significant enough to cause RO membrane fouling.

One example of a beach well desalination plant that has faced a problem with elevated iron in the source water is the 4500 m^3/day (1.2 mgd) Morro Bay SWRO facility located in northern California, in the United States. The plant source water is supplied by five beach wells, each with a production capacity of 1100 to 1900 m^3/day (0.3 to 0.5 mgd). The beach well intake water has an iron concentration of 5 to 17 mg/L. For comparison, open intake seawater typically has several orders of magnitude lower iron content.

The Morro Bay facility was originally designed without pretreatment filters, which resulted in plugging of the RO cartridge filters within half an hour of starting operations during an attempt to run the plant in 1996. The problem of high iron concentration was resolved by the installation of a pretreatment filter designed for a surface loading rate of 6.1 $m^3/m^2{\cdot}h$ (2.5 gpm/ft^2). For comparison, a typical open intake desalination plant is designed for pretreatment loading rates of 10 to 13.5 $m^3/m^2{\cdot}h$ (4.0 to 5.5 gpm/ft^2), and, therefore would require less pretreatment filtration capacity.

As indicated previously, the largest existing Pacific-coast seawater desalination plant—in Salina Cruz, Mexico—has also faced iron and manganese challenges, which have been resolved by the installation of pretreatment filters and chemical conditioning of the beach well water. The existing experience shows that the costs for pretreatment of seawater with high iron/manganese content collected by a beach well intake are typically comparable to or higher than the costs for pretreatment of seawater collected using an open ocean intake.

Source Water Quality Variations Open ocean intakes (especially if they are deeper than 10 m/30 ft) provide relatively consistent seawater quality in terms of concentration of total dissolved solids (TDS). The data on TDS concentration of intake source water that were collected for the development of the Huntington Beach seawater desalination project in Southern California, in the United States, indicate that the open intake salinity varied within 10 percent of its average value of 33.5 ppt.

Although beach wells in general produce source water of consistent salinity, they can also yield water of an unpredictably variable TDS concentration, with swings exceeding 30 percent of the average value. For example, the TDS concentrations of the two operational wells at the Salina Cruz water treatment plant vary in a wide range—between 16,800 and 21,800 mg/L for well number 2 and between 17,800 and 19,800 mg/L for well number 3. The wide range of source salinity concentration in this case is explained by the influence of fresh groundwater.

A similar trend was observed at the Morro Bay SWRO plant in California. During the plant's initial operation in 1992, the well water TDS was approximately 26,000 mg/L. In December 2001, the TDS concentration of the intake water was 6300 mg/L. The December 2002 data for the same plant indicate intake salinity of 22,000 mg/L.

The wide range of intake salinity over time in systems using beach wells would require the installation of variable frequency drives for efficient power use control, which would ultimately increase the construction cost of such systems and complicate their operation.

One important issue to consider when assessing the viability of using beach wells is the fact that intake well salinity can change unpredictably over time when the well operation is influenced by freshwater inflow to the well's source water aquifer. This uncertainty regarding intake water quality increases the risk of uncontrollable rises in the unit cost of water production over time and has to be taken in consideration when comparing the overall life-cycle costs of plant operations. Therefore, the quality of the beach well intake water has to be thoroughly characterized, by installing a set of test wells and collecting water quality samples under a variety of operational conditions.

Thorough year-round water quality characterization is of high importance for beach wells whose source water may be influenced by fresh groundwater aquifers with seasonal fluctuation of water quality.

Content of Contaminants That Are Difficult to Treat Usually open ocean intakes are considered a less viable source of water for desalination plants in areas located in close proximity to wastewater discharges or industrial and port activities. However, open intake seawater is typically free of endocrine disrupting or carcinogenic compounds such as MTBE, NDMA, and 1,4-dioxane. Long-term water quality data collected for the development of the Huntington Beach and Carlsbad SWRO projects in Southern California and a number of other desalination plants worldwide have confirmed this observation.

Beach well water, on the other hand, may contain compounds that are difficult to treat, especially when the wells are under the influence of contaminated groundwater. An example is the Morro Bay SWRO plant, where beach well intake water was contaminated by MTBE from a leak of an underground gasoline tank. (MTBE is a gasoline additive.) Similar problems were observed at California's 500 m^3/day (0.132 mgd) Santa Catalina Island seawater desalination plant, which uses beach well intakes.

The compounds of concern can be treated by a number of available technologies, including activated carbon filtration, ultraviolet (UV) irradiation, hydrogen peroxide oxidation, ozonation, etc. However, because these treatment systems will need to be constructed in addition to the SWRO system, this additional treatment may measurably increase the overall production cost of desalinated water.

Although beach wells have proven to be quite cost-competitive for plants of capacity smaller than 4000 m^3/day (1.1 mgd), open ocean intakes have found significantly wider application for large SWRO desalination plants. At present, worldwide there are less than one dozen operational SWRO facilities with capacity larger than 20,000 m^3/day (5.3 mgd) that use beach well intakes. The largest SWRO facility with beach wells is the 54,000 m^3/day (14.3 mgd) Pembroke plant in Malta. This plant has been in operation since 1991.

The largest SWRO plant in North America that obtains source water from beach wells is the 15,000 m^3/day (3.8 mgd) water supply facility for the Pemex Salina Cruz refinery in Mexico. This plant also has the largest existing seawater intake wells—three Ranney-type radial collectors, each with a capacity of 15,000 m^3/day (3.8 mgd).

Surface Water Intakes

Open intakes typically include the following key components: an inlet structure (forebay) with coarse bar screens, a source water conveyance pipeline or channel connecting the inlet structure to an onshore concrete screen chamber, and mechanical fine screens in the chamber. Depending on the location of the inlet structure, the intakes can be onshore or offshore. Offshore intakes with vertical inlet structures are the most commonly used for seawater desalination projects. The offshore inlet structure is usually a vertical concrete or steel well (vault) or pipe located at or above the ocean floor and submerged below the surface of the water (Fig. 4.3).

The open intake inlet system may include passive wedgewire screens (Fig. 4.4). The use of such screens eliminates the need for coarse and fine screens on shore. Wedgewire screens are cylindrical metal screens that have trapezoidal-shaped wedgewire slots with openings of 0.5 to 10 mm. They combine very low flow-through velocities, small slot size, and naturally occurring high sweeping velocities at the screen surface to minimize impingement and entrainment. These screens are designed

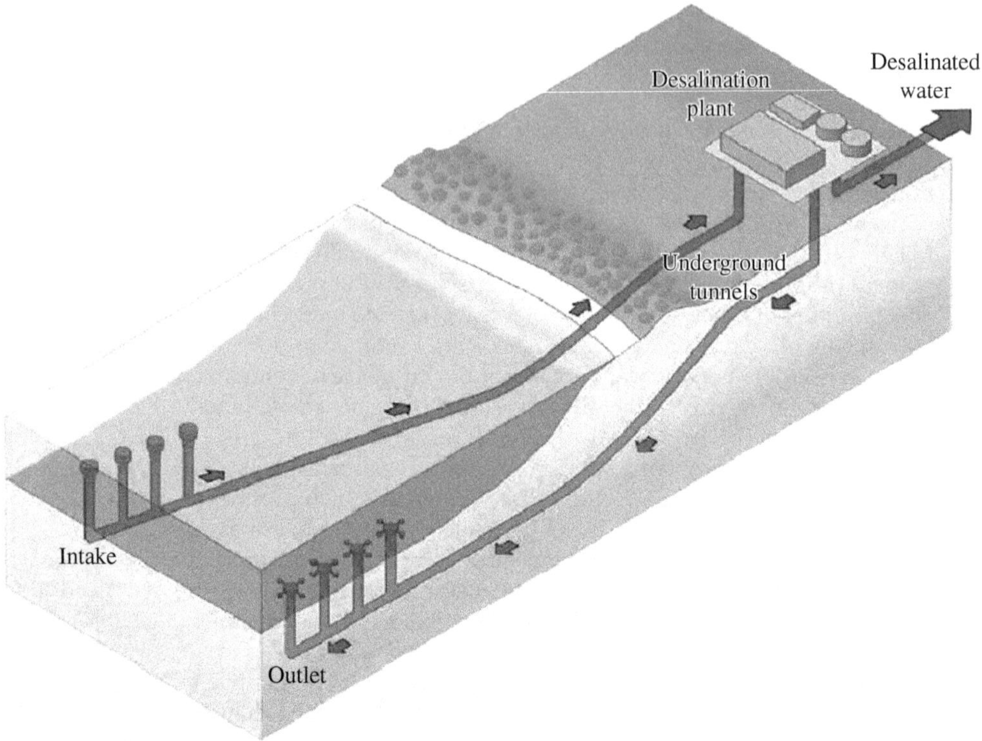

FIGURE 4.3 Desalination plant with offshore intake. (*Source: Sydney Water.*)

FIGURE 4.4 Wedgewire screen. (*Source: Acciona Agua.*)

to be placed in a water body where a significant prevailing ambient cross-flow current exists, with velocities higher than 0.3 m/s (1 ft/s). This high cross-flow velocity allows organisms that would otherwise be impinged on the wedgewire intake to be carried away with the flow.

An integral part of a typical wedgewire screen system is an airburst back-flush system, which directs a charge of compressed air to each screen unit to blow off debris back into the water body, where it is carried away from the screen unit by the ambient cross-flow currents.

Co-located intakes are a type of open intake for desalination plants co-sited with existing power generation stations that use seawater for once-through cooling purposes. Intake and/or discharge of collocated desalination plants is typically directly connected to the discharge outfall of a coastal power plant. The warmer cooling water discharged by the power plant is less viscous than the ambient ocean water, which reduces the energy needed for desalination by membrane separation. In addition, the use of co-located intakes eliminates in most cases the need to construct separate intake and outfall for the desalination plant, which reduces project's overall capital expenditures.

Open intakes face some of the same challenges that are associated with the siting and construction of beach wells—for example, beach erosion and impacts from large-magnitude earthquakes and storms. In addition, since open intakes collect water directly from the water column, the source water of the desalination plant could contain large quantities of debris, algae, silt, hydrocarbons, and other contaminants of anthropogenic or natural origin. Since RO membranes are easily fouled by such contaminants, the use of open intakes usually requires elaborate seawater pretreatment, as compared to the construction of beach wells.

Considerations for Selection of SWRO Plant Intake Type

At present, open ocean intakes are the most widely used type of intake technology worldwide, because they can be installed in practically any location and built in any size. While open intakes are suitable for all sizes of desalination plants, their cost-effectiveness depends on a number of location-related factors, such as plant size, depth and geology of the ocean floor, and performance impact of sources of contamination (e.g., wastewater and storm water outfalls, ship channel traffic, and large industrial port activities).

Mainly due to the fact that favorable hydrogeological conditions for subsurface intakes are often impossible to find in the vicinity of the plant site, the application of this type of intake technology to date has been limited to plants of relatively small capacity. In addition, densely populated coastal areas, where large desalination plants are needed, have very limited land availability for installation of numerous beach wells, which often is an important factor and potentially a fatal flaw for the construction of subsurface intakes in certain coastal communities.

Open and subsurface intakes offer different advantages and usually have different disadvantages in terms of capital, operation, and maintenance costs; construction complexity; environmental impact; operational considerations; and need for subsequent source water pretreatment and concentrate disposal. Therefore, the selection of the most suitable intake system for the site-specific conditions of a given desalination project should be completed based on a life-cycle cost-benefit analysis and environmental impact assessment including all key project components—intake, pretreatment, membrane salt separation, and concentrate disposal.

Intake selection should be based on a reasonable balance between the cost expenditures and environmental impacts associated with the production of desalinated water. Project proponents should not be burdened with the use of the most costly intake alternative if the environmental impacts associated with the construction and operation of a less-expensive type of intake are minimal and can be reasonably mitigated.

While thorough feasibility evaluation of intake alternatives is warranted, this evaluation should be initiated with prescreening for fatal flaws based on site-specific studies for the selected intake location. If the prescreening shows that certain intake alternatives have one or more fatal flaws that preclude their use, such intake systems should be removed from the evaluation process; their detailed feasibility assessment would be unproductive and would only cause unwarranted project delays and expenditures.

4.4 Source Water Quality

The selection of a saline water source and thorough analysis of its water quality are of critical importance for the successful planning, implementation, and long-term operation of desalination projects. Chapter 2 presents a detailed overview of the water quality of various saline sources and provides guidance regarding the parameters and methods commonly used for characterizing the quality of source water.

Typically, the content of total dissolved solids and the concentration of key ions (sodium, calcium, magnesium, bromide, boron, chloride, sulfate, carbonate, and bicarbonate) in the source water are of prime importance for planning both brackish and seawater desalination plants. These parameters, along with water temperature and pH, drive the design and configuration of the reverse osmosis system of most desalination projects.

Since desalination plant's RO facilities are usually associated with over 60 percent of plant construction costs and O&M expenditures, the mineral content, temperature, and pH of the saline source water are considered factors of prime importance in the process of planning and designing a desalination project. Therefore, if brackish water of adequate quality and yield is available in the service area, the construction of a brackish water rather than a seawater desalination plant would usually be less costly, and is almost always preferred. Often, however, such a choice is limited by the availability of suitable brackish water sources, especially for larger projects.

Saline groundwater collected via subsurface intakes could contain high concentrations of dissolved and colloidal iron and manganese in reduced form, colloidal silica, nitrates, ammonia, cyanide, and radionuclides, and could have a very low level of oxygen. In such cases, the actual concentrations of these constituents could significantly impact project planning and design.

Some brackish groundwater sources may have an elevated concentration of natural organic matter that causes discoloration of the source water, or may contain odorous gases such as hydrogen sulfide. The presence of such contaminants at levels above the thresholds presented in Chap. 2 typically would require additional treatment to produce finished water of drinking quality.

On the other hand, surface saline water sources (i.e., brackish lake or river waters or open ocean seawater) could periodically be exposed to algal blooms and could contain floating oil, grease, and hydrocarbons. In addition, surface source waters could have high levels of suspended solids and nutrients, which typically originate from surface water runoff and/or anthropogenic contamination. Such contaminants should be taken into consideration in project planning because they could have a great impact on plant design and costs.

Since water quality can vary significantly over time, source water characterization should encompass both typical water quality conditions and events which result in extremely low or high values of the previously discussed water quality parameters (e.g., heavy storms and ship traffic, dredging of the intake area, algal blooms, seasonal changes in the direction of underwater currents and near-shore wind patterns, periodic industrial discharges, etc.).

Source water quality has a measurable impact on the cost of producing desalinated water. In general, construction and O&M costs increase with an increase in source water concentration of TDS and with a decrease in water temperature. Source seawater TDS concentration is directly related to the RO system's design feed pressure and the overall plant design recovery and configuration. Therefore, the use of lower-salinity source water typically allows a reduction of the costs associated with construction and operation of the RO system and at the same time an increase in plant recovery.

However, it is important to note that the consistency of the source water quality is often almost equally important for successful desalination plant design and operation, as is the level of TDS in the source water. For example, construction of a seawater desalination plant intake near the confluence of a river and the ocean could reduce the overall source water salinity and therefore decrease the plant's total energy use. However, if the river water carries heavy loads of turbidity, organics, nutrients, and man-made pollutants, the removal of the contaminants contributed by the river water may require a more elaborate desalination pretreatment, which in turn may negate the cost savings from lower-salinity water.

4.5 Product Water Quality

Product water quality is one of the key factors that have significant impact on plant configuration and costs. The sections below address key product water quality issues that have to be taken under consideration when planning brackish and seawater desalination plants.

4.5.1 Water Quality of SWRO Desalination Plants

Mineral Content

Content of minerals in the permeate produced by SWRO desalination plants may vary over a wide range, depending on the ion composition and temperature of the source water, the configuration of the RO membrane system, and the salt rejection of the membranes used for desalination. Projections of permeate water quality produced by SWRO systems of different configurations (i.e., single-pass, full two-pass, and split-permeate second pass) from different sources of seawater are presented in greater detail in Chap. 14.

Concentrations of TDS (300 to 500 mg/L), chloride (150 to 240 mg/L) and sodium (90 to 180 mg/L) in the permeate generated by a single-pass SWRO system are typically within United States Environmental Protection Agency (US EPA) regulatory requirements and World Health Organization (WHO) drinking water quality guidelines. However, if the intended use is irrigation of salinity-sensitive crops (e.g., avocados, strawberries) and/or ornamental plants (e.g., some species of palm trees, flowers, or grasses), the introduction of this desalinated water into the distribution system may pose potential challenges unless the TDS, chloride, and sodium are diluted by the other water sources in the distribution system to below 250 mg/L, 120 mg/L, and 80 mg/L, respectively. Alternatively, seawater treatment by a two-pass RO system can produce water of suitable quality for all municipal, agricultural and horticultural uses.

In addition, the ratio of sodium (Na^+), calcium (Ca^{2+}), and magnesium (Mg^{2+}) ions in irrigation water—referred to as the sodium adsorption ratio (SAR)—could also have an impact on some crops. An excessively high SAR value (i.e., a high level of sodium and low levels of calcium and magnesium) in the irrigation water contributes to soil dispersion and structural breakdown, which in turn results in filling up of the soil pores with finer soil particles and ultimately in diminished water and nutrient infiltration rates and reduced crop yield.

The permeate produced by a single-pass RO system is relatively high in sodium and very low in calcium and magnesium, as compared to traditional water supply sources. As a result, the SAR value of this permeate is usually unacceptably high (8 to 12 meq/L) for direct agricultural irrigation of most crops. However, RO permeate post-treatment including calcium addition and, as needed, second-pass RO treatment allows to reduce SAR of desalinated water to acceptable levels of 4 to 6 meq/L or less.

Typically, RO permeate has significantly lower concentrations of potassium (<1 mg/L), calcium (0.3 to 0.5 mg/L), and magnesium (0.4 to 4 mg/L) than does water produced from conventional fresh water sources such as rivers, lakes and aquifers (1 to 3 mg/L, 4 to 30 mg/L, and 10 to 40 mg/L, respectively). At these low mineral levels, the desalinated water is of inferior taste and has higher corrosivity than conventional water resources. Usually these water quality challenges are addressed by the addition of calcium hardness and alkalinity to RO permeate at levels of 60 to 120 mg/L as calcium carbonate.

While it is still a subject of debate, there are mounting recommendations to establish a minimum limit for magnesium in public water supplies based on its benefits for human health and agriculture. Typically, desalinated water contains magnesium levels of less than 2 mg/L; mineral supplementation to enhance human health protection and nutrient value for agricultural applications is recommended at levels of 5 to 10 mg/L and 15 to 20 mg/L, respectively (Cotruvo et al., 2010, Lahav et al., 2012).

The levels of boron and bromide in the desalinated water are usually an order of magnitude higher than those in conventional freshwater sources. For example, typical river water has a boron concentration of 0.05 to 0.2 mg/L, while source seawater boron levels are usually between 4 and 6 mg/L. The boron content of desalinated seawater treated by a single-pass SWRO system is usually between 0.7 and 1.5 mg/L. Two-pass SWRO systems typically produce water with boron levels between 0.3 and 0.5 mg/L. Both single and two-pass SWRO desalination systems produce fresh water compliant with the boron level of 2.4 mg/L included in the 2011 World Health Organization's drinking water guidelines.

Neither the US EPA nor any state except California has established drinking water regulatory requirements for boron. The California Department of Public Health has a boron action level of 1 mg/L. Some countries, such as Israel, Cyprus, Qatar, Bahrain, and the United Arab Emirates, have boron limits in drinking water of 0.5 mg/L or less. Usually, the low boron limits in these countries are driven by the use of a large portion of the desalinated water for agricultural application on citrus fruits (oranges, lemons, limes, etc.) and by the need to improve the quality of traditional water sources that have naturally high boron levels. High boron concentrations are known to have a negative impact on the yield, size, and color of citrus fruit.

Bromide levels in freshwater sources are usually between 0.05 and 0.3 mg/L, whereas source seawater has a bromide concentration of 55 to 85 mg/L and the permeate produced by a single-pass SWRO system typically has a bromide content between 0.6 and 0.9 mg/L. Two-pass SWRO systems can produce bromide levels in a range of 0.2 to 0.4 mg/L.

Drinking water can exhibit unpleasant changes in taste and odor if desalinated water with a bromide concentration of 0.4 mg/L or more is blended with other water sources that contain phenols. The bromide concentration of desalinated seawater may also have a significant negative impact on the finished water quality if this water is disinfected using chloramines rather than chlorine, or if it is ozonated.

Disinfection of desalinated water with chlorine only (in the form of chlorine gas or sodium hypochlorite) creates a very stable chlorine residual that shows minimal decay over long periods of time (60 days or more). Therefore, when desalinated water is used as the main water supply in a given service area, chlorination (rather than chloramination) is the most commonly applied disinfection method.

Applying a combination of chlorine and ammonia to desalinated water in order to create chloramines (a practice widely used in the United States, for example) may yield an unstable total chlorine residual that decays rapidly (within several hours) to unacceptably low levels if this water has elevated bromide content. Bromide concentration above 4 mg/L has a pronounced destabilizing impact on chloramine residual. This impact can be mitigated either by producing desalinated water of higher quality or by super-chlorinating the water (i.e., applying initial chlorine at doses of 3.5 to 4.0 mg/L).

Ozonation of desalinated water with a bromide concentration of 0.4 mg/L or more may result in the formation of bromate concentration that exceeds most drinking water regulations worldwide, which stipulate a maximum bromate level of 10 μg/L (Cotruvo et al., 2010).

Organics

Desalinated seawater produced by SWRO systems usually contains organics at a level which is an order of magnitude lower than those of most traditional freshwater sources (rivers, lakes, and groundwater aquifers). Therefore, when the desalinated water is disinfected with chlorine, its content of disinfection by-products (DBPs) is very low.

Low molecular weight algal toxins such as domoic acid and saxitoxin are potential concern for the quality of desalinated water originating from surface water sources. Such toxins are generated during algal blooms, when the concentration of algae in the seawater may increase several hundred times.

Algal bloom events are accompanied by an overall deterioration of source seawater quality, including discoloration, oxygen depletion, and an elevated content of organics released from algal cell decay. Certain algae, such as *Pseudo-nitzschia seriata* have red pigmentation; their excessive growth during algal blooms results in reddish discoloration of the seawater—red tide (American Water Works Association, 2011; see Fig. 4-5).

While a number of other organic toxins—such as yessotoxin, okadaic acid, brevetoxin, microcystin, and nodularin—are generated during algal blooms, domoic acid and saxitoxin are two algal toxins of specific interest because their molecular weight and size are comparable with the average molecule rejection size (molecular weight cutoff) of SWRO membranes, and theoretically some of them might pass through the

FIGURE 4.5 Red tide near Carlsbad, California.

membranes. Domoic acid is of particular concern because it concentrates up to several hundred times in shellfish, and when ingested with the contaminated shellfish it could cause amnesic shellfish poisoning.

A common practice of health departments of coastal states in the United States, Australia, and other countries worldwide is to monitor concentration of domoic acid in shellfish tissue and issue advisories for temporary discontinuation of shellfish harvesting when the concentration of domoic acid in the tissue exceeds 80 μg/L (American Water Works Association, 2011).

It should be pointed out that despite the small molecular weight of some of the algal toxins, seawater reverse osmosis membranes can completely reject such organic compounds and thereby can produce safe drinking water even when the source water is exposed to heavy algal blooms. Rejection of domoic acid and saxitoxin by SWRO membranes was been studied at the West Basin Municipal Water District and Carlsbad pilot SWRO plants in Southern California in 2005, during a 50-year red-tide algal bloom. The test results at both facilities indicate that the two algal toxins are completely rejected by the SWRO membranes and that permeate produced by these membranes is safe for human consumption.

Pathogens

While SWRO membranes are not an absolute barrier for microbial contaminants, typically they are expected to achieve pathogen removal of 4 to 6 logs or more. A pretreatment filtration system upstream of the RO desalination membranes typically provides an additional 2- to 4-log pathogen removal.

A virus-challenge study completed by the US Bureau of Reclamation (2007) using state-of-the-art pretreatment systems and SWRO membranes clearly indicates that membrane seawater desalination plants can consistently achieve 6- to 12-log removal of microbial contaminants (Fig. 4.6).

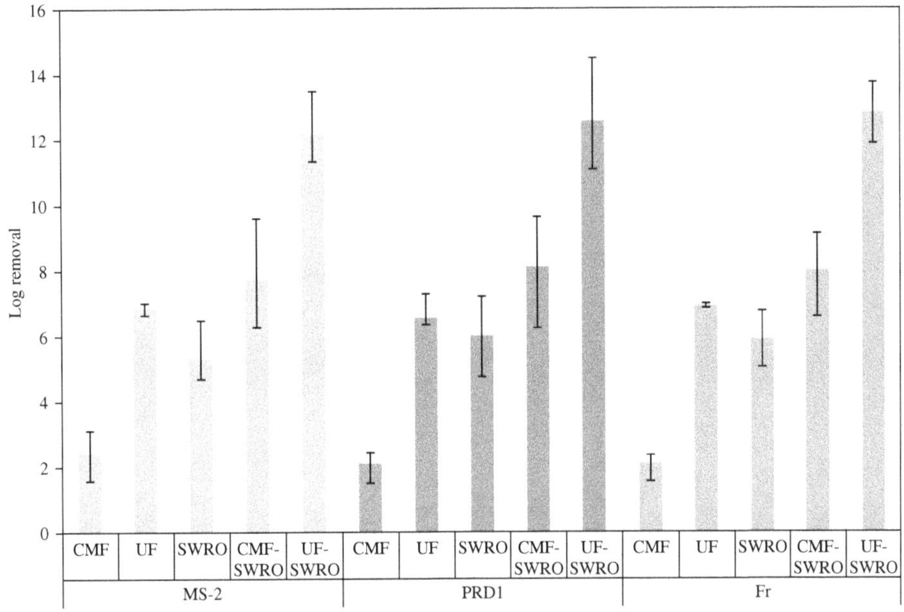

Figure 4.6 Pathogen log removal of seawater pretreatment and RO systems (*Source:* US Bureau of Reclamation, 2007).

This figure shows log removal of conventional media filtration (CMF) pretreatment, (i.e., single-stage sand/anthracite media filters), ultrafiltration (UF) membrane seawater pretreatment, seawater reverse osmosis membranes (SWRO), and combinations of conventional media filtration and SWRO treatment of seawater (CMF-SWRO) and ultrafiltration pretreatment and SWRO separation (UF-SWRO). Challenge testing in this study was completed using MS-2, PRD1, and Fr viruses.

US EPA regulations require drinking water production plants, including desalination facilities, to incorporate in their treatment process multiple barriers for removal or inactivation of pathogens. Table 4.2 summarizes the minimum and maximum reduction requirements and the credits given to typical treatment processes employed in seawater and brackish water desalination (American Water Works Association, 2011). Analysis of Table 4.2 indicates that a typical SWRO desalination plant including conventional pretreatment followed by an RO membrane system and chlorine disinfection could be assigned a total of 6-log virus removal credit, 5-log *Giardia* removal credit, and 4-log *Cryptosporidium* removal credit, which matches closely the maximum log-reduction requirements that may be imposed on a desalination project even under the worst-case scenario of source water quality.

While SWRO membranes can consistently provide over 4-log (99.99 percent) pathogen rejection, due to the lack of standard procedures for RO membrane integrity testing at present they are often credited with only 2-log pathogen removal by the regulatory agencies involved in public health protection. The 2-log removal credit of SWRO systems is assigned based on the continuous monitoring of the actual membrane's TDS log removal (measured as conductivity log removal).

Since SWRO membrane systems typically remove at least 2 log (99 percent) of the source water salinity, TDS removal in this case is used as a conservative surrogate

Pathogen	Log Reduction Requirement*†		Log Reduction Credits Allocated for Treatment Processes by State Regulatory Agencies						
	Min	Max	Slow Sand Filtration	Conventional Pretreatment	MF	UF	RO‡	UV	Chemical Disinfection§
Viruses	4	6	1	2	0.5–1¶	1–4¶	2	0	2
Giardia	3	5.5	2	2.5	4	4	2	2	0.5
Cryptosporidium	2	4	2	2	4	4	2	2	0

*Maximum reduction requirements are based on impaired water sources as determined by watershed sanitary survey monitoring of total coliforms and *Cryptosporidium* monitoring.

†States may require a minimum 0.5-log *Giardia* removal or 2-log virus disinfection beyond filtration to provide a multi-barrier treatment approach.

‡Credits for RO are based on California Department of Public Health guidelines and may differ as more states adopt policies on desalination.

§Chemical disinfection refers to the use of free chlorine following the RO process.

¶The virus removal credit for MF and UF will depend on the specific system used and the state in which the system is permitted

Table 4.2 Pathogen Log Reduction Credits Assigned to Typical Treatment Processes

measure of pathogen removal. As the desalination industry evolves, it is anticipated that alternative membrane integrity test procedures will be developed in the future that will allow assignment of a significantly higher pathogen removal credit to SWRO membranes, reflective of their actual ability to provide very high levels of pathogen removal.

4.5.2 Water Quality of BWRO Desalination Plants

Mineral Content

The wide variability in brackish source waters requires site-specific analysis of the ability of a product water to meet drinking water standards for various constituents. As compared to desalinated seawater, the product water from inland BWRO facilities typically has lower levels of sodium and chloride and higher content of other ions (typically calcium and/or magnesium cations and sulfate and/or bicarbonate anions). Chapter 14 presents examples of the water quality of permeate produced by high-salinity and low-salinity BWRO desalination plants.

Organics

Similar to SWRO membranes, BWRO membranes are capable of removing over 90 percent of most organics contained in the source water. Because of their higher molecular weight cutoff, however, BWRO membranes typically have a lower rejection of organic compounds characterized by small molecular weight.

Pathogens

Brackish water RO membranes can provide over 4 log of pathogen rejection. However, similar to SWRO membranes they are often credited with only 1 or 2 log of pathogen removal due to the lack of a standard online testing method that allows continuous monitoring of their actual pathogen removal and integrity.

4.5.3 Disinfection By-Products in Desalinated Water

Disinfection by-products include a range of compounds—such as trihalomethanes, bromine, iodine, bromates, and haloacetic acids—that are formed through the interaction of chlorine (and, to a lesser degree, chloramines) with organic matter and bromide in the source water or in the distribution system. The organic content of saline source water is typically high in inland surface water and variable in seawater and groundwater.

The organic content of desalinated water is usually an order of magnitude lower than that of most fresh surface water sources, and desalinated water thus has a significantly lower potential to form organics-related DBPs than do traditional freshwater supplies. While RO membranes reject most organics in the source water, the process is not as efficient for removing DBPs, which are formed when chlorine is used for source water pretreatment.

Brackish water membranes are less efficient in terms of DBP removal than are seawater RO membranes. In addition, because the BWRO permeate is more often blended with source water, this blend may require enhanced post-treatment to reduce DBPs.

4.5.4 Blending of Desalinated Water in the Distribution System

In projects where desalinated water is not the main source of water supply, blending it with other source waters of inferior quality (such as surface water or groundwater of elevated salinity) usually has a very positive effect on the quality of the water blend, and therefore it is highly desirable.

Blending low-DBP desalinated seawater with surface water of high DBP content can reduce the overall DBP concentration of the drinking water. However, as indicated previously, when desalinated water has high content of some unwanted minerals such as bromide, boron, sodium, and chloride, mixing this water with drinking water originating from other sources (river, lake, or groundwater aquifer) may have a negative impact on the quality of the blended water. Therefore, the compatibility of the blended water sources must be taken into consideration.

Potential differences in bromide and TOC levels in the blended waters may have an effect on the DBP concentration of the blend. The types of disinfection used for the various water sources may impact the formation of DBPs and the stability of the chlorine residual.

Before blending, desalinated water usually has significantly lower levels of calcium and magnesium ions, as well as low alkalinity concentration, compared with fresh surface water sources. Blending desalinated water with drinking water of high hardness and high alkalinity may be sufficient to provide the needed chemical stability, if the blended water meets target water quality requirements for corrosion control.

4.5.5 Wastewater Treatment and Water Reuse Considerations

Impact of High Boron Concentration on the Quality of Reclaimed Water

As previously indicated, elevated boron concentrations (above 1 mg/L), while safe for human consumption, may have an impact on water use for agricultural and horticultural irrigation. Municipal activities and household detergents add approximately 0.2 to 0.3 mg/L of boron to drinking water during its conversion to wastewater. Desalination treatment operations should therefore consider the impact this addition has on water reuse applications. Additional treatment during desalination to further reduce boron concentration may be necessary if reclaimed water with elevated boron content is to be used for irrigation of sensitive plants and crops.

Impact of Low Alkalinity on Wastewater Treatment Plant Nitrification

Often, desalinated water has a lower alkalinity content than other water sources. In such cases, the introduction of desalinated water to the distribution system will lower the alkalinity of the influent to the wastewater treatment system processing such water. Wastewater alkalinity concentration is very important if the treatment plant has a biological nitrification system. Such systems consume 7.14 mg of alkalinity (as calcium carbonate) for every milligram of nitrified ammonia (as N) contained in the wastewater.

While alkalinity is added to the desalinated product water at a dosage of 40 to 100 mg/L for corrosion protection, such dosage often is inadequate to sustain the wastewater treatment plant's nitrification process, even though wastewater alkalinity is typically 100 to 150 mg/L higher than that of drinking water. Possible solutions include increasing the alkalinity of the desalinated product water, employing biological denitrification in the activated sludge system of the wastewater plant, and directly increasing the alkalinity of the wastewater treatment plant's influent by feeding a strongly basic conditioning chemical, such as sodium or calcium hydroxide.

4.5.6 Selection of Target Product Water Quality

At present, reverse osmosis desalination technology combined with other commercially available pre- and post-RO water treatment processes allows for the production of water of practically any quality. The target product water quality for a given desalination project typically is determined based on the requirements regulating the finished product water and the specific water quality needs of the largest water users in the plant's service area. If such needs are predominantly industrial, agricultural, or horticultural in nature, the required quality of the desalinated water may in some cases be higher than that of drinking water.

As discussed previously, two of the key factors that influence the selection of the target product water quality are (1) the content of specific minerals (e.g., sodium, chloride, boron, and bromides) in the water and (2) the overall water production costs. These two factors are interrelated—production of higher-quality desalinated water is possible at an incrementally (15 to 50 percent) higher cost.

In most municipal applications, desalination plants are designed at a minimum to produce water of a quality that is compliant with drinking water regulations, especially if this water is the main source of supply for the service area. However, many utilities and municipalities worldwide use desalinated water as a supplemental source of water supply only, and this water is blended in the distribution system with other existing water supplies.

Since the quality of the desalinated water can typically be adjusted to a target level more easily than can the quality of the other traditional water sources, often desalinated water is produced at a quality higher than that of the other sources and subsequently blended in order to improve the final product water delivered to the customers. While this approach to water quality improvement results in an elevated cost of production of desalinated water, it often is the most cost-effective overall strategy for improving the water quality in the entire distribution system, as compared to providing additional treatment processes to the individual conventional water treatment plants supplying the same service area.

Another approach to determining the target product water quality for a desalination project is to try to match it as close as possible with the quality of the other traditional water resources delivered to the same service area. This approach, while usually more costly, simplifies the decision-making process in terms of potential modifications that would have to be made to the existing distribution system's operations and water quality.

In addition to the municipal uses discussed previously, the target desalinated water quality may be driven to even higher levels of salt removal by the needs of some industrial applications, especially these where ultrapure water is necessary. Such applications may necessitate the enhanced removal of sodium, silica, specific ions, oxygen, and other constituents, which would require RO permeate treatment through one or more additional water quality polishing processes, such as ion exchange, activated carbon adsorption, advanced oxidation, etc. Such water quality polishing steps can sometimes double the costs of desalinated water over expenditures associated with producing drinking water for potable use.

4.6 Plant Discharge

Typically, both brackish and seawater RO desalination plants generate three key waste streams: (1) concentrate (brine), which usually has 1.5 to 5 times higher salinity than the saline source water; (2) spent filter backwash water from the plant's pretreatment facilities, which has the same salinity as the source water; and (3) spent chemicals and flush water from periodic RO membrane cleaning, which usually are of lower salinity than the source water. Of these three site streams, concentrate is by far the largest in volume and in potential to cause environmental impacts. Therefore, engineering practitioners sometimes refer to desalination plant discharge as *concentrate discharge*, although they are not synonymous.

Management of concentrate and other waste streams associated with the production of desalinated water is one of the key project planning factors that determine plant's location, size, and treatment processes. Usually, in order for a given desalination plant site to be feasible, it has to be located within a reasonable distance (typically 0.5 to 10 km, or 0.3 to 6.1 mi) of suitable sites for concentrate disposal. For a plant discharge disposal site to be suitable, it must have a physical configuration and receiving capacity that allow for the concentrate and, if possible, other plant waste streams to be continuously disposed of in an environmentally safe manner for the entire useful life of the desalination project.

The most common methods for disposal of concentrate and plant discharge are (1) surface water discharge; (2) discharge to sanitary sewer; (3) deep aquifer well injection (for brackish water concentrate); (4) beach well discharge (for seawater concentrate); and (5) evaporation ponds. Other concentrate management methods which are not as widely practiced are (1) spray irrigation; (2) zero liquid discharge (ZLD) by concentrate evaporation and salt crystallization; and (3) beneficial use. Such methods are either very costly (e.g., ZLD) or seasonal in nature (e.g., spray irrigation and some methods of beneficial reuse). It is important to point out that, depending on the size of the project (especially for larger inland desalination projects), it may not be possible to apply a single method for concentrate disposal. Often such projects rely on a combination of multiple disposal alternatives.

Key desalination project planning activities associated with concentrate disposal include (1) water quality characterization of concentrate and other waste streams generated by the desalination plant; (2) development of feasible alternatives for management of the desalination plant's waste streams; and (3) selection of the most viable alternative for desalination plant discharge management based on environmental impact and life-cycle cost analyses.

4.6.1 Concentrate

The volume of concentrate generated by seawater desalination plants is significant, because a typical SWRO separation process converts only 40 to 55 percent of the source water into desalinated freshwater, rejecting the remaining source water as concentrate. Seawater concentrate contains over 99 percent of all source seawater salts and dissolved constituents, and its mineral content is approximately 1.5 to 2 times higher than that of the source seawater.

BWRO plants usually convert 70 to 90 percent of the source water into freshwater, and therefore they generate relatively smaller volumes of concentrate than do SWRO plants with the same freshwater production capacity. However, the mineral content per unit volume of brackish water concentrate is typically 2.5 to 6.5 times higher than for the source water.

Concentrate water quality is largely determined by the quality of the source water and the design of the desalination plant and therefore, it can be projected based on a thorough characterization of the source water quality. Open ocean seawater quality is usually very consistent; over 98 percent of the seawater concentrate's salinity is attributed to five dissolved minerals: sodium, chloride, sulfate, magnesium, and calcium. Therefore, the characterization of seawater concentrate focuses on the measurement of the concentrations of these minerals; the total content of dissolved solids, conductivity, pH, temperature, turbidity, silt density index, total suspended solids, and oxygen; and concentration of organic and inorganic contaminants defined by the regulatory requirements pertinent to the discharge area.

However, water quality of the concentrate generated by SWRO desalination plants with subsurface (i.e., well) intakes is strongly dependent on whether the coastal source water aquifer is influenced by contaminants present in surrounding aquifers. For example, alluvial aquifers often contain elevated concentrations of colloidal iron and manganese and have very low levels of oxygen, which may have a dramatic impact on the quality of the source and product water and on the plant concentrate. Therefore, such aquifers should also be characterized thoroughly during the planning phase of the desalination project.

The water quality of brackish water desalination processes may vary significantly between locations and may contain additional constituents, such as colloidal iron, manganese, silica, nitrate, phosphate, arsenic, cyanide, ammonia, and organics. BWRO concentrate may be dominated by sodium or calcium cations and chloride, sulfate, or bicarbonate anions. Groundwater originating concentrate frequently has high levels of carbon dioxide and hydrogen sulfide, which require degasification prior to discharge. Low oxygen levels in concentrate resulting from groundwater sources may also require aeration or other means to increase dissolved oxygen prior to discharge. Therefore, these water quality parameters will have to be included in the source water quality characterization.

In addition to the previously discussed water quality parameters, the desalination plant concentrate should also be analyzed for acute and chronic whole effluent toxicity. These parameters allow planners to account for the potential synergistic environmental impacts of various contaminants contained in the concentrate. Whole effluent toxicity of the concentrate is difficult to predict based on chemical characterization of the saline source water only. While acute and chronic whole effluent toxicity thresholds of the concentrate could typically be correlated with the level of salinity in the source water, some aquatic species can also be impacted by the ion makeup of the concentrate (i.e., the relative ratios of such ions as calcium, magnesium, sodium, etc.). Therefore, the most reliable and thorough characterization of the concentrate water quality could be achieved by pilot testing using desalination system with configuration, design, and operational conditions similar to those planned for the full-scale desalination plant and then generating concentrate from the pilot plant and analyzing it for all government-regulated discharge water quality parameters, including whole effluent toxicity.

4.6.2 Spent Filter Backwash Water

Spent filter backwash water is a waste stream produced by the pretreatment filtration system, which serves to remove solid particulates and other compounds before the water stream can be treated by RO membranes. All SWRO processes require a pretreatment step, and thus produce backwash water. Pretreatment is less frequently required for BWRO systems, unless surface water or groundwater containing high levels of iron and/or manganese is used as source water for the desalination plant.

The amount of solids contained in the spent filter backwash water is dependent on the source water quality and the type of pretreatment system employed (granular or membrane filters). Typically, membrane-based pretreatment systems produce larger volumes of backwash water (1.5 to 2 times), but require less coagulant, if any, as compared to granular filters, which tend to generate a waste stream with a higher content of solid constituents. Depending on the pretreatment system, the waste stream may contain iron salts used as coagulants in addition to suspended solids (debris, silt, shell particles, etc.) naturally occurring in the source water.

Often, spent filter backwash water (with or without treatment) is blended and discharged along with the concentrate. The blended plant discharge may contain elevated turbidity, total suspended solids, color, organic content, iron and manganese, and biochemical oxygen demand. The concentration of each contaminant of concern in the blend can be calculated as a flow-weighted average of the concentrations of the same contaminant in the individual waste streams. Alternatively, if a desalination pilot plant is available, the water quality of the mixed plant discharge can be determined by direct sampling and laboratory analysis.

4.6.3 Spent Membrane Cleaning Chemicals

Waste streams generated from the chemical cleaning of UF and MF pretreatment membranes usually contribute less than 1.0 percent of the total plant discharge volume, whereas spent RO membrane cleaning solutions are typically less than 0.5 percent of the discharge. Spent membrane cleaning chemicals should be characterized for the same water quality parameters as the desalination plant's concentrate and spent filter backwash water.

Methods for determining the quality and quantity of waste streams generated from desalination plant operations are presented in Chap. 16. That chapter also contains a description, review, and feasibility analysis of commonly used concentrate disposal and plant discharge methods and their benefits, constraints, and costs.

4.7 Conceptual Plant Design

4.7.1 Scope

Once the desalination plant's service area, location, site, source water quality, product water quality, and concentrate water quality have been determined, and the intake and discharge type and configuration have been selected, the next step of the desalination project planning process is to complete the conceptual plant design. This design defines the type and sequence of the plant's water treatment processes and equipment, establishes key facility and equipment design criteria, and incorporates a preliminary plant site layout and hydraulic profile, estimates of project capital and O&M costs, and a project implementation schedule. The conceptual plant design also addresses the type of technology and equipment to be used for energy recovery from the plant concentrate, post-treatment of the RO permeate, handling and disposal of the solids and liquid waste streams generated during source water pretreatment and membrane cleaning, and product water storage and delivery systems.

The conceptual plant design process takes into consideration the physical, operational, and environmental constraints imposed on the project. It usually involves the initial development of several project alternatives followed by the selection of the most viable alternative, based on a set of criteria such as capital and O&M costs; the size of the overall plant site footprint; environmental impacts; the carbon footprint of plant construction and operation; ease of plant operation and maintenance; overall plant performance in terms of energy and chemical use; plant freshwater production reliability, redundancy, and spare capacity; plant expansion and phasing flexibility; and the ability of the proposed plant design and facility configuration to accommodate future technologies and equipment.

4.7.2 Selection of Key Treatment Processes

As indicated previously, a typical desalination plant includes processes for removal of debris, suspended and colloidal solids, and fine silt from the source water, such as screens and filters, followed by processes for removal of dissolved minerals, organics, and pathogens. The combination of these two types of treatment processes (pretreatment and RO membrane separation) produces freshwater with low mineral and pathogen content (permeate).

A typical third step of the desalination plant treatment process is remineralization of the RO permeate for health and corrosion protection, followed by finished water disinfection (if the water is destined for potable use). If the RO permeate contains dissolved gases that have a negative impact on the taste and odor of the desalinated water (e.g., hydrogen sulfide), such gases usually are removed through an additional post-RO treatment process (typically involving oxidation and/or water degassing). Figure 4.7 presents a schematic of a typical desalination plant and indicates key treatment processes.

It should be pointed out that actual desalination projects do not always include all of the treatment steps and processes identified in Fig. 4.7. This figure depicts practically all technologies a desalination plant might incorporate (except for degasing) under the worst-case scenario in terms of source water quality. The figure is representative of the configuration of a seawater desalination plant with open ocean intake exposed to difficult-to-treat water with high content of turbidity, silt, algae, and oil.

Brackish water desalination plants that use intake wells producing low-turbidity and low-silt source water often do not have elaborate pretreatment systems, and they blend a portion of their source water with desalinated water to add minerals to the finished water and reduce the overall costs of water production. Figure 4.8 depicts a general schematic of a typical BWRO plant with well intake.

Pretreatment

Since the main purpose of pretreatment is to reduce the content of suspended solids and silt in the source water, and this content may vary significantly from one project to another, some plants (e.g., plants with well intakes collecting water from pristine saline aquifers that are not affected by surface water contamination) could have minimal pretreatment, which could include only cartridge or bag filtration. However, surface water intakes collecting water from heavily contaminated areas (e.g., industrial ports, shallow bays prone to frequent algal blooms, or locations near a wastewater treatment plant and/or storm drain discharge) could be exposed to significant contamination and often require a series of primary and secondary treatment facilities, such as those shown in Fig. 4.7, in order to produce water with a low content of suspended solids and silt (total suspended solids of < 1 mg/L, turbidity of < 0.3 NTU, and silt density index of < 4) that is suitable for RO separation.

In addition to removal of suspended solids and silt, desalination plant pretreatment is intended to minimize membrane scaling, i.e., excessive precipitation and accumulation of minerals such as calcium and magnesium salts and silica on the RO membrane surface that over time may foul the membranes and hinder the productivity and efficiency of the

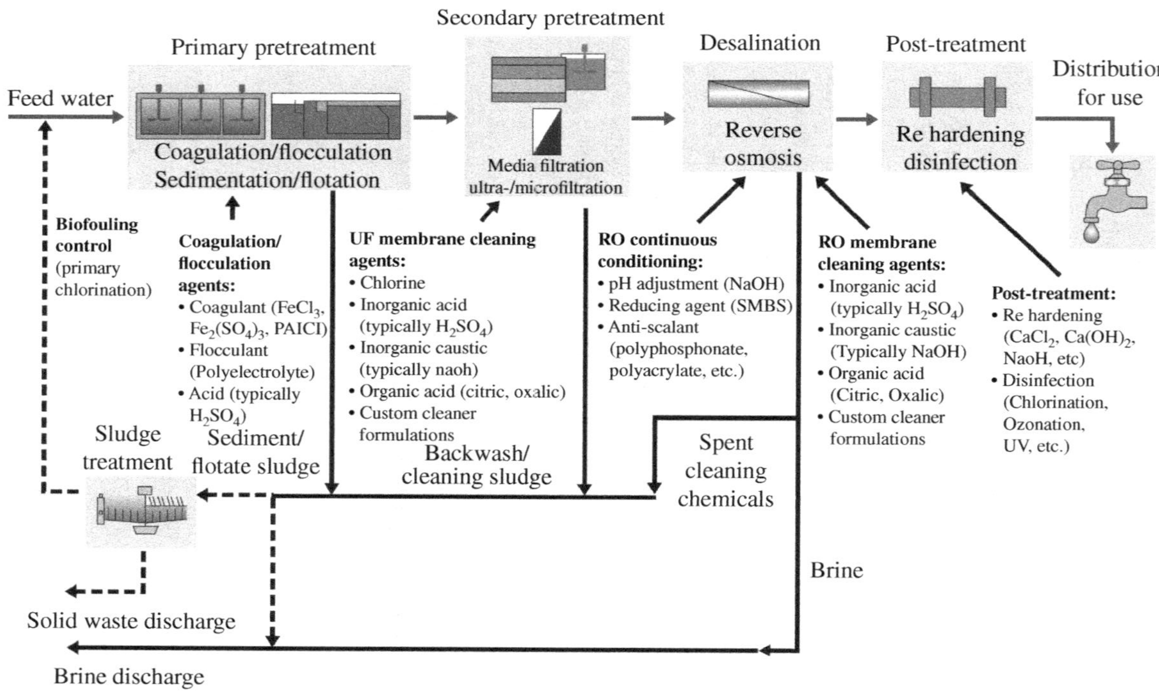

Figure 4.7 SWRO desalination plant schematic.

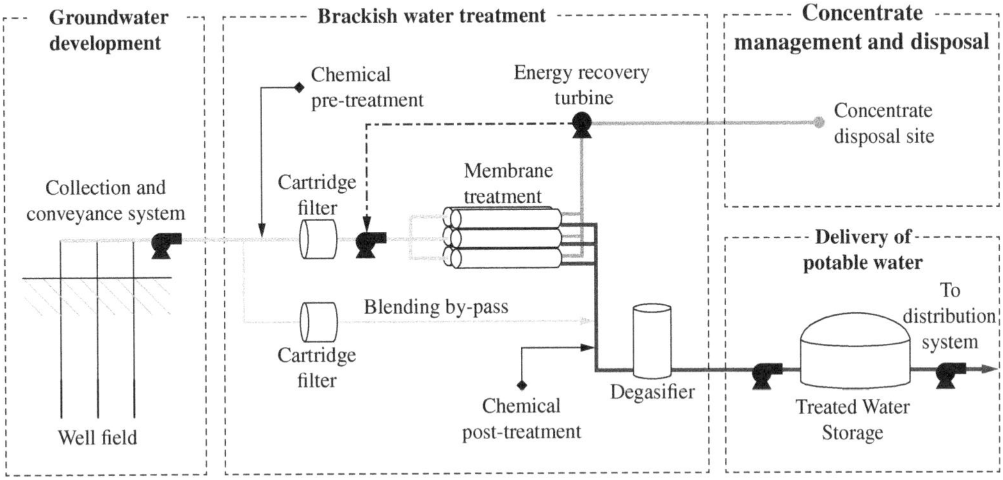

Figure 4.8 BWRO plant schematic. (*Source: NRS Consulting Engineers.*)

salt separation process. Membrane scaling is typically minimized by conditioning the source water with a specific class of chemicals termed antiscalants (scale inhibitors).

Besides productivity reduction caused by solids and minerals, RO membrane system performance can also be hindered by fouling from organic and microbial contaminants contained in the saline source water. Natural organics and particulate or colloidal fouling are commonly controlled by applying coagulants and flocculants to the source water in order to enlarge the particle size of these contaminants and, ultimately, remove them by sedimentation, dissolved air flotation, granular media filtration, UF or MF membrane filtration, or a combination of these processes (see Fig. 4.7). Chapters 8 through 13 discuss in greater detail the alternative processes that are widely used for saline water pretreatment and provide guidelines for their selection and design.

Membrane Salt Separation

At present, reverse osmosis is the salt separation process that is most commonly used for desalination. RO elements incorporating thin-film composite polyamide membranes in spiral-wound configuration are applied in over 90 percent of the municipal desalination projects built worldwide in the past two decades.

RO membrane elements have standard diameters and lengths and are typically installed in pressure vessels that house six to eight elements per vessel. The RO elements and pressure vessels are divided into brackish water and seawater types, depending on their application. Typically, seawater membrane elements and vessels are used to desalinate source water with a TDS concentration of 15,000 mg/L or higher. Brackish water RO elements and vessels are applied for source waters of lower salinity and for additional (second-pass) treatment of permeate generated by SWRO elements in order to produce desalinated water of very high quality (typically, concentrations of TDS, chloride, boron, and bromide lower than 100, 60, 0.5, and 0.4 mg/L, respectively). Fundamentals of the reverse osmosis process are provided in Chap. 3, while Chap. 14 contains detailed information regarding various RO system treatment configurations and their applications.

The RO system type and configuration are selected based on the source water quality of the desalination plant and the target product water quality. Since desalinated water of very similar quality could be produced from the same source water by a number of different RO system configurations and membrane products, usually the most viable RO system for a given project is determined based on a life-cycle cost-benefit analysis.

Post Treatment

As shown in Fig. 4.7, post-treatment of the desalinated water includes two types of processes: rehardening and disinfection. Rehardening is the addition of hardness and bicarbonate alkalinity to the RO permeate in order to provide corrosion protection for the distribution system conveying this water to the final users. The most common compounds used for the addition of hardness and alkalinity to desalinated water are calcium hydroxide (lime) and carbon dioxide. However, the use of calcite (limestone) in combination with carbon dioxide or sulfuric acid is becoming a more prevalent post-treatment technology for corrosion protection because of the lower-turbidity water it tends to produce. Issues and considerations involved in selecting the most viable RO permeate conditioning system for corrosion protection are presented in Chap. 15.

In addition to being rehardened, desalinated water produced for human consumption is also disinfected, through the addition of chlorine-based chemicals, such as chlorine gas, and sodium and calcium hypochlorite, or by UV irradiation. Ozonation is sometimes used for disinfecting finished water from BWRO desalination plants if the water has a low content of bromide. However, ozone is practically never applied for disinfecting desalinated seawater, because this water often has an elevated bromide content (i.e., bromide levels higher than 0.4 mg/L) and thus ozonation may result in excessive generation of bromate, which is a carcinogen.

4.7.3 Equipment Selection

The selection of equipment for a given desalination project is based on the type of the treatment process for which the equipment is intended, its efficiency in terms of energy use, its cost, its ease of operation and maintenance, the size and capacity of the individual equipment units available on the market, and its useful life and track record for similar applications. Another important factor for equipment selection is the quality of the materials from which the equipment is built and their suitability for the ambient environment to which the equipment is exposed.

Typically, plastic equipment and piping is preferred for low-pressure applications (i.e., for working pressures under 10 bar, or 145 lb/in^2). Except for plastic pressure vessels and RO membrane elements, and plastic or ceramic components of some of the available energy recovery systems, most of the other equipment used for high-pressure applications is usually made of high-grade (duplex or super duplex) stainless steel or is coated for corrosion protection.

Depending on the regulatory requirements for public health protection and the use of the finished water, in many countries (e.g., the United States, Australia, Canada, Switzerland, Germany) the quality and type of the materials selected for the desalination plant equipment and piping have to comply with regulatory requirements ensuring that they do not release into the finished drinking water any chemical compounds that are hazardous for human health. In the United States, for example, such requirements are stipulated in NSF/ANSI Standard 61 and are enforced by state and federal human health protection agencies.

4.7.4 Treatment Process Validation and Optimization by Pilot Testing

Often, the overall feasibility of the developed conceptual plant design—including type, performance design criteria, and configuration of the selected pretreatment filtration technology and RO membrane system, as well as chemicals for source water conditioning and membrane cleaning—is verified by pilot testing. Typically, pilot plants are facilities with a capacity between 1/100 and 1/2000 of the capacity of the actual desalination plant.

Key objectives of pilot testing are to collect project-specific data on source and concentrate water quality for project design and to evaluate plant performance at typical operational conditions as well as at maximum and minimum salinity, temperature, turbidity, colloidal contaminants (i.e., iron and manganese), and organic content in the source water. Therefore, especially for desalination plants with open intakes, it is critical to design the pilot testing schedule in such a manner that it allows the capture of events with potentially significant impact on plant operations, such as heavy storms and algal blooms, intense ship traffic, intake area dredging, and periodic waste discharges in the intake area from industrial facilities. For plants with subsurface intakes (wells, infiltration galleries, etc.), pilot testing will need to be of adequate length (at least 6 to 12 months, especially for larger projects) to determine the safe and reliable yield of the intake system and to account for water quality changes triggered by seasonal events (heavy rains and surface runoff), well fouling with silt and well biogrowth over time, and mobilization of contaminants in the source water from adjacent aquifers and/or sources of contamination (such as landfills, leaking underground gasoline, or fuel oil tanks and pipelines).

Pilot testing is the most viable method to generate technical data that are required for project's environmental review, such as the quality of the plant source water and waste streams (concentrate, spent filter backwash, spent membrane cleaning chemicals, and solids residuals) needed for assessment of the environmental impact of plant operations. In addition, side-by-side pilot testing is completed to assess the feasibility of alternative pretreatment technologies and new RO membrane elements, and configurations for the site-specific project conditions, and to optimize overall plant design. Pilot testing also creates opportunities for public outreach and education regarding the quality, benefits, and advantages of desalinated water as compared to alternative water supply sources.

4.7.5 Plant Configuration and Layout

Desalination plant configuration and layout are typically developed to maximize the flexibility of plant operations and minimize both the length of piping, and electrical conduits between the individual treatment facilities and equipment, and the overall footprint of the plant site. Another important consideration when developing plant configuration and determining the layout of buildings and interconnecting roads is the accessibility of key plant equipment (including pumps, motors, energy recovery equipment, pretreatment and RO membrane vessels, cartridge filters, etc.) for inspection, maintenance, and replacement.

The plant layout should be developed to simplify access for large trucks to plant areas designated for the storage of chemicals and of sludge (residuals). Roads to these facilities should be designed with turning radii adequate for the largest delivery and firefighting trucks. Such roads should be at least 6 m (18 ft) wide and should be paved and designed to withstand heavy trucks.

In addition to the facilities for water treatment, chemical storage, and solids handling; the electrical building; the plant motor control centers; the maintenance shop; and other areas for workers, storage, and administration, the plant layout should also incorporate adequate parking areas for employees and visitors, and landscaping.

In most urbanized coastal centers, the land available for the construction of desalination plants is limited and comes at a high cost. Therefore, plants in such areas are often designed with compact layouts, where some of the desalination equipment and facilities are installed in multistory buildings. If land is readily available, though, the least costly plant configuration locates facilities in single-story aboveground structures.

Often the shape of the available site determines the plant layout. For example, the site of the 462,000 m^3/day (132 mgd) Hadera seawater desalination plant in Israel is of very elongated shape (Fig. 4.9), which dictated locating all treatment facilities in one line following the plant treatment process sequence (i.e., intake, pretreatment, RO system, product water storage tanks, etc.).

The intake pump station and dual media gravity pretreatment filters are located closest to the ocean, followed by the RO building, the post-treatment limestone contactors, the circular product water storage tank, and the product water delivery pump station. All structures are built at grade. The main access road runs parallel to the plant buildings and provides access to all facilities, buildings, and storage areas. Chemical storage facilities are housed in the middle of the plant at approximately the same distance from all the main plant areas where chemicals are used continuously—the pretreatment filters and post-treatment facilities. This plant layout is fairly compact and functional.

Figure 4.10 depicts the layout of the largest SWRO facility presently in operation in the United States—the 95,000 m^3/day (25 mgd) Tampa Bay desalination

FIGURE 4.9 Layout of the Hadera SWRO plant, Israel. (*Source: IDE Technologies.*)

FIGURE 4.10 Layout of the Tampa Bay SWRO desalination plant, Florida. (*Tampa Bay Water.*)

plant. The plant site is approximately 3.4 ha (8.5 acres), and the layout is more rectangular. The rectangular building located in the center of the photo houses the filter effluent transfer pumps, cartridge filters, energy recovery equipment, and RO trains.

The two-stage filtration system (sand filters followed by diatomaceous filters) is located to the left of the RO building, while the post-treatment facilities for lime and carbon dioxide addition, the sodium hypochlorite disinfection system, the circular product water storage tank, and the pump station are shown in the upper right corner of the picture.

Chemical storage and feed facilities, as well as the solids handling system, are located in the center of the plant near the pretreatment filters. The empty area between the RO building and the post-treatment facilities is planned to be used for plant expansion to up to 132,000 m^3/day (35 mgd).

Figure 4.11 shows the 28,000 m^3/day (7.5 mgd) Southmost desalination plant in Brownsville, Texas. This is a typical BWRO desalination facility (see Fig. 4.8) using

Figure 4.11 Layout of the Southmost BWRO plant, Brownsville, Texas. (*Source: NRS Consulting Engineers.*)

groundwater with a TDS concentration in the range of 1800 to 2000 mg/L, collected by 20 supply wells (18 duty and 2 standby) with depths varying between 85 and 100 m (280 and 330 ft) and located approximately 25 km (15 mi) from the plant site.

The brackish source water has a very low content of particulates, organics, and silt. Approximately 5700 m^3/day (1.5 mgd) of the 34,000 m^3/day (9 mgd) of source water collected by the plant wells is bypassed and blended with 22,700 m^3/day (6 mgd) of permeate produced form the rest of the source water by BWRO desalination.

The desalination plant incorporates a product water tank that can store up to one day of the plant's production capacity. The pretreatment includes source water cartridge filtration and antiscalant addition only. The total plant site shown in Fig. 4.11 is 5.7 ha (17 acres). The layout was intentionally developed with additional room for a significant plant expansion.

4.7.6 Energy Use

Salt separation from saline water requires a significant amount of energy to overcome the naturally occurring osmotic pressure exerted on the reverse osmosis membranes. This in turns makes RO desalination several times more energy intensive than conventional treatment of freshwater resources. Table 4.3 presents the energy use associated with various water supply alternatives.

Analysis of Table 4.3 indicates that the energy needed for seawater desalination is approximately 8 to 10 times higher than that required for production of freshwater from conventional sources, such as rivers, lakes, and freshwater aquifers. Brackish water desalination typically requires significantly less energy. However, sources of low-salinity brackish water often are not readily available near urban centers.

Table 4.4 indicates typical ranges of energy use for medium and large seawater and brackish water desalination plants (i.e., plants with a freshwater production capacity of

Water Supply Alternative	Energy Use, kWh/m³
Conventional treatment of surface water	0.2–0.4
Water reclamation	0.5–1.0
Indirect potable reuse	1.5–2.0
Brackish water desalination	0.3–2.6
Seawater Desalination	2.5–4.0

*1 kWh/m³ = 3.785 kWh per 1000 gal

TABLE 4.3 Energy Use of Various Water Supply Alternatives*

20,000 m³/day or more). This table is based on actual data from over 40 SWRO and BWRO plants constructed between 2005 and 2011.

As shown in Table 4.4, the SWRO systems of best-in-class seawater desalination plants use between 2.5 and 2.8 kWh of electricity to produce 1 m³ of fresh water (9.5 to 10.5 kWh per 1000 gal), while the industry average is approximately 3.1 kWh/m³ (11.7 kWh per 1000 gal). The industry-wide medium range of energy use for production of fresh drinking water from brackish water varies across a significantly wider bracket—0.6 to 2.1 kWh/m³ (2.3 to 8.0 kWh per 1000 gal)—averaging 0.8 kWh/m³ (3.0 kWh per 1000 gal) for low-salinity BWRO desalination plants and 1.4 kWh/m³ (5.3 kWh per 1000 gal) for high-salinity desalination plants.

4.7.7 Chemicals Used in Desalination Plants

Chemical consumption at desalination plants is highly variable from one project to another and is greatly influenced by the source water quality. In general, the more contaminated the saline source water is with particulate, organic, microbial, and mineral foulants, the greater the amount of chemicals that is needed to produce the same volume of freshwater.

Table 4.5 lists the most common chemicals used in seawater and brackish water desalination plants and their typical dosage, points of application, and purpose. The table does not include chemicals for periodic membrane cleaning; those chemicals are discussed in detail in Chap. 14.

Classification	Low-Salinity BWRO Energy Use, kWh/m³	High-Salinity BWRO Energy Use, kWh/m³	SWRO Energy Use, kWh/m³
Low-end bracket	0.3–0.5	0.6–0.8	2.5–2.8
Medium bracket	0.6–1.2	1.0–2.1	2.9–3.2
High-end bracket	1.5–2.0	2.2–2.6	3.3–4.0
Average	**0.8**	**1.4**	**3.1**

*1 kWh/m³ = 3.785 kWh per 1000 gal

TABLE 4.4 Typical Energy Use for Medium and Large SWRO and BWRO Systems*

Chemical	Dosage, mg/L	Point of Application and Purpose
Ferric chloride or ferric sulfate	0.5–30	• Upstream of pretreatment systems for enhanced removal of solids and silt
Sulfuric acid	30–100	• At intake forebay for control of shellfish growth control in open intakes • Upstream of pretreatment systems for enhanced removal of solids and silt • Upstream of RO system for scale inhibition • Into permeate for reduction of pH and enhanced dissolution of calcite in post-treatment contactors • Into permeate for adjustment of the final product water's pH
Polymer (flocculant)	0–2	• Upstream of pretreatment systems for enhanced removal of solids and silt
Sodium hypochlorite	0–15	• At intake forebay (for open intakes) or well heads (for well intakes) and in intake pump station wet well for control of biogrowth • Upstream of secondary pretreatment for control of biofouling
Sodium bisulfite	0–50	• Upstream of RO system for removal of oxidant residual
Antiscalant	0.5–2	• Downstream of the point of addition of sodium bisulfite and upstream of the RO system for inhibition of scaling
Sodium hydroxide	10–40	• Into feed water of first or second RO passes for enhanced removal of boron • Into finished water for adjustment of pH
Lime	50–100	• Into RO permeate for addition of hardness and alkalinity
Carbon dioxide	30–80	• Into RO permeate for addition of alkalinity and enhanced dissolution of lime and calcite

Table 4.5 Chemicals Commonly Used in Desalination Plants

4.8 Project Implementation Schedule and Phasing

4.8.1 Project Duration

A detailed project implementation schedule has to be developed during the design phase of the desalination project. The plant construction schedule should include, at a minimum, the following information:

- Total duration of the project implementation.
- Duration and start date of contractor mobilization and site preparation.
- Duration and start date of the project engineering and design.
- Duration and start date of procurement and installation of high-pressure RO pumps and energy recovery equipment, pressure vessels and high-pressure stainless steel piping, RO membrane elements, and any other significant items

with long lead times (i.e., procurement, installation, or start-up requires over 3 months).

- Duration and start date of construction of the intake facilities, intake and discharge interconnecting piping, pretreatment system, RO system, and post-treatment facilities.
- Duration and start date of plant commissioning and startup.
- Duration and start date of acceptance testing.

Table 4.6 presents typical lengths of project implementation as a function of plant size. The total duration of the design and construction may vary from the periods indicated in Table 4.6 depending on the site-specific project scope and conditions. Some construction activities may take longer than indicated in the table, especially if most of the construction has to be completed in adverse weather conditions; if the plant footprint is too compact, if the construction staging area is very limited, and/or if the access to the site and the allowable hours of the day and days of the week for construction are burdened with significant constraints due to regulatory requirements related to noise, traffic, air pollution, or other concerns. Some of the construction activities may be accelerated by working in multiple shifts and pre-purchasing some of the long-lead equipment and piping. However, such project acceleration activities usually result in an increase in the overall construction costs.

4.8.2 Project Phasing

The desalination projects with the highest and lowest costs have a very distinct difference in phasing strategy. While the large, high-cost projects incorporate single intake and discharge tunnel structures built for the ultimate plant capacity (which often equals double the capacity of the first project phase), the desalination projects on the low end of the cost spectrum use multiple-pipe intake systems constructed mainly from high-density polyethylene or glass-reinforced plastic that have a capacity commensurate with the production capacity of the desalination plant. Additional multiple-intake pipes and structures are installed as needed at the time of plant expansion for these facilities.

Plant Size, m³/day*	Design, Months	Construction, Months	Start-Up and Commissioning, Months	Total,† Months
Less than 1000	1–2	2–3	1–2	4–7
5000	2–3	4–6	1–2	7–11
10,000	2–4	6–8	1–2	9–14
20,000	3–5	8–10	2–3	13–18
40,000	3–6	14–16	2–3	19–25
100,000	5–8	18–20	3–4	26–32
200,000	6–10	20–24	3–4	29–38

*1 mgd = 3,785 m³/day

†Accelerated implementation of some of the activities is possible but is likely to result in a cost increase.

Table 4.6 Typical Lengths of Desalination Project Implementation

While single-phase construction of desalination plant intake and outfall structures dramatically reduces the environmental and public controversy associated with expanding the plant's capacity at a later date, this "ease-of-implementation" benefit typically comes with an overall cost penalty. The notion that the higher costs associated with building complex intake and outfall concrete tunnels in one phase will somehow be offset by economies of scale usually does not yield the expected overall project cost savings. The main reason is the fact that the cost of 100 m (300 ft) of deep concrete intake or discharge tunnel is over quadruple the cost of an intake or discharge of the same capacity constructed from multiple high-density polyethylene or glass-reinforced plastic pipes located on the ocean bottom—but the economy of scale from single-phase construction is usually less than 30 percent.

4.9 Project Economics

A detailed discussion of key project cost components and their typical ranges for low- and high-salinity BWRO projects and for SWRO projects is provided in Chap. 17. Costs for low-salinity BWRO projects vary in a range of \$0.20 to \$1.50 per m^3 (\$0.80 to \$5.70 per 1000 gal). Costs for high-salinity BWRO desalination plants are in a range of \$0.30 to \$1.80 per m^3 (\$1.10 to \$6.80 per 1000 gal). Costs for production of freshwater by SWRO desalination are between \$0.50 and \$3.0 per m^3 (\$1.9 to \$11.3 per 1000 gal). These energy and cost ranges are based on a comparative analysis of desalination projects constructed worldwide over the past 10 years.

The costs of project capital, O&M, and overall water production depend on a number of factors, most of which are specific to the project's location, size, and technical and socioeconomic circumstances. In general, there are two types of factors that strongly influence desalination project costs: (1) ones controlled by the decisions of the facility owner and (2) ones beyond the control of the facility owner, including those that result from regulatory requirements and market forces.

4.9.1 Effect of Plant Size on Project Costs

Project size has a significant influence on the overall production cost of desalinated water. As illustrated in Fig. 4.12, the unit cost of water production by desalination is reduced with increase of plant capacity. On this figure, the cost of water production of a typical 5000 m^3/day (1.3 mgd) desalination plant is used as baseline for comparison, and is assigned an economy-of-scale factor of 1, while the reduced cost of water production for larger plants is shown as a fraction of this cost. For example, as seen on Fig. 4.12, a plant of capacity of 200,000 m^3/day (53 mgd) will produce water at half of the cost of plant of capacity of 5000 m^3/day (1.3 mgd).

This economy of scale is mainly driven by the size of the individual treatment and pumping units, especially the reverse osmosis trains. Currently, the largest RO train that can be built using off-the shelf standard equipment (high-pressure pumps, energy recovery devices, and 8-in. RO membranes) has production capacity of approximately 21,000 m^3/day (5.5 mgd). Construction of larger individual trains is possible, but usually is not as cost effective because it requires the use of custom-made RO system equipment, which is significantly more costly. As a result, some of the economy-of-scale savings are negated by the additional equipment costs.

For plants larger than 200,000 m^3/day (53 mgd), the economy-of-scale benefits are very limited, mainly because of the added complexity of flow distribution, treatment,

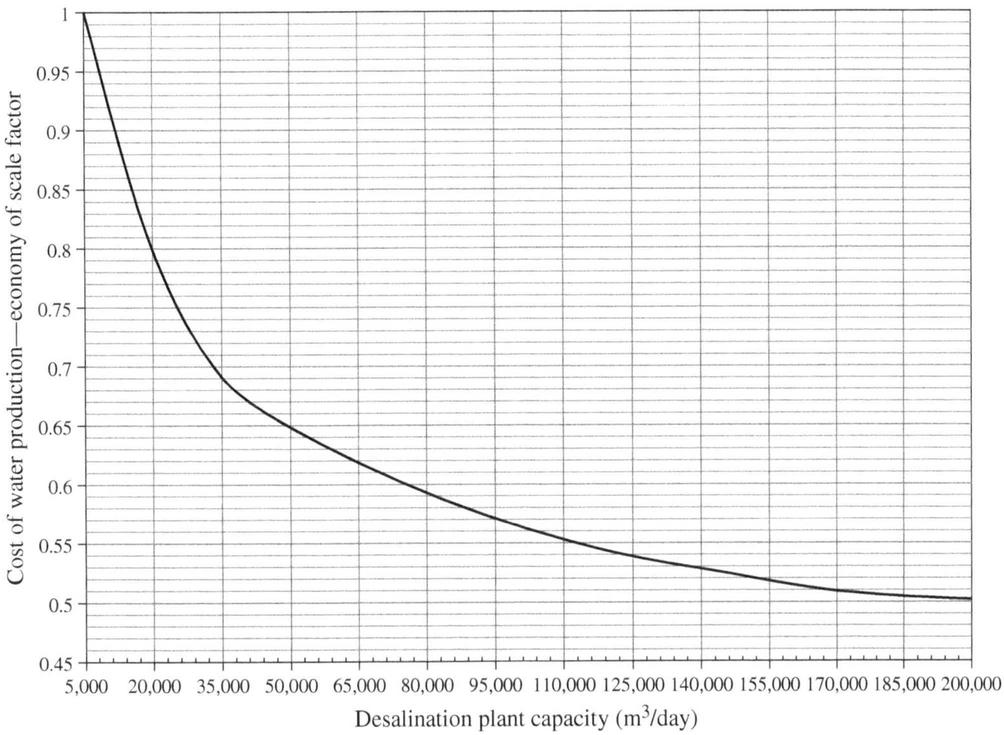

FIGURE 4.12 Relationship between plant size and cost of water production.

and operations. In fact, at present most plants with capacity larger than 200,000 m^3/day (53 mgd) are built as two identical parallel desalination systems that share a common intake and outfall.

As the maximum unit size of commercially available desalination plant equipment (pumps, membranes, pressure vessels, energy recovery systems, etc.) increases in the future, it is likely that the plant capacity at which economy of scale no longer yields measurable savings will shift to 400,000 m^3/day (106 mgd) or higher. A step in this direction is the introduction over the past 8 years of SWRO desalination elements with diameter 16 in. and larger.

4.9.2 Concentrate Disposal and Plant Costs

Depending on the site-specific conditions of a given project, concentrate disposal expenditures may have a measurable contribution to the total plant construction and O&M costs and to the overall cost of water. For small desalination plants with low-cost access to an existing wastewater collection system, concentrate disposal to that system usually is the most attractive disposal option. On the other hand, construction of long new discharge outfalls or a series of deep groundwater injection wells, although widely practiced for small desalination plants, is often costly and site-prohibitive for large projects because of the excessive length of beach area needed for construction of a large number of wells. Chapter 16 provides additional discussion of concentrate disposal methods and associated costs.

4.9.3 Energy Use and Project Costs

For seawater desalination plants, the cost of power is typically 20 to 35 percent of the total expenditures for production of desalinated water. Therefore, both unit power cost and plant power use have a profound effect on water production costs. For brackish water desalination plants, the cost of power is usually a much smaller percentage of the overall freshwater production costs. A detailed discussion of this topic is provided in Chap. 17.

4.9.4 Project Risks and Costs

Costs associated with project financing, development, and environmental review typically are 10 to 20 percent of the overall freshwater production costs. These costs are closely related to the potential risks associated with project implementation and operation.

Typically, financial institutions establish the interest rates of the funds they lend to the project and its acceptable financial structure, based on a thorough evaluation of the project risk profile. In order to provide low-interest funding for a given desalination project, financial institutions demand strong assurances that the project will be permitted and built in a timely and cost-effective manner; the power supply contract and tariff for the project will be reasonable; the operation and maintenance of the desalination plant will be professionally handled by an operations staff that has successful prior experience in desalination; and the project risks and costs related to environmental impact will be minimal and manageable.

In the case of build-own-operate-transfer (BOOT) projects, where a private contractor delvers water supply services rather than a physical asset to the agency using the desalinated water (see Sec. 4.10.3), financial institutions lending funds would also require a legally binding water purchase agreement (WPA) between the final user of the desalinated water (public agency or private end user) and the BOOT contractor that is fair and balanced and that apportions risks equitably between the two parties based on their practical ability to manage these risks.

The entity providing funding for a given project could be a combination of private sector commercial lenders, banks and multilateral agencies, and international financial institutions. Increasingly, funding for desalination projects is provided from the capital markets and from project bonds. Public sector bond underwriters and lenders and private sector lenders often have different approaches and requirements for evaluating and mitigating project risks.

As a general rule, lenders will only be willing and able to take risks that are quantifiable and manageable at reasonable costs. Typically, lenders are not involved with the construction, operation, or insurance activities related to project implementation. Therefore, they will not take risks associated with these activities, especially risks that they are not familiar with or that can be more appropriately managed by other parties involved in the project.

In order to mitigate risks at early stages, lenders may be involved in the key milestones of project development and implementation, including negotiation of project contracts, review of key design and construction activities, and review and approval of certification of project completion and acceptance testing for continuous commercial operation. Lenders generally exercise their review rights over project implementation with the assistance of an independent engineer.

The key project risks considered by the lending institutions when determining their interest and conditions (i.e., the cost of money) for funding desalination projects involve permitting (licensing), entitlements, power supply, construction, source water, technology, the regulatory environment, operations, demand for desalinated water, and finance.

Permitting (Licensing) Risks

Permitting (licensing) risks are risks associated with obtaining and maintaining all permits (licenses) required for all phases of the project implementation and for long-term plant operation, including environmental permits (such as the concentrate discharge permit and drinking water permit), construction permits, and operations permits. Because desalination projects are relatively new to most permitting agencies—and there is therefore a lack of precedent and experience in permitting them—the time and effort required for permitting of this type of projects are usually more extensive than for obtaining permits for conventional water and wastewater treatment plants.

Permitting of large desalination projects often requires long and costly environmental and engineering studies; in some cases, environmental opposition may impose significant political and legal pressures to delay and ultimately derail the project. As a result, permitting risk is considered by lending institutions and public agencies alike as one of the costliest and most significant risk exposures associated with the implementation of a desalination project.

For example, initial difficulties encountered in the permitting of the Tampa Bay seawater desalination project in Florida were one of the key reasons that the public utility that initiated the project (Tampa Bay Water) decided to proceed with project implementation under a BOOT method of delivery, which allows this risk and the associated permitting costs to be transferred to the private BOOT contractor. Experience with environmental review and permitting of the Carlsbad and Huntington Beach SWRO projects in California, which has spanned a period of more than 10 years, is also indicative of the permitting challenges and risks that large desalination projects can face, as well as of the complexity of the environmental review of desalination projects.

Entitlement Risks

Entitlement risks are mainly associated with the control and costs of using the site and infrastructure where the desalination plant and associated facilities will be located. In a case where the desalination plant will share existing intake and discharge infrastructure with other facilities, such as power plants or wastewater treatment plants, entitlement risks are mainly associated with potential changes of the technology and capacity of the existing host facilities in the future, which may require the desalination plant to build its own intake and discharge facilities or significantly modify its structures in order to accommodate the necessary changes implemented by the host facility.

For example, if a desalination plant uses an existing wastewater plant outfall and the owner of the wastewater treatment plant decides to expand its capacity and, therefore, decrease the allowable volume of concentrate discharge from the desalination plant through its outfall, then the desalination plant may face the need (and associated expenditures) to build its own outfall—unless it is contractually entitled to use a predetermined portion of the discharge capacity of the existing outfall throughout the useful life of the desalination plant. In this example, if the desalination plant does not have a contractual entitlement to use the wastewater plant outfall over the period for which a given lending institution would fund the project, the lending institution would consider this condition an entitlement risk and would increase the project's financing costs in order to provide adequate protection of the lender's investment. The size of the interest rate penalty of the borrowed funds will be commensurate with the additional expenditures needed to address this risk if loss of entitlement occurs in the future.

Power Supply Risks

Power supply risks are associated with the availability of power and the magnitude of the change in unit power cost over the useful life of the desalination project. Since the cost of power, especially for seawater desalination projects, can be over 30 percent of the total water production cost, the financial institution funding the project will require the plant's operation costs to be secured with a long-term power supply contract that allows the prediction of power tariffs and energy expenditures over the funding term of the project. Financial institutions typically expect the power tariff adjustments allotted in the power supply agreement to be reflected in and matched with the water tariff adjustments in the water purchase agreement.

Construction Risks

Construction risks stem from the potential increase in construction costs during the project implementation period due to unusual site subsurface conditions, delay of delivery and installation of key equipment, construction cost overruns, errors and omissions on the part of the designer or construction contractor, and performance and reliability risks related to plant start-up, commissioning, and acceptance testing.

Well-recognized construction companies with a proven track record of successful construction of desalination projects in similar settings and of similar size would greatly increase the confidence level of lenders involved in project financing. Usually, construction companies that are newcomers to the desalination industry are considered to have a higher construction risk profile. Similarly, companies with significant cost and schedule overruns and/or ongoing litigation on projects of similar size and complexity will be accepted less favorably by project lenders.

Typically, turnkey fixed price and fixed schedule contracts that allow the owner to hold key contractors fiscally accountable for their performance obligations are favored by project lenders. As a proven mechanism to mitigate construction risks, the financial community prefers construction contract completion guarantees with clauses that require that performance and payment bonds of 10 to 30 percent of the turnkey construction price be available to the lenders to rectify construction problems. Typically, the size of the performance and payment bond is commensurate with the probability of the contractor's default, which in turn is related to the contractor's previous track record with similar projects and experience with key technologies and equipment that are proposed for the desalination project.

Source Water Quality Related Risks

Risks related to source water quality are associated with the potential impacts of the quality of the project source water on the desalination plant's operation and performance, and with the effect that potential changes in source water quality over the useful life of the desalination project would have on the cost of water production. For example, an increase in source water turbidity, organics, or other compounds that may result in accelerated fouling of the membrane elements or in the need for more elaborate pretreatment is typically of concern for the financial community. These risks related to water quality uncertainty can be addressed by not locating the desalination plant intake in the vicinity of discharges of existing wastewater treatment plants, near industrial outfalls, or in large industrial or commercial ports and shipping channels.

In BOOT projects, the risks related to source water quality are contractually addressed by including a source water quality specification in the water purchase agreement with the public or private entities purchasing the water and in the agreements for turnkey engineering, procurement, and construction (EPC) and O&M services. These agreements should also contain provisions for adjustments to the cost of water when the actual source water quality is outside of the contractual specifications and when unpredictable deviations from the source water quality specifications have a material impact on plant performance and costs.

Technology Risks

Technology risks are related to the potential downsides of using new and unproven technologies with limited or no track record at large-scale desalination plants. Although the use of new technologies typically has performance benefits such as reduced construction costs, consumption of power and/or chemicals, and expenditures, there are also potential downsides: an inability to meet contractual obligations regarding product water quantity and/or quality, as well as increased plant downtime due to process underperformance, equipment failure, or malfunction of key system components.

While project engineers typically tend to focus on the cost and performance advantages, project lenders always take into consideration both potential upsides and downsides on a life-cycle cost basis when evaluating the risks and benefits associated with using new technology for a given project. If potential downsides outweigh the cost savings over the useful life of the project or the lending period, then the technology is considered higher risk, and financing terms typically penalize rather than reward the use of the technology. Usually, the project lender turns this risk into a cost overrun amortized over the term of lender investment and then into an incremental increase in the interest rate of the funds that the lender commits to the project.

The use of new technology without a proven track record for similar applications, although attractive from an engineering point of view, may not always be beneficial for reducing the overall project costs, and in reality it may increase the cost of water production through increased financing costs. In general, if a new technology is introduced and lacks a full-scale track record of actual availability (downtime), an assumption of 5 to 10 percent downtime for the new equipment is commonly used by the financial community to evaluate technology risks. This corresponds to the fact that new technology used for the first time on a given project usually goes through two or three generations of improvements before it reaches the typical reliability of a well-proven and mature technology (i.e., technology with downtime of less than 1 percent and a full-scale track record of 5 years).

Regulatory Risks

Regulatory risks are associated with the effects that changes in environmental, engineering, construction, or other government regulations that are pertinent to a given project may have on construction and/or O&M costs. Regulatory changes may occur during the period of desalination plant construction (for example, changes in electrical or building codes) and/or during the period of plant operations (i.e., new regulatory requirements for product water quality or more stringent regulations for concentrate and waste stream disposal).

The financial community typically looks for flexibility features in desalination plant design that will allow for accommodation of future regulation-driven technology changes, and for contractual provisions that permit regulatory risks to be mitigated through cost-of-water tariff adjustments.

Operational Risks

Operational risks are associated with plant operation and maintenance over the useful life of the facility or the term of the lender's investment. Consistent and reliable plant operation and maintenance is the key to generating the adequate and steady revenue stream required to meet financial obligations. If the project owner or finished water purveyor does not have in-house experience with operating desalination plants of a similar size, contracting the O&M services to an experienced and well-established specialized private contractor with proven track record typically results in lower financing costs. As the desalination market matures, O&M challenges and risks associated with a shortage of local skilled labor are resolved over time, and the importance of this risk diminishes.

Desalinated Water Demand Risk

Desalinated water demand risk is closely related to the need for high-quality water in the service area of the desalination plant and the affordability of this water as compared to available existing water supply sources. Typically, in a public-private partnership, the project lender will look for a "take-or-pay" provision in the BOOT contract that ascertains that a predetermined minimum volume of desalinated water is purchased by the final user under all circumstances or else the final user pays for this minimum amount of desalinated water regardless of use.

The lending community considers the water demand risk of desalination projects to be relatively high in conditions where the costs of alternative freshwater supplies (i.e., groundwater and surface water) are significantly lower than the cost of desalinated water, and where the need for water is driven by temporary drought or seasonal shortages of freshwater.

Concerns of the financing community associated with desalinated water demand may be mitigated by putting in place a water cost structure that provides a temporary subsidy for the use of desalinated water, equal to the difference between the cost of desalinated water and the cost of water from other existing sources.

Examples of such subsidies are the \$0.32 per m^3 (\$1.20 per 1000 gal) credit given to Tampa Bay Water by the Southwest Florida Water Management District (SWFWMD)for the potable water produced at the Tampa Bay Water seawater desalination plant and the \$0.20 per m^3 (\$0.80 per 1000 gal) credit that the SWFWMD has committed to providing its customers for the use of desalinated seawater.

Similar direct or indirect mechanisms of reducing the water demand risk are used at state or local government levels throughout the world. In many countries, the desalination cost subsidy is implicitly provided at a governmental level, often by the state or local government taking on a number of the risks presented previously by providing payment guarantees and thereby indirectly subsidizing desalinated water costs.

Financial Risks

Financial risks are directly related to the financial strength (credit) of the entity that will be the final user of the desalinated water and will be responsible for all payment obligations associated with the project financing, as well as to the fiscal stability of the parties involved in construction and operation. Project lenders favor financial agreements with entities that have a proven track record in servicing and repaying debt and equity obligations on similar projects and that do not carry an excessive amount of previous fiscal obligations.

Other financial risks are those associated with the political stability of the country in which the desalination project is planned and the country's currency stability (currency risk). Many of the financial risks can be addressed cost effectively by involving the private sector in project financing.

Before financial institutions commit to fund a given project, they carefully quantify the risks and typically address the outstanding ones that are not already adequately mitigated by contractual and technical means through incremental increases in the interest rate of the funds they lend. The project's delivery and financing method has a significant effect on the cost of desalinated water. Although desalination projects worldwide have been delivered under a number of different methods and financial arrangements, most breakthroughs in reducing the cost of water to date have been achieved under a BOOT method of project delivery.

4.10 Contractor Procurement for Project Implementation

Desalination projects can be implemented using a number of contracting methods, which can be grouped into three key categories: design-bid-build (DBB), design-build-operate (DBO), and build-own-operate-transfer (BOOT). Because of the more mature market and longer experience, most of the brackish desalination projects in the United States and elsewhere are delivered as DBB projects. The DBB method has also been commonly used for procurement of small and medium seawater desalination plants in Europe, the United States, and Israel, and for large-scale desalination projects in the Middle East. For comparison, large seawater desalination projects in Europe, Israel, Asia, the Caribbean, and the United States are typically implemented using the BOOT method of delivery.

Which contracting method is selected mainly depends on the type of owner (public agency or private entity), the project's risk profile and the owner's experience with similar projects, and the source of project funding—loans, grants, bonds, equity, or a mixture of these. The project contracting method often has a significant influence on project costs, and therefore it deserves considerable attention.

4.10.1 Design-Bid-Build

Project Parties and Their Roles

Under the traditional DBB method of project delivery, the desalination plant's owner is typically a public entity (municipality or utility), which is responsible for the overall project implementation as well as for the financing and long-term plant operation and maintenance.

In most cases, under this method of project delivery the owner retains a consulting engineer to prepare detailed technical specifications for the desalination project, which are then used to procure the construction contractor or contractors to build the project. The construction contractors complete their work under the supervision of the owner and the consulting engineer; their main responsibility is to implement the requirements indicated in the specifications.

Key Advantages and Challenges

The key advantage of this delivery method for the owner is that the owner retains complete control over the plant ownership, design, and implementation. Because the owner most often operates the desalination plant with an in-house staff, it also retains all opportunities to take advantage of the cost savings that membrane technology advancements can yield over the long term.

The key challenge for the owner under this method of project delivery is that the owner takes practically all risks associated with project development (permitting and permit compliance, site availability and underground conditions, future power tariff changes, potential environmental damages and associated mitigation efforts), project implementation (faulty design, blunders in technology and equipment selection, contractor deviations from engineering specifications, start up and commissioning risks and delays), and project financing. If the owner decides to operate the desalination plant with its own staff, it takes all risks associated with long-term project operation and performance, such as the risks that the desalination plant may not be capable of producing desalinated water at or above the design capacity; operating at or below the projected use of power, cartridge filters, membranes, and chemicals; and meeting all applicable regulations regarding product water quality and concentrate discharge. Since the owner is responsible for the project financing, it also carries the financial burden associated with the project, including reduction of the owner's available bonding capacity for the implementation of future projects.

The DBB project delivery method is most suitable for owners that have prior experience with the permitting and implementation of desalination projects and operation of desalination plants. For owners lacking such experience, the use of the DBB method is advisable for the implementation of small desalination projects with a low risk profile, which would allow them to gain the necessary experience and develop in-house desalination plant O&M capabilities.

4.10.2 Design-Build-Operate

Project Parties and Their Roles

Similar to the DBB method of project delivery, the DBO approach also involves asset ownership by a public entity (utility or municipality). Under this method of delivery, the owner is responsible for project development, permitting, and financing.

The owner's consulting engineer typically develops detailed performance specifications and a preliminary project design, which are then used to prepare a tender and retain a DBO contractor. This contractor is responsible for the final process design, and for the detailed design, construction, start-up, and commissioning, as well as for the long-term operation of the desalination plant. Usually, the DBO contracting team consists of an engineer, a contractor, and a private operations company (operator).

Key Advantages and Challenges

One key advantage of the DBO method of delivery over the DBB approach is that the early coordination of the facility planning and design with key construction activities and plant O&M requirements allows optimization of the plant design and reduction of life-cycle water production costs. Another advantage for the public entity (utility or municipality) that will use the desalinated water is that it retains ultimate ownership of the desalination plant. In addition, under this method of delivery the owner transfers most of the plant's O&M risks to a private operator that has the experience and skills to manage these risks more cost effectively.

A modified DBO approach used in Australia is the alliance contracting concept. Under this delivery method, the owner (the public partner) and the private DBO contractor share responsibilities, risks, and rewards for project delivery and performance. The alliance method gives the public agency an opportunity to be more actively involved throughout the project implementation and to exercise more control over the final product. These benefits are traded for the assumption of some of the project design and construction risks that are traditionally apportioned to the private DBO or BOOT contractor.

4.10.3 Build-Own-Operate-Transfer

Project Parties and Their Roles

The main difference between BOOT and the other two methods of delivery is that the public entity purchases water (a commodity) rather than a physical asset (the desalination plant). The project ownership is retained by the BOOT contractor.

The BOOT contractor is responsible for all aspects of project implementation, including environmental and construction permitting, design, equipment procurement, construction, start-up, commissioning, long-term operations and permit compliance, and project finance. As indicated previously, BOOT projects are usually financed with a combination of equity and debt.

The repayment obligations on the debt bonds or commercial construction loan for this type of project are typically revenue based. They are "nonrecourse" to the private company that delivers the project and the public agency purchasing the desalinated water, because the net worth of the owners of the project company and the public agency does not have to be used to provide security for debt repayment.

The public or private entity that is the final user of the desalinated water procures a turnkey BOOT contractor based on a performance specification developed by the owner's engineer. The BOOT contractor sells product water at a guaranteed price, quality, quantity, and point of delivery under a water purchase agreement (WPA).

Once the terms for payment of services are set by the WPA, the BOOT project owner or developer usually retains a turnkey contractor to provide all engineering, procurement, and construction (EPC) services needed to build and commission the desalination plant, as well as a private O&M contractor to operate the plant over the entire term of the WPA. Often, the BOOT project owner or developer also serves as an EPC and/or O&M contractor and provides some or all of the equity needed to finance the project.

The WPA, EPC, and O&M contracts—in combination with other entitlements, such as environmental and construction permits, the land purchase or lease agreement, the power purchase agreement, the agreement for access to source water, and the agreement for concentrate and waste disposal services—are used as a proof of control of the BOOT contractor over the project cash flow, which is necessary to secure private financing for the BOOT project.

The financing costs associated with a BOOT project are a direct function of the strength of its contracts and the financial and operating strength of both the entity purchasing the water and the EPC and O&M contractors. A well-structured BOOT project with good WPA, EPC, and O&M contracts and willing participants typically can be financed with 80 percent debt and 20 percent equity. If the project structure is strong and the risk profile is favorable, an even lower percentage of equity may be allowed.

The WPA guarantees water delivery to the user of the desalinated water (public or private entity) at a predetermined quantity, quality, and availability over its entire term. It also guarantees a predetermined payment to the BOOT contractor for the delivered water, and thereby secures a revenue stream that the BOOT contractor can pledge to obtain project financing. The key provisions recommended for incorporation in a well-structured water purchase agreement—in order to minimize the project financing cost and therefore the overall cost of water production—are a take-or-pay clause, firm water purchase obligations, provisions to assign the water contract to lenders, a firm and transparent water structure, change-in law-provisions, clearly defined water quality standards, and liability for third-party claims.

A take-or-pay clause in the contract commits the water purchasing entity to agree to purchase a minimum amount of water at any given time in order to pay for the fixed costs of water incurred by the BOOT contractor if the desalination facility is put on standby. The contract should not contain provisions that allow the purchasing entity to unilaterally terminate or substantially revise the contract in the future. The contract should provide the financial institutions that will contribute equity and debt funds for project implementation with a right and ample opportunity to cure project default if the BOOT contractor fails to perform its obligations. The WPA should also have a water tariff structure that provides adequate coverage of the fixed water production costs and includes water cost escalation factors tied to price adjustment indexes for third-party commodities (power, chemicals, labor, etc.) and foreign exchange fluctuations.

A well-written WPA should have a change-in-law clause that allows the BOOT contractor to adjust the water tariff in order to reflect the additional costs it will actually incur in order to comply with future environmental and/or other regulations that have a material impact on the costs of water production. The WPA should contain clear specifications for product water flowrate and quality, the plant's capacity availability factor, the locations of water delivery, and the procedures for measuring the delivered water flow and monitoring the quality of the desalinated water.

The WPA should have provisions equally protecting both the BOOT contractor and the water purchaser from claims by the ultimate water consumers. In most cases, the BOOT contractor sells the water to a wholesale water supply agency, which in turn conveys and distributes the product water to the actual consumers. The BOOT contractor can only be required to be liable for the product water quality at the point of delivery to the wholesale agency; unless a problem in the distribution system is caused by deviations from the guaranteed water quality specifications, the BOOT contractor should not be required to take responsibility for changes in the water quality caused by malfunctions of the wholesale supplier's distribution system or conveyance facilities. On the other hand, the BOOT contractor should carry liability for impacts on the wholesale supplier's distribution system in case the BOOT contractor supplies inferior out-of-spec product water that causes such impacts.

Water purchase agreements have a number of other provisions that aim to define the contractual division of responsibilities and risks between the BOOT contractor and the water purchaser. These provisions may vary from project to project, but in general they have to be such that the project risks are apportioned between the BOOT contractor and the water purchaser commensurate with the parties' ability to control and mitigate the risks and to deliver water to the ultimate consumer at the lowest overall cost and a competitive market price.

Key Advantages and Challenges

Most of the large seawater desalination facilities built over the past 10 years (or currently under construction) have been delivered under public-private partnership arrangements using the BOOT method of project implementation. The BOOT method is preferred by municipalities and public utilities worldwide because it allows for cost-effective transfer to the private sector of the risks associated with the number of variables affecting the cost of desalinated water, such as intake water quality and its effects (sometimes difficult to predict) on plant performance, permitting challenges, start-up and commissioning difficulties, the fast-changing membrane technology and equipment market, and limited public sector experience with the operation of large seawater desalination facilities.

4.11 Project Funding

The most common methods of financing desalination projects are government funding, conventional (bond or construction loan) financing, and private project financing.

Under the government funding scenario, a local or state government or public agency directly lends funds or provides grants, subsidies, or guarantees for repayment of the funds required to build the desalination plant.

Under conventional (bond or construction loan) financing, long-term funds are raised by the issuance of bonds or the provision by a private lender to a public agency, private utility, or business enterprise of a long-term construction loan against an independent credit risk rating and/or ongoing revenues from water sales or other assets.

Under private project financing, one or more private lenders fund the desalination project via a special project company, relying only on the future cash flow from the project for repayment of the investment, with no recourse to the project owner, developer, and/or product water purchaser (nonrecourse financing).

Government financing of an entire desalination project is not very common at present and is usually available only for the construction of small projects and under emergency conditions. However, in many countries—such as the United States, Australia, Israel, Spain, and some Caribbean and Middle Eastern states—the government directly or indirectly subsidizes costs associated with desalination in order to close the gap between the cost of water from traditionally available surface and/or groundwater sources and the cost of desalinated water.

Often, the state government provides sovereign guarantees for payment for water supply services under a BOOT contract with a private company, especially in circumstances where the direct purchaser of desalinated water is a public agency under the fiscal and administrative control of the state government. A sovereign government guarantee is critical for privately financed projects when the contracting public agency does not have fiscal autonomy and/or is not rated for credit risk.

Conventional financing is based on the issuance of long-term debt in the form of general obligation or revenue bonds or a commercial bank loan for a given project. General obligation bonds are used for financing publicly owned projects. In order to issue this type of bond, the entity seeking funding (government, public utility, municipality, etc.) has to have taxing powers to support the payments of debt obligations.

The key advantage of general obligation bonds is that they are backed by the full taxing capacity of the governmental entity or public agency; consequently, this credit is considered the strongest security pledge available to a lender, and it therefore comes at the lowest available net interest rate. In addition, the issuance of general obligation bonds is usually simpler and frequently less costly than raising other types of debt.

However, the use of general obligation bonds for funding desalination projects has a number of constraints. In order to issue such bonds, most jurisdictions require prior legislative or voter approval of the bond issue and limit the amount of tax-supported debt that can be issued by a legal administrative entity (utility, municipality, authority, etc.). As a result, financing large desalination projects with general obligation bonds may reduce the government agency's ability to issue debt for future projects and may have a negative impact on the agency's credit rating. This type of bond cannot be issued by private entities or businesses. The interest rates for general obligation bonds typically vary from 2.5 to 4 percent.

The second option for conventional project financing is the use of public or private activity revenue bonds. The interest and principal of the long-term debt raised through revenue bonds are payable solely through the revenue generated from the specific utility and/or the project owner. Revenue bonds are generally tax exempt and are typically issued at interest rates lower than those of taxable debt and bonds and construction loans, but higher than those of general obligation bonds. Typically, tax-exempt revenue bonds have interest rates of 3.5 to 6 percent. Taxable debt and bonds usually have interest rates of 4.5 to 8 percent.

Bonds are typically used to finance medium and large projects (i.e., 20,000 m^3/day (5.3 mgd) or higher). Smaller projects are often funded by construction loans issued by commercial banks or lenders that specialize in such financing. Fixed-rate commercial loans are widely used for this purpose; these loans have a constant interest rate and payment for their full term. The term of such a loan depends on the project size and risk profile, and typically is between 5 and 20 years. The interest rate for commercial loans is usually set at a spread ranging from 150 to 275 points (i.e., 1.50 to 2.75 percent) over internationally accepted and established interbank interest rates, such as the London Interbank Offered Rate (LIBOR). LIBOR is the rate that most creditworthy international banks charge each other for large loans.

Private project financing is widely used for implementing large BOOT desalination projects. Under this method of financing, the source of funds is private lenders—most often the BOOT project developer, private banks, and institutional investors such as pension and insurance funds. Private project financing is usually nonrecourse financing. This means that the purchaser and consumer of desalinated water (the public or private water supply entity and its customers) do not have any direct liability for repayment of the funds used for project development and construction, and therefore do not need to pledge any assets for fulfillment of obligations related to the project funding. The desalinated water user (the public or private entity purchasing the water from the private developer) only pays for water services and does not carry project payment obligations on its balance sheet.

The sole source of repayment of the funds invested in the project is the revenue generated from the sales of desalinated water. The responsibility for repaying funds for the development and implementation of a privately financed project lies within the special project company established by the private BOOT contractor. The assets of this company are owned by the project investors providing equity for the project.

Privately financed projects are usually funded by a combination of debt and equity. In some cases, funding can be obtained from multilateral lending agencies (i.e., the European Investment Bank, the World Bank, the Asian Development Bank, the European Bank for Reconstruction and Development) or national "export promoting" agencies. Debt may be in the form of bonds, commercial construction loans, and/or other financial instruments, with long-term or short-term repayment periods. The equity portion of the project funds is typically provided at the request of—and in accordance with the conditions of—the financial institution issuing the project debt.

Commercial banks, financial corporations, and project finance funds are typical sources of debt for desalination projects. Equity for a given desalination project is usually provided by the BOOT contractor and/or outside equity fund (i.e., private equity fund, insurance or pension fund). If the BOOT project is properly structured and priced, the BOOT contractor's equity contribution can be a direct cash payment and/or an indirect contribution of the funds it actually expends for project development ("sweat equity").

Revenue-based (nonrecourse) project financing is typically more complicated and costly to structure than asset-based debt. Transaction costs normally include financial advisory fees, bank fees, legal fees, and independent bank engineer fees. As a result, private nonrecourse financing may not be practical or cost competitive for relatively small desalination projects (projects with capital costs of less than $10 million), unless the transaction costs can be streamlined or multiple projects can be combined into one financial package.

When the project is operational, the revenue generated from the sale of desalinated water is used to (1) pay for the plant's O&M expenditures, (2) repay debt obligations, and (3) pay the return on the equity investment in this order of priority—i.e., O&M expenditures are paid first, debt is paid next, etc. Because project equity investors get paid last, after all other project-related payment obligations are met, and because plant revenue is the only source of repayment for all of the project's fiscal obligations, the equity investors are exposed to the highest risk of not achieving their return-on-investment goals.

Typically, project debt investors are protected by a take-or-pay clause in the water sales agreement between the BOOT contractor and the entity purchasing the desalinated water. However, project equity investors usually do not have such protection for their investment, and therefore their return-on-investment expectations are higher than those of the debt investors. In general, equity investors have expectations of returns commensurate with the returns yielded by financial stock markets trading securities of a comparable risk profile. However, these investors also take on the highest risk related to project performance.

Annual interest at a preset rate is charged for the use of the funds that lenders provide under any of the forms of project financing. For a given public utility, the cost of funds required to finance a desalination project will depend mainly on the utility's credit rating and on the restrictions that apply to the utility in relation to assuming new debt obligations. Public utilities with a relatively low credit rating and/or limited capacity to borrow adequate funds or issue bonds may often be able to obtain more favorable terms by using private sources of financing.

In addition to lowering the overall cost of project funding and the project risk profile, private sector involvement in the financing also has the benefits of keeping such financing off the balance sheet of the public utility embarking on the project and of distributing the implementation and performance risks and costs. Many public utilities that are newcomers to the desalination market prefer to minimize their project-related risks and fiscal exposure by opting to transfer key project risks and funding responsibilities to private companies and lending institutions that specialize in delivering desalination projects. Therefore, many of the recent large seawater desalination projects worldwide have been funded with a BOOT project delivery structure and nonrecourse private project financing.

4.12 References

American Water Works Association, *Desalination of Seawater, Manual of Water Supply Practices M 61*, American Water Works Assocation, Denver, Colorado, 2011.

Cotruvo, J., N. Voutchkov, J. Fawell, P. Payment, D. Cunliffe, and S. Lattemann, *Desalination Technology, Health and Environmental Impacts*, CRC Press and IWA Publishing, Boca Raton, Florida, 2010.

Lahav, O., N. Voutchkov, and L. Birnhack, *Post-Treatment of Desalinated Water*, Balaban Desalination Publications, L'Aquila, Italy, 2012.

US Bureau of Reclamation, *Evaluation of Desalination on Water Under the Influence of Surface Water Runoff for Pretreatment, Water Quality and Pathogen Removal Performance, Report No. 113*, US Bureau of Reclamation, Denver, Colorado, 2007.

CHAPTER 5

Environmental Review and Permitting

5.1 Introduction

The purpose of this chapter is to provide an overview of the potential environmental impacts associated with the construction and operation of desalination projects and to present alternatives for their minimization and mitigation. In addition, this chapter discusses the scope and nature of supporting studies that are typically completed to facilitate the environmental review and permitting (licensing) of desalination projects.

The environmental impacts of operating a seawater or brackish water desalination plant have many similarities to those of operating a conventional water treatment plant. Both conventional water treatment facilities and desalination plants have source water intake and waste stream discharge facilities whose operation may alter the aquatic environment or groundwater aquifers in which they are located. Desalination facilities and conventional water treatment plants use many of the same chemicals for source water conditioning, and therefore generate similar waste streams. In addition, desalination facilities, like conventional water treatment plants, incorporate equipment (i.e., pumps, motors, air compressors, valves, etc.) and treatment processes that generate noise pollution, consume electricity, and are directly or indirectly emit greenhouse gases.

Similar to conventional water supply projects, the construction of desalination plants generates traffic, noise, and other auxiliary environmental impacts. Such impacts are only temporary in nature and typically are minimized by detailed project planning and coordination with local agencies and residents of the areas impacted by construction-related activities.

When desalination projects are evaluated as freshwater supply alternatives to conventional water sources, the environmental impact of plant operations should be assessed in the context of the environmental impacts of water supply alternatives that may be used instead of desalination. Desalination projects are typically driven by the limited availability of alternative lower-cost water supply resources such as fresh groundwater aquifers or surface water (rivers, lakes, etc.). However, damaging long-term environmental impacts may also result from continued overdepletion of those conventional water supplies, including through interbasin water transfers.

For example, overpumping of freshwater aquifers over the years in a number of areas worldwide (e.g., the San Francisco-San Joaquin Bay Delta in northern California; wetlands in the Tampa Bay region of Florida; and freshwater aquifers and rivers and lakes in northern Israel and Spain, which supply water to sustain agricultural and urban centers in the southern regions of these countries), has resulted in substantial environmental impacts from use of the traditional freshwater resources in these regions. Such long-term transfers have affected the ecobalance in the freshwater resources to the extent that the long-term continuation of current water supply practices may result in significant and irreversible damage to the ecosystems and cause the intrusion of saline water into the freshwater aquifers, as in California's Monterey County and Salinas Valley. In such instances, the environmental impacts of construction and operation of new seawater desalination projects should be weighed against the environmentally damaging consequences from the continuation or expansion of the existing freshwater supply practices.

A rational approach to water supply management must ensure that sustainable and droughtproof local supplies are available and that long-term reliance on conventional water supply sources (i.e., surface water, groundwater) is reconsidered in favor of a well-balanced and diversified water supply portfolio combining surface water, groundwater, recycled water, water conservation, and desalination. The overall goal of a well-balanced and sustainable water supply portfolio is to identify a combination of alternative water supply sources that as a whole has the lowest overall environmental impact.

Despite many of the similarities of their environmental impacts, desalination plants have several distinctive differences from conventional water treatment plants: (1) they typically use more source water to produce the same volume of fresh water; (2) they generate discharge of elevated salinity—typically 1.5 to 10 times higher total dissolved solids (TDS) concentration than the source water; and (3) they use 2 to 10 times more electricity to produce the same volume of freshwater. Therefore, this chapter focuses on the three key environmental impact aspects that differentiate desalination projects from other water supply alternatives: (1) intake impingement and entrainment (I&E), (2) the concentrate's impact on the aquatic environment, and (3) the carbon footprint of plant operations.

5.2 Intakes—Environmental Impacts and Mitigation Measures

Desalination plant intakes collect saline water either directly from the water column of surface water sources (open intakes) or from underground aquifers (subsurface intakes). The types, configurations, and applications of various intakes are discussed in detail in Chap. 6. This chapter mainly focuses on their potential environmental impacts and on mitigation measures.

5.2.1 Open Intakes

Environmental Challenges

As with any other natural surface water source currently used for freshwater supply around the globe, surface brackish water and seawater contain aquatic organisms (algae, plankton, fish, bacteria, etc.). *Impingement* occurs when such organisms are trapped against the intake screens by the force of the flowing source water. Loss of marine life through the plant's treatment facilities (pumps, filters, etc.) is typically referred to as *entrainment*.

A third term, *entrapment*, describes impacts associated with offshore intake structures. These structures typically comprise an offshore pipe riser covered with a velocity cap, a long intake tunnel, and an onshore screening facility. Organisms that pass through the offshore velocity cap and are unable to escape the intake velocity in the intake tunnel are often referred to as *entrapped*. They have technically been entrained into the intake system, but their ultimate fate has not yet been determined. Depending on the mesh size of the screens at the onshore screening facility, these organisms can impinge on or entrain through the final screen mesh. It should be pointed out that impingement typically involves adult aquatic organisms (fish, crabs, etc.) that are large enough to actually be retained by the intake screens, while entrainment mainly affects aquatic organisms small enough to pass through the particular size and shape of the intake screen mesh.

Impingement and entrainment of aquatic organisms are not unique to open intakes of desalination plants. Conventional open freshwater intakes from surface water sources (i.e., rivers, lakes, estuaries) may also cause measurable impingement and entrainment.

Attention to intake impingement and entrainment issues associated with desalination plant operation is prompted by Sec. 316(b) of the 1972 US Clean Water Act, which regulates intake of cooling water for the steam electric industry and by the environmental scrutiny associated with the public review process of desalination projects in California. The magnitude of environmental impacts on aquatic organisms caused by impingement and entrainment from desalination plant intakes is site specific and varies significantly from one project to another.

Open ocean intakes are typically equipped with coarse bar screens followed by fine screens (Figs. 5.1 and 5.2), which prevent the majority of adult and juvenile aquatic organisms (fish, crabs, etc.) from entering the desalination plants. Most aquatic organisms collected with the saline source water used for production of desalinated water are removed by screening and filtration before this water enters the reverse osmosis desalination membranes for salt separation.

FIGURE 5.1 Intake bar screen. (*Source:* GHD.)

Figure 5.2 Fine intake screens.

A comprehensive multiyear assessment study of impingement and entrainment at the open ocean intakes of 19 power generation plants, completed by the California State Water Resources Control Board in 2010, provides an insight into the magnitude of these intake-related environmental impacts (State Water Resources Control Board, 2010). In this study, the estimated total average annual impingement of fish by seawater intakes varied between 0.12 g per year per m^3·day of intake flow (0.31 lb per year/mgd of intake flow) for Diablo Canyon Power Plant and 6.27 g per year/m^3·day (52.29 lb/mgd) for Harbor Generating Station; for all 19 plants, it averaged 0.8 g per year /m^3·day (6.63 lb/mgd). Taking into consideration that this amount is the total annual impact, the average daily impingement rate is estimated at 0.002 g per year /m^3·day (0.018 lb per year/mgd) of intake flow (0.8 g per year/m^3·day divided by 365 days = 0.002 g per day/m^3·day = 0.018 lb per day/mgd).

Using the California study results as a baseline, for a medium-size desalination plant with a production capacity of 40,000 m^3/day (10.6 mgd) collecting 100,000 m^3/day (26 mgd) of intake flow, the daily impingement impact is projected to be 0.2 kg/day (0.002 g × 100,000 m^3/day = 200 g/day = 0.2 kg/day = 0.44 lb/day), which is minimal.

The California report referenced above also gives a baseline for assessment of the entrainment impact of seawater intakes. The results indicate that the annual magnitude of such impact on larval fish can vary in a wide range—from 20 larval fish per cubic meter per day (0.08 million larval fish per million gallons per day) for the Contra Costa Power Plant to 1530 larval fish per cubic meter per day (5.8 million larval fish per million gallons per day) for the Encina Power Plant—and illustrate the fact that the entrainment impact is very site specific.

The average annual entrainment is estimated at 565 larval fish per cubic meter per day (2.14 million larval fish per million gallons per day) of intake flow. Prorated for a 100,000 m^3/day (26 mgd) intake of a 40,000 m^3/day (10.6 mgd) seawater desalination plant, this annual entrainment impact is 56.5 million larval fish per year.

While this number seems large, expert evaluation and research indicates that large entrainment numbers for larval fish do not necessarily equate to a measurable impact on adult fish population. Due to the high natural attrition mortality rate of larval fish and limitations of the availability of food, very few larval fish actually develop to the juvenile and adult stages in the natural environment. The majority of larvae is lost to predation and exposure to destructive forces of nature, such as wind and wave action. The impact of such forces on fish populations is several orders of magnitude higher than that of seawater intakes.

Potential I&E Reduction Solutions

While impingement and entrainment associated with open intake operations are not expected to create biologically significant impacts under most circumstances, it is prudent to use the best available site, design, technology, and—when needed—mitigation measures for minimizing the loss of marine life and maintaining the productivity and vitality of the aquatic environment in the vicinity of the intake.

Deep Offshore Intakes Intakes in enclosed bays and estuaries have the greatest potential to cause elevated impingement and entrainment impacts. Since typically, the number of marine species per unit volume of water decreases with depth, intakes at least 300 m (1000 ft) from the shore and at least 6 m (20 ft) below the surface usually result in significantly lower environmental impacts. As indicated previously, open intakes may also exhibit an entrapment effect—fish and other aquatic organisms that are drawn into the offshore conduit cannot return back to the open ocean because they are stranded between the intake forebay and fine screens. The use of velocity caps and low forebay through-screen velocity can reduce this entrapment effect.

Low Through-Screen Velocity Impingement occurs when the intake through-screen velocity is so high that the marine species cannot swim away and are retained at the screens. The US Environmental Protection Agency (US EPA) has identified a velocity threshold of 0.15 m/s (0.5 ft/s), below which impingement is practically nonexistent. Therefore, designing intake screening facilities to always operate at or below this velocity would address impingement impacts.

Small Bar Screen Openings The use of bar screens with a distance between the exclusion bars of no greater than 23 cm (9 in.) is recommended for preventing large organisms from entering the seawater intake (WateReuse Association, 2011).

Suitable Fine Screen Mesh Size After entering the bar screen, the saline water has to pass through fine screens to prevent debris from interfering with the treatment process at the downstream desalination plant facilities. The mesh size of the fine screen is a very important design parameter and should be selected so that it is fitted to the size of a majority of the larval organisms it is targeted to protect. Typically, the openings of most fine screens are 9.5 mm (3/8 in.) or smaller, because most adult and juvenile fish are larger than 10 mm in size. Many fish larvae are larger than 2 mm (1/16 in), and so a mesh of that size or smaller (i.e., 0.5 to 2 mm) could be an effective barrier and entrainment reduction measure.

Design Enhancements for Collection of Minimum Intake Flow Membrane reverse osmosis desalination plants typically collect saline water for one or more of the following purposes: (1) to use it as a source water for freshwater production, (2) to apply it as a backwash water for the source water pretreatment system, or (3) to predilute concentrate generated during the salt separation process down to environmentally safe salinity levels before it is discharged.

Most desalination plants that incorporate filtration for pretreatment of source water collect 4 to 10 percent of additional water to wash their pretreatment filtration systems and discharge the spent filter backwash. A design approach that may allow significant reduction of this water use is treatment and reuse of the backwash water. Such backwash treatment and reuse has cost implications but is a prudent practice aimed at reducing overall plant intake flow and associated impingement and entrainment.

Collecting additional source water for concentrate predilution may be needed when existing wastewater intake or power plant outfalls are used for concentrate discharge and the existing outfall volume is not sufficient to produce adequate dilution of the saline discharge. This additional flow intake could be eliminated by designing facilities for storing concentrate during periods of low outfall flows when adequate dilution is not available, or by installing a discharge diffuser system that allows enhancement of concentrate dissipation into the ambient marine environment without additional dilution.

If the desalination plant's production has to vary diurnally, the design and installation of variable-frequency drives on the intake pumps could also allow decreased impingement and entrainment of the plant intake through close matching of collected source water volume and production needs.

Use of Low-Impact Intake Technologies Impingement and entrainment of aquatic organisms can be minimized through the use of various subsurface and open intake technologies. Currently, there are no federal or state regulations that specifically define requirements for reduction of impingement and entrainment caused by desalination plant intakes. However, the US EPA Sec. 316(b) of the Clean Water Act federal regulations have stipulated national performance standards for intake impacts at power generation plants that require an 80 to 95 percent reduction of impingement and a 60 to 90 percent reduction of entrainment as compared to uncontrolled intake conditions (US EPA, 2008). Technologies that can meet these performance standards are defined by the US EPA as the best technology available.

Subsurface Intakes Subsurface intakes are considered a low-impact technology in terms of impingement and entrainment. However, to date there have been no studies that document the actual level of entrainment reduction that can be achieved by these intakes.

Wedgewire Screen Intakes Wedgewire screens are cylindrical metal screens with trapezoidal "wedgewire" slots with openings of 0.5 to 10 mm. They combine very low flow-through velocities, small slot size, and naturally occurring high sweeping velocities at the screen surface to minimize impingement and entrainment. This is the only technology directly approved by the US EPA as a best technology available, provided that (1) sufficient ambient conditions exist to promote cleaning of the screen face; (2) the through-screen design intake velocity is 0.15 m/s (0.5 ft/s) or less; and (3) the slot size is appropriate for the size of eggs, larvae, and juveniles of any fish and shellfish to be protected at the plant intake site (US EPA, 2011).

Wedgewire screens are designed to be placed in a water body where significant prevailing ambient cross-flow current velocities exist, of 0.3 m/s (1 ft/s) or more. This high cross-flow velocity allows organisms that would otherwise be impinged on the wedgewire intake to be carried away with the flow.

An integral part of a typical intake with wedgewire screens is an airburst back-flush system, which directs a charge of compressed air to each screen unit to blow off debris back into the water body, where it is carried away from the screen unit by the ambient cross-flow currents.

Figure 5.3 presents a schematic of the wedgewire screen intake used at the 150,000 m^3/day (40 mgd) Beckton desalination plant in London, England. This plant is equipped with seven 3-mm wedgewire screens installed on the suction pipe of each intake pump. The total screen length is 3500 mm (11.5 ft) and the screen diameter is 1100 mm (3.6 ft). The plant intake is under significant influence of tidal exchange of river water and seawater.

Offshore Intake Velocity Cap A velocity cap is a configuration of the open intake structure that is designed to change the main direction of water withdrawal from vertical to horizontal (Fig. 5.4). This configuration is beneficial for two main reasons: (1) it eliminates vertical vortices and avoids withdrawal from the more productive aquatic habitat, which usually is located closer to the surface of the water; and (2) it creates a horizontal velocity pattern which gives juvenile and adult fish an indication of danger—most fish have receptors along the length of their bodies that sense horizontal movement, because in nature such movement is associated with unusual conditions. This natural indication provides fish in the area of the intake ample warning and opportunity to swim away from the intake.

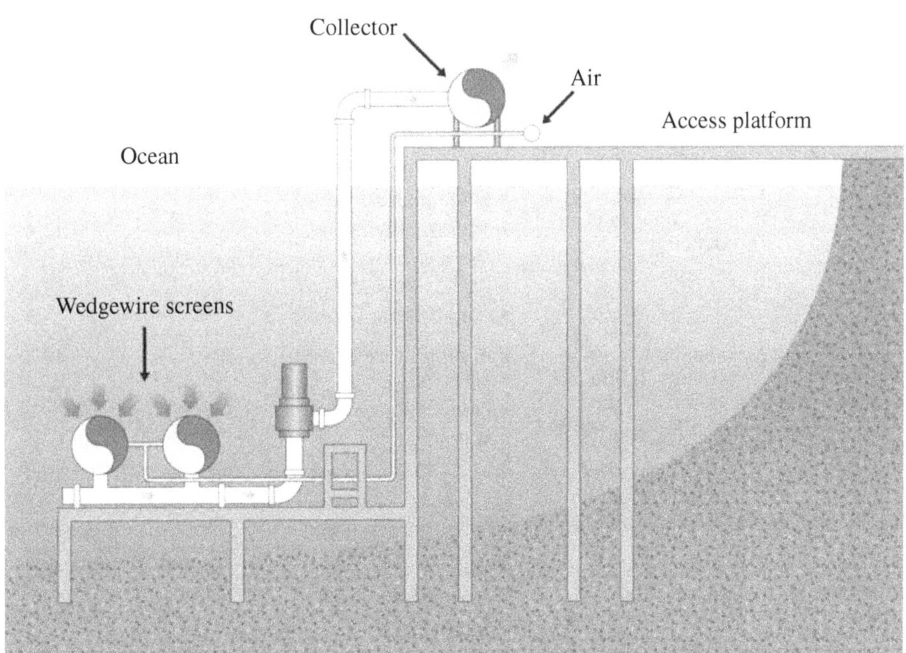

Figure 5.3 Wedgewire screen intake of Beckton desalination plant. (*Acciona Agua*.)

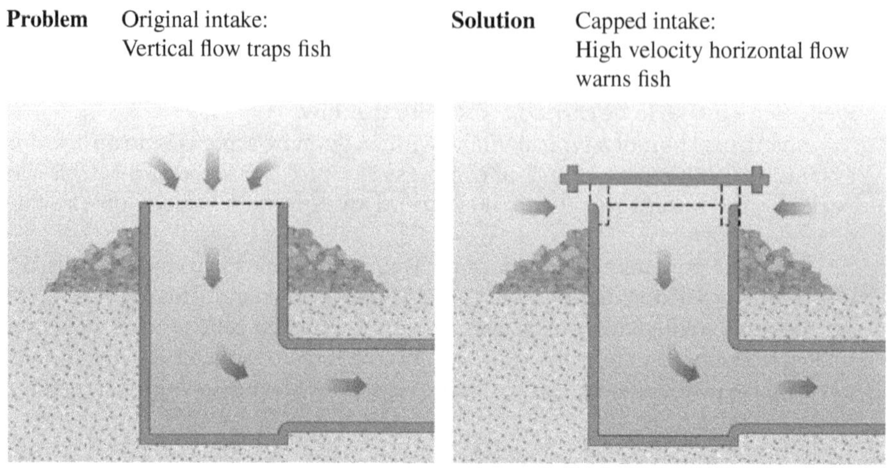

Figure 5.4 Velocity cap for entrainment reduction. (*Source:* US EPA.)

The velocity cap intake configuration has a long track record and is widely used worldwide. This is the original configuration of many power plant intakes in Southern California and of all large seawater desalination plants in Australia, Spain, and Israel constructed over the last 5 years. Based on a US EPA technology efficacy assessment, velocity caps can provide over 50 percent impingement reduction and can minimize

Type of I&E Reduction Measure	How It Works	Technologies	Impact Reduction Potential	
			Impingement	Entrainment
Physical barriers	Fish are blocked from passage and intake velocity is reduced	• Wedgewire screens • Fine mesh screens • Microscreening systems • Barrier nets • Aquatic filter barriers	Yes	Yes
Collection and return systems	Equipment is installed on fine screens for fish collection and return	• Ristroph traveling screens • Fine mesh traveling screens	Yes	No
Diversion systems	Fish are diverted from the screens and returned	• Angled screens with louvers • Inclined screens	Yes	Yes
Behavioral deterrent devices	Organisms are repulsed from the intake by introduced changes that alert them	• Velocity caps • Acoustic barriers • Strobe lights • Air bubble curtains	Yes	No

Table 5-1 Potential Impingement and Entrainment Reduction Technologies

entrainment and entrapment of marine species between the inlet structure and the plant's fine screens (US EPA, 2011).

Other Impingement and Entrainment Reduction Technologies In addition to the intake technologies described above, there are a number of other technologies that have been demonstrated to reduce the impingement and entrainment of open intake operations, mainly based on testing at existing power plant intakes. Table 5.1 provides a summary of such technologies. Not all of the technologies listed in the table can meet the US EPA performance targets under all conditions and circumstances or deliver both impingement and entrainment benefits. However, if needed, these technologies could be used in synergistic combination to achieve project-specific environmental impact reduction targets.

Fine mesh screens are one of the technologies that are equally popular for both desalination and power plant intakes. One type of fine mesh screen associated with the operations of the 95,000 m^3/day (25 mgd) Tampa Bay seawater desalination plant is shown in Fig. 5.5. This desalination plant is collocated with the 1200-MW Big Bend Power Station and uses cooling water from that plant as source seawater for desalination. The Tampa Bay desalination plant does not have a separate seawater intake. However, the intake of the power plant is equipped with 0.5-mm Ristroph fine mesh screens, which have been

Figure 5.5 Fine mesh screens (0.5 mm) of the Tampa Bay Big Bend Power Station's intake.

proven to reduce impingement and entrainment of fish eggs and larvae through the downstream conventional bar and fine screens of the power plant intake by over 80 percent (US EPA, 2011).

Unfortunately for the desalination plant, these screens are periodically bypassed (as allowed by permit) and screenings are conveyed to the power plant discharge outfall from which the desalination plant collects source seawater. As a result, the screenings can find their way to the desalination plant intake and can affect the performance of the plant's pretreatment system. This challenge necessitated the need for the remediated desalination plant to be equipped in 2005 with another set of fine screens located just upstream of the pretreatment facilities.

I&E Mitigation Measures

Environmental impact mitigation is typically applied if the site, design, and technology measures described previously do not provide adequate impingement and entrainment reduction to sustain the biological balance of the marine habitat in the area of the intake. Examples of types of activities that may be implemented by desalination facilities to provide environmental impact mitigation include (1) wetland restoration, (2) coastal lagoon restoration, (3) restoration of historic sediment elevations to promote reestablishment of eelgrass beds, (4) marine fish hatchery enhancement, (5) contribution to a marine fish hatchery stocking program, (6) artificial reef development, and (6) kelp bed enhancement.

The type and size of the mitigation alternative or combination of alternatives that is most suitable for a given project are typically selected to create a new habitat capable of sustaining types of species and levels of biological productivity comparable to those lost as a result of the intake operations. Since coastal wetlands are often the nursery areas for many of the species impacted by desalination intakes, wetland restoration is frequently the mitigation measure of choice. For example, development of new coastal wetlands is the preferred impingement and entrainment mitigation alternative for the 189,000 m^3/day (50 mgd) Carlsbad seawater desalination project in California.

The time and cost expenditures involved in the permitting, implementation, maintenance, and monitoring of such mitigation measures are significant, and such habitat-restorative measures are typically used when the impingement and entrainment reduction measures described in the previous sections are not readily available or viable for a given project.

Some environmental groups do not consider mitigation as an acceptable I&E management alternative, and have challenged the legality of the use of I&E mitigation measures for both power plant and desalination plant intakes. Court resolutions to recent legal challenges associated with the permitting of the Carlsbad and Huntington Beach seawater reverse osmosis (SWRO) projects in California, however, indicate that mitigation by environmental restoration is an acceptable I&E management solution as long as it is applied along with the best technology available and is suitable for the site-specific project conditions.

5.2.2 Subsurface Intakes

Because subsurface intakes naturally filter the collected saline water through the granular formations of the aquifer in which they operate, their use minimizes entrainment of aquatic organisms into the desalination plant. The source saline water collected by this type of intake typically does not require mechanical screening, and therefore subsurface intakes do not cause impingement impacts on the aquatic organisms in the area of the intake sections of this chapter.

Subsurface intakes can have a number of environmental impacts, such as loss of coastal habitat during construction and visual and aesthetic impacts, and can affect nearby coastal wetlands, depending upon the method of construction and the design for well completion. The magnitude of these impacts and potential mitigation measures are discussed in the following sections of this chapter.

Loss of Local Habitat during Construction

Small desalination plants [i.e., facilities with a freshwater production capacity of 20,000 m^3/day (5.3 MGD) or less] typically require a limited number of intake wells, and their impact on the natural habitat near the wells during construction is generally minimal. The individual intake wells for such installations are usually of capacity between 400 and 4000 m^3/day (0.1 and 1.1 mgd) and can often be constructed as low-profile or completely buried structures to minimize visual impacts. Because of the higher number of wells needed to supply adequate amounts of water for a large desalination plant, construction of these facilities may result in impacts over a larger area of local habitat, and since large wells are often constructed as above-grade structures, they have visual and aesthetic impacts (Fig. 5.6).

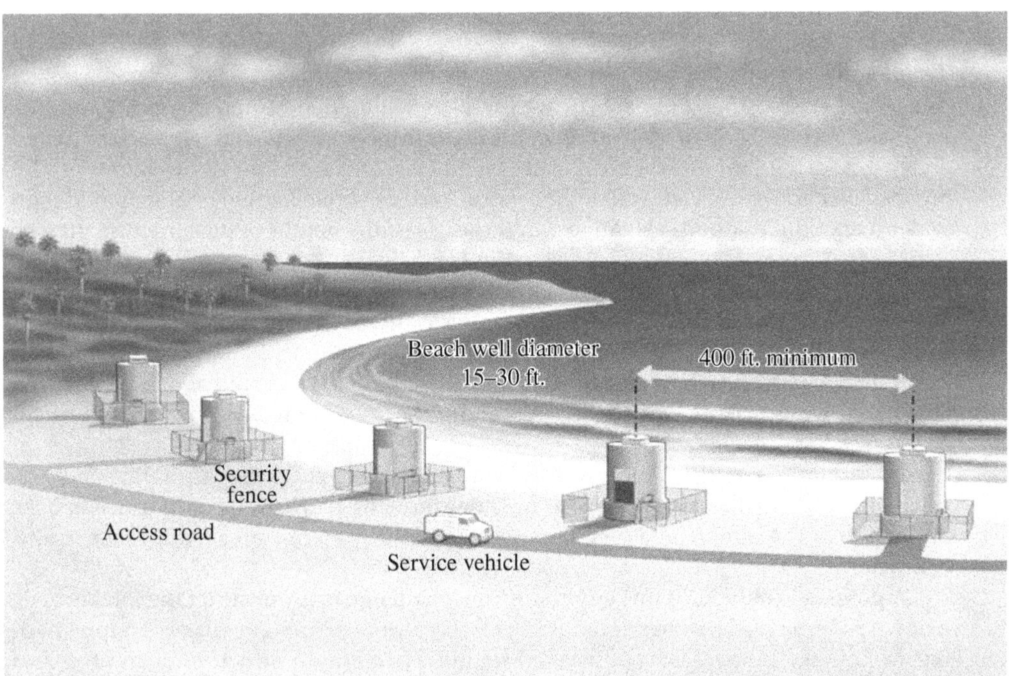

FIGURE 5.6 General schematic of beach well intake system.

If the site-specific geological conditions allow, SWRO plants collecting source water from coastal aquifers and brackish water reverse osmosis (BWRO) plants using brackish aquifers can employ high-production horizontal directionally drilled and/or radial or collector wells to reduce the number of individual wells needed and thereby to minimize environmental impacts associated with their installation. This type of high-productivity well would typically yield 7500 to 20,000 m^3/day (2 to 5 mgd) or more of source water.

For example, a hypothetical 40,000 m^3/day (10.6 mgd) seawater desalination plant that requires 100,000 m^3/day (26 mgd) of intake flow would need the construction of up to five operational beach wells (and one standby) with a capacity of 20,000 m^3/day (5.3 mgd) each to provide an adequate amount of source water. If radial or HDD wells are used, the minimum distance between the individual wells is 150 m (450 ft), and the area impacted by their installation would be spread over up to 600 m (1800 ft) of the shoreline. Therefore, assuming a typical 30-meter (100 ft) width of construction area for each well, the minimum terrestrial area (i.e., seashore) impacted during the construction of beach wells for a 40,000 m^3/day (10.6 mgd) desalination plant would be up to $30 \text{ m} \times 600 \text{ m} = 18{,}000 \text{ m}^2 = 1.8$ ha (4.5 acres).

For comparison, construction of a new open ocean intake would have a significantly smaller area of impact on the beach. For a 40,000 m^3/day (10.6 mgd) seawater desalination plant this beach area would be less than 2 acres.

If the desalination plant is co-located with an existing power plant, and the power plant discharge and/or outfall are used for source water collection, then no additional intake would be required, and therefore the beach zone and environment would not be disturbed with the installation of additional structures, equipment, and associated service infrastructure (access roads, electrical supply equipment, etc.).

Visual and Aesthetic Impacts

The visual and aesthetic impacts of well intakes will depend upon the location of the wellhead and the style of well completion used. If the well intake must be constructed above-grade, the pumps, electrical controls, motors, and auxiliary equipment typically are installed above the wet well of the caisson and/or above known or anticipated high water (e.g., tidal or flood) elevations. In these cases, the height of the structure may be 3 m (10 ft) or more above ground, as seen in Fig. 5.7.

The above-grade pump house facility can be designed in virtually any architectural style; however, this facility and its access provisions will change the visual landscape of the area in the vicinity of the intake. Taking into consideration that the desalination plant source water has to be protected from acts of vandalism if built above ground, the individual wells may, need to be fenced off or otherwise protected from unauthorized access (see Fig. 5.6).

If the intake wells are located in visually sensitive areas (e.g., public beaches), the installation of above-grade wells may degrade the recreational and tourism uses and value of the intake area (i.e., seashore), and may change area's appearance and character.

A potential solution to this environmental challenge is to construct the intake wells below-grade, at grade, or near grade. The electrical controls and auxiliary equipment of the well intake system can be installed within a watertight structure or located in a remote area near the intake for protection, if needed. In these cases, there may be little or no visual and aesthetic impacts from this type of intake. However, the costs associated

FIGURE 5.7 Radial intake well of a large seawater desalination plant.

with such well intakes and their support structures (pump control facilities, electrical substation, power supply conduits, etc.) will increase measurably.

Another alternative solution is to use open intakes, which—similar to near- and below-grade well intakes—are typically lower-profile structures that may blend better with the surrounding environment. The open intake piping can be directionally drilled under the seashore and/or ocean bottom to minimize both visual impact and impact on other coastal uses (recreation, water sports, fishing, etc.).

If the desalination plant is co-located with an existing power plant, construction of new onshore structures or facilities is typically not required; therefore, this type of intake is more favorable in terms of additional negative visual and aesthetic impact on the coastal environment and landscape.

Potential Impact on Wetlands and Groundwater Supplies

Operation of intake wells may have a negative impact on other local groundwater resources (e.g., fresh drinking water aquifers) or water bodies (e.g., perched water, wetlands, or interfaces between salt water and freshwater) that are hydraulically connected to the well extraction aquifer and within the radius of influence of the intake wells. Special attention should be given to intake well sites in the vicinity of existing freshwater supply well fields. Beach wells whose area of influence or source water collection extends to a nearby fresh or brackish groundwater aquifer may have a negative impact on the aquifer's capacity and water quality, and in some cases their operation may result in enhanced seawater intrusion (Fig. 5.8).

The operation of large intake wells located adjacent to wetlands may result in a drawdown of the water table that could affect (e.g., dry up) the wetlands, degrade local groundwater quality (e.g., increase its salinity), and cause other environmental impacts. Year-round study of the interaction between the aquifer from which water is extracted and nearby wetlands and underlying groundwater resources is warranted under such circumstances.

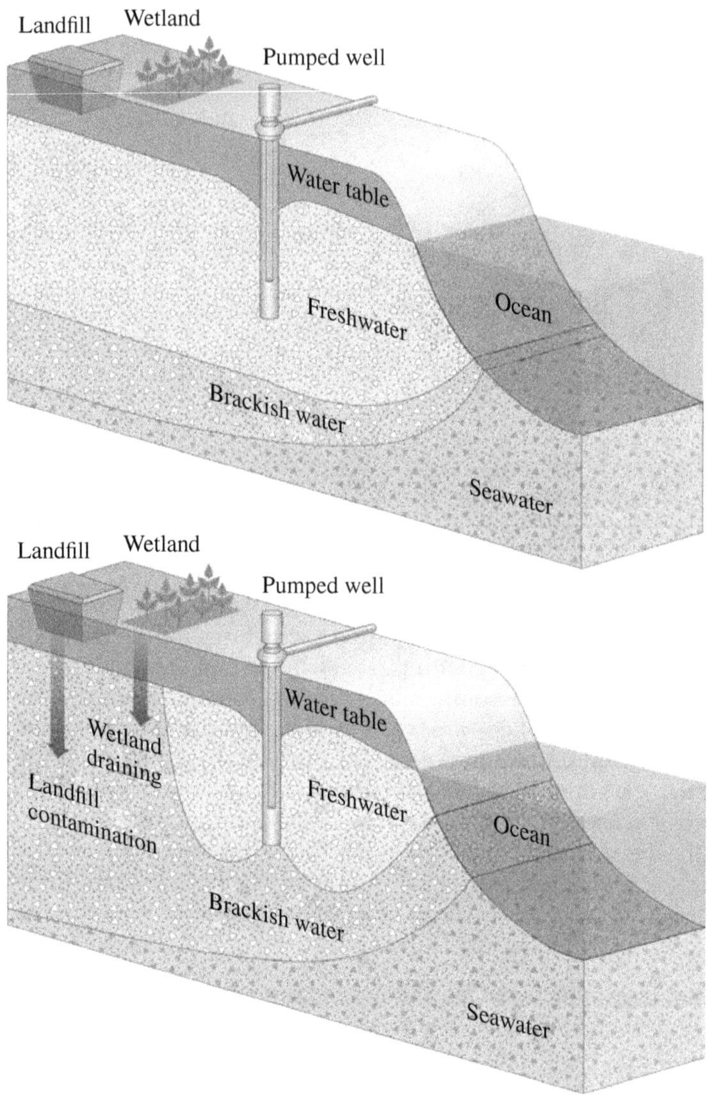

FIGURE 5.8 Seawater intrusion caused by beach well operation.

5.3 Discharge—Environmental Challenges and Solutions

One of the key limiting factors for the construction of new desalination plants is the availability of suitable conditions and locations for disposal of the high-salinity side stream, commonly referred to as *concentrate* or *brine*. Concentrate is generated as a by-product of the separation of the minerals from the source water used for desalination. This liquid stream contains most of the minerals and contaminants of the source water and pretreatment additives in concentrated form. If chemical pretreatment is used, the discharge may contain other compounds such as coagulants, antiscalants, polymers, or disinfectants.

The quantity of the concentrate is largely a function of the plant recovery, which in turn is highly dependent on the TDS concentration of the source water. Concentrate quality is determined by the content of minerals and other contaminants in the saline source water. Chapter 16 discusses concentrate water quantity and quality in greater detail.

Because the most prevalent method of concentrate disposal for both seawater and brackish water desalination plants is surface water discharge, this chapter focuses mainly on the environmental impacts of discharges to surface waters. Chapter 16 provides a thorough discussion of environmental considerations for other concentrate management alternatives (e.g., deep injection wells, land application, evaporation ponds, etc.).

Key environmental issues and considerations associated with concentrate disposal to surface waters include (1) salinity increase beyond the tolerance thresholds of the aquatic species in the area of the discharge, (2) concentration of source water constituents (i.e., metals, nutrients, radioactive ions, etc.) to harmful levels, and (3) discharge discoloration and low oxygen content.

5.3.1 Salinity Increase beyond the Tolerance Thresholds of Aquatic Species

The main environmental impact of concentrate on aquatic life in the vicinity of desalination plant discharge is typically associated with the salinity of this discharge and the ability of the native habitat to tolerate this salinity.

The maximum concentration of total dissolved solids that can be tolerated by marine organisms living in the desalination plant's outfall area is referred to as the *salinity tolerance threshold* and depends on the types of aquatic organisms inhabiting the area of the discharge and the period of time during which these organisms are exposed to elevated salinity (Mickley, 2006). These conditions are very site specific for the area of each desalination outfall, and therefore a rule of thumb for determining the salinity tolerance threshold is practically impossible to develop.

Marine organisms have varying sensitivities to elevated salinity. Some organisms are *osmotic conformers*, i.e., they have no mechanism to control osmosis, and therefore their cells conform to the same salinity as their environment. A large increase in the salinity of the surrounding marine environment due to concentrate discharge causes water to leave the cells of these organisms, which could lead to cell dehydration and ultimately to death.

Marine organisms that can naturally control the salt content and hence the osmotic potential within their cells despite variations in external salinity are known as *osmotic regulators*. Most marine fish, reptiles, birds, and mammals are osmotic regulators, employing a variety of mechanisms to control cellular osmosis. Salinity tolerances of marine organisms vary, but few shellfish (scallops, clams, oysters, mussels, or crabs) and reef-building corals are able to tolerate very high salinities.

Many marine organisms are naturally adapted to changes in seawater salinity. These changes occur seasonally and are mostly driven by water evaporation from the ocean surface, by rain and snow deposition and runoff events, and by surface water discharges.

The natural range of salinity fluctuations in the surface waters receiving concentrate from a given desalination plant can be determined based on information from sampling stations located in the vicinity of the discharge and operated by national, state, or local agencies and research centers responsible for surface water quality monitoring. In open ocean waters, the typical range of natural salinity fluctuation is at least ±10 percent of the average annual ambient seawater salinity concentration. This 10 percent threshold is a conservative measure of aquatic life's tolerance of elevated salinity. The actual salinity tolerance of most marine organisms is usually significantly higher than this level and often exceeds 40 ppt (Cotruvo et al., 2010).

Salinity in brackish surface waters and bays and estuaries can vary in a significantly wider range, and therefore most species inhabiting such waters are more easily adaptive to high-salinity discharges.

Seawater desalination plants usually produce concentrate with a salinity approximately 1.5 to 2 times higher than that of the ambient seawater. Since ocean water salinity in US open ocean coastal waters typically varies between 33 to 35 ppt, concentrate salinity is usually in the range of 50 to 70 ppt. While many marine organisms can adapt to this salinity range, some species are less tolerant to elevated salinity concentrations than others. For example, gobies, which are one of the most common species inhabiting California coastal waters, are tolerant to relatively high salinity concentrations and are known to inhabit the Salton Sea of California, which currently has an ambient salinity of 45 ppt. However, other common organisms such as abalone and sea urchins have a lower salinity tolerance.

The nature, magnitude, and significance of the impacts of elevated concentrate salinity mainly depend upon the types of marine organisms inhabiting the discharge area and the length of time of their exposure. A salinity tolerance study implemented in 2005 as part of the environmental impact review of the 189,000 m^3/day (50 mgd) Carlsbad seawater desalination project, and completed based on testing of over two dozen marine species frequently encountered along the California coast, indicates that based on tests of whole effluent toxicity, these marine species can safely tolerate salinity of 40 ppt (19.4 percent above ambient salinity; Poseidon Resources, 2007).

For this case in point, it is important to note that subsequent acute toxicity bioassay testing was completed, using standard topsmelt test organisms (*Atherinops affinis*), in conformance with the National Pollutant Discharge Elimination System (NPDES) permit requirements for the Carlsbad desalination project. This bioassay testing identified the following: (1) the No Observed Effect Concentration occurred at 42 ppt of concentrate salinity; (2) the Lowest Observed Effect Concentration was found to be 44 ppt; (3) the plant was well below the applicable toxicity limit for salinity of 46 ppt; (4) the No Observed Effect Time for a 60 ppt concentration was 2 hours; (5) the Lowest Observed Effect Time for the 60 ppt concentration was 4 hours. This means that for a short period of time, the species may be exposed to salinity as high as 60 ppt without any observed effect (Poseidon Resources, 2007).

A site investigation of a number of existing full-scale seawater desalination plants operating in the Caribbean, completed by scientists from the University of South Florida and the South Florida Water Management District (Hammond et al., 1998), concluded that salinity levels of 45 to 57 ppt have not caused statistically significant changes in the aquatic environment in the area of the discharge.

5.3.2 Concentration of Source Water Constituents to Harmful Levels

As indicated previously, salinity-related toxicity to aquatic life is the prime source of environmental impacts associated with surface water discharges. However, besides salinity, the reverse osmosis (RO) membrane separation process also removes over 90 percent of most of the other constituents in the source water, concentrating in the discharge by a factor of 1.5 to 10, depending on the desalination plant recovery.

Therefore, some contaminants in the saline source water (i.e., heavy metals, arsenic, cyanide, nitrates, toxins, etc.) that are regulated due to their potentially harmful impacts on the environment may be concentrated to levels that exceed acceptable regulatory thresholds.

In order to assess the potential environmental impacts of regulated water constituents, concentrate water quality should be tested for such constituents, and their actual levels should be compared to pertinent numeric regulatory standards. Practical experience shows that in most cases, BWRO and SWRO concentrate water quality meets regulatory standards associated with most surface waters. However, depending on the site-specific conditions and the discharge configuration and location, some source water constituents other than TDS could potentially reach harmful levels.

For example, because metal content in ocean water is naturally low, compliance with numeric standards for toxic metals usually does not present a challenge. However, concentrate co-discharged with effluent from a wastewater treatment plant may occasionally present a concern, because wastewater plant effluent contains metal concentrations that may be higher than those in the ambient surface source water. Similar attention to the metal levels in the combined discharge should be given to co-disposal of power plant cooling water and concentrate, especially if the power plant equipment leaches metals such as copper and nickel that may then be concentrated in the desalination plant discharge.

If the desalination plant has a pretreatment system that uses coagulant (such as ferric sulfate or ferric chloride), the waste discharges from the source water pretreatment may contain elevated concentrations of iron and turbidity that must be accounted for when assessing their total discharge concentrations.

Radionuclide levels in the ocean water often exceed effluent water quality regulatory standards, and SWRO plant concentrate is likely to contain elevated gross alpha radioactivity. This condition is not unusual for Pacific and Atlantic Ocean waters and must be well documented with adequate water quality sampling in order to avoid potential regulatory challenges.

Toxins such as domoic acid and saxitoxin, which are released by decaying algae during red tides and other algal bloom events, could potentially be harmful to human health and/or the marine environment, and are known to cause shellfish poisoning. However, practical experience shows that even under severe algal bloom conditions, such toxins typically occur at levels that do not present a threat to human health through direct injection of the desalinated water or concentrate. These toxins could cause shellfish poisoning, because they concentrate in shellfish tissue several hundred times, at which level they exceed toxicity levels for human health. For comparison, SWRO plants concentrate algal toxins only 1.5 to 2 times; at such concentrations, these toxins are below the human toxicity threshold. A practical solution to such a challenge is to use desalination plants with either deep open intakes or subsurface intakes in order to minimize the collection of algae and the toxins in their cells.

5.3.3 Discharge Discoloration and Low Oxygen Content

Concentrate from Plants with Open Intakes

Typically, concentrate from desalination plants with open surface water (ocean, river) intakes has the same color, odor, oxygen content, and transparency as the source water from which it was produced, and an increase or decrease in salinity will not change its physical characteristics or aesthetic impact on the environment. Usually, there is no relation between the level of salinity and biological or chemical oxygen demand of the desalination plant concentrate from open intakes. Therefore, concentrate generated by desalination plants with open intakes typically does not pose significant environmental challenges in terms of color and oxygen content. In fact, in some cases, such plant discharge may have a higher content of oxygen than the surface waters to which it is discharged, and actually may improve the quality of the receiving water body in terms of dissolved oxygen content.

Acids and scale inhibitors are often added to the desalination plant source water to facilitate the salt separation process. Typically, these additives are rejected by the reverse osmosis membranes and are collected in the concentrate. However, such source water conditioning compounds are applied at very low concentrations, and their content does not significantly alter the quality or quantity of the concentrate. The environmental implications of the use of such additives are typically well tested before their use, and only additives that are proven harmless for the environment and approved by pertinent regulatory agencies are actually applied for water treatment. All chemical additives used at desalination plants are typically of high purity and are approved for human consumption.

One condition that may cause a reduction and ultimately a depletion of the naturally high level of oxygen in the concentrate from desalination plants with open intakes is the overdosing of reducing chemical (i.e., sodium bisulfite or sulfur dioxide), which is added to remove chlorine in the saline water fed to the desalination plant's RO membrane system.

Typically, reducing chemical is applied at a dosage proportional to the chlorine content in the source water, such that the total chlorine residual in the water is reduced to less than 0.05 mg/L in order to protect the RO membranes from oxidation. However, sometimes due to operator error or monitoring instrument malfunction, the concentration of sodium bisulfite may exceed the dosage needed for removal of chlorine from the RO system feed water. In such cases, the excess content of reducing chemical left after dechlorination will react with the oxygen in the source water and reduce its content. As a result, both the desalination plant concentrate and the product water will have dissolved oxygen (DO) levels lower than those of the saline source water.

This potential environmental challenge is typically addressed by the installation of multiple instruments for monitoring of the chlorine content and oxidation-reduction potential (ORP) of the water treated with reducing chemical and of the concentrate—typically two ORP meters and one chlorine residual analyzer are installed in series on the pipeline feeding the RO system with source water. The ORP of the source water is an indirect indication of its oxygen content. In addition, ORP is measured in the desalination plant's source water and concentrate. If the ORP of the water treated with reducing chemical decreases below 10 percent of the ORP of the source water, then the dosage of the reducing chemical is decreased.

One condition in which the concentrate from a surface water source could be discolored is when it is blended with untreated spent filter backwash water from the desalination plant pretreatment facilities, especially if such backwash water contains iron-based coagulant (ferric hydroxide). Since ferric hydroxide (which also is commonly known as rust) has a red color, when it is blended with the colorless concentrate it will discolor the desalination plant discharge and may degrade the quality of the receiving surface waters. If such discharge is directed to a groundwater aquifer via deep well injection, it will degrade the aquifer's water quality and over time will decrease well discharge capacity.

A commonly applied solution to such environmental challenges is treatment of spent filter backwash water in solids handling facilities, including lamella sedimentation with subsequent dewatering of the sludge collected in the sedimentation tanks by mechanical dewatering equipment (centrifuges or belt filter presses). The dewatered sludge, which typically contains over 95 percent of the coagulant, usually is disposed to a landfill in solid form. For smaller desalination plants, the spent filter backwash water and other pretreatment conditioning chemicals are discharged to the nearby sanitary sewer for further treatment in a wastewater treatment plant. Most membrane pretreatment systems do not use coagulant and therefore are not challenged with the discharge discoloration issue.

Concentrate from Plants with Subsurface Intakes

The dissolved oxygen concentration of source water collected by intake wells is usually less than 2 mg/L, and it often varies between 0.2 and 1.5 mg/L. Desalination plant processes do not add an appreciable amount of DO to the intake water. Therefore, desalination plant product water and concentrate typically have a DO concentration similar to that of the source water. A low-DO concentration of the product water will either require re-aeration or result in significant use of chlorine.

If the low-DO concentrate from a desalination plant is to be discharged to an open water body (an ocean or a river) it typically will not be in compliance with the United States Environmental Protection Agency's daily average and minimum requirements for DO concentration—5 and 4 mg/L—respectively. Because large desalination plants using intake wells will discharge a significant volume of low-DO concentrate, this discharge can cause oxygen depletion and stress to aquatic life. Therefore, the concentrate from a beach well desalination plant has to be re-aerated before being discharged to surface water.

For a large desalination plant, the amount of air and energy necessary to increase the DO concentration of the discharge from near zero to 4 mg/L is significant and would have an effect on the production costs of freshwater. Discharge of this low-DO concentrate to a wastewater treatment plant outfall would also result in significant additional power use to aerate it prior to discharge.

For comparison, the concentrate from desalination plants with open intakes will typically have a DO concentration of 5 to 8 mg/L, which is adequate for disposal to a surface water body without re-aeration.

An alternative solution to the challenge of low-discharge DO concentration is to direct the concentrate for discharge to an aquifer of lower oxygen content, if such an aquifer is available in the vicinity of the desalination plant and if the discharge water quality will not degrade the aquifer in terms of other water quality parameters, such as salinity, solids, silt, etc.

5.4 Greenhouse Gas Emissions Associated with Plant Operations

5.4.1 Introduction

Gases that trap heat in the atmosphere are referred to greenhouse gases. Some greenhouse gases such as carbon dioxide occur naturally and are emitted to the atmosphere through natural processes and human activities. Other greenhouse gases (e.g., fluorinated gases) are created and emitted solely through human activities. The principal greenhouse gases that enter the atmosphere because of human activities are carbon dioxide, methane, nitrous oxide, and fluorinated gases.

Carbon dioxide enters the atmosphere through the burning of fossil fuels (oil, natural gas, and coal), solid waste, and trees and wood products, and also as a result of other chemical reactions (e.g., manufacturing of cement). Carbon dioxide is removed from the atmosphere (or *sequestered*) when it is absorbed by plants as part of the biological carbon cycle.

Methane is emitted during the production and transport of coal, natural gas, and oil. Methane emissions also result from livestock and other agricultural practices and from the decay of organic waste in municipal solid waste landfills. Nitrous oxide is emitted during agricultural and industrial activities as well as during combustion of fossil fuels and solid waste.

Hydrofluorocarbons, perfluorocarbons, and sulfur hexafluoride are synthetic, powerful greenhouse gases that are emitted from a variety of industrial processes. These gases are typically emitted in smaller quantities, but because they are potent greenhouse gases, they are sometimes referred to as *high global-warming potential gases*.

Changes in the atmospheric concentrations of these greenhouse gases can alter the balance of energy transfers between the atmosphere, space, land, and the oceans, and can ultimately result in global and local climate variability and permanent changes (National Research Council, 2001). Many elements of human society and the environment are sensitive to climate variability and change. Human health, agriculture, natural ecosystems, coastal areas, and heating and cooling requirements are examples of climate-sensitive systems. The extent of climate change effects, and whether they prove harmful or beneficial, will vary by region, over time, and with the ability of different societal and environmental systems to adapt to or cope with them.

Rising average temperatures are already affecting the environment. Some observed changes include shrinking of glaciers, thawing of permafrost, later freezing and earlier breaking up of ice on rivers and lakes, lengthening of growing seasons, shifts in plant and animal ranges, and earlier flowering of trees (Intergovernmental Panel on Climate Change, 2007).

Global temperatures are expected to continue to rise as human activities continue to add carbon dioxide, methane, nitrous oxide, and other greenhouse (or heat-trapping) gases to the atmosphere. Most of the United States is expected to experience an increase in average temperature as a result of increase in greenhouse gas emissions (Intergovernmental Panel on Climate Change, 2007).

According to a recent US EPA inventory of greenhouse gas (GHG) emission, the primary greenhouse gas emitted by human activities in the United States in 2006 was carbon dioxide, representing approximately 84.8 percent of total greenhouse gas emissions (US EPA, 20008). The largest source of carbon dioxide and of overall greenhouse gas emissions is fossil-fuel-based production of electricity; the second largest source is transportation. Despite the disproportional attention of some environmental groups to GHG emissions associated with water production, neither conventional nor membrane water treatment plants are major sources of GHGs.

5.4.2 Greenhouse Gas Emission Management

GHG management for seawater desalination plants is relatively new in the United States. The key step in such management is the development of a climate action plan

(CAP), which defines the carbon footprint of desalination plant operations as well as identifies a portfolio of alternative technologies and measures to achieve carbon neutrality for the project, from the use of state-of-the-art energy reduction technologies to the implementation of renewable energy projects. It also identifies carbon dioxide sequestration initiatives, including onsite carbon dioxide use, reforestation, and coastal wetland restoration.

An example of the key steps and approaches in the development of a CAP is presented in a case study for the 189,000 m^3/day (50 mgd) Carlsbad seawater desalination plant. This project is collocated with the Encina coastal power generation station, which currently uses seawater for once-through cooling (Fig. 5.9).

To address the challenge of rising global greenhouse gas emissions, California in 2006 enacted the Global Warming Solutions Act, which aims to reduce the GHG emissions of the state to 1990 levels by the year 2020. In response to this legislation, the Carlsbad project proponent, Poseidon Resources, developed a CAP to completely offset the carbon footprint associated with desalination plant operations. The key components of the climate action plan are described in the following.

Assessing a Project's Gross Carbon Footprint

The carbon footprint of a seawater desalination plant is the amount of greenhouse gases that will be released into the air from the power generation sources that will supply electricity for the plant. Usually, carbon footprint is measured in pounds or metric tons of carbon dioxide emitted per year. The total plant carbon footprint is dependent on two key factors: (1) how much electricity is used by the plant and (2) what sources (fossil

FIGURE 5.9 Aerial view of Carlsbad desalination project site.

fuels, wind, sunlight, etc.) are used to generate that electricity. Both of these factors can be variable over time, and therefore the climate action plan has to have the flexibility to incorporate changes.

The Carlsbad seawater desalination plant is planned to be operated continuously, 24 hours a day and 365 days per year, and to produce an average annual drinking water flow of 189,000 m^3/day (50 mgd). Average annual flow is typically expressed in cubic meters per day and calculated as the total flow produced by the desalination plant per year divided by 365 days.

When the plant was originally conceived, its total baseline power demand was projected at 31.3 MW, which corresponds to unit energy use of 3.96 kWh/m^3 (15.03 kWh per 1000 gal) of produced drinking water. This unit energy use incorporates both production of fresh drinking water and conveyance and delivery of this water to the distribution systems of the individual utilities and municipalities served by the plant.

However, over the lengthy period of the project's environmental review, seawater desalination technology evolved. By taking advantage of the most recently available state-of-the art technology for energy recovery and by advancing the design to accommodate the latest high-efficiency reverse osmosis system feed pumps and membranes, the actual project power use was reduced to 3.56 kWh/m^3 (13.48 kWh per 1000 gal) of drinking water. As a result, the total annual energy consumption for the Carlsbad seawater desalination project used to determine the plant's carbon footprint is 246,000 MWh/year. This energy use is determined for an annual average plant production capacity of 189,000 m^3/day (50 mgd). As actual production capacity may vary from year to year, so will the total energy use.

Next, in order to convert the desalination plant's annual energy use into the carbon footprint, the use is multiplied by the electric grid emission factor, which is the amount of greenhouse gases emitted during the production of unit electricity consumed from the power transmission and distribution system:

$$\text{Carbon footprint (lb CO}_2\text{/year)} = \text{Annual plant electricity use (MWh/year)} \times \text{Emission factor (lb CO}_2\text{/MWh)}$$

The actual value of the emission factor is specific to the supplier of electricity for the project. In the case of the Carlsbad seawater desalination project, this is San Diego Gas and Electric (SDG&E). Similar to other power suppliers in California, SDG&E determines its emission factor based on a standard protocol developed by the California Climate Action Registry, which was created by the California legislature (SB 1771) in 2001 as a nonprofit voluntary registry for GHG emissions. It is the authority in California that sets forth the rules by which GHG emissions are determined and accounted for. The CCAR is updated based on information submitted voluntarily by power generators and users. Based on information provided in its 2008 emissions report (California Climate Action Registry, 2008), the SDG&E emission factor was 546.46 pounds of carbon dioxide per megawatt-hour of delivered electricity. At present this factor is unknown since the emission factor has not been updated since 2009. At 246,000 MWh/year of energy use and an emission factor of 546.46 lb CO_2/MWh, the total carbon footprint for the Carlsbad seawater desalination project is calculated at 134.4 million pounds of carbon dioxide per year (61,100 metric tons of carbon dioxide per year). This carbon footprint is reflective of the latest energy-efficient design of the desalination plant. A more conventional desalination plant design (274,000 MWh/year of energy use) would have a carbon footprint of 149.7 million pounds of carbon dioxide per year.

If information on the emission factor is not available through CCAR, usually such factor can be obtained directly from the utility supplying power for the desalination project. It is important to note that the value of the emission factor is reduced with an increase of the portion of renewable power sources in the power supplier's energy resource portfolio. Because of the statewide initiatives and legislation to expand the use of renewable sources of electricity, the emission factors of all power suppliers are expected to decrease measurably in the future.

Carbon Emissions Reduction Due to Reduced Water Imports

In many parts of the world—such as Spain, Israel, Singapore, Australia, and California—most seawater desalination plants are built to replace in- or out-of-state water transfers. Long-distance water transfers are often very energy intensive, and the carbon footprint of such water supply alternatives may be comparable to that of a desalination plant of similar capacity. Offsetting the carbon footprint of such long-distance water transfers by building a desalination plant can be counted as a carbon-footprint reduction measure for the desalination plant.

For example, San Diego County imports approximately 90 percent of its water from two sources—the Sacramento –San Joaquin River Delta, traditionally known as the Bay-Delta, and the Colorado River. This imported water is captured, released, and conveyed via a complex system of intakes, dams, reservoirs, aqueducts, and pump stations (the State Water Project); it is then treated in conventional water treatment plants prior to its introduction to the water distribution system. The total amount of electricity needed to deliver this water to San Diego County via the State Water Project is 2.76 kWh/m^3 (10.45 kWh per 1000 gal), which includes 2.62 kWh/m^3 (9.93 kWh per 1000 gal) for delivery, 0.06 kWh/m^3 (0.21 kWh per 1000 gal) for evaporation losses, and 0.08 kWh/m^3 (0.31 kWh per 1000 gal) for treatment.

Over the past decade, the availability of imported water from the State Water Project has been in steady decline due to prolonged drought, climate change patterns, and environmental and population growth pressures. One of the key reasons for the development of the Carlsbad seawater desalination project is to replace 189,000 m^3/day (50 mgd) of the total volume of water imported via the State Water Project to San Diego County with fresh drinking water produced locally by tapping the ocean as an alternative drought-proof source of water supply.

Since the desalination project will offset the import of 189,000 m^3/day (50 mgd) of water via the State Water Project, once in operation it will also offset the electricity consumption of 2.76 kWh/m^3 (10.45 kWh 1000 gal) for conveyance of this volume of water and the GHG emissions associated with pumping, treating, and distributing this imported water. Based on the energy consumption of 2.76 kWh/m^3, the annual energy use for importing 189,000 m^3/day (50 mgd) of State Water Project water is 190,700 MWh/year. At 546.46 pounds of carbon dioxide per megawatt-hour, the total carbon footprint of the water imports that will be offset by desalinated water is therefore 104.2 million pounds of carbon dioxide per year (47,400 metric tons of carbon dioxide per year).

Taking into consideration that the gross carbon footprint of the desalination plant is 61,100 metric tons of carbon dioxide per year and that 47,400 metric tons of carbon dioxide per year (77.6 percent) of these GHG emissions would be offset by the reduction of 189,000 m^3/day (50 mgd) in water imports to San Diego County, the Carlsbad desalination plant's net carbon footprint is estimated at 13,700 metric tons of carbon dioxide per year. Lines 1–3 of Table 5.2 summarize the total annual power use and emissions,

Source	Total Annual Power Use, MWh/Year	Total Annual Emissions, Tons of CO_2 Per Year
Carbon Dioxide Emission Generation		
1. Seawater desalination and product water delivery—high energy efficiency design	246,000	61,100
2. Carbon emissions reduction due to reduced water imports	190,700	47,400
3. Total net power use and carbon emissions (line 1–line 2)	55,300	13,700
On-Site Carbon Dioxide Emission Reductions		
4. Energy-efficient plant design	Accounted for in line 1	Accounted for in line 1
5. Use of warm intake water	(12,300)*	(3100)
6. Green building design	(500)	(124)
7. On-site solar power generation	(777)	(193)
8. Use of CO_2 for water production	NA†	2100
9. Reduced energy for water reclamation	1950	(484)
10. Subtotal of on-site power/GHG emission reductions (sum of lines 4 through 9)	(15,527)	(6001)
Off-Site Carbon Dioxide Emission Mitigation		
11. CO_2 sequestration by revegetation of wildfire zones	NA	(166)
12. CO_2 sequestration in coastal wetlands	NA	(304)
13. Investment in renewable energy projects	(2260)	(561)
14. Other carbon offset projects and purchase of renewable credits	(37,513)	(6668)
15. Subtotal of off-site power/GHG mitigation reduction (sum of lines 11 through 14)	(39,773)	(7699)
16. Net CHG emission balance (line 3–line 9–line 14)		0

*Numbers in parentheses indicate reduction.
†NA = not applicable.

Table 5.2 Desalination Project Net GHG Emission Zero Balance

the power and emission reductions attributable to reduced water imports, and the net power use and net annual emissions.

Climate Action Plan for Net Carbon Footprint Reduction

The main purpose of the climate action plan for a given desalination project is to eliminate the plant's net carbon footprint by implementing a combination of as many of the following measures as are practically viable: energy-efficient facility design and operations; green building design; use of carbon dioxide for water production; on-site solar power generation; carbon dioxide sequestration through creation of coastal wetlands and reforestation; funding of renewable power generation projects; and acquisition of renewable energy credits. Carbon neutrality typically is achieved by a balanced combination of these measures.

The size and priority of the individual projects included in the CAP should be determined based on a life-cycle cost-benefit analysis and overall benefit for the local community. Implementation of energy efficiency measures for water production, green building design, and carbon dioxide sequestration projects in the vicinity of the project site should be given the highest priority.

The CAP is a living document that has to be updated periodically in order to reflect the dynamics of development of desalination and green energy generation technologies, as well as the efficiency and cost-effectiveness of various alternatives for carbon footprint reduction and offset. For example, once the Carlsbad seawater desalination plant is operational, the actual carbon footprint will be verified at the time of plant start-up and will be updated periodically to account for changes in power supplier's emission factor and for the actual performance of the already implemented carbon footprint reduction initiatives. Periodic assessment and reprioritization of activities that keep the desalination plant operations "green" is an essential component of the CAP, because both desalination technology and green power generation (e.g., solar, wind, and biofuel-based power) are expected to undergo accelerated development over the next decade as they evolve from marginal to mainstream sources of water supply and power supply, respectively. The specific carbon footprint reduction measures incorporated in the Carlsbad CAP, and their key benefits and constraints, are discussed below.

Energy-Efficient Plant Design

Over 50 percent of the energy used at seawater desalination plants is applied for separation of salts and freshwater by reverse osmosis. The seawater desalination project design should incorporate technologies and equipment to minimize the plant's energy consumption. One such option is the use of a state-of-the art energy recovery system based on a pressure exchanger, which typically allows recovery and reuse of over 30 percent of the total initial energy applied for salt separation. After membrane separation, most of the energy applied for desalination is retained in the concentrated stream (brine) that also contains the salts removed from the seawater. This energy-bearing stream is applied to the back side of the pistons of cylindrical isobaric chambers known as *pressure exchangers*. These pistons pump approximately 50 to 55 percent of the seawater fed into the reverse osmosis membranes for desalination. Since a small amount of energy (4 to 6 percent) is lost during the energy transfer from the concentrate to the feed water, this energy is added back to the feed flow by small booster pumps. The reminder (45 to 50 percent) of the feed flow is pumped by high-pressure centrifugal pumps equipped with high-efficiency motors. Pressure-exchanger-type energy recovery systems are described in greater detail in Chap. 14.

For example, the pressure exchanger energy recovery system for the Carlsbad seawater desalination plant is projected to recover 10,200 hp (7.6 MW) of power and yield 2650 hp (1.98 MW) of additional power savings as compared to the energy that could be recovered using older-generation Pelton wheel energy recovery equipment, which is common in the United States and worldwide.

In addition to the state-of-the-art pressure exchanger energy recovery technology, the Carlsbad desalination plant's design incorporates variable-frequency drives on seawater intake pumps, filter effluent transfer pumps, and product water pumps as well as premium-efficiency motors for all pumps in continuous operation that use power of

500 hp or more. Installation of premium-efficiency motors and variable-frequency drives on large pumps would result in an additional power savings of 1.26 MW (4 percent).

Harnessing, transferring, and reusing the energy applied for salt separation at very high efficiency by the pressure exchangers for the Carlsbad project allows a reduction of the overall amount of electric power used for seawater desalination by over 11.5 percent (3.24 MW) as compared to standard designs of similar facilities. These savings correspond to a total reduction in annual electricity use of 28,380 MWh/year and a carbon footprint reduction of 7000 tons of carbon dioxide per year; as shown in line 4 of Table 5.2, these savings are already accounted for in the high energy efficiency design figures in line 1 of Table 5.2.

Over 80 percent of the desalination plant's piping will be made of low-friction fiberglass reinforced plastic and high-density polyethylene materials, which in turn will yield additional energy savings for seawater conveyance. Plant operations will be fully automated, which will allow a reduction in plant staffing requirements and associated GHG emissions for staff transportation and services.

Use of Warm Intake Water

The viscosity of seawater decreases with an increase in seawater temperature; as a result, desalination of warmer seawater requires less energy. The Carlsbad seawater desalination plant is co-located with the Encina power plant, and its intake will be connected to the cooling water canal to take advantage of the warmer seawater discharged by the power plant. The difference between the average annual temperatures of the ambient seawater and the warm seawater that will be used as source water for the desalination plant is 5.5°C. Based on pilot testing results, this temperature increment is projected to result in 5 percent of additional energy savings and carbon footprint reduction (12,300 MWh/year and 3100 metric tons of carbon dioxide per year) as compared to desalinating cold seawater of ambient temperature. These savings in power and emissions are shown in line 5 of Table 5.2.

There are no additional capital or operations expenditures for using warm water from the power plant once-through cooling system. Therefore, when the power plant is operational, the desalination plant will use only warm cooling water. When the power plant is down, the desalination plant intake is designed to collect cold seawater from the same intake.

Green Building Design

As indicated in Chap. 4, whenever practical and viable, the desalination plant should be located on a site of little current value or public use. Reclaiming low-value land will reduce the project's imprint on the environment as compared to using new undisturbed site. For example, the Carlsbad seawater desalination plant will be located on a site occupied by a dilapidated fuel oil tank. This tank and its contents will be removed and the site will be reclaimed and reused to construct the desalination plant.

Another approach to reducing a desalination plant's physical imprint on the environment is minimizing the plant's site footprint. For example, a key green feature of the Carlsbad seawater desalination plant's design is its compactness (see Fig. 5.9). The plant facilities will be configured as series of structures sharing common walls, roofs, and equipment, which will allow a significant reduction of its physical footprint. The total area occupied by the desalination plant facilities will be approximately 2 ha (5 acres).

When built, this will be one of the smallest desalination plant site footprints in the world, by unit production capacity: 2 ha per 189,000 m^3/day (5 acres per 50 mgd).

For comparison, the 95,000 m^3/day (25 mgd) Tampa Bay seawater desalination plant (see Fig. 4.9) occupies 3.4 ha (8.5 acres), and the 330,000 m^3/day (86 mgd) SWRO plant in Ashkelon, Israel—which currently is the largest operating SWRO facility in the world—occupies 9.7 ha (24 acres). A plant with a smaller physical footprint will also yield a smaller construction-related carbon footprint: lower construction material expenditures and GHG emissions from construction equipment due to the smaller volume of excavation and concrete works. In addition, reduced construction site footprint generates less dust emissions and requires less water for dust control.

Whenever economically viable and practical, building design should follow the principles of the Leadership in Energy and Environmental Design (LEED) program. This is a program of the US Green Building Council, developed to promote the construction of sustainable buildings that reduce the overall impact of building construction and functions on the environment through (1) sustainable site selection and development, (2) energy efficiency, (3) materials selection, (4) indoor environmental quality, and (5) water savings.

Consistent with the principles of the LEED program, the desalination plant's buildings should include features and materials that allow minimal energy use for lighting, air-conditioning, and ventilation. For example, a portion of the walls of the main plant building of the Carlsbad seawater desalination plant will be equipped with translucent panels to maximize daylight and views to the outside. Nonemergency interior lighting will be automatically controlled to turn off in unoccupied rooms and facilities. A monitoring system will ensure that the ventilation in the individual working areas in the building is maintained at its design minimum requirements. In addition, the building's design incorporates water-conserving fixtures (lavatory faucets, showers, water closets, urinals, etc.) for plant staff service facilities and for landscape irrigation.

The green desalination plant buildings should incorporate low-emitting materials and thus pose less risk to the natural environment and the buildings' occupants. Low-emitting paints, coatings, adhesives, sealants, and carpeting should be used in the interiors of the buildings whenever possible. The building design team should include professional engineers who have achieved the LEED Accredited Professional designation and are experienced with the design and construction of green buildings.

For the Carlsbad seawater desalination project, for example, the additional costs associated with the implementation of the green building design as compared to the costs for a standard building are estimated at $5 million, and the potential energy savings are approximately 500 MWh/year. The potential carbon footprint reduction associated with this design is 124 tons of carbon dioxide per year (0.9 percent of the net power plant footprint). These figures are presented in line 6 of Table 5.2.

The unit cost of the carbon footprint reduction associated with the green building design was estimated for a project life of 30 years and an annual debt service of 6.5 percent (a capital recovery factor of 0.07657). At capital costs of $5 million, the annualized cost of this capital investment is $382,850/year. Because of the higher level of complexity and automation of the green building design as compared to a conventional design, the additional operations and maintenance costs associated with the green system of the building for the Carlsbad project are $34,650/year. Therefore, the total annual costs associated with this design are estimated at $417,500/year. At 124 tons of carbon dioxide reduction per year, this annualized cost corresponds to a unit carbon footprint reduction cost of just under $3400 per ton of carbon dioxide.

The total actual energy reduction that would result from green building design should be verified during plant commissioning, which will incorporate a LEED compliance review process. The LEED review process should be completed by an independent third-party consultant certified to complete such reviews.

On-Site Solar Power Generation

One enhancement of the green building design is the installation of a rooftop photovoltaic system for generating solar power. For example, the main desalination plant building is planned to have a roof surface of 4,645 m^2 (50,000 ft^2), which would be adequate to house a solar panel system that could generate approximately 777 MWh/year of electricity and reduce the net carbon footprint of the desalination plant by 193 metric tons of carbon dioxide per year, which is approximately 1.4 percent of the plant's net carbon footprint of 13,700 tons of carbon dioxide per year (see line 7 of Table 5.2).

The construction cost of the rooftop solar power system for the Carlsbad project is estimated at $4.1 million. The annual cost of power generation using this alternative is $313,937/year (at 30 years and 6.5 percent interest). In addition, the annual operation and maintenance costs for this system are estimated at $52,763/year. Therefore, the total annual costs for operation of this system are $366,700/year, which corresponds to unit cost of generated electricity of 47.2 cents per kilowatt-hour ($366,700/year divided by 777,000 kWh/year = $0.4719/kWh). This unit cost is approximately five times higher than the cost of power supply from the electric grid. The unit cost of carbon footprint reduction for this alternative is $1900 per ton of carbon dioxide.

Use of Carbon Dioxide for Water Production

Approximately 2100 tons of carbon dioxide per year is planned to be used at the desalination plant for post-treatment of the freshwater (permeate) produced by the reverse osmosis system. Carbon dioxide in a gaseous form will be added to the RO permeate in combination with calcium hydroxide or calcium carbonate to form soluble calcium bicarbonate, which adds hardness and alkalinity to the drinking water to protect the distribution system from corrosion.

In this post-treatment process of RO permeate stabilization, gaseous carbon dioxide is sequestered into a soluble form of calcium bicarbonate. Because the pH of the drinking water distributed for potable use is in a range of 8.3 to 8.5, at which carbon dioxide is in a soluble bicarbonate form, the carbon dioxide introduced in the RO permeate will remain permanently sequestered in this form and ultimately will be consumed with the drinking water.

A small quantity of carbon dioxide used in the desalination plant post-treatment process is sequestered directly from the air when the pH of the source seawater is adjusted by addition of sulfuric acid in order to prevent RO membrane scaling. However, a large amount is typically delivered to the desalination plant site by a commercial supplier. Depending on the supplier, carbon dioxide is of one of two origins: (1) a CO_2 generating plant or (2) a CO_2 recovery plant. CO_2 generating plants use various fossil fuels (natural gas, kerosene, diesel oil, etc.) to produce the gas by fuel combustion. CO_2 recovery plants produce carbon dioxide by recovering it from the waste streams of other industrial production facilities that emit CO_2-rich gases: breweries, commercial alcohol (i.e., ethanol) plants, hydrogen and ammonia plants, etc. Typically, if these gases are not collected via a CO_2 recovery plant and used in other facilities, such as a desalination plant, they are emitted to the atmosphere and therefore constitute a GHG release.

The Carlsbad desalination plant is planned to use only carbon dioxide produced in a CO_2 recovery plant. This requirement will be enforced by requiring the commercial supplier of carbon dioxide for the plant operations to provide a certificate of origin of each load of the water treatment chemical delivered to the plant site. This will encourage and incentivize the commercial suppliers and manufacturers of CO_2 to recover it from industrial waste streams rather than to generate new gas by combustion, and thereby to prevent its release to the atmosphere. Sequestration of CO_2 at the desalination plant by its conversion from a gaseous to a chemically bonded soluble form is therefore considered a carbon footprint reduction alternative. Through sequestration of 2100 tons of carbon dioxide per year in the desalination plant post-treatment process (see line 8 of Table 5.2), the net carbon footprint of the plant (13,700 tons of carbon dioxide per year) would be reduced by 15.3 percent. At an annual expenditure for carbon dioxide supply of approximately \$147,000/year, this carbon footprint reduction alternative is very cost competitive (\$70 per ton of carbon dioxide).

Reduced Energy for Water Reclamation

Often, water reclamation plants are equipped with brackish water reverse osmosis systems in order to reduce the salinity of the reused water. If the seawater desalination plant is designed to significantly reduce the salinity of the drinking water and therefore of the reclaimed water, then the operation of the brackish desalination system at the water reclamation plant could be discontinued, thereby saving energy and reducing GHG emissions.

For example, the Carlsbad Municipal Water District owns and operates a 15,000 m^3/day (4 mgd) water reclamation plant, which consists of advanced tertiary treatment facilities for the entire flow and of a 3785 m^3/day (1 mgd) brackish water reverse osmosis desalination system, which at present uses 1950 MWh of electricity per year. The purpose of the brackish water desalination plant is to reduce the salinity of the treated effluent from 1400 mg/L to below 1000 mg/L in order to make the effluent suitable for irrigation. The current high level of salinity of the reclaimed water is mainly due to the relatively high salinity of the city's drinking water, which can reach 1000 mg/L at times.

Once the Carlsbad seawater desalination plant is in operation and has completely replaced the existing high-salinity drinking water, the salinity of the city's reclaimed water is projected to be reduced by half. Therefore, the replacement of the city's existing high-salinity imported water supply with desalinated water would eliminate the need for operation of the brackish water desalination plant at the Carlsbad Water Recycling Facility. This in turn would reduce the carbon footprint of the Carlsbad Water Recycling Facility, which is otherwise 1950 MWh/year $\times$ 546.46 pounds of carbon dioxide per megawatt-hour = 1,065,957 pounds of carbon dioxide per year (484 tons of carbon dioxide per year). Since this GHG reduction is directly credited to the seawater desalination plant's operations, the Carlsbad desalination plant's carbon footprint could be reduced by 3.5 percent. The carbon footprint credit associated with the reduced energy for water reclamation is presented in line 9 of Table 5.2.

Carbon Dioxide Sequestration by Revegetation in Wildfire Zones

Almost every year, many dry parts of the world, such as California, are exposed to measurable loss of forest, urban, and suburban trees due to large wildfires. A desalination project could participate in a locally administered carbon offset program aimed at the revegetation of areas impacted by wildfires or the planting of new trees and other vegetation to combat GHG emissions.

For example, in response to a reforestation program administered by the California Coastal Commission, the Carlsbad project's proponent committed to investing $1 million in reforestation activities. More specifically, when the project developer updates the plant's net carbon footprint for the preceding year, it will calculate the cost of offsetting that year's net carbon emissions at a rate equal to the purchase of such carbon offsets. The project proponent will then pay the amount resulting from this calculation to either the San Diego County Air Pollution Control District or another entity identified by the California Coastal Commission as responsible for administering a San Diego area wildfire revegetation program. The project's proponent will continue making its annual offset payments to the revegetation program until the cumulative total of such payments equals $1 million, at which time the project proponent may elect to direct annual offset payments to other projects so long as those projects satisfy accepted standards for offsetting carbon emissions.

According to the *Tree Guidelines for Coastal Southern California Communities* issued by the USDA Forest Service (McPherson et al., 2000), the average annual costs for tree planting and care increase with the size of the mature tree and for medium-size trees range between $16 and $23 per tree (with an average of $19.50 per tree). Average annual maintenance costs for trees are estimated at $3 to $5 per tree (average $4 per tree). Updated for inflation, the 2008 average tree-planting cost was $26.70 per tree and the average annual maintenance cost was $5.50 per tree.

Assuming tree maintenance costs for 25 years at $5.50 per tree, the total life-cycle expenditure per tree is $137.50. When this is added to the tree-planting cost of $26.70, the total cost for planting and maintenance of the trees included in the reforestation project would be $164.20 per tree. At commitment of $1 million and total costs of $164.20 per tree, the total number of trees to be replanted is 6090. At an annual tree sequestration rate of 27 kg (60 lb) per tree over the 25-year period of the desalination plant's operations, the total annual carbon footprint reduction associated with the tree sequestration project is estimated at 365,400 lb (166 metric tons) of CO_2 per year, as shown in line 11 of Table 5.2. This is approximately a 1.2 percent reduction of the desalination plant's net footprint. At an annual expenditure for tree reforestation of approximately $33,500/year, the unit carbon footprint reduction cost for this alternative would be $200 per ton of CO_2.

Carbon Dioxide Sequestration in Coastal Wetlands

In addition to the benefit of marine habitat restoration and enhancement, coastal wetlands also act as a sink of carbon dioxide. Tidal wetlands are very productive habitats that remove significant amounts of carbon from the atmosphere, storing a large portion of it in the wetland soils. While freshwater wetlands also sequester CO_2, they are often a measurable source of methane emissions. For comparison, coastal wetlands and salt marshes release negligible amounts of greenhouse gases, and therefore their carbon sequestration capacity is not measurably reduced by methane production.

Based on a detailed study completed in a coastal lagoon in Southern California (Brevik & Homburg, 2004), the average annual rate of sequestration of carbon in coastal wetland soils is estimated at 0.03 kilograms of carbon per square meter per year. Another source (Trujillo, 2007) indicates that in addition to accumulating CO_2 in the soils, central and Southern California tidal marshes could also sequester 0.45 kilograms of carbon per square meter per year in the macrophytes growing in the marshes, and 0.34 to 0.63 kilograms of carbon per square meter per year in the algal biomass.

For example, as part of the Carlsbad seawater desalination project, the project developer is planning to develop 37 acres of new coastal wetlands in San Diego County. These wetlands will be designed to create habitat for marine species similar to those found in the Agua Hedionda Lagoon (see Fig. 5.9), from which source seawater is collected for the power plant and desalination plant operations. Once the wetlands are fully developed, they will be maintained and monitored over the life of the desalination plant operations. The cost of the wetland restoration project is estimated at $3 million.

Taking into consideration that the total area of the proposed wetland project is 37 acres (149,739 m^2) and the maximum sequestration capacity of the coastal wetlands is 1.11 kilograms of carbon per square meter per year, then the wetlands carbon sequestration capacity would be up to 83 tons of carbon per year. With a conversion factor from carbon to carbon dioxide of 3.664, the offset of the desalination plant's carbon footprint by the wetlands project is estimated at 304 tons of CO_2 per year as shown in line 12 of Table 5.2 (a 2.2 percent reduction of the net carbon footprint). At a total annual cost for wetlands development and maintenance of approximately $120,000/year, the unit carbon footprint reduction cost for this alternative would be $400 per ton of CO_2.

Site-specific research is planned in order to quantify the actual carbon sequestration capacity of the new wetlands system proposed for development as a part of the Carlsbad seawater desalination project, once the wetlands project is completed and is fully functional. Typically it takes 3 to 5 years for a coastal wetlands project to be fully functional and begin to yield enhanced habitat and GHG sequestration benefits.

Investment in Renewable Energy Projects

An alternative approach to offsetting the GHG emissions of a given desalination project is to invest in renewable energy projects located in the service area of the desalination plant. For example, the project developer plans to invest in a number of green power projects (rooftop photovoltaic systems, diesel bus conversion to clean natural gas vehicles, etc.) with its public partners who will be receiving desalinated water from the Carlsbad seawater desalination plant. The total carbon footprint offset for the desalination plant is projected at 2260 MWh/year, or 561 tons of CO_2 per year (4.1 percent of net carbon footprint). These credits are shown in line 13 of Table 5.2.

Other Carbon Offset Projects and Renewable Credits

The remainder of the project's carbon emissions could be offset by the purchase of a combination of carbon offset projects and renewable energy credits (RECs). Contracts for offset projects provide more price stability and are typically established for longer terms (10 to 20 years) than RECs (1 to 3 years).

For example, about 1.5 to 2 years before operations begin, the project developer will create and issue a request for proposals for carbon offset projects and renewable energy credits. The developer will then select the best mix from the responses and contract for their acquisition or development. The exact nature and cost of the offset projects and RECs will be known once the request process is complete. The offset purchases and RECs will be 6668 tons of CO_2 per year, as shown in line 14 of Table 5.2 If the project developer were to purchase local offset projects, each with 10-year contract terms, the initial solicitation would cost an estimated $1.2 million in offset projects for the first decade and over $5 million over the project's 30-year life. Offsets or RECs would be used as the swing mitigation option to true up annual changes to the project's net carbon footprint.

Project Annual Net Zero Carbon Emission Balance

Table 5.2 summarizes the total and net carbon footprint estimate of the Carlsbad seawater desalination project and quantifies GHG emission reduction and mitigation options that are planned for implementation in order to reduce the plant's net carbon emission footprint to zero. Analysis of that data presented in Table 5.2 indicates that for this example case study, up to 43.8 percent of the GHG emissions associated with seawater desalination and drinking water delivery will be reduced by on-site reduction measures and the remainder will be mitigated by off-site mitigation projects and purchase of renewable energy credits. It should be noted that the contribution of on-site GHG reduction activities is expected to increase over the useful life of the project (i.e., in the next 30 years) because of the following key reasons: (1) in the near future, most power suppliers countrywide are planning to significantly increase the percentage of green power sources in their electricity supply portfolios, which in turn will reduce their emission factors and the desalination plant's net carbon footprint; and (2) advances in desalination technology are expected to yield further energy savings and carbon footprint reductions. The mitigation costs of the various alternatives are summarized in Table 5.3.

The lowest unit cost of carbon footprint reduction can be achieved by using carbon dioxide for post-treatment of the desalinated water ($70 per ton of CO_2). The costliest carbon footprint reduction options are green building design ($3400 per of ton CO_2) and installation of rooftop solar power generation system ($1900 per ton of CO_2). Development of new coastal wetlands is a very promising carbon footprint reduction option ($400 per ton of CO_2) that could cost only a fraction of what it would cost for construction of a solar panel generation system. Similarly, reforestation could also be a cost-competitive GHG reduction alternative ($200 per ton of CO_2). As compared to green power generation alternatives (solar and wind power), reforestation and wetland mitigation have added environmental benefits. For example, the new coastal wetlands developed in relation to a seawater desalination project could be designed to create a habitat for species that are impacted by the intake operations of the desalination plant via impingement and entrainment on the intake screens.

Alternative	Unit Cost, $ per ton of CO_2 reduced
Green building design	3400
On-site solar power generation	1900
CO_2 sequestration in coastal wetlands	400
CO_2 sequestration by revegetation of wildfire zones	200
Use of CO_2 for water production	70

Table 5.3 Unit Costs of Carbon Footprint Reduction Alternatives

5.5 Traffic, Noise, and Other Auxiliary Impacts

Noise, air pollution, and traffic associated with the construction and operation of desalination plants are similar to those associated with the implementation of conventional water treatment plant projects.

Noise is of specific importance because the high-pressure reverse osmosis feed pumps operate at very high rotation speed and are usually a significant source of noise pollution (60 to 96 dB). These pumps feed flow to the reverse osmosis treatment trains and interconnected energy recovery devices and are the key sources of noise. They should be located in the reverse osmosis building, which will contain the generated noise.

Usually, desalination plants are equipped with large intake seawater pumps, pretreatment filter transfer pumps, and product water transfer pumps, which are often located outdoors. Potential noise mitigation measures for these pumps are as follows: (1) use of centrifugal pumps, (2) use of water-cooled motors, (3) installation of acoustic enclosures, (4) installation of sound curtains, and (5) installation of pumps and motors in an enclosed building.

Centrifugal pumps, which have relatively low noise levels, should be used when possible as an alternative to noisier piston pumps. The main sources of noise in a centrifugal pump station are the pump motors. Water-cooled motors are recommended instead of standard air-cooled motors to reduce noise levels. Commercially available acoustic enclosures can be installed around the pump motors or the entire pump station to contain and dissipate the noise from the outdoor mechanical equipment. Industrially sewn sound curtains can be installed around the pump stations using floor-mounted hardware. If it is required to comply with stringent acoustic attenuation requirements, all large pumps and motors can be installed in an enclosed building designed to reduce noise emissions. Often, the noise in the main desalination building, which houses the high-pressure pumps and energy recovery devices, is attenuated by acoustic control panels installed on the walls of the building.

5.6 Framework of Environmental Impact Assessment

5.6.1 Introduction

The purposes of the environmental impact assessment (EIA) for a given desalination project are to (1) review the project's purpose, need, scope, site, and supporting infrastructure; (2) analyze and quantify potential environmental impacts from the project's implementation; and (3) identify feasible mitigation measures aimed at eliminating or reducing all significant adverse environmental effects of the proposed desalination project. The EIA encompasses both temporary construction-related effects and long-term effects associated with project operation.

Typically, an EIA addresses three key areas:

- Impacts on biological (terrestrial and marine) resources in natural habitats (surface and groundwaters, soils, sediments, air) on and around the project site and supporting infrastructure.
- Impacts on socioeconomic and cultural resources, such as effects on public well-being, services, and utilities; changes in existing and planned land uses; traffic and circulation; aesthetics, light, and glare; recreation; natural scenery; and population growth.

- Impacts related to public health associated with the quality of the product water delivered to the project service area, the integrity and function of the existing water distribution system, air quality, noise pollution, and disposal of the waste streams generated during the production of desalinated water.

The environmental review process of a given project includes the following key steps: (1) project EIA scoping, (2) project definition, (3) environmental analysis, and (4) identification of significant environmental impacts and development of mitigation measures. The scope of these project review efforts is described next.

5.6.2 Project EIA Scoping

The initial desalination project planning efforts described in Chap. 4 allow for the definition of several alternatives for project location, size, and implementation, typically summarized in a project master plan. This master plan usually serves as a starting point for the project's environmental impact assessment.

The first step of the EIA scoping is to evaluate and prescreen the initial alternatives defined in the master plan based on preliminary determination of their environmental impacts and compliance with pertinent regulatory requirements. A key activity associated with this phase is to gain a more thorough understanding of the potential socioeconomic and regulatory constraints associated the implementation of the project alternatives defined in the master plan. Such an understanding is built through meetings and discussions with public and regulatory agencies with jurisdiction over the project's implementation and through public scoping meetings conducted with citizens located in the project service area and near the potential project plant sites. The public scoping meetings aim at identification of key concerns, issues, preferences, and expectations of the general public in terms of project scope, appearance, location, and size.

Based on a thorough analysis of the collected information, project alternatives are reviewed for fatal flaws, and usually one preferred project and one or two alternative projects and/or project components are selected as an outcome of the project EIA scoping effort. This project step also defines the scope of the detailed engineering and environmental review studies needed to evaluate the potential environmental impacts of the preferred project.

5.6.3 Project Definition

Once the scope of the project environmental review effort is determined, the next step is to gain detailed site-specific information on the preferred project and the selected project alternatives, including project size, location, needs and objectives; project phasing; environmental setting; desalination plant intake, discharge, and treatment alternatives; product water delivery system configuration and pipeline routes; and entitlements or agreements, permits, and approvals needed for project implementation.

Project definition also includes completion of geotechnical investigation and utility, biological, and archeological surveys of the sites of the plant intake, treatment facility, and concentrate disposal and water distribution pipeline routes, as well as characterization of the intake and discharge water quality and ambient physical and hydrodynamic conditions of the surface water body (for open intakes and discharges) or intake and discharge aquifers (for subsurface intakes and deep injection wells).

In addition, project definition encompasses an evaluation of the potential impacts the introduction of desalinated water may have on the physical integrity and water quality of the water distribution system. Special attention is often given to the impact of the desalinated water on industrial, agricultural, and horticultural users located in the desalination plant's service area.

The typical engineering documentation produced during the project definition phase of the EIA includes the following:

1. A visual impact report, with desalination plant site renderings from key local viewpoints.
2. An air quality assessment, reflecting impacts from air emissions due to construction traffic and plant operations
3. A climate action plan, which defines measures that will be needed to mitigate, reduce, or eliminate greenhouse gas emissions associated with project construction and operations.
4. A biological resources report, which identifies terrestrial life inhabiting the project site and routes of product water delivery pipelines and aquatic life in the intake and discharge areas—this report focuses on the presence of endangered species and protected habitats.
5. A cultural resources report, discussing archeological sites, artifacts, or areas of cultural significance that may be impacted by the project construction and operations.
6. A geotechnical report, which provides information on soil types and conditions including soil load-bearing capacity and soil and groundwater contamination with hazardous substances.
7. A bathymetry report, which describes the geophysical conditions in the intake and discharge areas for project alternatives that include open intakes and discharges
8. A hydrology report, which identifies ocean currents, winds, beach erosion, seismicity, and other factors that may impact the performance and integrity of the plant intake and discharge systems (for open intakes).
9. An impingement and entrainment assessment, quantifying the impact of construction of open intakes on the aquatic life near their locations.
10. A hydrodynamic study, which identifies the size, shape, location, and salinity distribution of the zone of dilution of the desalination plant concentrate around the location of its discharge.
11. An acoustic assessment, which defines potential impacts from noise generated by desalination plant equipment in the vicinity of the plant site.
12. A traffic impact analysis, providing insights into the changes of traffic patterns and intensity in the vicinity of the project site as a result of project construction and normal operation.
13. A water supply assessment, which addresses issues such as impacts of the desalinated water on the distribution system's water quality, corrosion, and hydraulics.

The studies listed here are typical for most desalination projects. However, depending on the project size, site-specific conditions, and regulatory requirements, some studies may not be necessary, or additional site-specific studies may be needed. This observation holds true especially for projects that may incorporate the use of existing infrastructure (e.g., existing outfalls and/or discharges) or are proposed to be co-sited with other existing facilities, such as power plants or wastewater treatment plants.

5.6.4 Environmental Analysis

The background information (data, studies, and investigations) collected during the project definition phase is analyzed to identify and quantify the potential environmental impacts of the preferred project and potential project alternatives. This analysis focuses on the following key impact areas:

1. Land use and relevant planning
2. Geology, soils, and seismicity
3. Hydrology, drainage, and storm water runoff.
4. Air quality
5. Noise
6. Public services and utilities
7. Aesthetics, light, and glare
8. Hazards and hazardous materials
9. Construction-related impacts
10. Surface water quality and aquatic biological resources (for surface intakes and discharges) or groundwater quality of the saline intake and discharge aquifers (for subsurface intakes and deep injection well discharges)
11. Product water quality and impacts on the water distribution system

The land use impact assessment focuses on the compatibility of the desalination project with the land zoning requirements pertinent to the project site and water distribution infrastructure. This assessment also addresses the consistency of the project's needs, objectives, size, scope, service area, and points of interconnection to the water distribution system with all applicable local, regional, and state water resource management plans, policies, and other requirements.

Assessments of geology, soils, and seismicity target the suitability of the site's topography, geology, and location for project construction and long-term operation. In seismic areas, special attention is given to the project's proximity to major active faults and potential for soil liquefaction and subsidence, lateral spread, and landslides; the projected magnitude of seismic force; and the potential impact on the plant from tsunamis and/or seiche waves. This information is also used to determine the likelihood of long-term operational wind- and water-driven erosion impacts on plant intake and discharge structures and other project-related facilities.

Hydrology, drainage, and storm water runoff issues under assessment include the likelihood that the project will violate applicable storm water discharge requirements; substantially deplete groundwater supplies or interfere with groundwater recharge and lower the local groundwater table level; alter the existing drainage pattern of the project site area; create or contribute surface runoff that cannot be handled by the existing or planned water drainage systems; and expose the project structures and staff to flood risks. A particular attention is given to the location of structures holding and handling hazardous waste. Such structures are expected to be located outside of the 100-year flood hazard area in the location.

Air quality impacts of the project are evaluated in terms of the project's potential to violate any air quality standards established for the project area by pertinent regulatory agencies, to release significant quantities of greenhouse and hazardous gases, or to create objectionable odors affecting the nearby residents. The environmental review includes mobile emission sources (e.g., employee vehicles and chemical delivery trucks) and stationary emission sources (e.g., emissions from chemical storage and handling facilities) as well as indirect GHG emissions associated with the on-site and off-site power generation facilities used to supply the desalination plant with electricity.

Desalination plants employ large motors, pumps, and energy recovery devices that can be significant sources of noise pollution. The environmental review in terms of noise identifies the location of noise-sensitive receptors in the vicinity of the plant site (e.g., hospitals, schools, residential buildings, etc.) and evaluates the potential for project-related noise emissions to exceed applicable local standards, ordinances, and other pertinent regulations and thus have a negative impact on the well-being of the neighboring community.

The construction and operation of desalination plants involves the use of a number of public services and utilities, such as fire and police services, roadways, wastewater collection services (typically for the disposal of spent membrane cleaning solutions and sanitary wastewater), solids waste collection services, storm water, drinking water and gas utilities, and electrical power supply, telephone, and cable utilities and services. The project environmental review assesses the potential for overloading of such services and utilities and for noncompliance with service-related pertinent regulatory requirements.

Potential effects related to aesthetics, light, and glare encompass negative impact on the site's character or on a scenic vista in the vicinity of the plant; substantial damage to scenic resources such as trees, rock outcroppings, historic buildings, scenic highways; and creation of significant light and glare affecting the daytime or nighttime views of the plant area.

Desalination plants use some water treatment chemicals (i.e., sulfuric acid, sodium hypochlorite, ferric coagulants, etc.) that, depending on the size of their on-site storage facilities, may be classified as hazardous materials. Therefore, the environmental review addresses the potential of the transportation, storage, use, and disposal of such chemicals to create significant public and/or environmental hazards.

The environmental review also focuses on construction-related impacts, such as effects on adjacent land uses and traffic, biological resources (vegetation, special-status wildlife habitats, and endangered or protected species), historical/archeological and paleontological resources, hydrology and water quality, air quality, noise, public services and utilities, aesthetics, and hazards and hazardous materials.

As discussed previously, desalination plants with open intakes and discharge outfalls may have potential negative impacts on surface water quality (ocean, sea, river estuary, etc.) and aquatic biological resources. The main desalination plant intake impacts are associated with impingement and entrainment of aquatic organisms and potential loss of natural habitat occupied by the intake structures and piping, while the discharge impacts are mainly related to salinity or to toxicity triggered by mineral-ion imbalance. Such impacts are discussed in detail in previous sections of this chapter.

Product water quality is one of the key environmental review topics for desalination projects with service areas supplied by a number of different sources of drinking water. This is especially true when the concentration of mineral constituents such as boron, bromide, sodium, and chloride in desalinated water are at higher levels than those in the other water sources with which the desalinated water is mixed in the water distribution system.

The environmental impacts associated with such differences can potentially be reduction of the growth rate and yield of some citrus crops, strawberries, avocados, and other crops or ornamental plants irrigated with such water, as well as the formation of excessive amounts of bromates, bromamines, and disinfection byproducts which may have potential human health impacts. In addition, desalinated water often has very low hardness and alkalinity; therefore, if its water quality is not properly adjusted, such water can cause corrosion in the water distribution system and household plumbing.

Another aspect of desalinated water quality which is evaluated during project EIA is the potential of the water to contain algal toxins that could be a human health concern if they exceed certain level. Other potential product water quality parameters of concern could be constituents in the saline source water that are not fully rejected by the RO membranes, such as gases (e.g., hydrogen sulfide) or nitrates. Source water quality characterization and RO membrane performance analysis allow prediction of the potential for such effects.

5.6.5 Identification of Significant Environmental Impacts and Development of Mitigation Measures

The project environmental review described in the previous section allows to determine the magnitude of the potential environmental impacts associated with the construction and operation of the proposed desalination project, and the identification of mitigation measures for the impacts that are found to be significant. Such mitigation measures may involve changes in project scope, site, location, design, methods of source water collection and concentrate discharge, scope, service area, or potential water uses.

In some cases, while environmental impacts may be found to be significant, they may be unavoidable and may be found acceptable if the project benefits outweigh them or the impacts are temporary in nature and have a short duration. If such significant impacts are found to be unavoidable, the EIA usually identifies minimization and/or remediation activities that could address them.

The project mitigation measures identified by the EIA are usually incorporated in the desalination plant's construction and operation permits and licenses, and are enforced by the respective regulatory agencies that issue such permits/licenses and have responsibility for independent project monitoring and oversight.

All of the project's significant environmental impacts and their proposed mitigation measures are incorporated into a project environmental management plan, which typically is integrated into the overall desalination project implementation plan and schedule. Besides identifying key environmental impact mitigation activities, this plan also defines monitoring requirements that allow documentation and confirmation of the effectiveness of these activities.

A more detailed discussion of the key components and activities associated with the preparation of a typical desalination project environmental impact assessment is presented elsewhere (Cotruvo et al., 2010).

5.7 Overview of the Desalination Project Permitting Process

One of the key activities related to environmental impact assessment for a given desalination project is to identify all applicable regulatory requirements associated with project planning, design, construction, and operation, and to develop a plan to obtain project permits and licenses stipulating such regulatory requirements (i.e., a project permitting plan).

The number and type of permits, as well as the permit requirements and regulatory agencies responsible for issuing and enforcing such permits, varies significantly from project to project, country to country, state to state, and even on a regional or local agency level. Therefore, the permitting process and plans are always project specific.

The *Guidelines for Implementing Seawater and Brackish Water Desalination Facilities* developed by the Water Research Foundation in cooperation with the WateReuse Research Foundation, the US Bureau of Reclamation, and the California Department of Water Resources (Water Research Foundation, 2010) provide a general overview of permitting and regulatory requirements and challenges in the United States. Texas and California have state-specific general guidelines for desalination project environmental planning, review, and permitting (R. W. Beck, 2004; California Department of Water Resources, 2008).

In general, permits related to desalination projects can be divided in three categories: (1) those related to source water intake, (2) those associated with plant discharge, and (3) those pertaining to product water. The main issues associated with such permits and typical permitting support studies are discussed in the following sections.

5.8 Permits Related to Source Water Intake

5.8.1 Key Permitting Issues

Permits and issues related to source water intake vary significantly depending on the type of desalination project (seawater vs. brackish water) and the type of intake (subsurface vs. open intake). As indicated previously, the main potential environmental impacts associated with open intakes are impingement and entrainment of aquatic organisms. As a result, most open-intake related permitting activities are associated with addressing these issues. Other permit-related issues for such intakes address intake structure impact on naval traffic and other coastal zone uses. Subsurface intakes (i.e., wells and infiltration galleries) do not cause any impingent and the rate of their entrainment impacts is considered significantly smaller than that of open intakes. Therefore, most permits for subsurface intakes focus on regulating the source water collection capacity and quality.

Requirements for Plants with Subsurface Intakes

For seawater projects with subsurface intakes, the main drinking water permit-related issues are associated with determining whether the intake is under the influence of surface water contamination, which in turn is used by regulators to establish source water treatment requirements. In California, construction of subsurface intakes within

8 km (5 mi) from the shore is also under the jurisdiction of the California Coastal Commission and requires a coastal development permit.

For brackish water projects with subsurface intakes the key permit approvals encompass a water test well permit, a water production well registration permit, a production well construction and alteration permit, and a production well operation permit. In addition, an important requirement associated with well permitting is the possession of water rights for take (pumping) of groundwater at a target intake capacity from a specific aquifer and/or groundwater basin.

Requirements for Plants with Open Intakes

The main permitting requirements for brackish and seawater desalination plants with open intakes relate to the environmental impacts of these intakes in terms of impingement and entrainment of aquatic organisms.

In addition, desalination plants in the United States need permits under the federal Clean Water Act, Sec. 404 (dredge and fill permit) and the Rivers and Harbors Act (Sec. 10 permit) for the construction of new intake forebay structure and pipes in navigable waterways. Such permits are administered by the US Army Corps of Engineers and usually require the concurrence of other federal agencies regulating coastal development, such as the National Oceanic and Atmospheric Administration, and—for some states—state regulatory agencies, such as the California Coastal Commission and the California State Lands Commission. However, if the desalination plant is co-located with a power plant and therefore does not require the construction of new intake, a Sec. 404 (dredge and fill) permit is not needed.

Desalination projects with open intake/outfall structures may require review by the National Marine Fisheries Service and/or US Fish and Wildlife Service for potential impacts on endangered, threatened, or sensitive species and may need an incidental take permit for protected marine mammals and migratory birds that may be impacted by open intake operations. In some states, depending on the location of the intake, an erosion prevention permit may also be needed for intake construction.

Some desalination projects include intake, discharge, and product water delivery pipelines that cross roads, highways, and utilities under the jurisdiction of state and local agencies. Construction of such project components typically requires encroachment permits, easements, or utility permits and authorizations.

As compared to subsurface intakes, water rights are not needed for desalination projects with open seawater intakes. However, in some states, such water rights are required if the open intake is located in enclosed or semienclosed water bodies such as bays and estuaries or in a river.

5.8.2 Permitting Support Studies

Subsurface Intakes

For subsurface intakes, the main type of permitting-related study needed is a test well pumping study, which allows for a determination of the production capacity of the intake wells and source water quality. In addition, a hydrogeological study is necessary to assess the type and productivity of the source water aquifer and to determine whether the well water quality is under the influence of surface or subsurface contamination sources and whether the well pumping may mobilize pollutants and minerals from adjacent aquifers and introduce them into the source water aquifer.

Open Intakes

Impingement and Entrainment Study The environmental permitting and assessment of open intakes requires the completion of a 12-month study, which involves collection of source water samples in the vicinity of the intake usually two to four times per month and allows a determination of the annual and daily amounts of marine organisms that could potentially be impinged on the source water intake and entrained into the desalination plant. Impingement is very much dependent on the specific intake type, configuration, and through-screen velocity. As indicated previously, entrainment is typically considered proportional to the plant flow and the abundance of marine organisms in the intake area.

The results from the impingement and entrainment studies are used to project the potential reduction of the quantity and type of aquatic species in the area of influence of the desalination plant intake as a result of its continuous operation. Once the impacted aquatic species are projected and their potential losses quantified, that impact is evaluated against the productivity of the intake area to determine whether it is significant and whether the impacted aquatic species are of critical importance to sustain the local aquatic ecosystem or are of measurable commercial or recreational fishing value.

Flow, Impingement, and Entrainment Minimization Plan If the regulatory agencies with jurisdiction over the source water resources determine that the open intake of the desalination plant at its initial configuration and size will cause significant environmental impacts, then the desalination project proponent (permit applicant) will have to prepare and submit for agency review a plan indicating how the proponent proposes to minimize intake flow, impingement, and entrainment by the selection of alternative intake and location and by the use of a combination of best technologies available, and design and operational measures. Such measures are summarized in a flow, impingement, and entrainment plan, which is submitted to the pertinent regulatory agencies for review and approval.

Impingement and Entrainment (I&E) Mitigation Plan While intake impingement can be reduced significantly through the use of low through-screen velocity, entrainment cannot be fully eliminated; so depending on the local environmental setting, the permitting agencies with jurisdiction over the intake area may require the project proponent to develop and implement an I&E mitigation plan. The purpose of such a plan is to implement an aquatic life restoration program that produces environment (e.g., wetlands, artificial coral reefs, etc.) and aquatic species in kind to those lost as a result of impingement and entrainment. Potential I&E mitigation projects are discussed in previous sections of this chapter.

5.9 Permits Related to Plant Discharge

5.9.1 Key Permitting Issues

Permitting of desalination plants with surface discharges is mainly focused on the environmental impacts of plant concentrate on aquatic life of the receiving waters. Groundwater aquifer discharges are typically regulated to protect their water quality and to maintain and enhance their other beneficial uses.

Overview of Regulatory Framework in the US

Desalination plant discharges are classified by the US EPA as industrial waste despite the fact that they are distinctly different from most industrial discharges. Several regulatory programs in the United States address the disposal of desalination plant discharges: (1) the Clean Water Act; (2) the Underground Injection Control Program, ordinances that protect groundwater; and (3) the Resource Conservation and Recovery Act, which regulates solid waste residuals. The permits required for concentrate discharge in the United States vary depending on the type of discharge (surface or subsurface).

Disposal to surface water discharges requires a National Pollutant Discharge Elimination System (NPDES) permit, which in most states is issued by state regulatory agencies that are delegated such responsibility by the US EPA. Besides numeric limits for specific contaminants and whole effluent toxicity, NPDES permits typically contain receiving water quality antidegradation provisions that require the plant discharge to be within 10 percent of the ambient levels of naturally occurring contaminants and to prevent impairment of the receiving water quality in terms of color, odor, and visual appearance.

Wastewater collection system discharges require a permit issued by the local wastewater collection and management agency to meet its sewer ordinance and the Clean Water Act industrial pretreatment program requirements, as stipulated in the agency's NPDES permit.

Concentrate disposal by land application (percolation ponds, rapid infiltration basins, landscape and crop irrigation, etc.) has to comply with federal and state regulations to protect groundwater, public health, and crops and vegetation. Land application also requires permits from state agencies.

Concentrate disposal by deep well injection is regulated by the US EPA under the Underground Injection Control Program of the Safe Drinking Water Act. The related construction, monitoring, and other permits are issued and enforced by the US EPA region or state agency that has jurisdiction over the desalination plant's location.

The Resource Conservation and Recovery Act regulates the disposal of solid waste generated by desalination plants, such as precipitated salts and sludge. If a given plant generates solids that contain arsenic or other toxins above levels that classify them as hazardous waste, and such sludge does not pass the toxic characteristic leaching procedure test, then the sludge is considered a hazardous waste and must be handled accordingly.

It should be pointed out that sludge generated from typical seawater desalination plants with open intakes is usually nonhazardous and can be disposed to a sanitary landfill without further treatment. One exception is the sludge generated by saline water pretreatment with diatomaceous media filters, since the diatomaceous media is considered a hazardous material in the United States. For comparison, sludge from some brackish water sources can sometimes contain high levels of naturally occurring toxic compounds such as arsenic and cyanide, which may require disposal to a hazardous waste landfill.

Salinity and Whole Effluent Toxicity Requirements for Surface Discharges

At present, most countries worldwide do not have numeric standards for salinity content in the concentrate discharge; the discharge limit for this water quality parameter is usually established for the site-specific conditions of a given project (Eniav and Lokiec, 2002; Mauguin and Corsin, 2005; Sadhwani et al., 2005). The pertinent federal and state laws in the United States regulate the salinity of desalination plant concentrate discharges by establishing project-specific acute and chronic whole effluent toxicity (WET) objectives. WET is a more comprehensive measure of the environmental impact of concentrate than a salinity limit because WET water quality objectives also account for potential synergistic environmental impacts of concentrate with other constituents in the discharge.

According to current regulations in the United States, if a desalination plant discharge meets all water quality objectives defined in the applicable federal and state regulations as well as acute and chronic WET objectives, then the proposed discharge does not present a threat to aquatic life—regardless of what the actual salinity level of the discharge is or what increase above ambient salinity it may cause—because WET accounts for the salinity-related environmental impacts of concentrate.

The California Ocean Plan (California State Water Resources Control Board, 2009) establishes a daily maximum for acute toxicity in receiving water quality of 0.3 acute toxicity units. Requirement III.C.4(b) of the California Ocean Plan specifies that this maximum applies to ocean waters outside the acute toxicity mixing zone. The acute toxicity mixing zone is defined as follows:

> The mixing zone for the acute toxicity objective shall be 10 percent (10%) of the distance from the edge of the outfall structure to the edge of the chronic mixing zone (zone of initial dilution).

The California Ocean Plan defines the zone of initial dilution as the zone in which the process of initial dilution is completed. Initial dilution is defined within App. I of the California Ocean Plan as follows:

> Initial dilution is the process which results in the rapid and irreversible turbulent mixing of wastewater with ocean water around the point of discharge. For a submerged buoyant discharge, characteristic of most municipal and industrial wastes that are released from the submarine outfalls, the momentum of the discharge and its initial buoyancy act together to produce turbulent mixing. Initial dilution in this case is completed when the diluting wastewater ceases to rise in the water column and first begins to spread horizontally.

Despite the fact that environmental impacts associated with concentrate salinity are indirectly regulated through site-specific acute and chronic WET objectives, the discharge permits for some of the existing seawater desalination plants in the United States also contain specific numeric salinity limits (Table 5.4).

The Carlsbad project's NPDES discharge permit, for example, contains an effluent limitation for chronic toxicity at the edge of the zone of initial dilution in combination with numeric limitations for average daily and average hourly total dissolved solids (salinity) concentrations of 40 parts per thousand (ppt) and 44 ppt, respectively. These salinity limits were established based on a site-specific salinity tolerance study and chronic and acute toxicity testing completed for this project (City of Carlsbad, 2005). The referenced limits are applicable to the point of discharge and are both reflective of protective against the acute toxicity effect of the proposed discharge.

The 189,000 m^3/day (50 mgd) Huntington Beach SWRO project's NPDES permit also contains a limit for chronic toxicity but does not contain numeric limits for salinity. Instead, the potential acute toxicity effect of the discharge is limited by a ratio of the daily discharge flow from the desalination plant and the power plant intake cooling water flow, which provides dilution to the concentrate. This dilution ratio requirement

Desalination Plant	Total Flow, mgd	Average Daily TDS, ppt*	Maximum Hourly TDS, ppt*	Acute Toxicity, Acute Toxicity units	Chronic Toxicity, Chronic Toxicity units	Flow Ratio
Carlsbad—50 mgd • 33.5 ppt TDS (source) • 67.0 ppt (concentrate)	54/60.3 (conventional pretreatment) 57/64.5 (membrane pretreatment)	40 (19.4% above ambient)	44 (31.3% above ambient)	0.765	16.5	Mixing zone—15.1:1
Huntington Beach—50 mgd • 33.5 ppt TDS (source) • 67.0 ppt (concentrate)	56.59 (conventional pretreatment)	None	None	None	8.5	Mixing zone—7.5:1 Minimum dilution—2.24:1
Tampa Bay—25 mgd 26 ppt TDS (source) 43 ppt (concentrate)	22.8 (conventional pretreatment)	35.8 (38% above ambient)	35.8 (38% above ambient)	None	None	Dilution—28:1 (minimum 20:1)

*1 ppt (part per thousand) = 1000 mg/L.

Table 5.4 Examples of Desalination Plant Discharge Limits

effectively provides a limit for the 40 ppt salinity discharge from the desalination plant and is derived from site-specific analysis.

Concentrate disposal to surface water bodies (oceans, bays, rivers) may have impacts other than direct changes in salinity. In some circumstances, concentrate plume density may lead to increased stratification, reducing vertical mixing. This stratification may in turn reduce dissolved oxygen levels in the water column or at the bottom of the ocean in the area of the discharge, which may have ecological implications.

5.9.2 Permitting Support Studies

Support Studies for BWRO Plants with Subsurface Discharges

Deep well injection is one of the most commonly used methods for discharge of concentrate from brackish water desalination plants. In the United States, such discharges at a federal level are regulated under the Safe Drinking Water Act's Underground Injection Control Program. A typical permitting-related study for such discharges is the water quality characterization of the concentrate and of the hypersaline aquifer that is planned to receive the concentrate, in order to confirm that the receiving aquifer will not be impaired by the discharge. In addition, a hydrogeological study is completed to determine whether the receiving hypersaline aquifer is confined and can prevent the delivered concentrate from migrating to other adjacent aquifers and contaminating them. Additional details of the requirements associated with the construction of deep injection wells for concentrate disposal are presented in Chap. 16.

Support Studies for SWRO Plants with Subsurface Discharges

Shallow exfiltration beach wells and galleries are sometimes used for disposal of concentrate from small desalination plants into the coastal aquifer closest to the ocean bottom. Such discharges are typically regulated by an NPDES permit as surface discharges.

Permitting-related studies for such wells mainly focus on determining the concentrate's water quality and ability to meet applicable surface water quality discharge requirements, since this concentrate is ultimately dispersed by the receiving surface water body (ocean, river, etc.).

Support Studies for BWRO and SWRO Plants with Surface Water Discharges

Discharge Salinity Dispersion Modeling The main purpose of the evaluation of the concentrate dispersion rate from the point of discharge is to establish the size of the zone of initial dilution (ZID) required to dissipate the discharge salinity plume down to within 10 percent of the ambient water's TDS levels; and to determine the TDS concentrations at the surface, the midlevel of the water column, and the bottom in the ZID.

The TDS concentration of the saline plume at these three levels is then compared to the salinity tolerance of the aquatic organisms inhabiting the surface (mostly plankton), the water column (predominantly invertebrates), and the bottom sediments of the receiving water body in order to determine the impact of the concentrate salinity discharge on these organisms.

The discharge salinity field in the ZID and the ZID boundaries are established using hydrodynamic modeling. This modeling allows the determination of the most suitable location, design configuration, and size of the discharge outfall and diffusers if a new outfall is needed, or the assessment of the feasibility of using existing wastewater or power plant outfall facilities.

The model selected for identifying the boundaries of the desalination plant discharge should be used to define the concentrate plume dissipation boundaries under a variety of outfall and diffuser configurations and operational conditions.

Evaluation of concentrate dispersion and recirculation for large desalination plants usually requires sophisticated discharge plume analysis and is completed using various computational fluid dynamics software packages tailor-made for a given application (Bleninger and Jirka, 2008; Cotruvo et al., 2010). The models most widely used for salinity plume analysis are CORMIX and Visual Plumes. Both models allow depiction of the concentrate plume dissipation under a variety of outfall and diffuser configurations and operational conditions. These models have been developed for and approved by the US EPA for analysis of the mixing zone and establishment of total maximum discharge limits. However, CORMIX and Visual Plumes are near-field models that do not account for the far-field mixing and advective processes associated with shoaling waves and coastal current systems. Therefore, discharge modeling is extended beyond the near-field ZID using various computational fluid dynamics software packages tailor-made for a given application.

Discharge Whole Effluent Toxicity Study Whole effluent toxicity testing is an important component of the comprehensive evaluation of the concentrate discharge's effect on aquatic life. Completion of both acute and chronic toxicity testing is recommended for the salinity levels that may occur under the worst-case combination of conditions in the discharge. Use of at least one species indigenous to the targeted discharge area is desirable.

In the case of concentrate discharge through an existing wastewater treatment plant outfall, at least one species of the echinoderm taxa (e.g., urchins, starfish, sand dollars, or serpent stars) is recommended to be tested with a worst-case scenario blend of concentrate and wastewater effluent (typically, maximum wastewater effluent flow discharge combined with average concentrate flow).

In the United States and Australia, the discharge permit (license) issued by the government regulatory agency in charge of the surface discharges typically includes limits of whole effluent toxicity. Performance of WET bioassay testing in the United States is required to conform to protocols approved by the US EPA for assessing acute, chronic, and bioaccumulative toxicity to receiving water biota. The standard bioassay tests use approved pollutant-sensitive species.

Some brackish water RO plants in the United States have produced concentrates that fail WET limits tests. Most of the cases in Florida were associated with high calcium levels, and some were complicated by toxicity from high fluoride levels (Mickley, 2006). Toxicity caused by high levels of major ions is a correctable chemical imbalance, as opposed to toxic contamination from heavy metals or pesticides (Mickley, 2000). For this reason, Florida has exceptions for major ion toxicity when it is the only toxicity present in a concentrate.

Concentrate Water Quality Characterization Study A characterization study of the concentrate water quality involves the collection of concentrate samples from a pilot desalination plant and laboratory analysis of these samples for the discharge water quality parameters established by the pertinent regulatory agencies with jurisdiction over concentrate disposal. At a minimum, concentrate samples are recommended to be collected under near-average source water quality conditions (i.e., annual average salinity, temperature, and turbidity) as well as at extreme conditions, such as heavy rain events, algal blooms, dredging near the intake area, seasonal agricultural runoff, and very low and high source water temperatures and salinities.

The pilot plant used for generation of concentrate samples should be operated at the same recovery, flux, and product water quality targets as would the planned full-scale desalination plant, in order to collect representative samples. If possible, the same type of RO membrane elements should be used as well.

The concentrate water quality data collected from the sampling events should be compared against the numeric limits of the applicable regulatory requirements. Key parameters that should be given attention in assessing concentrate compliance with applicable numeric water quality standards for effluent discharge are TDS, metals, turbidity, and radionuclides.

One important issue with all concentrate water quality analyses is that most of the laboratory analysis guidelines worldwide were developed for testing freshwater rather than high-salinity concentrate. The elevated salt content of the concentrate samples can interfere with the standard analytical procedures and can often produce erroneous results. Therefore, concentrate analysis must be completed by an analytical laboratory experienced with and properly equipped for brackish and seawater analysis. The same recommendation applies for the laboratory retained to complete the whole effluent toxicity testing and source water quality characterization.

If pilot testing is not possible for a given project, the mineral content of the concentrate can be projected by characterizing the desalination plant source water quality for typical operational conditions (plant recovery rate, membrane flux, product water quality, and source water salinity and temperature) and analyzing these data with RO membrane performance projection software available from all key manufacturers of membrane elements (e.g., Hydranautics, Toray, Dow). This software calculates the content of key ions in the concentrate based on the content of the same ions in the source water, the type of the RO elements, and the main design criteria of the RO desalination system, such as recovery, membrane flux, and membrane age. It should be pointed out, however, that this concentrate water quality characterization method is less desirable than pilot testing, because the currently available membrane projection software does not have provisions for calculating the concentrations of most of the regulated metals, organics, and pathogens in the discharge.

Salinity Tolerance Evaluation Study Determining the tolerance of aquatic organisms to the actual desalination plant concentrate is very beneficial, because it can allow minimization of the complexity of the plant outfall structure, especially if the discharge area is inhabited by salinity-tolerant species. A novel method to identify the salinity tolerance of the aquatic life in the area of a desalination plant discharge was developed at the Carlsbad seawater desalination demonstration plant. This method includes the following four key steps: (1) determining the test salinity range, (2) identifying site-specific test species inhabiting the discharge area, (3) testing biometrics at average discharge salinity, and (4) testing salinity tolerance at varying concentrate dilution levels.

Determining Test Salinity Range The first step of the salinity tolerance evaluation (STE) method is to define the minimum and maximum TDS concentrations that are projected to occur in the area of the discharge after the start-up of desalination plant operations. This salinity range should be established with consideration of the effect of mixing and the associated dilution in the area of the discharge as a result of the site-specific natural hydrodynamic forces in the receiving water body (currents, winds, tidal movements, temperature differences, etc.) as well as the mixing energy introduced with the desalination plant's discharge diffuser system. If the desalination plant concentrate is diluted with other discharge (e.g., cooling water from power plant or wastewater treatment plant effluent) prior to its exit from the outfall into the surface water body, this additional dilution should also be accounted for in establishing the salinity range in which the salinity tolerance of the aquatic species is assessed.

Because of the complexity of the various factors that impact mixing and dilution of desalination plant concentrate with the ambient surface water, especially for medium and large projects [i.e., projects with a discharge volume of 20,000 m^3/day (5.3 mgd) or higher], the actual salinity range that would occur in the area of the discharge should be determined based on hydrodynamic modeling (Jenkins and Wasyl, 2001; Einav and Lokiec, 2003).

At a minimum, the salinity test concentrations should range from that at the middle of the water column and the middle of the zone of initial dilution to the maximum bottom salinity concentration at the edge of the ZID (Jenkins and Wasyl, 2001).

The ZID is often defined as the area of the ocean within a 300 mm (1000-ft) radius from the point of the desalination plant discharge. However, the ZID boundary in terms of TDS should be established as the distance from the point of discharge at which the salinity of the concentrate is within 10 percent of the ambient salinity of the receiving water body. Depending on the naturally occurring mixing intensity at the point of discharge, this distance may be shorter or longer than 300 m (1000 ft).

Identifying Test Species The purpose of the second step of the STE method is to identify the most sensitive site-specific species that would be indicative of the salinity tolerance of the aquatic flora and fauna in the area of the desalination plant discharge. These species are used for the biometrics and salinity tolerance tests. At least three species should be selected for the tests: one representative for the fish population in the area, one for the invertebrate population, and one for the macroalgal population (i.e., kelp, red algae, etc.), if such species are present and occur in significant numbers (Chapman et al., 1995; California State Water Board, 1996; Graham, 2004).

The selection of the specific test species should be completed by an expert aquatic biologist who is very familiar with the site-specific flora and fauna in the area of the desalination plant discharge. The test species should be selected based on (1) presence and abundance in the area, (2) environmental sensitivity (i.e., endangered and protected marine species are the first priority), (3) sensitivity to salinity in the range projected to occur in the discharge, and (4) significance in terms of commercial and recreational harvesting and fishing.

Testing Biometrics The purpose of the biometrics test is to track how well the indicative test species could handle a long-term steady exposure to the elevated average discharge salinity that would occur in the middle of the zone of initial dilution after the desalination plant goes into operation. The biometrics test should be completed in a large aquarium (test tank) in which the desalination plant concentrate is blended with ambient water from the receiving surface water body (ocean, river, etc.) to obtain a salinity that is not to be exceeded in the middle of the ZID in the ocean for at least 95 percent of the time. This salinity level should be maintained in the aquarium for the duration of the biometrics test.

In addition, a second aquarium (control tank) of the same size and with the same number and type of test aquatic organisms should be employed, with the main difference that this tank should be filled up with ambient water from the receiving water body collected from the area of the discharge. The control tank should be operated in parallel with the test tank; observations from this tank will be used as a baseline for comparison and statistical analysis.

Once the salinity in the aquariums is set to target levels, they are populated with the selected test species. Key biometric parameters (appearance, willingness to feed, activity, weight gain or loss, and gonad production) of these species are monitored frequently (a minimum of every 2 days) by an expert marine biologist over a prolonged period of time (a minimum of 3 months, preferably 5 or more months). Percent weight gain or loss and fertilization for one or more of the test and control organisms should be measured as well. At the end of the test, the qualitative and quantitative biometric parameters of the aquatic species in the test and control tanks should be compared to identify whether the species exhibit statistically significant differences—especially in terms of weight gain or loss and fertilization capabilities.

Salinity Tolerance Test The main purpose of the salinity tolerance test is to establish whether the selected test species will survive the extreme salinity conditions that may occur within the ZID and on the edge of the ZID, and whether these test organisms will be able to retain their capacity to reproduce after exposure to these conditions for a length of time that is expected to occur in the worst-case scenario for full-scale plant operations. The test species should be exposed to several blends of concentrate and ambient receiving surface water that may occur within the range of the discharge salinities. The low end of the range should be the average salinity in the ZID (middle depth), and the high end should be the maximum salinity above the seabed at the boundary of the ZID. In general, discharge salinity is expected to decrease with an increase in the distance from the point of concentrate discharge and to increase with depth. The rate of decrease of concentrate salinity from the point of discharge depends on the hydrodynamic conditions in the vicinity of the discharge.

Similar to the biometrics test, this experiment requires one aquarium that contains all test species at ambient salinity (control tank) and separate aquaria for each test salinity concentration. The duration of the salinity tolerance test should be determined by the length of occurrence of the worst-case discharge salinity scenario. This duration should be established based on the results of the hydrodynamic modeling of the desalination plant discharge.

Usually, extreme salinity discharge conditions are not expected to continue for more than 1 to 2 weeks. However, if this is likely in specific circumstances, then the length of the study should be extended accordingly. Starting from the low end of the salinity concentration, individual test tanks should be set for salinity increments of 1 to 2 ppt until the maximum test salinity concentration is reached.

Case Study—Application of the STE Test for the Carlsbad Desalination Project The STE procedure described here was applied to assess the discharge impact of the 189,000 m^3/day (50 mgd) Carlsbad seawater desalination project in Southern California. This project includes direct connection of the desalination plant intake and discharge facilities to the discharge outfall of an adjacent coastal power generation plant using seawater for once-through cooling (Fig. 5.10). The power plant has a total of five electricity generators, and depending on the number of units in operation it pumps between 76,000 and 2,271,000 m^3/day (20 and 600 mgd) of cooling water through the condensers. The warm cooling water from all condensers is directed to a common discharge tunnel and lagoon leading to the ocean. The full-scale desalination facility is planned to tap into this discharge

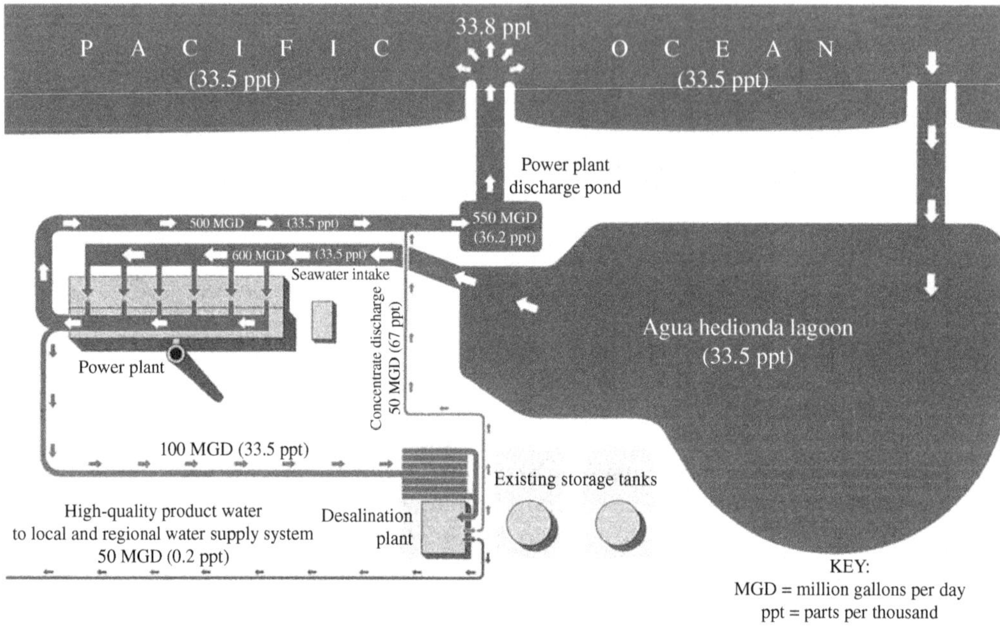

Figure 5.10 Schematic of Carlsbad seawater desalination plant.

tunnel for both feed water and discharge of high-salinity concentrate downstream of the intake area.

Water collected from one end of the power plant discharge canal will be conveyed to the desalination plant to produce freshwater, and the concentrate from the desalination plant will be returned into the same discharge canal, approximately 270 m (810 ft) downstream of the point of intake. The desalination plant concentrate, containing approximately twice the salinity of the source seawater (67 ppt vs. 33.5 ppt), will be blended with the remaining cooling water discharge from the power plant and conveyed to the ocean for disposal.

The salinity range of the mixed discharge from the Carlsbad seawater desalination plant and the power plant will be between 35 and 40 ppt. The average salinity in the middle of the ZID is projected to be 36.2 ppt. Therefore, the biometrics test was completed for this salinity, while the test range for the salinity tolerance test covered 37 to 40 ppt in 1 ppt increments. Both tests were executed by a marine biologist very familiar with the local flora and fauna in the area of the future desalination plant discharge (Le Page, 2004).

A list of the 20 marine species selected for the biometrics test for the Carlsbad project is presented in Table 5.5. The salinity tolerance test was completed using three local species that are known to have highest susceptibility to stress caused by elevated salinity (Le Page, 2004): (1) the purple sea urchin (*Strongylocentrotus purpuratus*, Fig. 5.11), (2) the sand dollar (*Dendraster excentricus*, Fig. 5.12), and (3) the red abalone (*Haliotis rufescens*, Fig. 5.13).

The biometrics test continued for a period of 5.5 months. The results of this test, summarized in Table 5.6, indicate that all organisms remained healthy throughout the test period. No mortality was encountered, and all species showed normal activity and

Scientific Name	Common Name	Number of Individuals
Paralichthys californicus	California halibut	5 juveniles
Paralabrax clathratus	Kelp bass	3 juveniles
Paralabrax nebulifer	Barred sand bass	3 juveniles
Hypsoblennius gentilis	Bay blenny	5
Strongylocentrotus franciscanus	Red sea urchin	4
Strongylocentrotus purpuratus	Purple sea urchin	14
Pisaster ochraceus	Ochre sea star	3
Asterina miniata	Bat star	3
Parastichopus californicus	California sea cucumber	2
Cancer productus	Red rock crab	2
Crassadoma gigantea	Giant rock scallop	3
Haliotis fulgens	Green abalone	3
Megathura crenulata	Giant keyhole limpet	3
Megastraea undosa	Wavy turban snail	3
Cypraea spadicea	Chestnut cowrie	3
Phragmatopoma californica	Sandcastle worm	1 colony
Anthopleura elegantissima	Aggregating anemone	4
Muricea fruticosa	Brown gorgonian	1 colony
Haliotis rufescens	Red abalone	5
Dendraster excentricus	Sand dollar	5

TABLE 5.5 Marine Species Used for the Carlsbad Biometrics Test

FIGURE 5.11 Purple sea urchin (*Strongylocentrotus purpuratus*).

Figure 5.12 Sand dollars (*Dendraster excentricus*).

Figure 5.13 Red abalone (*Haliotis rufescens*).

Scientific Name	Common Name	Avg. % Weight Change, g	% Weight Change (Control Group)*	Appearance and Feeding
Paralichthys californicus	California halibut	91.3	96.9	Strong
Paralabrax clathratus	Kelp bass	114.3	104.8	Strong
Paralabrax nebulifer	Barred sand bass	106.8	113.5	Strong
Hypsoblennius gentilis	Bay blenny	120.0	107.1	Strong
Strongylocentrotus franciscanus	Red sea urchin	2.8	2.4	Strong
Strongylocentrotus purpuratus	Purple sea urchin	7.9	7.2	Strong
Pisaster ochraceus	Ochre sea star	3.8	4.6	Strong
Asterina miniata	Bat star	2.8	3.1	Strong
Parastichopus californicus	Sea cucumber	–2.2	2.3	Strong
Haliotis fulgens	Green abalone	9.6	7.7	Strong
Megathura crenulata	Giant keyhole limpet	5.1	4.7	Strong
Lithopoma undosum	Wavy turban snail	3.9	2.4	Strong
Cypraea spadicea	Chestnut cowrie	0.6	1.0	Strong
Anthopleura elegantissima	Aggregating anemone	115.9	48.9	Strong
Haliotis rufescens	Red abalone	9.2	7.8	Strong
Dendraster excentricus	Sand dollar	3.5	4.5	Strong

*None of the weight changes are statistically significant.

Table 5.6 Overall Condition of Biometrics Test Species

feeding behavior. The appearance of the individuals remained good, with no changes in coloration or development of marks or lesions.

The duration of the salinity tolerance test for the Carlsbad project was 19 days. The results of this test, given in Table 5.7, show that both sand dollars and red abalones had 100 percent survival in all test tanks and in the control tank. One individual of the purple sea urchins died in each of the test tanks and one died in the control tank. Therefore, the adjusted survival rate for the purple sea urchins was also 100 percent.

These test results confirm that the marine organisms in the discharge zone would have adequate salinity tolerance to the desalination plant discharge in the entire range of operations of the desalination plant (i.e., up to 40 ppt). All individuals of the three tested species behaved normally during the test, exhibiting active feeding and moving habits. The biometrics and salinity tolerance tests were completed in 420-L (110-gal) marine aquariums (Fig. 5.14).

Organism Observed	Salinity, ppt	Mortality	Elapsed Time to First Mortality, days
Red abalone	33.5 (control tank)	0	N/A*
Red abalone	37	0	N/A
Red abalone	38	0	N/A
Red abalone	39	0	N/A
Red abalone	40	0	N/A
Sand dollar	33.5 (control tank)	0	N/A
Sand dollar	37	0	N/A
Sand dollar	38	0	N/A
Sand dollar	39	0	N/A
Sand dollar	40	0	N/A
Purple sea urchin	33.5 (control tank)	1	1
Purple sea urchin	37	1	1
Purple sea urchin	38	1	4
Purple sea urchin	39	1	4
Purple sea urchin	40	1	6

*N/A = not applicable.

Table 5.7 Results of Carlsbad Desalination Project Salinity Tolerance Test

Figure 5.14 Carlsbad desalination project test tank.

In summary, the salinity tolerance evaluation method applied to the Carlsbad seawater desalination project confirmed that the elevated salinity in the vicinity of the plant discharge would not have a measurable impact on the marine organisms there and that those organisms can tolerate the maximum salinity of 40 ppt that could occur in the discharge area under extreme conditions.

Additional acute and chronic toxicity studies completed subsequently for this project using the United States Environmental Protection Agency's standard whole effluent toxicity test (Weber et al., 1998) have confirmed the validity of the STE method. WET testing using red abalone (*Haliotis rufescens*) showed that the chronic toxicity threshold for this species occurs at TDS concentrations of over 40 ppt. An acute toxicity test completed using another standard WET species, the top smelt (*Atherinops affinis*), indicated that the salinity in the discharge could reach over 50 ppt on a short-term basis (1 day or more) without impacting this otherwise salinity-sensitive species.

5.10 Permits Related to Product Water

5.10.1 Key Permitting Issues

In the United States, product water quality for potable use is regulated by the Safe Drinking Water Act. Compliance with the act and its subsequent regulations is enforced by the US EPA and in most states delegated to the state authorities responsible for public health protection. Desalinated water has to comply with all regulatory requirements for potable water. From that prospective, permits and associated studies related to drinking water are very similar to those required for conventional freshwater treatment plants. As discussed previously, the main differences with respect to potentially negative impacts on water quality originate from sometimes significantly different contents of some minerals (i.e., bromide, boron, sodium, and chloride) and the typically lower alkalinity and hardness of desalinated water.

Additional concerns associated specifically with desalinated brackish water are the potentially high content of odorous gases, nitrates, and sometimes disinfection byproducts. A frequent concern is also the content of algal toxins in desalinated seawater.

5.10.2 Permitting Support Studies

The supporting studies that are commonly completed as a part of the process for obtaining drinking water permits for a given desalination project include (1) a watershed sanitary survey or source water assessment, (2) a process performance and operations monitoring plan, and (3) a plan for product water integration in the distribution system.

Watershed Sanitary Survey or Source Water Assessment

Watershed Sanitary Survey For desalination plants using open intakes or subsurface intakes under the influence of surface or subsurface contamination, the US EPA requires the completion of a watershed sanitary survey, which characterizes the source water quality in terms of contaminants regulated by the Safe Drinking Water Act and additional water quality parameters specified by applicable state regulatory agencies. This survey also identifies potential sources of water quality contamination in terms of pathogens and other anthropogenic pollutants located within a 1-mi radius of the intake.

The US EPA Ground Water Rule requires a sanitary survey of the wellhead facilities and well construction for all plants (including desalination facilities) with groundwater well intakes. The US EPA Long Term 2 Enhanced Surface Water Treatment Rule defines intake siting requirements, setbacks, depth or laterals, and other intake configuration details. In addition, it requires 12 to 24 months of *Cryptosporidium* monitoring to determine the bin classification of both open intakes and subsurface intakes that are determined to involve groundwater under direct influence of surface water contamination.

The scope and content of a typical sanitary survey of the desalination plant intake area was discussed in greater detail in Chap. 2.

Source Water Assessment If the desalination plant uses a subsurface intake that is not under the influence of surface water contamination (e.g., confined brackish water aquifers), current regulations allow for a more simplified source water quality characterization: a source water assessment. However, conclusively determining if water collected from a subsurface intake is or is not under the influence of surface water contamination can prove challenging, since current regulations do not specify a clear path to such a determination. This determination is more straightforward if the source of water is a deep confined brackish water aquifer, but less so if the facility uses subsurface intakes such as vertical beach wells, infiltration galleries, or horizontal and slant wells, which collect water from unconfined shallow aquifers. This issue is further complicated by the fact that beach erosion and storm impacts can change the depth of the filtration layer separating the intake from the surface water over the useful life of the desalination project.

In general, a watershed sanitary survey is a much more detailed evaluation of a drinking water source than is a source water assessment. One challenge in the case of seawater desalination is the selection of an appropriate approach to delineating the source area. Delineating the entire watershed as the source area is consistent with the guidance for watershed sanitary surveys. However, in the case of an ocean intake, the appropriate method to delineate the source area is unclear, because the boundaries will be affected by issues such as tides and currents. In the absence of specific regulatory requirements, the source water area for the desalination project is recommended to encompass zone with a radius of at least 1.64 km (1 mi) from the point of plant intake.

Process Performance and Operations Monitoring Plan

This plan identifies the key treatment processes that are intended to be employed for source water desalination, defines the projected removal of contaminants and inactivation of pathogens in the source water, and provides an overview of the overall plant operation approach. In addition, it identifies what monitoring equipment, procedures, and control measures which are planned to be incorporated in the desalination plant design and operations to ensure that the plant is producing water that meets applicable regulatory requirements under both normal and extreme source water quality conditions (e.g., heavy storms and algal blooms for surface intakes). The plan also discusses monitoring that will be implemented in the product water distribution system and at the points of water delivery and final use.

Plan for Product Water Integration in the Distribution System

This plan is of critical importance when the distribution system will receive desalinated water of varying quality and quantity that will be blended with multiple other water sources of diverse quality. The plan typically identifies the diurnal and monthly fluctuations of the flow and quality of the mix of desalinated water and other water sources, as well as potential water quality impacts in terms of disinfection byproducts, bromides, boron, sodium, chloride, odor, taste, nitrification, salinity, and temperature.

If different disinfectants are used for the various water sources delivering in the same distribution system, a desktop and/or laboratory analysis is typically completed to gain a better understanding of the overall stability of the disinfection residual in the possible water blends as well as the potential changes of disinfection by-products in the water.

Similarly, if waters of different hardness, pH, temperature, and alkalinity are blended in the distribution system, at a minimum a desktop water quality analysis should be completed to confirm the compatibility of the post-treatment methods and chemicals applied to the desalinated water with these of the other water sources and to ascertain that the introduction of desalinated water will not trigger corrosion of the distribution system and household plumbing, loss of chlorine residual, and deterioration of water quality in terms of disinfection byproducts, taste odor, color and temperature.

Other Studies Depending on the site-specific contaminants encountered in the source water, the regulatory agencies involved in issuing drinking water permits may require the completion of additional studies, such as algal toxin rejection and control analysis and source water characterization for endocrine disruptors and/or specific carcinogens, if sources of such pollutants may be present in the source water intake area.

5.11 Other Permits

Depending on the type and size of the desalination plant, and the plant intake and outfall type and location, the construction and operation of a given project may require a number of other permits. For example, in California a special agency is dedicated to permitting facilities located in the 8 km (5 mi) coastal zone—the California Coastal Commission. This agency regulates project construction and operation through a coastal development permit, the obtainment of which involves the completion of most of the studies discussed in the previous sections. Other permits, such as the stream alteration permit issued by the California Department of Fish and Game, regulate activities within inland waters, bays, and estuaries. If the desalination plant has its own power generation system or a standby diesel generator, in most states the operation of that equipment will require an air emission permit.

As seen from the overview presented in this chapter, numerous state and local agencies may be stakeholders in the desalination project permitting process. Therefore, the very first steps in that process are to identify all permits and authorizations needed for the planning, construction, and operation of the desalination project; determine the permitting jurisdictions and their requirements; and define a clear sequence of actions to minimize the time, efforts, and expenditures associated with project implementation.

5.12 References

Brevik, E. C., and J. A. Homburg, "A 5000 Year Record of Carbon Sequestration from a Coastal Lagoon and Wetland Complex, Southern California, USA," *Catena*, vol. 57 (X): 221–322, 2004. T.

Bleninger, G.H. *Jirka*, "Modeling and Environmentally Sound Management of Brine Discharges for Desalination Plants *Desalination*, vol. 221: pp. 585–597, 2008.

California Climate Action Registry, *Annual Emissions Report, San Diego Gas and Electric, 2006*, (http://www.climateregistry.org/tools/carrot/carrot-public-reports.html, accessed August 2012).

California Department of Water Resources, *California Desalination Planning Handbook*, California Department of Water Resources, Sacramento, 2008.

California State Water Board, *Procedures Manual for Conducting Toxicity Tests Developed by the Marine Bioassay Project*, 96-1WQ, California State Water Board, Sacramento, 1996.

California State Water Resources Control Board, *California Ocean Plan—Water Quality Control Plan, Ocean Waters of California*, California State Water Resources Board Sacramento, CA, 2009.

Chapman, G. A., D. L. Denton, and J. M. Lazorchak, *Short-term Methods for Estimating the Chronic Toxicity of Effluents and Receiving Waters to West Coast Marine and Estuarine Organisms*, USEPA Report No. EPA/600/R-95/136, US EPA, Cincinnati, Ohio, 1995.

City of Carlsbad, California, *Environmental Impact Report for Precise Development Plan and Desalination Plant EIR SCH# 2004041081*, Poseidon Resources, Carlsbad, 2005.

Cotruvo, J., N. Voutchkov, J. Fawell, P. Payment, D. Cunliffe, and S. Lattemann, *Desalination Technology, Health and Environmental Impacts*, CRC Press and IWA Publishing, Boca Raton, Florida, 2010.

Einav, R., and F. Lokiec, "Environmental Aspects of a Desalination Plant in Ashkelon," *Desalination*, vol. 156, pp. 79–85, 2003.

Graham, J., "Marine Biological Considerations Related to the Reverse Osmosis Desalination Project at the Encina Power Plant, Carlsbad, CA," *Environmental Impact Report for Carlsbad Seawater Desalination Plant, City of Carlsbad*, Poseidon Resources, Carlsbad, 2004.

Hammond, M., N. Blake, P. Hallock-Muller, M. Luther, D. Tomasko, and G. Vargo, *Effects of Disposal of Seawater Desalination Discharges on Near Shore Benthic Communities*, Southwest Florida Water Management District and University of South Florida, Tampa, 1998.

Intergovernmental Panel on Climate Change, "Climate Change 2007: Impacts, Adaptation, and Vulnerability," *Third Assessment Report of the Intergovernmental Panel on Climate Change* (M. L. Parry, O. F. Canziani, J. P. Palutikof, P. J. van der Linden, and C. E. Hanson, eds.). Cambridge University Press, Cambridge, United Kingdom, 2007.

Jenkins, S. A., and J. Wasyl, "Hydrodynamic Modeling of Dispersion and Dilution of Concentrated Seawater Produced by the Ocean Desalination Project at the Encina Power Plant, Carlsbad, California," *Environmental Impact Report for Carlsbad Seawater Desalination Plant, City of Carlsbad*, Poseidon Resources, Carlsbad, Place, 2001.

Le Page, S., "Salinity Tolerance Investigations: A Supplemental Report for the Carlsbad, CA Desalination Project," *Environmental Impact Report for Carlsbad Seawater Desalination Plant, City of Carlsbad*, Poseidon Resources, Carlsbad, 2004.

McPherson, G. E., K. I. Scott, J. R. Simpson, Q. Xiao, and P. J. Peper, *Tree Guidelines for Coastal Southern California Communities* (http://www.ufei.org/files/pubs/cufr_48.pdf), USDA Forest Service, Pacific Southwest Research Station, Davis, CA, 2000.

Mickley, M. C., *Membrane Concentrate Disposal: Practices and Regulation*, 2nd ed., Desalination and Water Purification Research and Development Program Report N. 123, U.S. Department of Interior, Bureau of Reclamation, Denver, Colorado 2006.

Mauguin, G. and P. Corsin, "Concentrate and Other Waste Disposals from SWRO Plants: Characterization and Reduction of Their Environmental Impact," *Desalination*, vol. 182, pp. 355–364, 2005.

National Research Council, *Climate Change Science: An Analysis of Some Key Questions* (http://www.nap.edu/catalog/10139.html), Committee of the Science of Climate Change, National Research Council, Washington, DC 2001.

Poseidon Resources, *Flow Entrainment and Impingement Minimization Plan*, 2007.

R. W. Beck, *Guidance Manual for Permitting Requirements in Texas for Desalination Facilities Using Reverse Osmosis Processes*, Texas Water Development Board, Austin, Texas 2004.

Sadhwani, J. J., J. M. Veza, and C. Santana, "Case Studies on Environmental Impact of Seawater Desalination," *Desalination*, 185, 1, 2005.

Truilio, L., *Notes on Carbon Sequestration and Tidal Salt Marsh Restoration*, 2007. (www.sfbayjv.org/tools/climate/CarbonWtlandsSummary_07_Trulio.pdf accessed August, 2012)

United States Environmental Protection Agency, *Technological Development Document for Proposed Section 316 (b) Phase II Rule*, USEPA-821-R-11-001, March 28, 2011.

Water Research Foundation, *Guidelines for Implementing Seawater and Brackish Water Desalination Facilities*, Water Research Foundation, Alexandria, Virginia, 2010.

WateReuse Association, *Desalination Plant Intakes—Impingement and Entrainment Impacts and Solutions*, White Paper, 2011. (http://www.watereuse.org/sites/default/files/u8/IE_White_Paper.pdf accessed August 2012.)

Weber, C. I., W. B. Horning, II, D. J. Klemm, T. W. Nieheisel, P. A. Lewis, E. L. Robinson, J. Menkedick, and F. Kessler, *Short-term Methods for Estimating the Chronic Toxicity of Effluents and Receiving Waters to Marine and Estuarine Organisms*, EPA/600/4-87/028, National Information Service, Springfield, VA, 1998.

CHAPTER 6

Intakes for Source Water Collection

6.1 Introduction

The main purpose of saline water intakes is to collect source water of adequate and consistent flow and quality over the entire useful life of the desalination plant. The type and configuration of intake selected for a given desalination project has a significant impact on the nature and quantity of foulants contained in the source water and on the complexity of the pretreatment system needed to control reverse osmosis membrane fouling.

Desalination plant intakes can be divided in two main categories: surface (open) and subsurface (groundwater) intakes. Open intakes receive ambient water directly from a saline water body via a submerged inlet structure. The water collected by open intakes contains the same amount of solids, organics, and other contaminants as the surface water body from which it originates.

Subsurface intakes collect groundwater from a saline (brackish water or coastal seawater) aquifer. This source water aquifer can be either a deep, confined brackish aquifer, which typically is not connected to a surface water body, or a shallow, unconfined aquifer, which may or may not be hydraulically connected to a surface water body. Deep brackish water aquifers are the most commonly used sources of water for brackish desalination plants. For example, the majority of the existing brackish water desalination plants in Florida and Texas tap deep brackish water aquifers, which usually have salinity between 800 and 9000 mg/L. Unconfined shallow coastal aquifers are typically used as water sources for seawater desalination plants. Their source water is filtered through the aquifer soils, and as a result it has relatively low particulate content and high water quality.

Planning considerations for the selection of intake type and location are discussed in detail in Chap. 4, while environmental impacts of intake operations and potential alternatives for their minimization and mitigation are described in Chap. 5. This chapter focuses on intake design considerations.

6.2 Open Intakes

6.2.1 Types and Configurations

Based on the location of their inlet structure, open intakes are commonly classified as onshore and offshore. The inlet structure of onshore intakes is constructed on the banks of the source water body, while the inlet structure of offshore intakes is located several hundred to several thousand meters away from the shore.

Onshore Open Intakes

To date, onshore intakes have found application mainly for very large thermal or hybrid seawater desalination plants. Such intakes typically consist of a large, deep intake canal ending in a concrete forebay structure equipped with coarse bar screens followed by fine screens and the intake pump station.

The largest seawater reverse osmosis desalination plant with onshore intake is the 109,000 m^3/day (29 mgd) Point Lisas desalination plant in Trinidad. In this case, the intake consists of a concrete forebay structure located on the shore of the industrial port's ship-turning basin (the Gulf of Paria) and is equipped with coarse and fine screens and vertical turbine pumps.

Another desalination plant that uses source water from an onshore open intake is the Tampa Bay desalination plant. This plant collects most of its source water from the discharge of the power plant with which it is co-located. A portion of the source water, however, is collected directly from an onshore intake located on one side of an intake canal (Fig. 6.1) and mixed with cooling water from the power plant to maintain a desalination plant feed water temperature below 38°C (100°F) at all times.

Offshore Open Intakes

These intakes typically consist of a velocity-cap-type inlet structure (Fig. 6.2), one or more intake water conduits (pipelines or an intake tunnel), an onshore intake chamber, trash racks, fine screens, and a source water intake pump station.

The inlet structure of offshore open intakes is usually either a vertical well (vault) made of concrete, copper-nickel, or steel; or a wedgewire screen, located 4 to 10 m (13 to 33 ft) above the floor of the water body and submerged between 4 and 20 m (13 and 66 ft) below the surface. Open intakes are the most commonly used type of source water collection system for medium and large desalination plants. Table 6.1 presents examples of open offshore intakes for large seawater desalination plants (Baudish et al., 2011).

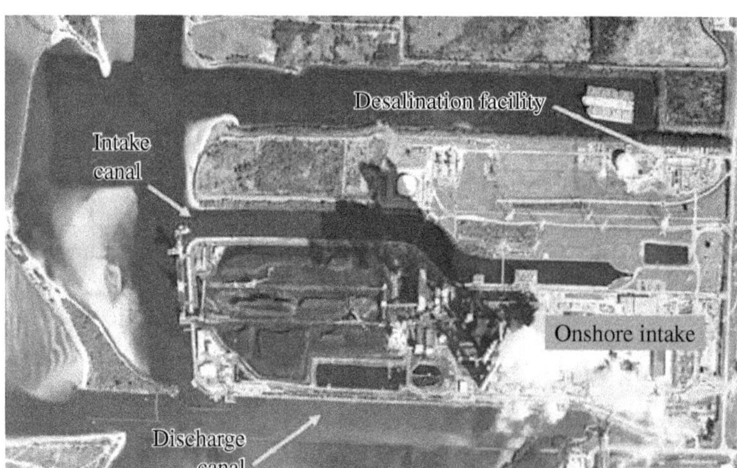

Figure 6.1 Tampa Bay desalination plant intake.

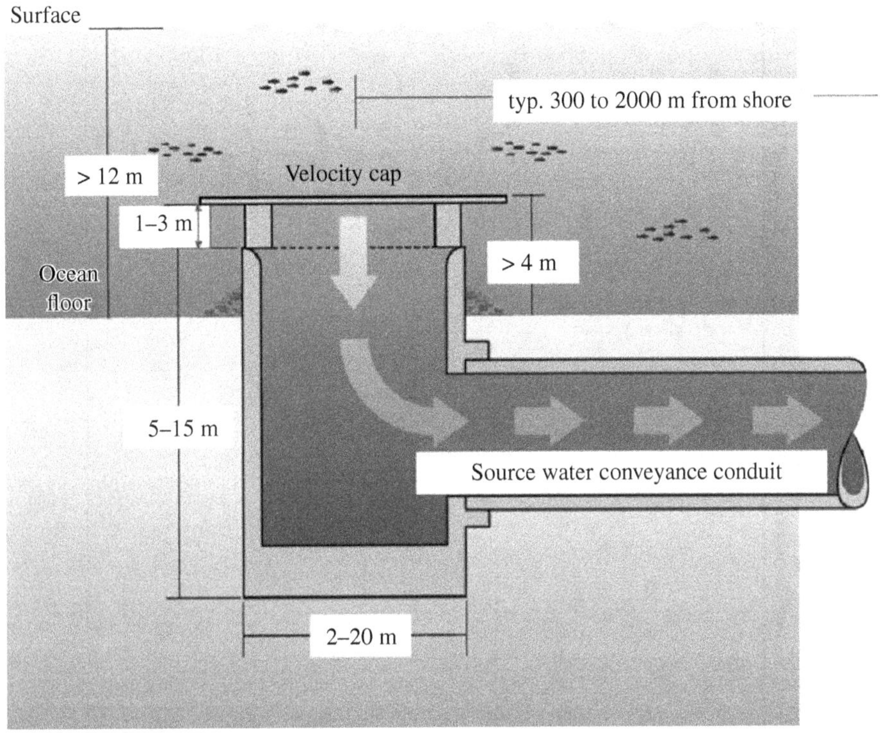

Figure 6.2 Velocity cap intake inlet structure.

In Table 6.1, the maximum entrance velocity is the through-screen velocity at the intake bars. The depth below the water surface is the distance between the mean water level and the top of the velocity cap of the intake. The distance from the bottom is the distance between the bottom of the ocean and the top of the velocity cap.

Offshore intakes, which usually extend several hundred meters away from the shoreline and sit 8 to 20 m below the water surface, are typically not influenced by the fresher coastal aquifers. Therefore, such intakes usually yield saline water with the same TDS content as the ambient ocean water.

One exception is offshore intakes located near the entrance of a large river or other freshwater body into the ocean. Since old river beds and associated alluvial aquifers can often extend far beyond the tidal zone, the water quality of such offshore intakes may be influenced by the groundwater quality in the alluvial aquifers, which often is inferior to that of open ocean seawater in terms of content of iron, manganese, and other undesirable contaminants.

Co-located Intakes

Configuration Co-located intakes are connected to the discharge of an adjacent power generation station that uses saline water for once-through cooling (Fig. 6.3). At present, this type of intake configuration has found application mainly for seawater

Desalination Plant and Production Capacity	Maximum Entrance Velocity, m/s (ft/s)	Depth below Water Surface, m (ft)	Distance from the Bottom, m (ft)	Number of Inlet Structures and Conduits	Inlet Structure Diameter, m (ft), and Screen Size, mm (in.)	Conduit Diameter, m (ft),Material, and Distance from Shore, m (ft)
Adelaide, Australia 300,000 m^3/day (79 mgd)	0.15 (0.50)	18 (59)	5.0 (16.4)	1/1	9.5 (31.2) 100 (4)	2.8 (9.2) Concrete (Tunnel) 1000 (3300)
Sydney, Australia 500,000 m^3/day (132 mgd)	0.15 (0.50)	24 (79)	6.0 (20.0)	4 inlet structures on a common tunnel	8.5 (27.9) 340 (13)	3.4 (11.2) Concrete (Tunnel) 300 (980)
Gold Coast, Australia 136,000 m^3/day (36 mgd)	0.05 (0.16)	22 (72)	4.4 (14.4)	1/1	5.8 (19.0) 140 (5.5)	2.8 (9.2) Concrete (Tunnel) 1400 (4600)
Perth I, Australia 143,000 m^3/day (38 mgd)	0.10 (0.33)	8 (26)	4.0 (13.0)	1/1	8.0/(26.4) 100 (4)	2.8 (9.2) GRP pipes* 300 (1000)
Perth II, Australia, 300,000 m^3/day (79 mgd)	0.15 (0.50)	10 (33)	4.0 (13.0)	2/2	7.0 (23) 100 (4)	2.4 (9.1) HDPE pipes† 500 (1600)
Fujairah I, United Arab Emirates 170,000 m^3/day (45 mgd)	0.10 (0.33)	10 (33)	6.0 (19.7)	3/3	3.0 (9.8) 80 (3)	2.0 (6.6) GRP pipes 380 (1250)
Al Dur, Bahrain 240,000 m^3/day (63 mgd)	0.10 (0.33)	4 (13)	2.3 (7.5)	4/4	7.2 (23.6) 80 (3)	2.4 (7.9) GRP pipes 1500 (4920)

*GRP = glass-reinforced plastic.
†HDPE = high-density polyethylene.

TABLE 6.1 Examples of Open Offshore Intakes for Seawater Desalination Plants

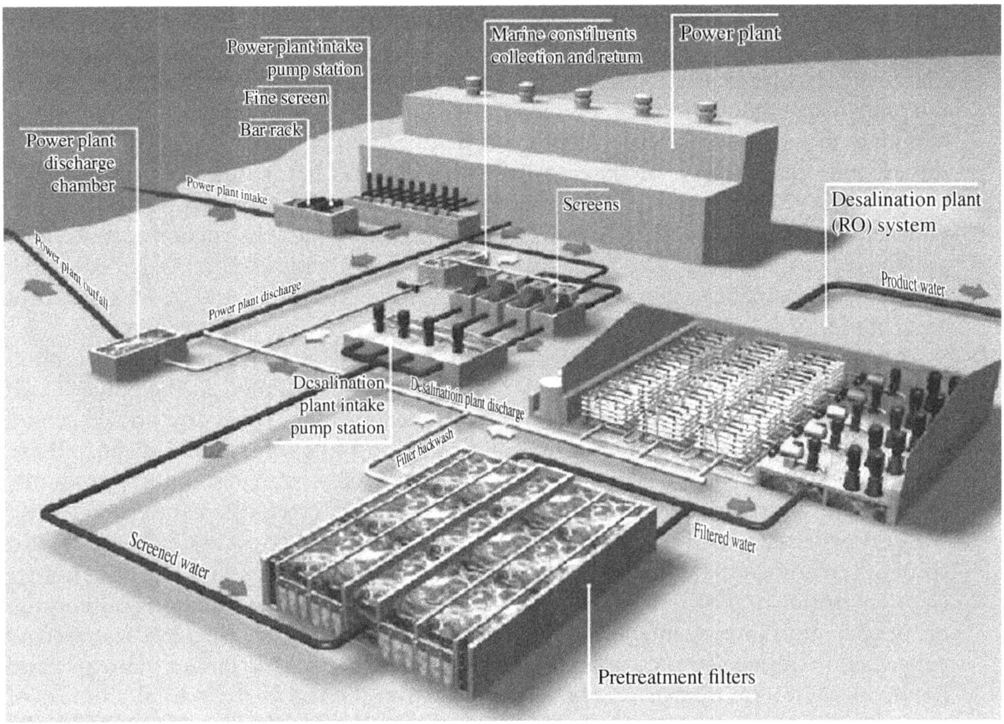

Figure 6.3 Co-location of desalination plant and power generation facility. (*Source: Kennedy Jenks*)

desalination facilities; however, it could also be applied for inland brackish water desalination plants where the host power plant uses brackish intake and discharge wells for once-through cooling.

The Co-location configuration allows the use of the power plant's cooling water as both saline source water for the desalination plant and blending water for salinity reduction of the desalination plant's concentrate. The Co-location implications associated with concentrate discharge are discussed in greater detail in Chap. 16.

The desalination plant intake is connected at a location between the point of exit of the warm cooling water from the power plant and the point of discharge of this water to the receiving water body (ocean, river, saline aquifer, etc.); Under this configuration the intake can collect a portion of the cooling water for desalination.

The desalination plant concentrate is conveyed to the power plant discharge downstream of the point of connection for the desalination plant intake. As a result, co-located desalination plants do not have their own intake and discharge.

Potential Co-location Benefits Sharing intake infrastructure also has environmental benefits, because it avoids the need for new construction in the area of open water body near the desalination plant. The construction of a separate new open intake structure and pipeline for the desalination plant could cause significant disturbance of the benthic marine organisms inhabiting the water body.

Usually, coastal power plants with once-through cooling systems use large volumes of seawater. Because the power plant's intake, seawater has to pass through the small-diameter tubes [typically 10 mm (3/8 in.) or less] of the plant condensers to cool them, the plant discharge cooling water is already screened through bar racks and fine screens similar to those used at surface water intake desalination plants (see Fig. 6.3). Therefore, desalination plant whose intake is connected to the discharge outfall of power plant usually does not require the construction of a separate intake structure, intake pipeline, and screening facilities (bar racks and fine screens). Since the construction cost of a new surface water intake structure for a desalination plant is typically 5 to 30 percent of the total plant construction expenditure, power plant co-location could yield significant cost savings.

The need for installation of additional fine screening facilities for the desalination plant intake is driven by the screenings disposal practice adopted by the power plant and the type of desalination plant pretreatment system. Typically, power plants remove the screenings retained at their bar racks and fine screens and dispose these waste debris to a landfill or return them to the ocean. However, in some power plants, the screenings collected at the mechanical screens are discharged into the cooling water downstream of the condensers. In this case, the power plant discharge would contain screenings that need to be removed at the desalination plant intake. Another clear environmental benefit of the co-location of power and desalination plants is the reduced overall impingement and entrapment of marine organisms as compared to construction of two separate open intake structures (one for the power plant and one for the desalination plant). This benefit stems from the fact that the total biomass of the impacted marine organisms is typically proportional to the volume of the intake saline source water. Because the same intake saline water is used twice (once for cooling and once for desalination), the net intake inflow of saline water and aquatic organisms is minimized.

6.2.2 Selection of Open Intake Type

Onshore versus Offshore Intake

Onshore intakes have one key advantage—they are usually the lowest-cost type of intake, especially for large desalination plants. However, such intakes typically produce the worst source water quality, because in most cases they are designed to collect water from the entire depth of the water column and are located in the surf zone, where breaking waves continuously lift particles from the bottom into suspension and thereby significantly increase water turbidity as compared to deeper waters.

The first 8 to 10 m (26 to 33 ft) of water from the surface in the surf zone typically contains several times higher levels of turbidity, algae, hydrocarbon contamination, silt, and organics than do deeper waters. The quality of water collected by onshore intakes can vary significantly, because in many cases it is influenced by wind, tides, currents, ship traffic, storms, and freshwater runoff. Onshore intakes can also be exposed to beach erosion and direct wave action with irreversibly damaging consequences.

Because of the lower-quality source water, onshore open intakes have found very limited application for membrane desalination plants. Unless the specific site location or costs dictate the need to use this type of intake, they are less desirable for reverse osmosis desalination plants. However, onshore intakes are often the prime choice for thermal desalination plants, where source water quality in terms of suspended solids, organic, and algal content is of secondary importance and has minimal influence on the desalination process.

Wedgewire Screens versus Conventional Inlet Structure

Wedgewire screens are preferred over conventional offshore velocity-cap-type intakes when suitable site conditions exist. According to one source (Gille, 2003), they can be used for desalination plants with an intake capacity of up to 20,000 m^3/h (127 mgd). Three key criteria for the suitability of the site are: (1) the sweeping velocity of currents naturally occurring in the vicinity of the intake should be at least 0.3 m/s (1.0 ft/s); (2) the minimum depth of the water above the screens at all times (including low-tide conditions) should be at least 50 percent of the screen diameter; and (3) the available distance between the bottom of the water body and the screens should be at least 50 percent of the screen diameter or 1.0 m (whichever is higher).

Additional design considerations for wedgewire screens are provided in Chap. 8. The key benefits of wedgewire screens as compared to conventional intakes are (1) lower costs of construction and operation and maintenance, (2) reduced entrainment of aquatic species, and (3) faster and simpler installation.

6.2.3 Selection of Open Intake Location

Besides the location of the desalination plant site, other key factors that control the selection of the intake location are:

- Potential for beach erosion in the intake area
- Location and direction of underwater currents
- Presence and location of active seismic faults
- Topography and geology of the floor of the water body
- Location of environmentally sensitive habitats along the intake pipe route and near the intake inlet
- Location and size of municipal and industrial wastewater discharges within a 1-km (0.6-mi) radius from the intake
- Size of waves and depth of wave impacts; ship and boat traffic
- Tide and wind characteristics in the intake area

Intakes should be located away from areas of active beach erosion and seismic faults; high waves; strong underwater currents carrying debris, silt, plankton, sea grass, weeds, and other stringy materials; and locations with heavy ship and boat traffic. If sensitive marine habitats are encountered, either the intake route should be modified or the intake conduits should be installed via directional drilling or tunneling under the sensitive area instead of via open trench construction or laying the conduit on the bottom of the water body.

6.2.4 Minimization of Impingement and Entrainment Impacts

Open ocean intakes will result in some entrainment of aquatic organisms (as compared to subsurface intakes), because they take source water directly from the saline water body rather than water that has been prefiltered through the coastal aquifer formations. In addition, some marine organisms may be impinged on the screening facilities and ultimately could be entrained in the plant's source water flow. Impingement and entrainment of marine organisms are not unique to saline water intakes. The same phenomena and extent of environmental impact may be observed at the intakes of conventional water treatment plants collecting surface water from rivers, lakes, and estuaries. Potential impingement and entrainment impacts of open intake operations, along with commonly applied alternatives for their reduction and mitigation, are discussed in detail in Chap. 5.

6.2.5 Design Considerations

Onshore Intakes

As indicated previously, onshore intakes have found limited application for the collection of source water for reverse osmosis desalination plants. However, depending on the site-specific conditions of the source water body, they may offer a low-construction-cost alternative to offshore intakes. It should be noted that because of the inferior quality of the water they collect, the use of onshore instead of offshore intakes is likely to result in increased complexity and costs of the pretreatment system and may require the construction of multiple pretreatment facilities. Therefore, the most suitable combination of intake and pretreatment system should be determined based on a detailed source water quality evaluation and lifecycle cost-benefit analysis.

Depending on the coastal conditions, onshore intakes could be installed on a sandy coast with low gradient, on a rocky coast, or in a natural or artificial enclosure (e.g., ship-turning basin, marina, industrial port, or lagoon). Of the three coastal environments, rocky bottom conditions are the most favorable for constructing onshore intakes. Key factors associated with the design of such intakes are wind and swell regimes, water level variations, the tidal regime, bathymetry, and coastal currents.

Onshore intakes on sandy coast with a low gradient are usually constructed with a long entrance canal, which is designed to protect the intake from littoral sediment transport by prevailing near-shore currents, winds, swells, and tides. The intake canal is constructed with jetties that are oriented to create a protective shield for the intake against the prevailing current. Without the jetties, the canal would fill up with sand and would need to be dredged frequently. The jetties are constructed of stone blocks and usually extend to an elevation of 2 to 3 m (6.6 to 10 ft) above the mean water level. In order to avoid fish entrapment, the canal is designed for an average velocity of 0.3 m/s (1 ft/s) at mean water level and 0.15 m/s (0.5 ft/s) at high water level. The canal ends in an onshore inlet structure with trash racks.

Rocky coast onshore intakes typically are concrete or metal structures that open directly to the water body. Depending on the site-specific conditions, such intakes may be designed with jetty protection. If possible, it is preferable to design the water entrance at a depth of at least 2 m (6.6 ft) below the low-tide level and to protect the entrance with wave-breaking jetties. The main difference between rocky bottom and sandy bottom shoreline conditions is that wave action typically does not cause a significant stirring up of sediments and elevated source water turbidity with a rocky bottom.

Onshore intakes in natural or artificial enclosures are typically well protected against wave and wind action and therefore have more consistent water quality. However, one concern is that the embayment accumulates silt and sand, so unless the area in front of the intake is dredged periodically, the intake capacity decreases over time. Dredging operations have implications for both cost and source water quality.

Offshore Intakes

Selection of the location, depth, and configuration of offshore intakes is of critical importance for the source water quality and the performance of the desalination plant. These intake design parameters are established based on a series of on-site investigations of the intake area conditions, including a bathymetric profile, a geotechnical survey, wave and time survey, underwater current survey, and source water quality profile.

Bathymetric Profile The bathymetric profile provides a topographic representation of the floor of the water body and indicates the bottom slope and distance to the water surface and the shore. Usually, several initial topographic profiles are completed within the vicinity of the target intake location and one of these profiles is selected as a preferred alternative based on the floor topography. As a second step of the engineering investigation of the intake area, a geotechnical survey and water quality characterization is completed along the route of the preferred water conduit.

Geotechnical Survey This survey aims to determine the geological formations, the presence and location of seismic faults, and the seabed conditions along the route of the preferred intake conduit. This information is used to decide whether the intake pipeline should be installed directly on the bottom of the water body, laid in an open-cut trench, or directionally drilled in the bottom stratum.

Flat sandy bottom conditions usually allow installation of the intake pipeline directly on the bottom or in a trench. Pipeline installed directly on the bottom is typically anchored by concrete saddles or engineered backfill to protect it against destructive current and wave action.

For rocky bottom conditions, installing the intake pipeline in trenches is usually costly and laying it on the surface of the ocean bottom is unsuitable because of the inadequate support and risks of pipe abrasion and damage. In such conditions, the intake conduit is directionally drilled or tunneled under the ocean bottom.

Wave and Tide Survey The purpose of this survey is to determine the magnitude of wave heights that could occur in the area of the intake and identify the wave height with a 1000-year average recurrence interval (which is typically used for design purposes), if long-term records of wave heights are available. Such projection is important in assessing the impact of the horizontal currents that high waves could create in shallow water and the force that such currents could exert on the intake structure, as well as the depth of sediment resuspension that the horizontal currents could trigger. In addition, the survey would need to include a review of tidal fluctuation information in order to estimate the submergence of the intake inlet structure under minimum, maximum, and average tide elevations.

Underwater Current Survey Underwater currents can have a measurable impact on source water quality and the ability of wedgewire screens to function properly. The direction and velocity of prevailing underwater currents are also of critical importance for determining the location of intake and outfall facilities and the minimum distance between them. Underwater current information is typically available from navigation (nautical) maps.

Biological (Ecological) Survey The purpose of this survey is to identify the presence of environmentally sensitive habitats that may be located in the intake inlet area or along the pipeline conduit route. Habitats of high biological productivity—such as sea grass beds (meadows), kelp forests, algal mats, seaweed bays, coral reefs, salt marshes, and mangrove flats—should be avoided, because they are used as spawning grounds for many marine species and often are very fragile habitats.

Source Water Quality Profile The source water quality survey is typically completed at a minimum of 3 to 5 locations along the preferred route of the intake conduit—typically at distances of 300 m (1000 ft), 500 m (1640 ft), 1000 m (3280 ft), 1500 m (4920 ft), and 2000 m (3785 ft) from the shore. The exact location of each water quality sampling point represents a potential intake location and is selected such that it avoids strong underwater currents (unless wedgewire screens are used), bottom irregularities, environmentally sensitive habitats, and active beach erosion zones.

Once the sampling locations are determined, a sampling protocol is developed that identifies the sampling depth and frequency and the water quality parameters for which each sample will be tested. Some of the sampling can be automated by installing buoys equipped with water quality monitoring instruments for collection of data parameters such as conductivity, temperature, dissolved oxygen content, turbidity, and pH of the source water.

Most water sampling, however, is completed manually at intervals from once or twice per month to once every quarter for a period of one year to gain a thorough understanding of the seasonal water quality variations at the potential intake locations. Additional sampling is usually carried out to capture extreme water quality events such as heavy storms, algal blooms, intake area dredging, periods of elevated ship traffic, seasonal currents, and winds.

Selection of Inlet Location and Depth Once the topographic, water quality, and geologic data are collected and analyzed, the actual location of the desalination plant's offshore inlet structure is selected such that it yields the best source water quality (the lowest turbidity, silt density index, algal content, and seasonal variation in source water quality) and at the same time is closest to the shore.

The selected location and depth of the intake inlet structure should be such that the intake is adequately submerged at low tide, protected from the damaging wave motion of storms, and far enough offshore to avoid the near-shore surf zone, where storms can cause suspension of large quantities of silt and sediment and can ultimately damage the intake structure and the interconnecting piping.

For the selected intake location, the optimal intake depth and distance from the bottom are determined by preparation of depth profiles of turbidity, algal content, salinity, and temperature. The goal is to find a depth from the surface and distance from the bottom that yields the best source water quality (lowest silt density index, turbidity, and algal content, and highest temperature).

Such profiles should be prepared for worst-case water quality conditions, such as heavy storms or high waves, high-intensity winds, lowest water algal blooms with algal counts over 2000 cells per milliliter, seasonal underwater currents, intake area dredging, and heaviest ship traffic conditions. The optimum intake depth is determined as the shortest distance below the water surface where the various extreme environmental conditions that may occur in the intake area have minimal or no impact on the source water quality and plant operations.

Intake Inlet Structure—Design Considerations and Criteria

General guidelines for the depth of inlet structures for offshore intakes for medium and large size desalination plants are presented in Fig. 6.2 and Table 6.2.

Feature	Recommended Size	Notes
Number of inlet structures	2 minimum	Number of intake structures is dictated by the plant's design availability factor and the size of the intake conduit.
Diameter (size)	2–20 m (13–130 ft)	Most inlet structures have a circular shape. If the inlet is located in a current, the shape may be rectangular, with the shorter side against the current.
Distance from mean surface water level to top of velocity cap	8 m (26 ft) minimum at mean water level 4 m (13 ft) minimum at low water level 20 m (66 ft) maximum 12–20 m (40–66 ft) optimum	Shallower intake structures may be needed if the desirable depth cannot be achieved within 2000 m (6600 ft) of the shore. Structures deeper than 20 m (66 ft) may not be cost effective.
Distance from bottom to top of velocity cap	One diameter of the seabed pipe or 4 m (13 ft), minimum	This distance is determined by the depth of the sweeping current at the bottom.
Distance from the shore	300–2000 m (1000–6600 ft)	This distance is determined by the length of the tidal or active beach erosion zones.
Distance between coarse screen bars	50–300 mm (2 to 12 in.)	A wider distance is preferred for tropical environments with anticipated heavy growth of shellfish or coral.
Through-screen velocity	0.10–0.15 m/s (0.3–0.5 ft/s)	A through-screen velocity of 0.10 m/s is recommended for source waters with jellyfish content higher than 1 organism per cubic meter. Through-screen velocity should be determined for 50 percent of the area between the bar screens.

Table 6.2 Key Design Criteria for Inlet Structures of Offshore Intakes

Inlet Depth Typically, the optimum depth of the top of the intake velocity cap below the mean surface water level is 12 to 20 m (39 to 66 ft). It is recommended that the top of the inlet velocity cap is submerged at a minimum of 8 m (26 ft) and 4 m (13 ft) below the water surface at mean and lowest-tide water levels, respectively. Minimum water depth should be confirmed with the authorities responsible for regulating naval vessel navigation in the vicinity of the intake.

Intakes at depths shallower than 8 m (26 ft) from the mean water surface level usually produce source water of inferior quality due to accelerated wave- and wind-driven mixing near the ocean surface. Since water quality does not improve significantly beyond a depth of 20 m (65 ft), and water temperature decreases with depth, the construction of very deep intakes typically is not cost effective, taking under consideration that construction and maintenance costs of intake inlets increase with depth. In addition, deeper water is colder, and because of its higher viscosity, membrane salt separation by reverse osmosis requires more energy. Since the relationship between water temperature and viscosity becomes curvilinear for temperatures below 8 to 12°C (46 to 54°F), the use of colder source water would result in an almost exponential increase in energy consumption for desalination of source water with a temperature below 10°C (50°F).

One benefit of colder waters is their lower biofouling potential. However, this benefit is most pronounced for waters warmer than 20°C (68°F) and rich in organics (total organic carbon > 1.5 mg/L), and is practically negligible for waters colder than 12°C (54°F).

The distance between the bottom of the inlet mouth and the ocean floor should be at least 50 percent of the length of the bar screens but no less than 4 m (13 ft), in order to prevent excessive carryover of bottom sediments into the downstream intake facilities. If the intake area is affected by frequent naval traffic, then the minimum depth from the bottom should account for the depth of bottom sediment resuspension by large ships and boats.

Inlet Screens Inlet screens are typically coarse bar screens with a distance between bars of 50 to 300 mm (2 to 12 in.). The bar length is usually between 1 and 3 m (3 and 10 ft). The design maximum through-screen velocity varies depending on the content of jellyfish in the source water and is usually in a range of 0.10 to 0.15 m/s (0.3 to 0.5 ft/s). It is important to note that the design through-screen velocity should be calculated for 50 percent of the total area between the screen bars, to allow for growth of shellfish and accumulation of debris over time. It is anticipated that the inlet screening area will be reduced by 30 to 50 percent every 18 to 24 months, and therefore the screens will need to be cleaned periodically by divers.

A maximum through-screen velocity of 0.10 m/s or less is typically selected if the source water contains a jellyfish load of one organism per cubic meter of source water or more. This lower through-screen velocity allows minimization of the intake of jellyfish into the plant and thereby of their negative impact on downstream screening facilities. Jellyfish outbreaks are typical for warmer, shallower, and polluted waters, especially if the intake is located in the middle of an underwater current though which jellyfish travel frequently.

Typically, small jellyfish specimens that can enter the intake weigh between 200 and 400 g per individual; for large plants with few fine screens, jellyfish could add significant load to the screens, which in extreme conditions could cause equipment damage. In addition, jellyfish are very difficult to clean from the screens, and often the standard debris jet sprays with which fine screens are equipped cannot remove them. From this perspective, drum screens are usually more difficult to clean once the jellyfish attach on the screens, and therefore reducing jellyfish intake by reducing the inlet through-screen velocity is of critical importance, especially for plants with fine drum screens. If large quantities of jellyfish enter the intake, they could render the fine screens inoperable or significantly reduce the plant's source water capacity.

Inlet Materials Inlet shafts are typically built from corrosion-resistant materials such as copper-nickel alloy, concrete, or stainless steel. The screen bars are made of either 90/10 copper-nickel or super duplex stainless steel for seawater applications, and of duplex stainless steel for brackish water intakes.

Inlet Configuration Small plants are usually designed with a single inlet structure and conveyance pipeline. However, an intake configuration with two or more inlets connected to individual conduits (pipelines or tunnels) is preferable in order to avoid shutting down the entire plant when the intake system is taken out of service for cleaning. For plants with two or more inlets, it is important to note that the inlets should be located an adequate distance from each other to avoid mutual impacts on their flow patterns. As a rule of thumb, the individual inlet shafts (towers) should be installed at a distance of two to three inlet shaft diameters from each other. In addition, intake inlets should be located at an adequate distance—typically 400 to 1000 m (1300 to 3300 ft)—away from and always upstream up the discharge, so that prevailing currents can dissipate the discharge concentrate away from the intake and short-circuiting can be avoided. The minimum distance between the intake inlet and discharge diffuser field locations should be determined based on hydrodynamic modeling.

If the plant intake is designed with a single conduit, usually the conduit diameter is sized for 120 to 130 percent of the annual average intake flow, in order to accommodate biogrowth on the conduit walls and silt and debris accumulation on the bottom. If two or more conduits are installed, each of them is designed for 60 percent of the plant intake flow.

Typically, intake inlets and conduits have to be cleaned once every 1 to 2 years in order to maintain their conveyance capacity. Usually, intake and conduit cleaning takes 2 to 5 days. Therefore, desalination plants with a product water capacity larger than 150,000 m^3/day (40 mgd) are commonly designed with three to five inlets that can be serviced individually. An exception are inlets located in very deep water [15 m (49 ft) or more], where the process of aquatic growth on the inlet bar screens and conduits is significantly slower because of the colder temperatures and limited abundance of marine life and nutrients.

Intake Water Conduit Configuration Depending on the plant capacity, intake water conduits are either pipes or tunnels. Most reverse osmosis desalination plants with a freshwater production capacity smaller than 300,000 m^3/day (79 mgd) are designed with pipeline conduits. An intake pipeline can be constructed using one of three methods: (1) it can be directly installed on the bottom of the water body and anchored with concrete blocks, (2) it can be buried in trenches, or (3) it can be directionally drilled or micro-tunneled in the bottom sediments.

Pipelines are laid directly on the bottom of the water body if the bottom is sandy and flat and the intake is located on a low-energy shoreline (e.g., bays, semienclosed lagoons, deep quiescent waters) with minimal exposure to beach erosion, heavy storms, typhoons, tsunamis, or strong underwater currents. For example, the intakes of the desalination plant in Ashkelon, Israel, are constructed in this manner. Typically, HDPE pipes are used in such a configuration. The intake pipe is delivered to the site in long sections, welded onshore, and fitted with concrete blocks at regular intervals. A trench is dredged on the bottom to flatten the pipeline route before the pipe is installed. The pipe with the attached concrete blocks is closed on one end, towed, adjusted to fit the design route, and then sunk by being filled up with water. Finally the inlet is installed and connected to the pipe.

Pipelines are installed in trenches, typically at a depth of 1 to 5 m (3 to 16 ft), when the water body's bottom is uneven or exposed to strong currents and wind or tidal action. Usually, open trench excavation is mainly suitable for relatively shallow sandy or soft rock bottom. Examples of seawater desalination plants with such intakes are Perth I and II, Australia; Al Dur, Bahrain; Fujairah I in the United Arab Emirates; Barka III in Oman; and others. Directional drilling, or micro-tunneling, is the most viable method for installing of intake pipes in a rocky coastal environment.

Dredge-and-cover pipeline construction is an alternative to pipeline installation in trenches and is typically applied for deeper waters. A relatively shallow trench is constructed by dredging the bottom; the pipe is installed in the trench and then covered with several layers of suitable-size gravel and rock and then covered with the dredged material. Depending on the site-specific bottom conditions, a cover of rock armor may also be added on the top of the pipe. As in the installation of HDPE pipe directly on the bottom, the pipe is secured along its trench route by concrete weight blocks.

Intake Pipeline Materials Materials that have found wide application for desalination plant intake pipelines to date are glass-reinforced plastic (GRP), high-density polyethylene (HDPE), polyvinyl chloride, concrete, cement-lined ductile iron, and cement- or epoxy-lined mild steel. GRP and HDPE are the two most commonly used pipeline materials.

HDPE has found the widest application for intake pipelines installed directly on the bottom of the water body because it is very durable and resistant to corrosion, it can handle moderate wave and wind action without needing to be buried, it is more flexible than GRP, and it can be easily transported, installed, and fused in long sections. The two main disadvantages of HDPE as compared to GRP are that it is only available in limited diameters [up to 2000 mm (78 in.)] and it is costlier.

GRP piping is practically not restricted in terms of size—the largest diameter is 4000 mm (157 in.). However, it has lower flexibility than HDPE, it is lighter and thus tends to float, and it cannot carry high dynamic loads. Therefore, GRP pipe has to be buried in a trench and covered with armor rock (crushed stone) or other material to protect it against floating, currents, and other natural destructive forces.

While concrete and steel pipe are also employed for desalination applications, they are significantly more costly, less durable, and more difficult to install. Therefore, they are not commonly used in new desalination projects.

Intake conduits for large membrane desalination plants are typically concrete tunnels directionally drilled at a depth between 5 and 20 m (16 and 66 ft) below the bottom of the source water body and connected to the inlet screen structure via a vertical riser shaft. A tunneled intake can be one of two types—a tunnel with a single intake inlet structure at the end (Fig. 6.4) or a tunnel with multiple inlet structures located along its length (Fig. 6.5). Examples of tunnel intakes with a single inlet are the Gold Coast and Adelaide plants in Australia; the Sydney and Melbourne's Victoria desalination plants have intake tunnels with four inlet structures connected via vertical shafts.

Tunnel construction usually begins with the installation of the onshore shaft and launch chamber for the tunnel-boring machine. This machine allows for construction of the tunnel in concrete segments (rings). In one segment installation cycle, the machine drills space adequate to install one segment and then its rams are retracted and the new tunnel ring segment is installed. The tunnel ring is waterproofed and grouted before the drilling machine moves forward to drill space to install the next segment.

Once the horizontal tunnel is built, one or more riser shafts are constructed from an offshore platform: first an external caisson is installed into the seabed that acts like a protective shell, and then the actual shaft drilling and excavation are completed. Once the shaft reaches the tunnel level, it is sealed at the top and stub connection tunnels are constructed by drilling multiple horizontal holes to form the tunnel connector and then

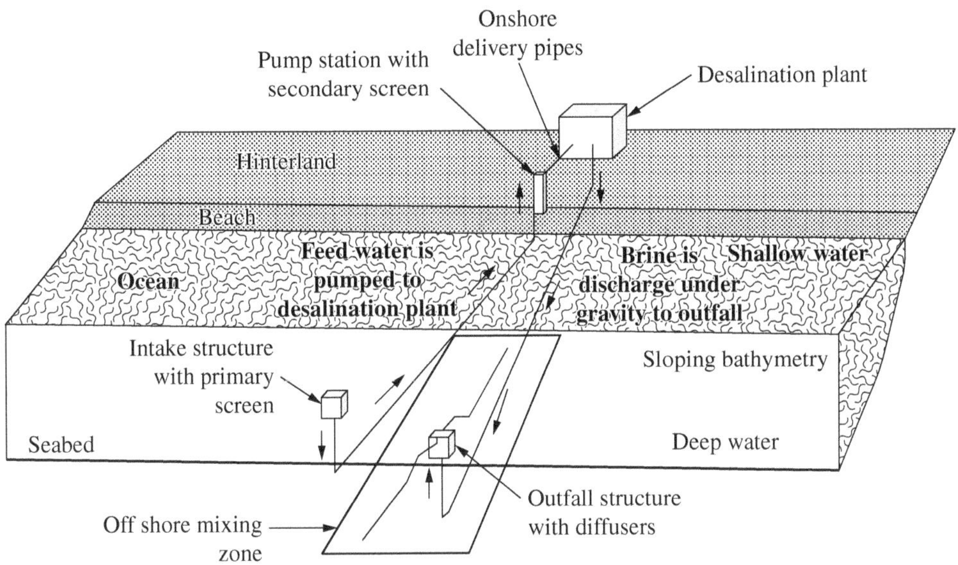

FIGURE 6.4 Offshore intake with a single inlet structure. (*Source:* WaterSecure.)

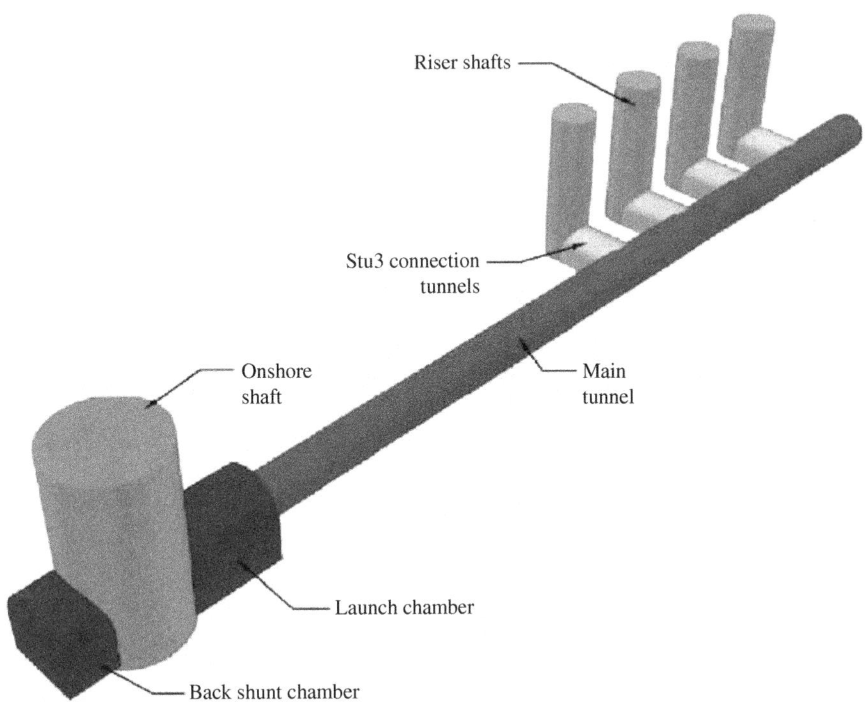

FIGURE 6.5 Intake with multiple inlet structures. (*Source:* Victoria DSE.)

grouting and waterproofing the tunnel connector walls. As a final step, the inlet structure is installed on the top of the riser shaft.

Design Example of Intake Inlet Structure This design example illustrates the sizing of the inlet for a reverse osmosis seawater desalination plant designed for an average production flow of 300,000 m^3/day, a maximum daily production flow (Q_{max}) of 315,000 m^3/day, a recovery (R) of 46 percent, and a volume of additional water uses (backwash water BW = 5 percent and other waters OW = 1 percent) of a total 6 percent of the intake water flow. The design example is illustrated in Fig. 6.6.

1. Calculate the plant intake flow.

$$Q_{in} = (Q_{max} \times 100/R) \times [1 + (\mathrm{BW} + \mathrm{OW})/100] \tag{6.1}$$

$$= (315{,}000 \times 100/0.46) \times [1 + (0.05 + 0.01)/100] = 725{,}870\ \mathrm{m^3/day} = 8.4\ \mathrm{m^3/s}$$

2. Select the depth from the ocean surface to the top of the velocity cap. $H_{s\text{-}vc}$ = 12.2 m is selected based on the bathymetric survey, water quality survey, and depth profile as discussed previously (see Fig. 6.6).
3. Determine the depth from the ocean surface to the ocean bottom. $H_{s\text{-}b}$ = 20.1 m based on the bathymetric survey for the selected inlet location.
4. Select the distance between the ocean bottom and the bottom of the bar screen. $H_{b\text{-}vc}$ = 4.1 m is selected based on the water quality (turbidity and silt density index) profile along the entire depth, which allows for identification of the distance from the bottom where the turbidity and silt density index (SDI) decrease significantly over the near-bottom turbidity and SDI.

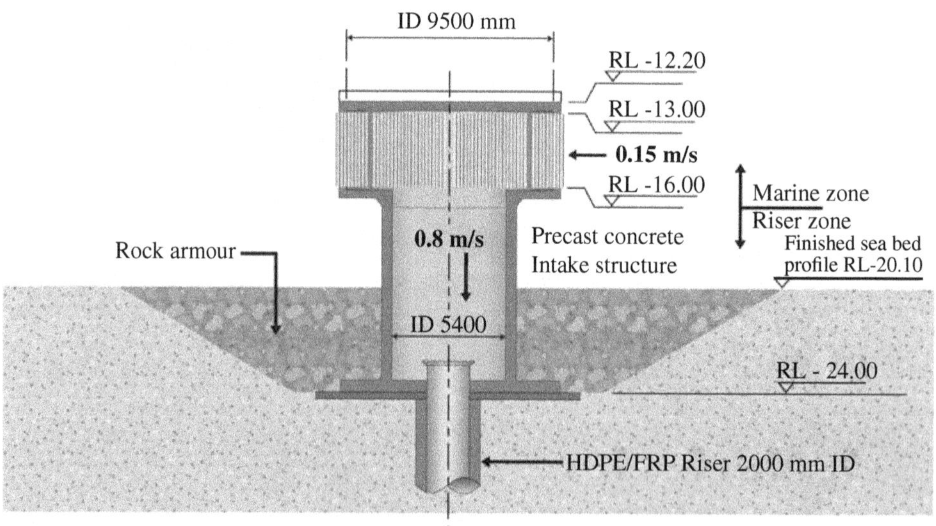

Figure 6.6 Design example of an intake inlet.

5. Calculate the length of the inlet bars.

$$L_b = H_{s\text{-}b} - H_{s\text{-}vc} - H_{b\text{-}vc} = 20.1 - 12.2 - 4.1 = 3.8 \text{ m}$$

6. Calculate the total design active surface area of the bar screen.

$$A_{sbs} = (Q_{max}/V_{ts}) \times (100/A_{\%}) \tag{6.2}$$

where V_{ts} = design through-screen velocity (selected to be 0.15 m/s) and $A_{\%}$ = available screening area expressed as a percentage of total through area of the screen openings (assumed to be 50 percent for design purposes).

$$A_{sbs} = (8.4/0.15) \times (100/50) = 112 \text{ m}^2 \text{ (1,206 ft}^2\text{)}$$

Taking into consideration that the total surface area of the intake inlet cylinder is

$$A_{sbs} = 2\pi r L_b \tag{6.3}$$

Then the radius of the inlet will be

$$r = A_{sbs}/2\pi L_b \tag{6.4}$$

$$= 112/(2 \times 3.1416 \times 3.8) = 4.7 \text{ m (15.4 ft)}$$

The diameter is rounded up to 9.5 m/31.2 ft (see Fig. 6.6).

The bottom portion of the intake shaft is designed for a velocity of 0.8 m/s (2.6 ft/s) and the diameter of the vertical shaft is oversized by 50 percent to accommodate potential accumulation of shellfish on the walls. The diameter at the fully open shaft is $[(4 \times 8.4)/(0.8 \times 3.1416)]^{1/2} = 3.66$ m (selected 3.65 m/12 ft). This diameter is increased by 50 percent to a design value of 5.475 m (selected 5.4 m/17.7 ft).

The intake conveyance pipeline is designed for a velocity of 2.5 m/s (8.2 ft/s) and a maximum flow of 8.4 m^3/s; $[(4\times8.4)/(2.5\times3.1416)]^{1/2} = 2.07$ m, rounded to 2000 mm (80 in).

6.2.6 Costs of Open Intakes

Construction Costs of Onshore Intakes

Figure 6.7 provides a budgetary estimate for the construction costs of onshore intakes as a function of the desalination plant's intake flow. Because of the significant impact of the site-specific condition of the actual intake costs, such costs may vary within 30 percent above or below the values indicated in this figure.

Construction Costs of Offshore Intakes

The construction costs are graphed in Fig. 6.8 for intake systems with offshore inlet structures and HDPE pipelines and for structures with concrete tunnels. These costs are presented as a function of the plant intake flow and are expressed in dollars per meter of intake conduit. Similar to the costs of onshore intake construction, these costs will vary from one location to another; based on site-specific conditions such as water depth, geology, and currents, they could be within a 30 percent envelope of the values indicated in the figure. Analysis of this figure indicates that the construction of desalination plant intakes with deep tunnels is typically several times costlier than the installation of HDPE pipeline on the bottom of the source water body.

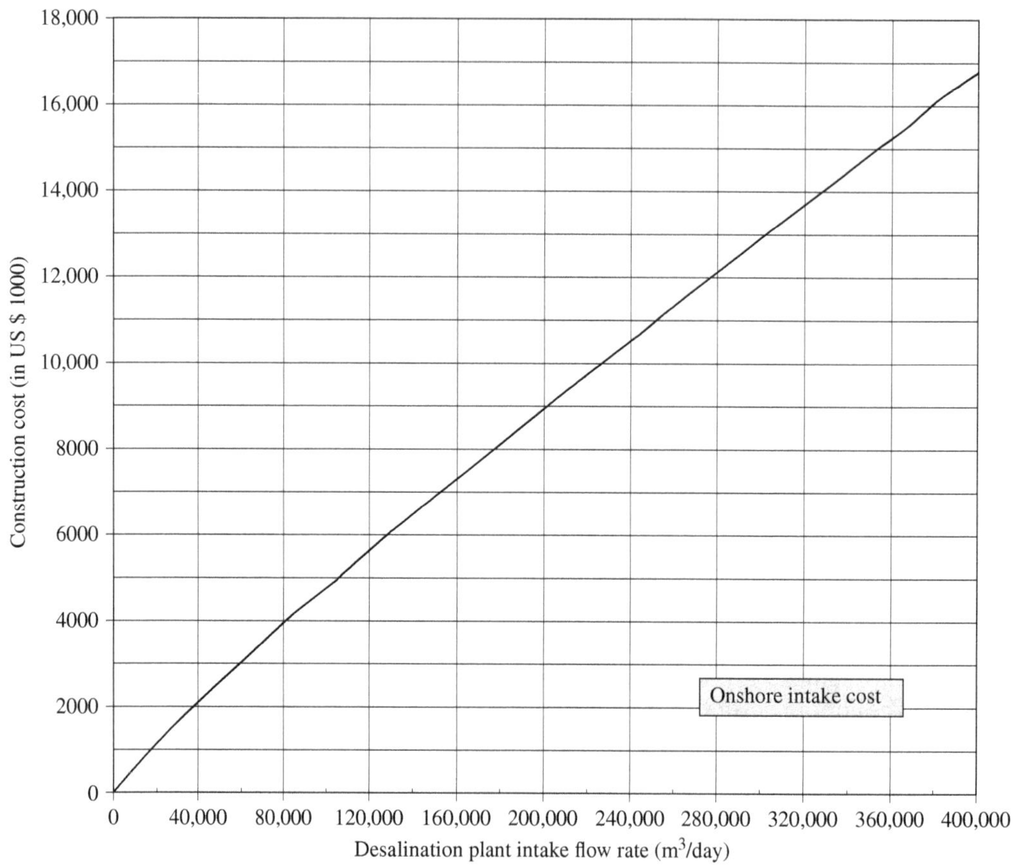

FIGURE 6.7 Onshore intake cost.

6.3 Subsurface Intakes

Subsurface intakes are the predominant type of source water collection facility for brackish desalination plants of all sizes. Source water for over 80 percent of the brackish desalination plants in the United States is supplied using wells. Most of these wells collect water from deep, confined aquifers of low to medium salinity (i.e., total dissolved solids of 600 to 3000 mg/L). Typically, deep brackish wells yield a source water of low turbidity (< 0.4 NTU) and silt content (silt density index < 1), which can be processed through the reverse osmosis system with minimal or no pretreatment (usually cartridge or bag filtration only).

Subsurface intakes for seawater desalination installations collect source water from either a saline near-shore (coastal) aquifer or an offshore aquifer under the ocean floor. The salinity of the coastal aquifers varies as a result of ocean water tidal movement and of changes in elevation and salinity of the fresh and/or brackish groundwater aquifers that are hydraulically connected to the coastal aquifer. As a result, coastal aquifers are

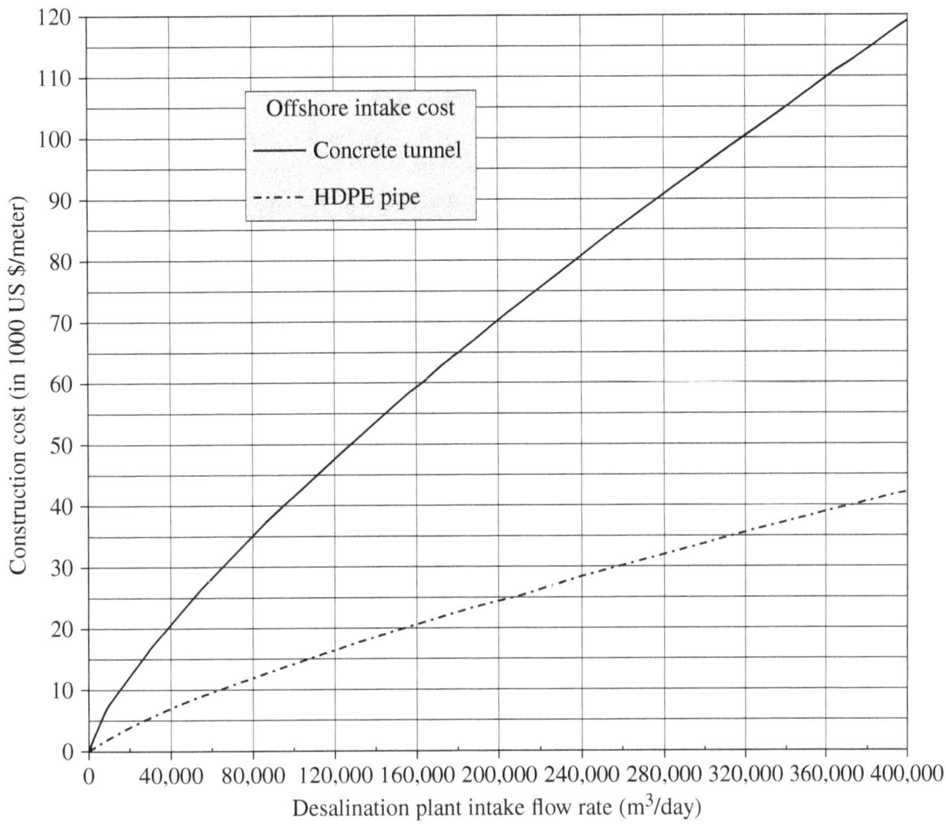

FIGURE 6.8 Offshore intake cost.

typically a source of saline water whose total dissolved solids concentration is lower than that of open ocean water. In contrast to brackish desalination plants, where wells are the intake type of choice, at present less than 10 percent of seawater desalination plants worldwide use well intakes.

The most common types of subsurface intakes for desalination plants are (1) vertical wells, (2) horizontal directionally drilled (HDD) wells, (3) horizontal Ranney-type wells, and (4) infiltration galleries. Vertical wells are used for both brackish and seawater desalination facilities. The other three types of subsurface intakes have found application mainly in seawater desalination projects.

The subsurface intake facilities are relatively simple to build, and the saline water they collect is pretreated via slow filtration through the subsurface soil formations in the area of source water extraction. Therefore, raw water collected using subsurface intakes is usually of better quality than that from open intakes in terms of solids, silt, oil and grease, algal content, natural organic contamination, and aquatic microorganisms. When seawater intake wells are located in coastal aquifers with significant freshwater influence (e.g., river estuaries), they also produce source water of lower salinity.

6.3.1 Types and Configurations

Vertical Wells

Vertical wells (Fig. 6.9) are the most commonly used type of subsurface intake. They consist of the following key components: casing well screen, filter pack, well seal, and surface seal. Vertical wells have a submersible or vertical turbine pump installed inside the well casting, which is a steel or nonmetallic (typically, fiberglass) pipe that lines the well borehole to protect the well against caving.

The diameter of the casing has to be adequate to house the well intake pump and provide ample room for pump service. While the diameter of the well casing is determined mainly by the well screen's size and yield, the diameter of the well borehole has to be at least 0.1 m (4 in.) greater than that of the well casing to accommodate the installation of the well seal. Usually, the well casing diameter is between 200 and 1200 mm (8 and 48 in.), and the well depth is typically less than 75 m (250 ft).

The well screen, the intake portion of the well, is a sievelike structure with slotted or perforated openings. It is located at a depth corresponding to the water-carrying zone of the aquifer. The screen's depth, opening size, diameter, and length are key design criteria for well performance. They are selected to maximize the well's safe yield, control the well entrance velocity, and avoid excessive entrance of sand and other particulates, which have a negative impact on the well's useful life and water quality.

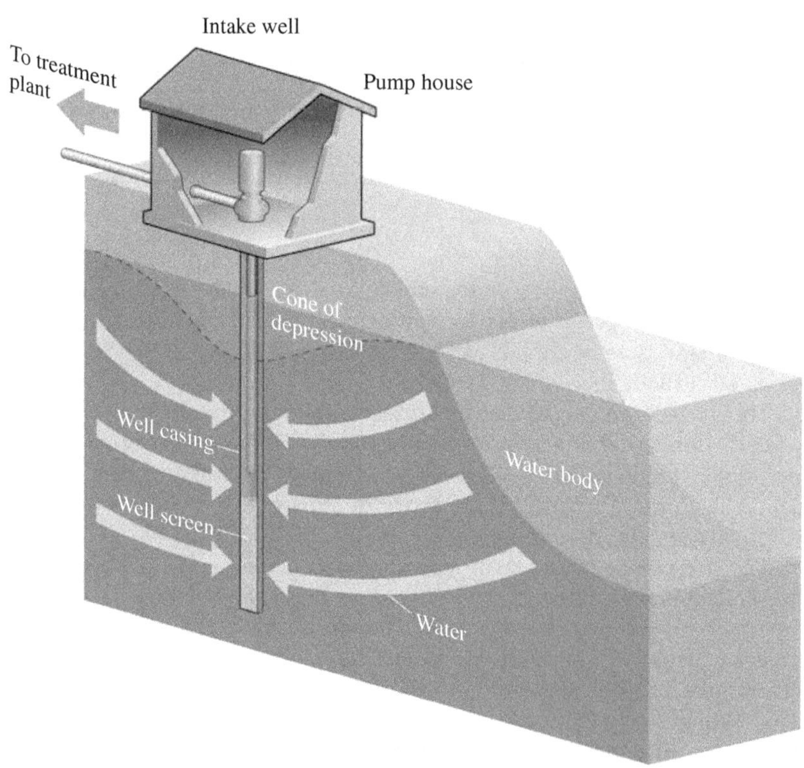

Figure 6.9 Vertical intake well.

Detailed guidelines for the selection of well casings and screens are provided elsewhere (Roscoe Moss Company, 2012).

The performance of the well screen is enhanced by a gravel (filter) pack, which consists of clean, uniform, and well-rounded gravel and sand placed between the borehole wall and the well screen to prefilter the groundwater entering the well. Typically, the gravel pack extends at least 1 m (3 ft) above the well screen.

A well seal is installed above the filter pack to prevent soil and contaminants from entering the well screen area. The well seal is a cylindrical layer of cement, bentonite, or clay placed in the annulus of the well between the well casting and the borehole. Typically the well seal extends at least 0.6 m (2 ft) above the top of the gravel pack, and usually through the elevation of the soil frost zone. The aboveground portion of the well is finished with a concrete surface seal. The surface and well seals protect the well from surface runoff contamination and support the casing. Once the vertical well is constructed, it has to be monitored frequently to secure its long-term performance and identify early signs of potential malfunction and failure. The most common causes of well failure are collapse of the borehole, corrosion of the casing, improper or defective construction techniques, growth of organisms within the borehole, and formation of mineral concentrations or crusts in the open-hole or screened section of the borehole.

The 80,200 m^3/day (21 mgd) Sur SWRO plant in Oman is the largest plant with vertical intake wells in operation at present. The intake area consists of high-yield karstic formations that have an average transmissivity of 7,000 m^3/day·m (David et al., 2009). The well field includes 33 (25 duty and 8 standby) beach wells capable of producing 70 to 100 L/s (1.6 to 2.3 mgd) each. Well depths are 80 to 100 m (260 to 330 ft), and the diameter of the wells is 14 in. Each well is equipped with 14-in.-diameter PVC casing and a screen with a slot size of 3 mm. The wells are surrounded by gravel packs. Each well is equipped with a duplex stainless steel submersible pump and has an average drawdown of 12 m (39 ft).

Horizontal Directionally Drilled Wells

Horizontal diretionally drilled (HDD) collector wells consist of a relatively shallow blank well casting with one or more horizontal perforated screens bored at an angle (typically inclined at 15° to 20°) and extending from the surface entry point underground past the mean tide line. This type of well has found application mainly in seawater desalination installations.

One of the most widely used HDD well intakes today is the Neodren well intake system. A general schematic of Neodren HDD collectors is shown in Fig. 6.10 (Peters and Pinto, 2008). The source water is collected at a relatively slow rate via a number of perforated HDPE pipes with 120-µm pore openings and is naturally filtered through the ocean bottom sediments before it reaches the desalination plant. A typical collector pipe size is 450 mm (18 in.), but collectors as large as 710 mm (28 in.) in diameter could be installed. Typically, the individual HDD collector pipes deliver the source water into a common wet well, from where it is pumped to the desalination plant for further processing. The individual HDD well collectors usually yield between 50 and 150 L/s (1.1 and 3.4 mgd).

Usually, collector pipes are installed at a depth between 5 and 10 m (16 to 33 ft) below the surface of the ocean bottom in separate boreholes by drilling under the ocean seabed 200 to 600 m (660 to 2000 ft) into the coastal aquifer, to a location that can yield

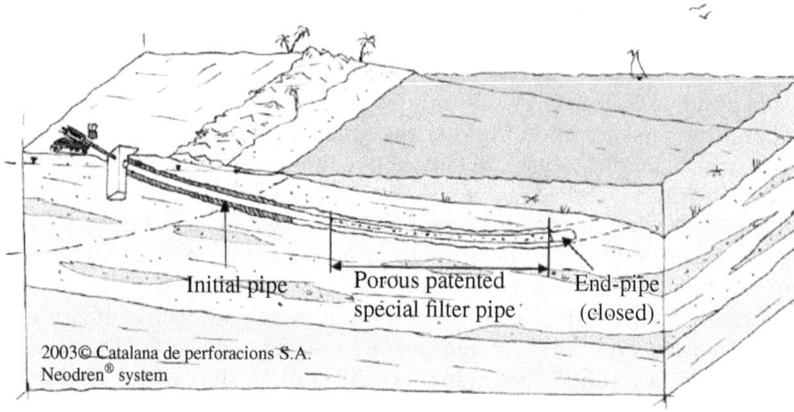

Figure 6.10 Neodren HDD intake. (*Source:* Peters and Pinto, 2008.)

seawater of ambient salinity and avoid water collection from fresh near-shore aquifers in the vicinity of the plant site. The drilling technology employed for the installation of the Neodren system is well proven and has found application for a number of other uses, such as laying fiber-optic cables and oil and gas pipelines.

The Neodren HDD intake technology is patented by the Spanish company Catalana de Perforacions. This technology has been used for over 10 years in several dozen small and medium-size seawater desalination plants in Spain, and currently it is under construction, consideration, and pilot testing at a number of other plants worldwide.

The collection capacity of source seawater depends on the number and diameter of the horizontally drilled perforated pipes, the length of the perforated portion of the pipes, and the transmissivity and depth of the seabed in which the collector pipes are drilled.

The natural seabed filtration process removes practically all coarse solids and particulates which are 50 μm or larger from the seawater and prevents marine organisms in all phases of development (adults, juveniles, and larvae) from entering the desalination plant (i.e., it protects marine life against impingement and entrainment). This system is also an effective barrier against the heavy solids loads generated during algal blooms and oil spills.

Data available to date (Peters et al., 2007), however, indicate that while the HDD system can successfully reduce source seawater turbidity and total organic carbon, this reduction typically is not adequate to directly apply the water collected by the HDD intakes to the SWRO membranes. The source water would need to be pretreated through a filtration system prior to membrane separation.

One of the largest seawater desalination plants that has HDD well intake, is located in Spain—the 65,000 m^3/day (17 mgd) Cartagena Canal (San Pedro del Pinatar) plant. The intake system consists of 20 HDD wells arranged in a fan shape—Fig. 6.11.

The individual intake wells are between 500 and 600 m (1600 and 2000 ft) long and have a diameter of 355 mm (14 in.). Each well produces between 100 and 140 L/s (2.3 and 3.1 mgd). The plant operates at 45 percent recovery. The source sea water is collected in a large wet well located underground and pumped to the plant using submersible pumps.

FIGURE 6.11 HDD intake of San Pedro del Pinatar SWRO plant.

Horizontal Ranney-Type Wells

This type of well consists of a concrete caisson that extends below the ground surface and has water well collector screens (laterals) projected out horizontally from inside the caisson into the surrounding aquifer (Fig. 6.12). Since the well screens in the collector wells are placed horizontally, a higher rate of source water collection is possible than with most vertical wells. This allows the same intake water quantity to be collected with fewer wells. Individual horizontal intake wells are typically designed to collect between 0.0044 and 1.75 m^3/s (0.1 and 40.0 mgd) of source water.

The caisson of a horizontal collector well is constructed of reinforced concrete with an inside diameter of 2.7 to 6.0 m (9 to 20 ft) and a wall thickness from approximately 0.5 to 1.0 m (1.5 to 3.0 ft). The caisson depth varies according to site-specific geologic conditions, ranging from approximately 10 m to over 45 m (30 to 150 ft).

The number, length, and location of the horizontal laterals are determined based on a detailed hydrogeological investigation. Typically, the diameter of the laterals ranges from 0.2 to 0.3 m (8 to 12 in.) and their length extends up to 60 m (200 ft). The size of the lateral screens is selected to accommodate the grain size of the underground soil formation. If necessary, an artificial gravel-pack filter can be installed around the screens to suit finer-grained deposits. Usually one well has 2 to 14 laterals oriented towards the source water body (ocean, river).

Horizontal Ranney-type wells are typically coupled with an intake pump station installed above the well caisson. The well intake pump station can be designed with submersible pumps to minimize noise levels. However, intakes that consist of mid- and larger-size wells most frequently employ horizontal or vertical turbine pumps, because

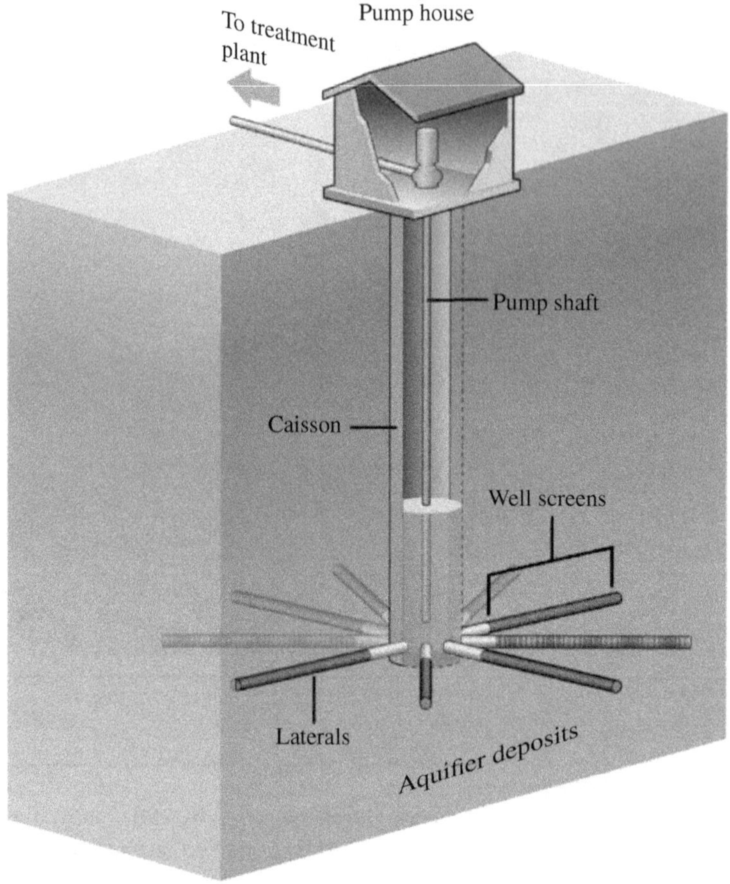

Figure 6.12 Horizontal Ranney-type intake well.

this type of pump usually has higher energy efficiency and requires less power than submersible pumps.

Ranney wells are not as commonly used for seawater intakes as vertical wells are. The largest installation of horizontal wells is located in Salina Cruz, Mexico (Fig. 6.13) and consists of three wells that are designed to deliver 14,500 m^3/day (3.8 mgd) of seawater each. The wells are located on the beach, and the water quality they deliver is not directly suitable for seawater desalination. The source seawater from these wells contains high levels of iron and manganese, and has to be treated in greensand filters prior to SWRO separation.

One potential challenge with such wells is that if the source water contains hydrogen sulfide. That compound is likely to be oxidized to elemental sulfur, which could subsequently cause reverse osmosis membrane fouling (Missimer et al., 2010).

Infiltration Galleries (Seabed or Riverbed Filtration Systems)

These subsurface intake systems consist of a submerged slow sand filtration bed located at the bottom of the source surface water body (i.e., ocean, lake, or river). The bottom of

FIGURE 6.13 Two horizontal wells for Salina Cruz SWRO plant, Mexico.

this engineered filtration bed contains a number of equidistant horizontal perforated pipes that convey filtered source water collected from the bed to the wet well of an intake pump station located on shore (Fig. 6.14). Infiltration galleries are typically implemented when conventional horizontal or vertical intake wells cannot be used due to unfavorable hydrogeological conditions. For example, they are suitable for intakes

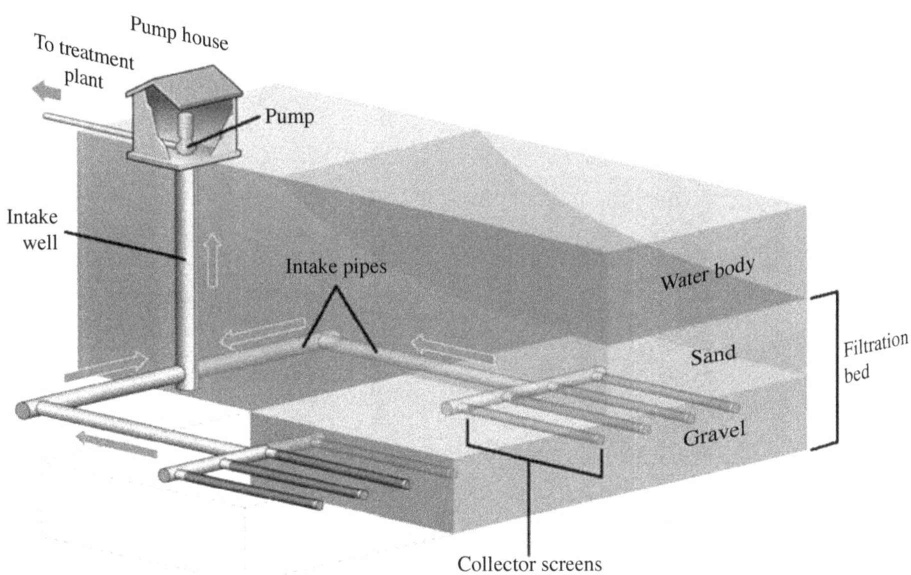

FIGURE 6.14 Infiltration gallery.

where the permeability of the underground soil formation is relatively low, or in the case of river or seashore filtration, where the thickness of the beach or the onshore sediments is insufficient to install conventional intake wells.

Filtration beds are sized and configured using the same design criteria as slow sand filters. The design surface loading rate of the filter media is 0.12 to 0.25 $m^3/m^2 \cdot hr$ (0.05 to 0.10 gpm/ft^2). Such well systems can be installed in areas with good natural ventilation because they rely mainly on wave and current action to remove the solids retained and accumulated on the surface of the well bed during the filtration process. One potential challenge with infiltration galleries is the biofouling of the filtration media, which could reduce their production capacity over time.

The largest seawater desalination plant with a seabed infiltration system in operation at present is the 50,000 m^3/day (13.2 mgd) Fukuoka District RO facility in Japan. This plant has been in operation since 2006. The infiltration bed is 313.6 m long and 64.2 m wide (2 ha/5 acres), and it is designed to collect a source seawater flow of 130,000 m^3/day (34 mgd). The design infiltration velocity is 0.25 $m^3/m^2 \cdot hr$ (0.10 gpm/ft^2).

The filtration media in the well is configured in three distinct layers (Fig. 6.15): at the bottom—2.3 m (7.5 ft) of graded gravel pack with stone sizes between 20 and 40 mm that surrounds the horizontal well collectors; in the middle—0.3 m (1.0 ft) of finer graded gravel of size 2.5 to 13 mm; and on the top—1.5 m (5.0 ft) of natural sand excavated from the ocean bottom. The filtration media is submerged at 11.5 m (37.7 ft) below the ocean surface.

FIGURE 6.15 Seabed filtration media configuration of the Fukuoka SWRO plant.

Figure 6.16 Segment of a 600-mm intake collector screen.

The filtered water collectors are polyethylene pipe screens 60 m (200 ft) long and 600 mm (24 in.) in diameter (Fig. 6.16) that are installed at a distance of 5 m (16 ft) from each other. The collector pipes are designed for an inflow velocity of 3 cm/s (0.1 ft/s). The screens collect the source water flow into a central pipe with a diameter of 1580 mm (62 in.) and length of 1178 m (3860 ft) that conveys it into a two-tank water collection well for pumping to the desalination plant. The collected water is pretreated with UF membrane filtration prior to desalination in the SWRO membrane system.

6.3.2 Design Considerations

A detailed discussion of key planning issues associated with the construction of subsurface intakes is presented in Chap. 4, while potential environmental impacts associated with well construction are analyzed in Chap. 5.

Key characteristics of well performance and design are yield, static and pumping water levels, and the cone of depression. Well yield indicates how much water can be withdrawn from a given well for a preset period of time. It is typically measured in cubic meters per second (m^3/s) or cubic feet per second (ft^3/s) for large wells and liters per second (L/s) or gallons per hour (gal/h) for small wells. Pumping and static water levels are the groundwater levels in the well when pumping from the well is on and off, respectively.

When the well is operational, the surface of the groundwater level in the aquifer takes an inverted cone shape due to directional water flow toward the well. This inverted shape is called the cone of depression (see Fig. 6.9).

Well type, size, and capacity are determined based on a hydrogeological investigation, which typically includes the following key steps:

1. Complete a preliminary geological survey to identify if the selected site is generally suitable for the construction of a subsurface water intake.
2. Drill-test to collect samples of the aquifer formation deposits for visual classification and grain-size distribution analysis.
3. Install one or more test wells and observation wells, and conduct a pumping test to determine the site-specific hydraulic characteristics of the aquifer that are necessary for subsurface system design and determination of the intake system yield. The test should be a minimum of 72 h. Usually a number of incremental tests are completed in sequence to determine the optimum yield. This initial testing is followed with a longer-term test that allows for determination of the aquifer's transmissivity.
4. Collect adequate amount of source water samples and analyze the water quality, with special emphasis on the content of iron, manganese, barium, strontium, silica, radon, carbon dioxide, arsenic, and hydrogen sulfide in the source water. If the aquifer water quality is under the influence of a surface water source (e.g., river, lake, ocean) whose quality and quantity vary seasonally, then complete year-round intake water quality sampling to determine seasonal fluctuations of source water quality.
5. If the subsurface intake system will require the installation of multiple collection facilities (wells, infiltration galleries, or river bank filtration facilities), then complete a computer model analysis to establish the response of the production aquifer to pumping and the potential impact of groundwater collection on adjacent fresh or saline water aquifers that could be in interaction with the water supply aquifer.

The key factors that determine if the use of subsurface intake is practical and/or economical are the type of the water source aquifer (confined or unconfined); the aquifer permeability (hydraulic conductivity), which is a measure of the velocity of water movement through the ground (typically measured in meters per second); the average specific yield (productivity) of the aquifer (in cubic meters per day per linear meter of riverbank or seashore along which the collector wells are located); the thickness of the production aquifer deposits; and the existence of nearby fresh or brackish water aquifers that could be negatively impacted by the intake well operations or have a measurable effect on intake well water quality.

A confined aquifer (also referred to as an *artesian aquifer*) is a water-saturated geological formation between two layers of low permeability (i.e., bedrock) that restrict the vertical movement of the groundwater in or out of the aquifer. Confined aquifers are often pressurized by the surrounding geological formations, and therefore collecting water from such aquifers may not require pumping. Unconfined aquifers are groundwater-saturated formations with a fluctuating water level or table. Such fluctuation is driven by recharge from surface runoff (rain or snowmelt) or changes in the water table of a surface water body (ocean, river, lake, etc.) hydraulically connected to the aquifer.

Coarse-grained porous and highly permeable geological formations (e.g., sandstones, beach sand and alluvial deposits, coarse-grained gravel and limestone) connected to a brackish riverbed (for brackish plant intakes) or to the ocean floor (for seawater intakes) that have a specific yield (transmissivity) in excess of 1000 m^3/day·m and a water carrying zone of at least 6 m (20 ft) are most suitable for subsurface intakes. The higher the aquifer permeability, transmissivity, and thickness, the larger the well yield that the aquifer can support. Such soil conditions often exist along coastal dunes, reefs, and alluvial deltas. Fractured rock formations an potentially deliver a high volume of flow, but often the collected water is of inferior quality. The worst subsurface substrates for installation of wells are those formed as a result of volcanic activity, such as basalt and lava, as well as granite and clay formations.

One key criterion for whether vertical wells can be used or horizontal wells need to be installed is the presence of faults along the coast in parallel to the ocean shore. If such faults exist, typically there is no hydraulic connection between the coastal aquifer and the sea, and therefore vertical wells will not be suitable.

Site Selection

The next step after an aquifer of suitable yield has been identified is to select the location of the actual well field. Well fields should be located perpendicular to the main direction of the groundwater flow (i.e., along the banks of a nearby brackish river or parallel to the seashore). In addition, production well fields should be sited away from and uphill of potential contamination sources, such as septic tanks, landfills, industrial plants, and underground fuel storage tanks.

It is preferable to install beach wells as close as possible to the shoreline—30 to 50 m (100 to 160 ft)—in order to avoid influence on freshwater wells located near the shore and to collect water with the same salinity as the ambient seawater, unadulterated with freshwater influence. Once the location of the well field is established, several alternative methods for drilling could be applied, depending on the size and type of wells. The methods are similar to those used for freshwater well construction and are discussed in greater detail in other sources (Roscoe Moss Company, 1990; Rodriguez-Estrella and Pulido-Bosh, 2009; Williams, 2011).

Vertical Well Yield

The production capacity of a vertical well can be determined based on the following formula:

$$Q_{\text{well}} = (T \times A_{\text{d}})/4.4 \tag{6.5}$$

where Q_{well} = well production capacity, m^3/day
T = aquifer transmissivity = $k \times h_0$, m^2/d
k = aquifer permeability, m/d
h_0 = thickness of the aquifer, m
A_{d} = design aquifer drawdown, m

The actual thickness of the aquifer (h_0) is established based on hydrogeological investigation. The aquifer permeability is determined from pumping tests completed in the target well field area. Figure 6.17 illustrates the calculation of the aquifer permeability and transmissivity based on the results from an actual pumping test completed for the development of well intake in the Al-Birk SWRO plant in Saudi Arabia (Jamaluddin et al., 2005).

The pumping test, which continued for 7 days (followed by a 12-h recovery period), allowed for the determination of the stable flow of the pump and the elevation of the

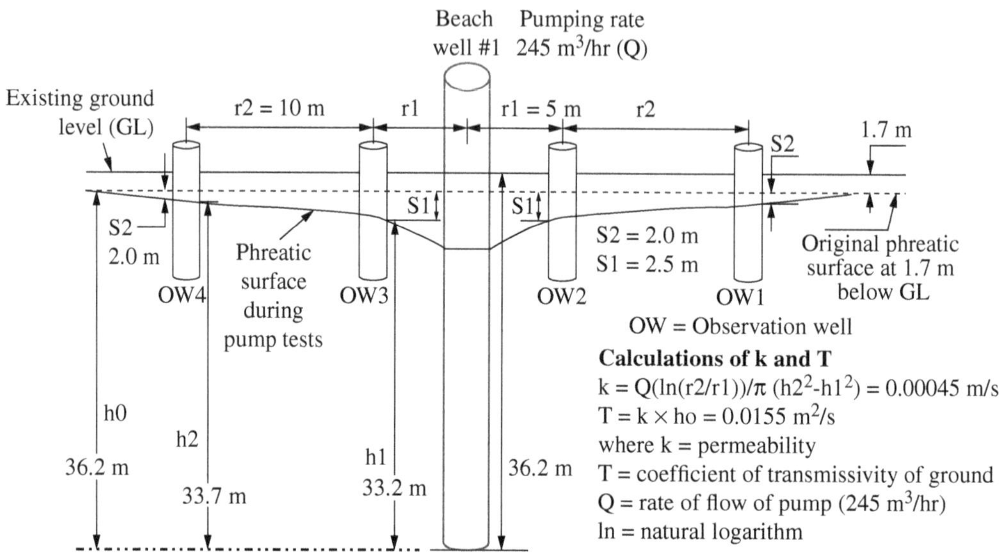

FIGURE 6.17 Example calculation of the transmissivity of a well intake aquifer. (*Source: Jamaluddin et al., 2005.*)

groundwater level at four observation wells (OW1 through OW4). In this particular example, the transmissivity T of the 36.2-m-thick (119 ft) aquifer was calculated at $0.0155\ m^2/s = 1339\ m^3/day \cdot m$. As indicated previously, aquifers with a transmissivity higher than $1000\ m^3/day \cdot m$ are considered suitable for installation of intake wells.

In order to determine the production capacity of the long well, the designer will need to first decide what will be the acceptable aquifer drawdown during normal facility operation. This drawdown is related to the safe yield and is determined based on the potential impact of the radius of influence of the intake well on other production wells in the area (if any), and on areas that could be negatively impacted by it (e.g., wetlands that could be dried up), and on areas that could cause negative impact on the quality of the collected well water such as sources of underground contamination (e.g., drawing water from the drainage area of a nearby landfill or cemetery). In this case, it was determined that the acceptable aquifer drawdown would be 10.0 m (8.2 ft). Based on this depth and the transmissivity determined from the pumping test, the beach well production capacity was calculated at

$$Q_{\text{well}} = (T \times A_d)/4.4 = (1339 \times 15.0)/4.4 = 4565\ \text{m}^3/\text{day} = 1.2\ \text{mgd}\ (837\ \text{gal/min})$$

Using Table 6.3 (HDR Engineering, 2001), which provides an initial recommendation for the nominal size of the pump bowls and the well casting, it can be determined that for this example, the nominal size of the pump bowl will be 250 to 300 mm (10 to 12 in.) and the optimum size of the well casting will be 350 to 400 mm (14 to 16 in.).

6.3.3 Costs of Subsurface Intakes

Vertical intake wells are usually less costly than horizontal wells, but their yield is relatively small—typically 0.004 to 0.044 m^3/s (0.1 to 1.0 mgd)—and thus they are typically

Well Capacity, m³/day (gal/min)	Nominal Size of Pump Bowl, mm (in.)	Optimum Size of Well Casting, mm (in.)
<545 (100)	100 (4)	150 (6)
409–954 (75–175)	125 (5)	200 (8)
818–1910 (150–350)	150 (6)	250 (10)
1640–3820 (300–700)	200 (8)	300 (12)
2730–5450 (500–1000)	250 (10)	350 (14)
4360–9810 (800–1800)	300 (12)	400 (16)
6540–16,400 (1200–3000)	350 (14)	500 (20)
10,900–20,700 (2000–3800)	400 (16)	600 (24)

Table 6.3 Well Pump Bowl Size and Casting Diameter as Functions of Projected Yield

Intake Well Production Capacity, m³/day	Construction Costs in 2012 US$ as a Function of Well Capacity *Q*, m³/day, and Well Depth *H*, m
1000–2000	$40Q + 700H + 25{,}000$
2000–4500	$50Q + 850H + 50{,}000$
4500–6500	$65Q + 1100H + 80{,}000$
6500–10,000	$76Q + 2000H + 150{,}000$
10,000–15,000	$85Q + 2100H + 190{,}000$
15,000–30,000	$90Q + 3300H + 260{,}000$

Table 6.4 Construction Costs of Vertical Intake Wells

used for supplying relatively small quantities of water, usually less than 20,000 m³/day (5.3 mgd). Table 6.4 provides construction costs for vertical intake wells as a function of well capacity and depth.

The costs listed in Table 6.4 do not include expenditures associated with the construction of groundwater monitoring wells for the well field or piping for delivery of the source water to the desalination plant. However, they do include the capital expenditures for the intake well pumps and auxiliary equipment associated with pump operations (i.e., electrical and instrumentation, civil works, etc.). Costs for HDD wells and Ranney wells are typically 30 to 50 percent higher than those of vertical wells, and infiltration galleries are usually the costliest type of subsurface intakes.

Cost Example

Consider the example of a 40,000 m³/day (10.6 mgd) seawater desalination plant with an intake capacity of 98,440 m³/day and 22 duty wells plus 3 standby wells, each with a diameter of 350 mm (14 in.), depth of 40 m (130 ft), and individual capacity of 4475 m³/day. The construction costs of such a deep well injection system can be calculated using Table 6.4: $(22 + 3) \times [(50 \times 4475) + (850 \times 40) + 50{,}000] = \$7.7$ million.

6.4 References

Baudish, P. A., H. A. Lavery, R. N. Burch, D. D. Pain, D. G. Franklin, and P. J. Banks, "Design Considerations and Interactions for Tunnelled Seawater Intake and Brine Outfall Systems," IDA World Congress, Perth, Western Australia, September 4–9, 2011. Ref: IDAWC/PER11-242.

David, B., J. P. Pinot, and M. Morrillon, "Beach Wells for Large-Scale Reverse Osmosis Plants: The Sur Case Study," IDA World Congress, Atlantis, the Palm-Dubai, United Arab Emirates, November 7–12, 2009. Ref: IDAWC/DB09-106.

Gille, D., "Seawater Intakes for Desalination Plants," *Desalination*, 156: 249–256, 2003.

HDR Engineering, *Handbook of Public Water Systems*, 2nd ed., John Wiley and Sons, Place, 2001.

Jamaluddin, A. T. M., A. M. Hassan, M.O. Saeed, L. M. Bakheet, A. Al-Reweli, M. M. Al-Amri, *Sub-Surface Intake to Salvage a SWRO Plant from Chronic Fouling Problems, IDA World Congress*, Singapore, September 11–16, 2005: Ref: SP05-174.

Missimer, T. M., R. G. Maliva, M. Thompson, W. S. Manahan, and K. P. Goodboy, "Reduction of Seawater Reverse Osmosis Treatment Costs by Improvement of Raw Water Quality: Innovative Intake Designs," *Desalination and Water Reuse*, 20 (3): 12–22, 2010.

Peters, T., and D. Pinto, "Seawater Intake and Pretreatment/Brine Discharge—Environmental Issues," *Desalination*, 221: 576–584, 2008.

Peters, T., D. Pinto, and E. Pinto, "Improved Seawater Intake and Pre-Treatment System Based on Neodren Technology," *Desalination*, 203: 134–140, 2007.

Rodriguez-Estrella, T., and A. Pulido-Bosch, "Methodologies for Abstraction from Coastal Aquifers for Supplying Desalination Plants in South-East Spain," *Desalination*, 249: 1088–1098, 2009.

Roscoe Moss Company, *Handbook of Ground Water Development*, J. Wiley and Sons, New York, 1990.

Roscoe Moss Company, Los Angeles, CA, 2012.

Williams, D. E., "Design and Construction of Slant and Vertical Wells for Desalination Plant Intake," IDA World Congress, Perth, Western Australia, September 4–9, 2011, Ref: IDAWC/PER11-050.

CHAPTER 7

Intake Pump Stations

7.1 Introduction

The main function of intake pump stations is to reliably deliver source water collected by the plant intake to the downstream pretreatment facilities at a target range of flow rates and pressures (the source water delivery regime). Intake pump stations include the following key components: the source water receiving well (wet well), pumps, service and auxiliary equipment and facilities, interconnecting piping, and chemical feed systems.

7.2 Types and Configurations

Source water pumps for subsurface intakes (i.e., vertical and horizontal wells) are either vertical turbine pumps or submersible centrifugal pumps located directly in the intake well. Typically, vertical turbine pumps are of two types: line-shaft and submersible. The line-shaft vertical turbine pumps are most suitable for wells with a depth of 90 m (300 ft) or less. Their suitability is marginal at depths over 300 m (1000 ft), because the long shaft stretch wears the impellers and bowl unless the thrust bearings are very carefully adjusted. Submersible vertical turbine pumps are most suitable for depths of 210 m (700 ft) or more.

In desalination plants with open intakes, the source water pumps and their motors can be located in or on a wet well, in a dry well, or in a metal encasement (can) designed to serve as a wet well. Canned pumps are wet-well pumps, but because of their different configuration they are often considered a separate category.

7.2.1 Wet-Well Pump Stations

Wet-well pump stations are the most commonly used type of desalination plant intake pump stations at present. In this type, the pumps are located in a wet well and supported on a concrete slab on the top of the well. Typically, vertical turbine pumps are used in such applications. Figure 7.1 depicts the intake pump station of the Gold Coast Desalination Plant in Australia.

Wet-well pump stations have found industry-wide application because of their simplicity and lower costs. This configuration does not require the construction of a separate structure for the pumps and the well and is therefore more compact. However, one key disadvantage of the wet-well configuration is that the pumps are continuously submerged and exposed to corrosion. Corrosion effects are usually aggravated by frequent changes in the water level in the well. In addition, the maintenance of vertical

Figure 7.1 Wet-well intake pump station of Gold Coast Desalination Plant.

pumps is often more difficult because it requires a more complex and taller crane structure. Therefore, it is critical to provide ventilation to the pump wet well in order to minimize corrosion.

These pumps are practically impossible to service in place, so the quality and design of their submerged bearings are of critical importance, because they are subject to overload and are not accessible from the surface.

7.2.2 Dry-Well Pump Stations

In a dry-well configuration, the intake pumps are located in a separate structure, and except for the intake suction header, they are readily accessible for inspection and maintenance. Since they are not as exposed to corrosion as wet-well pumps, they typically require less maintenance and operator attention.

The key disadvantages of this configuration are the greater construction costs and risk of outage due to flooding. If the pumps are installed on a floor above the suction well, they will require priming equipment. Construction of the pump floor at a depth allowing a flooded suction configuration typically results in significant excavation costs and construction complexity.

7.2.3 Canned Pump Stations

Canned pump stations are equipped with a metal suction can (barrel) that surrounds the pumping unit and is specially designed and hydraulically optimized to create near-uniform flow conditions (Fig. 7.2). Usually such pumps are as equally efficient as dry- or wet-well pump stations.

Figure 7.2
Canned pump.

The can may either have a closed bottom (as shown in Fig. 7.2) or be connected directly to the piping header. Most canned pumps have a closed-bottom configuration. The canned pump configuration is applied to vertical turbine pumps only. Because of the limited volume of liquid in the can (wet well), this configuration is sensitive to flow surges—therefore, it is critical that the can always be full and the velocity of flow between the can and the piping header does not exceed 1.2 m/s (4 ft/s). In addition, the distance between the suction nozzle and the discharge elbow opening of the can should be at least five can diameters.

One of the key advantages of this type of pump is its minimal space requirement—the footprint of such pumps is 25 to 30 percent smaller than of conventional wet-well designs and over 50 percent smaller than the footprint of dry-well pump stations. In addition, this pump configuration typically yields the lowest plant construction costs. These were the main reasons, for example, why a canned pump configuration was selected for the intake pump station of the 95,000 m^3/day (25 mgd) Tampa Bay desalination plant in Florida.

7.3 Planning and Design Considerations

The selection of the site location and depth of the intake pump station is typically driven by the location of the intake structure. The type, size, and number of pumps and pump motors, as well as the type of pump motor speed controls (constant or variable speed), are typically determined based on the source water delivery regime.

At a minimum, intake pump stations are designed with two duty pumps and one standby. For plants designed for a production capacity availability factor of 96 percent or more, if five or more duty pumps are selected to be used, installation of two standby units should be considered. Installation of a variable-frequency drive on at least one of the pump motors is recommended for greater flexibility and efficiency of plant operations.

7.3.1 Pump Station Location

The pump station should be sited at a location and grading that prevent flooding and tsunami damage. The pump motor elevation should be selected to be higher than the 100-year flood level, and the ground floor of the pump station structure should be at least 0.6 m (2 ft) higher than the adjacent ground.

Usually the intake pumps are installed downstream of the plant screens, in order to protect the pump impellers from abrasion and damage. However, for some plants with deep offshore open intakes, the intake pumps could be installed upstream of the plant's fine screens. Such is the case at the Gold Coast SWRO plant in Australia (see Fig. 8.7). At this plant, the intake tunnel is located over 67 m (220 ft) below the ground surface; and because of the significant depth, a decision was made to mount the vertical intake pumps directly into the intake tunnel shaft and to screen the source water downstream of the intake pump station. This configuration has resulted in a significant decrease in the costs and time for construction of the plant intake system (Baudish et al., 2011). When pumps are installed downstream of the coarse screens but upstream of the fine screens, they have to be equipped with strainers that can be periodically accessed and cleaned.

7.3.2 Pump Room Configuration

The pump room layout should be designed so that it allows easy access to all equipment, valves, instrumentation, etc. The motor control center, control cabinet, variable-frequency drives, and their associated equipment, transformers, and electrical switchgear should be located in ventilated rooms at or above ground level. The motor control center room should have at least one double door no less than 2.7 m (9 ft) high, with access to the site's roadway.

Split-case horizontal mixed-flow centrifugal pumps are most commonly used for dry-well pump stations, while wet-well pump stations are typically equipped with vertical turbine pumps. In both cases, the minimum clearance between adjacent items of equipment (pumps, motor control centers, instrumentation cabinets, etc.) should be at least 1 m (3.3 ft). The vertical distance between the tallest equipment (i.e., pump motors) and overhead obstructions should be at least 2.5 m (8.2 ft), or the manufacturer's recommended minimum maintenance clearance plus 0.3 m (1 ft). Clearance around large equipment—such as engines, pump connections larger than 500 mm (20 in.), and motors larger than 400 hp—should be at least 2.5 m (8.3 ft).

7.3.3 Inlet and Suction Well Configuration

The pump inlet to a wet-well or dry-well pump stations should be of the flange and flare elbow type if it is 500 mm (20 in.) or less. Inlets for larger pumps should be of the draft tube type, cast in the suction floor slab. The inlet should be sized on the basis of a maximum velocity of 1 m/s (3.3 ft/s). It is recommended that inlet suction piping be designed for a maximum velocity of 1.5 m/s (5 ft/s).

To prevent air pockets, suction piping should be leveled or sloped slightly upwards toward the intake pumps. It should be as short as possible but not shorter than its own diameter. Also to avoid air pockets, an eccentric side-down reducer should be used in reducing the suction piping to the suction opening of the pump and where operating on suction lift.

A gate valve should be installed on the suction piping to isolate the individual pump from the well. The valve stem can be installed horizontally to avoid air pockets. Butterfly valves should not be used on intake suction piping.

The suction well configuration (depth, width, distance between pump suction headers, submergence, etc.) for wet-well intake pump stations is determined by the pump size, type, and number. In general, suction wells should be designed such that they distribute flow evenly to all pumps in the well and have a uniform nonvortex flow pattern. In vertical pump wet-well systems, key design criteria—such as submergence of the suction head, distance from the back wall, bay width, length of approach floor channel, and floor clearance—are typically related to the diameter of the intake pump bell (Bhrends et al., 2011).

Suction bell velocity is one of the most critical parameters associated with pump performance and prevention of vortex formation in the wet well. For pumps with up to 4.6 m (15 ft) of head, the suction bell velocity should be held up to 0.8 m/s (2.5 ft/s); for pumps with up to 15 m (50 ft) of head, bell velocity should be below 1.2 m/s (4 ft/s); and above 15 m of head, the bell velocity can be up to 1.7 m/s (5.5 ft/s). Since desalination intake pumps are designed similarly to pumps for other water treatment applications, other references can be used for more comprehensive information on this topic (Karassik et al., 1986; Stewart 1986; Hicks and Edwards, 1987; Hydraulic Institute, 1994 and 1998; Nemdili and Hellmann, 1999).

Discharge Piping

If the discharge pipe is relatively short (less than 20 pipe diameters), the diameter of the pipe and the pump discharge nozzle should be the same. If the discharge piping is longer, its size should be increased by one diameter for lengths of up to 300 m (1000 ft) and by two diameters for longer distances.

A check valve and a gate or butterfly valve should be installed on the discharge piping; the check valve should be located between the pump and the butterfly or gate valve. The check valve protects the pump from excessive back pressure and prevents the intake flow from being conveyed backward through the pump when the pump is shut down. Gate or butterfly valves are used to isolate pumps for maintenance.

The intake flow meters most widely used for seawater applications are magnetic. They are installed on a straight pipe run that is at least 10 pipe diameters long in the upstream direction from the flow meters and 5 pipe diameters downstream. Some meters can operate at shorter distances; the manufacturer should be consulted to confirm the meter configuration and location.

7.3.4 Surge Analysis and Protection

If the pump station has a design flow rate greater than 5500 m^3/day (1000 gal/min) or a discharge pipe diameter greater than 250 mm (10 in.), or the distance between the intake pump station and the pretreatment facilities is longer than 300 m (1000 ft), the completion of a surge analysis is recommended in order to evaluate the potential for hydraulic transients and their negative impact on the source water delivery pipeline, pumps, and other equipment. Check valves, surge tanks, rupture disks, and air release/vacuum relief valves should be installed accordingly.

Surge control valves are installed to prevent buildup of surge pressure. They consist of hydraulically operated valves on the discharge, complete with a speed control device to allow independent timing of the closing and opening speeds. The valve controller has hydraulic and safety equipment wired to function in sequence with the pump motor starting gear. Pressure relief valves are typically diaphragm-activated globe valves or angle valves that are installed on the discharge piping for flow control and pressure regulation.

7.3.5 Corrosion Protection

The submerged portion of metal intake structures can be protected against corrosion by employing a cathodic corrosion protection system. Typically the corrosion protection system contains a sacrificial anode which is consumed instead of the equipment to which it is connected. In order to minimize corrosion and its negative impacts on intake operations, intake pump parts in direct contact with the saline water (i.e., pump casting, bell mouth, shaft, and impeller) are made of super duplex stainless steel for seawater installations and duplex stainless steel for high-salinity brackish desalination plants.

At most desalination plants, glass-reinforced plastic is used for the piping interconnecting the desalination plant intake pumps with the downstream pretreatment facilities. This material is suitable for practically all plant sizes and conditions, except for very high pressures.

Typically, intake pump station wells are concrete structures. However, for smaller plants, lined or coated steel or glass-fused steel could be used instead.

Bypass Connection to Discharge

Many desalination plants have direct connection of the desalination plant intake to the discharge, which allows them to discharge a portion or all of the intake water in case of source water contamination with excessive levels of hydrocarbons or algal bloom solids (Baudish et al., 2011). Installation of such bypass is strongly recommended for operational flexibility and accelerated plant commissioning.

7.3.6 Water Quality Monitoring and Controls

Monitoring and control of the intake pump station is typically incorporated into the plant's system control and data acquisition system, and includes a sonic level sensor, wet-well level indicator, pump controller, flow meter and flow recorder, control valves, temperature sensors and indicators, and oil and grease sensors.

Intake Pump Flow Metering

Source water intake pumps are equipped with magnetic flow meters, which measure and record the pumped flow continuously. If the intake flow is discontinued for any reason, including non-routine facility operations, the flow rate monitoring system triggers automatic shutdown of the intake pumps.

Intake Water Temperature Measurement

Desalination plant intake pump stations are typically equipped with instrumentation for continuous measurement of the intake water temperature. Any fluctuations of the intake temperature outside preset normal limits defined in the reverse osmosis (RO) or ultrafiltration pretreatment membrane warranties (whichever is lower) triggers the alarm and ultimately shutdown. This monitoring equipment provides additional protection against unusual intake water quality conditions.

Intake Water Salinity and Turbidity Measurement

Intake pump stations of desalination plants are usually equipped with instrumentation for continuous measurement of the intake water salinity. Any fluctuations of intake water salinity and/or turbidity outside preset normal operational minimum and maximum limits triggers an alarm. This monitoring equipment provides additional protection against unusual freshwater or surface water streams entering the desalination plant intake.

Intake Water Oil Spill and Leak Detection Monitoring

Intake pump stations of desalination plants with open intakes are usually equipped with instrumentation for oil spill and leak detection. In most plants, if the monitoring system detects total hydrocarbons in the intake water at concentrations higher than 0.04 mg/L, it activates an alarm to warn plant operations staff about the potential for oil contamination in the source water. Hydrocarbon levels above 0.1 mg/L usually automatically trigger intake pump shutdown. This monitoring equipment provides additional protection against unusual intake water quality conditions. Depending on the type of pretreatment system available upstream of the desalination membranes (e.g., dissolved air flotation systems), sometimes the maximum trigger concentration for hydrocarbons in the source water can be set at levels of up to 1 mg/L.

Wet-Well Level Monitoring

Intake wells of desalination plants are typically provided with transmitters to indicate the water level, with low-level switches. This level-monitoring instrumentation protects the intake pumps from running dry and overheating.

7.4 Chemical Feed Systems

7.4.1 Sodium Hypochlorite Feed System

Most desalination plants are equipped with chemical feed systems for adding sodium hypochlorite to the desalinated water. For desalination plants with offshore open intakes, sodium hypochlorite is usually injected at the intake structure in order to suppress biological growth in the pipeline connecting the intake structure with the onshore intake pump station. In addition, sodium hypochlorite is added to minimize biological growth in the pretreatment and RO systems and the interconnecting channels. The dosage of sodium hypochlorite in the source water is set so that the injected chlorine is consumed almost completely by the time the water exits the plant pretreatment system. Chapter 9 provides additional discussion of the use of sodium hypochlorite for biofouling control.

Seawater contains large numbers of marine organisms such as shellfish (e.g., clams, mussels, barnacles) and bacteria that tend to attach to and grow on the inner walls of the intake pipelines of desalination plants. Excessive biogrowth over time results in increased pipeline head loss and ultimately in reduced intake capacity. Practical experience shows that addition of sodium hypochlorite to the intake inlet structure and piping reduces but does not completely inhibit biogrowth on the pipe walls. Therefore, at present intake pipelines of most full-scale desalination plants are taken out of service once every 12 to 18 months for cleaning.

Typically, pipeline biogrowth is controlled by periodic "pigging," i.e., removal of marine species, sediments, and scale from the pipe walls with an abrasive scraping device ("pig")—made of plastic, light metal, or ice. These pipeline cleaning systems are often designed to to pig the pipeline in either direction. The pig, propelled by pressurized flow, removes and conveys the pipe debris to the well of the intake pump station. Full-scale pipeline cleaning usually takes 1 or 2 days per pipe—a downtime during which the desalination plant operates at reduced capacity or is completely shut down. Besides pigging, intake pipeline cleaning is accomplished manually by divers which enter the intake pipeline through manholes located every 200 to 500 meters along the intake length and remove scale, sediments and marine growth from the pipeline walls.

7.4.2 Sulfuric Acid Feed System

Since chlorination alone is not likely to completely eliminate shellfish growth in the intake, an innovative approach that can be used to achieve improved cleaning combines periodic use of sulfuric acid and sodium hypochlorite to minimize the attachment of shellfish to the walls of the intake pipelines.

This chemical biofouling control strategy is based on the fact that the colonization of the intake pipeline walls by shellfish is a gradual process over several weeks and that the binding and attachment process of shellfish plankton to the microbial biofilm and pipe walls is accomplished by calcium-rich compounds produced by the shellfish that can be dissolved at low pH. The colonization of the intake pipe walls is completed in two steps: (1) a microbial biofilm forms on the surface of the pipeline, and (2) shellfish plankton attaches to the biofilm by forming bonds between calcium and extracellular organics released by the shellfish and grows into fully developed organisms that feed on the organic matter conveyed with the intake water.

Sulfuric acid (or any other acid) that reduces the pH of the intake seawater to 4 or less can dissolve the calcium-organic bonds between the shellfish, biofilm, and pipe walls, thereby preventing shellfish from permanently attaching to the pipe surface. The addition of sodium hypochlorite suppresses the formation of bacterial biofilm on the pipeline walls and the growth of shellfish plankton and adult organisms.

Under this biogrowth management approach, sulfuric acid and sodium hypochlorite are added downstream of the point where the intake inlet structure connects to the intake pipeline. The sodium hypochlorite line has to be flushed after chemical addition, and then the same pipe can be used to feed the sulfuric acid. Alternatively, a separate line can be installed to add sulfuric acid in the same location.

Only one chemical (sodium hypochlorite or sulfuric acid) should be added at a time. The chemicals are delivered at the following dosages, frequencies, and durations:

Sodium hypochlorite:

- Dosage (summer, winter) = 6 mg/L, 4 mg/L
- Frequency (summer, winter) = every 48 h, weekly
- Duration = (summer, winter) = 6 h, 4 h

Sulfuric acid:

- Dosage (average, maximum) = 100 mg/L, 140 mg/L to the amount needed to reach a pH of 4

- Frequency (summer, winter) = weekly, biweekly
- Duration (summer, winter) = 6 h, 4 h

Sodium hypochlorite and sulfuric acid are added within 2 h of each other. The dosage of sulfuric acid is determined experimentally and is based on the goal of reducing the pH of the source seawater to 4 or less. At this pH, the organic bonds of the shellfish to the pipeline walls are typically dissolved. The periodic short-time pH reduction below 4 is expected to also have a beneficial impact on the RO membrane elements, because it loosens biofilm, iron residue, and scale attached to them.

Besides sodium hypochlorite and sulfuric acid, another chemical that may occasionally be added at the intake pump station is powdered activated carbon. This chemical is not routinely used for source water conditioning; in some desalination plants it is provided for potential addition only under conditions of large oil spills or heavy algal blooms.

7.5 Intake Pump Station Construction Costs

A cost graph for intake pump stations with wet-well and dry-well configurations is presented in Fig. 7.3. Construction costs are depicted as a function of the desalination plant intake flow. The cost estimate does not include the expenditures for intake screens,

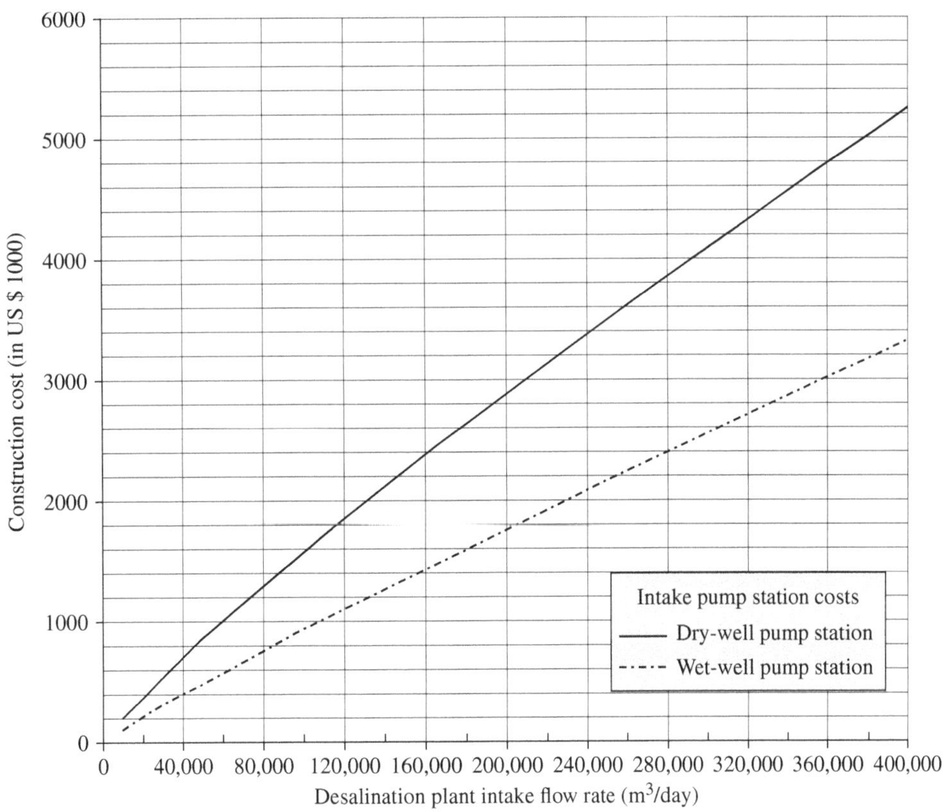

FIGURE 7.3 Intake pump station costs.

intake structure, or piping interconnecting the pump station and the downstream pre-treatment facilities.

7.6 References

Baudish, P. A., H. A. Lavery, R. N. Burch, D. D. Pain, D. G. Franklin, and P. J. Banks, "Design Considerations and Interactions for Tunnelled Seawater Intake and Brine Outfall Systems," IDA World Congress, Perth, Western Australia, September 4–9, 2011. Ref: IDAWC/PER11-242.

Bhrends, B., G. Witte, R. C. van Niekerk, and P. Springer, *Optimization of Intake Stations as a System Function*, IDA World Congress, Perth, Western Australia, September 4–9, 2011. Ref: IDAWC/PER11-086.

Hydraulic Institute, *American National Standards for Centrifugal Pumps*, 15th ed., Hydraulic Institute Parsippany, NJ, 1994.

Hydraulic Institute, *American National Standard for Pump Intake Design, ANSI/HI 9.8-1998*, Hydraulic Institute Parsippany, NJ, 1998.

Hicks, T. G., and T. W. Edwards, *Pump Application Engineering*, McGraw-Hill, New York, 1971.

Karassik, I. J., W. C. Krutzch, H. W. Fraser, and J. P. Messina, *Pump Handbook*, 2nd ed., McGraw-Hill, New York, 1986.

Nemdili, A., and D.-H. Hellmann, "The Requirements to Successful Centrifugal Pump Application for Desalination and Power Plant Processes," *Desalination*, 126: 199–205, 1999.

Stewart, H. L., *Pumps*, 2nd ed., Macmillan, New York, 1986.

CHAPTER 8

Source Water Screening

8.1 Introduction

Screening facilities are the first treatment step of every desalination plant. Depending on the type of intake and pretreatment, screening may be as simple as cartridge filtration or as sophisticated as a series of mechanical screens, designed to sequentially remove large debris and marine organisms, and microscreens, designed to retain silt, plankton, sand, shell particles, and other solid debris in the saline source water. The main purpose of the screens is to protect the downstream pretreatment or reverse osmosis (RO) facilities from equipment and structure damage, accelerated filter media clogging and fouling, and reduction of product water treatment capacity.

Open ocean intakes are typically equipped with coarse bar screens followed by smaller-size (fine) screens with openings of 1 to 10 mm that prevent the majority of adult and juvenile aquatic organisms (fish, crabs, etc.) from entering the desalination plant. While coarse screens are always stationary, fine screens can be of two types—stationary (passive) or periodically moving (i.e., rotating) screens.

Most aquatic organisms collected with the source water used for production of desalinated water are removed by screening and downstream filtration before the saline source water enters the reverse osmosis desalination membranes for salt separation. After screening, the water is typically processed by finer membrane or granular media pretreatment filters.

8.2 Bar, Band, and Drum Screens

A typical surface water intake system for a medium or large membrane desalination plant with open intake includes a set of manually or mechanically cleaned bar racks followed by automated traveling fine bar screens or fine mesh band or drum screens.

8.2.1 Coarse Bar Screens (Bar Racks)

Bar racks usually have a distance of 50 to 300 mm (2 to 12 in.) between the bars; their purpose is to prevent large debris and aquatic life from entering the plant intake. For offshore intakes, the screens are installed on the intake's vertical inlet tower (Fig. 8.1). The design flow-through velocity for clean screens is typically 3 to 4 cm/s (0.10 to 0.15 ft/s). Such low design velocity is selected not only to minimize impingement of aquatic life on the screens but also to account for the loss of flow-through surface as a result of shellfish growth and debris accumulation on the surface of the coarse bars. It is recommended that the screen bars be manufactured of super duplex stainless

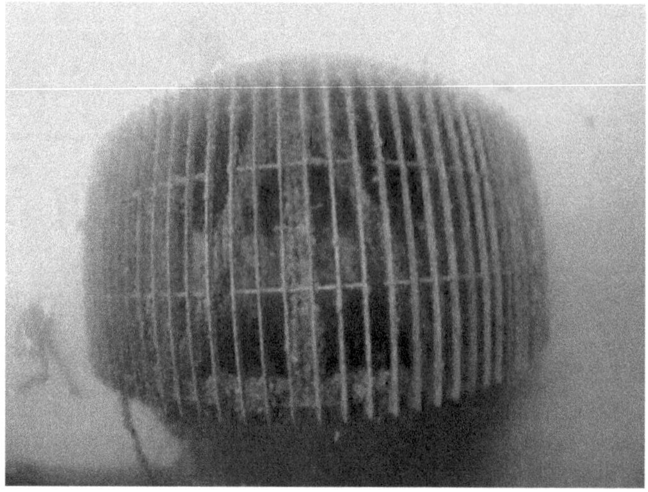

Figure 8.1 Coarse bar screen of open offshore intake.

steel or copper–nickel alloys (the latter are preferred) in order to suppress marine growth. Shellfish growth on the screens can narrow the open space between the bars by over 50 percent; as a result, either the bars have to be cleaned manually every several years.

Onshore bar screens (Fig. 8.2) are usually equipped with an automated raking mechanism that periodically removes the accumulated debris and allows maintenance

Figure 8.2 Coarse screen for onshore intake. (*Source*: Atlas.)

of the open space between the bars at approximately the same flow-through velocity over time. Therefore, the design flow-through velocity of these screens is typically twice that of the coarse screens of open intakes, i.e., 8 cm/s (0.3 ft/s) rather than 4 cm/s (0.15 ft/s). As discussed in Chap. 5, it is important to maintain the intake through-screen velocity below 15 cm/s (0.5 ft/s) at all times in order to minimize impingement of aquatic organisms on the screens.

8.2.2 Fine Screens

Rotating Screens

Fine self-cleaning rotating screens typically have 3- to 10-mm openings. They are installed vertically in water intake channels downstream of the coarse screens and are equipped with rotating cleaning equipment, often combined with water-spraying nozzles to remove the debris from the screen surface. These nozzles are supplied with cleaning water by pumps sized for a flow of 45 to 68 m^3/h (200 to 300 gal/min) and a pressure of 4 to 7 bar (60 to 100 lb/in^2). Because one of the main functions of the fine screens is to protect the intake pumps from damage, the actual screen openings should be smaller than the distance between the intake pump impellers.

Two types of rotating fine screens that have found wide application in desalination plants are band and drum screens. Typically, band screens are installed at small and medium plants, whereas drum screens have found wider implementation for some of the largest desalination facilities worldwide. Fine bar screens are sometimes also used in open intakes, but they are not as common as band and drum screens.

Band Screens These vertical travelling screens consist of individual screening panels with fine mesh openings that are attached on support roller chains, which in turn are installed on metal-framed guide tracks (Fig. 8.3). This figure depicts bar rack (course screen) followed by fine band screen.

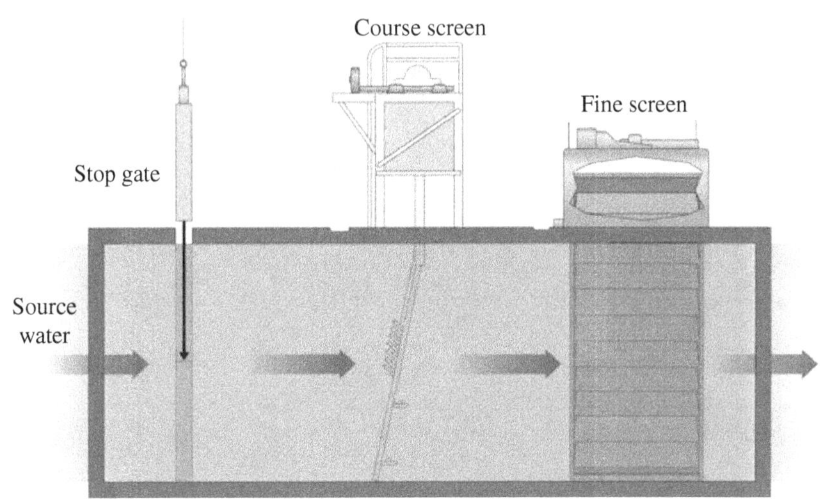

Figure 8.3 Intake structure with bar rack followed by band screen. (*Source: Traveni.*)

As the intake source water travels through the screens, the debris contained in the water is removed and accumulated on the screen panels. The screen panel mesh is typically made of polyamide, polyester, super duplex stainless steel for seawater applications, or duplex stainless steel for brackish water intakes. The debris accumulation causes a gradual increase in screen headloss. Once the headloss reaches a preset level (or after a preset time), the screening panels are rotated upward and the debris collected on the panels is moved to the deck level, where it is removed from the panels into collection troughs by low-pressure water sprays. The debris is either conveyed to collection bins and disposed of as solid waste or recycled back to the source water body.

Typically, the screens are designed to enter into a cleaning cycle at a water elevation differential of 0.1 to 0.2 m, which usually corresponds to approximately 30 percent reduction of the screening area. Most commercially available band screens travel at velocities of 2 to 10 m/min (6.6 to 33 ft/min). These screens are typically designed for two-speed or variable-speed operation. The design screen area efficiency factor for band screens is usually 0.5 to 0.6 (i.e., 50 to 60 percent of the screening area is active filtration area used to determine through-screen velocity). These fine screens are designed to maintain through-screen velocity below 15 cm/s (0.5 ft/s) at all times, and normally operate at velocities of 6 to 10 cm/s (0.2 to 0.3 ft/s).

Band screens can be configured in three flow patterns—through-flow, center-flow, and dual-flow (Fig. 8.4; Rogers, 2009).

While the through-flow pattern is commonly used, its main disadvantage is that if the backwash spray does not effectively remove the screenings from the surface of the

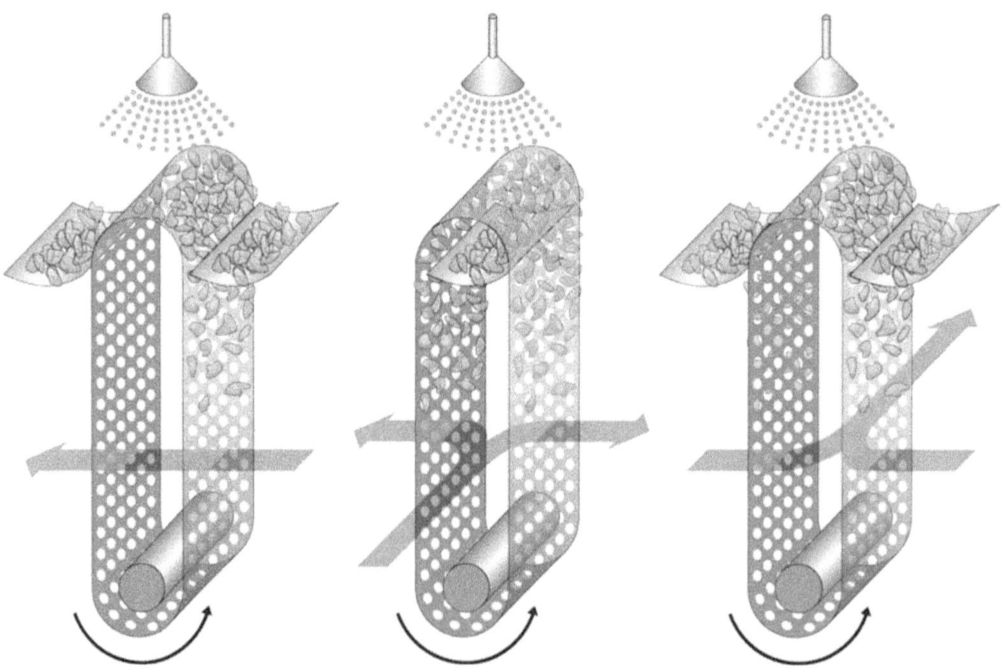

Figure 8.4 Through-flow, center-flow, and dual-flow screen patterns. (*Source: Ovivo.*)

panels, these screenings will be conveyed to the back (clean) side of the screen and released into the screened water.

In the center-flow configuration, the screen panels are oriented in parallel with the flow—the water enters in the space between the two screens and passes out through the screens at both sides simultaneously. This configuration allows retention of all debris on the inner side of the screens, thereby addressing the main challenge associated with the through-flow pattern. The key disadvantage of this configuration is that it produces divergent turbulent flow, which is not favorable for the intake pump hydraulics. In the dual-flow pattern, the feed water enters from the outer sides of both screens and is collected inside the screens, which improves channel flow hydraulics while retaining solids only on the outer side of the screens. Despite the benefits of the center-flow and dual-flow screen patterns, the most widely used configuration is the flow-through pattern, mainly because of the higher construction and installation complexity of the other two.

Newer installations are more commonly designed in a dual-flow in-to-out configuration. An example of the application of such screens is the 300,000 m^3/day (79 mgd). Adelaide desalination plant, where three units, each capable of treating 100 percent of the intake flow of 624,600 m^3/day (165 mgd), are installed in individual channels. The screens have an effective width of 2.8 m (9.2 ft), mesh openings of 3 mm (0.1 in.), and 50 mesh panels. The screen material is super duplex stainless steel.

Fine mesh screens are modified band screens that use finer screening panels—with screen openings of 0.5 to 1 mm (0.02 to 0.04 in.)—and are equipped with buckets that allow the capture of fish and some other aquatic organisms. Such screens have been found to significantly reduce entrainment of aquatic organisms (see Chap. 5). They sometimes are installed downstream of the conventional band screens, as in the intake at Tampa Electric's power plant, which also serves as a cold water intake for the Tampa Bay seawater desalination plant. The fish and other organisms captured in the screen buckets are conveyed to the source water through a low-pressure, low-speed pump system.

Besides fine mesh screens, there are other modified traveling band screens that have been specifically designed to reduce impingement and entrainment of marine species. Such screening technologies are discussed in greater detail elsewhere (Mackey et al., 2011).

Drum Screens Drum screens (Fig. 8.5) have found wide application for intakes of large seawater desalination plants in Australia, the Middle East, and Europe. For example, the Sydney Desalination Plant and the Gold Coast Desalination Plant in Australia are equipped with drum screens.

These screens consist of a rotating cylindrical frame covered with wire-mesh fabric. The frame is located in a screen structure and supported by a horizontal center shaft that rotates slowly on roller bearings. The screens are rotated by a drive located at shaft level.

The most commonly used water pattern for such screens is in-to-out (also referred to as *double-entry*), in which the source water enters the inner side of the cylinder and moves radially outward, creating a hydraulically beneficial converging flow pattern. Debris deposited on the inner surface of the screen is removed by a water spray jet located on the top of the screen and is collected in a water trough (Fig. 8.6).

Drum screens are available in unit capacities of up to 3,000,000 m^3/day (270 mgd). Similar to band screens, they are also available in single-entry, double-exit out-to-in,

Figure 8.5 Drum screen of the Sydney Desalination Plant.

and in-to-out configurations, as well as a double-entry, single-exit (out-to-in) flow pattern (Rogers, 2009).

Drum screens are more advantageous for applications where the source water debris and materials may fluctuate significantly, because they are less susceptible to overtorque from a large influx of solids to the screen over a short period of time, which can occur in onshore open intakes or shallow offshore intakes. In addition, drum screens typically create lower flow-through head loss at the same flow.

Besides hydraulic loading, drum and band screens are also frequently designed based on the solids load—especially jellyfish outbreaks, when the amount of these marine organisms in the water cans exceed 300 tons/h. In the case of jellyfish outbreaks, they can completely blind the screening surface of the screens, and their removal is very cumbersome. From this perspective, manual removal (scrubbing) of jellyfish from the screens is usually easier with band screens than with drum screens. If drum screens, however, are designed to handle large debris or jellyfish loads, they can perform with minimum maintenance during jellyfish outbreak episodes.

One important difference between drum screens and band screens is that the latter have an overall footprint that is approximately 30 to 50 percent smaller footprint. Therefore, if space is at a premium, band screens may be a preferred option. In addition, band screens are usually 30 to 40 percent less costly. However, they typically have higher maintenance (mainly equipment) costs.

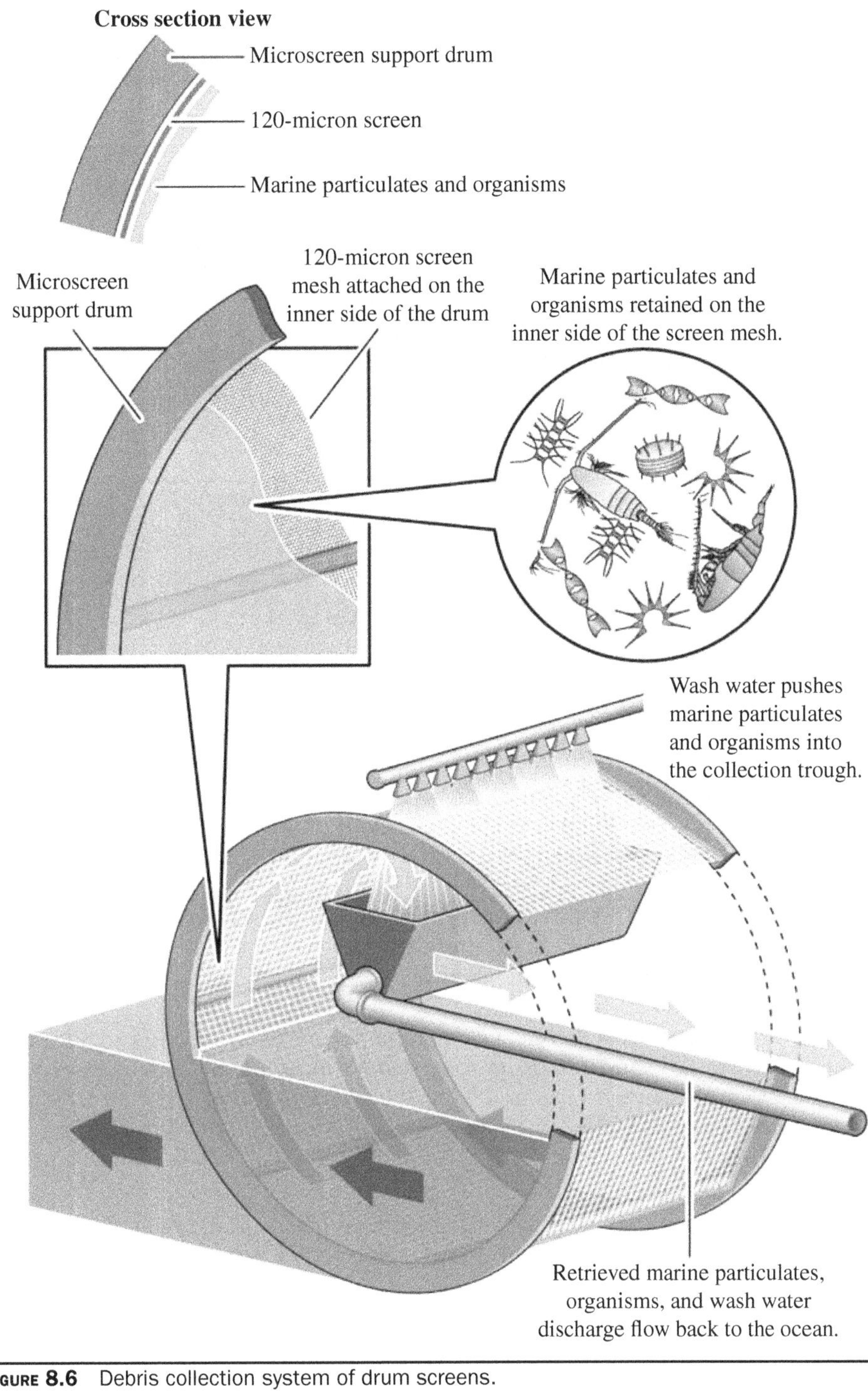

FIGURE 8.6 Debris collection system of drum screens.

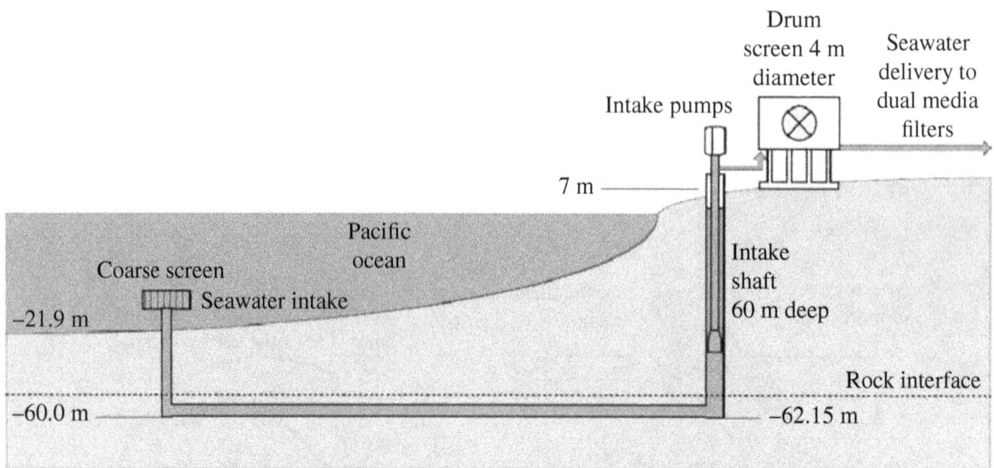

Figure 8.7 Schematic of the Gold Coast Desalination Plant's intake system. (*Source: SKM.*)

Figure 8.7 depicts a general schematic of the intake system of the 170,000 m^3/day (45 mgd) Gold Coast Desalination Plant (Baudish et al., 2011). This intake has shafts 67 m (221 ft) deep connecting the offshore intake riser to the plant intake pump station. The intake structure has a diameter of 5.8 m (19 ft), a height of 4.4 m (47.2 ft), and screens that are 2 m (6.6 ft) high with 140-mm (5.5-in.) bars made of copper-nickel alloy.

The intake entrance point is approximately 4 m (13.1 ft) from the ocean bottom and 1400 m (0.85 mi) offshore. The total depth of the intake structure is 14 m (46 ft). The maximum intake flow rate is 5 cm/s (0.16 ft/s). The intake pumps are vertical turbines and are located on the top of the intake shaft. The plant has one duty and one standby 3-mm mesh wire drum screen, located in an elevated aboveground structure (Fig. 8.8), which allows the water from the screens to be fed by gravity to the plant's single-stage dual media gravity filters.

Costs of Band and Drum Screens The graph presented in Fig. 8.9 provides budgetary cost estimates for band and drum screens as a function of the desalination plant's feed water flow. As shown in this graph, band screens are generally less costly than drum screens for projects of the same size.

Wedgewire Screens

Wedgewire screens are passive screens (with no mechanical moving parts) located offshore that are directly connected to the suction end of the intake pump piping, thereby eliminating the need for additional coarse and fine screening facilities. The desalination plant with the largest wedgewire screen intake in operation at present is the 150,000 m^3/day (40 mgd) Beckton plant in the United Kingdom (Fig. 8.10). The wedgewire screens are located 3 m (10 ft) above the bottom. They are made of copper–nickel alloy and have 3-mm (1/8-in.) openings. The through-screen flow velocity is 0.15 m/s (0.5 ft/s); Moore et al., 2009.

FIGURE 8.8 Drum screen intake structure of the Gold Coast Desalination Plant.

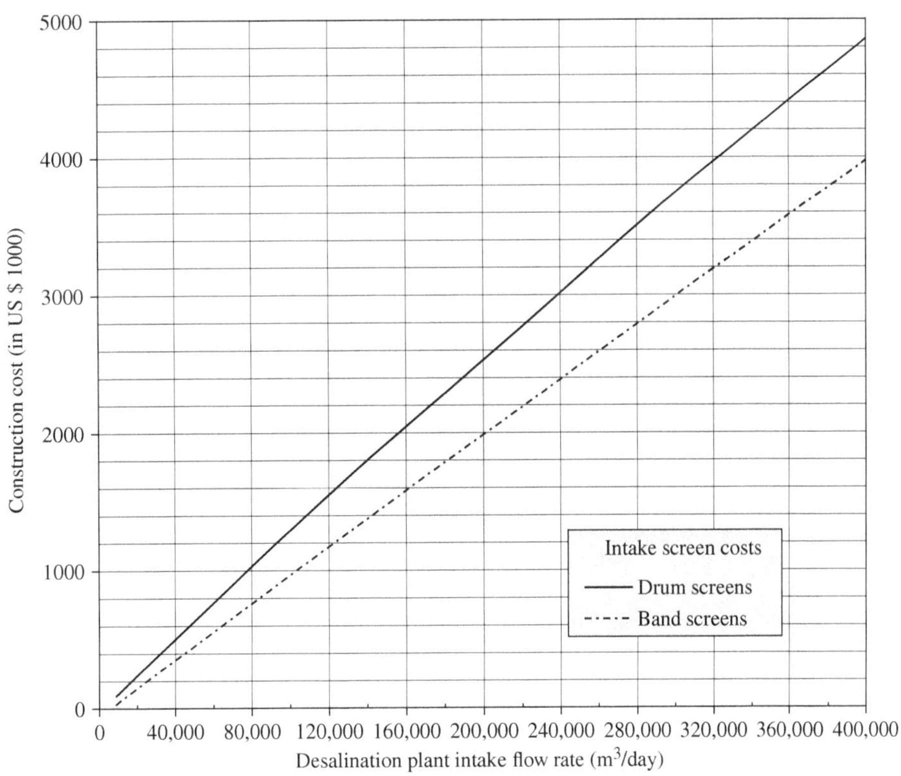

FIGURE 8.9 Drum and band screen costs.

Figure 8.10 Wedgewire screens of the Beckton desalination plant. (*Source: Ovivo.*)

Wedgewire screens are cylindrical metal screens with trapezoidal "wedgewire" slots that have openings of 0.5 to 10 mm. The screen size most commonly used for desalination plants at present is 3 mm.

Wedgewire screens combine very low flow-through velocities—10 to 15 cm/s (0.3 to 0.5 ft/s)—small slot size, and naturally occurring high sweeping velocities at the screen surface to minimize impingement and entrainment. These screens are designed to be placed in a water body where significant prevailing ambient cross-flow current velocities exist, at least 0.3 m/s (1 ft/s). These high cross-flow velocities allow organisms that would otherwise be impinged on the wedgewire intake to be carried away with the flow. Therefore, wedgewire screens are considered by the US EPA to be a best technology available for impingement and entrainment reduction (see Chap. 5).

An integral part of a typical wedgewire screen system is an airburst back-flush system, which directs a charge of compressed air to each screen unit to blow off debris back into the water body, where it is carried away by the ambient cross-flow currents.

The screens need to be installed a minimum of 1 m (3.3 ft) from the bottom to avoid the entrance of sand and silt into them. If the intake is located in a relatively shallow, tidally influenced area, the depth from the bottom should be increased to a minimum of 2 m (6.6 ft).

Typically the material used for such screens is 90/10 copper-nickel alloy, super duplex stainless steel, or titanium. Copper-nickel alloy usually offers the optimum combination of reasonable costs and resistance to corrosion and erosion. Figure 8.11 presents a graph indicating the costs of wedgewire screens as a function of the plant's source water flow.

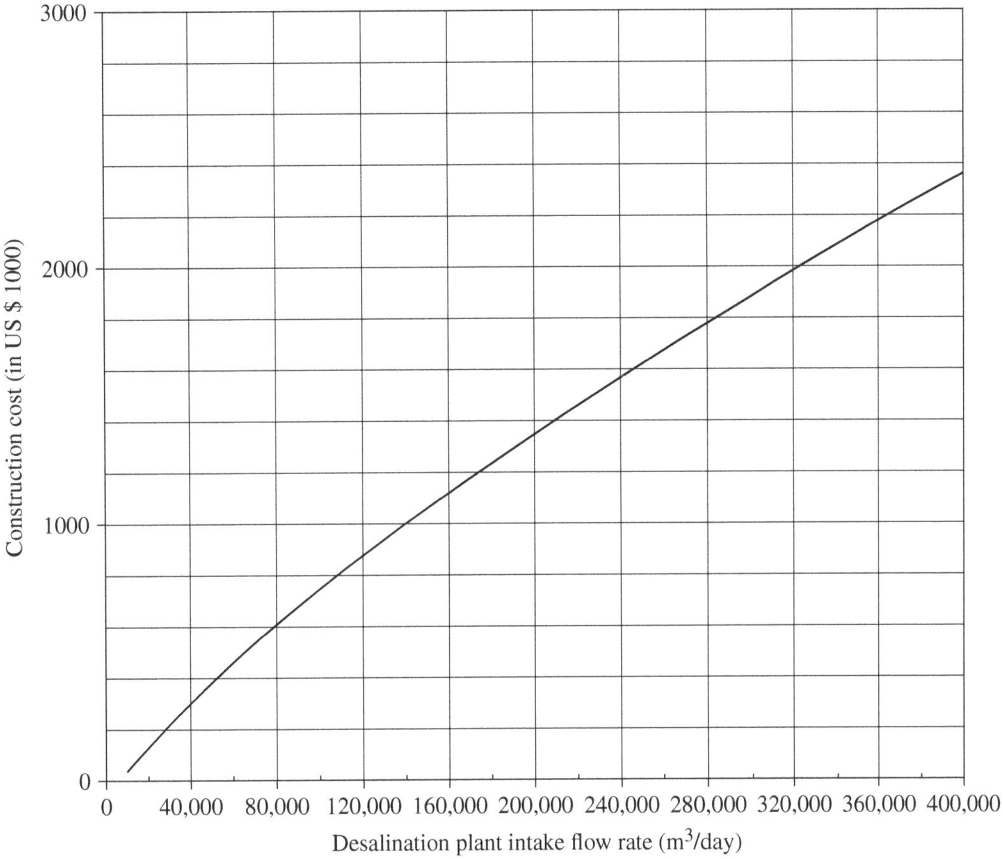

FIGURE 8.11 Wedgewire screen costs.

8.3 Microscreens

8.3.1 Types and Configurations

If the filtration pretreatment system selected for the RO desalination plant is of the membrane type, the fine screens described in the previous sections do not provide sufficient removal of fine source water particles to protect the integrity of the membrane pretreatment system. Typically, microscreens—microstrainers (Fig. 8.12) or disk filters (Fig. 8.13)—can be used for this application.

Most microstrainers consist of screen with small openings (80 to 400 microns) located inside a filtration chamber. The source water enters on the inner side of the strainer, moves radially outward through the screen, and exits through the outlet. The gradual buildup of solids on the inner surface of the screen creates a filter cake that increases the differential pressure between intake and filtered water over time. When the differential pressure reaches a preset value, the deposited solids are removed by a jet of backwash water. The self-cleaning process typically takes 30 to 40 s.

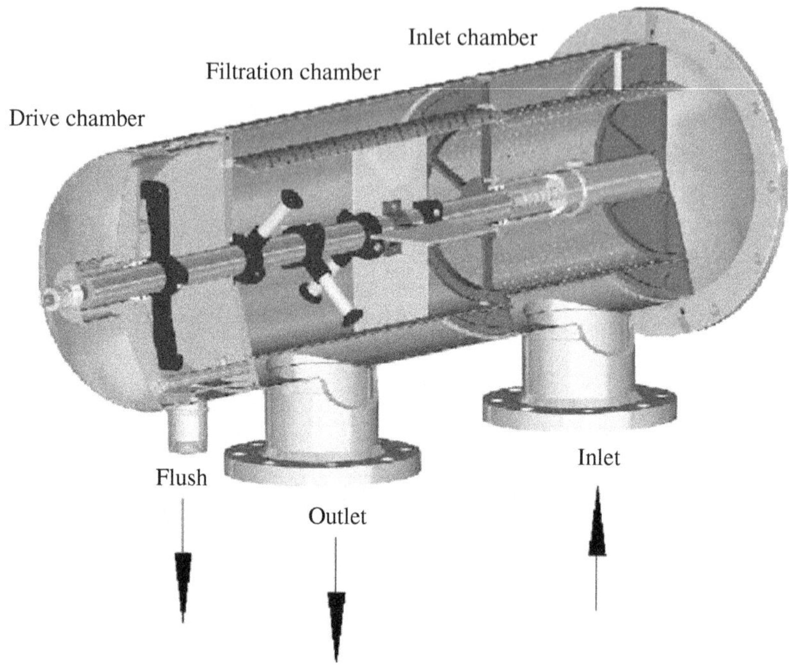

Figure 8.12 Self-cleaning microstrainer.

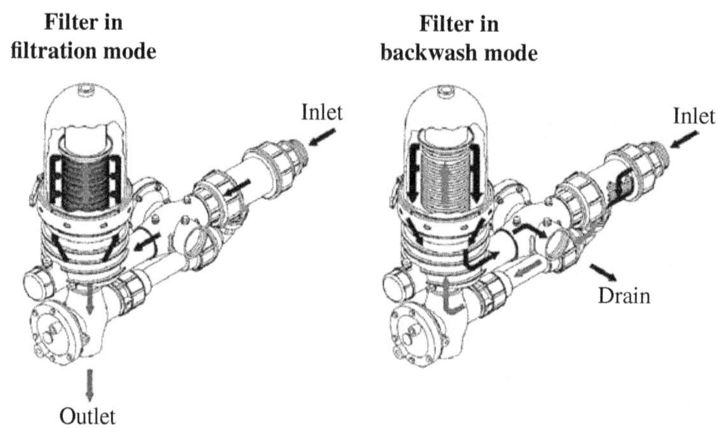

Figure 8.13 Disk filter modes of operation.

Disk filters are equipped with polypropylene disks that are diagonally grooved on both sides to a specific micrometer size. A series of these disks are stacked and compressed on a specially designed spine. The groove on the top of a disk runs opposite to the groove below, creating a filtration element with a series of valleys and traps for source water debris. The stack is enclosed in corrosion- and pressure-resistant housing.

Desalination plants with disk filter microscreens are usually also equipped with conventional coarse screens or a combination of coarse and fine screens that retain debris larger than 10 mm (0.4 in.). During the filtration process, the filtration discs are tightly compressed together, thus providing high filtration efficiency. Filtration occurs while water is percolating from the peripheral end to the core of the element (Fig. 8.13). Source water debris and aquatic organisms (mainly plankton) smaller than the size of the microscreens (80 to 120 μm) are retained and accumulated in the cavity between the filter disks and the outer shell of the filters, thereby increasing the head loss through the filters. Once the filter head loss reaches a preset level, typically 0.35 bar (5 lb/in^2) or less, the filters enter backwash mode. All debris retained on the outer side of the filters is then flushed by tangential water jets of filtered water flow under 0.15 to 0.2 bar (2 to 3 lb/in^2) of pressure; the flush water is directed to a pipe that returns the debris and aquatic organisms retained on the filters back to the surface water body from which the source water originated.

Because of the relatively low differential pressure these filters operate at, they are likely to minimize impingement of marine organisms in the source water. Since the disk filtration system could be equipped with an organism return pipe, aquatic organisms could be returned to the source water body, thereby reducing their entrainment.

One of the key issues associated with using membrane pretreatment is that the UF and MF membrane fibers can be punctured by sharp objects contained in the source water, such as broken shells or sharp sand particles. In addition, source water may contain barnacles, which in their embryonic phase of development are only 130 to 150 μm in size and can pass through the screen openings unless those openings are 120 μm or smaller.

If barnacle plankton passes through the screens, it could attach to the walls of downstream pretreatment facilities, grow on these walls, and ultimately interfere with pretreatment system operations. Once barnacles establish colonies in the pretreatment facilities and equipment, they are very difficult to remove; they can withstand chlorination, which is a very effective biocide for most other marine organisms. Therefore, the use of fine microscreens or disk filters (80 to 120 μm) is essential for reliable operation of desalination plants using membrane pretreatment. Microscreens are not needed for pretreatment systems using granular media filtration, because these systems effectively remove fine particulates and barnacles in all phases of their development.

8.3.2 Design Example

Disk filters have found wide application as microscreens for source seawater prior to membrane pretreatment. The following example illustrates the application of a disk filter microscreen for a 40,000 m^3/day (10.6 mgd) seawater RO desalination plant designed for 43 percent system recovery and a total plant seawater intake flow of 98,440 m^3/day (26 mgd).

Manufacturer	Arkal (or equal)
Model	Galaxy 6″ Spin Klin
Unit disk filter capacity	4320 m^3/day (1.14 mgd)
Number of arrays	2
Number of disk filter units per array	12
Number of disks (spines) per disk filter	8

Number of jets per spine	48
Filter size	100 μm
Array inlet and Outlet piping diameter	500 mm (20 in.)
Inlet and outlet diameter of disk filters	150 mm (6 in.)
Pressure loss (after filter cleaning)	0.15 bar (2.1 lb/in^2)
Pressure loss triggering backwash	0.3 bar (4.2 lb/in^2)
Average pressure loss during operation	0.22 bar
Number of filters washed at one time	12 (one array)
Backwash flow	16 m^3/min per array (4,230 gpm)
Backwash cycle length	6 min
Backwash frequency	14 washes per day
Total backwash volume	0.3 to 0.5 percent of intake flow
Backwash pumps (horizontal centrifugal)	1 duty and 1 standby
Backwash pump head	4 bar (58 lb/in^2)
Disk filter material	Polypropylene

This example is developed for a particular type of popular disk filter (Spin Klin, manufactured by Arkal, Israel; see Fig. 8.13). These filters have found application at a number of desalination plants with membrane pretreatment, such as the 300,000 m^3/day (79 mgd) Adelaide desalination plant and the Southern Seawater Desalination Plant (Perth II) in Australia, the 100,000 m^3/day (26 mgd) Chennai SWRO plant in India, and other plants in the Middle East and Europe.

A number of other manufacturers provide similar equipment, although the specific unit sizes and design criteria vary (Gille, 2003). Equipment manufacturers should be consulted to identify the microscreen system's design criteria for a specific desalination project.

8.3.3 Microscreen Costs

Figure 8.14 depicts a budgetary cost graph for microscreen systems as a function of the intake flow they are designed to process. These costs are presented in year 2012 US dollars.

8.4 Cartridge Filters

8.4.1 Types and Configurations

Cartridge filters are fine microfilters of nominal size from 1 to 25 μm made of thin plastic fibers or other fine filtration media that is installed around a central tube to form standard-size cartridges (Fig. 8.15). Often they are the only screening device between the intake wells and the RO system in brackish and seawater desalination plants with well intakes producing high-quality source water. Cartridge filters are RO membrane protection facilities rather than screening devices; the main purpose they serve is to capture particulates in the pretreated source water that may have passed through the

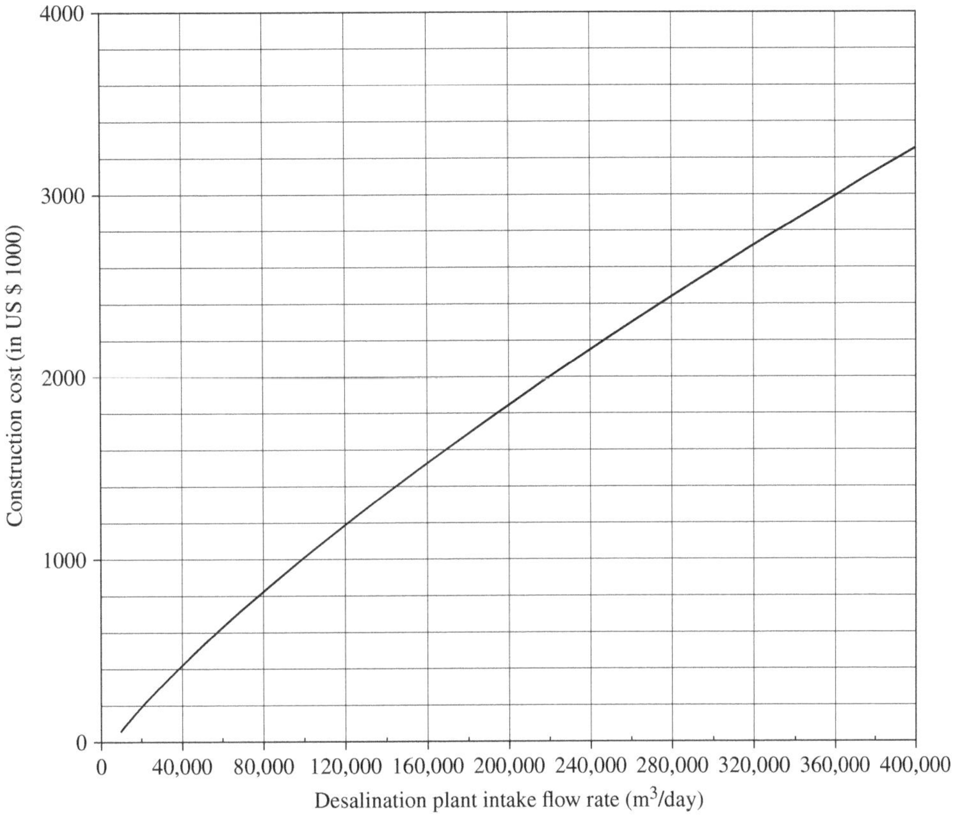

Figure 8.14 Microscreen costs.

Figure 8.15 Cartridge filters installed in a horizontal vessel.

upstream pretreatment systems in order to prevent damage or premature fouling of the RO membranes.

Although wound (spun) polypropylene cartridges are most commonly used for seawater and brackish water applications, other types, such as melt-blown or pleated cartridges of other materials have also found application. Standard cartridge filters for RO desalination plants are typically 101.6 to 1524 cm (40 to 60 in.) long and are installed in horizontal or vertical pressure vessels (filter housings). Cartridges are rated for removal of particles of 1, 2, 5, 10, or 25 μm, with the most frequently used size being 5 μm.

8.4.2 Planning and Design Considerations

Cartridge filters are typically installed downstream of the granular media filtration system (if such a system is used for pretreatment) to capture fine sand, particles, and silt that may be contained in the pretreated water. When the source seawater is of very high quality—a silt density index (SDI) below 2—and does not need particulate removal by filtration prior to desalination, cartridge filters are used as the only pretreatment device, in this case serving as a barrier to capture fine silt and particulates that can occasionally enter the source water during the start-up of intake well pumps or due to failure of intake equipment or piping.

A typical indication of whether the pretreatment system of a given desalination plant operates properly is the SDI reduction through the cartridge filters. If the pretreatment system performs well, then the SDI of the source water upstream and downstream of the cartridge filters is approximately the same.

If the cartridge filters consistently reduce the SDI of the filtered source water by over 1 unit, this means that the upstream pretreatment system is not functioning properly. Sometimes the SDI of the source water increases when it passes through the cartridge filters. This almost always occurs because the cartridge filters have not been designed properly or are malfunctioning and providing conditions for growth of biofouling microorganisms on and within the filters.

A frequently debated question is whether cartridge filters are needed downstream of MF or UF membrane pretreatment systems, taking into consideration that the cartridge filter pores are one to two orders of magnitude larger than those of the membrane filters. The answer to this question is highly dependent on the quality of the pretreatment membrane's fiber material and the type of flow pattern through the pretreatment system.

For UF or MF filtration systems that have a direct flow-through pattern—where the desalination plant feed pumps convey water directly through the membrane pretreatment system without an interim pumping—the pretreatment membranes are more likely to be exposed to pressure surges. If the fiber material of the pretreatment membranes is weak and breaks easily under pressure surge conditions, the pretreatment system is more likely to experience fiber breaks. Broken membrane fibers will release small amounts of particles into the RO feed water, which could cause accelerated membrane fouling unless it is captured by cartridge filtration.

In addition, if the broken membrane fibers release sharp particles contained in the source water, these particles could also damage the RO membranes. Sharp broken-shell particles may find their way into the UF or MF pretreated water if shellfish plankton contained in the source water passes through the microscreens, grows to adult shellfish organisms (e.g., barnacles) on the walls of the pretreatment system feed pump station, and releases portions of shells that have been broken into small, sharp particles by the feed pumps. The shell particles will be pressurized onto the UF/MF membrane filter fibers, causing punctures and ultimately entering the filtered flow. In such cases, the use of cartridge filters downstream of the membrane pretreatment system is a prudent engineering practice.

Cartridge filters are operated under pressure, and the differential pressure across them is monitored to aid in determining when filter cartridges should be replaced. In addition, valved sample ports should be installed immediately upstream and downstream of the cartridge filter vessels for water quality sampling and monitoring (including SDI field testing).

Cartridge filtration systems are designed for hydraulic loading rates of 0.2 to 0.3 L/s per 250 mm (3 to 5 gal/min per 10 in.) of length. Additional filtration capacity is normally provided to allow replacement of cartridges without interruption of water production. Pressure vessels are typically constructed of duplex stainless steel for seawater RO installations.

The pressure drop across a clean cartridge filter is usually specified as less than 0.2 bar (2.8 lb/in^2). Commonly, cartridges are replaced when the filter differential pressure reaches 0.7 to 1.0 bar (10.1 to 14.5 lb/in^2). The operational time before replacement depends on the source water quality and the degree of pretreatment. Typically, a cartridge filter replacement is needed once every 6 to 8 weeks. However, if the source seawater is of very good quality cartridge filters may not need replacement for 6 months or more.

For RO systems where sand in the feed water might be anticipated, rigid melt-blown cartridges or cartridge filters with single open ends and dual O-rings on the insertion nipple (rather than conventional cartridges with dual open ends) are commonly used. The single-open-end insertion filters have positive seating and an insertion plate, which does not allow deformation of the filter cartridge under pressure caused by sand packing. Double-open-end cartridge filters are held in place by a spring-loaded pressure plate.

8.4.3 Design Example

This example presents the sizing and configuration of the cartridge filtration system for a 40,000 m^3/day (10.6 mgd) seawater desalination plant with a total plant seawater intake flow of 98,440 m^3/day (26 mgd).

Design feed flow, Q_{in}	98,440 m^3/day = 1140 L/s
Cartridge filter material	Pleated polypropylene
Cartridge filter size	5 μm
Cartridge filter length, L_{cf}	1016 mm (40 in.)
Selected design loading rate, DLR	0.25 L/s per 250 mm
Number of cartridge filters needed	Q_{in}/[DLR × (L_{cf}/250)] (8.1) = 1140/[0.25 × (1016/250)] = 1122
Number of cartridge vessels	6 (selected to match RO trains)
Cartridge vessel material	Glass-reinforced plastic
Number of cartridges per vessel	1122/6 = 187 (selected 180)
Actual cartridge filter loading rate	1140/[180 × 6 × (1016/250)] = 0.26 L/s per 250 mm (4.2 gal/min per 10 in.)

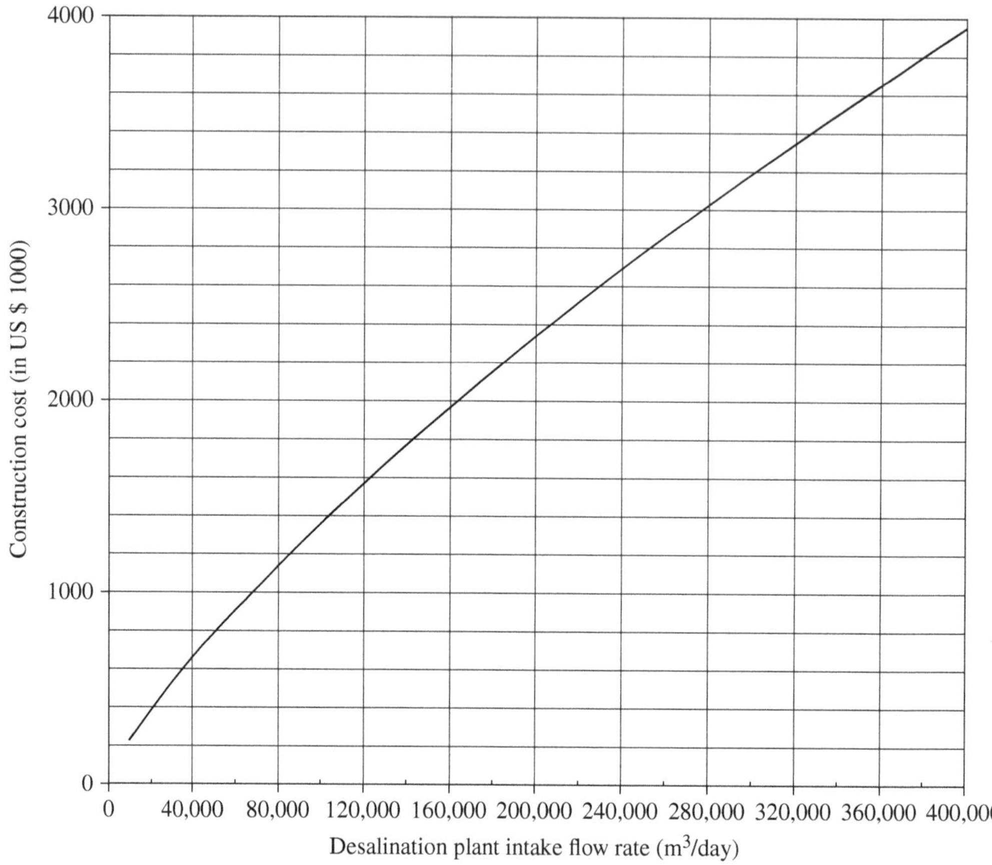

FIGURE 8.16 Cartridge filter system costs.

In summary, the cartridge filtration system for the 40,000 m^3/day (10.6 mgd) desalination plant will consist of six cartridge vessels, each of which will contain 180 cartridge filters of size 5 μm and length 40 in.

8.4.4 Cartridge Filter System Costs

A budgetary cost graph of cartridge system construction costs is presented in Fig. 8.16. Actual desalination project costs may vary in a range of 30 percent above or below the values presented in the figure.

8.5 References

Baudish, P. A., H. A. Lavery, R. N. Burch, D. D. Pain, D. G. Franklin, and P. J. Banks, "Design Considerations and Interactions for Tunnelled Seawater Intake and Brine Outfall Systems," IDA World Congress, Perth, Western Australia, September 4–9, 2011. Ref: IDAWC/PER11-242.

Gille, D., "Seawater Intakes for Desalination Plants," *Desalination*, 156: 249–256, 2003.

Mackey E. D., N. Pozos, W. James, T. Seacord, H. Hunt, and D. L. Mayer, *Assessing Seawater Intake Systems for Desalination Plants: Joint Report of the Water Research Foundation, WateReuse Research Foundation, the US Bureau of Reclamation and the California Department of Water Resources*, WateReuse Research Foundation, Alexandria, Virginia, Place, 2011.

Moore B. J., J. J. Malfeito, and J. Zorrilla, "An Overview of the 150,000 m^3/day Beckton Desalination Plant in London," IDA World Congress, Atlantis, the Palm-Dubai, United Arab Emirates, November 7–12, 2009. Ref: IDAWC/DB09-089.

Rogers, N., "A Review of Onshore Screening Technologies for Fine Screening of Large Intakes," IDA World Congress, Atlantis, the Palm-Dubai, United Arab Emirates, November 7–12, 2009. Ref: IDAWC/DB09-070.

CHAPTER 9

Source Water Conditioning

9.1 Introduction

In order to reduce its fouling potential, saline source water is conditioned prior to reverse osmosis (RO) separation using various chemicals: coagulants, flocculants, scale inhibitors, oxidants (i.e., chlorine, chlorine dioxide, etc.), and oxidant reduction compounds (i.e., sodium bisulfite and sulfuric acid). Coagulants and flocculants are added to enhance removal of particulate and colloidal foulants in saline water pretreatment facilities. Scale inhibitors are introduced in the saline source water after pretreatment filtration to suppress crystallization of mineral-scaling foulants on the surface of the RO membranes.

Oxidants (typically sodium hypochlorite or chlorine dioxide) are fed to the source water to minimize pretreatment and RO membrane biofouling and excessive growth of aquatic organisms (e.g., shellfish) on the inner surface of the intake piping, equipment, and structures. Sodium bisulfite or other reducing chemicals are added to the pretreated source water to remove residual chlorine and/or other oxidants prior to the water's introduction into the RO membranes.

9.2 Coagulation

Coagulant addition is accomplished ahead of the pretreatment sedimentation tanks, dissolved air flotation units, or filters. The coagulants most frequently used for membrane plant source water conditioning prior to sedimentation or filtration are ferric salts (ferric sulfate and ferric chloride). Aluminum salts (such as alum or polyaluminum chloride) are not typically used, because it is difficult to maintain aluminum concentrations at low levels in dissolved form, and small amounts of aluminum may cause irreversible mineral fouling of the downstream RO membrane elements.

The optimum coagulant dosage depends on the pH and should be established based on an on-site jar or pilot test for the site-specific conditions of a given application. Practical experience indicates that the optimum pH for coagulation of particles in saline waters is highly temperature dependent. As the temperature decreases, the optimum pH for coagulation increases, and vice versa. For example, the optimum pH at a temperature of 10°C (50°F) is 8.2, while at a source water temperature of 35°C (95°F), the optimum pH decreases to 7.4 (Edzwald and Haarhoff, 2012).

The use of coagulant is critical for effective and consistent performance of granular media pretreatment filtration systems. However, it should be pointed out that the needed amount of coagulant is dependent on the size and charge of the particles dominating in the source water. Coagulation allows the granular filtration process to also remove finer particulate debris and microplankton from the source water. Properly operating filters can remove particles as small as 0.5 μm. However, if the source water contains low turbidity (usually < 0.5 NTU) and the prevailing size of particles is less than 5 μm (which is common for deep intakes with low algal content), coagulant addition does not yield a significant improvement in the granular media filtration process. In this case, the addition of a minimal amount of coagulant (i.e., 0.5 mg/L or less) or even no coagulant addition at all is viable. In such conditions, however, it is critical to have a prolonged period of coagulation and flocculation (i.e., coagulation and flocculation times of 10 min or more), because for these particles the main mechanism for floc formation is physical contact rather than charge attraction.

Coagulation is critical for source waters of high turbidity, especially if that turbidity is caused by surface runoff (e.g., rain events, river water, or wastewater discharge influence) and resuspension of bottom sediments (e.g., frequent boat traffic, dredging of the source water area, periodic strong currents near the intake, or strong wind events in shallow intake areas). In this case, a rule of thumb is to add a coagulant dosage that is approximately 2 times higher than the source water turbidity.

Membrane pretreatment can remove particles as fine as 0.04 μm (microfiltration membranes) or 0.01 μm (ultrafiltration membranes) without coagulation. Therefore, for these systems coagulation is typically applied when the saline source water contains particles of natural organic matter with a high negative charge that can be coagulated easily and removed via filtration, when heavy algal blooms or oil spill events occur.

9.2.1 Types of Coagulation Chemicals and Feed Systems

This chemical conditioning of source water includes three key components: the chemical feed system, coagulation tanks, and flocculation tanks. The purpose of coagulation tanks is to achieve an accelerated mixing of the coagulant and the source water and to neutralize the electrical charge of the source water particles and colloids. The subsequent agglomeration of the coagulated particles into larger, easy-to-remove flocs is completed in flocculation tanks.

It should be noted that flocculation tanks are always installed downstream of the coagulation tanks, independent of whether additional flocculant chemicals are fed to the source water. While coagulation is a relatively rapid chemical reaction, flocculation is a much slower process and typically requires longer contact time and mixing conditions. Therefore, design requirements for coagulation and flocculation systems differ.

The main purpose of the coagulant feed system is to achieve uniform mixing of the added coagulant with the source water, which promotes accelerated attraction of the coagulant particles to the solid particles in the source water (i.e., to facilitate efficient coagulation). The two types of coagulant mixing systems most widely used in desalination plants are in-line static mixers (Fig. 9.1) and mechanical (flash) mixers installed in coagulation tanks (Fig. 9.2).

In-line static mixers have lower energy and maintenance requirements and are relatively easy to install. They typically operate at a velocity range of 0.3 to 2.4 m/s (1 to 8 ft/s) and are designed to run in a plug-flow hydraulic mode in order to provide

Figure 9.1 In-line static mixer.

Figure 9.2 Flash mixers in coagulation tank.

uniform mixing within the entire pipe cross section. The velocity gradient × contact time for such mixers can be determined by the formula

$$G \times T = (1212 \times d) \times \left(D_p \times \frac{L}{Q} \right)^{0.5} \tag{9.1}$$

where G = velocity gradient, sec^{-1}
T = time, sec
d = diameter of static mixer, in.
D_p = differential pressure through mixer, lb/in^2
L = length of static mixer, in.
Q = flow, gpm

Although in-line static mixers are simple to install and significantly less costly, they have two disadvantages: (1) their mixing efficiency is a function of the flow rate, because the mixing energy originates from the flow turbulence; and (2) they are proprietary equipment—a project designer must rely on the equipment manufacturer for performance projections. Static mixers also create additional head losses of 0.3 to 1 m (1 to 3.3 ft) that need to be accounted for in the design of the intake pump station. Another important issue is provision of an adequate length of pipeline (at least 20 times the pipe diameter) between the static mixer and the entrance to the pretreatment filters in order to achieve adequate flocculation.

Mechanical flash mixing systems consist of coagulation tank with one or more mechanical mixers and chambers. The coagulation tank is designed for a velocity gradient multiplied by time G × T = 4000 to 6000. The power requirement for the mechanical mixers is 2.2 to 2.5 hp per 10,000 m^3/day. This type of mixing usually provides a more reliable and consistent coagulation, especially for desalination plants designed for significant differences in minimum and maximum plant production (i.e., a disparity of more than 1:10).

9.2.2 Planning Considerations

Overdosing of coagulants and their inadequate mixing with the source water are some of the most frequent causes of RO membrane mineral fouling. When overdosed, coagulant accumulates on the downstream facilities and can cause fast fouling of the downstream cartridge filters (Fig. 9.3) following the pretreatment step, as well as iron fouling of the RO membranes (Fig. 9.4).

The effect of coagulant (iron salt) overdosing on the silt density index (SDI) can be recognized by visually inspecting the SDI test filter paper. In Fig. 9.5, the first two SDI test pads are discolored as a result of coagulant overdosing. The numbers below the pads are the SDI readings.

In such situations, a significant improvement of source water SDI can be attained by reducing the coagulant feed dosage or, in a case of poor mixing, modifying the coagulant mixing system to eliminate the content of unreacted chemical in the filtered seawater fed to the RO membrane system.

9.2.3 Design Example

In-Line Static Mixer

Applying Eq. 9.1 for a desalination plant with a freshwater production capacity of 40,000 m^3/day (10.6 mgd), an intake flow of 98,440 m^3/day (26 mgd = 18,044 gal/min),

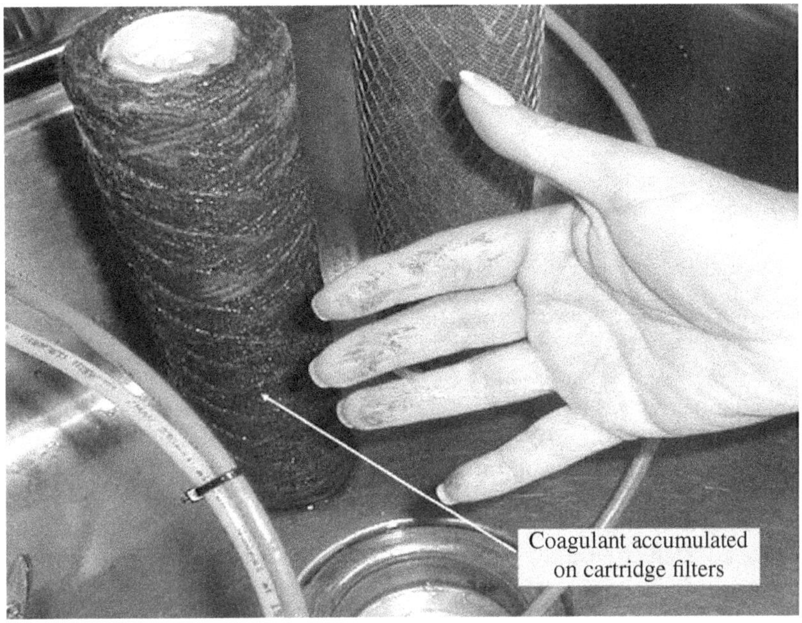

Figure 9.3 Coagulant accumulation on cartridge filters due to overdosing.

Figure 9.4 Coagulant residue on the RO membrane feed due to overdosing.

16.2 5 min. | 16.3 5 min. | 3.7 15 min.

Figure 9.5 Iron accumulation on SDI test pads due to coagulant overdosing.

a 30-in. mixer with a length of 5 ft, and differential pressure through the mixer of 0.5 lb/in^2, the G × T = (1212 × 30) x[0.5 × (5 × 12)/18,044]$^{1/2}$ = 1482. The optimum G × T mixing range is usually between 500 and 1600.

Coagulation Chamber with Flash Mixers

The key design criteria for such a flocculation system are listed here:

Number of coagulation tanks	4
Tank width × length × depth	1.2 m × 1.2 m × 1.4 m (3.9 ft × 3.9 ft × 4.6 ft)
Tank volume, *V*	2.0 m^3 (21.5 ft^3)
Contact time, *T*	(2.0 m^3 × 4 tanks × 24 h × 60 min)/98,440 m^3/day = 0.12 min = 7 s
Mixing energy per tank at 2.5 hp per 10,000 m^3/day, W	8.2 hp = 6120 W
Absolute viscosity of water, μ_a	0.00114 N·s/m^3
Velocity gradient multiplied by time, G × T	$W/(\mu_a \times V)^{0.5} \times T = [6120/(0.00114 \times 2.0 \times 4)]^{0.5} \times 7 = 5734$
Type of mixer	Vertical shaft with hydrofoil blades
Blade area as a percentage of tank area	0.15 percent
Shaft speed	40 to 80 r/min

It is important to point out that the velocity gradient multiplied by time G × T provided by the coagulation chambers (5734) is significantly higher than that provided by the static mixer (1482) and thus corresponds to a more robust mixing.

9.3 Flocculation

Flocculants (polymers) are sometimes applied in addition to coagulants to improve seawater pretreatment. However, polymer overdosing, even if only minimal, may also cause organic fouling on the RO membranes. Often, the potential for RO membrane fouling due to polymer overdosing is more significant than the benefit of polymer use. Therefore, many desalination plants do not condition the coagulated source water with polymers. If polymers are used, only nonionic or anionic polymers are usually applied, because most RO membrane elements carry a negative surface charge. If overdosed, cationic polymer is likely to form thin film on the membrane surface, and foul the RO membrane elements.

9.3.1 Types of Flocculation Chemicals and Feed Systems

The type and dosage of polymer (nonionic or anionic) that is most suitable for a given application should be determined by jar and/or pilot testing. Typically, polymer is added at a very low dosage (0.25 to 0.5 mg/L). Polymer dosages higher than 1 mg/L should be avoided, because they usually result in a high content of unused polymer in the filter effluent, which in turn plugs the cartridge filters and deposits on the membrane elements, thereby shortening the useful life of the cartridge filter and expediting the need for RO membrane cleaning.

9.3.2 Planning and Design Considerations

The formation of large flocs that can be removed easily by the downstream sedimentation, dissolved air flotation, or filtration processes is a slower process, and therefore it requires a longer retention time than coagulation. Most widely used flocculation systems in saline water pretreatment are mechanical flocculators with vertical mixers.

Key Design Criteria

The key design criteria for such flocculation systems are listed here:

Minimum number of tanks	4
Velocity gradient	30 to 120 s^{-1}
Contact time	10 to 40 min
Number of flocculation chambers in series	2 to 4
Water depth	3.5 to 4.5 m
Blade area as a percentage of tank area	0.1 to 0.2 percent
Shaft speed	2 to 6 r/min

9.3.3 Design Example

An example of a flocculation tank for a 40,000 m^3/day (10.6 mgd) seawater desalination plant, which is a part of dissolved air flotation clarifier system, is presented in Chap. 10.

9.4 Scale Inhibitors

The formation of mineral deposits (scaling) on the surface of the RO membranes is caused by the precipitation of low-solubility salts such as calcium carbonate, magnesium carbonate, barium sulfate, strontium sulfate, and silica. The stability of these compounds depends on their concentration in the concentrate flow stream, the water temperature, pH, desalination plant recovery, and other factors.

As plant recovery increases, more salts in the source water are likely to reach the point at which their solubility is exceeded and they begin forming crystals (referred to as the *solubility limit*) and to ultimately accumulate on the membrane surface and cause mineral membrane fouling (scaling). Therefore, for a given saline water mineral composition, RO desalination systems face a threshold of maximum recovery at which the scaling process destabilizes membrane performance due to excessive accumulation of salt crystals or amorphous scale on the membrane surface.

Determination of the scaling potential of saline source waters is rather complex, and therefore it usually is performed using computer software. Such software is available from key suppliers of antiscalants, such as Nalco, Avista, Genesys, etc. In their software, chemical manufacturers relate antiscalant requirements to the scaling potential of each salt that can be formed in the concentrate in enough quantity to create measurable scaling. The antiscalant dosages are then determined for each scale-forming mineral, and recommendations are made based on the salt that would require largest amount of antiscalant. It is important to point out that the results of this type of software are very sensitive to the accuracy and completeness of the source water quality analysis. Sometimes parameters such as barium, strontium, and fluoride are not measured in the source water, and assumed concentrations are used instead. Such practices may often cause the projection results to be inaccurate. General steps to determine the limiting salts and to manually calculate allowable permeate recovery for saline waters of given mineral content are illustrated elsewhere (American Water Works Association, 2007).

El-Manharawy and Hafez (2001) studied the use of molar ratio of sulfates and bicarbonates as a tool to predict the scaling potential of saline source waters. Table 9.1 summarizes their results. The table is indicative of the fact that in typical open ocean seawater, the main cause of scaling is sulfate, while in low-salinity brackish water, the predominant type of scaling is caused by carbonate salts. This does not, however, mean that seawaters cannot cause carbonate scale—it means that carbonate scale formation is at a relatively low rate and that typically not more than 10 percent (by weight) of the scale observed on seawater reverse osmosis (SWRO) membranes is contributed by calcium carbonate.

One of the key factors associated with the scaling potential of saline source waters is their ionic strength. Typically, the higher the ionic strength (i.e., concentration of total dissolved solids) of the source water, the higher the recovery threshold at which scaling would occur at the same temperature and mineral composition. Another important factor associated with the scaling potential of the source water is temperature—usually the scaling potential of calcium carbonate increases with an increase in source water temperature.

Silica, especially if it is in colloidal sate, is a compound that sometimes creates scaling challenges in brackish waters. Usually silica is considered to be of concern when its concentration in the RO concentrate exceeds 140 mg/L (Wilf et al., 2007). In typical

Molar Ratio of SO_4/HCO_3	Chloride Concentration, mg/L	Sulfate Scaling Potential	Carbonate Scaling Potential
> 15	≥ 20,000	High	Low
10–15	≥ 10,000	Medium to high	Medium
1–10	≥ 3000	Medium	Medium to high
< 1	< 3000	Low	High

Table 9.1 Sulfate-to-Bicarbonate Molar Ratios and Scaling Potential

open ocean seawaters, silica is usually below 20 mg/L and is not considered a compound of high scaling potential, because SWRO systems typically operate at relatively low recoveries (40 to 50 percent).

Barium and strontium concentrations in source waters correlate highly with their concentrations observed in membrane scales. Usually these compounds are at levels in seawater that are too low to make them a significant source of scale. However, in some brackish waters they can be at levels an order of magnitude higher than in seawater and may result in scaling, especially in brackish water plants operating at high recoveries.

Scales vary in texture and appearance; typically, calcium sulfate scales formed on RO membranes treating seawater have an orderly, prismatic, crystalline structure. Sulfate crystals can reach a length of 20 mm and width of 5 mm (El-Manharawy & Hafez, 2001). However, sulfate scale from high-salinity brackish waters can vary significantly from one water source to another. On the other hand, carbonate scales formed on brackish water membranes are typically fine, amorphous white deposits.

Scaling control depends on the particular mineral salts that precipitate on the membrane surface. For example, calcium carbonate scales can be prevented from forming by acidification of the source water. Acids convert carbonate ions (CO_3^{2-}) into soluble bicarbonate ions and carbonic acid, and ultimately into carbon dioxide.

For prevention of calcium carbonate and other scaling, commercially available scale inhibitor chemicals (antiscalants) are often added to the source water, or alternatively, scaling foulants are removed by softening or nanofiltration pretreatment facilities located upstream of the RO system.

Some of the compounds naturally contained in seawater (such as humic acids) serve as natural chelating agents and scale inhibitors. Therefore, acidification of seawater prior to membrane salt separation is not usually needed or commonly practiced.

When a high level of boron removal is targeted, seawater acidification is not advised; it will have a negative impact on boron rejection by the RO membranes. In addition, overdose of acid can cause corrosion of piping and equipment and create iron-based colloidal fouling on the RO membranes. Therefore, the benefits associated with acid addition and acid dosage will need to be weighed against the potential problems that acidification can cause.

Often, SWRO systems have to be designed to remove boron to levels below 1 mg/L. In this case, a most common practice for enhanced boron removal is to increase the source seawater's pH to a range between 8.8 and 11. At this high pH range, RO membrane scaling is very likely to occur, and scale inhibitors are therefore typically added to prevent it.

Some scale inhibitors prevent the formation of seed crystals, while others deform the seed crystals so they cannot grow and cause problems in the membrane system. In some cases, dispersants are added to the scale inhibitor formulations to aid in preventing deposition of colloidal material.

It is important to note that antiscalants are designed so that they do not pass through the membranes; they are therefore contained in the concentrate. This is an important issue in terms of their potential environmental impacts and toxicity.

9.4.1 Acids

A sulfuric acid feed upstream of RO membrane systems is commonly used for calcium carbonate control. Calcium carbonate is the most common scaling compound in brackish water (Wilf et al., 2007), and therefore BWRO desalination plants are usually equipped with acid addition systems.

The addition of acid lowers the carbonate concentration by converting bicarbonate to carbon dioxide. The carbon dioxide passes through the RO membranes and is removed or used in the post-treatment system. Sulfuric acid is only effective against carbonate scale, and because of the addition of sulfates with this chemical, it actually increases the calcium sulfate scaling potential of the source water.

Sometimes hydrochloric acid is used as a scale inhibitor instead of sulfuric acid. Sulfuric acid is usually preferred over hydrochloric acid for cost and safety reasons; however, hydrochloric acid may be used if sulfate introduced to the source seawater by the addition of sulfuric acid significantly affects the system design and cost.

9.4.2 Other Scale Inhibitors

Sodium hexametaphosphate has been one of the most commonly used scale inhibitors in the past, but in recent years it has frequently been replaced by proprietary chemical formulations because of their improved effectiveness, long storage life without loss of strength, resistance to microbial growth while in the feed tank, ease of handling, and other reasons. Sodium hexametaphosphate can serve as a bacterial nutrient, and because it contains phosphates, its use or overdose could result in concentrate discharge with high phosphorus content, which in turn could trigger algal blooms in the discharge area. Therefore, the use of this otherwise popular and effective scale inhibitor is limited.

Phosphonates such as aminotrismethylenephosphonic acid, 1-hydroxyethyllidene-1, 1-diphosphonic acid, and 2-phosphonobutane-1,2,4-tricarboxylic acid have found wide application for high-temperature waters. They are usually very suitable for prevention of calcium and barium sulfate scale and for inhibition of calcium carbonate scale formation.

Phosphonates decrease the precipitation rate of salts that have exceeded their solubility thresholds. These antiscalants are particularly efficient for SWRO systems with high pH operation for enhanced boron removal. Besides calcium scale, other important scaling compounds that impact system operations are magnesium carbonate and magnesium hydroxide. Phosphonates also can react with and remove low levels of iron in the source water and inhibit silica fouling.

Polymeric dispersants based on polyacrylic acid and maleic acid (also referred to as polyacrylates) are commonly applied as calcium carbonate scale inhibitors. However, they are incompatible with coagulants used for seawater pretreatment and are therefore not recommended for such applications. They distort the crystalline growth of the scales on the surface of the membranes.

It is important to select the correct scale inhibitor for the specific application. For example, the presence of iron in the source water can cause precipitation and membrane fouling with some types of antiscalants. Scale inhibitor feed systems typically include positive-displacement metering pumps (or centrifugal pumps for large systems) drawing from a day tank or other storage device (for small plants), such as 55-gallon drums or larger-capacity totes.

9.5 Biocides

Oxidants such as sodium hypochlorite and chlorine dioxide are often used to suppress the growth of aquatic organisms (e.g., shellfish, barnacles) on the inner surfaces of intake pipes, equipment, tanks, distribution channels, and other structures in contact with the source seawater, as well as to minimize biofouling of RO membranes.

9.5.1 Sodium Hypochlorite

Sodium hypochlorite (NaClO) is the most commonly used oxidant today. When added to water, sodium hypochlorite generates hypochlorous acid (HClO) and sodium hydroxide (NaOH):

$$NaClO + H_2O \rightarrow HClO$$

Hypochlorous acid in turn dissociates into hydrogen (H^+) and hypochlorite (ClO^-) ions:

$$HClO \leftrightarrow H^+ + ClO^-$$

The sum of sodium hypochlorite, hypochlorous acid, and hypochlorite ions is termed (and measured as) *free residual chlorine*. Chlorine in all of its forms is a toxicant that attacks all aquatic organisms and typically destroys them by oxidation of their tissue and cells. It should be pointed out, however, that the use of chlorine has several drawbacks. Chlorination cannot destroy all forms of biofouling organisms and therefore it is not an absolute barrier to RO membrane biofouling. Chlorine or other oxidants added to the source water will need to be removed before they reach the RO membranes, because they will cause permanent damage to the membranes' polymeric structure.

In addition, chlorine and other oxidants break down otherwise nonbiodegradable natural organic matter into biodegradable organic compounds and destroy the outer walls of bacterial cells. They thereby cause the release of intracellular material into the source water. Since the intracellular material released from algal and bacterial cells as a result of oxidation is rich in easily biodegradable organics, it serves as a food to bacteria that have already colonized the RO membranes or survived the chlorination process.

Long-term exposure to chlorine triggers the production of extracellular polysaccharides or DNA by some of the microorganisms in the source water as a defense mechanism, which in turn protects the biofilm-forming bacteria. As a result, while continuous use of chlorine may have a short-term benefit in controlling RO membrane biofouling, in the long term it usually does not solve this problem. Therefore, it is not recommended.

Intermittent chlorination has been found to be a more efficient method of RO biofouling control than continuous chlorination. In this case, chlorine or another oxidant is fed to the source seawater at very high dosages (usually 3 to 5 mg/L) one to four times per day. Sometimes, shock chlorination is applied less frequently (i.e., only one to three times a week). Since marine organisms are very adaptive to ambient conditions, usually a random schedule of intermittent shock chlorination works better than a preestablished chlorination schedule.

9.5.2 Chlorine Dioxide

Chlorine dioxide is a weaker oxidant than chlorine but it is fairly effective for most aquatic microorganisms while at the same time, it is not as aggressive in terms of RO membrane oxidation. Therefore, if used intermittently and in low dosages (0.2 to 0.5 mg/L) and at low pH, it can be applied without the need for dechlorination—it is weak enough to not cause permanent damage to the RO membranes' polymeric structure.

Recent studies have shown that the feasibility of using chlorine dioxide for biofouling control is pH dependent and that such use may not cause RO membrane degradation, if the pH of the source water is below 8 (Erikson and Dimotsis, 2012). Ambient seawater has a pH in a range of 7.8 to 8.2, and the use of chlorine dioxide without dechlorination may not always be suitable if the pH of the source water is not adjusted.

Reverse osmosis membrane manufacturers differ in their views regarding the use of chlorine dioxide without subsequent dechlorination, and they have to be consulted regarding application dosages and dechlorination if chlorine dioxide is chosen as a biocide.

Because of its short useful life, chlorine dioxide has to be generated at the desalination plant site. However, some of the chlorine dioxide generators available on the market also produce small amounts of chlorine in the form of HClO and ClO^-, which could oxidize the membrane elements over time. Therefore, the chlorine dioxide system selected for a given RO desalination project would have to be equipped with provisions to remove this residual amount of chlorine, or a dechlorination system would need to be installed.

Chlorites are another undesirable site product of the generation of chlorine dioxide. They are carcinogenic and are not removed by the pretreatment process. Therefore, the level of chlorites is recommended to be monitored in the desalinated product water when chlorine dioxide is applied.

9.5.3 Chloramines

Another type of oxidant that has found a wide application for water reclamation plants with RO membrane treatment are chloramines. Chloramines are created by the sequential addition of chlorine and ammonia to the source water, and have been found to be very efficient because they are weak enough not to cause oxidation of the RO membrane film.

Although chloramination is a very common and efficient practice for controlling biofouling of RO membranes that treat wastewater or brackish water with a low bromide content (i.e., bromide concentration below 0.05 mg/L), it is not recommended for desalination applications where the saline source water has high content of bromides, such as seawater.

As compared to wastewater, seawater contains an order of magnitude higher concentration of bromide. When mixed with ammonia, bromide creates bromamines, which are several times stronger oxidants than chloramines and can cause rapid and irreversible damage of the RO membrane elements. Therefore, chloramination is not commonly practiced for seawater desalination applications.

9.5.4 Nonoxidizing Biocides

Most nonoxidizing biocides are proprietary formulas that have toxic effects on marine species; they usually are low-molecular-weight compounds that can penetrate the cell walls of bacteria and inhibit their metabolism and enzymatic system. Experience with nonoxidizing biocides shows that they are efficient only when they are applied in large dosages over a short period of time. Long-term applications at small dosages usually have limited benefit, because most biofilm-forming bacteria can adapt to nonoxidizing biocides over time.

9.6 Dechlorination

Because RO membranes are damaged by exposure to oxidants, such as chlorine, when the saline source water is conditioned with chlorine or other strong oxidants, those oxidants will need to be removed prior to membrane separation. Typically,

RO membranes will degrade irreversibly after an exposure of 200 to 1000 h at a free chlorine dosage of 1 mg/L. Higher chlorine dosages will shorten this time to only several days. Usually, membrane degradation is expedited if the water is alkaline.

In order to protect membrane integrity, residual chlorine or other oxidants that are not consumed by the source water impurities are typically removed by the addition of reducing compounds (oxidant scavengers), which react with the oxidants in the source water and create nonoxidizing side products.

The most commonly applied reduction chemical is sodium metabisulfite ($Na_2S_2O_5$). When introduced to the source water, it creates sodium bisulfite ($NaHSO_3$), which reduces hypochlorous acid to sulfuric acid (H_2SO_4), hydrochloric acid (HCl), and sodium bisulfate (Na_2SO_4)—none of which is an oxidizing compound.

$$Na_2S_2O_5 + H_2O \rightarrow 2NaHSO_3$$

$$2NaHSO_3 + 2HClO \rightarrow H_2SO_4 + 2HCl + Na_2SO_4$$

Approximately 3.0 mg/L of sodium metabisulfite is needed to remove 1.0 mg/L of free chlorine. Typically, the sodium metabisulfite dosage is optimized based on a reading of the oxidation-reduction potential of the source water at the entrance to the RO system trains. This potential should be maintained at less than 200 mV in order to protect the RO membrane integrity.

Sodium metabisulfite is usually introduced to the feed water immediately after cartridge filtration on the suction side of the booster pumps that feed pretreated source water into the RO system's high-pressure feed pumps (Fig. 9.6).

FIGURE 9.6 Point of chemical addition to pretreated water.

Because the reaction between residual chlorine and sodium bisulfite is practically instantaneous, no separate mixing device is needed. Mixing provided by the pretreated water booster pumps and the high-pressure RO feed pumps is typically adequate to achieve practically complete removal of chlorine or other strong oxidants prior to membrane separation.

Overdosing of sodium bisulfite is not recommended for two reasons: (1) after consuming chlorine and other strong oxidants in the source water, this reducing compound will react with oxygen naturally occurring in the water and reduce the content of oxygen in the desalination plant concentrate, which in turn may have a negative impact on the marine environment receiving this concentrate; and (2) sodium bisulfite could serve as food to some of the biofouling bacteria growing on the RO membranes and thereby exacerbate membrane biofouling.

Another compound that can be used as a reducing agent is activated carbon. Activated carbon, however, is more costly, and the reaction is slower; thus it has not found wide use for dechlorination.

9.7 Planning and Design Considerations for Source Water Conditioning

All chemical feed systems for source water conditioning chemicals have two key components—storage and feed solution preparation tanks and chemical feed pumps. The chemicals are stored on-site in buildings or storage areas that allow their safe loading, containment, and handling.

In conditions where the ambient air temperatures may be reduced below freezing for portions of the year or frequently exceed 35°C, the chemical storage and feed facilities are installed in buildings. Otherwise, they are located under a shed providing protection from direct sunlight exposure.

Most chemicals are delivered to the site as liquid solutions, because they are easier to handle and store. Some chemicals, however (e.g., dry ferric chloride, dry ferric sulfate, and polymer), are sometimes delivered in powder form. Chemicals are stored in tanks of materials suitable for their safe containment. Usually, chemical storage tanks are designed for 15 to 30 days of supply.

Prior to their use, chemicals are diluted down from the concentration at which they are delivered to an application concentration, in order to simplify their pumping and mixing with the source water. Dilution can be completed either in-line or in a batch mode. In the second case, the diluted chemicals are stored in separate tanks often referred to as *day tanks* because they typically store one day of the needed volume of chemical at its application concentration. Both the chemical storage tanks and the day tanks are typically equipped with ultrasonic level transducers to monitor storage level. The chemicals are delivered at application solution to the point of their injection using diaphragm-type metering pumps. These pumps usually have an adjustable diaphragm positioner or stroke rate that allows control over the dosage of the delivered chemical.

9.7.1 Properties of Commonly Used Source Water Conditioning Chemicals

Table 9.2 provides a summary of the key characteristics of most commonly used saline source water conditioning chemicals. The typical product concentration in this table is the concentration of most commonly available commercial products. Users should

Chemical	Typical Application	Typical Product Concentration, %	Bulk Density, kg/L	Application Concentration, %
Liquid ferric chloride	Coagulation	40	1.42	5
Liquid ferric sulfate	Coagulation	40	1.55	5
Sulfuric acid	pH adjustment	98	1.83	20
Sodium hypochlorite	Biogrowth control	13	1.23	5
Sodium bisulfite	Dechlorination	99	1.48	20
Antiscalant	Scale control	99	1.0	20
Sodium hydroxide	pH adjustment	50	1.525	20

Table 9.2 Properties of Commonly Used Conditioning Chemicals

consult the chemical supplier for the properties of the specific product they are purchasing.

9.7.2 Example Calculations

This section presents the calculations associated with the design of the ferric chloride storage and feed system for a 40,000 m^3/day (10.6 mgd) seawater desalination plant designed for an intake flow of 98,440 m^3/day (26 mgd) and an average and maximum feed dosage of 15 mg/L and 50 mg/L, respectively.

Chemical Use

The daily amount of the needed chemical in kg/day is calculated using Eq. 9.2:

$$Q_{dc} = [\text{Concentration (mg/L)} \times \text{Flow (m}^3\text{/day)}]/1000 \tag{9.2}$$

For this specific example, the daily average and maximum chemical use are

$$\text{Qavg}_{dc} = (15 \times 98{,}440)/1000 = 1477 \text{ kg/day}$$

$$\text{Qmax}_{dc} = (50 \times 98{,}440)/1000 = 4922 \text{ kg/day}$$

Chemical Storage Tanks

The daily chemical use of 1477 kg/day is at 100 percent chemical concentration. Because the actual commercial product of liquid ferric chloride is delivered at 40 percent concentration, then the amount of chemical that will need to be stored on-site for 30 days is

$$\text{Ast} = (\text{Average daily amount/Storage concentration}) \times \text{Storage time} \tag{9.3}$$

$$= (1477/0.4) \times 30 = 110{,}775 \text{ kg of 40\% liquid ferric chloride for 30 days}$$

Taking into consideration that the bulk density D_d of this chemical is 1.42 kg/L = 1420 kg/m^3, the actual storage volume is

$$\text{Vst} = \text{Ast}/D_d = 110{,}775/1420 = 78 \text{ m}^3$$

Typically, the actual storage tank volume is selected to be 10 to 15 percent larger than necessary, in order to allow venting space or free board at the top and sediment accumulation at the bottom, which are inactive storage areas. As a result, the actual tank storage would be 78 × 1.15 = 90 m³. Assuming three individual storage tanks with a diameter of 3 m (10 ft) each, the depth of each tank will be (90 m³/3 tanks)/[π × (3/2)²] = 4.3 m (14 ft).

Water Dilution Flow

The average dilution flow (in L/h) needed to reduce the chemical concentration from its delivery concentration C_d to its application concentration C_a can be calculated by the following formula:

$$Qd_{avg} = Qavg_{dc} \times [(C_d/C_a) - (1/D_d)]/24 \tag{9.4}$$

$$= 1477 \times [(0.40/0.05) - (1/1.42)]/24 = 449 \text{ L/h (119 gal/h)}$$

The maximum dilution flow will be calculated for the maximum concentration of chemical that will have to be delivered at design flow.

$$Qd_{max} = 4922 \times [(0.40/0.05) - (1/1.42)]/24 = 1497 \text{ L/h (396 gal/h)}$$

Chemical Metering Pumps

The chemical metering pumps have to be designed for the maximum capacity of chemical they have to deliver:

$$Qc_{max} = Qmax_{dc}/(D_d \times 24 \text{ h}) \tag{9.5}$$

$$= 4922/(1.42 \times 24) = 144 \text{ L/h (38 gal/h)}$$

Typically, for plants of this size and chemical fluctuation, at least two operating pumps and one standby will be provided for chemical feed, i.e., the individual pumps will have a capacity of 72 L/h (19 gal/h).

9.8 References

American Water Works Association, *Manual of Water Supply Practices, M46, Reverse Osmosis and Nanofiltration*, 2nd ed., American Water Works Association, Denver, Colorado, 2007.

Edzwald, J. K., and J. Haarhoff, *Dissolved Air Flotation for Water Clarification*, American Water Works Association and McGraw-Hill, New York, 2012.

El-Manharawy, S., and A. Hafez, "Molar Ratios as a Useful Tool for Prediction of Scaling Potential inside RO Systems," *Desalination*, 136: 243–254, 2001.

Erikson, P., and G. Dimotsis, "Field Experience: Can Chlorine Dioxide be Used as a Biocide in RO Plants?," *IDA Journal*, 4 (1), 14–20, 2012.

Wilf, M., L. Awerbuch, C. Bartels, M. Mickley, G. Pearce, and N. Voutchkov, *The Guidebook to Membrane Desalination Technology: Reverse Osmosis, Nanofiltration and Hybrid Systems Process Design, Applications and Economics*, Balaban, Rehovot, Israel, 2007.

CHAPTER 10

Sand Removal, Sedimentation, and Dissolved Air Flotation

10.1 Introduction

The purpose of sand removal, sedimentation, and dissolved air flotation pretreatment systems is to minimize the content of coarse materials such as grit, debris, and suspended solids collected by the plant intake and to protect downstream filtration facilities from solids overloading. The source water collected by onshore intakes and shallow offshore open intakes usually does not contain large quantities of sand, but it can have elevated content of floating and suspended solids. Well intakes typically have a very low content of suspended solids, but depending on their design and the subsurface soil conditions, they can produce source water of elevated sand content, especially when they are brought into service after a long shutdown.

10.2 Sand Removal Systems

A well-designed desalination plant intake usually produces source water of low sand and silt content. Therefore, source water desalination plants typically are not designed to have separate sand removal facilities. Small quantities of sand and coarse silt contained in the source water are retained by the plant's sedimentation or filtration facilities. However, in locations where a desalination plant's open intake is located adjacent to an area of prolonged/seasonal wind or wave-driven turbulence, significant ship traffic, turbulent underwater currents, or frequent dredging activities, a large amount of sand and silt may enter the desalination plant continuously and may need to be removed in separate facilities. Sand removal facilities may or may not be followed by sedimentation basins. Often, the source water may contain low levels of turbidity but a large amount of fine silt and sand. In this case, construction of grit removal facilities instead of clarifiers is more appropriate and cost effective.

10.2.1 Settling Canals and Retention Basins

Some large onshore intakes are designed with long canals that deliver the source water into retention basins, where it is presettled and sand, silt and large debris are accumulated. The source water from the reservoir overflows into the forebay of the screening facilities or intake pump station, from where it is conveyed into the main desalination plant pretreatment system. Such canals and retention basins are dredged periodically or equipped with sediment removal or flushing systems to minimize solids accumulation over time.

While retention reservoirs are suitable for dampening the effect of heavy rain events, winds, currents, ship traffic, and other sources of elevated solids content in the source water, they may present problems such as excessive algae accumulation, especially if the flow velocity is relatively low and the water remains in the reservoirs for a long time.

10.2.2 Strainers

Depending on the size of the desalination plant, the grit removal facilities most widely used in practice are 200- to 500-μm strainers (Fig. 10.1). Strainers of this size can remove sand and silt particles of 0.10 mm or larger.

Strainers are typically used for small and medium desalination plants, i.e., plants with a capacity of 20,000 m^3/day (5.3 mgd) or less. Microstrainers and microscreens are discussed in greater detail in Chap. 8.

10.2.3 Cyclone Separators

Cyclone separators have found application in removal of sand from groundwater, especially for small desalination plants. In such systems, the inlet pressure from the intake well pumps drives the source water into the top of the separator chamber at a tangent, causing rotation and the formation of a vortex in the center of the separator. The vortex action forces the separation of heavy particles from the water. These particles accumulate at the bottom in a collection chamber, from where they are periodically removed. In most recent desalination plant designs, cyclone separators and strainers have been replaced by microscreens, described in Chap. 8.

FIGURE 10.1 Sand strainers.

10.3 Sedimentation Tanks

10.3.1 Introduction

Sedimentation is typically used upstream of granular media and membrane filters when the membrane plant's source water has a daily average turbidity higher than 30 NTU or experiences turbidity spikes of 50 NTU or more that continue for a period of several hours. If sedimentation basins are not provided, large turbidity spikes may cause the pretreatment filters to exceed their solids holding capacity (especially if granular media filters are used), which in turn may impact filter pretreatment capacity and reduce the duration of filter runs. If the high solids load continues, the pretreatment filters will enter a condition of continuous backwash, which in turn will render them out of service.

10.3.2 Planning and Design Considerations

Sedimentation basins for source water pretreatment should be designed to produce settled source water of turbidity less than 2.0 NTU and a measurable silt density index (SDI_{15}) below 6. To achieve this level of turbidity removal, sedimentation basins are typically equipped with both coagulant (most frequently iron salt) and flocculant (polymer) feed systems. The needed coagulant and flocculant dosages should be established based on jar and/or pilot testing.

If the source water turbidity exceeds 100 NTU, then conventional sedimentation basins are often inadequate to produce a turbidity of the desired target level (less than 2 NTU). Under these conditions, sedimentation basins should be designed for enhanced solids removal by the installation of lamella plates or the use of sedimentation technologies that combine lamella and fine granular media for enhanced solids removal.

Typically, the use of enhanced sedimentation technologies is needed for treating source water from open shallow intakes that are under a strong influence of high-velocity currents, river water, or wastewater discharges of elevated turbidity. This condition could occur when the desalination plant intake is located in a river delta area or is influenced by a seasonal surface water runoff, strong winds and currents. For example, during the rainy season, the intake of the Point Lisas water desalination plant in Trinidad is under the influence of the Orinoco River currents, which carry a large amount of alluvial solids. As a result, the desalination plant's intake turbidity can exceed 200 NTU (Irwin and Thompson, 2003). To handle this high solids load, the plant source water is settled in lamella sedimentation tanks prior to conventional single-stage dual-media filtration (Fig. 10.2). While this plant has lamella settlers, it does not incorporate separate sand removal facilities or strainers upstream of it.

Key Design Criteria

High-Rate (Lamella) Settlers To date, rectangular lamella settlers have found the widest application in pretreatment of saline water originating from open ocean intakes. Key design criteria for this type of sedimentation tanks are:

Minimum number of tanks	2
Water depth	3.5 to 5.0 m (11.5 to 16.4 ft)
Mean flow velocity	0.3 to 1.1 m/min (1.0 to 3.6 ft/min)
Detention time (in the lamella module)	10 to 20 min

FIGURE 10.2 Point Lisas desalination plant, Trinidad. (*Source: Desalcott.*)

Surface loading rate (lamella module)	1.0 to 2.0 $m^3/m^2{\cdot}h$ (0.4 to 0.8 gal/min per ft^2)
Launder weir loading	4.0 to 8.0 $m^3/m{\cdot}h$ (250 gal/day per ft)

Lamella modules (Fig. 10.3) used in high-rate settlers are proprietary products; the design engineer should consult the equipment manufacturers regarding the configuration, number, and size of lamella modules as well as the design surface loading rate and depth of the sedimentation tank.

10.3.3 Design Example

This example illustrates general design criteria for a lamella settler intended to provide pretreatment of the source water for a 40,000 m^3/day (10.6 mgd) seawater reverse osmosis (SWRO) desalination plant designed for 43 percent system recovery. The plant's source water turbidity reaches levels of 80 NTU during storm events (which may last several days). The plant is equipped with a combination of lamella settlers followed by a single-stage dual granular media filter. The source water is relatively low in terms of hydrocarbon content, with a maximum concentration of 0.04 mg/L or less. The source water is not frequently exposed to algal blooms, and when such events occur periodically, they are of low intensity, with an algal content of the source water lower than 20,000 cells per milliliter.

FIGURE 10.3 Lamella modules—Cape Preston desalination plant, Australia.

The plant filter backwash flow is 5 percent of the intake flow, and the lamella clarifier waste stream (sludge) flow is an additional 0.5 percent of the intake flow. The pretreatment system is designed to operate with addition of coagulant and flocculant and with pH adjustment of the source water flow.

The lamella settler system is designed to treat a total of 98,440 m^3/day (26 mgd): (40,000/0.43)/[(1 – (.05 + 0.005)]. Key design parameters of this system are shown in Table 10.1.

The dimensions (width, length, and depth) and the net surface area per lamella module presented in Table 10.1 are provided by the lamella supplier. The surface loading rate is calculated by dividing the total feed flow to the lamella settlers by the total surface area of all lamella modules. This loading rate should be comparable to the loading rate used for the design of conventional settlers, i.e., 1 to 2 $m^3/m^2{\cdot}h$ (0.4 to 0.8 gal/min·ft^2).

However, if the surface loading rate is calculated by dividing the feed flow by the total physical surface area of the lamella settlers, it is approximately 22 times as high in this example (Table 10.1). This comparison illustrates the fact that lamella settlers are significantly more space efficient and economical than conventional settling tanks. Therefore, they have found wider implementation for desalination plant pretreatment than have conventional clarifiers.

Component/Parameter	Specifications/Design Criteria
Feed Water	
Design Flow rate, m^3/day (mgd)	98,440 (26)
Turbidity, NTU	0.5–80
SDI	6–16
Algal content, cells per milliliter	< 20,000
Design Chemical Dosages	
Ferric chloride, mg/L	15 (0.5–50)
Cationic polymer, mg/L	0.5 (0–1)
Sulfuric acid, mg/L (target pH = 6.7)	8 (0–30)
Lamella Settlers	
Number of settler tanks	4
Number of lamella modules per tank	4
Width of lamella modules, m (ft)	1.24 (4.1)
Length of lamella modules, m (ft)	8.67 (28.4)
Depth of lamella modules, m (ft)	2.588 (9.8)
Net surface area per lamella module, m^2 (ft^2)	235 (2528)
Surface loading rate per module area, $m^3/m^2{\cdot}h$ (gal/min·ft^2)	1.09 (0.4)
Setter tank surface area, m^2 (ft^2)	43 (463)
Settler tank surface loading rate, $m^3/m^2{\cdot}h$ (gal/min·ft^2)	23.8 (9.8)
Water depth, m (ft)	5.5 (20.8)

TABLE 10.1 Example of a Lamella Settler Pretreatment System for a 40,000 m^3/day (10.6 mgd) Desalination Plant

10.4 Dissolved Air Flotation Clarifiers

10.4.1 Types and Configurations

Dissolved air flotation (DAF) technology is very suitable for the removal of floating particulate foulants such as algal cells, oil, grease, or other light solid contaminants that cannot be effectively removed by sedimentation or filtration. DAF systems can typically produce effluent turbidity of < 0.5 NTU and can be combined in one structure with dual-media gravity filters for sequential pretreatment of the source water.

DAF process uses very small air bubbles to float light particles and organic substances (oil, grease) contained in the source water (Fig. 10.4). The floated solids are collected at the top of the DAF tank and skimmed off for disposal, while the low-turbidity source water exits near the bottom of the tank.

The surface loading rate for removal of light particulates and floatable substances by DAF is approximately one-tenth of that needed for conventional sedimentation. Another benefit of DAF as compared to conventional sedimentation is the higher density of the formed residuals (sludge). While residuals collected at the bottom of sedimentation basins typically have a concentration of only 0.3 to 0.5 percent solids,

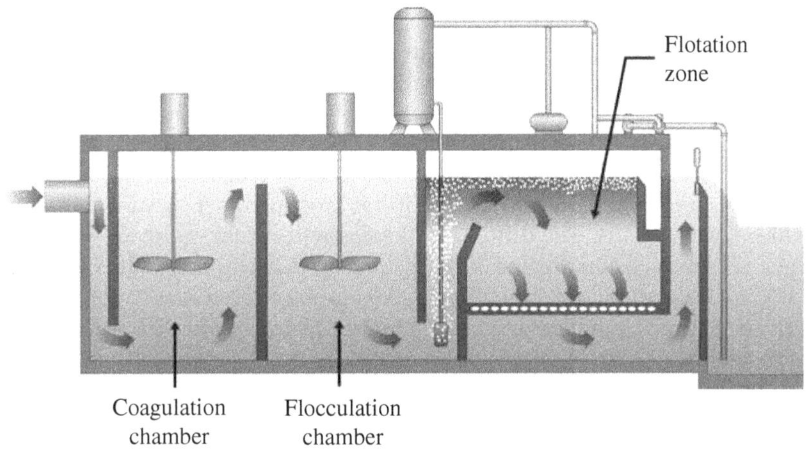

Figure 10.4 General schematic of a DAF clarifier.

DAF residuals (which are skimmed off the surface of the DAF tank) contain a solids concentration of 1 to 3 percent.

In some full-scale applications, the DAF process is combined with granular media filters to provide a compact and robust pretreatment of source water with high algal and/or oil and grease content. Although this combined DAF and filter configuration is very compact and cost competitive, it has three key disadvantages: (1) it complicates the design and operation of the pretreatment filters; (2) DAF loading is controlled by the filter loading rate, and therefore DAF tanks are typically oversized; and (3) flocculation tanks must be coupled with individual filter cells.

10.4.2 Planning and Design Considerations

The feasibility of DAF for pretreatment of saline surface waters is determined by the source water quality and governed by the turbidity concentration and overall life-cycle pretreatment costs. The DAF process can handle source water with turbidity of up to 50 NTU. Therefore, if the source water is impacted by high-turbidity spikes or heavy solids (usually related to seasonal river discharges or surface runoff), then DAF may not be a suitable pretreatment option. In most algal bloom events, however, source water turbidity almost never exceeds 50 NTU, so the DAF technology can handle practically any algal bloom.

Although DAF systems have a much smaller footprint than conventional flocculation and sedimentation facilities, they include a number of additional equipment associated with air saturation and diffusion and with recirculation of a portion of the treated flow, and therefore their construction costs are typically comparable to those of conventional sedimentation basins. Usually, the operation and maintenance costs of DAF systems are higher than those of gravity sedimentation tanks, due to the higher power use for the flocculation chamber mixers, air saturators, recycling pumps, and sludge skimmers. The total power use for DAF systems is usually 2.5 to 3 kWh per 10,000 m^3/day (1.0 to 1.2 kWh/mgd) of treated source water, which is significantly higher than that for sedimentation systems—0.5 to 0.7 kWh per 10,000 m^3/day (0.2 to 0.3 kWh/mgd) of treated water.

DAF clarifiers for seawater applications have several key differences from those for fresh surface waters: (1) they have to remove smaller algal cells and therefore have to have diffusers that create smaller bubbles; (2) seawater has a significantly higher density than freshwater and as a result requires operation at higher air pressures in order to provide adequate solids removal; and (3) seawater particles and algae usually have a lower electrical charge than freshwater solids, which makes them more difficult to coagulate and flocculate and thus requires larger contact chambers than those of freshwater DAF systems. The differences between seawater and freshwater applications of DAF are discussed in greater detail elsewhere (Edzwald and Haarhoff, 2012).

Practical experience shows that DAF system design that is not adopted to the specific water quality challenges of seawater pretreatment often does not meet performance expectations of high algal content removal, especially during normal (nonalgal bloom) source water conditions, when the content of algae in the water is low (< 500 cells per liter) and source water turbidity is < 5 NTU.

The smaller algal particles in ocean water require smaller air bubbles for effective removal. The optimum range of the size of the air bubbles is directly related to the predominant size of algal cells in the source water, which can be determined by the completion of algal profiles during algal blooms and non-algal bloom conditions.

Most existing commercially available DAF technologies have been created for wastewater and freshwater applications, and therefore the majority of the bubbles generated by their diffuser systems are in the range of 40 to 100 μm. Often, the type of algae dominating during algal bloom events in the Persian Gulf, for example, are an order of magnitude smaller than freshwater algae, i.e., they are picoplankton (0.2 to 2 μm) and nanoplankton (2 to 20 μm)—see Fig. 10.5. If such small plankton is the main cause of algal blooms, conventional DAF systems designed to remove larger (40 to 100 μm) freshwater algal cells are likely to have limited removal efficiency.

In addition, as indicated in a recent study (Zhu and Bates, 2012), commonly applied source water chlorination practices may result in algal cell destruction and may further diminish the benefits associated with DAF pretreatment.

Because of its higher density and viscosity, seawater requires 20 to 30 percent higher air saturation, and introduction of the air at higher pressures. As a result, while the

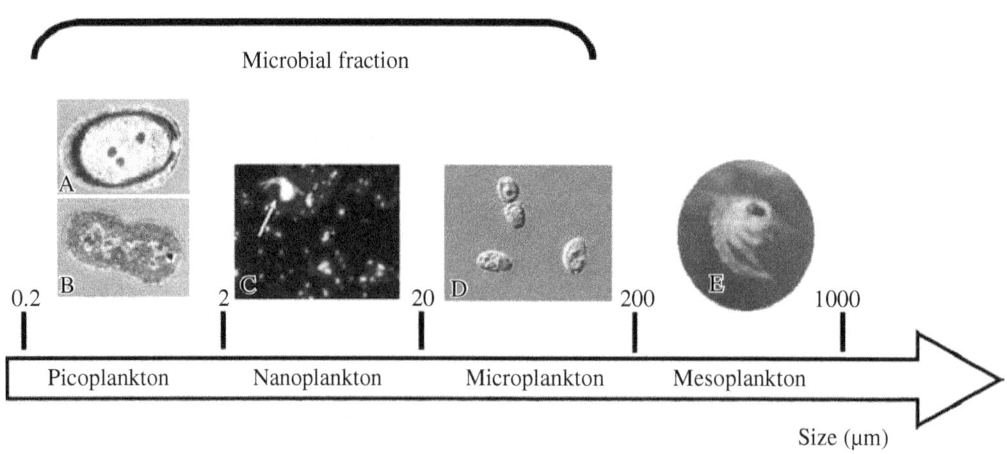

Figure 10.5 Plankton size classification.

required pressure of the feed water recycled to the DAF for fresh water is 4 to 6 bar (58 to 87 lb/in^2), the actual pressure needed for seawater DAF operations to form a large percentage of small bubbles commensurate with the size of seawater algae is typically 6 to 9 bar (87 to 130 lb/in^2).

Low-charge particles require longer contact time and better mixing in the coagulation and flocculation chambers in order to form large enough flocs for effective removal in the flotation zone of the DAF. With freshwater particles and algae that carry a strong negative charge, the addition of coagulant (ferric chloride or sulfate) that carries a positive charge will result in the creation of large flocs in a very short time because of strong opposite-charge attraction. With fine, uncharged seawater particles, the main mechanism of floc formation is direct physical contact with the coagulant particles, which requires more time, especially if the solids concentration is very low (e.g., at low feed water turbidity). As a result, the typical 5- to 7-min contact time used to design the flocculation chambers of DAF systems for freshwater applications will be insufficient for the formation of adequate flocs of seawater particles; a contact time of at least 10 min is needed for DAF systems processing seawater. One proprietary DAF system designed for seawater pretreatment applications addresses this challenge by installing a device referred to in Fig. 10.6 as the Turbomix, which increases particle collision and flocculation.

Key Design Criteria

DAF systems include three key components: coagulation chamber, flocculation chamber and flotation zone (Fig. 10.4). In addition, DAF systems also have clarified water recirculation system and air-saturation system that are used to introduce air into the feed water of the DAF clarifier. In some designs, the DAF system has only a flocculation chamber attached to the DAF clarifier, and coagulation is achieved through in-line static mixing.

The key design criteria for these DAF system components are as follows:

In-Line Static Mixer (or Coagulation Chamber)	
Velocity gradient, $G \times T$	500 to 1600
Flocculation Chamber	
Contact time	10 to 20 min
Number of flocculation chambers in series	2 to 4
Water depth	3.5 to 4.5 m (11.5 to 15.0 ft)
Type of mixer	Vertical shaft with hydrofoil blades
Blade area as a percentage of tank area	0.1 to 0.2 percent
Shaft speed	40 to 60 r/min
Flotation Chamber	
Minimum number of tanks	2 (same as filter cells, if combined with filters)
Tank width	3 to 10 m (10 to 33 ft)
Tank length	8 to 12 m (26 to 39 ft)
Tank depth	2.5 to 5 m (8 to 16 ft)
Surface loading rate	10 to 40 m^3/m^2·h (4 to 16 gpm/ft^2)
Hydraulic detention time	10 to 20 min

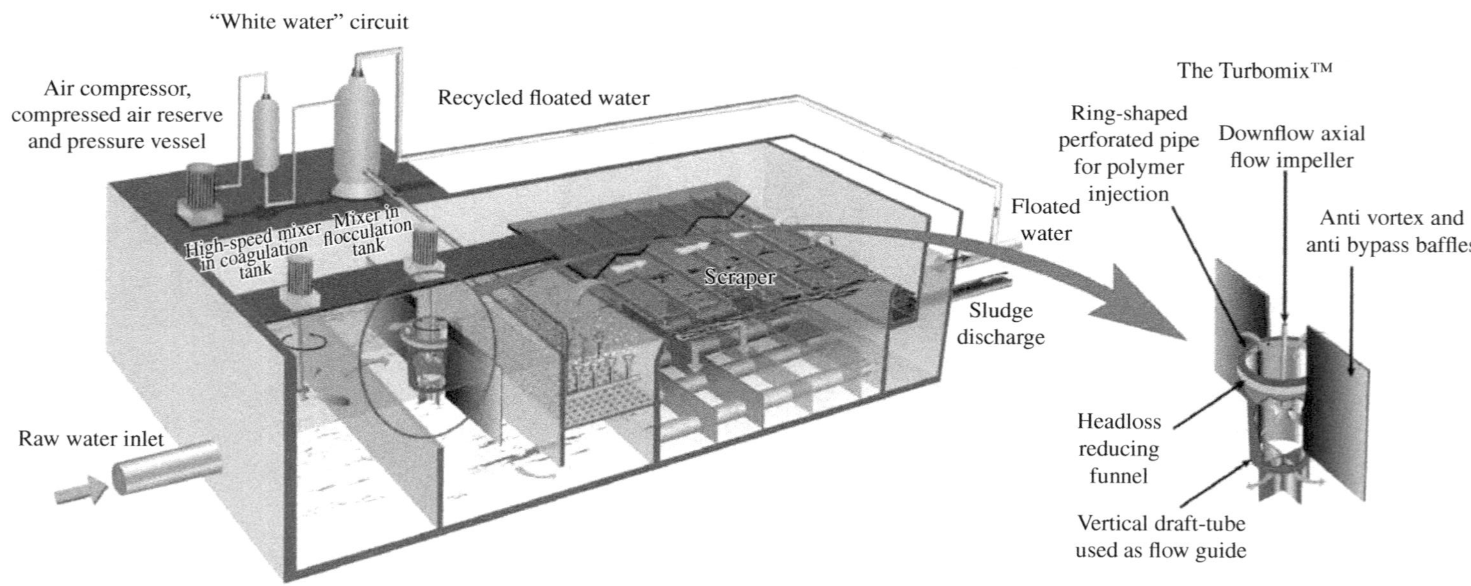

FIGURE 10.6 DAF flocculation chamber with Turbomix funnel. (*Source: Veolia.*)

Treated Water Recycling System

Recycling rate	8 to 15 percent of intake flow
Maximum air loading	10 g/m^3
Saturator loading rate	60 to 65 m^3/m^2·h (25 to 27 gpm/ft^2)
Operating pressure	6 to 9 bar (87 to 130 lb/in^2)

A DAF process with built-in filtration is used at the 136,000 m^3/day (36 mgd) Tuas seawater desalination plant in Singapore (Kiang et al., 2007). This pretreatment technology has been selected for this project to address the source water quality challenges associated with the location of the desalination plant's open intake in a large industrial port (e.g., oil spills) and the frequent occurrence of algal blooms in the area of the intake.

The source seawater has a total suspended solids concentration that can reach up to 60 mg/L at times, and oil and grease levels can be up to 10 mg/L. The facility uses 20 built-in filter DAF units, two of which are operated as standbys. Plastic covers shield the surface of the tanks to prevent the impact of rain and wind on DAF operations and to control algal growth. Each DAF unit is equipped with two mechanical flocculation tanks located within the same DAF vessel. Up to 12 percent of the filtered water is saturated with air and recirculated to the feed of the DAF units.

A combination of DAF followed by two-stage dual-media pressure filtration has been successfully used at the 45,400 m^3/day (12 mgd) El Coloso SWRO plant in Chile, which at present is the largest desalination plant in South America. The plant is located in the city of Antofagasta, where seawater is exposed to year-round red tide events, which have the capacity to create frequent particulate fouling and biofouling of the SWRO membranes (Petry et al., 2007). The DAF system at this plant is combined in one facility with a coagulation and flocculation chamber. The average and maximum flow rising velocities of the DAF system are 22 and 33 m^3/m^2·h (9 and 14 gal/min·ft^2), respectively. This DAF system can be bypassed during normal operations and is typically used during algal bloom events.

The downstream pressure filters are designed for surface loading rate of 25 m^3/m^2·h (10.2 gal/min·ft^2). Ferric chloride at a dosage of 10 mg/L is added ahead of the DAF system for source water coagulation. The DAF system reduces source seawater turbidity to between 0.5 and 1.5 NTU and removes approximately 30 to 40 percent of the source seawater organics.

Another example of a large seawater desalination plant incorporating a DAF system for pretreatment is the 200,000 m^3/day (53 mgd) Barcelona facility in Spain. The pretreatment system of this plant incorporates 10 high-rate AquaDAF units equipped with flocculation chambers, followed by 20 first-stage dual-media gravity filters and 24 second-stage pressurized dual-media filters. The purpose of the DAF system is mainly to remove algae and reduce source water organic content. Because the plant intake is located near a large port area, the DAF unit is also designed to handle potential oil contamination in the source water.

The intake of the desalination plant is located 2200 m 7200 ft) from the coast and 3 km (2 mi) away from the entrance of a large river (the Llobregat River) to the ocean, which carries significant amount of alluvial and natural organic matter. After coagulation with ferric chloride and flocculation in flash mixing chambers, over 30 percent of these organics are removed by the DAF system.

10.4.3 Design Example

This example DAF clarifier is designed to the same seawater desalination plant specifications described in Sec. 10.3.3, i.e., a 40,000 m^3/day (10.6 mgd) SWRO desalination plant with 43 percent system recovery. The plant source water turbidity reaches levels of 80 NTU during storm events and up to 40 NTU during algal blooms. This source water is planned to be treated by a combination of DAF clarifiers and granular dual-media filters.

The plant filter backwash flow is 5 percent of the intake flow, and the lamella clarifier waste stream (sludge) flow is another 0.5 percent of the intake flow. The maximum algal count in the source water is 60,000 cells per milliliter and the hydrocarbon levels can reach 0.5 to 1 mg/L. The pretreatment system is designed to operate with addition of coagulant and flocculant and adjustment of pH of the source water flow.

The pretreatment system will need to be designed to treat a total of 98,440 m^3/day (26 mgd): $(40{,}000/0.43)/[(1-(.05+0.005)]$. Source water coagulation will be completed by in-line static mixers. Design parameters of the DAF clarifier are summarized in Table 10.2.

10.5 Lamella Settler and DAF Clarifier Costs

Figure 10.7 provides a depiction of the construction costs of lamella settlers and DAF clarifiers.

As indicated in Fig. 12.5, lamella settlers are less costly than DAF clarifiers for the same volume of pretreatment source water. However, lamella settlers do not remove

Component/Parameter	Specifications/Design Criteria
Feed Water	
Design flow rate, m^3/day (mgd)	98,440 (26)
Turbidity, NTU	0.5–80
SDI	6–16
Design Chemical Dosages	
Ferric chloride, mg/L	15 (0.5–50)
Cationic polymer, mg/L	0.5 (0–1)
Sulfuric acid, mg/L (target pH = 6.7)	8 (0–30)
Flocculation Tanks	
Number per DAF tank	1
Total number	4
Width, m (ft)	4.85 (15.9)
Length, m (ft)	8.0 (26.4)
Depth, m (ft)	4.9 (16.1)
Number of mixers per tank	2
Total retention time, min	11
DAF Tanks	
Number	4
Width, m (ft)	4.85 (15.9)
Length, m (ft)	10.0 (32.8)
Depth, m (ft)	4.9 (16.1)
Total surface area, m^2 (ft^2)	194 (2,088)

Table 10.2 Example of a DAF Clarification System for a 40,000 m^3/day (10.6 mgd) Desalination Plant

Component/Parameter	Specifications/Design Criteria
Surface contact zone area, m^2 (ft^2)	38 (409)
Surface flotation area, m^2 (ft^2)	156 (1678)
Surface loading rate at 15% recycling, $m^3/m^2 \cdot h$ (gal/min·ft^2)	30.2 (12.3)
Circulation Pumps	
Number	4 active + 1 standby
Capacity, m^3/h (gal/min)	154 (680)
Delivery pressure, bar (lb/in^2)	7 (100)
Air Compressors	
Number	4 active + 1 standby
Capacity, m^3/h (gal/min)	15 (66)
Delivery pressure, bar (lb/in^2)	10 (142)
DAF Saturator Tanks	
Number	4
Capacity per tank, m^3/h (gal/min)	100 (440)
Net volume per tank, m^3 (gal)	4 (1060)

TABLE 10.2 Example of a DAF Clarification System for a 40,000 m^3/day (10.6 mgd) Desalination Plant (*Continued*)

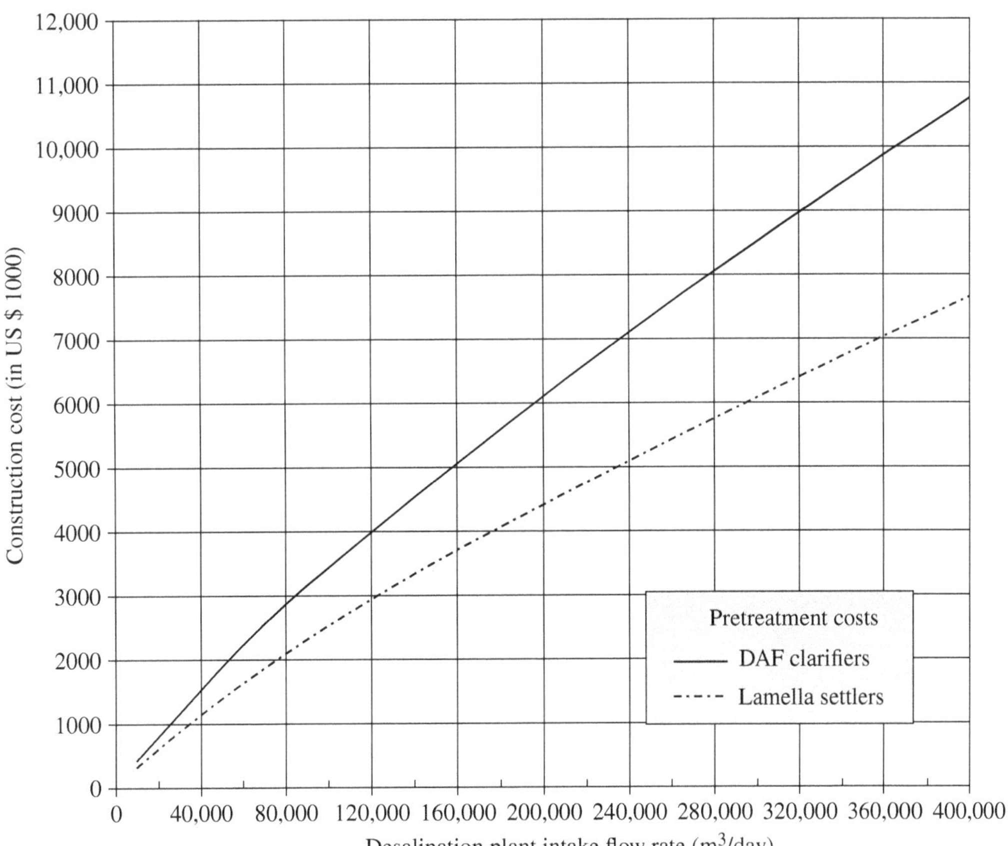

FIGURE 10.7 Construction costs of lamella settlers and DAF clarifiers.

algae and hydrocarbons well, and often DAF clarifiers are the preferred primary treatment step of choice.

Based on analysis of Fig. 12.5, the estimated costs of the example lamella settlers and DAF clarifiers for the 40,000 m^3/day (10.6 mgd) SWRO desalination plant described in Secs. 10.3 and 10.4 are \$1.2 and \$1.5 million, respectively.

10.6 References

Edzwald K. J., and J. Haarhoff, *Dissolved Air Flotation for Water Clarification*, American Water Works Association and McGraw-Hill, New York, 2012.

Irwin, K. J., and J. D. Thompson, "Trinidad SWRO—Orinoco Fluctuations Fail to Make Filters Falter," *Desalination and Water Reuse*, 13 (3) 12–16, 2003.

Kiang, F. H., W. W. L. Young, and D. D. Ratnayaka, "The Singapore Solutions," *Civil Engineering*, 77 (1): 62–69, 2007.

Petry, M., M. A. Sanz, C. Langlais, V. Bonnelye, J. P. Durand, D. Guevara, W. M. Nardes, et al., "The El Coloso (Chile) Reverse Osmosis Plant," *Desalination*, 203: 141–152, 2007.

Zhu, I. X., and B. J. Bates, "Seawater Desalination Pretreatment for Harmful Algae Blooms Using Dissolved-Air Flotation," *IDA Journal*, 4 (1): 34–37, 2012.

CHAPTER 11

Pretreatment by Granular Media Filtration

11.1 Introduction

Granular media (conventional) filtration is the most commonly used source water pretreatment process for reverse osmosis (RO) desalination plants today, other than cartridge filtration. This process includes filtration of the source water through one or more layers of granular media (e.g., anthracite coal, silica sand, garnet). Conventional filters used for saline water pretreatment are typically rapid single-stage dual-media (anthracite and sand) units. However, in some cases where the source water contains high levels of organics (total organic carbon concentration is higher than 6 mg/L) and suspended solids (monthly average turbidity exceeds 20 NTU), two-stage filtration systems are applied. In this configuration, the first filtration stage is mainly designed to remove coarse solids and organics in suspended form. The second-stage filters are configured to retain fine solids and silt and to remove a portion (20 to 40 percent) of the soluble organics contained in the saline water by biofiltration. All pretreatment systems are designed to achieve the filtered water quality specifications listed in Table 11.1.

11.2 The Filter Operation Cycle

Granular media filtration is a cyclical process that incorporates two sequential modes of operation: (1) source water processing (filtration) and (2) filter media backwash.

11.2.1 Source Water Processing (Filtration)

During the filtration cycle, the water moves in the direction of the size gradation of the media, and solids in the water are retained on and around the media grains.

As the feed water is filtered through the media, the content of solids and silt in the water decreases. Usually, properly operating filters remove 90 to 99 percent of the solids and silt contained in the source water. Some of the aquatic microorganisms in the source water are also retained on the filter media. These microorganisms consume a portion of the dissolved organics in the source water. The organic load removal efficiency of the filters is a function of three main factors: media depth, surface loading rate, and

Parameter	Concentration/Level
Turbidity (daily average/maximum), NTU	< 0.1/0.5
Silt density index	< 3 (at least 95 % of the time) < 5 (at all times)
Total organic carbon, mg/L	< 1
pH (minimum/maximum)	4.0/9.0
Oxidation-reduction potential, mV	< 200
Chlorine residual, mg/L	≤ 0.02
Total hydrocarbons, mg/L	≤ 0.04

Table 11.1 Target Minimum Pretreated Water Quality

temperature. Removal of organics by the filters increases with an increase in depth and temperature and with a decrease of surface loading rate.

The solids retained in the pore volume between the filter grains reduce this volume over time and create hydraulic losses through the filter media (filter bed resistance).

Most filters used for saline water pretreatment operate at a constant filtration rate, which means that the feed pressure of these filters increases over the filtration cycle to compensate for the head losses in the filter bed caused by accumulation of solids. Once the filter media head losses reach a preset maximum level, the filter is taken out of service and media backwash is activated. A typical filtration cycle continues for 24 to 48 h. Deeper filters with a larger surface area have a greater capacity to retain solids and therefore usually have longer filtration cycles.

11.2.2 Filter Media Backwash

Granular media filters are typically backwashed using filtered source water or concentrate from the RO membrane system. The backwash frequency of filter cells is usually once every 24 to 48 h; spent (waste) backwash volume is 2 to 6 percent of the intake source water. The use of RO concentrate instead of filtered effluent to backwash filter cells allows for a reduction in backwash volume and in the energy needed to pump source water to the desalination plant.

During backwash of downflow filters which are preferred for surface water pretreatment, the backwash water flows upward through the filters, scours the filter grains, removes the solids accumulated on them, expands the filter bed, and transports the removed solids toward the backwash troughs.

From experience, it is known that backwashing filter media grains smaller than 0.8 mm with water only is difficult. Therefore, at present a typical backwash includes a combination or sequence of air and water washing. Air creates greater turbulence and enhances particle scrubbing. The length of water and air backwashing cycles is a function of the solids content in the source water and typically is between 5 and 8 min.

The applied bed expansion depends on the size of the filter media—the smaller the media, the larger expansion is needed. For example, media with a diameter of 1.2 mm require an expansion of only 10 to 15 percent. Filter media with a size of 0.8 mm needs an expansion of 20 to 25 percent. For sand filter media of 0.4 to 0.6 mm, used frequently in pretreatment filters, the backwash rate should provide 30 to 50 percent media bed expansion for optimal filtration performance.

The number of filter cells and the individual production capacity of each cell are typically selected to allow full flow operation with one filter cell out of service in backwash and one out of service for maintenance. Additional information on the design of granular media filtration systems is provided elsewhere (American Water Works Association, 2007; Wilf et al., 2007).

11.3 Key Filtration System Components

11.3.1 Filter Cells

A typical granular media filtration system consists of number of individual units (cells or vessels) that operate in parallel. The number of filter cells is mainly dependent on the total flow the filters are designed to handle. The construction cost of the filtration system is usually reduced when fewer individual cells are used. However, the minimum number of filters is limited by the following key factors: (1) the practical maximum size of the individual filter bed [100 to 150 m^2 (1080 to 1610 ft^2)]—larger beds are likely to result in nonuniform backwash; (2) the increase in the filtration rate of the filters remaining in operation when one or two filters are in a backwash mode; and (3) the configuration of the RO system, i.e., the number of individual trains and the planned mode of operation of the desalination plant.

In order to maintain consistent, high-quality filter performance, the number of filter cells should be selected in such a manner that when one cell is out of service for backwash or maintenance, the hydraulic loading rate of the filters remaining in operation does not exceed 20 percent of the average loading rate with all units in service; and when two units are out of service, this rate should be less than 30 percent of the average loading rate.

In general, even for very small desalination plants, the minimum recommended number of individual pretreatment filters is four. For plants with a capacity higher than 5000 m^3/day (1.3 mgd), 6 to 8 filter units are preferable (Kawamura, 2000).

For desalination plants larger than 10,000 m^3/day (2.6 mgd), filter cells are usually divided into two groups that can be operated independently and paired with one-half of the desalination plant RO trains. In plants larger than 200,000 m^3/day (53 mgd) the desalination plant is typically divided into at least two sets of two filter groups, each with 8 to 32 individual filter cells.

11.3.2 Filter Media

The type, uniformity, size, and depth of filter media are of key importance for the performance of pretreatment filters. Dual-media filters have two layers of filtration media—a typical design includes 0.4 to 0.8 m (1.3 to 2.6 ft) of anthracite or pumice over 0.4 to 2 m (1.3 to 6.6 ft) of sand.

Deep dual-media filters are often used if the desalination plant's filtration system is designed to achieve enhanced removal of soluble organics from source water by biofiltration. In this case, the depth of the anthracite level is enhanced to 1.5 to 1.8 m (4.9 to 5.9 ft).

If the source water is relatively cold [i.e., the average annual temperature is below 15°C (59°F)] and at the same time is of high organic content, a layer of granular activated carbon (GAC) of the same depth is used instead of a deeper layer of anthracite, because the biofiltration removal efficiency will be hindered by the low temperature. During biofiltration a portion of the soluble organics in the source water is metabolized by the microorganisms that grow on a thin biofilm formed on the granular filter media. For comparison, the GAC media removes a portion of the source water organics mainly by adsorption.

Tri-media filters have 0.45 to 0.60 m (1.5 to 2 ft) of anthracite or pumice as the top layer, 0.2 to 0.4 m (0.7 to 1.3 ft) of sand as the middle layer, and 0.10 to 0.15 m (0.33 to 0.5 ft) of garnet or limonite as the bottom layer. These filters are used if the source water contains a large amount of fine silt or experiences algal blooms dominated by pico and microalgae (0.5 to 20 μm). Filter media density varies as shown in Table 11.2.

The effective size d_{10} of the medium is the size of the opening of the sieve for which 10 percent of the grains (by weight) are smaller in diameter. The uniformity coefficient (UC) is the ratio between the opening size d_{60} of a sieve for which 60 percent of the grains (by weight) are smaller and the effective size of the medium:

$$\mathrm{UC} = d_{60}/d_{10} \tag{11.1}$$

The uniformity coefficient is an important parameter because it indicates how similar the media particles are in size. In general, for media of the same size, a higher uniformity coefficient allows for an increased filter cycle length.

The d_{60} value of the filter media can also be used to determine the filter backwash rate at 20°C (68°F; Qasim et al., 2000) using the following formulas:

$$\text{For sand, } U_b = d_{60} \tag{11.2}$$

$$\text{For anthracite, } U_b = 0.47 \times d_{60} \tag{11.3}$$

For temperatures other than 20°C, the backwash time can be adjusted through the application of an adjustment coefficient for water viscosity:

$$U_{bt} = U_b \times K^{0.333} \tag{11.4}$$

where K = the absolute viscosity of water at the temperature (kg/m·s).

The size of the media and uniformity coefficient should always be configured to decrease along the direction of the flow, while the specific density should increase. This configuration prevents the intermixing of the different types of media during back-washing. Intermixing of the media results in shorter filter cycles and the need for more frequent backwashing.

The depth of the filter bed is typically a function of the media size and follows the general rule of thumb that the ratio l/d_e between the depth l of the filter bed (in millimeters) and the effective size d_e of the filter media (in millimeters) should be in a range of 1000 to 1500. For example, if the effective size of the anthracite media is selected to be 0.65 mm,

Medium	Typical Effective Grain Size, mm	Specific Density, tons/m³ (lb/ft³)	Uniformity Coefficient
Pumice	0.8–2.0	1.2 (75)	1.3–1.8
Anthracite	0.8–2.0	1.4–1.7 (87–104)	1.3–1.8
Silica sand	0.4–0.8	2.60–2.65 (162–165)	1.2–1.6
Garnet	0.2–0.6	3.5–4.3 (218–268)	1.5–1.8

TABLE 11.2 Typical Filter Media Characteristics

the depth of the anthracite bed should preferably be $0.65 \times 1500 = 975$ mm, or approximately 1 m (3.3 ft).

The depth of the GAC medium is estimated based on the average contact time in that medium, which is recommended to be 10 to 15 min. For example, if a filter is designed for a surface loading rate of 9 $m^3/m^2{\cdot}h$ (4 gal/min·ft^2), the depth of the GAC medium should be at least $9 \times [10 \text{ min}/(60 \text{ min/h})] = 1.5$ m (4.9 ft).

When each filter media layer is first placed in the filter cells, an additional 3 to 5 cm (1.2 to 2 in.) of medium should be added to the design depth of the layer to account for the removal or loss of fine particles from the newly installed bed after backwashing. It should also be pointed out that if the filters are designed to achieve removal of total organic carbon (TOC) by biofiltration, it will take at least 4 to 6 weeks to create a sustainable biofilm on the surface of the filter media that can yield steady and consistent filter performance and TOC removal. If the source water is relatively cold (i.e., below 20°C), then biofilm formation may take several weeks longer.

11.3.3 Media Support Layer and Filter Underdrain System

The filtration medium is typically supported by a layer of gravel bed. The gravel bed is graded in three to six layers (Kawamura, 2000) and is located on the top of a filter underdrain system. There are two types of filter underdrain systems widely used at present: block underdrainsand false bottom underdrains with nozzles.

The nozzle type of underdrain has found a wider application for desalination applications. In this system, nozzles penetrate the bottom of the underdrain and their main functions are to collect filtered water uniformly during the filtration cycle and to distribute backwash water more evenly during the backwash cycle. Block underdrain systems are constructed from light-weight moulded plastic (usually high density polyethylene) blocks. The filter cells are configured as boxes with a rough-finished, plain flat floor. The use of block underdrains does not require construction of a sub-structure such as supporting piers or beams. The modular underdrain blocks are set in base grout directly on the flat filter floor and are placed end to end in rows to form continuous laterals through the length of each row. The space between the rows is filled with grout, locking the blocks together and resulting in a flat tile floor that is an integral part of the filter box. The surface of the blocks has dispersion orifices for air and water backwash.

Typically, the flow velocity in the channel, pipes, and the false bottom below the underdrain system is designed to be relatively low—0.6 m/s (2 ft/s)—in order to provide a uniform flow pattern distribution.

11.3.4 Service Facilities and Equipment

At present, most filters used for saline water pretreatment have air and water backwash systems. As a result, each filtration system is equipped with air blowers (usually one or two duty and one standby) and water backwash pumps (typically two to three duty and one standby).

11.4 Filter Types and Configurations

11.4.1 Single-Medium and Dual- and Tri-Media Filters

Single-medium filters are not very commonly used for saline water pretreatment, because of their limited ability to perform under varying source water conditions. Typically, such filters can be applied for desalination plants with subsurface intakes producing water with turbidity of < 2 NTU, total suspended solids (TSS) of < 3 mg/L, and a silt density index (SDI) of < 6. The only large-scale application of single-medium (0.7-mm sand) filters is at the Tampa Bay desalination plant in Florida. Most other plants worldwide use dual-media filters with a top layer of pumice or anthracite and a bottom layer of sand.

Tri-media filters are not commonly used for pretreatment of saline waters. Typically, such filters are suitable for capturing small plankton and fine silt that cannot be well retained by the top two layers (anthracite and sand). Since the cost of filter cells increases with depth, often instead of a single deep tri-media gravity filter, a combination of a coarser media (anthracite or sand) gravity filter followed by a pressure filter containing finer (sand and garnet) media is used.

11.4.2 Single- and Two-Stage Filters

Two-stage filtration is typically used when the source water contains high levels of turbidity (usually above 20 NTU) and organics (TOC > 6 mg/L) for long periods of time (i.e., weeks or month). Such conditions occur in desalination plant intake areas exposed to prolonged algal bloom events (which sometimes can last for several months) or located in river estuaries that are exposed to elevated turbidity levels during the wet season of the year.

Two-stage filtration systems consist of coarse (roughing) filters and fine (polishing) filters operated in series. Usually the first-stage filter is a single-medium (e.g., coarse sand or anthracite) or dual-media type, while the second-stage filter is configured as a dual-media filter with the design criteria described in the previous section. The first (coarse media) filter typically removes 60 to 80 percent of the total amount of solids contained in the source water and is designed to retain all large debris and most of the coarser floating algal biomass. The second-stage filter removes over 99 percent of the remaining solids and fine silt as well as the microalgae contained in the source water, typically producing effluent turbidity of less than 0.05 NTU.

Two-stage filters have several advantages. The filtration process through the coarse media filters not only removes large particulate foulants but also enhances coagulation of the fine particulates contained in the source water, which makes their removal in the second-stage filters less difficult and allows the second-stage filters to be designed as shallow-bed filters rather than deep-bed ones, and to operate at higher surface loading rates. This benefit results in a reduced size of the dual-media filters and in a lower total amount of coagulant (ferric salt) needed to achieve the same final filter effluent water quality as compared to single-stage dual-media filters.

Two other benefits of two-stage filters are that: (1) they can handle larger fluctuations of intake source water turbidity because of the larger total filter media volume and solids retention capacity; and (2) when the second-stage filters are designed as deep-bed (rather than shallow-bed) ones, they can achieve enhanced TOC removal by biofiltration. While deep single-stage dual-media filters can typically remove 20 to 30 percent of the TOC contained in the source seawater, two-stage systems with deep second-stage filters can achieve 40 to 60 percent of TOC removal, mainly due to enhanced fine particle coagulation and biofiltration.

11.4.3 Downflow and Upflow Filters

Most filters used in pretreatment of seawater and brackish water are downflow filters. This flow direction allows large algal particles to be retained at the top of the filter media and removed with the backwash water with minimum breakage and consequent release of organics. If upflow filtration is used, algae contained in the source water are pressed against the filter media and unwanted dissolved organics are released from the broken algal cells into the filtered water.

11.4.4 Filters Combined with Dissolved Air Flotation Clarifiers

In cases where the saline source water contains a large amount of algal particulates and/or oil and grease, and space is at premium, dissolved air flotation (DAF) and granular media filtration processes can be combined in one structure that has the DAF clarifier located above the filter cell (see Fig. 11.1).

In this configuration, granular media filters are typically designed as dual-media (anthracite and sand) downflow filters. The design surface loading rate of these filters is usually two to three times that of single-stage dual-media filters, i.e., 15 to 35 $m^3/m^2{\cdot}h$ (6 to 14 $gal/min{\cdot}ft^2$).

Since the operations and maintenance costs of DAF clarifiers are relatively high, it is recommended that the filtration portion of the pretreatment system is designed for the lower end of that range—15 to 20 $m^3/m^2{\cdot}h$ (6 to 8 $gal/min{\cdot}ft^2$)—which will allow operation of only the filtration portion of the system when the source water quality is good and the levels of turbidity and organics in the water are low.

11.4.5 Gravity and Pressure Filters

Depending on the driving force for water filtration, granular media filters are classified as either gravity or pressure filters. The main differences between the two types of filters

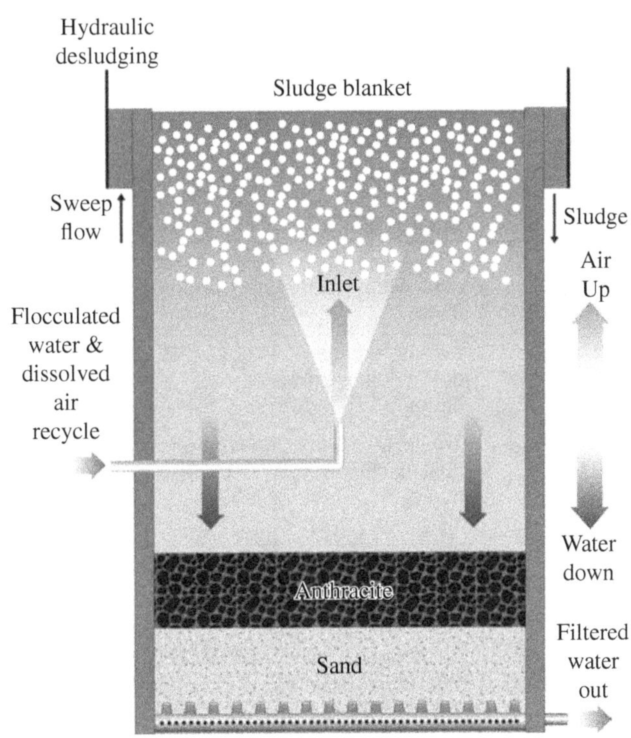

FIGURE 11.1 Combined DAF clarifier and granular media filter.

are the head required to convey the water through the media bed, the filtration rate, and the type of vessel used to contain the filter media. Because of the high cost of constructing large pressure vessels with proper wetted surfaces for corrosion resistance, pressure filters are typically used for small- and medium-capacity RO plants. Gravity pretreatment filters are used for both small and large RO desalination projects.

Gravity Filters—Description

Typically, gravity filters are reinforced concrete structures that operate at a water pressure drop through the media of between 1.8 and 3.0 m (6 and 10 ft). The hydrostatic pressure over the filter bed provides the force needed to overcome the head loss in the media. Single-stage dual-media downflow gravity filters are the predominant type of filtration pretreatment technology used in desalination plants with a capacity higher than 40,000 m^3/day (10.6 mgd). Table 11.3 provides examples of key design criteria for desalination plants of various sizes and water qualities.

Some of the largest seawater reverse osmosis (SWRO) desalination plants in the world in operation today, such as the 125,000 m^3/day Gold Coast SWRO plant (see Fig. 11.2) are equipped with dual-media single-stage gravity filters.

Gravity Filters—Key Advantages

Better Removal of Algal Material from the Seawater Surface saline water (e.g., seawater) always contains a measurable amount of algae, whose concentration usually increases several times during the summer period and may increase up to 10 times during periods of algal blooms (which may or may not exhibit themselves as red tides).

There is a large variety of algal species in the seawater. Some algal species that occur during red tide events have cells that are relatively easy to break under pressure as low as 0.3 to 0.6 bar (4 to 9 lb/in^2).

Desalination Plant Location and Capacity	Pretreatment System Configuration	Average and Maximum Filter Loading Rates, $m^3/m^2 \cdot h$ (gal/min·ft^2)	Notes
Ashkelon SWRO plant, Israel—330,000 m^3/day (86 MGD)	40 single-stage dual-media gravity filters	10/12 (25/30)	Open intake, 1000 m (3280 ft) from shore
Sydney SWRO plant, Australia—250,000 m^3/day (66 MGD)	24 single-stage dual-media gravity filters	8/12 (20/30)	Open intake, 300 m (984 ft) from shore
Fujairah SWRO plant, United Arab Emirates—170,000 m^3/day (45 MGD)	14 single-stage dual-media gravity filters	Filtration rate: 8.5/9/5 (21/23)	Shallow offshore open intake; high algal bloom potential
Gold Coast SWRO plant, Australia—136,000 m^3/day (36 MGD)	18 single-stage dual-media gravity filters	8/10 (20/25)	Open intake, 1500 m (4920 ft) from shore
Tuas desalination plant, Singapore—136,000 m^3/day (36 MGD)	20 combined DAF and sand (110 cm) media filters	6/10 (15/25)	Open intake in an industrial port

Table 11.3 Examples of Large Desalination Plants with Gravity Filters

FIGURE 11.2 Single-stage gravity filters at the Gold Coast SWRO desalination plant, Australia.

When algal cells break, they release cytoplasm into the source water that has a very high content of easily biodegradable polysaccharides. When the amount of polysaccharides released by the broken algal cells exceeds a certain level in the filtered water, it typically triggers accelerated biofouling on the RO membranes.

Two practical approaches to address this problem are: (1) the use of a dissolved air flotation facility ahead of the pretreatment filters to gently remove algal cells and prevent their breakage (which is preferable) and (2) installation of a granular activated carbon media layer (activated carbon layer/cap) on the surface of the filters to remove some of the polysaccharides and other organics in the source water.

Pressure filters usually operate at filtration pressure which is several times higher than that of gravity filters. Because the operating pressure of these filters is often higher than the threshold at which algal cells break, pressure filters have the disadvantage of causing accelerated biofouling when filtering source water with a very high algal content. This effect is likely to manifest itself mainly in the summer and during algal blooms when the level of TOC in the source water exceeds 2 mg/L.

Pressure filters are used in medium and large desalination plants in Spain, Algeria, and Australia. In most successful applications, the source water quality is very good (TOC < 1 mg/L, SDI < 4, and turbidity < 4 NTU). Most of the Spanish SWRO desalination plant intakes are relatively deep, and the algal content in the source water is fairly low. At a depth of 10 to 20 m (33 to 66 ft), the concentration of algae is significantly lower than that at the water surface—and therefore, as long as the desalination plant intake is fairly deep, biofouling caused by breakage and decay of algal biomass may not be as significant a problem as it would be for shallow intakes or intakes located at the surface of the water body (e.g., near-shore open intakes).

Longer Useful Life of the Filter Structure Typically, gravity filters are concrete structures with a useful life of 50 to 100 years. Pressure filters are steel structures with a life span of 25 years or less. The internal surface of the pressure filters used in desalination plants is typically lined with a rubber or epoxy coating that needs to be inspected occasionally and replaced every 5 to 10 years.

Lower Power Use Because pressure filters typically operate at several times higher feed pressures than gravity filters, the energy use for pressure filtration is proportionally higher.

Higher Solids Retention Capacity and Better Handling of Turbidity Spikes Gravity media filters have approximately two to three times the volume of filtration media and retention time that pressure filters have for the same water production capacity. Therefore, this type of filter can retain proportionally more solids, and as a result, pretreatment filter performance is less sensitive to occasional spikes in source water turbidity.

Pressure filters usually do not handle solids and turbidity spikes as well, because of their smaller solids retention capacity (i.e., smaller volume of pores that can store solids before the filter needs to be backwashed). If the source water is likely to experience occasional spikes of high turbidity (20 NTU or higher) due to rain events, algal blooms, naval traffic, ocean bottom dredging operations in the vicinity of the intake, seasonal change in the underwater current direction, or spring upwelling of water from the bottom to the surface, then pressure filters will produce effluent with inferior quality (higher SDI and turbidity) during such events. Therefore, use of pressure filters will likely result in more frequent RO cleaning.

Simpler Inspection and Maintenance Gravity filters are typically covered with light plastic panels that protect the filter cells from direct sunlight (Fig. 11.3) or are installed in buildings (see Fig. 11.2).

If covers are used, they can be easily removed and the filter cells can be inspected visually for irregularities—malfunctioning filter backwash nozzles, weir corrosion, poorly backwashed areas of filter media, formation of "mud balls," etc. Pressure filters are completely enclosed and very difficult to inspect for the same problems. As a result, they have to be designed with a higher contingency factor (reserve capacity). A 15 to 20 percent reserve capacity is recommended, to accommodate potential flow distribution problems and uneven backwash air and water distribution.

Easier Accommodation of Membrane Pretreatment in the Future As membrane pretreatment technology evolves and new membrane systems available on the market are designed to better handle challenges of algal blooms, it will be very advantageous to modify exiting conventional granular media filters into submersible membrane pretreatment filters. This upgrade will be possible as long as gravity filter cells are designed with adequate depth and configuration to accommodate submersible ultrafiltration and microfiltration membranes.

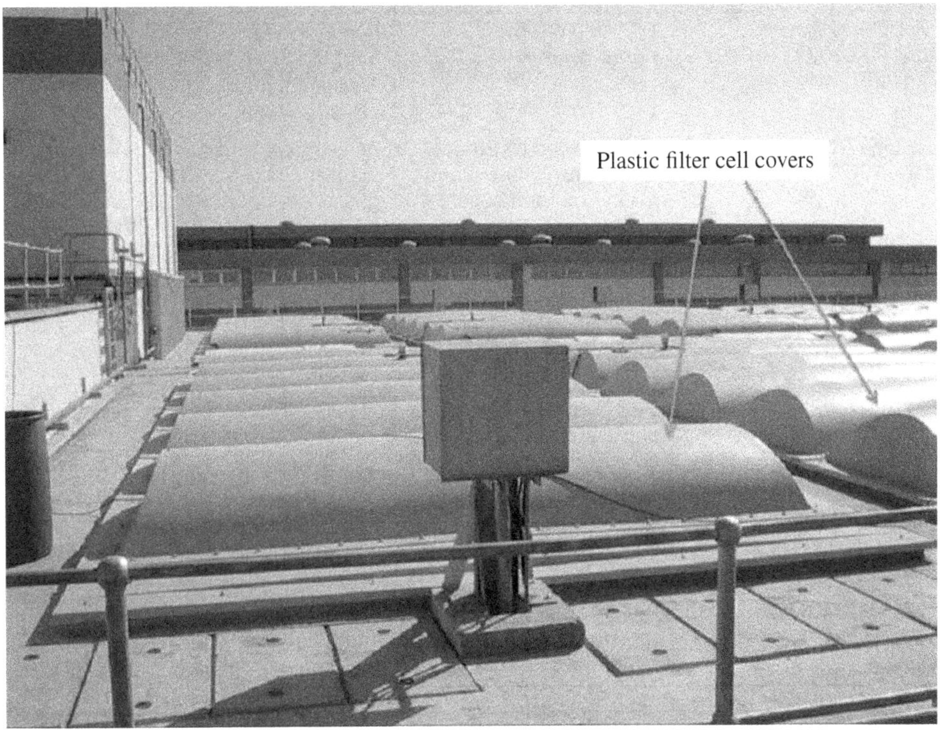

FIGURE 11.3 Gravity filters in Ashkelon, Israel, protected with plastic covers to control algal growth.

Pressure Filters—Description

Pressure filters have a filter bed configuration similar to those of gravity filters, except that the filter media is contained in a steel pressure vessel. They have found application mainly at small and medium seawater desalination plants, usually with a production capacity of less than 20,000 m^3/day (5.3 mgd). However, there are a number of installations worldwide where pressure filters are used for pretreatment of significantly larger volumes of water (Table 11.4).

In most cases, for good source water quality (SDI < 5 and turbidity less than 5 NTU) the pressure filters are designed as single-stage dual-media (anthracite and sand) units. Some plants with relatively poor water quality use two-stage pressure filtration systems. Pressure filters are available in two vessel configurations: vertical and horizontal.

Vertical pressure filters (Fig. 11.4) are customarily used in smaller plants; individual vessels have a maximum diameter of 3 m. Horizontal pressure filters (Fig. 11.5) are used more popular for medium and large facilities. The largest desalination plant using horizontal granular media pressure filters for seawater pretreatment is the 143,000 m^3/day (38 mgd) Perth I (Kwinana) SWRO facility in Perth, Australia (Fig. 11.6). Horizontal filters allow a larger filtration area per filter vessel than do vertical units. However, vertical vessels can usually be designed with deeper filter media, if deep filters are needed to handle spikes in source water turbidity.

Desalination Plant Location and Capacity	Pretreatment System Configuration	Average and Maximum Filter Loading Rates, $m^3/m^2 \cdot h$ ($gal/min \cdot ft^2$)	Notes
Al Dur SWRO plant, Bahrain—240,000 m^3/day (63 mgd)	DAF followed by horizontal pressure filters	DAF surface loading rate: 25–30 (61–73) Pressure filtration rate: 18–24 (44–59)	Shallow offshore open intake in an area prone to algal blooms
Barcelona SWRO plant—200,000 m^3/day (53 mgd)	DAF followed by 20 dual-media gravity filters and 20 horizontal pressure filters	DAF surface loading rate: 25–30 (61–73) Gravity filtration rate: 8–10 (20–25) Pressure filtration rate: 15–20 (37–49)	Deep offshore open intake in an industrial port and near a river estuary
Perth I SWRO plant, Perth, Australia—143,000 m^3/day (38 mgd)	24 single-stage dual-media pressure filters	14/18 (34/44)	Shallow open intake
Fujairah II SWRO plant, United Arab Emirates—140,000 m^3/day (37 mgd)	DAF followed by 16 single-stage dual-media pressure filters	15/20 (37/49)	Shallow offshore open intake; high algal bloom potential
Carboneras SWRO plant, Spain—120,000 m^3/day (32 mgd)	40 single-stage dual-media pressure filters	12/15 (30/37)	Offshore open intake
El Coloso SWRO plant, Chile—45,400 m^3/day (12 mgd)	DAF followed by 13 two-stage dual-media horizontal pressure filters	DAF surface loading rate: 22–33 (54–81) Filtration rate: 25 (61)	Open intake in an industrial port with frequent red tides

TABLE 11.4 Large Seawater Desalination Plants with Pressure Filters

FIGURE 11.4 Vertical pretreatment pressure filters.

FIGURE 11.5 Horizontal pretreatment pressure filters.

FIGURE 11.6 Single-stage horizontal pressure filters, Perth I SWRO plant. (*Source*: *Water Corporation.*)

Compared to gravity media filters, which operate under a maximum water level over the filter bed of up to 2.5 m (8.2 ft), pressure filters typically run at a feed pressure equivalent to 15 to 30 m (49 to 98 ft) of water column. The magnitude of the feed pressure is often driven by the suction pressure requirements of the high-pressure feed pumps for the downstream RO system.

One key advantage of using pressure filters is that they can avoid the need for intermittent pumping of the pretreated source water. A typical RO system with gravity pretreatment filters requires installation of filter effluent transfer pumps to convey the filtrate from the filter effluent well to the high-pressure RO feed pumps. The use of pressure filters can eliminate the need for such interim filter effluent transfer pumps, because the filtrate is already pressurized by the intake pumps, and the pretreatment filters do not break the hydraulic grade line.

Pressure Filters—Key Advantages

The key advantages of the pressure filters are discussed below.

Lower Construction Costs Pressure filters are prefabricated steel structures (see Figs. 11.4 through 11.6); their production costs per unit filtration capacity are lower than those of concrete gravity filters. Since pressure filters are designed with approximately 2 to 3 times the surface loading rates of gravity filters—25 to 45 $m^3/m^2{\cdot}h$ (10 to 18 $gal/min{\cdot}ft^2$) versus 8 to 15 $m^3/m^2{\cdot}h$ (3 to 6 $gal/min{\cdot}ft^2$)—their volume and size are smaller, and therefore they usually are less costly to build and install.

Smaller Footprint Because of their smaller volume and filtration area, pressure filters occupy a smaller footprint. If the available site has a limited footprint, this is an important factor to consider when selecting granular media filtration technology.

Simpler Installation Because pressure filter vessels are prefabricated, their installation time is approximately 20 to 30 percent shorter than that of gravity filters with concrete structures.

No Effect of Sunlight on Algal Growth in Filter Weirs Since pressure filters are completely enclosed, sunlight cannot reach the filter weirs, distribution system, and media to induce algal growth that would have a negative impact on filter performance. Gravity filters (especially if they are not located in a building or covered with nontranslucent panels) can grow algae on all wetted components exposed to direct sunlight.

11.5 Filter Performance

11.5.1 Removal of Solids

Since the purpose of the pretreatment filters for RO plants is not only to remove over 99 percent of all suspended solids in the source water but also to reduce the content of the much finer silt particles by several orders of magnitude, the design of these facilities is usually governed by the target SDI of the filter effluent rather than by target turbidity or pathogen removal rates.

Filter efficiency in terms of solids removal (reduction of turbidity and total suspended solids) is not directly related to its removal efficiency for silt and fine colloids

(SDI reduction capability). Dissolved organics and coagulant (iron salts) can absorb on or in the SDI filter test pad and result in increased SDI values. Full-scale experience at many granular media pretreatment filter installations indicates that filters can consistently reduce source water turbidity to less than 0.1 NTU while at the same time producing effluent with a SDI frequently exceeding 4. In many cases, granular media filters at RO desalination plants need to be designed more conservatively than similar filters at conventional surface water treatment plants in order to capture fine solids, silt, and colloidal organics contained in the saline source water.

11.5.2 Removal of Organics

Typical gravity and pressure dual-media filters with a conventional filter bed depth of 1.0 to 1.4 m (3.3 to 5.3 ft) have a relatively low organics removal rate—15 to 20 percent. This removal rate however, increases significantly with depth and can reach 25 to 35 percent for filters with a total filter depth of 2 m (6.6 ft) or more. If a carbon cap is installed on the top of the filter media (above the layer of anthracite), the filter's TOC removal rate can be increased to 40 to 50 percent.

11.5.3 Removal of Microorganisms

Algae

The rate of algae removal by the filters will depend mainly on the size of the algae and the size of the filter media. Most algae larger than 100 μm are typically retained on the surface of the top medium (anthracite/pumice). Practical observations indicate that the closer the desalination plant is located to the equator, the larger the percentage of micro- and picoalgae in the source seawater. Such algae are not well removed by conventional sand media with a size of 0.4 to 0.6 mm (400 to 600 μm) and require the installation of a third layer of finer filter medium. Depending on the size of the media and the algae dominating in the source water, algal removal can typically vary between 20 percent and 90 percent.

Bacteria and Viruses

Desalination pretreatment filters typically provide 99 percent (2-log) removal of pathogens, but sometimes may have lower removal rates for marine bacteria because these bacteria are typically of smaller size than human pathogens and may pass through the filters.

11.6 Source Water Pretreatment Prior to Granular Media Filtration

Most saline water particles and microorganisms have a slightly negative charge that has to be neutralized by coagulation. In addition, these neutralized particles need to be agglomerated in larger flocs that can be effectively retained within the filter media. Therefore, source water conditioning by coagulation and subsequent flocculation are necessary prior to granular media filtration. Coagulation and flocculation processes, and design criteria associated with source water conditioning prior to filtration, are discussed in Chap. 9.

Source water may need to undergo additional pretreatment prior to filtration (sand removal, sedimentation, DAF) depending on its quality. Alternative pre-filtration processes and configurations are presented in Chap. 10.

11.7 Planning and Design Considerations

The design criteria presented here should be used as guidelines only—it is recommended that media size, depth, and configuration, especially for medium and large desalination plants, be selected based on pilot testing for the site-specific conditions and water quality associated with the project over a period that encompasses worst-case scenario water quality (e.g., significant rain events of intensity higher than 15 mm, dredging near the intake area, red tide events, seasonal winds and currents, etc.).

11.7.1 Single-Stage Dual-Media Filters

Gravity Filters—Key Design Criteria

Key design criteria for single-stage dual-media gravity filters for medium and large desalination plants are presented below.

Filter type	Dual media, downflow
Backwash	Air–water
Average filter cell run Duration	24 h
Flow distribution to individual cells	Pipe (if a concrete channel is used, the channel depth should be tapered to keep velocity in the distribution channel above 2 m/s at all times).
Number of filter cells	8 to 18
Filter cell width	3 to 8 m (10 to 26 ft)
Filter cell depth	4.5 to 7.5 m/15 to 25 ft (typically 5 m 16 ft)
Filter cell length-to-width ratio	2:1 to 4:1 (typically 3:1)
Individual filter cell area	25 to 100 m^2 (270 to 1100 ft^2)
Maximum water depth above filter bed	2.5 m/8 ft (should be equal to or slightly higher than filter bed head loss, which usually is 1.8 to 2.4 m/6 to 8 ft)

Filtration Rate (at Desalination Plant Intake Design Flow)

With all filters in service	8 to 10 $m^3/m^2 \cdot h$ (3 to 4 gpm/ft^2)
With two filters out of service	15 $m^3/m^2 \cdot h$ (6 gpm/ft^2)

Filter Media

Top Layer: Anthracite or Pumice

Anthracite/pumice layer depth for deep bed filters	0.8 to 1.8 m (2.6 to 6.0 ft)
Anthracite/pumice layer depth for shallow bed filters	0.4 to 0.8 m/1.3 to 2.6 ft [used for source water of low turbidity (< 5 NTU) and low organics content(TOC < 2 mg/L)]

Anthracite/pumice effective size	0.8 to 2 mm (typically 1.2 mm)
Anthracite/pumice uniformity coefficient	1.3 to 1.7 (preferably < 1.4)
Anthracite specific gravity	1.5 to 1.6 tons/m^3
Anthracite bulk density	0.8 to 0.85 tons/m^3
Pumice specific gravity	1.1 to 1.2 tons/m^3
Pumice bulk density	0.4 to 0.55 tons/m^3
Bottom Layer: Sand	
Sand layer depth for deep bed filters	0.8 to 2 m/2.6 to 6.5 ft (recommended)
Sand layer depth for shallow bed filters	0.4 to 0.6 m (1.3 to 2.0 ft)
Sand effective size	0.4 to 0.6 mm
Sand uniformity coefficient	< 1.4
Sand specific gravity	2.65 tons/m^3
Sand bulk density	1.5 to 1.9 tons/m^3

Air–Water Filter Backwash System

Maximum backwash rate	55 $m^3/m^2 \cdot h$ (22 gpm/ft^2)
Average backwash rate	40 to 45 $m^3/m^2 \cdot h$ (16 to 18 gpm/ft^2)
Duration (total air plus water)	15 to 30 min (includes filter cell draining and fill-up)

Pressure Filters—Key Design Criteria

Key design criteria for single-stage dual-media pressure filters for small and medium desalination plants are very similar to those of gravity filters. Design criteria by which pressure filters differ from gravity filters are presented here.

Number of filter vessels	6 to 20
Filter vessel diameter	1.2 to 6 m/4 to 20 ft (typically 3 m/10 ft)
Filter vessel length	2.5 to 15 m/8 to 50 ft (typically 6 m/20 ft)
Depth of filter bed	0.6 to 0.9 m (2 to 3 ft)

Filtration Rate (at Desalination Plant Intake Design Flow)

With all filters in service	12 to 25 $m^3/m^2 \cdot h$ (5 to 10 gpm/ft^2)
With two filters out of service	30 $m^3/m^2 \cdot h$ (12 gpm/ft^2)

Head Loss across the Filter Vessel

Total headloss across the filter	15 to 30 m/45 to 90 ft (average 20 m/65 ft)
Net headloss available for filtration	7.5 to 15 m (25 to 50 ft)

11.7.2 Two-Stage Filters

Key Design Criteria

Key design criteria for the first, coarse media, stage of a two-stage media filtration system are as follows:

Filter type	Single or dual media, downflow, air–water backwash
Average filter cell run length	24 to 48 h

Filtration Rate (at Desalination Plant Intake Design Flow)

With all filters in service	12 to 25 $m^3/m^2{\cdot}h$ (5 to 10 gpm/ft^2)
With two filters out of service	30 $m^3/m^2{\cdot}h$ (12 gpm/ft^2)

Anthracite or Sand Filter Media

Anthracite layer depth	0.4 to 1 m (1.3 to 3.3 ft)
Anthracite effective size	1 to 2 mm (typically 1.5 mm)
Anthracite uniformity coefficient	< 1.5
Sand layer depth	0.4 to 1 m (1.3 to 3.3 ft)
Sand effective size	0.4 to 0.6 mm
Sand uniformity coefficient	< 1.5

Air–Water Filter Backwash System

Maximum backwash rate	60 $m^3/m^2{\cdot}h$ (25 gpm/ft^2)
Average backwash rate	45 to 55 $m^3/m^2{\cdot}h$ (18 to 22 gpm/ft^2)
Duration (total air plus water)	20 to 30 min (includes filter cell draining and fill-up)

All other filter design parameters are the same as for single-stage dual-media filters, which are described in the previous section. As indicated previously, the second stage (polishing filter) is typically designed as a dual-media shallow filter, unless enhanced organics removal is needed.

11.7.3 Design Examples

The design examples provided below were developed for a hypothetical 40,000 m^3/day (10.6 mgd) seawater desalination plant designed for a recovery of 43 percent. The pretreatment filtration system has to be designed to produce the quality of filtered water presented in Table 11.1.

Example of a Single-Stage Dual-Media Gravity Filter

In this example, the source water has a turbidity of 0.3 to 10 NTU (TSS of 0.5 to 15 mg/L) with occasional spikes of turbidity to 15 NTU (TSS = 20 mg/L) during rain events. The source water's SDI is in a range of 6 to 12. Maximum algal count is < 20,000 cells per milliliter, and total hydrocarbon levels are below 0.04 mg/L at all times. This water quality is typical for an open ocean intake at a medium depth (6 to 8 m/20 to 26 ft from the water surface).

Since open ocean water does not contain elevated content of silica, iron, or manganese, these fouling compounds will not have impact on filter design. Because of the relatively high level of turbidity in the water, the backwash volume is expected to be 5 percent of the intake source water. Since the plant is designed for a total 43 percent recovery, the total volume of filtered water that will need to be produced for operations is 40,000/0.43 = 93,023 m^3/day (24.6 mgd). In addition, the filtration system will have to be designed to produce backwash water for the filters at a volume that is approximately 5 percent of the source water flow. As a result, the total plant intake flow for which the filters will need to be designed is 93,023/(1–0.05) = 97,920 m^3/day (25.9 mgd).

Table 11.5 presents a summary of the key design criteria for the dual-media (sand and anthracite) gravity filtration system for seawater pretreatment. The system is designed for air–water backwash and filtration cycles of 24 to 48 h, depending on the source water turbidity. The desalination plant source water is preconditioned using ferric chloride for coagulation and polymer for flocculation. The chemical source water conditioning system also includes addition of sulfuric acid to maintain optimum pH for the coagulation process. The filtration system is designed with rinse-to-waste provisions that discharge the first 10 to 15 min of the flow produced by individual filtration cells immediately after their backwash, in order to avoid sending this higher-turbidity water to the RO membrane system.

Example of a Single-Stage Dual-Media Pressure Filter

Pressure filters are less costly than gravity filters in terms of construction costs. However, they apply significantly higher pressures to the algae in the source water, and therefore are preferably used as single-stage filters when the source water is collected by a very deep open ocean intake (typically deeper than 10 m) or by subsurface intakes (e.g., beach wells) that have prefiltered the algae contained in the water via slow sand filtration.

In this example, the source water has a turbidity of 0.2 to 2 NTU (TSS of 0.5 to 5 mg/L) and consistent water quality that is typically not affected significantly by algal blooms or rain events. The source water's SDI is in a range of 3 to 6. Maximum algal count is < 1000 cells per milliliter, and total hydrocarbon levels are below 0.04 mg/L at all times.

For the purposes of this example, it is assumed that the source water does not contain elevated content of silica, iron, or manganese and that therefore these fouling compounds will not have an impact on filter design. Because of the relatively low level of turbidity in the water, the average backwash volume is expected to be only 3 percent of the intake source water. Since the plant is designed for a total 43 percent recovery, the total volume of filtered water that will need to be produced for operations is 40,000/0.43 = 93,023 m^3/day (24.6 mgd). In addition, the filtration system will have to be designed to produce backwash water for the filters at a volume that is approximately 3 percent of the source water flow. As a result, the total plant intake flow for which the filters will need to be designed is 93,023/(1–0.03) = 95,900 m^3/day (25.3 mgd).

Key design criteria for the dual-media (sand and anthracite) pressure filtration system for seawater pretreatment are provided in Table 11.6. Similar to the previous example, the source water is preconditioned using ferric chloride for coagulation and polymer for flocculation. However, the chemical dosages are smaller, reflecting the better-quality source water. The chemical source water conditioning system also includes addition of sulfuric acid to maintain optimum pH for the coagulation process.

Component/Parameter	Specifications/Design Criteria
Feed Water	
Design flow rate, m^3/day (mgd)	97,920 (25.9)
Turbidity, NTU	0.3–10 (maximum 20)
SDI	6–10 (maximum 12)
Design Chemical Dosages	
Ferric chloride, mg/L	10 (0.5–30)
Polymer, mg/L	0.25 (0–1)
Sulfuric acid, mg/L (target pH = 6.7)	8 (0–30)
Filter Cells	
Number	10
Width, m (ft)	4 (13.2)
Length, m	15 (49.2)
Area, m^2 (ft^2)	60 (649)
Total filter cell depth, m (ft)	6.0 (19.7)
Total filter media depth, m (ft)	1.8 (5.3)
Water depth above the filter bed, m (ft)	2.0 (6.6)
Top Layer Filter Medium	
Type	Anthracite
Depth, m (ft)	1 (3.3)
Effective size, mm	1
Uniformity coefficient	1.4
Specific gravity, tons/m^3	1.55
Bottom Layer Filter Medium	
Type	Sand
Depth, m (ft)	0.8 (3.3)
Effective size, mm	0.6
Uniformity coefficient	1.3
Specific gravity, tons/m^3	2.65
Filter Performance Parameters	
Average surface loading rate, $m^3/m^2 \cdot h$ (gal/min·ft^2)	6.8 (2.8)
Maximum surface loading rate with two units out of service, $m^3/m^2 \cdot h$ (gal/min·ft^2)	8.5 (3.5)
Filtration cycle, h	24–48
Backwash Rate	
Average, $m^3/m^2 \cdot h$ (gal/min·ft^2)	40 (16)
Maximum, $m^3/m^2 \cdot h$ (gal/min·ft^2)	50 (20)

Table 11.5 Example of a Single-Stage Dual Granular Media Gravity Filtration System for a 40,000 m^3/day (10.6 mgd) Desalination Plant

Component/Parameter	Specifications/Design Criteria
Feed Water	
Design flow rate, m^3/day (mgd)	95,900 (25.3)
Turbidity, NTU	0.2–2 (maximum 5)
SDI	3–6 (maximum 8)
Design Chemical Dosages	
Ferric chloride, mg/L	1 (0.5–5)
Polymer, mg/L	0.15 (0–0.5)
Sulfuric acid, mg/L (target pH = 6.7)	8 (0–30)
Filter Cells	
Number	8
Diameter, m (ft)	3.5 (11.5)
Length, m (ft)	12 (39.4)
Filter vessel cross-section area, m^2 (ft^2)	42 (452)
Total filter media depth, m (ft)	1.2 (4)
Top Layer Filter Medium	
Type	Pumice
Depth, m (ft)	0.6 (2.0)
Effective size, mm	0.6
Uniformity coefficient	1.3
Specific gravity, tons/m^3	1.15
Bottom Layer Filter Medium	
Type	Sand
Depth, m (ft)	0.6 (3.3)
Effective size, mm	0.4
Uniformity coefficient	1.3
Specific gravity, tons/m^3	2.65
Filter Performance Parameters	
Average surface loading rate, $m^3/m^2 \cdot h$ (gal/min·ft^2)	11.9 (4.9)
Maximum surface loading rate with two units out of service, $m^3/m^2 \cdot h$ (gal/min·ft^2)	15.9 (6.5)
Filtration cycle, h	24–48
Backwash Rate	
Average, $m^3/m^2 \cdot h$ (gal/min·ft^2)	35 (14)
Maximum, $m^3/m^2 \cdot h$ (gal/min·ft^2)	40 (16)

Table 11.6 Example of a Single-Stage Dual Granular Media Pressure Filtration System for a 40,000 m^3/day (10.6 mgd) Desalination Plant

Example of a Two-Stage Gravity and Pressure Filter System

Two-stage filtration systems are usually applied when the source water is collected by a relatively shallow open intake (4 to 8 m from the water surface) or an onshore open intake that collects water from the entire depth of the water column. In this case, the purpose of the first-stage granular gravity media filter is to remove the larger particles captured from the surface water column, such as large algae, silt, and solids. Typically, the first-stage filter produces filtrate with a turbidity below 2 NTU and an SDI in a range of 6 to 8 or better. The main reason why the first-stage filters are selected to be gravity driven rather than pressure driven is to minimize the breakage of algae biomass and associated release of organics from it. The gravity filters operate at hydrostatic pressure, which practically eliminates such breakage by not allowing the algae to penetrate deep into the filters and instead gently retain them on the surface of the top layer of filtration media.

The second-stage filter receives filtrate from the first stage and is designed to polish this filtrate to levels acceptable for processing through the RO system (see Table 11.1). Because most of the algal mass in the source water has already been removed by the coarser first-stage filter, the possibility of algal breakage and release of easily biodegradable organics is reduced significantly. This allows the second-stage filters to be designed as pressure-driven units and the project economics to benefit from the lower costs associated with this type of filter.

The source water in this example is of worse quality than that in the previous two examples; it has a turbidity of 2 to 30 NTU (TSS of 5 to 50 mg/L) with occasional spikes of up to 50 NTU during the most severe period of algal blooms, which occur periodically in the intake area. The source water's SDI is in a range of 10 to 16. Maximum algal count is < 40,000 cells per milliliter, and total hydrocarbon levels are below 0.10 mg/L at all times.

Similar to the other two examples, in this example it is also assumed that the source water does not contain elevated content of silica, iron, or manganese, and therefore these fouling compounds will not have an impact on filter design. Because the pretreatment system consists of two-stage filtration, the overall volume of backwash water is expected to increase, with the first-stage filter's daily backwash water volume equal to 6 percent of the total daily plant intake flow and the second-stage pressure filters to use an additional 4 percent of the total flow.

Taking into consideration that the plant is designed for the same 43 percent recovery, the total volume of filtered water that will need to be produced for operations is still 40,000/0.43 = 93,023 m^3/day (24.6 mgd). However, in order to accommodate the 10 percent (6 + 4) of flow for backwashing of the two stages of filters, the intake source water volume will be 93,023/(1 − 0.10) = 103,360 m^3/day (27.3 mgd).

In this case, as with the other two examples, the source water is preconditioned using ferric chloride for coagulation, polymer for flocculation, and pH adjustment to optimize the use of coagulant. While provisions for addition of coagulant are usually incorporated ahead of each of the two filtration stages, most of the particle precipitation, coagulation, and flocculation occur in the coagulation and flocculation chambers upstream of the first-stage filters. Therefore, coagulant addition upstream of the second-stage filters to water that has already been treated with coagulants is not typically practiced. However, this flexibility to have facilities for in-line coagulation and flocculation of the feed to the second-stage filters is sometimes used when the source water quality is very good and the first-stage filtration can be bypassed.

It should be pointed out that two-stage filtration systems are not very commonly used, because in most existing desalination projects, the location and depth of the intake of the desalination plant has been selected such that the collected source water is of good quality and requires only a single-stage filtration. However, as larger reverse osmosis desalination projects are built or are co-located with power plants in hybrid configurations, the only practical low-cost intake solutions are either onshore intakes or relatively shallow near-shore open intakes, both of which will require the consideration of a more elaborate pretreatment system. An alternative solution for very large desalination plants—400,000 m^3/day (106 mgd) or more—would be to construct deep tunnels under the ocean bottom that extend outside the tidal zone at locations where an intake can be built at a level of 10 m (33 ft) or more below the water surface. Experience with construction of deep tunnel-type intakes in Australia indicates that the costs for such structures could be much higher than the costs of a two-stage pretreatment system. In such cases, two-stage filtration systems could become a more attractive and cost-effective solution than deeper intakes. Tables 11.7 and 11.8 present key design criteria of the first and second stages of a two-stage pretreatment system for a 40,000 m^3/day (10.6 mgd) plant with a shallow or onshore intake.

11.8 Pretreatment System Costs

Figure 11.7 depicts the 2012 construction costs for granular media gravity and pressure filters as a function of the desalination plant intake flow they pretreat. As seen from this figure, pressure filters have a lower cost than gravity filters for the same daily volume of pretreated saline source water.

For example, in the case of the 40,000 m^3/day (10.6 mgd) single-stage gravity filtration system presented in Table 11.5, the plant intake flow is 97,920 m^3/day (25.9 mgd). For this flow, the estimated construction cost of the pretreatment system is \$6.7 million. For the example of the pressure-driven filtration system in Table 11.6, which has a feed flow of 95,900 m^3/day (25.3 mgd), the estimated construction cost of the system is \$3.8 million.

Similarly, for the example of the two-stage gravity and pressure granular media filtration pretreatment system in Tables 11.7 and 11.8, the cost of the first-stage gravity filters—which have a feed flow of 103,360 m^3/day (27.3 mgd)—is \$7.0 million, and the cost of the second-stage pressure filters—at a feed flow of 96,900 m^3/day (25.6 mgd)—is \$3.9 million. As a result, the total construction cost for filtration pretreatment at for this plant is \$10.9 million. Comparison of the costs of these three examples underlines an important fact—the cost of the pretreatment system for the same plant can vary in a wide range (in this case between \$3.8 and \$10.9 million) depending on source water quality.

Chapter 17 presents an example cost estimate for a 40,000 m^3/day (10.6 mgd) desalination plant designed for 50 percent recovery. The intake source water flow of this plant with 5 percent backwash water is $(40{,}000/0.50)/(1 - 0.05) = 84{,}210$ m^3/day (22.3 mgd). At this intake flow, the total cost of a single-stage dual-media gravity filtration pretreatment system is \$5.85 million (see Fig. 11.7). Comparison of the pretreatment system construction costs of a plant with the same freshwater production capacity but a lower recovery (43 percent instead of 50 percent) indicates that the 7 percent difference in recovery results in approximately 15 percent higher capital expenditure for the pretreatment system (\$6.7 million instead of \$5.85 million). However, designing the desalination plant for a higher recovery yields a lower construction cost for the RO system. While the construction cost savings from the design of the RO system at higher recovery usually are greater than the additional expenditures for a larger pretreatment system, the key factors that define the optimum plant recovery are the

Component/Parameter	Specifications/Design Criteria
Feed Water	
Design flow rate, m^3/day (mgd)	103,360 (27.3)
Turbidity, NTU	2–30 (maximum 50)
SDI	8–12 (maximum 16)
Design Chemical Dosages	
Ferric chloride, mg/L	15 (2–40)
Polymer, mg/L	0.25 (0.25–1.5)
Sulfuric acid, mg/L (target pH = 6.7)	8 (0–30)
Filter Cells	
Number	8
Width, m (ft)	4 (13.2)
Length, m (ft)	15 (49.2)
Area, m^2 (ft^2)	60 (649)
Total filter cell depth, m (ft)	5.4 (19.7)
Total filter media depth, m (ft)	1.2 (5.3)
Water depth above the filter bed, m (ft)	2 (6.6)
Top Layer Filter Medium	
Type	Anthracite
Depth, m (ft)	0.6 (3.3)
Effective size, mm	1.4
Uniformity coefficient	1.4
Specific gravity, tons/m^3	1.55
Bottom Layer Filter Medium	
Type	Sand
Depth, m (ft)	0.6 (3.3)
Effective size, mm	0.8
Uniformity coefficient	1.4
Specific gravity, tons/m^3	2.65
Filter Performance Parameters	
Average surface loading rate, $m^3/m^2 \cdot h$ (gal/min·ft^2)	9 (3.7)
Maximum surface loading rate with two units out of service, $m^3/m^2 \cdot h$ (gal/min·ft^2)	12 (4.9)
Filtration cycle, h	24–48
Backwash Rate	
Average, $m^3/m^2 \cdot h$ (gal/min·ft^2)	45 (18)
Maximum, $m^3/m^2 \cdot h$ (gal/min·ft^2)	55 (23)

Table 11.7 Example of First-Stage Gravity Filters of a Two-Stage Filtration System for a 40,000 m^3/day (10.6 mgd) Desalination Plant

Component/Parameter	Specifications/Design Criteria
Feed Water	
Design flow rate, m^3/day (mgd)	96,900 (25.6)
Turbidity, NTU	0.2–5 (maximum 10)
SDI	4–8 (maximum 10)
Design Chemical Dosages	
Ferric chloride, mg/L	0 (0–8)
Polymer, mg/L	0 (0–0.5)
Sulfuric acid, mg/L (target pH = 6.7)	0 (0–30)
Filter Cells	
Number	8
Diameter, m (ft)	3.5 (11.5)
Length, m (ft)	12 (39.4)
Filter vessel area, m^2 (ft^2)	42 (452)
Total filter media depth, m (ft)	1.2 (4)
Top Layer Filter Medium	
Type	Sand
Depth, m (ft)	0.6 (3.3)
Effective size, mm	0.6
Uniformity coefficient	1.3
Specific gravity, tons/m^3	2.65
Bottom Layer Filter Medium	
Type	Garnet
Depth, m (ft)	0.6 (2.0)
Effective size, mm	0.3
Uniformity coefficient	1.3
Specific gravity, tons/m^3	4.10
Filter Performance Parameters	
Average surface loading rate, $m^3/m^2{\cdot}h$ (gal/min·ft^2)	12 (4.9)
Maximum surface loading rate with two units out of service, $m^3/m^2{\cdot}h$ (gal/min·ft^2)	16 (6.5)
Filtration cycle, h	24–48
Backwash Rate	
Average, $m^3/m^2{\cdot}h$ (gal/min·ft^2)	35 (14)
Maximum, $m^3/m^2{\cdot}h$ (gal/min·ft^2)	40 (16)

Table 11.8 Example of Second-Stage Pressure Filters of a Two-Stage Filtration System for a 40,000 m^3/day (10.6 mgd) Desalination Plant

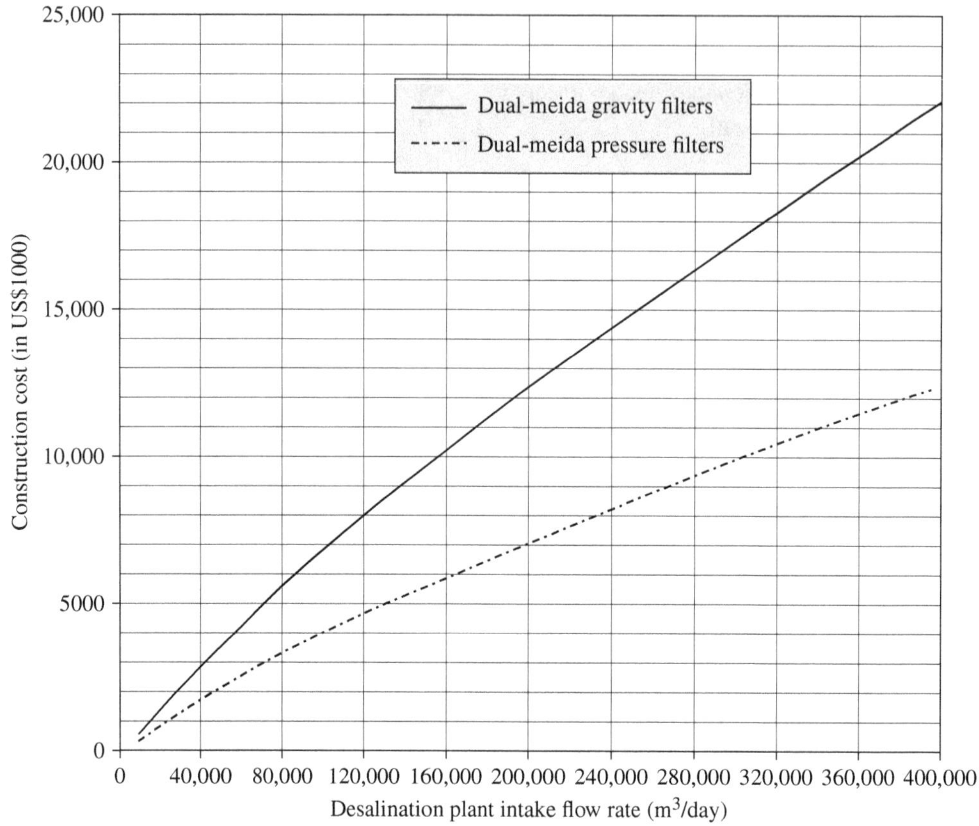

FIGURE 11.7 Construction costs of dual-media gravity and pressure filters.

cost of energy and the type of applied energy recovery system. Since designing the desalination plant for higher recovery results in elevated energy use (and sometimes also in more frequent membrane cleaning), if the cost of electricity is relatively high and the energy recovery system efficiency is sensitive to the plant recovery (as it is with pressure exchangers), design at a lower recovery may be more cost effective overall despite the elevated pretreatment costs.

11.9 References

American Water Works Association, *Manual of Water Supply Practices, M46, Reverse Osmosis and Nanofiltration*, 2nd ed., American Water Works Association, Denver, Colorado, 2007.

Kawamura, S., *Integrated Design and Operation of Water Treatment Facilities*, 2nd ed., John Wiley & Sons, New York 2000.

Qasim, S. R., E. Motley, and G. Zhu, *Water Works Engineering—Planning, Design & Operation*, Prentice Hall PTR, Upper Saddle River, New Jersey, 2000.

Wilf, M., L. Awerbuch, C. Bartels, M. Mickley, G. Pearce, and N. Voutchkov, *The Guidebook to Membrane Desalination Technology: Reverse Osmosis, Nanofiltration and Hybrid Systems Process Design, Applications and Economics*, Balaban, Rehovot, Israel, 2007.

CHAPTER 12

Pretreatment by Membrane Filtration

12.1 Introduction

Particulate, colloidal, and some organic foulants contained in water can be removed successfully using microfiltration (MF) or ultrafiltration (UF) pretreatment. Figure 12.1 depicts a general schematic of water desalination plant with membrane pretreatment. As indicated in this figure, saline water pretreatment includes several key components: (1) coarse and fine screens similar to those used for plants with conventional pretreatment; (2) microscreens to remove fine particulates and sharp objects from the water that could damage the membranes; and (3) a UF or MF membrane system.

Depending on the force (pressure or vacuum) driving the filtration process, membrane pretreatment systems are classified as pressurized (pressure driven) or submerged (vacuum driven). Depending on the size of the membrane pores, the membrane systems used for pretreatment are classified as microfiltration (pores of 0.04 μm) or ultrafiltration (pores of 0.02 μm). While earlier generations of MF elements had pore sizes of 0.1 to 0.2 μm, at present the difference between MF and UF element pores is only a factor of 2 to 3.

Table 12.1 provides a list of some of the large seawater desalination plants with membrane pretreatment in operation at present.

One of the largest pressurized membrane pretreatment systems in the world is located in Perth, Australia, at the 300,000 m^3/day (79 mgd) Southern Seawater Desalination Plant also known as Perth II plant (Fig. 12.2). The largest desalination plant with vacuum-driven membrane pretreatment is the 300,000 m^3/day (79 mgd) Adelaide desalination facility in Australia (Lazaredes and Broom, 2011).

The application of membrane filtration for saline water pretreatment is gaining a wider acceptance (Busch et al., 2009). The number of medium and large desalination plants with membrane pretreatment increased from less than half a dozen in 2002 to over 40 in 2011 (Gasia-Brush et al., 2011).

Practical experience at the 40,000 m^3/day (10.6 mgd) Al Dur seawater desalination plant in Bahrain (Burashid and Hussain, 2004) has shown that membrane pretreatment alone may not always provide a competitive solution, especially for challenging source saline waters of high organic content and biofouling potential. A number of desalination plants with relatively shallow intakes and source water of high turbidity and organic loads (e.g., Shuwaikh, Beckton, Kindasa) have an additional pretreatment step prior to the membrane filtration system in order to cope with these source water quality

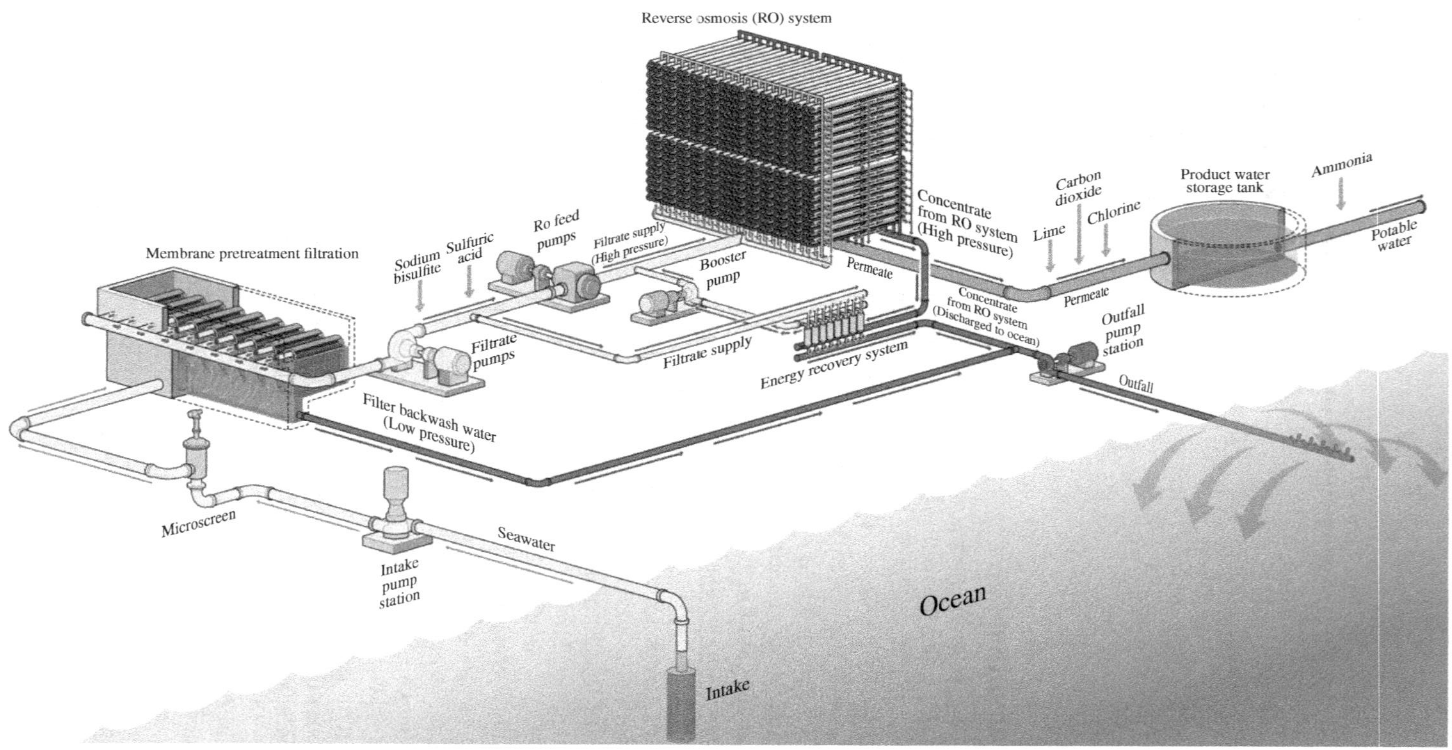

FIGURE 12.1 General schematic of a desalination plant with membrane pretreatment.

Desalination Plant Location and Capacity	Pretreatment System Type, Configuration, and Supplier	Hydraulic Loading Rate, L/m²·h/(gfd)	Notes
Shuwaikh desalination plant, Kuwait—350,000 m³/day (92 mgd)	Pressurized UF (Norit)	60–77 (35–45)	Shallow intake
Adelaide SWRO Plant, Australia—300,000 m³/day (79 mgd)	Submerged UF (Memcor)	52–64 (30–37)	Deep open intake
Southern Seawater Desalination Plant, Perth, Australia—300,000 m³/day (79 mgd)	Pressurized UF (Memcor)	40–50 (23–29)	Shallow open intake
Qingdao SWRO plant, China—232,000 m³/day (61 mgd)	Pressurized UF (Norit)	80–93 (47–54) (average 85)	Open intake
Palm Jumeirah SWRO plants (2), United Arab Emirates—96,000 m³/day (25 mgd) each	Pressurized UF (Norit-Pentair)	60–80 (35–54) (average 68)(40)	Open intake
Beckton desalination plant, London, United Kingdom—150,00 m³/day (40 mgd)	Pressurized UF (Norit-Pentair)	40–60 (23–35)	Saline river intake
Fukuoka SWRO plant, Japan—96,000 m³/day (25 mgd)	Pressurized UF (Hydranautics)	60–80 (35–54)	Subsurface intake
Kindasa SWRO plant, Saudi Arabia—90,000 m³/day (24 mgd)	Dual-media pressurized filtration followed by pressurized UF (Hydranautics)	Dual-media filtration rate: 15–20 m³/m².h (6–8 gpm/ft²) UF flux: 80–100 (54–59 l/m².hr)	Near-shore open intake in an industrial port
Yuhuan SWRO plant, China—36,000 m³/day (10 mgd)	Submerged UF (GE Zenon)	40–60 (23–35)	Open intake

TABLE 12.1 Large Desalination Plants with Membrane Pretreatment

challenges. However, in most installations worldwide, membrane filtration is the only pretreatment system prior to reverse osmosis desalination.

12.2 The Filtration Process

All membrane pretreatment processes have four operational modes: (1) processing, (2) backwash, (3) cleaning, and (4) integrity testing. These operational modes are typically monitored and controlled by a programmable logic controller.

FIGURE 12.2 Module of the UF pretreatment system at the Southern Seawater Desalination Plant in Perth, Australia.

12.2.1 Processing (Filtration)

Membrane filtration of the saline source water takes place during the processing phase. Depending on the specific membrane product and configuration, the filtration process can occur in either direct-flow or cross-flow mode. In direct-flow mode, all of the source water passes through the membranes. In cross-flow operations, only a portion of the source flow (typically 90 to 95 percent) passes through the membranes, while the remaining flow (reject) travels along the feed side of the membranes; its movement along the membrane surface generates shearing velocity that evacuates the solids removed from the saline water out of the membrane.

Usually in cross-flow mode, a portion of the reject stream is recirculated back to the feed system. The cross-flow pattern in these membranes is similar to that in RO membrane elements. The key benefit of such a flow pattern is that membranes can be operated continuously. The main problem with cross-flow elements, however, is that they have a relatively lower packing density, which limits their productivity and requires significant energy costs to maintain a flow tangential to the membrane surface. Therefore, the newest UF and MF membrane elements available on the market are designed to operate in a direct-flow configuration. Direct-flow membranes, however, cannot be operated continuously, because the solids are accumulated on the membrane surface and have to be removed periodically via intermittent backwash.

The two most important membrane performance parameters associated with the filtration cycle of membrane pretreatment systems are membrane flux and transmembrane pressure (TMP).

Membrane Flux

Membrane flux is the volume of pretreated water produced by a unit of membrane area. This parameter is most commonly measured in liters per square meter per hour ($L/m^2 \cdot h$, also often notated as Lmh or lmh) and gallons per day per square foot (gfd). The two flux measures relate as follows: 1 gfd = 1.705 lmh. Typically, pretreatment systems of desalination plants are designed for a flux of 40 to 80 lmh (24 to 47 gfd)—see Table 12.1.

Accumulation of solids on the surface of the membrane and in the membrane pores (membrane fouling) increases with an increase in membrane flux. In order to maintain a reasonably long membrane cycle (30 min or more), the operational flux has to be selected such that the fouling rate of the membranes is reasonably low and that within the time of one filtration cycle, the pressure loss created by the solids accumulated on the membranes stays below the maximum pressure the membrane feed pumps are designed to deliver. Such a flux is referred to as *sustainable flux*. Usually, the higher the solids content in the source water, the lower the design sustainable flux for the same filter cycle length.

Transmembrane Pressure

Transmembrane pressure (TMP) is the difference between the feed pressure and the filtrate pressure of the pretreatment system. This pressure drives the flow through the membranes and therefore is directly related to the membrane flux. The TMP also has an impact on membrane fouling and filtration cycle length.

For most UF and MF membrane systems used for saline water pretreatment, the TMP is usually reported in bars (bar) or pounds per square inch (lb/in^2). Sometimes it is reported in kilopascals (kPa), where 1 bar = 100 kPa = 14.5 lb/in^2. Typically, pretreatment systems operate at a TMP between 0.2 and 1.0 bar (2.9 to 14.5 lb/in^2). Pressure-driven systems can operate at TMPs higher than 1 bar, while vacuum-driven (submerged systems) systems are limited by the maximum vacuum of 1 bar. Because of the potential for excessive vacuum to collapse the membrane fibers, the maximum TMP of submerged membrane pretreatment systems is usually limited to 0.7 bar (10 lb/in^2).

Most pressure systems can operate at a TMP of up to 2.5 bar (36 lb/in^2). Usually, the greater the durability and flexibility of the membrane fibers, the higher the maximum pressure those fibers can handle. Based on this and other criteria, the maximum TMP varies by supplier.

Membrane Permeability

Similar to reverse osmosis membranes, another important performance parameter associated with the filtration cycle is membrane permeability, which is defined as the ratio between the membrane flux and the transmembrane pressure. Membrane permeability is measured in Lmh/bar or gfd/(lb/in^2) [1 gfd/(lb/in^2) = 25 Lmh/bar]. Most MF and UF membrane elements used for saline water pretreatment operate at a membrane permeability of 75 to 500 Lmh/bar [5 to 20 gfd/(lb/in^2)].

12.2.2 Backwash

During processing mode, solids filtered out of the source water accumulate on the feed side of the membrane surface. These solids are periodically removed from the filtration system by backwashing of the membranes with filtered water or concentrate. Backwash is usually triggered by timer and occurs every 30 to 120 min for approximately 30 to 60 s. Backwash can also be initiated when the TMP reaches a certain maximum threshold, beyond which the membrane system cannot perform at the target flux and filtered water quality. If the threshold TMP is exceeded, typically the membrane system production capacity (flux) is decreased, the filtered water quality deteriorates, and the membranes could be exposed to irreversible fouling.

Membrane backwash is a multistep process that usually applies a combination of filtered water and air in a sequence and at rates designed to maximize the removal of particulates that have accumulated in the membrane system during the processing cycle. Backwash plays a very important role in the normal operation of membrane systems, because membranes have a significantly smaller volume or capacity available to store solids within the system than do granular media filters. This smaller solids retention capacity of membrane systems is the main reason that membrane modules have to be backwashed 30 to 50 times as frequently as filter cells (i.e., typically once every 30 min vs. once every 24 h).

Air–water backwash is mainly intended to remove particulates from the membrane pretreatment system, and does not involve the use of any cleaning chemicals. However, over time the membrane surface also accumulates organic deposits and biofilm. This type of membrane fouling is controlled by chemically enhanced backwash (CEB), also referred to as *maintenance wash*, which is typically practiced once or twice per day. During CEB, the membranes are soaked for several minutes in chlorine and sometimes other cleaning chemicals (acids, alkalies, or sodium bisulfite) and then backwashed. The needed chemical dosages are a function of the predominant types of foulant in the source water and on the membranes, and of the type of membrane material and configuration.

12.2.3 Cleaning

Periodic membrane backwash and CEB do not completely eliminate membrane fouling, and therefore the TMP needed to produce filtered saline water of the target volume (flux) and quality increases over time. Once the TMP reaches a preset level (typically 0.7 to 0.8 bar (10 to 12 lb/in^2) for submerged systems and 1.5 to 2.5 bar (22 to 36 lb/in^2) for pressurized systems), the membrane modules have to be taken off-line and cleaned with chemicals that aim to reduce the TMP to a reasonable level.

Membrane cleaning is typically needed every 1 to 3 months and is performed using a combination of low-pH solution of citric or sulfuric acid followed by a high-pH solution of sodium hydroxide and sodium hypochlorite. The cleaning chemicals are recirculated through the membranes for a period of 8 to 24 h and then the membranes are flushed and returned to normal operation. Depending of the nature of the fouling, sometimes other cleaning chemicals (biocides) are used to address specific fouling compounds (e.g., oil and grease, excessive biogrowth, etc.).

Typically, inorganic foulants such as iron and manganese solids accumulated on the membrane surface are cleaned with citric acid, sulfuric acid, or hydrochloric acid. Fouling caused mainly by organic materials is treated by a base such as sodium hydroxide, while biological and algal fouling is cleaned using disinfectants such as sodium hypochlorite, peracetic acid, or surfactants. In all cases, the clean-in-place (CIP) is completed with heated cleaning solutions and RO permeate or filtrate.

12.2.4 Integrity Testing

All membrane pretreatment systems are equipped with integrity testing features that allow the detection of occasional breaks or punctures in the membrane fibers or leafs; cracks in membrane modules, piping, and connectors; and other problems that could occur during membrane production, installation, or operation. The most widely used membrane system integrity test is a pressure hold and visual test performed while the system is off-line.

During the pressure hold test, water is purged from the system using filtrate and then air is applied under a pressure of 0.3 to 1 bar (4.2 to 14.2 lb/in^2) and the decay of the air pressure is monitored over time. Typically, the membrane module integrity is adequate when the pressure loss over a 5-min period is less than 10 percent of the initial pressure applied to the membranes. The pressure hold tests and conditions can vary for the various commercially available membrane products and configurations, and therefore the membrane integrity testing system and conditions have to be coordinated with the membrane system supplier or equipment manufacturer.

Besides the pressure hold test, other off-line membrane integrity tests used include the vacuum hold test, the bubble point test, diffusive airflow tests, etc. (American Water Works Association, 2007). In addition, membrane integrity is monitored online by counting particle passage, measuring filter effluent turbidity for the individual membrane modules, or acoustic sensing.

The most popular method for online integrity monitoring is continuous measurement of the effluent turbidity for the individual membrane modules (trains) comprising the membrane system. Usually a breach of integrity in a given train or module in a membrane pretreatment system is identified by comparing the effluent turbidity of that train or module to the average turbidity of the other modules.

12.3 Key Filtration System Components

12.3.1 Filter Vessels and Modules

MF and UF filtration membranes are configured in individual functional filtration units referred to as *modules*. The most commonly used membrane module configurations are hollow fiber, tubular, flat plate, and spiral. The membrane modules are contained in housings, shells, or cassettes that are assembled into larger membrane filtration system components—vessels and racks.

12.3.2 Membrane Filtration Media

Membrane Materials

Membranes used for saline water pretreatment are typically made of polyethersulfone (PES), polyvinylidene diflouride (PVDF), or polysulfone. All of the membrane products made of these materials are hydrophilic, with PES being the most hydrophilic of all of these materials. Hydrophilic materials have two key advantages: (1) they wet easily, which makes them more permeable for a given pore size, and (2) they have higher resistance to attachment of organic materials on their surface (i.e., to biofouling). Table 12.2 presents a summary of key membrane suppliers and the types and materials of their membrane products.

An overview of the information in Table 12.2 indicates that the most commonly used membrane materials at present at PVDF and PES. Pearce (2011) provides detailed information on the benefits of alternative membrane materials. In general, PES has higher permeability but lower durability and chlorine resistance. Conversely, PVDF membranes have higher strength and flexibility. Such material features are very important if the membranes will be exposed to high pressures or pressure surges and have to withstand significant mechanical stress. These membrane characteristics, especially flexibility, are of critical importance if the membranes will be cleaned by air backwash, which typically creates significant stress on the membrane fibers and potting interface. This also is one of the main reasons why PES membrane systems are backwashed with water only. Accidental release of large amounts of air in the feed to PES systems could cause fiber breakages and potting interface challenges.

Conditions of elevated fiber breakage could occur, for example, in a pretreatment and RO system configuration without break tanks between the intake pump station, pretreatment system, and RO feed piping. In this pump-through configuration, the desalination plant's intake pumps convey source water directly through the UF/MF system into the suction of the booster pump of the RO feed system. Because the required booster pump suction pressure is usually in a range of 2.5 to 4 bar (36 to 58 lb/in^2), the UF/MF system has to be designed to convey water at this pressure. This direct pumping configuration results in the feed pressure to the UF/MF system in a range of 2.7 to 5 bar (39 to 73 lb/in^2) , which is relatively close to the maximum pressures most pressure-driven membrane pretreatment systems are designed to operate at (i.e., around 6 to 8 bars/87 to 116 lb/in^2). If a hydraulic surge is triggered by the abrupt shutdown of the intake pumps, and the surge protection of the pipeline connecting the

Membrane Manufacturer	Type of Membrane	Membrane Material	Direction of Flow
Norit (Pentair)	Pressure-driven UF	PES	Inside-out
Memcor (Siemens)	Pressure- and vacuum-driven MF	PVDF	Outside-in
Hydranautics—(Nitto Denko)	Pressure-driven UF	PES	Inside-out
Hyflux	Pressure-driven UF	PES & PVDF	Outside-in
GE Zenon	Vacuum-driven UF	PVDF	Outside-in
Dow	Pressure-driven UF	PVDF	Outside-in
Toray	UF and MF Pressure-driven	PVDF	Outside-in
Pall/Asahi	Pressure-driven MF	PVDF	Outside-in
Inge (BASF)	Pressure-driven UF	PES	Inside-out
Koch	Pressure-driven UF	polysulfone and PES	Inside-out

Table 12.2 Materials of MF/UF Membrane Products Used for Saline Water Pretreatment

intake pump station to the feed of the UF system is inadequate, this surge could create pressures along the water flow path well above double the operating pressure (i.e., over 10 bar/145 lb/in^2), which the membrane material and potting interface may not be able to withstand if their strength is inadequate.

In general, PVDF membrane systems are better suited to handle pressure surges, and have enhanced durability and chlorine resistance. However, this type of membrane product is also more expensive and has a lower permeability and base resistance than PES. In addition, PES membranes have a narrower pore size distribution, which can be beneficial in terms of the filtered water quality.

Membrane Geometries

The most widely used membranes in saline water pretreatment have hollow-fiber, tubular, or spiral-wound geometry. Hollow-fiber membranes typically consist of several hundred to several thousand membrane fibers bounded at each end by epoxy or urethane resin and encased in individual modules. Typically the internal diameter of the membrane fibers is 0.4 to 1.5 mm (0.02 to 0.06 in).

Depending on the membrane manufacturer, the hollow-fiber (capillary) membrane elements may be operated in an inside-out or outside-in flow pattern. The inside-out mode of operation provides a better control of flow and more uniform flow distribution. But the outside-in flow pattern usually results in lower head losses through the module, and operation under this pattern is less sensitive to the amount of solids in the source saline water.

Tubular membranes have inner tube diameters that are an order of magnitude larger than those of hollow fiber membranes (i.e., 1 to 2.5 cm/0.4 to 1.0 in). The individual membrane tubes are placed inside a fiberglass-reinforced plastic or stainless steel tube, and the two ends of the tube are sealed with a gasket or other clamp-type device. The typical flow pattern for these membranes is inside-out, i.e., the source saline water is introduced into the tube lumen under pressure and flows through the walls of the membrane tubes into the outside shell of the module.

The key advantages of hollow-fiber membrane elements are as follows: (1) the high ratio of surface area to volume (packing density) allows for a reduction in the overall footprint of the filtration system; (2) the fibers can be easily backwashed; (3) filtration can be completed at a low pressure—TMP is typically 0.2 to 1 bar (2.9 to 14.5 lb/in^2); and (4) the pressure drop across the membrane modules is low (0.1 to 1 bar/1.45 to 14.5 lb/in^2).

Tubular membranes have the following advantages: (1) large channel diameters allow them to treat waters of higher solids content as compared to hollow-fiber membranes operated in an inside-out mode (this advantage is not significant if an outside-in hollow-fiber membrane is compared with an inside-out tubular membrane); and (2) they can be operated at approximately double the cross-flow velocity, which is beneficial in terms of biofouling control.

12.3.3 Service Facilities and Equipment

All membrane pretreatment systems have three types of service support facilities and equipment: (1) backwash system, (2) CIP system, and (3) cleaning chemical feed system. The backwash system typically includes a filtered water storage tank, backwash pumps, and (depending on the membrane system) air compressors for air backwash. Some membrane pretreatment systems (e.g., Norit X-Flow) use only water backwash.

The CIP system for the membrane pretreatment facility is very similar in configuration to that of the RO system. Sometimes, the same CIP system is used for both pretreatment and RO membrane cleaning, although this is not desirable, especially for larger desalination plants. The chemical feed system usually includes acid, base, sodium hypochlorite, and sodium bisulfite storage and feed systems to service the CEB and CIP membrane maintenance activities. In addition, some of the membrane pretreatment systems are designed for enhanced performance by addition of conditioning chemicals to the feed water (coagulants, flocculants, powdered activated carbon, and pH adjusters). However, most saline water pretreatment systems are designed to operate without source water conditioning under normal operational conditions.

12.4 Filter Types and Configurations

Depending on the type of driving filtration force, membrane pretreatment filters are divided in two categories—pressurized (pressure driven) and submerged (vacuum driven).

12.4.1 Pressurized Membrane Systems

Pressurized UF and MF systems consist of membrane elements installed in pressure vessels, which are grouped in racks (trains) similar to those of RO systems (Fig. 12.3).

FIGURE 12.3 Pressurized MF system with vertical membrane elements.

At present, practically all key membrane suppliers offer pressurized UF and MF systems. Depending on the direction of the feed flow through the membrane, pressure-driven systems are divided into outside-in (also referred to as pressure-driven outside, or PDO, feed) systems and inside-out (or pressure-driven inside, or PDI, feed). In PDO systems the source water is distributed around the filter fibers, and after passing through the membrane, the filtered water is collected through the fiber lumen. In PDI systems the source water is fed into the filter lumen and is collected on the outside of the fibers.

In general, PDO systems are more difficult to clean with water backwash only, requiring air backwash to achieve the same level of productivity recovery as PDI systems. In addition, PDO systems operate either at higher feed pressures for the same design flux as PDI systems or at lower fluxes at the same design TMP.

A general schematic indicating the key components of a pressurized system is presented in Fig. 12.4. As shown in the figure, the source water conveyed by the intake pumps passes through a microscreen into a wet well, from where it is pumped into the UF system. The filtered water is collected from the system and directed to a storage tank, from where it is pumped into the RO system.

12.4.2 Submerged Membrane Systems

Submerged UF and MF systems consist of membrane modules installed in open tanks (Fig. 12.5).

A general schematic indicating the key components of a submerged pretreatment system is presented in Fig. 12.6. All submerged systems are outside-in systems, in which the filtered water is conveyed into the fiber lumens by the vacuum applied on the lumens. The membrane modules are typically installed in concrete or metal tanks and are designed so they can be removed relatively easily for inspection. Each of the tanks can be operated individually and taken out of service for cleaning, inspection, and maintenance. Usually the tanks are open to the air; they can be installed under a light shed for direct protection of the membranes and equipment against sunlight. In some existing plants, the submerged source water pretreatment system is installed in a building.

12.4.3 Comparison of Pressurized and Submerged Systems

Pressurized membrane systems use membrane elements installed in pressure vessels or housings, and the membrane separation process in these systems is driven by of pressure. Submerged systems use membrane modules or cassettes that are immersed in tanks and operate under a slight negative pressure (vacuum). It is recommended that the following issues be considered in choosing between a submerged and a pressurized membrane pretreatment system.

Handling Source Water Quality Variations

Submerged membrane systems are usually more advantageous for treating source saline water of variable turbidity, such as intake surface waters that experience frequent turbidity fluctuations of 20 NTU or more. Because these membrane systems are located in tanks (vessels) with a relatively large holding volume, they can retain and equalize the source water solids load in the tanks and thereby reduce the impact of water quality fluctuations on pretreatment system performance. Since shallow open intakes often yield source water with wide turbidity fluctuations, submerged pretreatment systems are usually more suitable for such applications.

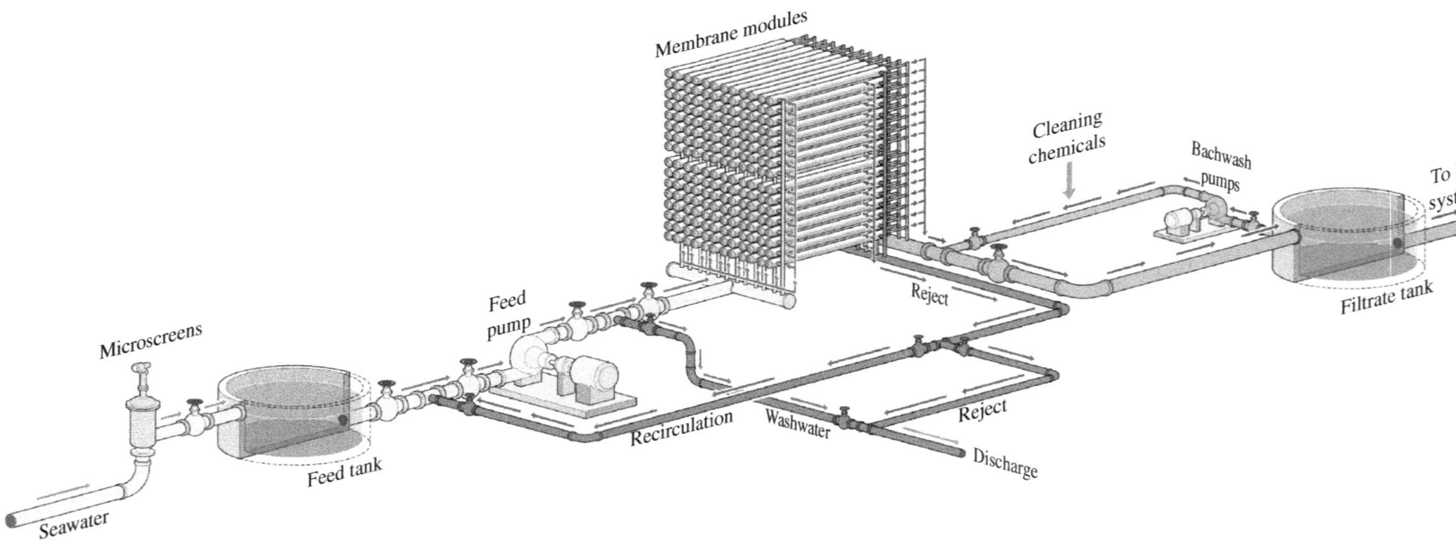

FIGURE 12.4 General schematic of a pressurized membrane pretreatment system.

FIGURE 12.5 Submerged MF system.

Pressurized membrane systems have a limited capacity to retain solids, due to the fact that the individual membrane elements are located in a tight membrane vessel with a very small retention volume. Therefore, if a pressurized system is exposed to a large amount of solids, the membrane elements and vessels will fill up with solids very quickly, which in turn will trigger frequent membrane backwash and result in destabilization of the membrane system's performance. If a pressurized membrane system is overloaded with solids, it will have to be derated to its capacity to hold solids. Otherwise, it will experience very frequent backwashes and ultimately interruption of normal operations.

To address this deficiency of pressurized membrane systems, some membrane manufacturers offer membrane modules with adjustable fiber density, which allows customization of membrane system design to more challenging water quality. Typically, these customized membrane elements have fewer fibers and more empty space within the membrane elements, thereby providing more volume to retain higher influent solids loads. However, this customization is usually at the expense of installing more membrane elements and enlarging the overall size and cost of the membrane system.

Typically, the tanks in which submerged membrane elements are installed provide a minimum hydraulic retention time of 10 to 15 min and have an order of magnitude

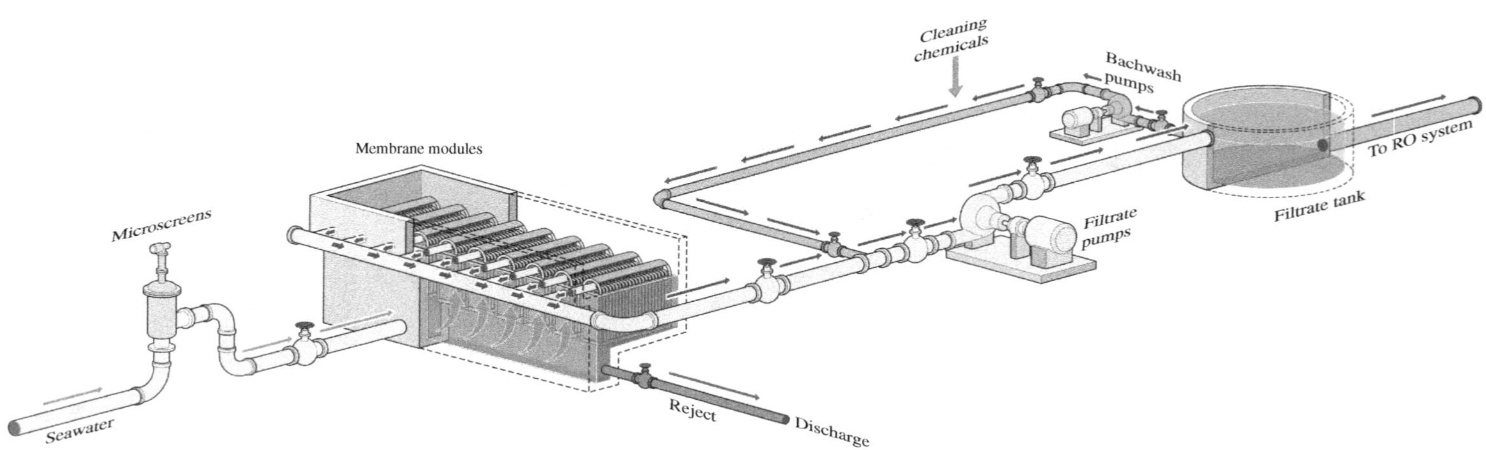

Figure 12.6 General schematic of a submerged membrane pretreatment system.

higher volume and capacity available to handle saline source water of elevated turbidity and to store solids than pressure-driven membrane elements. This renders submerged membrane pretreatment systems more suitable for high-turbidity water applications. The aeration scouring that submerged systems which is mainly applied for backwashing also improves their tolerance to high solids loads. In addition, since submerged systems usually operate at a lower transmembrane pressure, their fouling rate is lower and they have more stable operation during transient solids load conditions.

Pressurized membrane pretreatment systems, however, are often more suitable for cold saline source waters (i.e., saline waters with a monthly average temperature for the coldest month of the year of 15°C or less). The productivity of submerged systems is more sensitive to source water temperature and viscosity. The maximum operational transmembrane pressure available for submerged membrane systems is limited to 1 bar of vacuum, although in practical terms such systems operate at a lower maximum TMP (0.7 to 0.8 bar/10 to 12 lb/in^2).

Usually, pressurized membrane systems are very suitable and cost competitive for source waters collected by deep open intakes and subsurface wells, since such waters have relatively limited turbidity fluctuation. Because pressure-driven systems can operate at higher fluxes for the same TMP, they can be designed more aggressively and be more cost competitive in such applications.

System Footprint

If the source water has high solids content that limits the design flux of pressurized systems, submerged membrane pretreatment systems are usually more space efficient, because they permit the installation of larger membrane surface area per unit facility footprint. Significant space reduction is achieved by the fact that the submerged membrane elements do not need to be installed in individual membrane vessels. In addition, submerged systems typically have only one pipeline system for permeate collection. The distribution of the feed water and the collection and evacuation of the spent filter backwash water are completed at tank level, which allows the system design to be simplified and the number of valves, pipes, and auxiliary service facilities to be reduced.

The smaller footprint of submerged membrane pretreatment systems renders them more beneficial for large water treatment plants. Typically, submerged membrane systems occupy 10 to 20 percent less space than pressurized membrane installations for pretreatment systems with the same design flux. However, for colder source water of very high quality—which is typical for desalination plants with deep open intakes or subsurface intakes—because pressurized systems can be designed at 20 to 30 percent higher fluxes, they yield pretreatment installations of a smaller overall site footprint.

Submerged pretreatment systems have a clear advantage when it comes to retrofitting existing granular media pretreatment systems. For example, submerged membrane tank modules can potentially be installed in existing granular media filter cells, filter backwash tanks, disinfection contact tanks, or other existing structures. Typically, a conservatively designed granular media filter structure can house a submerged membrane filtration system with 1.5 to 2 times the production capacity of the original filtration system, with only moderate structural modifications.

Equipment and Construction Costs and Energy Requirements

Depending on the size of the system and the quality of the intake saline water, site-specific conditions may favor the use of either a pressurized or a submerged membrane pretreatment system. Pressurized systems are typically very cost competitive at small and medium installations because they can be manufactured and assembled in a factory off-site and shipped as packaged installations without the need of significant site preparation or construction of separate structures.

As pressure-driven pretreatment technology evolves, the construction costs of these membrane systems are reduced and larger individual modules become available on the market. Most recent projects indicate that pressure-driven membrane systems are becoming very cost competitive for all plant sizes (Bush et al., 2009).

Equipment and construction costs of larger plants with more challenging saline source water quality are typically lower when using submerged membrane systems, especially for plant retrofits. An exception to this rule of thumb is the treatment of low-temperature source waters.

Because for the same water quality, submerged systems of the same type (MF or UF) usually operate at a lower pressure than pressurized systems, their total power use is slightly lower. Typically, submerged systems may use 10 to 30 percent less energy than pressurized systems for water sources of medium to high turbidity and temperature between 18°C and 35°C.

Commoditization Potential

Currently, submerged and pressurized membrane systems differ in the type and size of the individual membrane elements, the configuration of the membrane modules, the type of membrane element backwash, and the type of membrane integrity testing method. However, submerged systems are easier to standardize due to their simplified configuration.

The lack of membrane system unification, standardization, and commoditization makes the membrane plant owner dependent on the membrane manufacturer supplying the system to continue to provide membrane elements for the system and to improve existing technology in order to stay competitive and match the performance of other membrane manufacturers in the future. As a result, the owner of the membrane water treatment plant takes the risk that the membrane technology used at the time of plant construction will become obsolete and out-of-date in the near future due to the accelerated dynamics in the development of new membrane technologies and products or due to the original system manufacturer's exiting the membrane market.

The inherent risks associated with the incompatibility of the membrane technologies available today can be partially mitigated by selecting and configuring a membrane pretreatment system such that it can be designed to accommodate its replacement with at least one other existing system or membrane element of similar type. From this point of view, conservatively designed submerged membrane systems offer a better opportunity to handle future changes.

Currently, the submerged systems available on the market have many more similarities than differences, as compared to pressurized systems. Typically, all existing submerged systems use tanks of a similar size and depth to house their membrane modules or cassettes, and have comparable membrane CIP and backwash systems.

Submerged systems can be designed around the use of a particular membrane technology, but because of their similarities, their tanks and auxiliary facilities can be sized to accommodate the replacement of the initially selected membrane system with an alternative membrane system from other manufacturer, if needed in the future. For comparison, pressurized systems are more difficult to commoditize because of the major differences in size, diameter, type of pressure vessels, and type of backwashing system.

12.5 Filter Performance

12.5.1 Removal of Solids

MF and UF membrane systems have been shown to be very effective for turbidity removal as well as for reduction of nonsoluble and colloidal organics from saline source waters. Turbidity can be lowered consistently below 0.1 NTU and filter effluent silt density index (SDI_{15}) levels can be reduced below 3 over 90 percent of the time.

12.5.2 Removal of Organics

Membrane pretreatment does not remove a significant amount of dissolved organics or aquatic bacteria, which cause RO membrane biofouling. Because of the very short water retention time of the membrane pretreatment systems, they do not provide a measurable biofiltration effect unless they are designed as membrane bioreactors. For comparison, granular media filters—depending on their configuration, loading rate, and depth—can remove 20 to 40 percent of soluble organics from source water.

12.5.3 Removal of Microorganisms

Algae

UF and MF membranes can remove most algae. Their operation will typically not be affected by mild or moderate algal blooms, when algal content in the saline source water is less than 20,000 cells per milliliter. Depending on the membrane type, product, and operating TMP, higher algal content could cause varying degrees of reduction of membrane productivity, resulting in shorter filtration cycles and more frequent CEB and CIP membrane cleanings.

Bacteria, Viruses, and Protozoa

Both MF and UF systems can remove 4 or more logs of pathogens such as *Giardia* and *Cryptosporidium*. The newest-generation membranes can also effectively remove viruses. Typical UF membrane elements with a pore size of 0.01 to 0.02 μm can remove over 4 logs (99.99 percent) of viruses. MF elements with pore sizes of 0.03 μm or less can achieve 3-log virus removal. Older MF membranes (with pore openings of 0.1 to 0.2 μm) do not provide effective virus removal.

In general, neither MF nor UF membranes remove marine bacteria completely and they cannot be considered effective barriers for preventing biofouling of the downstream SWRO elements.

12.6 Planning and Design Considerations

To date, UF membranes have found wider application for saline water pretreatment than MF membranes. Results from a comparative study of the two (Kumar et al., 2006) indicate that "tight" (20,000 Da) UF membranes can produce filtrate with lower RO membrane fouling potential than can 0.1-μm MF membranes. For comparison, these MF membranes produce filter effluent of water quality similar to that of 100,000-u UF membranes.

Under conditions in which large amounts of silt particles of size comparable to that of the MF membrane pores are brought into suspension by naval ship traffic or ocean bottom dredging near the area of the intake, the silt particles may lodge in side the MF membrane pores during the filtration process and ultimately may cause irreversible membrane fouling. Since UF membranes have smaller pores and a different membrane fiber structure than MF membranes, they typically do not face this problem. Potential problem of this nature can usually be identified through side-by-side pilot testing of MF and UF membrane systems during periods of elevated silt content.

The two most important parameters associated with the design of any membrane pretreatment system are the design flux and feed water recovery. Membrane flux determines the amount of total membrane area and the number of modules or elements needed to pretreat a certain volume of saline water. Feed water recovery indicates the fraction of the source saline water that is converted into filtrate suitable for saline water desalination. A number of factors can impact the selection of these two parameters and ultimately influence the size and configuration of the membrane pretreatment system. These factors are discussed in the following sections.

12.6.1 Source Water Turbidity

The quality of the source saline water has a significant impact on the configuration and design of the membrane pretreatment system. Saline water with a higher turbidity will result in the need for a system designed around lower membrane flux and will usually yield lower overall recovery due to higher backwashing frequency requirements. Since the decrease of design membrane flux results in a proportional increase in the total needed membrane surface area (i.e., requires additional membrane modules and equipment), and reduced system recovery means that additional saline water will need to be collected to produce the same volume of filtered water, then depending on the solids content of the saline source water, it may be more economical to remove a portion of the solids prior to membrane pretreatment.

Typically, saline source water with an annual average turbidity lower than 20 NTU can be treated economically without upstream solids removal. If the source water has a consistently higher turbidity, or if the water intake experiences frequent and extensive algal blooms and/or turbidity spikes, then solids removal by dissolved air flotation, sedimentation, or coarse media filtration may be warranted and economical.

The most prudent approach to determining the effect of saline source water on the design of the membrane pretreatment system is to complete pilot testing during times of the year when algal blooms and/or frequent rain events occur. If turbidity spikes in the source water related to rain and algal blooms occur during different periods of the year, pilot testing should encompass both of these periods.

12.6.2 Source Water Organic Content

Similar to turbidity, a high organic content in the source saline water will result in a lower design flux and sometimes lower recovery. Depending on the nature of the organic compounds in the saline water, pretreatment membranes can experience biofouling similar to that of SWRO membranes. If the organics in the source saline water are natural organic matter that can be coagulated easily, such as humic acids, coagulation and flocculation upstream of the membrane filtration may result in significant improvement of membrane flux and performance.

However, neither MF nor UF membranes are very effective in removing organics and marine bacteria associated with algal bloom events, even if the source saline water is conditioned with coagulant. Therefore, if the source saline water is exposed to frequent and intensive algal bloom events in which the source water's total organic carbon is higher than 2 mg/L for prolonged periods of time (i.e., a week or more), the membrane pretreatment system will need to be designed for a conservatively low flux and recovery.

The most prudent approach to determining the effect of biofouling related to algal bloom on the design of the membrane pretreatment system is to complete pilot testing during the warmest month of the year, when algal blooms are most likely to occur and are most intense.

12.6.3 Source Water Temperature

Saline water viscosity increases with a decrease in temperature. Viscosity affects a membrane's ability to produce filtered water, as more pressure (or vacuum, for submerged systems) is required to overcome the resistance associated with flow across the membrane surface area when operating at a constant flux (i.e., producing the same filtered flow). Typically, average design membrane flux is established for the average annual temperature, flow, and turbidity and adjusted down for the minimum monthly average temperature using the correction factor shown in Table 12.3.

For example, if the flux determined for average annual conditions and temperature is 80 Lmh (47 gfd), and the average annual temperature is 20°C but the minimum monthly average temperature is 15°C (59°F), then the design flux should be reduced by approximately 15 percent (i.e., down to 68 Lmh/40 gfd) in order for the membrane pretreatment system to be able to produce the same filtrate flow during all months of the year at approximately the same recovery and power demand.

Minimum Monthly Average Temperature, °C	Flux Correction, %
5	55
10	30
15	15
20	0
25	−10

Table 12.3 Temperature Correction Factor

The correction factor presented in Table 12.3 is a rule of thumb based on practical experience—and it may vary from one membrane product to another. Most membrane manufacturers have recommended capacity compensation factors for their systems; they should be consulted when such a factor is selected for the site-specific conditions of a given project. The most prudent approach to determining the effect of temperature and increased viscosity on the design of the membrane pretreatment system is to complete pilot testing during the coldest month of the year.

Experience with Existing Installations

While UF and MF membrane experience for saline water applications is fairly limited at present, over 10 years of such full-scale experience exists for freshwater applications. An operations survey completed at 10 fresh drinking water treatment plants in the United States (Atassi et al., 2007) identified a number of challenges these plants experienced during their start-up, acceptance testing, and full-scale operations (Table 12.4). While the MF and UF systems were used for freshwater treatment, most of the lessons learned from these applications are relevant to saline water applications as well.

12.7 Overview of Membrane Products Used for Saline Water Pretreatment

12.7.1 Norit (Pentair X-Flow)

Norit (now Pentair X-Flow) has two main membrane products that have found wide application for desalination plant pretreatment: Seaguard (Fig. 12.7) and Seaflex (Fig. 12.8).

The Seaguard system consists of horizontal glass-reinforced plastic vessels (housings) to which source water is delivered by common distribution piping. Each housing contains four UF membrane elements (modules), model SXL 225, connected in series. The Seaflex pretreatment system consists of single vertical membrane elements (modules) fed from a common source water line. The single module of the Seaflex HP (high-pressure) pretreatment system is installed in a glass-reinforced plastic pressure vessel. The vertical module of the Seaflex LP (low-pressure) system is directly attached to the feed lines and is not contained in a separate vessel. Table 12.5 provides a summary of the key design parameters for these systems.

Both the Seaguard and Seaflex membrane pretreatment systems employ PES UF membranes with an inside-out flow pattern and pore sizes of 0.02 to 0.025 μm. The horizontal Seaguard systems have some similarities with RO systems in terms of vessel diameter and general configuration. However, these systems have only four membrane elements (instead of the seven or eight used in RO vessels) and the UF vessels fed from both ends with a dead end at the center of the vessel. This configuration is driven by the fact that because of the high fluxes at which these systems operate, flow distribution within the vessels is even worse than in RO systems, and the fourth element in a sequential UF configuration will be used very inefficiently (i.e., will have very low productivity). In order to address the uneven hydraulic distribution within the vessels (housings), Norit (Pentair-X Flow) has incorporated bypass tubes within the vessels.

The horizontal (Seaguard) design allows for a more efficient use of the height of existing buildings, since a number of vessels can be added vertically in the same basic footprint. This system also has fewer valves and overall shorter interconnecting piping.

Utility Number	Primary Membrane System Problems	Root Causes of Problems	Lessons Learned
1	• Inability to meet design capacity	• Lower achievable flux than projected • Pilot testing did not address extreme water quality conditions	• Pilot test during extreme water quality conditions • Use a conservative safety factor when up-scaling pilot testing results
2	• High CIP frequencies • Excessive downtime and operations and maintenance costs	• Excessive membrane fouling • Pilot testing did not address extreme water quality conditions	• Pilot test during extreme water quality conditions • Consider additional pretreatment to address extreme water quality conditions
3	• Inability to meet design capacity	• Undersized membrane ancillary support systems	• Ancillary support systems can be a significant bottleneck if undersized
4	• Higher membrane replacement costs	• Shorter membrane life than projected • Potential membrane fouling and lack of previous data from suppliers	• Additional pretreatment may be needed to obtain the useful membrane life indicated by the membrane supplier
5	• Excessive downtime and maintenance • Lower water quality than projected	• Excessive fiber breakage • Fouling or poor water quality putting higher stress on the fibers than expected	• Lack of experience with use of membranes for a given water quality may require a change in membrane chemistry and durability
6	• Inability to meet design capacity	• More downtime than anticipated • Manufacturer failure to include valve opening and closing time in integrity tests • Insufficient installed membrane capacity	• Complete a thorough review of the downtime for all MF and UF system operational steps under the worst-case operations scenario
7	• Higher operations and maintenance costs than expected	• More frequent chemical cleaning needed than initially projected	• Pilot test during extreme water quality conditions • Use a conservative safety factor when up-scaling pilot testing results
8	• Excessive system downtime	• Failures in membrane potting • System failure to handle water pressure and potting materials not tested previously	• Never use a membrane that has components or materials that have never been tested previously
9	• Difficult system operation	• Insufficient system training for staff	• Plan for additional staff training beyond the minimum offered by the manufacturer
10	• Excessive downtime • Failure to meet product water quality targets	• Frequent failing of membrane integrity testing • Air leakage from gaskets and valves	• Make sure that replacement of failed gaskets, valves, and seals is included in the manufacturer's membrane system warranty

*Based on Atassi et al, 2007.

Table 12.4 UF/MF Membrane System Survey—Lessons Learned*

FIGURE 12.7 Norit (Pentair X-Flow) Seaguard UF pretreatment system.

FIGURE 12.8 Norit (Pentair X-Flow) Seaflex UF pretreatment system.

Parameter	Seaguard	Seaflex HP	Seaflex LP
Membrane elements (modules)	SXL 225	Aquaflex	Aquaflex
Module configuration	Horizontal	Vertical	Vertical
Number of modules per pressure vessel (housing)	4	1	1 (no separate housing)
Module length, m (in.)	1.5 (60)	1.68 (66)	1.68 (66)
Module diameter, mm (in.)	200 (8)	200 (8)	225 (9)
Membrane area, m^2 (ft^2)	40 (430)	55 (592)	40 (430)
Fiber inner diameter, mm, (in)	0.8 (0.0315)	0.8 (0.0315)	0.8 (0.0315)
Fiber outer diameter, mm (in)	1.3 (0.0512)	1.3 (0.0512)	1.3 (0.0512)
Maximum transmembrane pressure, bar (lb/in^2)	3 (43.5)	3 (43.5)	3 (43.5)
pH operating range	1.5–13	1.5–13	1.5–13
Weight per module, kg (lb)	35 (77)	44 (100)	44 (100)
Suitable for Pump-through configuration?	Yes	Yes	No

Table 12.5 Norit (Pentair X-Flow) Pretreatment Systems

The Seaguard and Seaflex HP systems are designed to handle a maximum design pressure of 8 bar (116 lb/in^2), which allows them to be operated in a pump-through mode, i.e., without interim pumping between the plant intake and RO system. The Seaflex LP system is designed for a maximum flow-through pressure of 5 bar (73 lb/in^2) and therefore is not suitable for operation as a direct-flow pass-through system.

The vertical (Seaflex) configuration has a better flow distribution within its single vessel, which allows 10 to 12 percent higher design flux at a comparable membrane surface area. The vertical membrane configuration also allows for easier and better membrane cleaning and flushing.

12.7.2 Memcor (Siemens)

As indicated in Table 12.1, Memcor (Siemens) has both vacuum- and pressure-driven membrane systems. These are divided into two lines of products—Xpress or X for small package systems and Component or C for engineered systems designed to serve plants with a production capacity of over 4000 m^3/day (1.1 mgd). Both the X and C modules are available in pressure-driven (XP and CP) and submerged configurations (XS and CS). All membranes are made of PVDF with a pore size of 0.03 μm.

Despite the fact that Memcor (Siemens) membranes are classified as MF type, their pore size is adequately small to remove up to 3 logs (99.9 percent) of viruses. Table 12.6 provides a summary of the key performance parameters of the Memcor (Siemens) membrane elements that have found application for seawater pretreatment.

Both submerged and pressurized Memcor (Siemens) units are designed for a combination of air and water backwash. Submerged systems have a slightly higher backwash volume (reject flow) than pressure-driven systems—approximately 7 to 8 percent of the total intake flow, as compared to 4 to 5 percent of the total intake flow.

Parameter	CS:10V (Submerged)	CP:L10V (Pressurized)	CP:L20V (Pressurized)
Membrane elements (modules)	S10V	L10V	L20V
Module configuration	Vertical	Vertical	Vertical
Number of modules per unit (maximum)	900	960	960
Module length, m (in.)	1.19 (47)	1.19 (47)	1.80 (71)
Module diameter, mm (in.)	130 (5.2)	150 (5.9)	120 (4.7)
Membrane area, m^2 (ft^2)	27.9 (300)	55 (252)	38.1 (410)
Fiber inner diameter, mm (in.)	0.5 (0.002)	0.25 (0.001)	0.25 (0.001)
Fiber outer diameter, mm (in)	0.8 (0.0315)	0.5 (0.0197)	0.5 (0.0197)
Maximum transmembrane pressure, bar (lb/in^2)	1.2 (17)	1.5 (22)	1.5 (22)
pH operating range	1–10	1–13	1–13
Weight per module, kg (lb)	6.5 (14)	6.5 (14)	9 (20)
Suitable for pump-through configuration?	No	Yes	Yes

TABLE 12.6 Memcor (Siemens) Membrane Pretreatment Systems

12.7.3 Hydranautics

Hydranautics pretreatment membranes are low-pressure UF modules that have found application at a number of saline water desalination plants (see Table 12.1). Figure 12.9 shows a Hydranautics HYDRAcap UF membrane module.

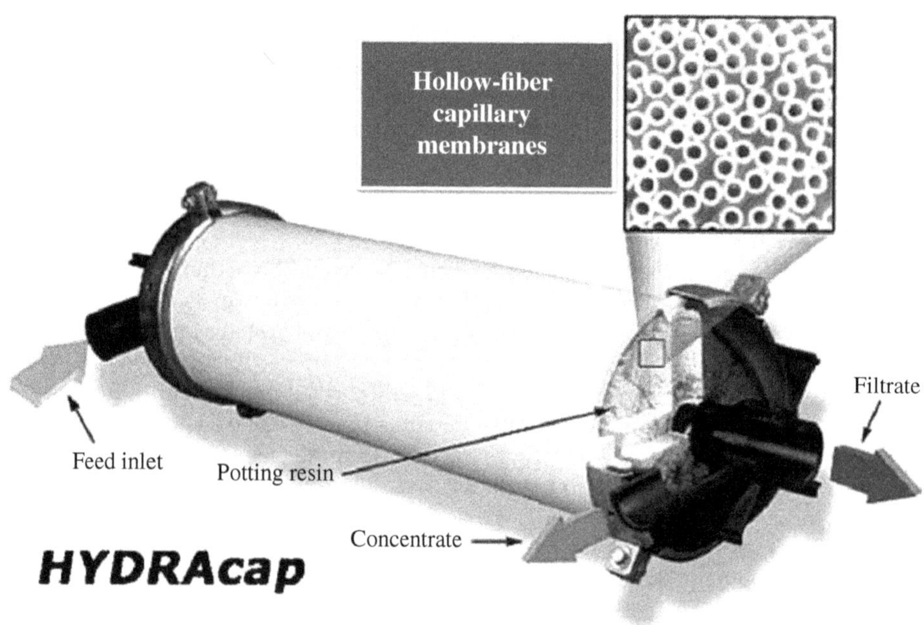

FIGURE 12.9 HYDRAcap UF membrane module.

The key membrane performance parameters are presented here.

Membrane type	Pressure-driven inside-out UF
Models	HYDRAcap 60 and 60-LD
Typical operating pressure	0.2 to 0.5 bar (3–7.0)
Backwash-triggering TMP	Typically 1.1 to 1.4 bar (16 to 20 lb/in^2)
Filtration cycle length	15 to 60 min
Backwash duration	30 to 60 s
Membrane material	Hydrophilic polyethersulfone (PES)
Nominal pore size	150,000 Da
Module diameter	225 mm (8.8 in)
Module length	1000 and 1500 mm (40 and 60 in)
Membrane module filtration area	30 m^2 (1000-mm module) and 46 m^2 (323 ft^2 and 495 ft^2) (1500-mm module)
Design flux	60 to 100 lmh (35 to 59 gfd)
Number of modules needed to produce 1000 m^3/day of filtrate	10 to 16 HYDRAcap 60 1500-mm modules

12.7.4 GE Zenon

General Electric (GE) Zenon's ZeeWeed submerged UF pretreatment system is based on hollow-fiber membrane modules and operated in an outside-in-mode under low suction pressure. The UF fibers are combined into bundles that are installed in standard-size modules (Fig. 12.10).

Up to 96 membrane modules can be installed in a membrane cassette; these cassettes are immersed in tanks fed with source water. An air scouring system is usually installed in the tanks to loosen and release the solids retained within the hollow-fiber bundles during the backwash cycle of system operation.

Although GE Zenon has three baseline submerged membrane UF products (ZW 500, ZW 1000, and ZW1500), at present the membranes most widely used for desalination are ZeeWeed 1000. The original ZW1000 membranes have been modified over the past 5 years, and currently three versions of the original product are available on the market: ZW1000-V2, -V3, and -V4. In addition, the ZW1000-V4 version is offered in two options with a high surface area specifically designed to handle high solids loads. This V-4 version is most suitable for desalination plants with shallow to medium-depth open intakes and high solids loads.

The main differences between the three versions of the original ZW1000 are the total membrane cassette surface area and the reduced internal fiber diameter. The V2 version has a smaller surface area than the other versions because it is designed to handle higher solids loads by having a relatively smaller fiber packing density—the design is tailored for wastewater applications. The V3 version of the ZW1000 has fibers with a greater inner diameter than the V2 version (0.5 mm vs. 0.4 mm), which allows for reduced TMP and increased membrane productivity. These performance benefits make the ZW1000-V3 membranes likely the most suitable product

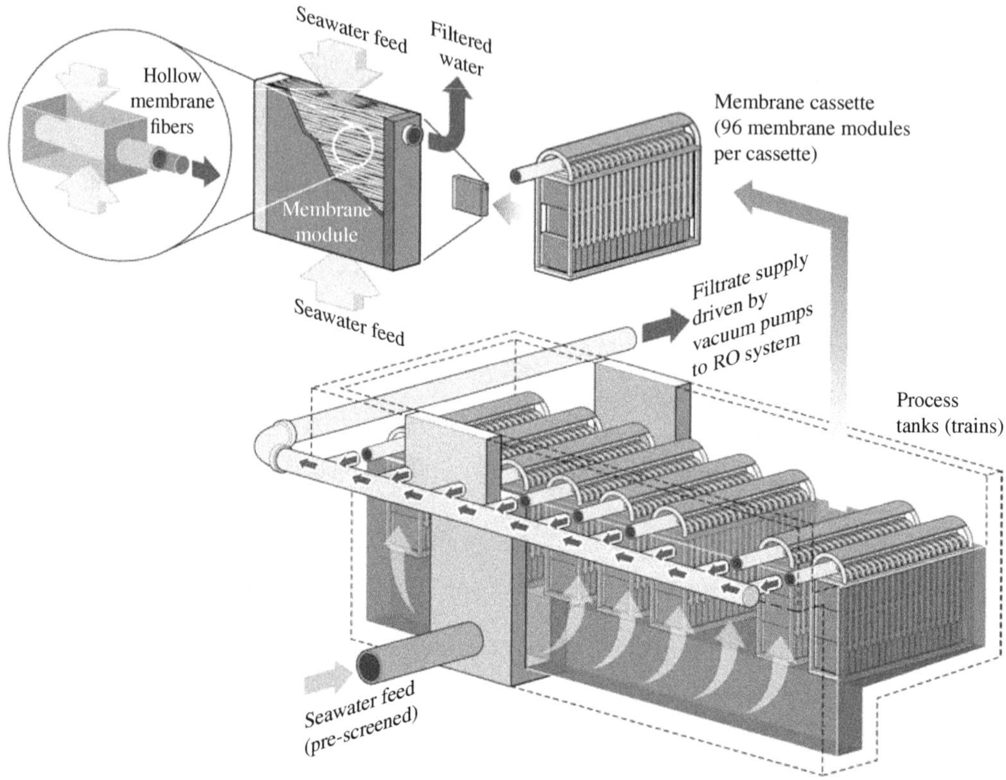

Figure 12.10 GE Zenon ZeeWeed vacuum-driven UF membrane system.

for waters of lower solids content, such as seawater collected using relatively deeper open intakes.

GE Zenon also a offers pressurized outside-in membrane product, the ZW1500. This product has a similar membrane fiber material to the ZW1000-V4; however, the fibers are thicker and better suited for intensive air and water backwash. The ZW1500 has a significantly greater fiber length than the ZW1000 (1.7 m vs. 0.6 m/5.5 ft vs. 2.0 ft) and is designed to operate at a maximum TMP of 2.75 bar (40 lb/in^2) . It has a diameter of 180 mm (7 in.) and a total membrane area of 51.1 m^2 (550 ft^2). Table 12.7 presents the key performance parameters of the ZW1000-V2, -V3, and -4 elements that are typically offered for desalination plant pretreatment applications.

12.7.5 Other UF and MF Membrane Products

Other commercially available membrane products used for pilot and full-scale desalination projects include Hyflux and Inge pressure-driven PES UF membranes, Toray pressure-driven modified PVDF UF and MF membranes, Dow pressure-driven PVDF UF membranes, Koch pressure-driven polysulfone UF and Pall/Asahi Aria pressure-driven MF membranes.

Parameter/Membrane Module	ZW1000-V2	ZW1000-V3	ZW1000-4
Module configuration	Horizontal, installed in cassettes	Horizontal, installed in cassettes	Horizontal, installed in cassettes
Number of modules per cassette (maximum)	96	96	96
Cassette width, m (ft)	2.11 (6.9)	1.82 (6)	1.82 (6)
Cassette height, m (ft)	2.54 (8.3)	2.6 (8.5)	2.6 (8.5)
Membrane area, m^2 (ft^2)	37.2 (400)	46.5 and 55.2 (500 and 594)	48.1 and 51.1 (517 and 550)
Fiber inside diameter, mm	0.4	0.5	0.47
Fiber outside diameter, mm	0.7	0.8	0.95
Maximum transmembrane pressure, bar (lb/in^2)	0.69 (10)	0.69 (10)	0.9 (13)
pH operating range	2–10	2–10	2–10
Weight per cassette, kg (lb)	1007 (2215)	1007 (2215)	1100 (2420)
Suitable for pump-through configuration?	No	No	No

Table 12.7 GE Zenon Seawater Pretreatment Systems

12.8 Design Examples

12.8.1 Submerged UF Pretreatment System

This design example illustrates the determination of the configuration of a vacuum-driven UF membrane pretreatment system for a 40,000 m^3/day (10.6 mgd) seawater desalination plant designed for 43 percent SWRO system recovery and 5 percent reject flow (i.e., 95 percent UF system recovery). The source water turbidity varies between 0.3 and 10 NTU (a total suspended solids concentration of 5 to 15 mg/L), with occasional spikes of up to 15 NTU (total suspended solids = 20 mg/L). The maximum algal count in the source water is 20,000 cells per milliliter, and the hydrocarbon levels are below 0.04 mg/L. The pretreatment system is designed to operate without addition of coagulant or flocculant and without pH adjustment of the source water flow.

The pretreatment system will need to be designed to treat a total of 97,920 m^3/day [(40,000/0.43)/(1 – 0.05) = 97,920]. Other key design parameters are as follows:

Source Water

Total flow	97,920 m^3/day (25.9 mgd)
Temperature (average annual/ minimum monthly average)	20°C/15°C
Turbidity (average annual/daily maximum)	1.5 NTU/10 NTU
Algal count (average annual/daily maximum)	200 cells per milliliter/20,000 cells per milliliter

Target Quality of Pretreated Saline Water

Turbidity (average/maximum)	0.05 NTU/0.3 NTU
SDI_{15}	< 3 (95 percent of the time); < 5 at all times

Vacuum-Driven UF Membrane Pretreatment System

Membrane module	GE Zenon ZeeWeed 1000-V3
Average flux (at average annual temperature of 20°C)	40 Lmh/23 gfd (based on pilot test)
Temperature correction factor for minimum monthly average temperature of 15°C	1.15
Design flux at minimum monthly average temperature of 15°C	45/1.15 = 39.1 Lmh (22.9 gfd)
Total membrane area required	(97,920 × 1000)/(39.1 × 24) = 104,348 m^2 (1,123,202 ft^2)
Number of membrane modules at 55.2 m^2 per module	104,348/55.2 = 1890 modules.
Number of membrane cassettes at 48 modules per cassette	1890/48 = 40 cassettes

Each tank is designed to house one spare cassette, i.e., the tank structure's dimensions are determined for six cassettes (five plus one). The final UF system configuration includes eight tanks sized to house six cassettes per tank, with five cassettes installed per tank along with connections to a sixth cassette. With one tank in backwash and one tank out of service for cleaning, the plant operating flux will be 97,920 × 1000/(8 × 5 × 48 × 55.2 × 24) = 38.5 Lmh (23 gfd). This flux is within the acceptable range of 30 to 45 lmh (18 to 26 gfd) determined by pilot testing.

12.8.2 Pressure-Driven UF Pretreatment System

This example illustrates the key design steps and criteria for a pressure-driven Norit Seaguard UF pretreatment system for a desalination plant of the same size—40,000 m^3/day (10.6 mgd)—and source water quality as described in Sec. 12.8.1. The pressure-driven UF pretreatment system has the same recovery rate (95 percent) and therefore the same design feed flow. Similar to the vacuum-driven system in the previous example, this pretreatment system is designed to produce filtered water of target quality without the need to condition the source water with chemicals, i.e., no coagulant, flocculant, or acid is added to the feed water.

Source Water

Total feed flow	97,920 m^3/day (25.9 mgd)
Temperature (average annual/minimum monthly average)	20°C/15°C

Turbidity (average annual/daily maximum)	1.5 NTU/10 NTU
Algal count (average annual/daily maximum)	200 cells per milliliter/20,000 cells per milliliter
Target Quality of Pretreated Saline Water	
Turbidity (average/maximum)	0.05 NTU/0.3 NTU
SDI_{15}	< 3 (95 percent of the time); < 5 at all times
Pressure-Driven UF Membrane Pretreatment System	
Membrane module (8-in. element)	Norit (Pentair X-Flow) Seaguard SXL 225
Average flux (at average annual temperature of 20°C)	65 Lmh/38 gfd (based on pilot test)
Temperature correction factor for minimum monthly average temperature of 15°C	1.15
Design flux at minimum monthly average temperature of 15°C	65/1.15 = 56.5 Lmh (33 gfd)
Total membrane area required	(97,920 × 1000)/(56.5 × 24) = 72,212 m^2 (777,290 ft^2)
Number of membrane modules at 40 m^2 per module	72,212/40 = 1805 modules
Number of membrane vessels at four modules per vessel	1805/4 = 451 vessels

As indicated previously, the vessels are configured in trains. The proposed design would have 10 trains at 48 vessels per train for a total of 480 vessels. This configuration incorporates 6 percent standby capacity (480/451 = 1.06). The total number of installed modules (8-in. UF elements) is 10 × 48 × 4 = 1920. With one UF train in backwash and one train out of service for cleaning, the plant operating flux will be (97,920 × 1000)/(8 × 48 × 4 × 40 × 24) = 66.4 Lmh (39 gfd), which is within the acceptable range of 50 to 75 Lmh (29 to 44 gfd) determined by pilot testing.

12.9 Pretreatment System Costs

The construction costs for membrane pretreatment are difficult to determine based on existing projects because of the relatively limited track record of this type of pretreatment as compared to granular media filtration, and also because of the diversity of membrane products and configurations presently available on the market. Therefore, rather than a single cost curve, Fig. 12.11 presents a range of year-2012 costs of membrane pretreatment for desalination plants as a function of the plant intake flow rate.

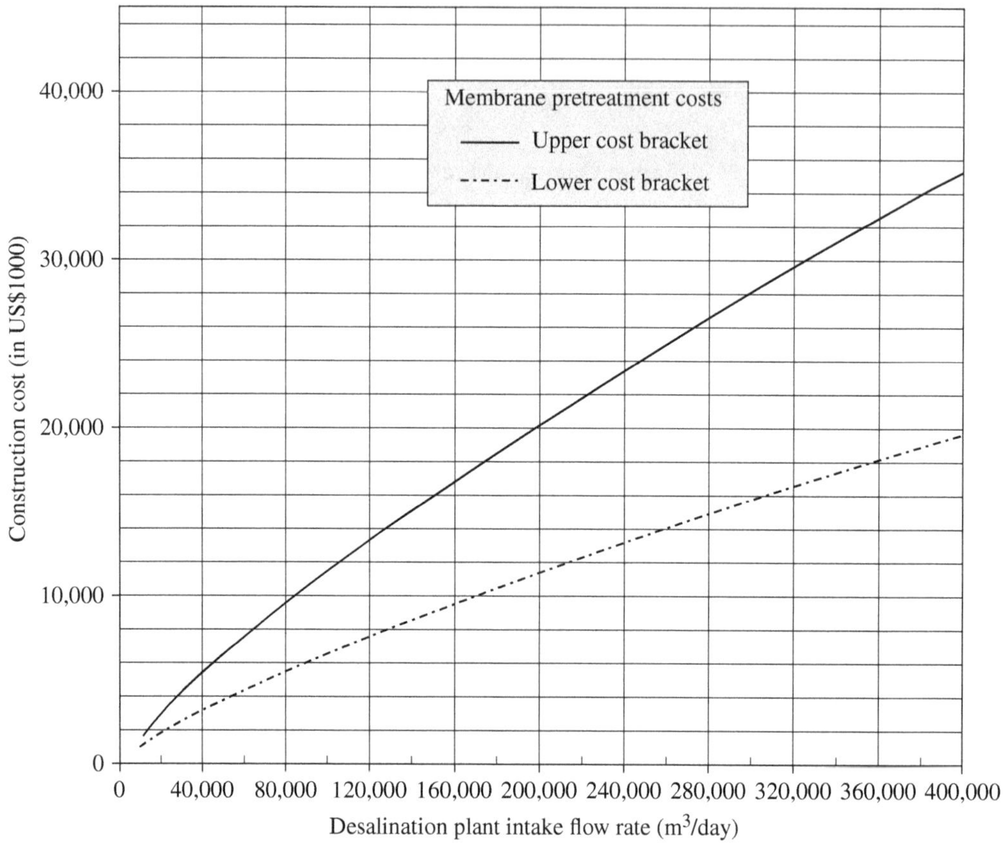

FIGURE 12.11 Construction costs of membrane pretreatment systems.

For the example of the 40,000 m^3/day (10.6 mgd) desalination plant described in the previous sections of this chapter, which has an intake flow of 97,920 m^3/day (25.9 mgd), the construction cost of the intake pretreatment system is estimated in a range of $6.5 million to $11.3 million (average of $8.9 million) based on the cost curves depicted in the figure.

12.10 References

American Water Works Association, *Manual of Water Supply Practices, M46, Reverse Osmosis and Nanofiltration*, 2nd ed., American Water Works Association, Denver, Colorado, 2007.

Atassi A., M. White, and L. M. Rago, "Membrane Failure Headlines: Facts behind the Hype Regarding Initial Problems Experienced at Low Pressure Membrane Installations," *Proceedings of the 2007 AWWA Membrane Technology Conference*, Tampa, Florida, USA, March 18–21, 2007.

Burashid, K., and A. R. Hussain, Seawater RO Plant Operation and Maintenance Experience: Addur Desalination Plant Operation Assessment, Desalination, 165: 11–22, 2004.

Busch, M., R. Chu, and S. Rosenberg, "Novel Trends in Dual Membrane Systems for Seawater Desalination: Minimum Primary Pretreatment and Low Environmental Impact Treatment Schemes," IDA World Congress, Atlantis, the Palm-Dubai, United Arab Emirates, November 7–12, 2009. Ref: IDAWC/DB09-019.

Gasia-Bruch, E., M. Busch, V. Garica-Molina, and U. Kolbe, "Dual Membrane Systems in Seawater Desalination: Drivers for Selection and Field Experiences," IDA World Congress, Perth, Western Australia, September 4–9, 2011. Ref: IDAWC/PER11-244.

Kumar, S. A., S. Adam, and W. Pearce, "Investigation of Seawater Reverse Osmosis Fouling and Its Relationship to Pre-treatment Type," *Environmental Science and Technology*, 40 : 2037–2044, 2006.

Lazaredes, H., and G. Broom, "Review of the Pre-treatment Selection/Design for Australia's Big Six Desalination Projects," IDA World Congress, Perth, Western Australia, September 4–9, 2011. Ref: IDAWC/PER11-244.

Pearce, G. K., *UF/MF Membrane Water Treatment—Principles and Design*, Water Treatment Academy, TechnoBiz Communications, Bangkok, Thailand, 2011.

CHAPTER 13

Comparison of Granular Media and Membrane Pretreatment

13.1 Introduction

Membrane filtration technologies have a number of advantages over conventional granular media filtration systems. Granular media filtration however, is a well-understood and widely used pretreatment technology with a proven track record, which has a number of features that may render it cost competitive under specific circumstances. Therefore, the selection of filtration technology for pretreatment of saline water should be based on a thorough life-cycle cost-benefit analysis.

Side-by-side pilot testing of the two types of systems is also highly recommended to develop background information on system performance for objective evaluation and selection of technology. The following issues have to be taken into consideration in choosing between granular media and membrane pretreatment filtration for a specific application.

13.2 Effect of Source Water Quality on Performance

Membrane filtration has a wider spectrum of particle removal capabilities than conventional media filtration. Because the particulate separation process is based on filtration through a membrane with a fairly uniform pore size, removal efficiency is higher and more consistent than with the more randomly porous granular media filtration bed. Single- or dual-media filters usually have lower removal efficiency in terms of raw source water organics in suspended form, disinfection by-product precursors, fine particles, silt, and pathogens.

Membrane filtration technologies are less prone to upsets caused by seasonal changes in source water turbidity, color, pathogen contamination, and size and type of water particles, because their primary treatment mechanism is mechanical particle removal through fine-pore membranes (Pearce, 2011). Therefore, the upstream chemical coagulation and flocculation of the source water particles is of lesser importance for their consistent and efficient performance.

In contrast, the pretreatment performance of granular media filtration systems is very dependent on how efficient the chemical coagulation and flocculation of the source water is ahead of the filtration process. Therefore, for applications where intake source water quality experiences significant seasonal variations and presents a challenge in terms of high level of pathogens and elevated concentration of fine particles and particulate organics, membrane filtration technologies are likely to offer performance benefits. However, if the source water for the desalination plant is collected from an open intake located far from the tidally influenced zone and at an adequate depth to be exposed to only limited seasonal variations (typically 10 m or deeper), granular media filtration may offer a very cost-effective pretreatment alternative to membrane filtration.

Source water temperature is a very important factor in selecting a pretreatment system. Vacuum-driven membrane pretreatment systems are usually less cost effective than conventional granular media filtration systems for source water with a temperature lower than 15°C, because the productivity (flux) of vacuum-driven membrane filtration is dramatically reduced by the significant increase in unit weight of source water at low temperatures (American Water Works Association, 2007).

Another condition under which the use of granular media filtration may have certain additional benefits is when the source water is exposed to sudden and unpredictable changes of specific contaminants, such as very high- or low-pH chemical spills, oil and grease spills, frequent exposures to very high source water temperature, or contaminants that may damage the microfiltration (MF) or ultrafiltration (UF) pretreatment membranes irreversibly. If the membrane elements are permanently damaged, the cost of their replacement could be significant, especially for large reverse osmosis (RO) desalination plants.

As indicated in a previous chapter, source water may naturally contain sharp particles that can damage MF and UF membranes upon contact. To remove these sharp source water particles, the RO plant intake system should incorporate a microscreening system with a mesh size of 120 μm or less ahead of the membrane pretreatment system (see Chap. 8 for more details). The performance and reliability of conventional granular media pretreatment systems are not sensitive to the content of sharp objects in the source water and do not require elaborate and costly microscreening ahead of the filters. Typically, mechanical traveling screens with openings of 3 to 10 mm provide adequate protection for conventional granular media pretreatment systems.

A particular challenge for seawater pretreatment systems with open intakes could be the source water's content of barnacle plankton small enough to pass through the microscreening system, which could grow to adult barnacles on the walls of the wet well of the pumps feeding the pretreatment system. When some of the barnacles enter the membrane pretreatment feed pumps, their shells are broken by the pumps into small, sharp particles that are then pumped (or drawn by vacuum) against the membrane fibers, causing occasional punctures and over time resulting in a loss of membrane integrity and performance. Typically, such challenge can be solved by matching the size of the microscreens upstream of the membrane pretreatment system with the size of the smallest barnacle (or other shellfish) plankton species that may occur in the saline source water. This issue warrants detailed year-round investigation, because the size and type of plankton species contained in the source water typically change seasonally. If such a challenge is likely to occur for a particular project, it may be prudent to consider the installation of cartridge filters downstream of the membrane pretreatment system that would be able to capture particles and other impurities that may pass through the punctured pretreatment fibers.

Since installation and operation of a microscreening system downstream of the source water intake screens is only needed if membrane filtration is used for pretreatment, the cost of microscreening of the source water should be taken into consideration in comparing conventional and membrane filtration pretreatment.

On the other hand, the use of membrane pretreatment eliminates the need and costs for installation and operation of a cartridge filter system ahead of the RO feed pumps. Cartridge filters are needed when a granular filtration system is used for pretreatment in order to protect the downstream RO membranes from damage caused by fine sand particles that may be conveyed occasionally with the pretreated water.

The occurrence of frequent and prolonged red tides or other algal blooms in the area of the source water intake is another important factor to consider when selecting the type and configuration of source water pretreatment system. As indicated in a previous chapter, many of the marine microalgae that grow excessively during algal blooms cannot withstand an external pressure of more than 0.3 to 0.6 bar (4 to 8 lb/in^2), and their cells could break when exposed to pressure- or vacuum-driven MF or UF filtration.

When algal cells break, they release easily biodegradable organic compounds that can trigger accelerated growth and formation of a biofilm of marine bacteria on the RO membranes. In turn, this accelerated biofilm formation can foul the RO membranes and result in significant reduction of desalination plant's production capacity within several weeks of the beginning of the algal bloom. In such source water conditions, gravity down-flow granular media filtration may be preferable over membrane pretreatment, because it allows removal of the algae from the source water with minimum breakage of their cells.

13.3 Surface Area Requirements

Membrane technologies are typically more space efficient than granular media filtration ones. The smaller footprint of membrane filtration is usually of greater importance in upgrading existing desalination plants with limited site area availability or where the cost of new land acquisition is very high. Typically, the footprint of a conventional single-stage dual-media filtration system is 30 to 50 m^2 per 1000 m^3/day (1200 to 2100 ft^2/mgd) of desalination plant production capacity.

Depending on the type and size of the membrane modules and the intake water quality characteristics, a membrane filtration system may have a footprint 20 to 50 percent smaller than that of conventional filtration system. The space benefits of membrane filtration are more significant for high-turbidity source water, where two-stage granular media filtration may be required to achieve comparable performance to a single-stage membrane pretreatment system. For source water that is more difficult to treat, which necessitates the granular media filtration system to be designed for surface loading rates of less than 10 $m^3/m^2{\cdot}h$ (4 gpm/ft^2) or requires a two-stage granular media filtration to produce comparable filter effluent, a membrane filtration system's footprint may be up to 50 percent smaller.

As a rule of thumb, under typical surface water quality conditions, the footprint of granular media filters designed for a surface loading rate of 8.5 to 12 $m^3/m^2.h$ (3.5 to 5 gpm/ft^2) is approximately 30 to 50 percent larger than that of a UF or MF system producing similar filtered water quality. For better-than-average source water quality—with

Waste Stream	Granular Media Filtration, % of Plant Feed Volume	Membrane Filtration, % of Plant Feed Volume
Intake bar screens wash water	0.1–0.2	0.1–0.2
Microscreen wash water	0 (not needed)	0.5–1.5
Spent Membrane backwash water (reject)	2–6	5–10
Chemically enhanced backwash	0 (not needed)	0.2–0.4
Spent CIP chemicals	0 (not needed)	0.03–0.05
Total	**2.1–6.2**	**5.83–12.15**

Table 13.1 Comparison of Waste Streams from Granular Media and Membrane Pretreatment

a silt density index (SDI) of < 4—where granular media filters can perform adequately at surface loading rates of 15 to 20 $m^3/m^2{\cdot}h$ (6 to 8 gpm/ft^2), the total footprint difference is usually 20 to 30 percent, in favor of membrane pretreatment.

13.4 Quantity and Quality of Generated Residuals

Conventional and membrane pretreatment systems differ significantly by the type, quality, and quantity of residuals generated during the filtration process (Table 13.1).

Typically, granular media filtration systems generate only one large liquid waste stream—spent filter backwash water. The volume of this stream in a well-designed plant varies between 2 and 6 percent of the total plant feed source water volume. In addition to the particulate solids and colloids that are contained in the source water, this waste stream also contains coagulant (typically iron salt).

Membrane pretreatment systems generate two large liquid residual streams: (1) spent membrane backwash water (reject) and (2) membrane cleaning solution from daily chemically enhanced backwash (CEB) which also is referenced as maintenance wash. The volume of the spent membrane backwash water is typically 5 to 10 percent of the plant's intake source volume, i.e., approximately double the spent filter backwash water volume of granular media pretreatment systems.

The difference in total liquid residual volume generated by membrane pretreatment systems is even larger, taking into account that the microscreens needed to protect the membrane pretreatment filters will be a source of an additional waste discharge from their intermittent cleaning. While conventional traveling fine bar screens use 0.1 to 0.2 percent of the intake source water for cleaning, microscreens generate waste screen-wash volume that equals 0.5 to 1.5 percent of the intake flow. The relatively larger waste stream volume of the membrane pretreatment system will require a proportionally larger intake source water volume, which in turn will result in increased size and construction costs for desalination plant intake facilities and pump station, and in higher operation and maintenance (O&M) costs for pumping source water to the pretreatment facilities.

In addition to the daily membrane washing and monthly membrane cleaning, cost-competitive design and operation of membrane pretreatment systems requires daily chemically enhanced membrane backwash using large dosages of chlorine (typically 200 to 1000 mg/L) and strong bases and/or acids over a short period of time. This performance-enhancing CEB adds to the volume of the waste streams generated at the RO membrane plant and to the overall cost of source water pretreatment. The daily volume of waste stream generated during CEB is usually 0.2 to 0.4 percent of the volume of the intake source water.

Another waste stream that is associated only with membrane pretreatment is generated during the periodic chemical cleaning of the pretreatment membranes. Extended off-line chemical cleaning, often referred to as *clean-in-place* (CIP)—during which membranes are soaked in a solution of hydrochloric and/or citric acid, sodium hydroxide, and surfactants—is critical for maintaining steady membrane performance and productivity and it is usually needed once every one to three months. The membrane CIP generates an additional waste stream that is equal to 0.03 to 0.05 percent of the source water volume.

One key advantage of membrane pretreatment systems is that the spent filter backwash water they generate contains less source water conditioning chemicals (coagulant and polymer) and therefore is more environmentally friendly than the spent filter backwash stream generated by conventional pretreatment facilities. This benefit stems from the fact that the coagulant dosage for source water pretreatment by membrane filtration is typically one-third to one-half that for granular media filtration.

In some cases, the source water may not need to be conditioned with coagulant before membrane pretreatment, and this spent filter backwash can then be disposed of along with the RO concentrate without further treatment. For comparison, due to the high content of iron, the spent filter backwash from granular media filtration pretreatment would need to be treated by sedimentation, and the settled solids would need to be dewatered and disposed to a sanitary landfill. Otherwise, the high content of iron salt in the backwash water would cause the desalination plant's discharge to have a red color due to the high content of ferric hydroxide in the discharge.

The waste streams generated during the CEB and the CIP membrane cleanings should be pretreated on-site in a neutralization tank prior to discharge. The additional treatment and disposal costs of the waste membrane cleaning chemicals should be taken into consideration when comparing membrane and granular media pretreatment systems.

13.5 Chemical Use

Typically, the cost of chemical conditioning for source water with granular media filtration is in a range of 4 to 6 percent of the total annual O&M costs for production of desalinated water. Granular media pretreatment systems use 50 to 100 percent more source water conditioning chemicals (iron salts and sometimes polymer) than membrane pretreatment systems to remove the same amount of particulate and colloidal foulants. In the case of vacuum-driven membrane pretreatment, coagulants can be omitted completely or applied only intermittently, during periods of severe algal blooms.

Typically, granular media pretreatment systems, do not use any chemicals for filtration media cleaning (except for an occasional addition of chlorine). In contrast, membrane pretreatment systems use a significant quantity of membrane cleaning chemicals for CEB and CIP, which in terms of total annual chemical costs may be comparable to the total costs of source water conditioning chemicals used by granular media filters. The cost of these cleaning chemicals should be considered in the cost-benefit analysis of the plant's pretreatment system. Another factor that should be accounted for in the analysis of overall plant chemical use and costs is the RO system cleaning frequency, and therefore the RO membrane cleaning costs. These costs may be reduced with membrane pretreatment, due to its typically better capabilities for removal of solids and silt. However, if microbial fouling is the predominant type of RO fouling that occurs at a given desalination plant, membrane filtration typically does not offer any significant advantages to granular media pretreatment, and in some cases may accelerate the biofouling rate due to enhanced breakage of algal cells and release of easily biodegradable organics under the high pressure or vacuum needed for filtration.

A significant difference between the two types of pretreatment systems is the amount of chlorine used for filtration media maintenance. In order to control the pretreatment filtration rate and RO biofouling, granular media filters are occasionally fed with chlorinated source water that contains 1.5 to 5.0 mg/L of chlorine. This so-called shock chlorination is typically completed once to several times per month for 4 to 8 h at a time.

For comparison, chemically enhanced backwash with chlorine dosages of 20 to 1000 mg/L is performed on all pretreatment membranes once to two times per day for a period of 20 to 30 min. Since the CEB process of most pretreatment membranes involves air–water backwash, some of the applied chlorine is stripped into the surrounding air and can cause corrosion of nearby equipment and unprotected concrete and metal structures. Therefore, the use of a protective filter cell structure and equipment coatings and of suitable corrosion-resistant materials is of critical importance. In addition, typically some of the applied chlorine is soaked into the membranes, which may leach chlorine into the RO system feed for the first 20 to 40 min after CEB. Therefore, dechlorination of the filtered source water with sodium bisulfite after CEB cleaning is very important, and the cost of dechlorination chemicals should be taken into consideration when comparing granular media and membrane source water pretreatment.

13.6 Power Use

Granular media pretreatment systems use a limited amount of power to separate particulates in the source water. As mentioned in a previous chapter, large RO desalination plants typically include a single-stage gravity granular filtration pretreatment process, which has minimal power requirements—typically less than 0.05 kWh/m^3 (0.2 kWh per 1000 gal). On the other hand, depending on the type of membrane system (pressure or vacuum), membrane systems use approximately 4 to 6 times as much power—0.2 to 0.4 kWh/m^3 (0.75 to 1.5 kWh per 1000 gal)—to remove particulates from the source water. More power is used not only to create a flow-driving pressure through the membranes but also to perform membrane backwashing and pump source or filtered water. The total power use has to be taken into consideration in completing a life-cycle cost comparison of conventional versus membrane pretreatment for a given application.

While the power demand difference holds true for comparison of single-stage gravity granular media filters and single-stage pressure- or vacuum-driven UF or MF filters, the difference is negligible if pressure granular media filters are used for pretreatment. Single-stage pressure granular media filters operate at comparable feed pressures and power demand to membrane pretreatment systems. Two-stage pressure filtration systems typically use more electricity than a single-stage MF or UF system producing comparable filtered water quality.

For example, a comparative cost analysis between a two-stage pressure filtration system and a single-stage pressure-driven membrane pretreatment system completed during the planning phase of the Perth II SWRO project in Australia revealed that the two-stage pressure filter system would use 20 percent more electricity than the membrane pretreatment system (Molina et al., 2009). For this project, the need to consider two-stage granular media pretreatment was driven by the significant fluctuation in source water turbidity (5 to 50 mg/L of total suspended solids), attributed to the relatively shallow plant intake.

13.7 Economy of Scale

Membrane and granular media pretreatment systems may yield different economies of scale depending on the water treatment plant's capacity. Usually, both technologies have a comparable economy of scale for plant capacity up to 40,000 m^3/day (10.6 mgd). For large desalination plants granular media filtration systems typically yield more economy-of-scale benefits. The anticipated reduction of construction costs for membrane plant capacity increase from 40,000 to 200,000 m^3/day (10.6 to 52.8 mgd) is in a range of 3 to 5 percent, whereas for granular media filtration plants the economy of scale for the same capacity increase is 8 to 10 percent.

The main reason for the smaller economy-of-scale benefits of membrane pretreatment technologies for large-capacity plants is the maximum size of membrane modules currently available on the market. Typically, depending on the manufacturer and the membrane technology, the largest membrane modules available at present have a water production capacity between 2000 and 4000 m^3/day (0.5 and 1 mgd), although recently some manufacturers of immersed membrane systems have begun offering membrane modules with a production capacity of up to 20,000 m^3/day (5.3 mgd). In comparison, the maximum size of the individual granular media filter cells can reach 32,000 m^3/day (8.5 mgd) or more, thereby allowing greater overall reduction of construction costs due to the fewer filter cells and less service equipment and piping.

One of the current trends in source water pretreatment worldwide is to use membrane technologies for large plant applications. As the number and type of large plant membrane application opportunities increase in the future, it is likely that membrane manufacturers will develop larger-scale individual membrane modules, which in turn will improve economy of scale and competitiveness of membrane pretreatment systems.

13.8 Filtration Media Replacement Costs

Properly operating granular media filters lose 5 to 10 percent of their media per year, which has to be replaced to maintain consistent performance. The costs of granular media replacement are usually predictable and relatively low. At present, a pretreatment membrane element's useful life typically varies in a range of 3 to 5 years. Assuming 5 years of membrane useful life, approximately 20 percent of the membrane elements would need to be replaced per year to maintain the pretreatment system's production capacity and performance. Taking into consideration that the annual costs of MF and UF membrane replacement are comparable to those for annual replacement of RO elements they are intended to protect, the use of membrane pretreatment would result an order-of-magnitude higher annual expenditures for media replacement than would the use of granular media filtration.

An additional factor that may contribute to the need for more frequent replacement of membrane elements is failure of membrane element integrity. Typically, the main reason triggering the need for early membrane element replacement is loss of integrity rather than loss of production capacity. The limited track record of long-term use of membrane systems and the uncertainty related to the factors triggering the need for their replacement have to be taken into consideration in selecting between granular media and membrane pretreatment technology for RO plants. The risk of loss of membrane integrity should be handled accordingly in the membrane element's useful life warranty provided by the manufacturer or supplier.

As indicated in the previous chapter, in most cases the use of membrane pretreatment will result in filtered source water with a lower capacity for particulate and colloidal RO membrane fouling. As a result, the use of membrane pretreatment instead of granular filtration should theoretically reduce the frequency of RO membrane cleaning and replacement. This should hold true especially for source waters with low microbial fouling potential. However, because of the limited full-scale performance track record to prove this assumption—and the fact that for seawaters with high microbial fouling potential, membrane pretreatment would make very little difference in terms of frequency of RO cleaning—at present, most RO membrane suppliers are reluctant to provide warranties for longer useful life or lower cleaning frequency for their RO membrane products. As a result, the potential benefits of membrane pretreatment often cannot be easily accounted for in an actual cost-benefit analysis for full-scale desalination projects.

13.9 Commoditization

Currently, all UF and MF membrane manufacturers offer their own design, size, and configuration of membrane elements and pretreatment systems. The membrane pretreatment systems differ by the filtration driving force (pressure or vacuum), the size of the individual membrane elements, the size of the membrane vessels, the configuration and size of the membrane modules, the type of membrane element backwash, and the type of membrane integrity testing method.

The absence of product uniformity and commoditization in the membrane market is a sign of a fast-growing field in the water equipment industry, and carries some benefits and some disadvantages. The availability of multiple membrane suppliers and systems allows for better accommodation of the site-specific needs of a given membrane application, thereby increasing the potential for use of membrane source water pretreatment. In addition, the lack of commoditization of the MF and UF membrane market, along with the increase in membrane applications in recent years, has spurred the interest of many manufacturers that traditionally have not produced membranes to enter the membrane market with new products. This in turn results in increased competition and accelerated development of new membrane technologies, products, and equipment.

Five to ten years ago, there were less than half a dozen membrane manufacturers that offered MF and UF membranes and membrane systems to the municipal market. That number has increased dramatically over the past five years, and today practically all large and many medium-size equipment manufacturers offer their own unique MF or UF membrane system.

The absence of standardization of membrane size, vessels, and configuration, however, also has a number of disadvantages that may hinder the use of membrane pretreatment, especially for large source water desalination plants. As the membrane market gets oversaturated with manufacturers offering similar membrane products, the market growth is likely to exceed the demand, which will trigger the exit of some of the current membrane manufacturers from the market. As a result, the manufacturers exiting the membrane market will no longer produce membrane elements and provide maintenance and support for their existing systems. Since their system configurations, membrane elements, and vessel types will be unique, the owners of such membrane systems will have to invest significant funds and efforts to modify their installations in order to accommodate alternative membrane equipment.

The current diversity of membrane element sizes and configurations and lack of standardization and commoditization may have a number of disadvantages for membrane plant owners in the long run. If an existing membrane manufacturer discontinues the production of membrane elements or a given type of membrane system (for example, abandoning production of pressure membrane systems in favor of submerged systems), membrane plant owners will incur additional costs to procure and install new pretreatment systems, because the other available membrane systems will be incompatible with their existing systems. While replacing or retrofitting the existing pretreatment systems to accommodate new membranes, the desalination plant owners will likely face reduced plant production capacity due to the downtime needed for replacement of the membrane system and to the fact that the productivity of old membrane elements (which cannot be replaced with alternative membrane products when needed) will decrease over time. Membrane plant owners are likely to also incur additional costs to train their staff in operating and maintaining the new membrane pretreatment systems. In addition, owners may experience a potential increase in unit membrane element and vessel costs over time, because the membrane elements will have to be purchased from a sole-source manufacturer rather than being competitively procured at market price and warranty conditions. Taking into consideration that MF and UF membrane element costs have been reduced dramatically over the past ten years, this disadvantage may have significant cost consequences.

Installation of nonstandardized membrane elements and vessels limits desalination plant owners' opportunities to benefit from the use of new and improved membrane pretreatment technologies and of elements that might be readily available in the future. In today's highly diversified membrane technology market, membrane plant owners are very dependent on the commitment of the pretreatment membrane manufacturers whose systems they use to excel in their existing technology and to develop competitive and compatible membrane elements and technologies in the future.

An example that illustrates these concerns was observed in the source water desalination membrane market in the recent past, when one of the key manufacturers of hollow-fiber RO membrane elements—DuPont's subsidiary Permasep—decided to exit the market for these membranes. In the 1990s, Permasep had a dominant portion in the source water membrane desalination market, supplying hollow-fiber RO membrane elements to several thousand installations worldwide. The hollow-fiber membrane elements and vessels produced by Permasep were different from and incompatible with those used by other hollow-fiber membrane manufacturers. Permasep's exit of the source water membrane desalination market triggered the need for significant modifications and expenditures by membrane plant owners using their RO membranes, in order to accommodate the necessary changes.

Standardization of membrane systems, elements, and vessels has another advantage to owner of membrane facilities that has been proven by the desalination membrane market—the significant reduction of membrane costs. Currently, RO desalination membranes and vessels are standardized in size and can be used interchangeably. The commoditization of the source water desalination market over the past 15 years contributed to the reduction of desalination membrane element costs by one-half to two-thirds, which consequently spurred the development of new large source water desalination plants worldwide.

Another often forgotten benefit of membrane technology standartization is the potential reduction in the cost of membrane plant funding and therefore in the overall cost of water production. The capital cost of a given source water desalination project consists of two key elements: (1) the cost of construction and (2) the cost of the capital needed to finance this construction. Since the cost of capital is typically 20 to 30 percent of the overall project costs, using commoditized membrane pretreatment systems could yield cost benefits sometimes higher than the savings that would result from the implementation of new and unique advanced technologies or equipment. A membrane pretreatment system that can accommodate a number of different membrane elements, vessels, and equipment is considered a lower investment risk and therefore a system with a lower cost of capital. With all other conditions being equal, the cost of capital (for example, the bond interest rate) for funding a project using standardized membranes or a well-proven conventional granular media pretreatment system will typically be lower than that for funding a desalination plant that uses a unique membrane pretreatment system configuration and elements that cannot be supplied competitively from alternative manufacturers.

Although a new, advanced membrane pretreatment system that has unique features may yield appreciable near-term savings in construction and operation costs, these savings may be compromised over the useful life of the project—typically 30 years or more—if the pretreatment system design is not flexible enough to accommodate the benefits of future membrane technologies. Based on the current status and diversity of MF and UF technologies, a sound approach to reducing risks associated with the funding and implementation of a membrane pretreatment system is to design the system configuration in a manner that can accommodate the replacement of the system or membrane elements with at least one other existing system or membrane element of a similar type. For example, if the preliminary engineering analysis and subsequent pilot testing indicate that a submerged vacuum-driven membrane pretreatment system is more suitable for a particular application, this desalination plant should be designed to accommodate at least one or two other submerged membrane systems currently available on the market. The additional expenditures in construction and installation costs to provide a flexible pretreatment system configuration that allows future modifications and use of alternative suppliers of the same type of membrane elements at minimal expenditure or replacement are very likely to be compensated by a lowering of the funding costs (costs of capital) for the project and a minimization of the overall life-cycle costs of the membrane plant.

13.10 Water Production Costs

At present, the overall cost of production of desalinated water using membrane pretreatment is typically 5 to 10 percent higher than that for freshwater produced by desalination plants with conventional source water pretreatment. However, when the source water quality is highly variable and/or the cost and availability of land are at premium, membrane pretreatment may be more cost advantageous. Also, when source water quality is fairly high, a membrane pretreatment system can be designed quite aggressively and can have a clear capital and life-cycle cost advantage over a conventional granular media filtration system.

Key factors that are often underestimated or omitted in comparing granular media and membrane pretreatment systems are (1) the additional capital and O&M costs of the microscreening system needed to protect the integrity of the pretreatment membranes; (2) the actual chemical costs and frequency of pretreatment membrane cleaning and chemically enhanced backwash; (3) the useful life and replacement costs of the pretreatment membranes—most analyses assume five years, while actual operational data shows that membranes need to be replaced in approximately three years due to loss of integrity; and (4) the 5 to 10 percent higher cost of project financing associated with the use of membrane pretreatment because of the long-term risk associated with the use of technology with a limited full-scale track record and less commoditization, especially for large-scale desalination plants.

Table 13.2 presents an example of a cost comparison for a conventional gravity dual-media filtration system and a UF vacuum-driven pretreatment system for a seawater desalination plant with a production capacity of 40,000 m^3/day (10.6 mgd). This example assumes conventional pretreatment that consists of single-stage dual-media filters and membrane pretreatment employing vacuum-driven UF membranes. The conventional pretreatment system is assumed to use 5 mg/L of ferric chloride for source water coagulation, whereas the membrane pretreatment system is designed to operate without coagulant addition. As a result, the design with conventional pretreatment incorporates a solids-handling system for treatment of spent filter backwash water, while the membrane pretreatment design does not include solids-handling facilities, and the spent membrane wash water is assumed to be disposed to the ocean with the RO system concentrate without further treatment. This assumption represents a best-case solids-handling scenario for desalination plant systems with membrane pretreatment.

Other assumptions in this example that favor membrane pretreatment are (1) the relatively high land costs of the desalination plant site; (2) a 12.5 percent higher design flux (and therefore smaller size) of the RO system using membrane pretreatment; (3) the avoidance of cartridge filter installation upstream of the RO system with membrane pretreatment; (4) the relatively long useful life of the membrane pretreatment filters (5 years); and (5) the reduction of RO membrane cleaning costs due to membrane pretreatment, which typically would not be the case if the main type of fouling that RO membranes experience is microbial biofouling.

A review of Table 13.2 indicates that both capital and O&M costs for conventional pretreatment are lower than the corresponding costs for membrane pretreatment. If the total construction costs for both plants are amortized using the capital recovery factor estimated for an amortization rate of 5.7 percent over a period of 20 years (11.752—see Chap. 17), the capital cost component of the seawater desalination costs will be \$0.43/m^3 (\$1.63 per 1000 gal) for the plant with conventional pretreatment and \$0.45/m^3 (\$1.70 per 1000 gal) for the plant with membrane pretreatment. These costs are determined by dividing the construction costs in Table 13.2 by the capital recovery factor and the annual production capacity of the plant. For example, for the plant with a conventional pretreatment system, this cost is calculated as follows: $74{,}000{,}000/(11.752 \times 40{,}000 \times 365) = \$0.43/\text{m}^3$.

The annual O&M costs for the conventional pretreatment system in this example are estimated at \$6,860,000/year (Table 13.2). When converted to unit costs their value is $\$6{,}860{,}000/(40{,}000 \times 365) = \$0.47/\text{m}^3$ (\$1.78 per 1000 gal). Similarly, the

Item	Granular Media Pretreatment	Membrane Pretreatment
Capital Costs, $1000		
Open ocean intake	2000	2100
Intake pump station	980	1050
Coarse and fine screens	500	540
Microscreens	0	1200
Coagulation and flocculation system	340	0
Cartridge filters	960	0
Source water chlorination system	150	160
Pretreatment membrane cleaning system	0	720
Filter tanks (excluding media or membranes)	3540	2530
Filtration media (sand and anthracite or UF membranes)	400	3100
Membrane pretreatment system service equipment	0	1800
Filter backwash system	380	640
Dechlorination system	80	140
Land	1000	720
Seawater reverse osmosis system	25,600	22,400
Post treatment system	1460	1460
Solids-handling facilities	1100	60
Discharge outfall	1830	1950
Other facilities and systems	7180	7180
Engineering and construction management	7350	8650
Start-up and commissioning	1460	1750
Other costs	17,690	19,450
Total capital costs	**74,000**	**77,600**
Amortized capital costs, $/m³ ($/1000 gal)	**0.43 (1.63)**	**0.45 (1.70)**
Operations and Maintenance (O&M) Costs, $1000		
Labor	420	500
Chemicals for coagulation and flocculation	220	0
Chemicals for pretreatment membrane cleaning	0	190
Chemicals for CEB of pretreatment membranes	0	120
Chemicals for RO membrane cleaning	120	80
Other chemicals	100	120
Microscreen maintenance and spare parts	0	30
Cartridge filter replacement	130	0
Pretreatment membrane replacement	0	220

Table 13.2 Cost Comparison for Granular Media and Membrane Pretreatment Systems

Item	Granular Media Pretreatment	Membrane Pretreatment
RO membrane replacement	650	460
Granular media addition	20	0
Other maintenance and spare part costs	680	800
Solids handling and sludge disposal	300	0
Disposal of spent membrane cleaning solution to sewer	30	80
Power use for seawater pretreatment	44	307
Power use by RO and other systems	3326	3326
Other O&M costs	820	820
Total annual O&M costs	**6860**	**7053**
Annual O&M costs, $/m³ ($/1000 gal)	**0.47 (1.78)**	**0.48 (1.82)**
Water production cost, $/m³ ($/1000 gal)	**0.90 (3.41)**	**0.93 (3.52)**

Table 13.2 Cost Comparison for Granular Media and Membrane Pretreatment Systems (*Continued*)

O&M costs for the desalination plant with a membrane pretreatment system are calculated as $0.48/m^3 ($1.82 per 1000 gal). Based on these capital and O&M cost estimates, the total water production costs for an RO plant with conventional and membrane pretreatment are $0.90/m^3 ($3.41 per 1000 gal) and $0.93/m^3 ($3.52 per 1000 gal), respectively.

For the example shown in Table 13.2, the cost of seawater desalination using membrane pretreatment is slightly higher (3.2 percent), even when the water quality and site-specific conditions favor the use of this type of pretreatment. The main items where the construction costs of the two systems differ significantly are the costs of filtration media, the pretreatment system, the RO system, and the solids-handling facilities. The intake costs for the desalination plant with a membrane pretreatment system are higher because this system would require the collection of 8 percent more source seawater than the conventional pretreatment system. As explained previously, this additional intake water is needed for the washing of the microscreens and the backwashing of the pretreatment membranes.

The costs associated with the RO system are lower for the membrane pretreatment system because this system is assumed to be designed at a 12.5 percent higher flux (15.3 Lmh vs. 13.6 Lmh). The high design flux allowance for RO systems with membrane pretreatment stems from the expectation that membrane filtration provides superior pretreatment.

The main differences in the O&M costs of the two systems are related to the higher use of power for the membrane pretreatment process and to the costs for pretreatment system maintenance and membrane replacement. It should be pointed out that, depending on the membrane pretreatment technology applied, the annual cost for replacement of pretreatment membranes can be comparable to that for replacement of RO membrane elements. On the other hand, the use of membrane pretreatment is expected to eliminate or significantly reduce the sludge disposal costs and to decrease the RO membrane replacement rate and the frequency and costs of cleaning.

Although the design assumptions used in Table 13.2 favor membrane pretreatment, in many cases not all of the benefits of this type of pretreatment will be applicable to the site-specific conditions of a given RO project; the difference in the cost of water between membrane and conventional pretreatment can exceed 10 percent in favor of conventional pretreatment. As membrane filtration technologies evolve and next generations of membrane products are more closely tailored to fit the specific challenges of saline water pretreatment, it is very likely that membrane pretreatment will become cost competitive for the majority of source water allocations.

13.11 Guidelines for Selecting a Pretreatment System

The most suitable pretreatment system type and configuration mainly depends on the source water quality, and more specifically on the type of foulants present in the source water. Table 13.3 provides a guideline for a combination of treatment processes that could be used for cost-effective pretreatment of saline source water as a function of its content of particulate and colloidal foulants (turbidity and SDI levels) and organic and microbial foulants (total organic carbon—TOC).

Source Water Quality	Recommended Combination of Pretreatment Technologies prior to RO Treatment	Notes
Turbidity < 0.1 NTU; SDI_{15} < 2; TOC (year-round) < 1 mg/L	Cartridge or bag filters only	Grit removal may be needed if intake wells are used
Turbidity ≥ 0.1 and < 5 NTU; SDI_{15} < 5; TOC (year-round) < 1 mg/L	Single-stage dual-media filters plus cartridge filters MF/UF pretreatment may be cost competitive if a 7- to 10-year RO membrane useful life is guaranteed	Coagulant addition may not needed if an MF/UF system is used
Turbidity ≥ 5 and < 30 NTU; SDI_{15} > 5; TOC (moderate algal blooms) < 4 mg/L	Single-stage dual-media filters plus cartridge filters, or MF/UF pretreatment	Coagulant addition may be needed
Turbidity ≥ 30 and < 50 NTU; SDI_{15} > 5; TOC (severe algal blooms) ≥ 4 mg/L and/or high oil spill potential	Sedimentation or DAF plus single-stage dual-media filters plus cartridge filters, or sedimentation or DAF plus MF/UF pretreatment	
Turbidity ≥ 50 NTU; SDI_{15} > 5; TOC (severe algal blooms) ≥ 4 mg/L and/or high oil spill potential	High-rate sedimentation or DAF plus two-stage dual-media filters plus cartridge filters, or high-rate sedimentation or DAF plus MF/UF pretreatment	DAF ahead of filtration may not be needed if algal blooms in the intake area are moderate (TOC < 2 mg/L) or oil contamination is not an issue

Table 13.3 Alternative Source Water Pretreatment System Configurations

The pretreatment configurations shown in Table 13.3 should be used as a guideline only. Often with saline groundwater and some saline surface water sources, other water quality parameters such as silica, iron, and manganese may dictate the selection of the most viable pretreatment system. The selection guidelines presented in Table 13.3 mainly apply for saline source waters that do not contain significant concentration of iron, or manganese, and have silica levels below 20 mg/L.

Thorough water quality analysis and pilot testing are recommended to define an optimum pretreatment system configuration for the site-specific source water quality of a given project, especially if that water originates from a source where silica, iron, and manganese are in measurable concentration in the source water.

It should be pointed out that in some cases, the pretreatment configurations listed in Table 13.3 could be modified, or additional pretreatment or source water conditioning could be needed for removal of scaling compounds (such as calcium and magnesium salts), colloidal foulants (i.e., iron and manganese), natural organic matter from nearby river estuaries, or pathogen contamination.

13.12 References

American Water Works Association, *Manual of Water Supply Practices, M46, Reverse Osmosis and Nanofiltration*, 2nd ed., American Water Works Association, Denver, Colorado, 2007.

Molina G. V., M. A. Marzal, and K. U. Hoehn, "Designing Membrane Systems for Coming Future: Perth II Desalination Plant," *Proceedings of IDA World Congress*, Dubai, United Arab Emirates, November 7–12, 2009. Ref: IDAWC/DB09-168.

Pearce, G. K., *UF/MF Membrane Water Treatment—Principles and Design*, Water Treatment Academy, TechnoBiz Communications, Bangkok, Thailand, 2011.

CHAPTER 14

Reverse Osmosis Separation

14.1 Introduction

This chapter provides an overview of the key components and configuration of membrane systems for brackish and seawater desaliantion applying spiral-wound polyamide thin-film composite reverse osmosis (RO) membane elements (modules), which are the most commonly used type of reverse osmosis separation systems at present. As indicated in Chap. 3, membrane separation systems may apply other types of membrane materials and configurations. However, they are not included in this book due to their relatively more limited areas of application. Additional information of alernative membrane materials, modules, and systems for liquid and gas separation is presented elsewhere (Li et al., 2008).

Figure 14.1 depicts a typical configuration of RO system. The filtered water produced by the plant's pretreatment system is conveyed by transfer pumps from a filtrated water storage tank through cartridge filters and into the suction pipe of the high-pressure RO feed pumps, which, in turn, deliver this water to the RO pressure vessels that contain the membrane elements where the actual desalination process occurs.

It should be pointed out that some of the components of the RO system depicted in Fig. 14.1 might differ from one desalination plant to another depending on the type of intake, source water quality, energy-recovery system, and design configuration. For example, for desalination plants with well intakes that produce source water of quality adequate to be treated directly by RO separation, the transfer pump and filtrate storage tank shown in Fig. 14.1 may be eliminated, and the source water may be pumped by the intake pumps through the cartridge filters directly into the suction header of the high-pressure pump. Such configuration is common for many BWRO desalination plants with deep well intakes (see Fig. 4.8). For BWRO desalination plants, it is also common to bypass some of the feed water and, after pretreatment via separate cartridge filtration units, to blend it with the desalinated water produced by the RO system.

Figure 14.1 also shows that the entire volume of the pretreated water is introduced into the RO system via the feed water pump and the high-pressure pump. This RO vessel feed configuration is used for BRWO and SWRO desalination plants where energy in the RO system concentrate is recovered and reused by its conversion into energy to drive the high-pressure RO pump. If isobaric energy-recovery systems are used, the energy contained in the RO concentrate is applied to directly pump a portion of the filter effluent into the RO system membrane vessels while the high-pressure pump conveys the rest of the flow. Such variations of the general RO system configuration presented in Fig. 14.1 are discussed in greater detail further in this chapter.

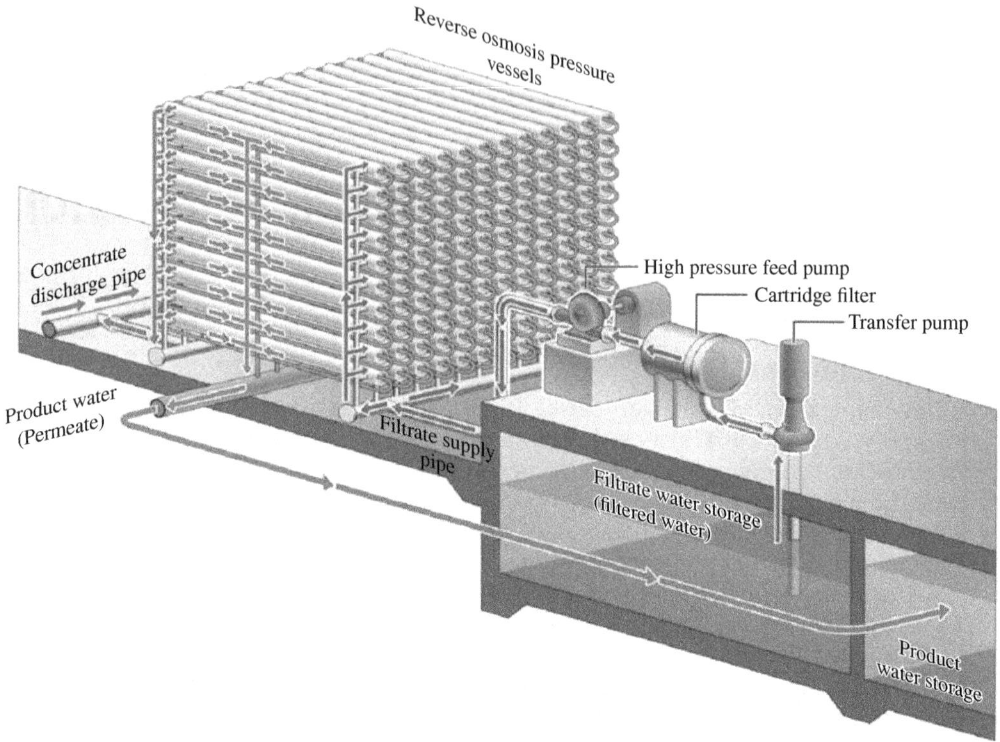

FIGURE 14.1 General RO membrane system configuration.

While cartridge filters are shown in Fig. 14.1, they and the filter effluent chemical-feed systems for sodium bisulfite, bactericide, and antiscalant, which are located upstream of the RO membrane skids, are often considered to be a part of the sfacilities water pretreatment system rather than components of the RO system because these facilities are not directly engaged into the actual RO membrane-separation process and serve to further condition the filter effluent for more stable and efficient separation process.

The main purpose of the cartridge filters is to protect the RO membranes from damage by particles that may have accidentally found their way into the pretreated water. The configuration and design of cartridge filters is discussed in greater detail in Chap. 8. Overview of systems for addition of conditioning chemicals (sodium bisulfide, biocide, and antiscalant) to the pretreated source water is presented in Chap. 9. This chapter focuses on the following RO system components: filtered water transfer pumps, high-pressure pumps, RO membranes, pressure vessels, interconnecting piping, RO skids (trains), energy-recovery systems, membrane-cleaning systems, and instrumentation and controls.

14.2 Filtered Water Transfer Pumps

Filtered water transfer pumps are typcially vertical turbine or horizontal centrifugal pumps designed to convey filtered water to the RO system. As indicated previously, filtered water from the plant pretreatment system or plant intake (if it is of adequate quality)

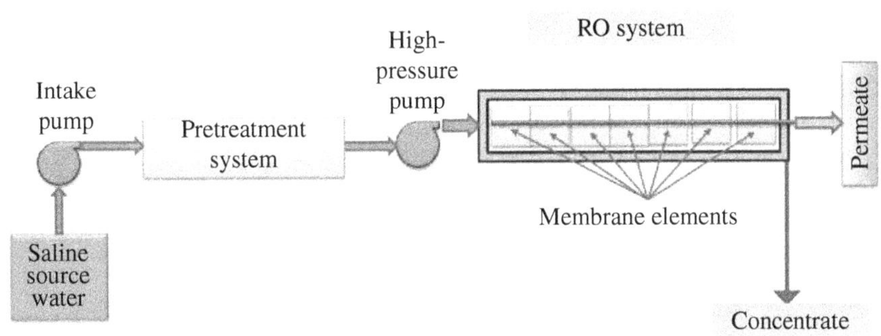

FIGURE 14.2 Direct flow-through desalination system.

could follow two alterantive flow patterns: direct flow-through desalination system (see Fig. 14.2) or desalination system with interim pumping (Fig. 14.3).

In the direct flow-through desaliantion system, the intake pump station is sized to deliver the suction pressure needed for the efficient operation of the high-pressure RO feed pump. In this case the pretreatment system is configured such that it does not break the pressure, which is accomplished by the use of either pressure granular media filters or pressure-driven membrane pretreatment filters. As a result, the pretreatment system will have to be designed to withstand the additional pressure needed for the suction of the high-pressure RO pump. For SWRO desalination systems this suction pressure could be 2 to 6 bars (29 to 87 lb/in^2). For BWRO plants the suction pressure is usually below 1 bar (14.5 lb/in^2).

In the case of a desalination plant with an interim filtered water transfer (Fig. 14.3), a separate pump station is installed to boost the filtered source water to the suction pressure needed for the efficient operation of the high-pressure RO pumps.

In state-of-the-art designs of SWRO systems, the filtered water transfer pumps are often equipped with variable frequency drives (VFDs) to allow for the feed pressure of the RO system to be cost-effectvely controlled by the feed pressure of the filtered water transfer pumps. Such control is often needed because seasonal (and sometimes diurnal)

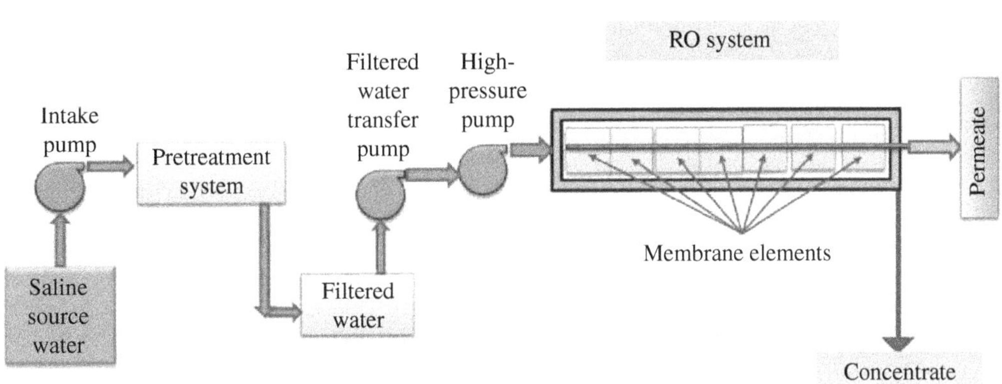

FIGURE 14.3 Desalination system with interim pumping.

changes in source water temperature and salinity have an impact on the osmotic pressure and net driving pressure (NDP) needed for desalination which in turn require the adjustment of the RO membrane feed pressure. As source water temperature decreases and/or salinity increases, the NDP and feed pressure needed to produce the same volume of permeate increase and vice versa. In order to maintian the high-pressure RO feed pumps at their maximum performance efficiency and constant feed flow at all times, the pressure they deliver has to be adjusted to match the changes in osmotic pressure (and related NDP) triggered by source water-quality fluctuations. Installation of VFDs results in overall reduction of plant energy use.

While pressure delivered to the RO vessels could alternatively be controlled by installing VFDs on the high-pressure RO pumps, because the cost of the VDFs is proportional to the size of the pump motor—and usually the size of the motor of the transfer pumps is an order of mangitude smaller than the size of the motor of the high-pressure pumps—signficant equipment cost savings can be obtained by installing the VFDs on the filter effluent transter pump motors instead of on the RO feed high-pressure pump motors. This holds especially true for SWRO desaliantion systems. Operating pressures in BWRO plants are signficanly lower, and energy savings from installation of VFDs do not always yield the same life-cycle cost benefits.

14.3 High-Pressure Feed Pumps

14.3.1 Overview

High-pressure feed pumps are designed to deliver source water to the RO membranes at pressure required for membrane separation of the fresh water from the salts, which typically is 5 to 25 bars (73 to 363 lb/in^2) for BWRO desalination and 55 to 70 bars (798 to 1015 lb/in^2) for seawater desalination. The actual required feed pressure is project and water-quality specific and is mainly determined by the source water salinity, temperature, target product water quality, and the configuration of the RO system. The pumps are sized based on required flow and operating pressures using standard performance curves supplied by pump manufacturers. All wetted pump materials should be of adequate-quality stainless steel, which is a function of the salinity of the water they process. Typically nanofiltration and low-salinity BWRO applications require the use of 316 L or greater quality stainless steel. Duplex and super-duplex stainless steel is recommended for high-salinity BWRO and SWRO applications, respectively.

Variable frequency drives are sometimes installed on the high-pressure pump motors to adjust motor speed in order to maintain optimum pump efficiency with changing feed pressure requirements driven by natural fluctuations in source water salinity and temperature. In addition, VFDs allow for the pumps to maintain optimum performance when membranes foul or scale and loose permeability over time.

If VFDs are not installed on the filtered water transfer pump or the high-pressure pump motors, then the feed flow and pressure of the centrifugal high-pressure pumps is adjusted via a pressure control valve (Fig. 14.4). This valve is throttled along with the flow control valve installed on the concentrate pipe to set RO system operations at target recovery, feed flow, and pressure. As RO membranes age over time, their permeability, and therefore their productivity decreases irreversibly. Typically, an RO system loses 8 to 15 percent of its initial productivity over a period of three to five years. In order

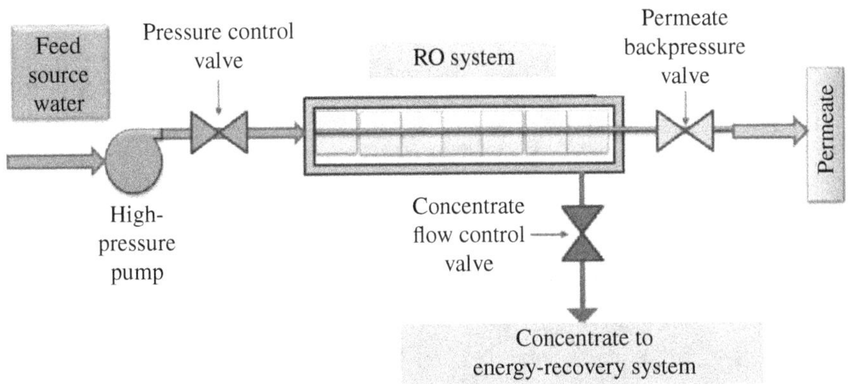

FIGURE 14.4 Main RO system control valves.

to compensate for this loss of productivity, usually the RO high-pressure pumps are oversized to deliver 10 to 15 percent higher pressure and flow than their initial design levels for new RO membranes.

Permeate backpressure valve (Fig. 14.4) is installed to control, within certain limits, the Beta factor (i.e., reduce concentrate polarization) in order to improve product water quality and reduce membrane fouling rate. The ability to control flux and fouling by back pressure is limited due to potential for thin-film delamination if permeate backpressure exceeds 0.3 bar (4.3 psi) above the concentrate pressure.

Because of the high pressures at which RO feed pumps operate, when these pumps are started their feed pressure has to be increased gradually [not more than 0.7 bars (10 lb/in^2) per second] until it reaches its design level in order to avoid hydraulic surge in the feed line. Depending on their severity, hydraulic surges could dislodge the membranes within the vessels, cause O-ring breaks and membrane compaction, and result in membrane leaf telescoping and cracking of the outer shell of the membrane elements. For larger desalination plants, the RO pump motors usually are designed with "soft start" provisions, which control the motor speed at start-up in order to avoid hydraulic surges. In addition, motorized valves on the RO system feed and concentrate are typically designed to have adequately long actuation time for pressure surge prevention.

14.3.2 Types of High-Pressure Pumps

The two most common types of high-pressure pumps used for RO systems are reciprocating (piston) pumps and centrifugal pumps.

Reciprocating High-Pressure Pumps

Reciprocating (positive displacement or piston) pumps (Fig. 14.5) are typically used for small size plants of capacity of 4000 m^3/day (1.1 mgd), or less.

In such pumps, the rotating motion of the motor is converted into reciprocating motion that drives the pump pistons. The water flow delivered by this type of pump changes with the number of pistons, piston surface area, and pump stroke length.

The two main advantages of these types of pumps as compared with centrifugal pumps are that they typically have higher efficiency (90 to 95 percent versus 80 to

FIGURE 14.5 Triplex reciprocating pump. (*Source: Calder.*)

88 percent) and a very flat pump curve (i.e., they can deliver near constant feed flow rate with practically the same efficiency as pressure changes).

One key disadvantage of reciprocating pumps is that they deliver pulsating flow, which, unless attenuated, may cause a pressure surge in the RO vessels and result in membrane connector and element damage. Pump flow pulsation (i.e., the difference between the minimum and maximum flow pumps deliver with every stroke) is mainly the function of the number of pump pistons. Typically, pumps with two pistons have flow pulsation of 46 percent, which is reduced down to 23 percent for pumps with three pistons, to 4 percent for pumps with seven pistons, and to 2 percent for pumps with nine pistons.

Typically, flow pulsation acceptable for reliable operation of the RO system is 20 percent or less—the lower the better. Because the costs of piston pumps increase with the number of pistons available for the same pump capacity, desalination plants usually use reciprocating pumps with a minimum of three pistons (also referenced as triplex pumps) or with four pistons (quadruplex pumps). In order to further reduce the flow pulsations below 20 percent, triplex pumps are equipped with suction stabilizers and pulsation dampeners.

Two key reasons why standard reciprocating pumps have not found wider application for high-pressure RO feed applications are: (1) pulsations increase with the system size and additional vibration-related challenges occur when multiple pumps are connected to a common header, and (2) pump operation at maximum speed (which yields a superior efficiency as compared with centrifugal pumps) results in elevated maintenance costs.

Because of the high maintenance expenditures often in practice, reciprocating pumps are operated at close to half of design speed, which reduces their efficiency down to the levels of centrifugal pumps (i.e., 80 to 85 percent).

Centrifugal Pumps

Centrifugal high-pressure pumps are used for all size desalination plants. One disadvantage of centrifugal pumps over reciprocating pumps is that the pressure they deliver changes with flow and vice versa (i.e., the pump curve is not "flat"). As a result, if the RO feed pump would need to retain efficiency over a wide range of operating pressures, it would need to be equipped with a VFD, or the filtered water transfer pump conveying water to its suction header has to be designed with VFD, as discussed previously. Since pump curve "flattens" with the number of pump stages, and therefore centrifugal pumps used for high-pressure RO system feed applications are usually a multistage type.

The efficiency of centrifugal pumps (Pump_{eff} in %) is proportional to their size

$$\text{Pump}_{\text{eff}} \sim n \times (Q/H)^{0.5} \times (1/H)^{0.25} \tag{14.1}$$

where n is the pump speed (min^{-1}), Q is nominal pump capacity (m^3/s), and H is pump head (m). According to this formula, for the same delivery pressure and speed, pump efficiency is directly proportional to the square of the pump flow capacity. In practical terms this means that, in general, two smaller centrifugal pumps will be less efficient than one large pump delivering the same flow as the sum of the flows of the two pumps. This particular feature of centrifugal pumps explains one of the main reasons for a recent trend of using fewer larger pumps supplying two or more RO trains via common manifold instead of applying a standard RO feed pump-train configuration, where each RO train is supplied by an individual high-pressure RO pump. For example, a 9500 m^3/day (2.5 mgd) SWRO train would typically be served by a single high-pressure centrifugal RO feed pump of efficiency of 83 percent. If, however, one pump is designed to serve two SWRO trains of the same capacity [i.e., the pump delivers 19,000 m^3/day (5 mgd)], its efficiency at the same delivery pressure will increase to 85 percent. If one pump serves 16 rather than two RO trains of the same size, the pump efficiency would reach 88 to 90 percent. Taking into consideration that SWRO high-pressure pumps typically consume 70 to 85 percent of the energy used for the entire desalination plant, using fewer, larger pumps could yield significant energy savings.

The two main reasons why sometimes the single-pump/single-RO system configuration is preferred over configuration with fewer, larger pumps are: (1) reliability concerns–if the main and standby pumps fail, at the same time the plant loses a large portion of its production capacity, (2) efficiency limitations–if the RO plant has to be operated in a wide range of production flows maintaining maximum pump efficiency may not be viable even if VFDs are installed. For example, if the desalination plant would need to cost-effectively produce only 10 to 20 percent of its design flow at times, such large capacity downturn could not be achieved with a configuration using only few RO feed pumps, because even if these pumps are equipped with VFDs they will not be able to maintain maximum pump efficiency over such a large range of flow rates.

As indicated by Eq. (14.1), the pump efficiency increases with the square root of its speed for the same flow and pressure. This is the main reason why desalination pump motors are designed to run at very high speeds—typically 3000 to 3500 rotations per minute (rpm). While technically these pumps can be designed to operate at higher rotational speeds (up to 12,000 rpm), such designs are not practical and cost effective because the pump shaft size and required net positive suction head (NSPH) increase with speed.

The four types of centrifugal pumps most frequently used for desalination applications are vertical turbine pumps, split-case multistage pumps, segmental ring (ring-section) multistage pumps, and high-speed single-stage pumps.

Vertical Turbine Pumps

Single and multistage vertical turbine pumps such as those depicted in Fig. 14.6 are most commonly used in brackish water reverse osmosis desalination plants. As shown in this figure such pumps are mounted in a stainless-steel can and depending on the operating pressure, they could be single or multistage units.

While single-stage horizontal centrifugal pumps with radially split casing (sometimes referred to as ANSI pumps) are also used for nanofiltration (NF) and brackish water desalination (BWRO) plants, they usually have lower efficiency and need higher NPSH than vertical pumps. Therefore, they have not found as widespread application in NF and BWRO plants as vertical turbine centrifugal pumps.

Split-Case Multistage Centrifugal Pumps

At present, horizontal spit-case multistage centrifugal pumps are most commonly used as high-pressure feed pumps for medium- and large-size SWRO desalination plants (Fig. 14.7). These pumps usually yield high efficiency (80 to 88 percent). The feed water inside the pumps is guided from stage to stage by a set of volute passageways.

FIGURE 14.6 Vertical turbine high-pressure BWRO pumps.

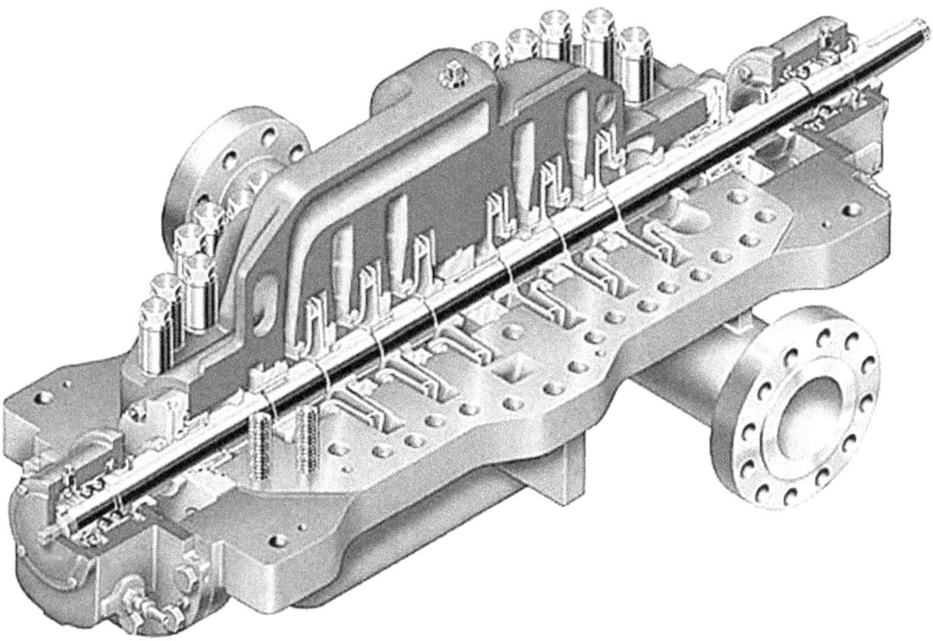

FIGURE 14.7 Horizontal split-case multistage pump. (*Source: Flowserve.*)

The pump impeller has an opposing surface direction design, which allows reducing a pump's net axial trust. Usually, large-size pumps have at least two stages, while smaller pumps have four to six stages.

Capacity of individual pumps commercially available for large installations ranges between 600 and 3000 m^3/h (2500 to 13,000 gpm). For example, the Ashkelon SWRO desalination plant in Israel has two sets of three duty and one standby horizontal split-case centrifugal pumps with capacities of 2500 m^3/h (11,000 gpm) each and are equipped with 7000-hp pump motors.

Over the past five years, radially split-case multistage centrifugal pumps have found application for a number of medium- and large-size SWRO desalination projects such as the 40,000 m^3/day (10.5 mgd) Dhekelia SWRO plant in Cyprus and the 250,000 m^3/day (66 mgd) Sydney Water SWRO plant in Australia (Fig. 14.8).

The key benefits of radially split versus horizontally split-case pumps are that they occupy less space and are easier to maintain because they have only one mechanical seal at the drive end (versus two seals for horizontally split case pumps). In addition, they have internal fiber-composite bearings, which are water lubricated as compared with horizontal split-case pumps, which have two sets of external grease-lubricated bearings.

The Sydney Water SWRO plant is equipped with 12 duty and one standby radially split-case pumps with motor sizes of 2800 hp each. Both Sydney and Dhekelia plants have the same type of pumps, and their efficiency ranges between 85 and 87 percent.

Segmental-Ring (Ring-Section) Multistage Pumps

These pumps are used for high-pressure RO applications in small- and medium-size desalination plants [i.e., plants with capacity of 1500 to 10,000 m^3/day (0.4 to 2.6 mgd)].

FIGURE 14.8 Radially split-case centrifugal pump of Dhekelia SWRO plant.

The key advantage of such pumps are their 20 to 30 percent lower equipment costs at the tradeoff of lower but reasonably good efficiency (70 to 80 percent versus 83 to 88 percent). The segmental ring pumps consist of individual pump stages located between the pump suction and discharge castings (Fig. 14.9).

The impellers of the segmental ring pumps are mounted on a common shaft. Such pumps typically have smaller diameter and lighter construction than the split-case multistage pumps, and, in smaller projects, where unit electricity costs are not high, and the main emphasis of the project design is to reduce plant construction costs, such type of pumps may be a cost-attractive alternative to the higher efficiency/higher cost split-case pumps.

High-Speed Single-Stage Pumps

Such pumps are typically used for small- and medium-size desalination plants and are combined with a specific type of energy-recovery device (turbocharger), which operate on a common shaft with the RO feed pumps and boost their pressure by applying energy recovered from the RO system concentrate. The combination of turbocharger and high-speed single-stage pumps is discussed in the next sections of this chapter.

The improved energy efficiency of these pumps is achieved based on the fact that, as seen from Eq. 14.1, pump efficiency is reversely proportional to the pump head. This means that two pumps in series operating at the same flow but at lower head will have

FIGURE 14.9 Segmental ring multistage pump.

higher efficiency (i.e., use less energy) than a single pump of the same type delivering the same total head and flow.

This principle applies for single- and multistage pumps. This is one of the reasons why the configuration of a filter effluent transfer pump in series with a high-pressure RO pump shown in Fig. 14.3 would typically be more energy efficient than the installation of a single high-pressure pump directly conveying water from the filtered water tank to the RO system.

14.4 Spiral-Wound Polyamide Membrane Elements

14.4.1 Overview

Standard size spiral-wound polyamide membrane reverse osmosis elements are commercially available from over two dozen membrane suppliers, including Hydranuatics–Nitto Denko, Dow Filmtec, Toray Membranes, Koch Membrane Systems, Woongjin Chemical, GE Water and Process Technologies, TriSep Corporation, Sempro Membranes, and others. However, at present over 90 percent of the municipal water sector is supplied with RO membranes from the first three manufacturers listed above. Many of the other manufacturers dominate specific industrial and point-of-use applications, which are not included in the scope of this book.

Spiral-wound RO membrane elements could be classified in three main categories by the type of water they are configured and designed to desalinate: (1) nanofiltration (water softening) (NF) elements, (2) brackish water desalination (BWRO) elements, and (3) seawater (SWRO) membrane elements. All three types of membrane elements have a similar configuration, which is described in greater detail in Chap. 3. However, they differ by the type of membrane material, salt rejection, permeability, and operating pressure range. The NF, BWRO, and SWRO membrane elements are specifically suited to process source waters in different salinity ranges at minimum membrane production costs, energy use, and maximum productivity. Table 14.1 provides a general overview of key performance parameters of the three types of membrane elements for standard size 8-in (200 mm) membranes.

14.4.2 Nanofiltration (Water Softening) Elements

These elements are designed to process waters of very low salinities (typically with TDS concentration < 1000 mg/L) and to mainly remove divalent ions, which cause water hardness (i.e., calcium and magnesium). Therefore, they are also referred to as softening membranes. Typically, NF membrane elements have higher permeability than brackish and seawater elements and comparable rejection of bivalent ions (i.e., Ca, Mg, Mn, Fe, SO_4), large organic molecules [i.e., trihalomethane (THM) precursors], natural pigments (NOM), and of pathogens (bacteria, protozoa, and viruses).

However, NF membranes have a significantly lower rejection of monovalent ions such as sodium, chloride, and boron as compared with BWRO and SWRO elements. While their "looser" membrane structure limits the ability of these elements to reject most monovalent salts, it yields higher permeability, which in turn allows the NF elements to be operated at relatively higher specific flux and significantly lower feed pressure than the BWRO and SWRO elements. Table 14.2 presents examples of commonly used commercially available 8-in (200-mm) diameter; 40-inch (1.0-m) long NF membrane elements and their key performance characteristics. Detailed technical specifications of these and other nanofiltration elements are available from the respective membrane manufacturers.

It should be pointed out that the rejection of commercially available NF membrane elements varies depending on their use—some products are designed specifically to remove color, THM precursors, and other large-molecule NOM. Such membranes may not be as efficient for membrane softening. There are also "tighter" NF elements that can remove not only hardness and organics but also up to 30 percent of the TDS in the source water. Therefore, it is recommended to consult membrane manufacturers as to which of their commercially available products would best fit a particular application.

Performance Parameter	Type of Membrane Element		
	NF	BWRO	SWRO
Typical source water salinity range, mg/L	400–1000	800–10,000	15,000–47,000
Operating feed pressure bars (lb/in^2)	5–8 (70–120)	10–15 (150–220)	55–70 (800–1000)
Average flux rate, Lmh (gfd)	20–40 (12–24)	20–40 (12–24)	14–16 (8–9.4)
Specific flux rate, Lmh/bar (gfd/lb/in^2)	3.5–7.0 (0.17–0.34)	2.0–4.0 (0.08–0.16)	1.0–1.5 (0.04–0.06)
Nominal salt rejection, %	70.0–95.0	99.0–99.7	99.5–99.8
Average product water flow rate per element, m^3/day (gpd)	20–25 (5000–6600)	20–25 (5000–6600)	12–15 (3200–4000)

TABLE 14.1 Key Performance Parameters of NF, BWRO, and SWRO Membrane Elements

Parameter	Commercial Membrane Element Model ESNA-LF2 Hydranautics	NF270-400 Dow Filmtec	SU620F Toray
Product water flow rate, m^3/day (gpd)	39.7 (10,500)	47.3 (12,500)	22.0 (5800)
Nominal $CaCl_2$ rejection, %	86	85–95	55
Test feed pressure, bars (lb/in^2)	5.2 (75)	4.8 (70)	3.4 (50)
Specific flux, Lmh/bar (gfd/lb/in^2)	7.6 (0.31)	15.8 (0.63)	8.6 (0.35)
Membrane surface area, m^2 (ft^2)	37.1 (400)	35.0 (380)	37.1 (400)
Maximum applied pressure, bars (lb/in^2)	42 (600)	42 (600)	42 (600)
Feed/brine spacer, mm (mil)	0.71 (28)	0.71 (28)	0.71 (28)

Table 14.2 Examples of Commonly Used NF Membrane Elements

14.4.3 Brackish Water Desalination Elements

BWRO elements are designed to treat source waters of salinity above 500 mg/L and as high as 10,000 to 15,000 mg/L. As shown in Table 14.1, their optimal operation range is typically up to 10,000 mg/L, but they could also treat higher salinities in multistage membrane configurations. The range of source water salinities beyond 15,000 mg/L is typically processed using SWRO membranes.

At present, there is no standard set of test conditions for all commercially available BWRO membrane elements. Standard 8-in (200-mm) diameter and 40-in (1.0-m) long BWRO membrane elements are tested at water salinity between 500 and 1500 mg/L, recovery rate of 15 percent; flux rate between 43.5 and 51.4 Lmh (25.6 and 30.2 gfd), and feed pressure of 6.7 to 10.3 bars (100 to 150 lb/in^2).

Depending on their key performance parameters, brackish water reverse osmosis elements can be subdivided in the following main groups: (1) high-rejection membranes, (2) low-energy membranes, (3) low-fouling membranes, and (4) high-productivity membranes.

High-Rejection BWRO Membranes

This type of membrane has several-tenths of a percent higher rejection (i.e., 99.5 to 99.7 percent) than standard BWRO elements, which reject 99.0 to 99.3 percent of the salts in the source water. Table 14.3 presents examples of 8-in high-rejection brackish water RO elements.

Low-Energy BWRO Membranes

These membrane elements are designed to produce approximately the same volume of water but at lower feed pressure (i.e., higher specific flux). Such membranes trade-off

Parameter	Commercial Membrane Element Model		
	ESPA 2+ Hydranautics	BW30-400 Dow Filmtec	TM720-400 Toray
Product water flow rate, m^3/day (gpd)	41.6 (11,000)	40.0 (10,500)	39 (10,200)
Nominal NaCl rejection, %	99.6	99.5	99.7
Test feed pressure, bars (lb/in^2)	10.3 (150)	15.8 (225)	15.8 (225)
Specific Flux, Lmh/bar (gfd/lb/in^2)	4.9 (0.20)	5.9 (0.24)	4.2 (0.17)
Membrane surface area, m^2 (ft^2)	39.5 (430)	35.0 (380)	37.1 (400)
Maximum applied pressure, bars (lb/in^2)	42 (600)	42 (600)	42 (600)
Feed/brine spacer, mm (mil)	0.71 (28)	0.71 (28)	0.71 (28)

Table 14.3 Examples of High-Rejection BWRO Membrane Elements

Parameter	Commercial Membrane Element Model		
	ESPA 4 Hydranautics	BW30 XLE-440 Dow Filmtec	TMH20-400 Toray
Product water flow rate, m^3/day (gpd)	49.2 (13,000)	48.1 (12,700)	49.2 (13,000)
Nominal NaCl rejection, %	99.2	99.0	99.4
Test feed pressure, bars (lb/in^2)	6.7 (100)	6.7 (100)	6.7 (100)
Specific flux, Lmh/bar (gfd/psi)	8.2 (0.33)	7.8 (0.24)	8.2 (0.35)
Membrane surface area, m^2 (ft^2)	37.1 (400)	40.8 (440)	37.1 (400)
Maximum applied pressure, bars (lb/in^2)	42 (600)	42 (600)	42 (600)
Feed/brine spacer, mm (mil)	0.71 (28)	0.71 (28)	0.71 (28)

Table 14.4 Examples of Low-Energy BWRO Membrane Elements

lower energy for lower rejection and are typically used if the source water salinity is relatively low, the unit energy costs are high, and/or if the source water is fairly cold. Examples of such membrane elements are shown in Table 14.4. As seen from this table, some of the elements also have a high membrane area.

Low-Fouling BWRO Membrane Elements

Low-fouling features of membrane elements are either obtained by changes of membrane surface chemistry (charge) or by the use of wider feed/brine spacers [31 or 34 mil (0.79 or 0.86 mm) versus standard 28 mil (0.71 mm)]. For example, the Hydranautics CPA3 BWRO membrane element incorporates membranes with a neutral surface charge (as compared with negatively charged conventional membranes), which reduces its fouling with positively charged particles such as ferric salts. In addition, this membrane has a 31-mil (0.79 mm) versus 28-mil (0.71 mm) spacer, which allows solids to pass more easily through the membrane elements.

Hydranautics ESPA4-LD is also a low-fouling type of membrane, but the lower-fouling feature of this membrane model is achieved mainly by increasing the membrane element spacer to 34 mil (0.86 mm). Similar BWRO membrane elements are available from Dow Filmtec (i.e., BW30-400/34i-FR), and Toray (TM720D-400). Table 14.5 provides examples of low-fouling BWRO membranes.

Figure 14.10 indicates that the use of RO elements with a wider brine spacer for treatment of high-fouling waters has a significant positive impact on the frequency of membrane cleaning (Bates et al., 2008). Membrane cleaning events are denoted as CIP (clean in place) in this figure.

Figure 14.8, illustrates the benefits of wider spacer RO elements in terms of membrane cleaning. For this example, the use of RO elements with 31-mil spacer instead of elements of standard 28-mil spacer has decreased the frequency of membrane cleaning from an average of once per month to once every six months. Application of membranes with an even wider, 34-mil, feed/brine spacer has extended the CIP frequency beyond six months.

	Commercial Membrane Element Model		
Parameter	**ESPA 4-LD Hydranautics**	**BW30 XLE-400/34i Dow Filmtec**	**TM720D-400 Toray**
Product water flow rate, m^3/day (gpd)	45.4 (12,000)	48.1 (10,500)	49.2 (8900)
Nominal NaCl rejection, %	99.2	99.5	99.8
Test feed pressure, bars (lb/in^2)	6.7 (100)	15.8 (225)	15.8 (225)
Specific flux, Lmh/bar (gfd/psi)	7.7 (0.23)	6.7 (0.20)	5.7 (0.17)
Membrane surface area, m^2 (ft^2)	37.1 (400)	37.1 (400)	37.1 (400)
Maximum applied pressure, bars (lb/in^2)	42 (600)	42 (600)	42 (600)
Feed/brine spacer, mm (mil)	0.86 (34)	0.86 (34)	0.86 (34)

Table 14.5 Examples of Low-Fouling BWRO Membrane Elements

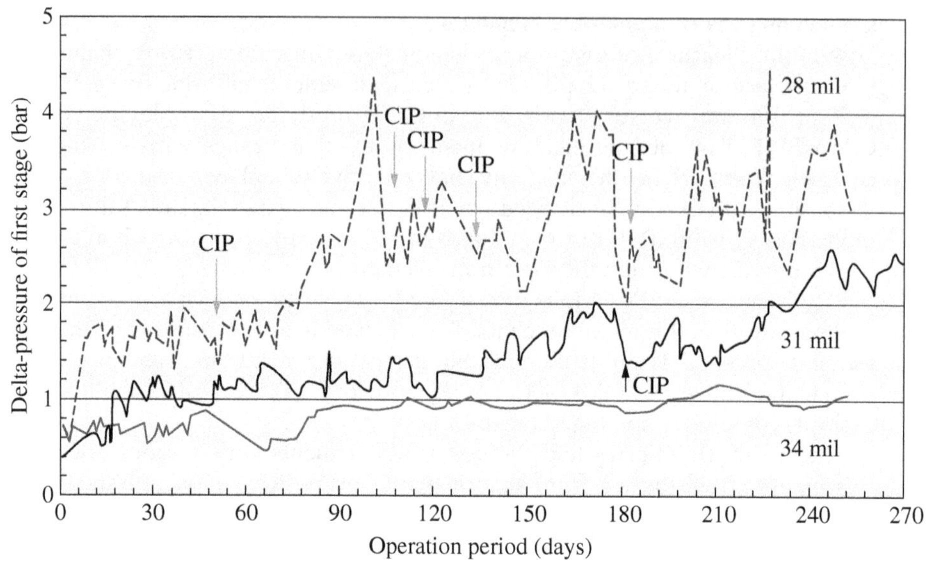

Figure 14.10 Impact of membrane brine spacer width on cleaning frequency (*Source: Hydranautics*).

High-Productivity BWRO Membranes

Increased productivity of membrane elements is typically achieved by increasing their total surface area which, in turn, is accomplished by incorporating one additional membrane leaf. At present, all main membrane manufacturers offer BWRO membrane elements of total surface area of 40.8 m^2 (440 ft^2). Examples of such commercial membrane elements are shown in Table 14.6.

14.4.4 Seawater Desalination Elements

Similar to BWRO membrane elements, SWRO membranes can also be classified in four main groups based on their performance: (1) high-rejection, (2) low-energy, (3) low-fouling, and (4) high-productivity. Standard-rejection membrane elements are designed to remove up to 99.6 percent of the salts in the source seawater. These membrane elements are most widely used today and have found applications in variety of RO system configurations. Compared with NF and BWRO elements, which vary significantly in terms of standard test conditions, all membrane manufacturers have adopted the same standard test feed salinity and pressure conditions for SWRO elements: 32,000 mg/L of NaCl and 55.2 bars (800 lb/in^2), respectively. However, membrane manufacturers have slight differences in the applied SWRO membrane test recovery (8 to 10 percent), and test flux rate (27.6 to 38.3 Lmh/16.3 to 22.5 gfd) between various products.

High-Rejection SWRO Membranes

High-rejection membrane elements are designed with tighter membrane structure, which allows to increase the mass of rejected ions and to reject smaller size ions, such as boron, for example. The higher-rejection membrane capabilities of 99.75 to 99.85 percent

Parameter	Commercial Membrane Element Model		
	ESPA 4-MAX Hydranautics	BW30 LE-440 Dow Filmtec	TM720D-440 Toray
Product water flow rate, m^3/day (gpd)	50.0 (13,200)	48.1 (12,700)	45.8 (12,100)
Nominal NaCl rejection, %	99.2	99.3	99.8
Test Feed Pressure, bars (lb/in^2)	10.3 (150)	10.3 (150)	15.8 (225)
Specific flux, Lmh/bar (gfd/lb/in^2)	7.7 (0.30)	7.8 (0.31)	7.0 (0.27)
Membrane surface area, m^2 (ft^2)	40.8 (440)	40.8 (440)	40.8 (440)
Maximum applied pressure, bars (lb/in^2)	42 (600)	42 (600)	42 (600)
Feed/brine spacer, mm (mil)	0.71 (28)	0.71 (28)	0.71 (28)

Table 14.6 Examples of High-Productivity BWRO Membrane Elements

come at a price—10 to 20 percent higher operating pressure. In general, these membrane elements are also more prone to fouling as compared with standard-rejection SWRO membrane elements, and their use requires more elaborate seawater pretreatment in terms of particulate, colloidal, and microbial foulants. Table 14.7 provides examples of high salinity-rejection SWRO elements.

Parameter	Commercial Membrane Element Model		
	SWC4+ Hydranautics	SW30 HR-380 Dow Filmtec	TM820K-440 Toray
Product water flow rate, m^3/day (gpd)	24.6 (6500)	22.7 (6000)	24.2 (6400)
Nominal NaCl rejection, %	99.80	99.70	99.86
Nominal boron rejection, %	91	90	96
Test feed pressure, bars (psi)	10.3 (800)	10.3 (800)	10.3 (800)
Specific flux, Lmh/bar (gfd/lb/in^2)	1.0 (0.040)	0.92 (0.037)	0.98 (0.039)
Membrane surface area, m^2 (ft^2)	37.1 (400)	35.0 (380)	40.8 (440)
Maximum applied pressure, bars (lb/in^2)	83 (1200)	83 (1200)	83 (1200)
Feed/brine spacer, mm (mil)	0.71 (28)	0.71 (28)	0.71 (28)

Table 14.7 Examples of High Salinity-Rejection SWRO Membrane Elements

Parameter	Commercial Membrane Element Model		
	SWC4B Max Hydranautics	SW30 XHR-400i Dow Filmtec	TM820K-440 Toray
Product water flow rate, m^3/day (gpd)	27.3 (7200)	22.7 (6000)	24.2 (6400)
Nominal NaCl rejection, %	99.80	99.80	99.86
Nominal boron rejection, %	95	93	96
Test feed pressure, bars (lb/in^2)	10.3 (800)	10.3 (800)	10.3 (800)
Specific flux, Lmh/bar (gfd/lb/in^2)	1.11 (0.044)	0.93 (0.038)	0.98 (0.039)
Membrane surface area, m^2 (ft^2)	40.8 (440)	37.1 (400)	40.8 (440)
Maximum applied pressure, bars (lb/in^2)	83 (1200)	83 (1200)	83 (1200)
Feed/brine spacer, mm (mil)	0.71 (28)	0.71 (28)	0.71 (28)

Table 14.8 Examples of High-Boron-Rejection SWRO Membrane Elements

Table 14.8 presents high-boron rejection SWRO elements provided by the same manufacturers. As seen, high-boron rejection membranes usually have high salt rejection as well. In the case of Toray's TM820K-440, this element combines highest salinity and boron rejection and high surface area (40.8 m^2/440 ft^2).

Low-Energy (High-Productivity) SWRO Membranes

Low-energy (high-productivity) SWRO membrane elements are designed with features to operate at lower feed pressure or yield more product water per membrane element, – namely: higher permeability and higher surface area. Increasing the total active membrane leaf surface area allows gaining significant productivity for the same size (diameter) membrane element. Active surface area of the membrane leaf is typically increased by adding additional membrane leaf/s and automating the membrane production process to minimize the membrane area occupied by the membrane envelope glue strip. High-productivity elements have a standard yield of 34 to 45 m^3/day (9000 to 12,000 gpd) and reasonably high salt rejection of 99.6 to 99.7 percent but typically lower than standard boron rejection.

The total active surface area of a membrane element can also be increased by increasing membrane size/diameter. Although 8-in (200 mm SWRO membrane elements are still the "standard" size for most widely used in large full-scale applications, larger 16-in (400 mm),18-in (460 mm), and 19-in (480 mm) diameter SWRO membrane elements are currently available. The key features of these large-size RO elements are discussed in the next section of this chapter.

The dynamics of the high-productivity (or low-energy) membrane element development is illustrated by an example of the development of seawater membranes. In the second half of the 1990s, the typical 8-in (200 mm) SWRO membrane element had a standard productivity of 19 to 23 m^3/day (5000 to 6000 gpd) at salt rejection of 99.6 percent. In 2003, several membrane manufacturers introduced high-productivity seawater membrane elements that are capable of producing 28 m^3/day (7500 gpd) at salt rejection of 99.75 percent. Just one year later, even higher productivity (34 m^3/day/9000 gpd at 99.7 percent rejection) seawater membrane elements were released on the market. Some of the newest high-productivity SWRO membrane elements have unit production capacity of 51 m^3/day (13,500 gpd), provide flexibility and choice, and allow us to trade productivity and pressure/power costs. Table 14.9 illustrates the performance characteristics of some of the high-productivity/low-pressure membrane elements available on the market.

Low-Fouling SWRO Membranes

The low-fouling [also referenced as "fouling-resistant" or "low-differential pressure (LD)" feature of most commercially available SWRO membranes at present is obtained by incorporating a wider (typically 34 mil/0.86 mm) feed/brine spacer in the membrane

	Commercial Membrane Element Model		
Parameter	**SWC6 Max Hydranautics**	**SW30 ULE-400i Dow Filmtec**	**TM820L-440 Toray**
Product water flow rate, m^3/day (gpd)	50.0 (13,200)	45.4 (12,000)	51.1 (13,500)
Nominal NaCl rejection, %	99.80	99.70	99.80
Nominal boron rejection, %	91	89	92
Test feed pressure, bars (lb/in^2)	10.3 (800)	10.3 (800)	10.3 (800)
Specific flux, Lmh/bar (gfd/lb/in^2)	2.03 (0.080)	1.85 (0.073)	2.08 (0.082)
Membrane surface area, m^2 (ft^2)	40.8 (440)	40.8 (440)	40.8 (440)
Maximum applied pressure, bars (lb/in^2)	83 (1200)	83 (1200)	83 (1200)
Feed/brine spacer, mm (mil)	0.71 (28)	0.71 (28)	0.71 (28)

Table 14.9 Examples of Low-Energy/High-Productivity SWRO Membrane Elements

element configuration at the expense of reducing the number of membrane leafs in the elements. Practical experience at seawater desalination plants to date shows that the use of wider spacer SWRO elements could be beneficial for high-fouling source waters such as those of the Persian Gulf and the Red Sea. Examples of low-fouling SWRO membranes are shown in Table 14.10.

14.5 Pressure Vessels

14.5.1 Description

As shown in Fig. 14.11, RO membrane elements are installed inside pressure vessels (housings) in a series of six to eight membranes per vessel. Membrane element interconnection within the vessels is typically accomplished by short plastic spool pipe segments with O-rings (interconnectors) or via specially designed interlocking devices (see Chap. 3).

Each pressure vessel is enclosed on its sides with closely fitting enclosures referenced as "end caps." The end caps are designed to withstand the membrane operating pressures and to restrict the movement of the membrane elements within the vessels.

Typically, one pressure vessel houses six to eight RO membrane elements. A recent design trend in SWRO plants is to install eight elements per vessel. Based on detailed cost-benefit analysis (Brusilovsky and Faigon, 2005; Liberman and Wilf, 2005;

	Commercial Membrane Element Model		
Parameter	**SWC4-LD Hydranautics**	**SWC5-LD Hydranautics**	**SW30 HRLE-370/34i Dow Filmtec**
Product water flow rate, m^3/day (gpd)	24.6 (6500)	34.1 (9000)	25.0 (6700)
Nominal NaCl rejection, %	99.8	99.8	99.8
Nominal boron rejection, %	93	92	92
Test feed pressure, bars (lb/in^2)	10.3 (800)	10.3 (800)	10.3 (800)
Specific flux, Lmh/bar (gfd/lb/in^2)	1.0 (0.040)	1.4 (0.056)	1.1 (0.044)
Membrane surface area, m^2 (ft^2)	37.1 (400)	37.1 (400)	34.4 (370)
Maximum applied pressure, bars (lb/in^2)	83 (1200)	83 (1200)	83 (1200)
Feed/brine spacer, mm (mil)	0.86 (34)	0.86 (34)	0.86 (34)

Table 14.10 Examples of Low-Fouling SWRO Membrane Elements

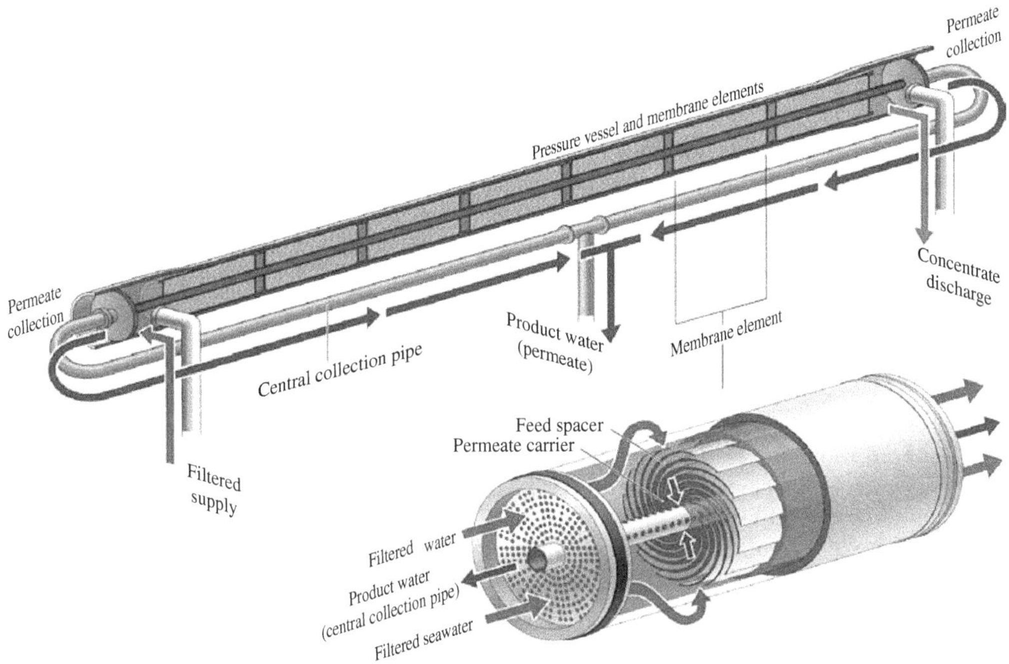

Figure 14.11 Spiralwound- RO membrane elements in vessels.

Wilf et al., 2008), the installation of eight instead of seven or six elements may be beneficial, especially for medium- and large-size desalination plants, because of the equipment cost reduction associated with the use of fewer vessels.

Besides capital cost reduction, the use of an eight-element vessel configuration could also lower the overall concentration polarization factor for the RO membranes due to higher feed/brine velocity and reduced recovery of the individual elements, which are beneficial in terms of fouling. However, the higher number of membrane elements in the vessel, the higher the differential pressure within the vessels, and the closest the vessels would operate to the maximum limit of pressure drop (also referred to as differential pressure) recommended by the membrane manufacturers of 4 bars (58 lb/in^2) beyond which permanent damage and compaction of the elements may occur. In addition, the use of eight elements will result in a slightly higher feed pressure.

In order to prevent movement of the membrane elements within the vessels, the end connection of the permeate side of the RO vessels is shimmed (supplemented axially). Membrane vessels are connected with steel pipe sections (ports) to the feed and concentrate lines of the RO train and with plastic ports to the permeate line. The minimum and maximum feed flows per individual 8-in vessels are recommended to be 10 and 17 m^3/h (44 and 75 gpm), respectively. The minimum recommended flow of concentrate per vessel is 2.7 m^3/h (12 gpm).

The entire volume of feed water to be processed by a given RO vessel is introduced into the front end of the vessel and applied onto the first membrane element. In most standard pressure-vessel configurations, permeate and concentrate are collected from the last element. However, as shown in Fig. 14.9, most-recent SWRO system designs incorporate permeate collection from both the front and back end of the membrane vessels. Such configurations and their benefits are discussed further in the next sections of this chapter.

14.5.2 Membrane Vessel Classification

Membrane vessels differ by their pressure class (i.e., their maximum operating pressure), by their diameters, the material from which they are produced, and by the location of their feed port. In general, pressure vessels can be produced to house from a single element to up to eight elements in series.

By Pressure Class

When the RO system is in operation, its pressure vessels are completely enclosed and pressurized at the operating pressure of this system. Based on their maximum pressure rating, pressure vessels are divided into three classes: (1) water softening (nanofiltration) pressure vessels designed for operation in a range of 3.5 to 10.5 bars (50 to 150 lb/in^2)—pressure rating of 150 lb/in^2, (2) BWRO pressure vessels designed to handle operating pressures of 10.5 to 42 bars (150 to 600 lb/in^2)—pressure ratings 450 and 600 lb/in^2, and (3) SWRO pressure vessels with operating pressures of 42 to 105 bars (600 to 1500 lb/in^2)—pressure ratings of 1000, 1200, and 1500 lb/in^2. Higher-pressure-rating vessels are also available.

By Diameter

Pressure vessels are designed to house a specific standard diameter membrane. Therefore, they are produced in standard membrane diameter sizes of 63 mm (2.5 in), 102 mm (4 in), 200 mm (8 in), and 400 mm (16 in).

By Material

The most common pressure vessel material is fiberglass-reinforced plastic (FRP). For specific industrial applications, where the pressure vessels have to be sanitized and/or operated at high temperatures (i.e., 65°C or higher), stainless-steel pressure vessels are more suitable. While stainless-steel pressure vessels can also be used for municipal water treatment, they are more expensive, heavier, and more difficult to handle, and therefore they have not found widespread application.

By Feed Port Location

Depending on the location of the feed ports, pressure vessels could be classified as end-port (end-entry), side-port (side-entry), and multiple-port vessels. Standard designs usually have end-entry and end-exit vessels. With side-entry vessels such as those shown in Fig. 14.11, the feed water enters from the side of the vessel, which often is preferred to the front entry because of the shorter length of distribution piping and the simpler disassembly and access to the membranes within the vessel.

Multiple-port vessels (Fig. 14.12) allow minimizing significantly the length of the feed water distribution piping to the vessels. With RO systems with multiple-port configurations, each vessel contains two side ports for the saline feed water flow and two side ports for the concentrate flow, and the pressure vessels are directly interconnected through Victaulic connections. Plugs/plates are installed between two adjacent stages to direct flow. One critical issue of the design of these systems is the uniform flow distribution, which may be impacted significantly by the headlosses in the side ports.

FIGURE 14.12 Multiple-port pressure vessels.

Previous studies indicate that the use of multiple-port vessels could yield piping cost savings of nearly 50 percent as compared with end-port connection configuration (Sachaf and Haarburger, 2008). However, multiple-port configurations also result in changes in the hydraulic regime and of flow distribution within and between the vessels, which, in turn, makes the design more complex (Verhuelsdonk et al., 2009).

14.5.3 Alternative Membrane Configurations within the Vessels

Standard Configuration

In standard membrane configuration, all membrane elements within the same vessel are identical, and their flux (fresh water productivity) decreases in the direction of the flow (see Figs. 3.18 and 14.11). As explained previously, this configuration results in the first two elements producing over 35 to 40 percent of the total plant flow, expanding feed energy pressure too fast and hindering the performance of the remaining downstream elements.

Usually, in RO systems using standard spiral-wound RO membrane elements, all of the feed water is introduced at the front of the membrane vessel, and all permeate and concentrate are collected at the back end. As a result, the first (front) membrane element is exposed to the entire vessel feed flow and pressure and operates at productivity per square meter of element (flux) significantly higher than that of the subsequent membrane elements.

With a typical configuration of seven elements per vessel and ideal uniform flow distribution to all RO elements, each membrane element would produce one-seventh (14.3 percent) of the total permeate flow of the vessel. However, in actual conventional RO systems, the flow distribution in a vessel is uneven, and the first membrane element usually produces over 25 percent of the total vessel permeate flow, while the last element only yields 6 to 8 percent of the total vessel permeate.

The decline of permeate production (flux) along the length of the membrane vessel is mainly due to the increase in feed salinity and associated osmotic pressure as the permeate is removed from the vessel, while the concentrate rejected from all elements remains in the vessel until it exits the last element. In addition, as the first element produces over 25 percent of the permeate flow, it also uses over 25 percent of the pressure/energy available for desalination. This energy is lost with permeate generated by the first RO element, instead of being available to obtain maximum performance efficiency of the remaining six RO elements in the pressure vessel.

Since a disproportionately large amount of energy is expended too early in the desalination process and the remaining six downstream RO elements are underworked, the overall energy efficiency of permeate production by the pressure vessels in conventional SWRO systems is not at optimum level.

Internally Staged Configuration with Different Membranes

Innovative hybrid membrane configuration combining SWRO elements of different productivity and rejection within the same vessel allows optimizing the use of energy introduced with the feed water to the desalination vessels (Fig. 14.13).

Ideally, redistributing and evening out the feed pressure and flux of all seven RO elements in the vessel to near-equal level can achieve the most energy-efficient desalination process with lowest fouling within the RO vessels. A novel membrane

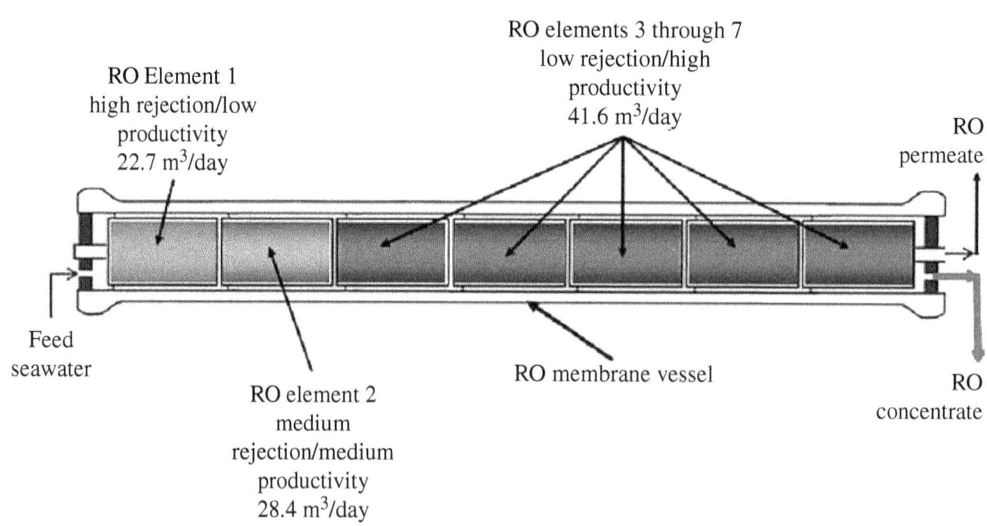

FIGURE 14.13 Interstage SWRO membrane configuration.

configuration design to achieve a more even flux distribution combines three different models of membranes with different permeability within the same vessel instead of using the same model of RO elements throughout the vessel (which is a typical configuration for conventional SWRO systems). This design is often referenced as Inter-Stage Design (ISD) and has been adopted in many recently constructed SWRO desalination plants worldwide (Mickols et al., 2005).

In the example shown in Fig. 14.13, the first (lead) element in ISD configuration, which receives the entire seawater feed flow of the vessel, is a low-permeability/high-salt rejection element (i.e., Dow Filmtec SW30 XHR-400i). Because of its low permeability, this element produces only 14 to 18 percent (instead of 25 percent) of the permeate flow produced by the entire vessel, thereby preserving the feed energy for more effective separation by the downstream RO membrane elements in the vessel.

The second RO element in the pressure vessel is of a standard (average) permeability (i.e., Dow Filmtec SW30 XLE-400i) and salt rejection and produces approximately 14 to 16 percent of the total flow, while the remaining five elements in the vessel are of the same high-permeability/low-rejection model (i.e., Dow Filmtec SW30 ULE-400i). This 1-1-5 combination of low-permeability/high-rejection and high-permeability/low-rejection elements results in a more even distribution of flux and pressure along the vessel and typically yields 5 to 15 percent energy savings, reducing the fouling rate of all membrane elements.

14.6 RO System Piping

High-quality stainless steel is typically used for high-pressure feed and concentrate piping of NF and RO systems. The higher the source water salinity and brine concentration, the higher the quality stainless steel is required to prevent the RO system piping from corrosion and to maintain its longevity. Besides stainless steel, copper-nickel (Cu-Ni) alloys have also found applications for brackish and seawater intake screens and other facilities. FRP and HDPE piping is used for low-pressure applications as well.

A commonly used measure for the quality of stainless steel in terms of corrosion is the parameter referenced as pitting resistance equivalent number (PREN), which is a function of the percent content of chromium (Cr), molybdenum (Mo), and nitrogen (N) contained in the steel:

$$\text{PREN} = \%\,\text{Cr} + 3.3 \times \%\,\text{Mo} + 16 \times \%\,\text{N} \tag{14.2}$$

Table 14.11 summarizes the piping materials recommended to be used for key RO system components. Typically, relatively low-quality stainless steel (i.e., 316 L) with PREN of 25 to 30 is suitable for conveyance of brackish water of low salinity (i.e., TDS < 1000 mg/L) and second-pass (brackish water) RO feed and permeate piping and valves. Duplex stainless steel of PREN of 30 to 40 is to be used for high-salinity BWRO applications. Piping and valves in contact with seawater or seawater concentrate are recommended to have PREN of more than 40.

Stainless-steel piping sections are typically welded together to various fittings. All welds are butt-type with 100-percent penetration. All stainless-steel assemblies are pickled and passivated following welding and then electro-polished until a homogeneous, polished finish is attained. Connections from the pressure vessel feed/concentrate ports to the pipe manifolds are typically accomplished via 90° ell weldments with grooved pipe couplings on both ends.

Steel–Common Reference Name	Areas of Application	PREN Number	Notes
316 L and LDX 2101–stainless steel	Permeate, low-salinity brackish water, second-pass RO	25	Not suitable for source seawater of seawater concentrate
2205–Duplex stainless steel	Permeate, high-salinity brackish water, second-pass RO	35	Can be used for source seawater with low DO level and temp. < 14°C
904 L–Duplex stainless steel	Permeate, high-salinity brackish water, second-pass RO	36	Can be used for source seawater with low DO level and temp. < 20°C
254 SMO	All applications	43	Super Austenitic stainless steel
SAF2507 & Zeron 100	All applications	42	Super duplex stainless steel
AL-6XN	All applications	47	Super duplex stainless steel

Table 14.11 Recommended Quality of Steel Piping and Equipment in Contact with Source Brackish Water, Seawater, and Concentrate

Welding of stainless steel is much more complex than that of regular construction steel and should be performed by highly qualified welders experienced with welding duplex and super-duplex stainless steel.

Reinforced flexible tubing could be used for NF and low-salinity BWRO systems. However, the useful life of such tubing is significantly shorter than that of steel, and it is not desirable because of the overall higher life-cycle replacement costs.

PVC material, schedule 80, is most commonly used for low-pressure permeate piping and valves. Connections of the permeate ports to the end caps of the RO pressure vessels are often made of low-pressure (and low-cost) flexible tubing to simplify RO membrane inspection and maintenance and reduce overall equipment costs. If flexible tubing is used, this tubing should be covered with UV-resistant coating for RO systems installed outdoors because exposure to sunlight would damage the piping, and this piping would need to be replaced every 24 to 48 months. Often, permeate manifold connections to the pressure vessel permeate ports are made using schedule 80 PVC U-bends, with a union at the connection to the vessel and a grooved pipe coupling at the connection to the manifold (Fig. 14.14).

The recommended RO distribution pipe velocities vary with the flow rate and material. General velocity recommendations for different pipe materials are as follows (Watson, 2006)–(1) stainless steel: 2.5 to 3.5 m/s (8 to 12 fps), (2) schedule 80 PVC piping: 1.5 to 2.0 m/s (5 to 7 fps), and (3) schedule 40 PVC piping: 1.0 to 1.5 m/s (3 to 5 fps), and FRP: 1.5 to 2.0 m/s (5 to 7 fps).

With the exception of control valves, valve sizes typically match the diameter of connected piping. Manual operators on 5-cm (2 in) and larger valves usually are gear-type. When valves are located above the trench grating, they are typically furnished with hand-wheel-type operators. Valves located beneath a trench grating are equipped with square nut operators. Openings are provided in the grating to allow access to the operating nut with a valve key.

Figure 14.14 Permeate manifolds of Tampa Bay plant RO train.

Manual valves located more 2 m (6.5 ft) above finished floors or grade levels are usually provided with chain-wheel duplex stainless-steel operators with duplex stainless steel. Buried valves are enclosed in concrete vaults or in valve boxes with cast-iron frames and covers. Automatic flow and control valves are furnished with electric motors. All automated valve actuators are typically provided with OPEN/CLOSED limit switches, with feedback to the control system for status indication. Modulating actuators usually are provided with position feedback. Local control stations have LOCAL/OFF/REMOTE control selector switches and OPEN/STOP/CLOSE pushbuttons for manual valve control in LOCAL mode.

All pressure piping is typically rated at a minimum of 150 percent of its design maximum operating pressure and is fully restrained. Pipe-coating and/or cathodic protection is used to protect buried metallic pipes from corrosion. Dissimilar metals are isolated for protection against electrolysis. Piping beneath concrete slabs or structures is usually encased in concrete with a minimum of 0.2 m (0.7 ft) of cover. Buried gravity piping typically is sloped uniformly without sags or crests. Minimum cover over buried pipe is 1.0 m (3.3 ft) or more. All process piping is usually equipped with vents and drains at pipe high and low points, respectively. Valves and hose connections are installed to match the service.

14.7 RO Skids and Trains

As indicated in the previous section, RO membrane elements are installed in pressure vessels that usually house six to eight elements per vessel. Multiple pressure vessels are arranged on support structures (referred to as skids or racks). The skids are typically made of powder-coated structural steel, plastic-coated steel, or plastic.

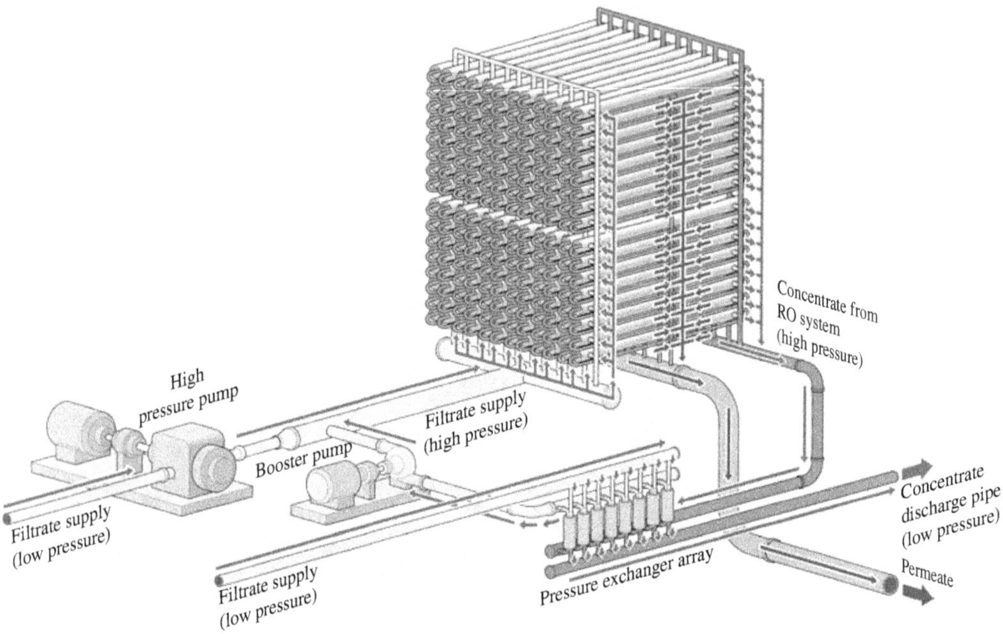

FIGURE 14.15 SWRO train with an isobaric energy-recovery system.

The combination of RO feed pump, pressure vessels, feed, concentrate and permeate piping, valves, couplings, and other fittings (energy-recovery system and instrumentation and controls) installed on a separate support structure (skid/rack), which can function independently, is referred to as RO train. Each RO train is typically designed to produce between 10 and 20 percent of the total amount of the membrane desalination product water flow. Figure 14.15 depicts one SWRO train equipped with a pressure-exchanger-type energy-recovery system.

The RO trains are configured and designed such that each individual train is capable of independently controlling total permeate and concentrate flows.

14.8 Energy Recovery Systems

14.8.1 Overview

A large portion (40 to 50 percent) of the energy applied for desalination of seawater is contained in the concentrate produced by the RO system. The maximum amount of energy that can be recovered from the concentrate, expressed as percentage of the total amount of energy introduced with the RO feed flow, can be calculated as follows:

$$ER_{max} = [(F_p - P_d) \times (1 - R)]/F_p \qquad (14.3)$$

where ER_{max} is maximum energy that can be recovered from concentrate expressed as percentage of the energy entering the RO system with the feed flow (%), F_p is the applied RO feed pressure (bars); P_d is the pressure drop across the membranes (bars), and R is the RO system recovery (%). For example, for a SWRO system operating at feed pressure of 65.0 bars (925 lb/in^2), 43 percent recovery, and differential pressure of 1.5 bars (21.3 lb/in^2), the maximum energy that can be recovered from the concentrate is [(65 bars – 1.5 bars) × (1 – 0.43)]/65 bars = 56 percent. This means that if the energy-recovery equipment has 100 percent efficiency, it could recover 56 percent of the energy introduced into the RO system.

For a brackish water desalination plant operating at feed pressure of 15 bars (217 lb/in^2), recovery of 75 percent, and differential pressure of 2.5 bars (36 lb/in^2) the maximum energy that can be recovered from concentrate is [(15.0 – 2.5 bars) × (1 – 0.75)]/15.0 bars = 21 percent. Since 21 percent of 15 bars is 3.15 bars (45.7 lb/in^2) of pressure energy, while 65 percent of 65 bars (943 lb/in^2) is 42.3 bars (613 lb/in^2); in practical terms, the amount of energy recovered by the SWRO system is 42.3 bars/3.15 = 13.4 times more (i.e., the rate of return on investment is over 13 times higher).

For NF systems operating at feed pressure of 8.0 bars, recovery of 90 percent and differential pressure of three bars, the maximum energy that could be recovered from the concentrate is [(8 – 3 bars) × (1 – 0.9)]/8 bars = 6.3 percent. Since 6.3 percent of eight bars is only 0.5 bars (7.3 lb/in^2), for practical comparison the amount of energy recovered by a typical SWRO system will be over (42.3/0.5 bars) = 84.6 times higher.

This analysis shows that in the BWRO and NF plants, the volume of concentrate and energy contained in it are typically one to two orders of magnitude lower than that of SWRO desalination plants. Therefore, energy-recovery equipment is usually not cost effective to install on NF systems. Low-cost energy-recovery equipment coupled with the high-pressure RO feed pumps however, have found application at high-salinity BWRO plants where unit energy costs are relatively high.

This energy can be recovered and reused for pumping of new saline source water by equipment specifically designed for this purpose—referred to as an energy-recovery device (ERD). Since energy used for seawater desalination contributes 50 to 70 percent of the total plant annual O&M costs and 25 to 35 percent of the total costs of fresh water production, reuse of this energy is beneficial and cost effective. The payback of equipment costs for installation of ER systems in SWRO plants through energy savings is usually less than five years. Advances in the technology and equipment allowing the recovery and reuse of the energy applied for seawater desalination have resulted in a reduction of 80 percent of the energy used for water production over the past 20 years.

Energy recovery equipment could be divided into two main groups based on the principle of its operation: centrifugal and isobaric ERDs. The key features, advantages, and disadvantages of these devices are discussed below.

14.8.2 Centrifugal Energy-Recovery Systems

In centrifugal energy-recovery devices, the pressure contained in the concentrate is applied to an impeller that converts this energy into rotational energy. This rotational energy is then used to reduce the energy needed to run the high-pressure pump. The three types of centrifugal energy recovery devices that have found wide application are Pelton wheel, hydraulic turbocharger, and Francis turbine (reverse running pump).

Pelton Wheel

Pelton wheel (turbine) is nineteenth-century technology originally developed for generation of hydroelectric power. This technology was adopted for desalination plant energy recovery over 20 years ago and consists of an enclosed turbine in which concentrate is applied through a high-velocity nozzle onto spoon-shaped buckets located on the periphery of the wheel (Fig. 14.16). The concentrate pressure is converted into rotational kinetic energy, which is applied to the shaft of the wheel. Since the Pelton wheel shaft is directly connected to the shaft of the high-pressure pump feeding source water to the RO system, the rotational energy of the wheel is directly applied for pumping of RO feed water.

After transferring its energy to the wheel, the concentrate discharges by gravity from it and is conveyed for disposal. Gravity conditions of concentrate disposal are of critical importance for the energy-recovery efficiency of this ERD. The efficiency of conversion of the concentrate energy into energy for pumping RO feed water is 80 to 90 percent and usually increases with the RO system recovery.

Since Pelton wheel is directly coupled with the high-pressure pump motor, the maximum size of RO train equipped with this type of ERD is dictated by the maximum size of commercially available Pelton wheels, which at present is 21,000 m^3/day (5.5 mgd). This size Pelton wheel is installed for the RO trains of the Point Lisas SWRO desalination Plant in Trinidad and, compared with conventional wheels, has two (rather than one) turbines attached on the shaft (Fig. 14.17). Other large plants with Pelton wheels include the 150,000 m^3/day (40 mgd) Beckton desalination plant in London and the 95,000 m^3/day (25 mgd) Tampa SWRO plant in Florida.

The key benefits of this ERD are that it is relatively simple to operate, and it is more compact and less costly than isobaric energy-recovery devices. It also is more energy efficient than Francis turbines. However, typically Pelton wheels are more costly and less efficient than turbocharges for small size installations.

Turbocharger

As previously discussed, the hydraulic turbo booster (HTB) (turbocharger) consists of a turbine and centrifugal pump connected on the same shaft. The HTB pump-turbine assembly is installed in series with a single-stage medium-pressure centrifugal pump, which is driven by an electric motor (Fig. 14.18). The system is equipped with a concentrate bypass to reduce this flow when more than the flow is needed to boost the feed pressure to target level.

The HTB is driven by the energy contained in the RO concentrate. The medium-pressure pump only delivers 50 to 75 percent of the total RO feed pressure (usually 35 to 46 bars/500 to 650 lb/in^2) needed for SWRO desalination. The rest of this pressure (25 to 37 bars/350 to 525 lb/in^2) for up to a total of 56 to 70 bars (800 to 1000 lb/in^2) is

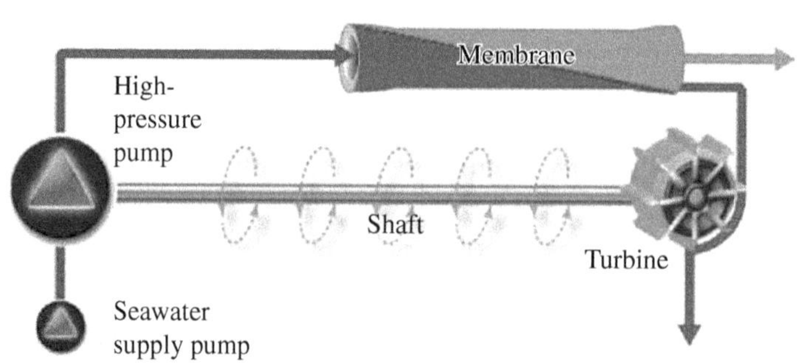

FIGURE 14.16 Pelton wheel ERD. (*Source: ERI.*)

FIGURE 14.17 Double-barreled Pelton wheel of Trinidad SWRO plant.

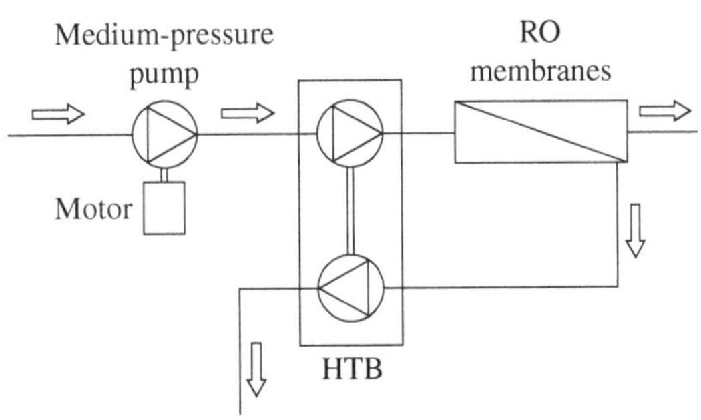

FIGURE 14.18 Turbocharger.

provided by the HTB. The energy efficiency of the HTB is 90 to 92 percent. Taking into consideration that the maximum pump efficiency of the single-stage medium-pressure pump is 85 to 90 percent, the total energy efficiency of the system is between 70 and 80 percent.

As shown by Eq. 14.1, pump efficiency is reversely proportional to the delivered pressure (i.e., the lower the pump feed pressure, the higher its efficiency for the same delivered flow and motor rotational speed). As a result, the efficiency of the medium-pressure pump is higher than a configuration in which a single-stage high-pressure pump has to deliver the same pressure and flow.

Since small size pumps have significantly lower efficiency than larger units, the use of turbochargers for small desalination plants usually yields greater overall energy efficiency. In addition, the HTB system is also less costly and more space efficient than all other ERDs. Figure 14.19 shows a turbocharger installed on a 12,500 m^3/day (3.3 mgd) RO train. The largest plant with turbochargers is the 500,000 m^3/day (132 mgd) Magtaa SWRO facility in Algeria.

The key advantages of a turbocharger are its low equipment cost, minimum space requirements, and simple operation and maintenance. The key drawbacks are its relatively lower efficiency for large-plant-size applications and higher sensitivity of energy-recovery efficiency to actual operating flow and pressure fluctuations.

A turbocharger has found applications for SWRO and BWRO. In the latter, a turbocharger could be used for interstage boosting (Fig. 14.20).

Francis Turbine (Reverse Running Pump)

Similar to the Pelton wheel, the Francis turbine is a device for conversion of the energy of the concentrate into the kinetic energy of the feed pump motor. The turbine is directly connected to the motor shaft. The two ERDs differ by the configuration of the turbine and the concentrate flow-path in it. Francis turbine is more sensitive to

Figure 14.19 Turbocharger on 12,500 m^3/day RO train.

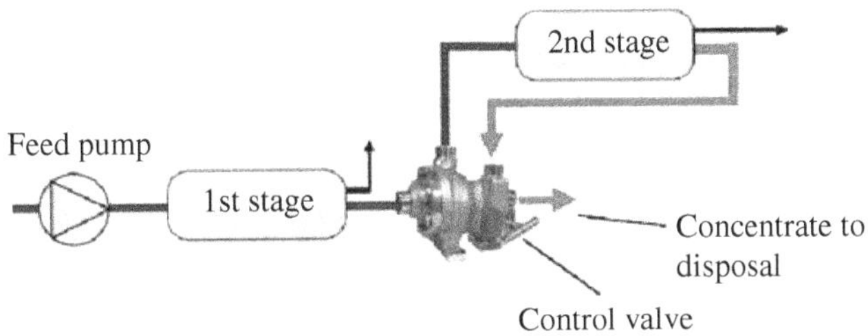

FIGURE 14.20 Turbocharger application for interstage boosting (*Source: FEDCO*).

flow changes and begins to turn only after the plant flow reaches a minimum of 40 percent of its design level. In addition, the energy-recovery efficiency is relatively lower, and its maximum level is achieved only for a very narrow range of RO feed pressure and flow conditions. Therefore, while this ERD device has been used in the past (until the early 1980s), and at present its use is limited mainly for brackish water RO desalination plants-some pump suppliers offer vertical high-pressure pumps with built-in Francis turbines.

14.8.3 Isobaric Energy Recovery Systems

Energy recovery systems working on the pressure-exchange principle (also referred to as isobaric chambers) deliver the energy of the concentrate via piston and directly pump new source water into the RO system (i.e., they transfer concentrate pressure directly into RO feed pressure).

Figure 14.15 depicts the configuration of a typical pressure-exchanger-based energy-recovery system. After membrane separation, most of the energy applied for desalination is contained in the concentrated stream (brine) that also carries the salts removed from the seawater. This energy-bearing stream (shown with arrows in Fig. 14.15) is applied to the backside of pistons of cylindrical isobaric chambers, also known as pressure exchangers (shown as vertical cylinders in Fig. 14.15). These piston-pump-type ERDs convey approximately 45 to 50 percent of the total volume of seawater fed into the RO membranes for salt separation. Since a small amount of energy (4 to 6 percent) is lost during the energy transfer from the concentrate to the feed water, this energy is added back to feed flow by small booster pumps to cover for the energy loss. The remainder (45 to 50 percent) of the feed flow is handled by high-pressure centrifugal pumps.

Harnessing, transferring, and reusing the energy applied for salt separation at very high efficiency (93 to 96 percent) by the pressure exchangers allows a dramatic reduction of the overall amount of electric power used for seawater desalination. In most applications, a separate energy-recovery system is dedicated to each individual SWRO train. However, some recent designs include configurations where two or more RO trains are serviced by a single energy-recovery unit.

One important feature of isobaric chamber ERDs (as compared with centrifugal ERDs such as the Pelton wheel) is that they are decoupled from the high-pressure pumps, and their individual size does not limit the total size of the SWRO trains. Since the general trend in SWRO system design is to build larger-size RO trains in order to benefit from the economy-of-scale related to shorter stainless-steel manifolds, larger size (16- and 18-in) RO elements, and fewer trains, decoupling of the ERD from the high-pressure RO pumps eliminates limitation on the RO train size and the use of larger SWRO feed pumps, which currently exists with centrifugal-type ERDs.

Isobaric chamber-type ERDs have found widespread application over the past 10 years, and the use of these systems has reduced the desalination power costs approximately 10 to 15 percent as compared with the last generation of energy-recovery technologies dominating the market before 2005 (Subramani et al., 2011). Pressure exchangers transfer the high pressure of the concentrated seawater directly into the RO feed water with 93 to 95 percent efficiency.

Pressure Exchanger

The pressure-exchanger (PX) technology developed by Energy Recovery International (ERI) consist of individual fiberglass vessels connected to common feed and concentrate manifolds, each of which contains a ceramic rotor with a number of cylinders (rotor chambers). In sequential process, the rotor chambers are filled up with low-pressure pretreated seawater, rotated by the flow of water itself and then exposed to high-pressure concentrate, which pressurizes this water out of the rotor chamber and delivers it into the RO feed line. As the concentrate pushes out the fresh water from the cylinder, it transfers its energy to it, and, at the end the cycle, the same cylinder contains only low-pressure concentrate. This cylinder is then rotated and exposed to filtered water with pressure higher than that of the low-pressure concentrate, which pushes it out of the cylinder and restarts the cyclic process (Fig. 14.21). Since there are no physical pistons to separate the concentrate from the feed water, some mixing of the two occurs in the contact zone.

At present, ERI pressure exchangers are the most widely used ERDs in medium and large SWRO plants worldwide. Figure 14.22 shows an example of the flow distribution of the ERI system for the Jeddah SWRO desalination plant in Saudi Arabia. It is advisable to contact the ERD supplier for assistance with the design and configuration of the ERD system for the site-specific conditions of a given project.

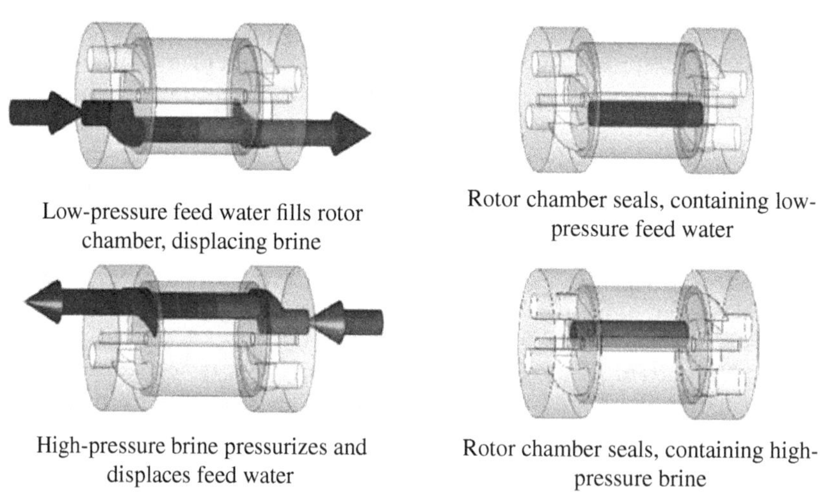

FIGURE 14.21 Working sequence of ERI pressure exchanger (*Source: ERI*).

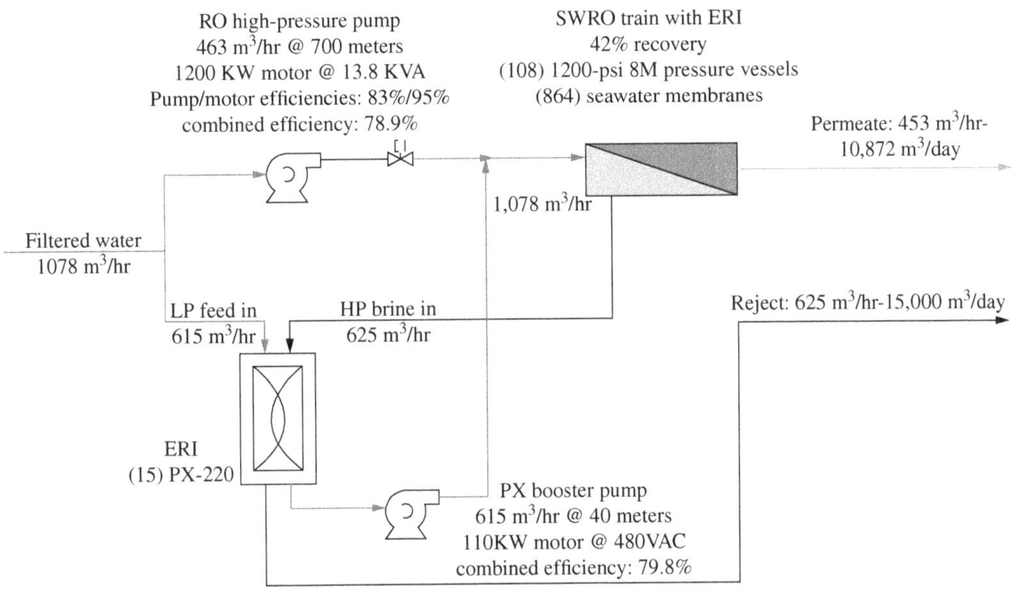

FIGURE 14.22 ERI system for the Jeddah SWRO plant, Saudi Arabia (*Source: ERI*).

The 462,000 m³/day (122 mgd) Hadera SWRO plant in Israel and the 300,000 m³/day (79 mgd) Adelaide and Southern (Perth) SWRO plants in Australia are among the largest facilities with ERI energy-recovery systems.

The key advantage of the ERI system as compared with most other ERDs is its high-energy recovery efficiency and reliability. This system is more compact as compared with the dual work exchanger energy-recovery (DWEER) system, which also applies the same direct pressure-exchange principle. The ERI system has fewer moving parts than DWEER, and its key components are made of low-cost, noncorrosive materials (FRP and ceramics).

Dual Work Exchanger Energy-Recovery System

The DWEER ERD is also an isobaric-type pump for direct conversion of concentrate pressure into RO feed water pressure via piston. This ERD system consists of two stainless-steel or plastic cylinders operating as a group with an actual piston in each cylinder, and two valves (one at each end) which control the feed and discharge of concentrate and RO water in and out of the cylinders, respectively. While one of the cylinders is filling up with filtered water from the pretreatment system, a piston is pushing out the low-pressure concentrate on the other side of the piston. Meanwhile, in the other cylinder, which is already filled up with filtered water from the previous cycle the high-pressure concentrate from the RO system pumps out this filtered water via the piston that separates the two. By the time the piston reaches the opposite end of the cylinder, the energy contained in the concentrate is completely converted into RO feed water pressure. This cycle reverses at the end of each stroke (Fig. 14.23).

Figure 14.24 presents a flow and pressure diagram of the DWEER system used at the 136,000 m³/day (36 mgd) Tuas SWRO Plant in Singapore. At present, the largest plant with DWEER ERDs in operation is located in Ashkelon, Israel. This plant has

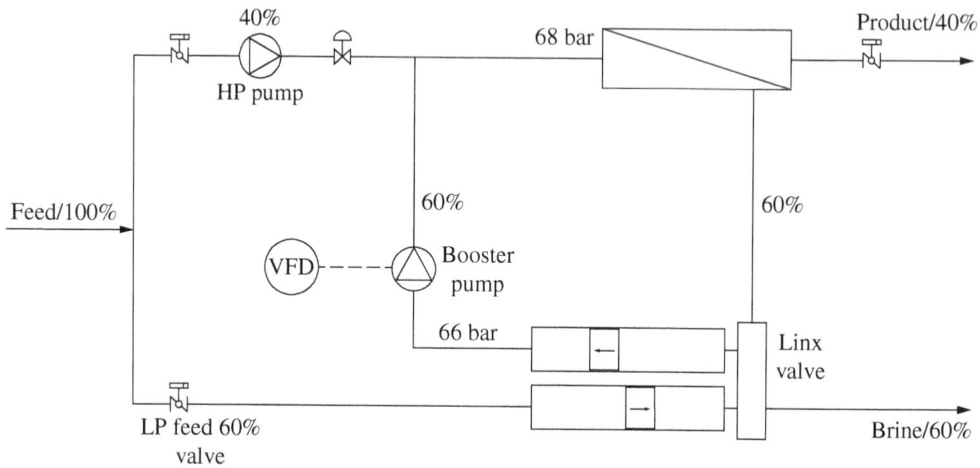

FIGURE 14.23 General schematic of DWEER ERD system. (*Source: DWEER*)

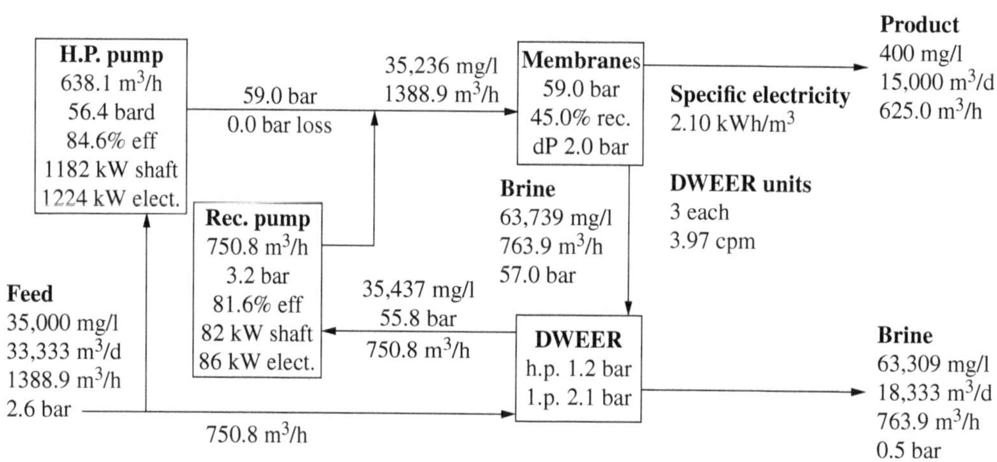

FIGURE 14.24 Flow and pressure diagram for the Tuas SWRO plant, Singapore. (*Source: DWEER*)

capacity of 330,000 m^3/day (86 mgd) and uses larger size DWEER units than those applied for the Tuas project. Other large plants with DWEER ERDs include the 250,000 m^3/day (66 mgd) Sidney Water SWRO plant and the 136,000 m^3/day (36 mgd) Gold Coast SWRO plant in Australia.

As compared with the ERI system, the DWEER system has lower-concentrate/filtered-water mixing which, depending on the model, may provide slightly higher overall energy efficiency. Older-generation DWEER systems are made of steel, which usually renders them more costly than the ERI systems. The latest generation of DWEER systems has addressed these challenges, and it is comparably competitive. Some of the disadvantages of the DWEER systems over the ERI ERDs are that they are more complex, have more moving parts, take longer to commission, and are more maintenance intensive.

14.9 Membrane Flushing and Cleaning Systems

14.9.1 Membrane Flushing System

RO systems are typically equipped with a permanently piped membrane flushing system to automatically flush vessels in the RO trains on shutdown in order to remove residual concentrate and prevent RO membranes from fouling and degradation. The flushing is accomplished using RO system permeate free from disinfectants or other chemicals, or, in the case of shorter RO system shutdowns, the RO vessels could be flushed using nonchlorinated and chemically conditioned filtered water. Flush water is typically stored in an on-site tank of sufficient capacity to flush all installed trains without concurrent refill.

14.9.2 Membrane Cleaning System

As indicated previously, all RO membranes accumulate foulants in the feed/brine spacer cavities over time, which cause differential pressure increase and have to be cleaned periodically in order to maintain their performance and useful life. The purpose of membrane cleaning is to dissolve and remove inorganic scales, dislodge and remove particulate and colloidal foulants, and break down and remove biological film accumulated in the feed/brine spacers.

Typical criteria for membrane cleaning applied in practice include: (1) 10 to 15 percent increase in normalized differential pressure (i.e., difference between feed and concentrate pressures), (2) 10 to 15 percent decrease in normalized permeate flow, (3) 10 to 15 percent increase in normalized permeate TDS concentration, and (4) before and after long-term RO train shutdown.

Normalized differential pressure, permeate flow, and permeate TDS concentration are calculated based on the difference between the initial values of these parameters, measured at the time of the start-up of the RO system, and their ongoing values for which difference is adjusted with a temperature correction factor in order to distinguish between changes in membrane performance caused by fouling and changes caused by temperature. The temperature correction factor used for data normalization is manufacturer and membrane model specific. In practice, actual desalination plant operations data is normalized using proprietary membrane supplier software, which is available on a membrane manufacturer's website.

Depending on the actual membrane fouling rate, RO trains may need to be cleaned as often as once per month or, for plants treating water with very low-fouling propensity, only once per year. In most well operating desalination plants, RO membrane trains are cleaned once every four to six months.

Figure 14.25 depicts a general schematic of clean-in-place (CIP) system for RO membrane cleaning. This system consists of one or more CIP tanks (typically equipped with mixers and cleaning solution heater), cleaning pump(s), cartridge filter, feed and recirculation piping, instrumentation and power supply, and control equipment. Non-chlorinated permeate is used to supply the CIP tank for cleaning solution makeup. Powdered or liquid chemicals are fed directly into the CIP tank, separate small dilution tank, or a CIP feed-recycle line is used for chemical mixing. Cleaning solution is pumped through the vessels of the train being cleaned via dedicated solution feed and return pipe headers.

The capacity of the installed cleaning solution storage tank(s) sized such that they are sufficient to clean all vessels within the largest single RO train from a single

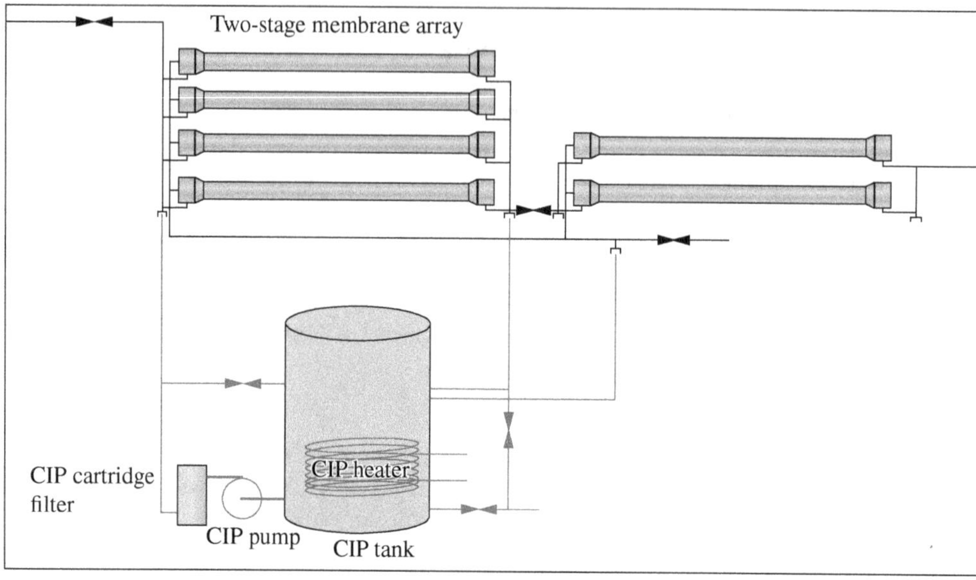

FIGURE 14.25 Schematic of a typical RO membrane cleaning system.

batch of prepared solution. In addition to the cleaning solution feed and return connections on the feed/concentrate manifolds of each train, individually isolated return connections are provided on the permeate header of each train to recycle permeate created during cleaning back to the cleaning solution storage tank(s). The CIP system is designed to mix and recirculate a range of alternate cleaning chemicals made up with RO permeate or dechlorinated potable water.

Usually hoses are used to connect small RO membrane systems to the cleaning feed line, cleaning concentrate return, and cleaning permeate return lines of the membrane system to the CIP tank. For larger systems, "hard" piping is commonly installed between the membrane trains and the CIP system. Membrane cleaning is completed with warm solution with temperature of up to 104°F (40°C). For systems treating cold water (with cold permeate for solution makeup), a heater is installed in the CIP tank. For NF or BWRO systems desalinating groundwater with a temperature of 25°C (77°F) or greater, cleaning is sometimes completed without preheating of the cleaning solution. For RO systems with 8 in (200 mm) vessels, the CIP flow rate needed for effective membrane cleaning typically is 8 to 10 m^3/h (35 to 44 gpm). The CIP pump discharge pressure is usually in a range of 4.5 to 4.8 bars (65 to 70 lb/in^2).

CIP system components are designed based on the number of pressure vessels that will be cleaned in each step. For multistage systems, the vessels included in each stage are cleaned in a separate step to prevent forcing foulants into the subsequent stages. Therefore, in tapered multistage RO systems (Fig. 14.25), the number of first-stage pressure vessels determines the size of the CIP system. Table 14.12 presents a list of typical membrane-cleaning solutions used in brackish and seawater desalination plants.

While many of the membrane-cleaning chemicals are generic products that could be procured through a number of chemical vendors, RO membrane suppliers often

Foulant Type	Cleaning Solution(s)
Inorganic salts (e.g., $CaCO_3$, $CaSO_4$, $BaSO_4$)	0.2% HCl; 0.5% H_3PO_4; 2% Citric acid
Metal oxides	2% Citric acid; 1% $Na_2S_2O_4$
Inorganic colloids (silt and particulates)	0.1% NaOH/0.05% Na dodecyl benzene sulfonate/pH 12
Silica (and metal silicates)	Ammonium bifluoride; 0.1% NaOH/0.05% Na dodecyl benzene sulfonate/pH 12
Biofilms and organics	Hypochlorite, hydrogen peroxide, 0.1% NaOH/0.05% Na dodecyl benzene sulfonate/pH 12, 1% sodium tripolyphosphate/1% trisodium phosphate/1% sodium EDTA

Notes: $BaSO_4$ = barium sulfate; $CaCO_3$ = calcium carbonate; $CaSO_4$ = calcium sulfate; HCL = hydrochloric acid; H_3PO_4 = phosphoric acid; $Na_2S_2O_4$ = sodium hydrosulfite; NaOH = sodium hydroxide; EDTA = ethylenediaminetetraacetic acid.

Table 14.12 Typical Membrane Cleaning Solutions

offer proprietary membrane-cleaning formulations based on laboratory and field experience with their own membrane products, which may be more suitable than generic cleaning chemicals. Therefore it is recommended to consult the plant membrane element supplier when making decisions regarding chemicals and cleaning procedures to be applied for a specific project. Key factors that influence the efficiency of membrane cleaning, besides selecting the appropriate chemicals, are temperature of the cleaning solution, chemical strength, cleaning flow, and length of time intervals of membrane soaking and chemical circulation.

The typical sequence of activities during CIP cleaning of RO train includes: (1) RO train flushing, (2) membrane disinfection and partial removal of iron and calcium using sodium bisulfide (SBS) or 2,2, dibromo-3-nitrilo-propionamide (DBNPA), (3) membrane flushing and high pH cleaning, (4) membrane flushing and evaluation of the high pH cleaning effect, (5) low pH cleaning, (6) membrane flushing and final disinfection with SBS or DBNPA, and (7) final membrane flushing and evaluation of overall cleaning effect. Discussion of waste streams generated during CIP cleaning is presented in Chap. 16.

14.10 Instrumentation and Controls

14.10.1 Overview

Instrumentation and control systems can be as basic as a manual control with automatic shutdown features for pump and membrane protection or as complicated as a supervisory control and data acquisition (SCADA) system. SCADA systems are often based on programmable logic controllers and remote telemetry units that are supervised by a host computer located in a control room near the membrane skids and high-pressure pumps.

The use of personal computers or industrial-grade, human-machine interface is now common (Fig. 14.26).

Systems designed for automatic control can monitor chemical feed systems, and they have alarm and report-generation capabilities. In many facilities, a personal computer is also used for membrane train performance normalization calculations and graph preparation, which facilitates plant performance monitoring and making of decisions on when to clean the membranes.

14.10.2 SCADA System

The plant SCADA system consists of information and the control networks. At a minimum, the information network is composed of personal computers, server/workstations, printers, hubs, and switches, etc. located at the central control room. The control network consists of programmable logic controllers (PLCs), remote input/output (RIO) panels, fiber optic or serial data cables, and execution components associated with key processes and equipment such as motorized valves and variable frequency drives.

Programmable Logic Controllers

Fully automated desalination plants include programmable logic controller (PLC) control panels for key plant components equipped with a redundant PLC and an operator interface. The PLC control panels are connected to the RIO panels via a fiber optic or data cable ring. All main plant processes and equipment including intake, pretreatment, RO system, post-treatment, product water storage and conveyance system, chemical feed and storage system, and solids handling system are equipped with PLCs,

FIGURE 14.26 Central control room of Gold Coast SWRO plant, Australia.

which allow monitoring of normal plant operations, equipment status, and alarm conditions in real time. The PLC network is typically connected to the SCADA system via Ethernet fiber optic or data cable. Fiber-optic cable minimum requirements are four-channel multimode.

Human-Machine Interface

Human-machine interface (HMI) includes graphic control screens, alarm functions, trend functions, data presentation, incident recording, etc. to monitor and control the entire plant. The HMI software is selected to be compatible with the PLC and the PC server/workstations. Currently, the most widely used HMI software is Intellution and Wonderware. Other proprietary packages have also found applications in recent projects. For consistency in the graphics, the same HMI software is used for the PLCs and at the workstations.

Local Control Panels

Local input/output control panels are used for medium and large plants. Local control panels are provided adjacent to all key facilities—and especially in remote locations, which are not within a walking distance from the main control panel (such as the intake pump station, product water storage tanks, etc.). All panels are typically designed to comply with the appropriate classification for the specific panel location. In the latest plant designs, local control panels are replaced by portable wireless tablets, which allow plant operators to have the same monitoring information and control over the individual treatment processes as they would in the central control room (Fig. 14.27).

FIGURE 14.27 Operator with wireless tablet for process monitoring and control.

14.10.3 Plant Monitoring and Control

The plant control system usually has manual and automatic control provisions: in the manual mode locally by the operator and in the automatic mode by the PLC with operator confirmation input. At a minimum, manual/auto controls for the following facilities are typically provided:

- The entire plant: normal start-up and shutdown and emergency shutdown
- Source water intake pump station
- Pretreatment processes
- High-pressure membrane feed pumps
- RO membrane racks
- Energy-recovery system
- Clean-in-place and flush systems
- Spent cleaning solution system
- Degasifier and scrubber system
- Product water transfer pumps
- Concentrate management system
- Chemical feed systems

Alarm and Monitoring Parameters

At a minimum, the following parameters are typically monitored and used for controlling desalination plant operations:

- Source water pH, conductivity, turbidity, and temperature
- Source water flow and pressure
- Train permeate and concentrate flow and hydraulic recovery
- Product transfer pump and finished water flow
- Train permeate and concentrate and plant finished water conductivity
- Degasifier effluent pH
- Finished water pH and chlorine residual
- Total plant waste concentrate flow
- Membrane control valve status
- Membrane feed, interstage, and concentrate pressures
- Product water clear-well/storage tank level
- Finished water storage tank level
- Product water pump flow and discharge pressure

Alarms are automatically activated when these parameters reach preset low and/or high values.

Shutdown of All RO Membrane Trains

The instrumentation and control system is typically designed with provisions to shutdown all membrane trains when the operator initiates a total plant shutdown manually or automatically under the following conditions:

- Pump "run" failure of source water intake pumps, critical chemical feed pumps, product transfer pumps
- Source water pressure below a preset minimum
- Feed water pH or turbidity excursion
- Water level in product water clear-well or finished water storage tank reaches preset high level
- Loss of power

Shutdown of Individual RO Membrane Trains

In addition to the total plant shutdown events listed above, the instrumentation and control system is typically designed with provisions to shut down individual RO trains when the operator initiates a train shutdown manually or automatically under the following conditions:

- Membrane feed pump low suction or high discharge pressure
- High permeate pressure
- Train recovery rate is out of range
- Train concentrate flow low
- Train permeate flow out of range
- Train permeate pressure high
- Water level in product water clear well or finished water storage tank reaches preset high level
- Loss of individual train power

Start/Stop Stations

For large- and medium-size plants, all motors servicing key equipment should have local control stations with "start/stop" buttons, local "running/stopped" indicator lights, local "lock-out," and local disconnect switches. The local start/stop stations have to be installed in addition to the provisions for remote control of the motor operations from the motor control center (MCC), PLC panel, or the plant control room workstations.

Uninterruptible Power Supply

Uninterruptible power supply (UPS) system is typically installed to provide backup power for plant instrumentation monitoring and control system. One or more UPS units are usually provided for the PLC panel and the workstations at the central control room. The UPS is sized to supply backup power for a minimum period of 30 minutes.

14.10.4 Instrumentation

The basic instrumentation required to monitor and control any RO system consists of control valves and devices for measuring flow, pressure, conductivity, pH, temperature, and liquid levels.

Instrument location is selected such that the instruments are subjected to significant vibrations and undue turbulence. Instruments should be easily accessible because they require frequent calibration and repairs.

Magnetic Flow Meters

Magnetic flow meters are popular flow-measurement devices in large-capacity membrane plants. This equipment can be used on most water streams encountered in RO plants with the possible exception of some permeate streams. Low conductivity (i.e., 20 mS/cm or lower) water will not give an accurate flow reading, so these meters need to be carefully selected. Vortex shedding meters can be used for low conductivity applications.

Typically, flow meters used at membrane plants are pulsed DC electromagnetic induction-type, which provide a signal proportional to the liquid flow rate. The recommended meter accuracy is plus or minus 0.1 percent of reading. All flow meters that are used at the desalination plant have to be factory calibrated. Magnetic meters have to be grounded per manufacturer's recommendations. A NFMA 4X flow converter/transmitter matched to the flow meter is typically provided for meters generating remote data signals. The output should be 4-20 mA into 0-1000 ohms. A local indication of actual flow rate and totalization display should be provided. The key advantages of magnetic flow meters, as compared with other alternative types of meters, are that the flow stream is completely free from obstacles, and, as a result, the headlosses through the meter are minimal. In addition, magnetic flow meters usually have a wider measuring range than most other types of flow meters. However, sometimes, when pipeline headloss is not a limiting factor, Venturi meters are used for measuring large plant intake flows. Venturi meters are lower-cost/lower-accuracy flow meters, which also could be equipped with remote data transfer.

Rotameters

These flow measurement devices are suitable only for small package plants and are acceptable low-cost equipment for local indication of small-volume chemical feed flows. Although rotameters can be supplied with a flow signal transmitter, this configuration is uncommon. Rotameter accuracy is sensitive to the viscosity and density of the measured liquid and to the concentration of particulates in the measured stream.

Electronic Pressure Transmitters

Electronic pressure transmitters can provide reading accuracy of 0.1 percent of their span, which is important when measuring differential pressures in critical locations at the desalination plant. These pressure transmitters can be electronically zeroed. The differential pressure transmitters are usually diaphragm actuated, microprocessor-based type. They are equipped with loop-powered units with a 4-20 mA output. Each transmitter is typically provided with stainless-steel mounting hardware and a five-way manifold. Differential pressure transmitters are used to measure the pressure drop across the following key treatment plant facilities:

- Pretreatment filters (if pressure type pretreatment filters are used)
- Cartridge filters
- Membrane train stage feed, interstage, concentrate, and/or stage and overall train differential pressure (feed to concentrate)

Conductivity Analyzers

The quality of the source and product water is typically monitored by conductivity analyzers. The conductivity sensor is an in-line-type sensor unit with a local indication and a transmitter for remote accurate continuous monitoring, indicating, and recording. Although conductivity meters are usually installed on-line, valved sample points for measuring conductivity/salinity using portable apparatus should also be provided at key locations, such as the source water intake pump station, feed to the RO membrane system, concentrate discharge, and the product water lines from the individual RO trains. Conductivity is measured in μS/cm (micro-Siemens per centimeter). Typically, high conductivity readings from the analyzer located on the permeate lines from the individual RO membrane trains trigger alarm locally and remotely at the central control room.

Temperature and pH Analyzers

Online electronic pH and temperature analyzers and transmitters are widely used in RO desalination plants. Online temperature analyzers are recommended to be installed on the feed line to the RO system, if the water temperature is expected to vary significantly (more than 50°C/10°F from the annual average temperature). A pH analyzer is typically installed as a minimum on the product water line.

Liquid-Level Sensors

The type and operational parameters of the liquid level sensors are the same as those used in conventional water treatment plants. Ultrasonic-type level sensors/transmitters are commonly used for water tanks/wells with a relatively quiescent surface, such as product water storage and chemical feed tanks. Usually, the liquid level sensors are potted/encapsulated in corrosion-resistance housing. These sensors are provided with automatic air temperature and density compensation. A microprocessor-based transmitter/converter converts the sensor output signal to level.

Level measurement accuracy of most sensors is plus or minus 1.0 percent. The output is an isolated 4-20 mA signal. For outdoor mounted units NEMA 4X enclosures with sunshields are provided. Liquid level signals for all key tanks and pump wet wells are transmitted to the PLC and the desalination plant control room workstations for continuous monitoring and alarm generation.

14.11 RO System Types and Configurations

14.11.1 Overview

Based on the source water salinity they process, desalination plants can be divided into three main groups: nanofiltration (softening) plants, brackish water desalination plants, and seawater desalination plants. In addition, depending on the number of sequential RO systems for treatment of permeate and concentrate, RO system configurations could be divided into two main categories: (1) single- and multipass RO systems, and

(2) single- and multistage RO systems. In all types of desalination plants, multipass and multistage RO systems could also be combined into configurations that allow to achieve target RO system recovery and product water quality at optimum life-cycle cost of water production. The various types of systems and their practical application are discussed in the following sections.

Single and Multipass RO Systems

An NF or RO system where the saline source water is desalinated (i.e., treated by reverse osmosis) only one time is referenced as a single-pass RO system (Fig. 14.28). RO systems that are designed to re-treat permeate multiple times are termed multipass RO systems.

Figure 14.28 shows general schematics of single-, two-, and three-pass RO systems. Since each RO pass provides additional treatment of permeate produced by the previous RO pass, the overall system permeate water quality improves with each pass. Therefore, multipass RO systems are applied when source water salinity is relatively high and the target product water quality cannot be achieved by treating the saline source water by reverse osmosis only once. While multiple RO pass systems allow generating product water of very high quality, such systems are also more costly and produce less water than a single-pass RO system processing the same volume of source water flow. Therefore, multipass RO systems are typically applied when a single-pass RO system cannot produce source water of desired quality. Another reason, multipass (usually two-pass) RO systems are employed is when the source water has a very high-fouling potential.

As discussed in Chap. 3, the rate of RO membrane fouling is exponentially proportional to the membrane flux, which, in turn, (for a given RO system) is proportional to the difference between RO system feed and permeate pressures. By using a two-pass

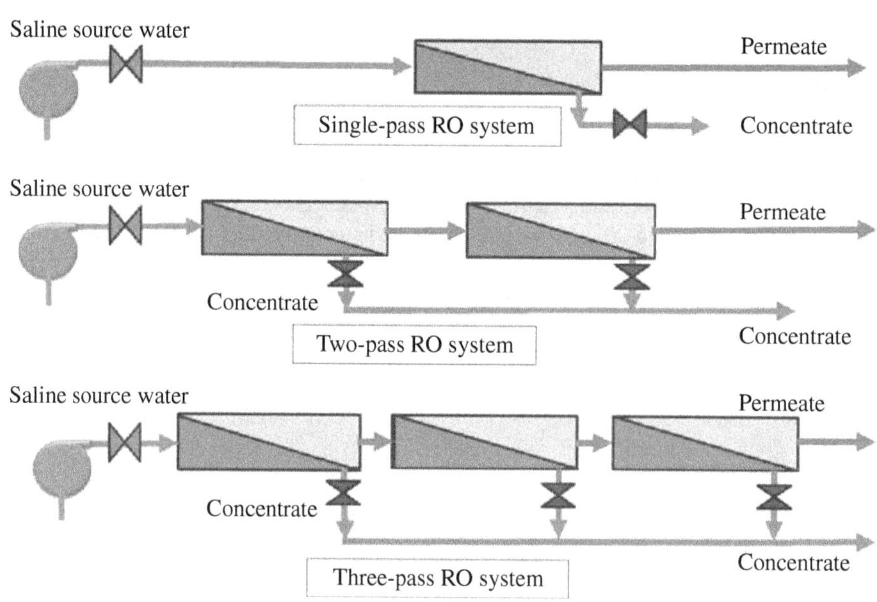

FIGURE 14.28 Single and multipass RO systems.

versus single-pass RO system, the feed pressure to the first pass could be reduced significantly (typically down to 65 to 75 percent of the total RO feed pressure) thereby diminishing membrane flux and fouling rate. This reduction could result in a measurable improvement of RO system operations (lower RO membrane cleaning frequency and differential/feed RO pressure). In this case, the use of a two-pass RO system, however, would require the installation of interpass RO pump to deliver the remaining (25 to 35 percent) of the total RO feed pressure needed for production of the target RO system water quality. The pump, which boosts the permeate pressure from the first pass to the second pass, is referred to as a booster (or second-pass) RO feed pump.

Single and Multiple RO Stage Systems

A key challenge associated with the use of multipass RO systems is that the overall recovery of such systems decreases with the number of the installed RO passes because a portion of the saline source water is converted into concentrate at each pass (Fig. 14.28).

In order to reduce the total volume of concentrate (i.e., increase the overall recovery/produce more fresh water) from the same volume of source water, the concentrate generated by the individual RO passes can be treated by a separate RO system, referenced as "stage." Typically, the membrane vessels in the BWRO system stages are grouped in parallel and are referred to as "arrays." The stages are arranged in series such that adequate cross-flow and minimum concentrate flow are maintained in each stage. Often, the ratio between the number of vessels in the first stage and second stage or NF and BWRO systems is selected at 2:1, and such configuration is referenced to as "2:1 Array." In three-stage NF and BWRO systems the typical array configuration is 3:2:1 (i.e., the number of vessels in the first stage is three times higher than those in the third stage). The second stage has two times higher number of vessels than the third stage.

Figure 14.29 depicts single-, two-, and three-stage RO systems. Use of multiple stages allows improvement of the overall recovery of the entire RO system. However, the additional concentrate treatment steps also increase RO system costs. The optimum number of stages and passes for a given RO system depends on many factors—such as source water quality, target permeate water quality and fresh water production flow, and cost of equipment and energy—and should be determined based on the site-specific conditions of each project.

It is important to note that the single-pass and single-stage RO systems shown in Figs. 14.28 and 14.29 are identical in configuration (i.e., an RO system in which the source water is treated only once can be referenced both as single-pass and single-stage). The difference of nomenclature can be explained by the main difference in configuration of SWRO and BWRO systems (i.e., in SWRO desalination plants it is more common that the source water undergoes RO treatment multiple times; while in NF and BWRO plants, the concentrate rather than the permeate generated by the first RO treatment undergoes additional processing by reverse osmosis). As a result, an SWRO system, where the source water and permeate are processed through reverse osmosis in sequence is referred to as a two-pass SWRO system; while a BWRO system, where the source water and the concentrate are processed through reverse osmosis in sequence, is referred to as a two-stage BWRO system.

14.11.2 NF System Configurations

Nanofiltration plants usually treat relatively low salinity water (typically less than 1000 mg/L), which has high elevated hardness, organic content, color, iron, or other divalent ions such as nitrates. Typically, NF plants are equipped with two-stage RO

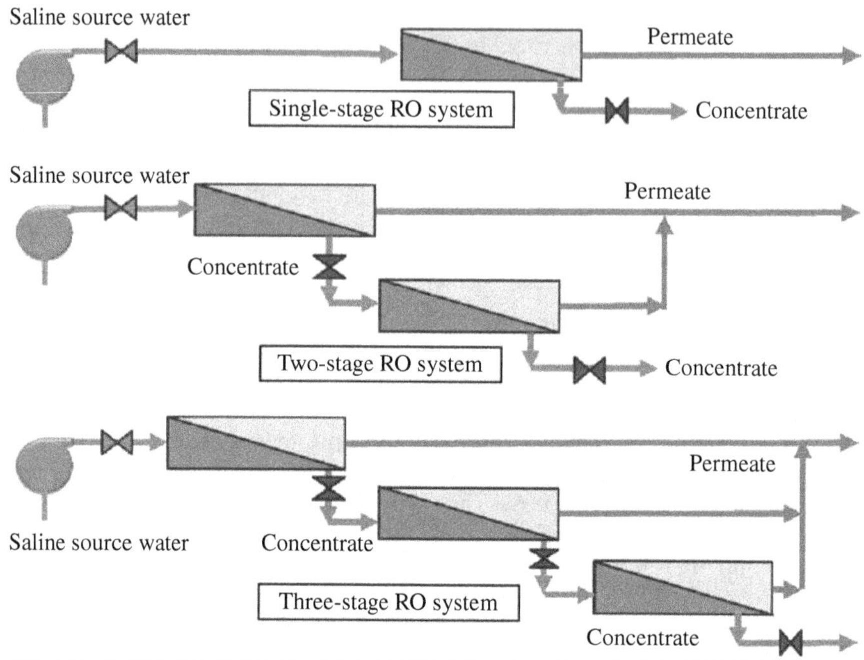

Figure 14.29 Single and multistage RO systems.

systems and are designed to operate at 80 to 90 percent recovery. The most commonly used NF membrane configuration includes pressure vessels with seven elements per vessel and a 2:1 membrane array.

Figure 14.30 depicts the general process schematic of the largest operational NF plant in the United States, the 151,000 m^3/day (40 mgd) Boca Raton plant in Florida. This plant processes well water of relatively low salinity (466 mg/L) but has high hardness (265 mg/L), alkalinity (265 mg/L); iron (0.3 mg/L) and NOM content (TOC = 12 mg/L). The plant is designed to operate for total recovery of 85 percent, and it has 10 two-stage NF trains with 2:1 arrays (72 vessels in the first stage and 36 vessels in the second stage). The plant uses Hydranuatics ESNA1-LF2 elements (see Table 14.2) operating at an average flux of 20.7 Lmh (12.2 gfd) and producing an average of 20.1 m^3/day (5300 gpd) per element. The plant permeate has total hardness of 50 to 80 mg/L, iron content below 0.04 mg/L, and TOC of less than 1 mg/L (Bartels et al., 2007).

14.11.3 BWRO System Configurations

Source water TDS concentration of BWRO plants typically ranges between 500 mg/L and 10,000 mg/L. Plants processing source water with salinity between 500 and 2500 mg/L and in a range of 2500 to 10,000 mg/L (or above) are referred to as low-salinity and high-salinity brackish water reverse osmosis (BWRO) desalination facilities, respectively.

Figure 14.31 illustrates a typical schematic of a low-salinity BWRO desalination plant. For such plants blending a portion (5 to 30 percent) of the source water flow (indicated in Fig. 14.31 as "Bypass Source Water") with RO permeate is common practice for

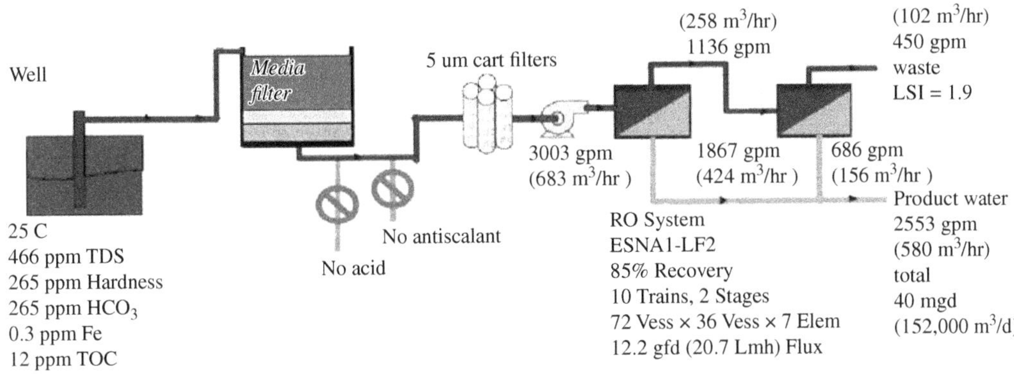

FIGURE 14.30 Schematic of Boca Raton NF plant, Florida. (*Source: Hydranautics.*)

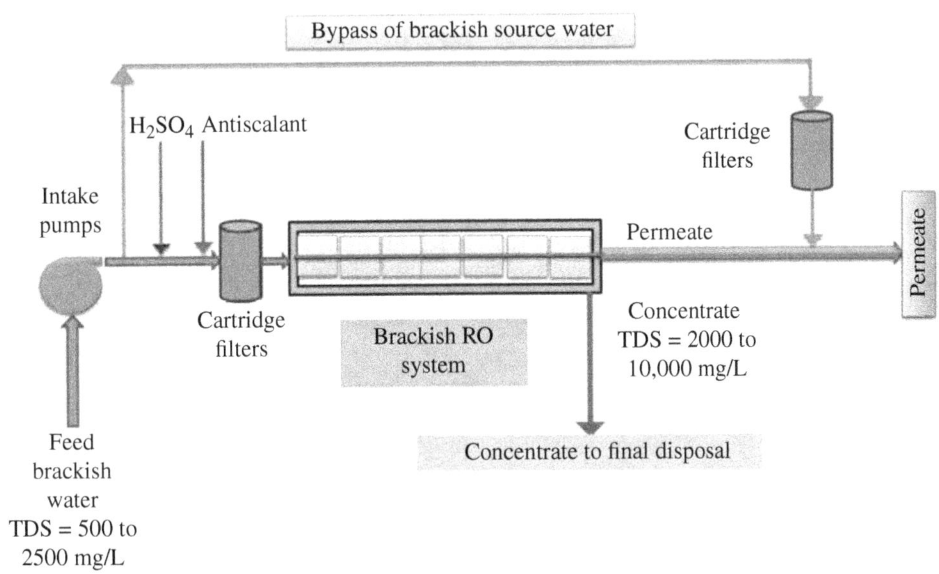

FIGURE 14.31 Schematic of typical low-salinity BWRO plant.

remineralization of the desalinated water. Low-salinity BWRO plants often process the source water through a single RO stage (pass) only. However, two-stage BWRO plants configured with 2:1 arrays are also common.

Table 14.13 provides an illustrative hypothetical example of the permeate water quality produced by a low-salinity BWRO plant operating at blending ratio of 28.6 percent and permeate recovery of 85 percent (Wilf et al., 2007). In this specific example, the TDS of the source seawater and RO permeate are 647.3 and 215 mg/L, respectively.

Figure 14.32 depicts a typical schematic of a high-salinity BWRO desalination plant. As indicated in this figure, the brackish source water in this case is usually treated by a two-stage RO system, where concentrate generated by the first RO pass is processed

Water Quality Parameter	Source Water Quality	Blended Permeate Water Quality
Temperature, °C	25	25
pH	7.0	6.6
Ca^{2+}, mg/L	96	29
Mg^{2+}, mg/L	11.7	3.5
Na^{+}, mg/L	90	32.1
K^{+}, mg/L	6.5	2.4
HCO_3^{-}, mg/L	72.6	30.4
SO_4^{2-}, mg/L	158.4	47.2
Cl^{-}, mg/L	190.7	61
F^{-}, mg/L	0.2	0.1
SiO_2, mg/L	24.3	9.3
TDS, mg/L	647.3	215

Table 14.13 Example of Product Water Quality Generated by Low-Salinity BWRO Plant

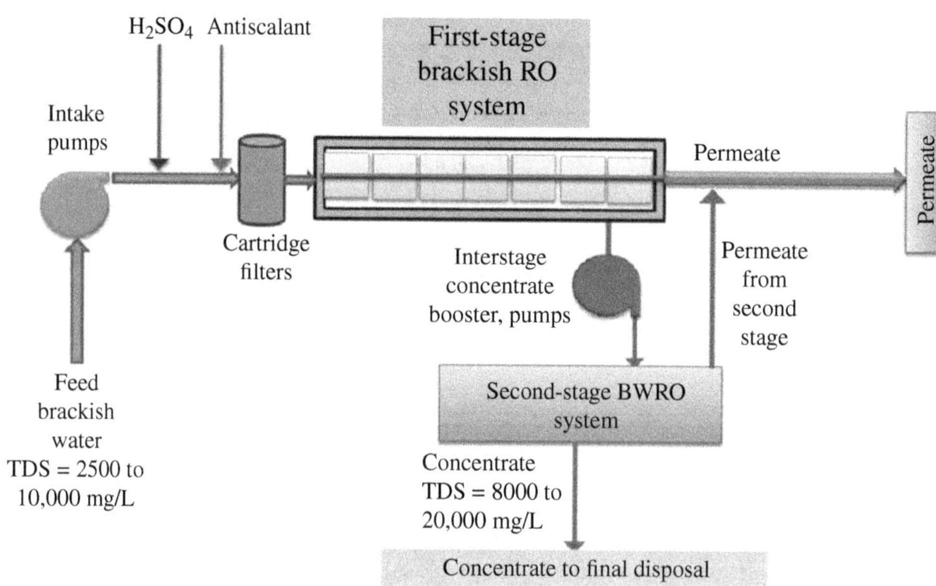

Figure 14.32 Schematic of typical high-salinity brackish water RO plant.

through a second RO system after increasing (boosting) the concentrate pressure via interstage booster pumps. Under this configuration the first-stage permeate usually contributes 75 to 85 percent of the total permeate flow, while the RO system, which processes concentrate from the first RO pass (typically referred to as "second stage"), produces the remaining 15 to 25 percent of the total RO system permeate flow.

A hypothetical example of the permeate water quality produced by a high-salinity BWRO plant is presented in Table 14.14 (Wilf et al., 2007). In this example, the desalination plant uses a brackish well water source of salinity of 5881 mg/L and processes this source through a two-stage BWRO system similar to that illustrated in Fig. 14.32. The overall plant recovery is 80 percent (i.e., 80 percent of the source water is converted into low-salinity permeate). The system is equipped with an interstage booster pump, which directs the concentrate generated by the first SWRO stage to the second RO stage. The recovery of the first RO stage for the example presented in Table 14.14 is 64 percent. The second stage recovers approximately 44 percent of the concentrate produced by the first-stage RO system.

14.11.4 Seawater System Configurations

The SWRO system configurations most widely applied at present include single-pass treatment, where the source water is processed by RO only once (Fig. 14.33), and two-pass RO treatment, where the seawater is first processed through a SWRO system, and then permeate produced by this system is reprocessed by brackish RO membranes (see Figs. 14.34 and 14.35).

Single-Pass SWRO Systems

Single-stage SWRO systems are designed to produce desalinated seawater (permeate) in one step using only a single set of RO trains operating in parallel. In general, between 800 and 900 SWRO membrane elements installed in 100 to 150 vessels are needed to

Water Quality Parameter	Source Water Quality	Blended Permeate Water Quality
Temperature, °C	28	28
pH	7.0	6.0
Ca^{2+}, mg/L	105	1.0
Mg^{2+}, mg/L	130	1.3
Na^{+}, mg/L	1837	84
K^{+}, mg/L	85	4.8
CO_3^{2-}, mg/L	0.6	0.0
HCO_3^{-}, mg/L	250	18.4
SO_4^{2-}, mg/L	479	5.5
Cl^{-}, mg/L	2970	123
F^{-}, mg/L	1.4	0.1
NO_3^{-}, mg/L	5.0	1.4
SiO_2, mg/L	17.0	0.5
TDS, mg/L	5880	240

TABLE 14.14 Example of High-Salinity BWRO Plant Permeate Water Quality

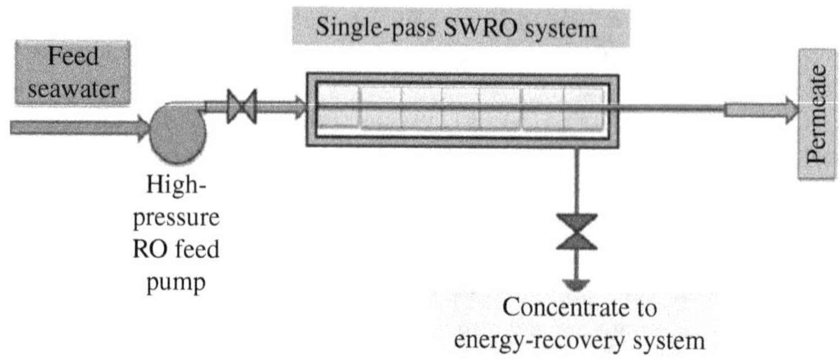

FIGURE 14.33 Single-pass SWRO system.

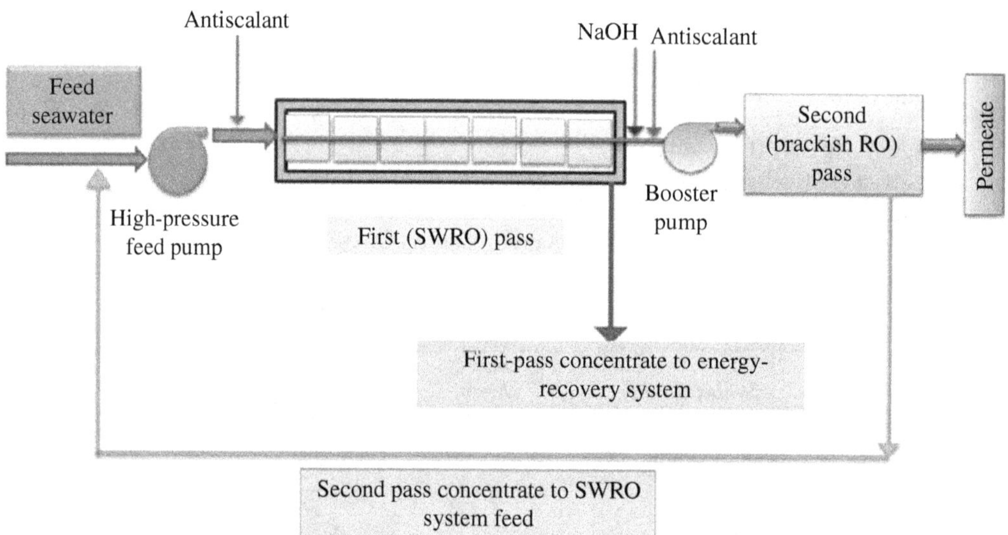

FIGURE 14.34 Conventional full two-pass SWRO system.

produce 10,000 m^3/day (2.6 mgd) of permeate suitable for potable use in a single-stage SWRO system.

Under a typical single-stage SWRO system configuration, each RO train has a dedicated system of transfer pump for pretreated seawater followed by a high-pressure RO feed pump. The high-pressure feed pump motor/operation is coupled with that of energy-recovery equipment (see Fig. 14.31).

Single-stage SWRO systems are widely used for production of drinking water. However, these systems have found limited industrial application mainly because of the water quality limitations of the produced permeate. Even if using the highest-rejection RO membrane elements commercially available today (nominal minimum rejection of 99.85 percent), the single-stage SWRO desalination systems typically cannot consistently yield permeate with TDS concentration lower than 200 mg/L, chloride level of less than

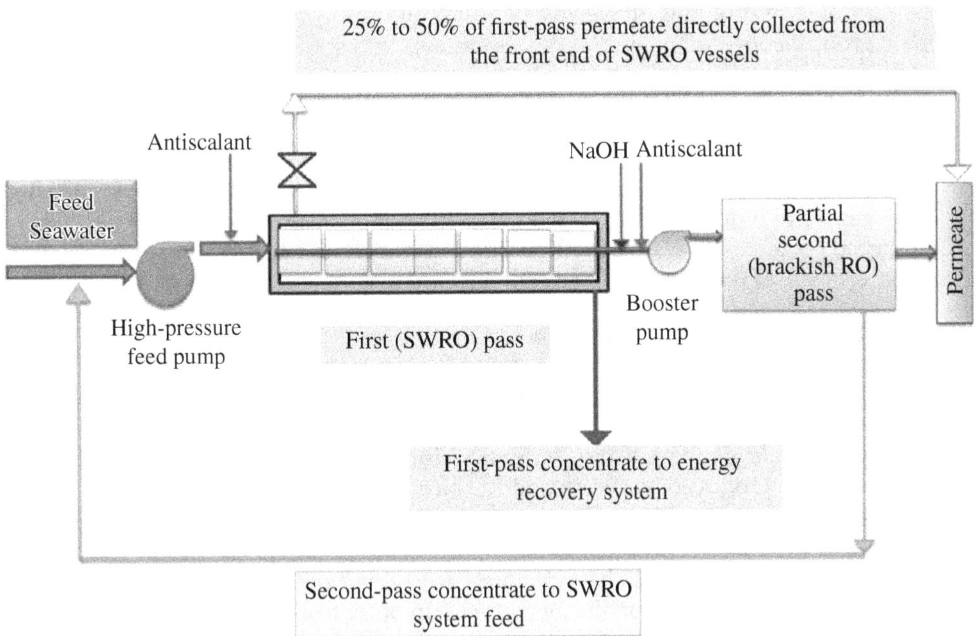

FIGURE 14.35 Split-partial two-pass SWRO system.

100 mg/L, and boron concentration lower than 0.5 mg/L, especially when source water temperatures exceed 18 to 20°C (64 to 68°F).

If enhanced boron removal is needed in such systems, high boron rejection membranes (see Table 14.8) are used, and/or sodium hydroxide and antiscalant are added to the RO system feed water to increase pH to 8.8 or more, which, in turn, improves boron rejection.

Two-Pass SWRO Systems

Two-pass SWRO systems are typically used when either the source seawater salinity is relatively high (i.e., higher than 35,000 mg/L) and/or the product water quality requirements are very stringent. For example, if high-salinity/high-temperature source water (such as Red Sea and Persian Gulf seawater, for example) is used in combination with standard-rejection (99.6 percent) SWRO membranes, then single-stage SWRO systems may not be able to produce permeate suitable for drinking water use. In this case, two-pass SWRO systems have proven to be an efficient and cost-effective configuration for potable water production. RO systems with two or more passes are also widely used for production of high-purity industrial water.

The two-pass SWRO systems typically consist of a combination of a single-pass SWRO system and a single- or multipass brackish water RO (BWRO) system connected in series (see Figs. 14.34 and 14.35). Permeate from the SWRO system (i.e., first pass) is directed for further treatment to the BWRO system (i.e., second pass) to produce a high-quality TDS permeate. The concentrate from the pass-two BWRO system is returned to the feed of the one-pass SWRO system to maximize the overall desalination system production capacity and efficiency. Two-pass SWRO systems are classified into two main groups: conventional full two-pass systems and split-partial two-pass systems.

Conventional Full Two-Pass SWRO Systems

In conventional full two-pass SWRO membrane systems (see Fig. 14.34), the source seawater is first treated by a set of SWRO membrane trains (referred to as first RO pass), and then the entire volume of desalinated water from the first pass is processed through a second set of brackish water desalination membrane trains (Greenlee et al., 2009). If enhanced boron removal is needed, sodium hydroxide and antiscalant are added to the feed permeate of the second-pass RO to increase pH and improve boron rejection.

Split-Partial Two-Pass SWRO Systems

In split-partial two-pass systems the second RO pass typically processes only a portion (50 to 75 percent) of permeate generated by the first pass. The rest of the low-salinity permeate is produced by the front (feed) SWRO elements of the first pass. This low-salinity permeate is collected and, without additional desalination, it is directly blended with permeate produced by the second RO pass (see Fig. 14.35).

As indicated in Fig. 14.35, the second-pass concentrate is returned to the feed of the first RO system pass. When the desalination system is designed for enhanced boron removal, this concentrate will have pH of 9.5 to 11.0 and potentially could cause precipitation of calcium carbonate on the membranes. In order to avoid this challenge, typically antiscalant is added to the feed to the partial second pass (brackish RO) system. Long-term experience with such configuration indicates that this solution is effective in preventing scaling of the first-pass RO membranes by the recycled second-pass concentrate. Because boron level in the front permeate stream is usually between 0.25 and 0.50 mg/L, no additional treatment of this stream is needed even if the plant is designed and operated for enhanced boron removal.

While the recycling of the second-pass concentrate returns a small portion of the source water salinity, and therefore it slightly increases the salinity of the seawater fed to the first RO pass, the energy use associated with this incremental salinity increase is significantly smaller than the energy savings of processing the entire volume of the first-pass permeate through the second pass. Under the split-partial two-pass configuration, the volume of permeate pumped to the second RO pass, and the size of this pass is typically 25 to 50 percent smaller than the volume pumped to the second RO pass under conventional once-through operation.

Since pumping energy is directly proportional to flow, the energy costs for the second-pass feed pumps are reduced proportionally (i.e., with 25 to 50 percent). For a SWRO system operating at 45 percent recovery, such savings will amount to 14 to 22 percent of the energy of the first-pass RO pump. The concentrate returned from the second pass carries only 1 to 2 percent of additional salinity to the first-pass RO feed, which reduces the energy benefit from such recovery proportionally (i.e., by 1 to 2 percent only). As a result, the overall energy savings resulting from the use of split-partial two-pass RO system as compared with a conventional two-pass RO system are between 12 and 20 percent. Practical experience with large SWRO desalination plants indicates that the average total RO system life-cycle cost savings associated with applying such SWRO system configuration are typically between 14 to 16 percent.

At present, most new SWRO desalination systems are designed with a split-partial second-pass configuration because this configuration allows reducing the size of the second-pass RO system and the overall fresh-water production costs. It should be pointed out that split-partial second-pass RO systems can be configured in several alternatives, which may involve the use of the same or different type of membrane elements within the first-pass SWRO system.

An example of a plant with partial-second-pass configuration is the 95,000 m^3/day (25 mgd) Tampa Bay seawater desalination facility in Florida. The second pass at this facility is designed to treat up to 30 percent of the permeate produced by the first-pass SWRO system as needed in order to maintain the concentration of chlorides in the plant product water always below 100 mg/L. The partial second pass at the Tampa Bay water seawater desalination plant was installed to provide operational flexibility and to accommodate the wide fluctuations of source water salinity (16,000 to 32,000 mg/L) and temperature (18 to 40°C/64 to 104°F).

Typically, the product water-quality target chloride concentration of 100 mg/L at this plant is achieved by only operating the first pass of the system. However, when source water TDS concentration exceeds 28,000 mg/L and/or the source water temperature exceeds 35°C (95°F), the second pass is activated to maintain adequate product water quality. The percent of first-pass permeate directed for additional treatment through the second-pass is a function of the actual combination of source water TDS and temperature and is adjusted based on the plant product water chloride level.

Product Water Quality of Single- and Two-Pass SWRO Systems

Tables 14.15 through 14.19 present a summary of the range of permeate water quality produced by typical single-pass and partial two-pass seawater desalination systems

Seawater Source: Pacific Ocean			
		Permeate Water Quality	
Water Quality Parameter	Pacific Ocean Source Seawater Quality	Single-Pass SWRO System	Split-Partial Two-Pass RO System
Temperature, °C	14–28	15–29	16–30
pH	8.0	6.3–7.2	7.6–7.8
Ca^{2+}, mg/L	358	0.6–1.1	0.2–0.5
Mg^{2+}, mg/L	1720	1.8–2.8	0.07–0.10
Na^{+}, mg/L	9900	78–134	9–20
K^{+}, mg/L	600	3.0–6.0	0.43–0.60
CO_3^{2-}, mg/L	2.0	0.0	0.0
HCO_3^-, mg/L	170	1.8–2.5	0.4–0.7
SO_4^{2-}, mg/L	2570	2.6–5.3	0.7–1.3
Cl^-, mg/L	18,100	130–195	13–20
F^-, mg/L	2.1	0.9–1.2	0.7–0.9
NO_3^-, mg/L	0.00	0.00	0.00
B^-, mg/L	4.5	0.7–1.2	0.3–0.5
Br^-, mg/L	73	0.6–0.9	0.2–0.4
TDS, mg/L	33,500	220–350	25–45

TABLE 14.15 Reverse-Osmosis Permeate Water Quality

Seawater Source: Atlantic Ocean			
		Permeate Water Quality	
Water Quality Parameter	Atlantic Ocean Source Seawater Quality	Single-Pass SWRO System	Split-Partial Two-Pass RO System
Temperature, °C	16–30	17–31	18–32
pH	8.0	6.3–7.2	7.7–8.0
Ca^{2+}, mg/L	410	0.7–1.4	0.3–0.5
Mg^{2+}, mg/L	1302	1.6–2.4	0.35–0.8
Na^{+}, mg/L	10,506	83–160	11–25
K^{+}, mg/L	390	2.4–4.5	0.35–0.50
CO_3^{2-}, mg/L	2.0	0.0	0.0
HCO_3^{-} mg/L	145	1.4–2.0	0.5–0.8
SO_4^{2-}, mg/L	2720	2.4–5.8	1.1–1.2
Cl^{-}, mg/L	19,440	146–220	20–34
F^{-}, mg/L	2.5	1.0–1.6	0.8–1.2
NO_3^{-}, mg/L	0.00	0.00	0.00
B^{-}, mg/L	4.5	0.7–1.2	0.3–0.5
Br^{-}, mg/L	78	0.8–1.1	0.3–0.5
TDS, mg/L	35,000	240–400	35–65

Table 14.16 Reverse-Osmosis Permeate Water Quality

Seawater Source: Mediterranean Sea			
		Permeate Water Quality	
Water Quality Parameter	Mediterranean Source Seawater Quality	Single-Pass SWRO System	Split-Partial Two-Pass RO System
Temperature, °C	1–28	17–29	18–30
pH	8.1	6.3–7.2	7.9–8.1
Ca^{2+}, mg/L	480	1.0–2.0	0.35–0.45
Mg^{2+}, mg/L	1558	1.9–2.8	0.5–1.0
Na^{+}, mg/L	12,200	98–196	15–34
K^{+}, mg/L	480	3.0–5.5	0.8–1.8
CO_3^{2-}, mg/L	5.6	0.0	0.0
HCO_3^{-} mg/L	160	1.7–2.4	0.5–0.8
SO_4^{2-}, mg/L	3190	2.9–6.3	1.4–2.95
Cl^{-}, mg/L	22,340	169–260	25–52
F^{-}, mg/L	1.4	0.7–1.1	0.5–0.8
NO_3^{-}, mg/L	0.00	0.00	0.00
B^{-}, mg/L	5.0	0.9–1.5	0.4–0.6
Br^{-}, mg/L	80	0.9–1.3	0.35–0.6
TDS, mg/L	40,500	280–480	45–95

Table 14.17 Reverse-Osmosis Permeate Water Quality

Seawater Source: Red Sea			
		Permeate Water Quality	
Water Quality Parameter	Red Sea Source Seawater Quality	Single-Pass SWRO System	Split-Partial Two-Pass RO System
Temperature, °C	22–33	23–34	24–35
pH	7.0–8.0	6.8–7.8	7.6–8.0
Ca^{2+}, mg/L	500	1.1–2.1	0.5–0.7
Mg^{2+}, mg/L	1540	1.8–3.4	0.7–1.0
Na^{+}, mg/L	13,300	142–220	20–38
K^{+}, mg/L	489	3.2–6.5	1.2–1.8
CO_3^{2-}, mg/L	2.4	0.0	0.0
HCO_3^{-}, mg/L	142.4	1.4–2.0	0.5–1.0
SO_4^{2-}, mg/L	3100	2.8–6.2	1.9–2.6
Cl^{-}, mg/L	22,840	195–276	29–58
F^{-}, mg/L	0.9	0.5–0.7	0.3–0.5
NO_3^{-}, mg/L	0.00	0.00	0.00
B^{-}, mg/L	5.3	1.2–1.7	0.45–0.80
Br^{-}, mg/L	80	1.0–1.4	0.45–0.60
TDS, mg/L	42,000	350–520	55–105

TABLE 14.18 Reverse-Osmosis Permeate Water Quality

Seawater Source: Persian Gulf			
		Permeate Water Quality	
Water Quality Parameter	Persian Gulf Source Seawater Quality	Single-Pass SWRO System	Split-Partial Two-Pass RO System
Temperature, °C	18–35	19–36	20–37
pH	6.0–7.0	5.1–6.0	5.1–6.0
Ca^{2+}, mg/L	570	1.4–2.6	0.6–0.8
Mg^{2+}, mg/L	1600	2.0–3.6	0.9–1.3
Na^{+}, mg/L	14,100	142–228	25–45
K^{+}, mg/L	530	4.3–6.8	1.5–2.2
CO_3^{2-}, mg/L	4.2	0.0	0.0
HCO_3^{-}, mg/L	155	1.8–2.3	0.6–0.9
SO_4^{2-}, mg/L	3300	3.1–6.5	2.1–3.2
Cl^{-}, mg/L	24,650	222–305	37.5–64
F^{-}, mg/L	1.5	0.9–1.2	0.5–0.8
NO_3^{-}, mg/L	0.00	0.00	0.00
B^{-}, mg/L	6.3	1.3–2.5	0.7–1.0
Br^{-}, mg/L	83	1.2–1.5	0.60–0.80
TDS, mg/L	45,000	380–520	70–120

TABLE 14.19 Reverse-Osmosis Permeate Water Quality

processing five different seawater sources: Pacific Ocean water, Atlantic Ocean water, Mediterranean seawater, Persian Gulf seawater, and Red Sea water.

Four-Stage SWRO Systems

In addition to the single- and two-pass SWRO systems described above, another reverse osmosis configuration that has found practical application in a number of large SWRO desalination plants that have to meet a stringent boron limit of 0.4 mg/L or less (i.e., Ashkelon, Hadera, and Sorek SWRO plants in Israel) is the four-stage SWRO system depicted in Fig. 14.36 (Gorenflo et al., 2007).

Table 14.20 presents a summary of the source water quality and the design and actual product water quality of the Ashkelon SWRO plant in Israel, which was the first facility where such reverse osmosis configuration was used and therefore has the longest track record of successful performance (Dreizin et al., 2008).

Two-Stage SWRO Systems

Two-stage SWRO membrane systems are mainly used to maximize the overall desalination plant recovery and reduce the volume of concentrate discharged by the desalination plant. A general schematic of a two-stage RO system is shown in Fig. 14.37. In these SWRO systems, typically the entire volume of the concentrate generated by the first-stage SWRO system is directed to a second-stage SWRO system for further treatment and enhanced recovery. Permeate from both systems is blended prior to final use.

The main advantage of such SWRO system configuration is that it allows achieving a high level of use (recovery) of the available source seawater and the energy used by the first-stage RO system. For example, while a single-stage SWRO system configuration typically allows turning 35 to 50 percent of the source seawater to potable water, the two-stage SWRO system recovery may reach 60 to 65 percent. Designing the SWRO plant around higher recovery allows minimizing the size of the plant intake and pretreatment facilities and the capital expenditures for their construction and operation.

However, because of the high salinity of the first-stage concentrate (typically above 55,000 mg/L), the practical implementation of two-stage SWRO systems requires use

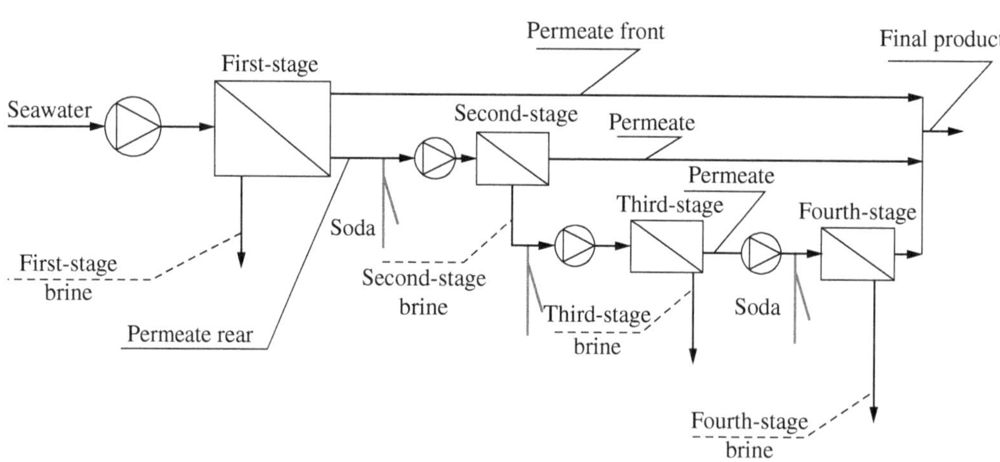

Figure 14.36 Ashkelon four-stage SWRO system (Gorenflo et al., 2007).

Ashkelon Seawater Desalination Plant			
Water Quality Parameter	Mediterranean Source Seawater Quality	Fresh Water Quality	
		Permeate	Finished Water
Temperature, °C	16–28	17–29	18–30
pH	8.1	7.5–8.5	8–8.5
Ca^{2+}, mg/L	483	0.2–0.5	90–110
Mg^{2+}, mg/L	1557	0.5–0.8	0.5–0.8
Na^{+}, mg/L	12,200	6–30	30–39
K^{+}, mg/L	481	1.5–1.8	1.5–1.8
HCO_3^{-}, mg/L	162	0.6–0.8	45–50
SO_4^{2-}, mg/L	3186	1.0–1.4	1.0–1.8
Cl^{-}, mg/L	22,599	10–15	10–15
F^{-}, mg/L	1.4	0.1–0.2	0.8–1.0
NO_3^{-}, mg/L	0.00	0.00	0.00
B^{-}, mg/L	5.3	0.2–0.3	0.2–0.3
Br^{-}, mg/L	80	0.2–0.3	0.2–0.3
TDS, mg/L	40,679	20–40	180–220

TABLE 14.20 Reverse-Osmosis Permeate Water Quality

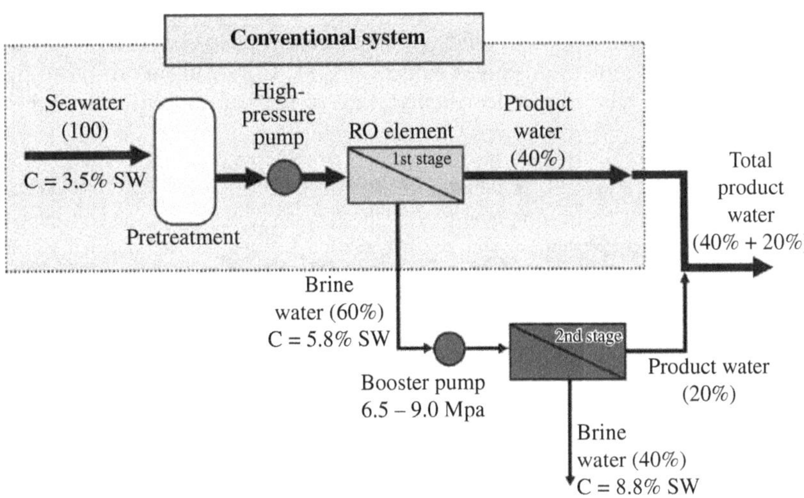

FIGURE 14.37 Two-stage SWRO system.

of high-pressure SWRO membranes, membrane vessels, and piping and auxiliary equipment that can withstand and perform well at very high pressures (up to 98 bars/1421 lb/in^2). The cost of this equipment is usually higher than the cost of the same size and type of equipment built to operate at more "standard" pressures (i.e., below 70 bars/1015 lb/in^2). Therefore, the viability of using a two-stage SWRO system has to be carefully assessed based on a comprehensive life-cycle cost analysis for the site-specific conditions of a given project. To date, two-stage systems have been mainly used to upgrade the capacity and improve energy use of existing conventional single-stage SWRO plants.

Sometimes, two-stage SWRO systems are designed to be operated with lower feed pressure to the first stage because of the high-fouling nature of the feed water. Since operation at lower feed pressure allows us to reduce fouling of the first-stage/pass elements, this configuration could be beneficial if the pretreatment system is not very robust and the water quality from the SWRO system is of high alluvial organic or particular content.

Hybrid SWRO Systems with Multiple Passes and Stages

The two-pass and two-stage RO system configurations may be combined to achieve an optimum plant design and tailor desalination plant operation to the site-specific water source water quality conditions and product water quality goals.

An example of a full-scale two-pass, two-stage SWRO system application is the 170,000 m^3/day (45 mgd) Fujairah seawater desalination plant (Rovel, 2005). A general treatment process schematic of this plant is depicted in Fig. 14.38. The plant uses Gulf of Oman source seawater.

The first pass of the Fujairah plant consists of 17 duty and one standby RO trains using standard rejection SWRO membranes producing permeate of TDS concentration of 400 to 500 mg/L. The overall recovery of the first-pass SWRO system is 43 percent. The second pass has eight BWRO trains with a total recovery rate of 90 percent. The second-pass BWRO trains have two stages and treat approximately 80 percent of the first-pass permeate to TDS concentration of 10 to 20 mg/L. The rest (i.e., 20 percent) of the first-pass permeate is blended with the second-pass permeate to produce finished water of TDS concentration of less than 120 mg/L. Concentrate produced by the second-pass RO system has salinity lower than that of the source seawater and is recycled to

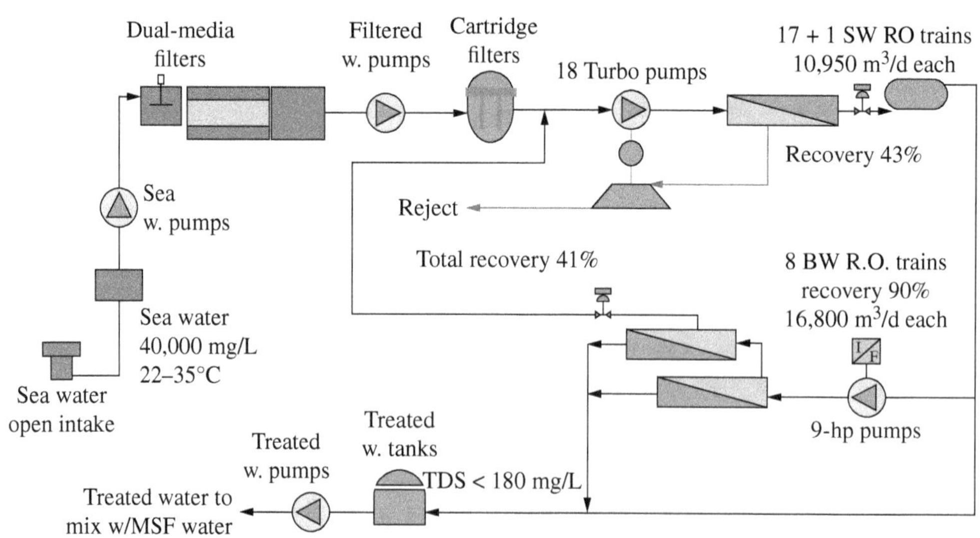

Figure 14.38 Fujairah seawater desalination plant schematic.

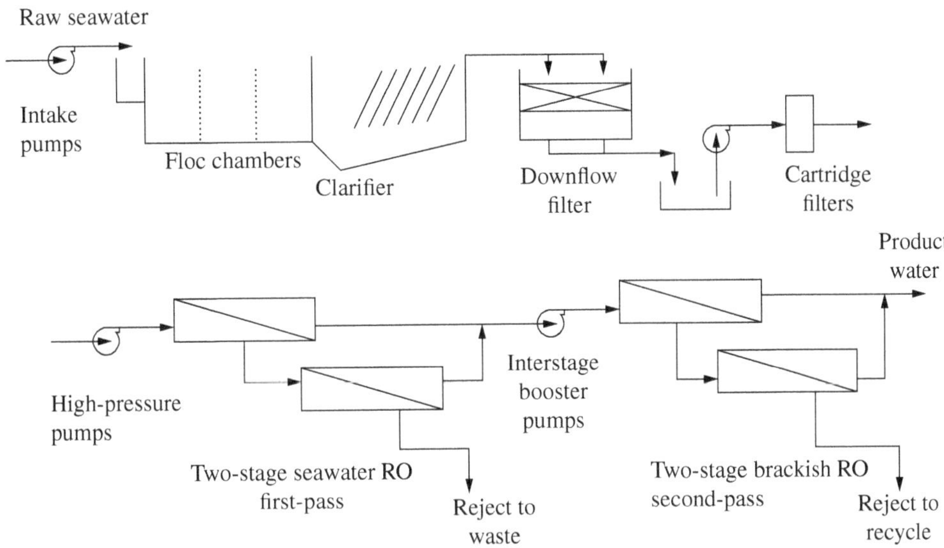

FIGURE 14.39 Point Lisas seawater desalination plant schematic.

the feed of the first-pass RO system (Fig. 14.38). Despite its complexity, the two-pass, two-stage RO system in Fujairah performs reliably.

An example of a two-pass, two-stage SWRO plant is the 146,000 m^3/day (38 mgd) Point Lisas facility in Trinidad (Fig. 14.39). This plant produces high-quality desalinated seawater of TDS concentration of 85 mg/L or less, which is predominantly used for industrial applications. The first pass of the Point Lisas SWRO system consists of six two-stage RO units. Each of the first-stage RO trains uses SWRO membranes and is coupled with Pelton wheel energy recovery device. The second-stage trains of the first pass are equipped with brackish water RO elements.

The entire volume of permeate from the first pass of the Point Lisas SWRO system is further treated in a second-pass RO system to meet the final product water quality specifications. The second-pass system also consists of two stages, each equipped with BWRO membranes. The Point Lisas seawater desalination plant has the same number of first- and second-pass RO membrane trains.

Because of the high-fouling potential of its source seawater, this two-pass stage configuration has proven to be beneficial. The high-pressure pumps of the first stage are designed for 80 percent lower feed pressure than those of a typical single-stage system, which allows us to significantly reduce the fouling on the RO elements of this pass. The permeate from the first pass is treated through a second-pass system, and an interstage booster pump is installed to deliver the remaining 20 percent of the feed pressure needed to produce water similar to that in a single-pass RO. Overall, this configuration produces better water quality, higher recovery, and results in less membrane fouling than a typical single-pass system for the same source water quality.

Three-Center RO System Configuration

As indicated previously, a typical conventional SWRO system is configured in individual equally sized RO trains, each of which is serviced by a separate transfer pump, cartridge filter vessel, high-pressure feed pump, and energy-recovery equipment dedicated to this RO train (see Fig. 14.15). The size of the individual RO trains depends of the overall production capacity of the SWRO plant and typically varies between 2000 m^3/day (0.5 mgd) and 21,000 m^3/day (5.5 mgd). The main advantage of this RO train-based configuration is that it is modular and allows for relatively easy flow distribution and service of the individual trains. Since typically the size of the individual RO trains does not exceed 10 to 20 percent of the total plant production capacity, train shutdown for maintenance (membrane cleaning and replacement and equipment service) is handled either by using a standby RO train or by temporary increase of the production capacity of the RO trains remaining in service.

The RO-train-based configuration is suitable when the SWRO plant is designed and intended to operate at a constant production output. At present, most of the existing large municipal SWRO plants worldwide are designed to supplement existing conventional water supply sources rather than to be the primary or the only source of water supply for a given area. Therefore, the operation of these SWRO plants does not need to have the flexibility to follow the actual diurnal and monthly fluctuations in product water demand, and most of the existing plants are designed to operate at constant production capacity.

In the future, the SWRO is likely to become a prime rather than a supplemental source of water supply for many coastal communities pressured by population growth, especially for large urbanized or industrial areas with limited traditional local sources of fresh water (i.e., groundwater or river or lake water). The SWRO systems servicing such areas have to be designed to have the operational flexibility of matching desalination plant production with the product water demand fluctuations.

Shift of the SWRO plant operational paradigm from constant to variable production flow requires a change of the typical SWRO configuration from one that is most suitable for constant production output to one that is most cost-effective for delivery of varying permeate production. A response to such shift of the desalination plant operational paradigm is the three-center RO system configuration in Ashkelon, Israel (see Fig. 14.40). Under this configuration, the RO membrane vessels, high-pressure pumps, and energy-recovery equipment are no longer separated in individual RO trains, but are rather combined in three functional centers: a high-pressure RO feed pumping center, a membrane center, and an energy recovery center (Liberman, 2002). The three functional centers are interconnected via service piping.

The RO feed pumping center includes only a few large-capacity high-pressure pumps that deliver seawater to the RO membrane center. The main benefit of using few large-capacity high-pressure pumps rather than a large number of small-capacity units is the gain in overall pumping efficiency.

Typically, the smaller the ratio between the pressure and the flow delivered by a given pump, the better the pump efficiency and the "flatter" the pump curve (i.e. the pump efficiency is less dependent of the variation of the delivered flow). Therefore pump efficiency can be improved by either reducing the pressure delivered by the pump or by increasing pump flow. Since the pump operating-pressure decrease is limited by the RO system target salt separation performance, the main approach to improve pump efficiency is to increase unit pump flow. While a conventional-size high-pressure RO feed pump of small capacity would typically have maximum total energy use efficiency of 80 to 85 percent, the use of a 10 times larger size pump may allow us to increase the pump efficiency to 88 to 92 percent, especially for large SWRO plants. This feature of the three-center design is valuable in the case of systems delivering varying flow.

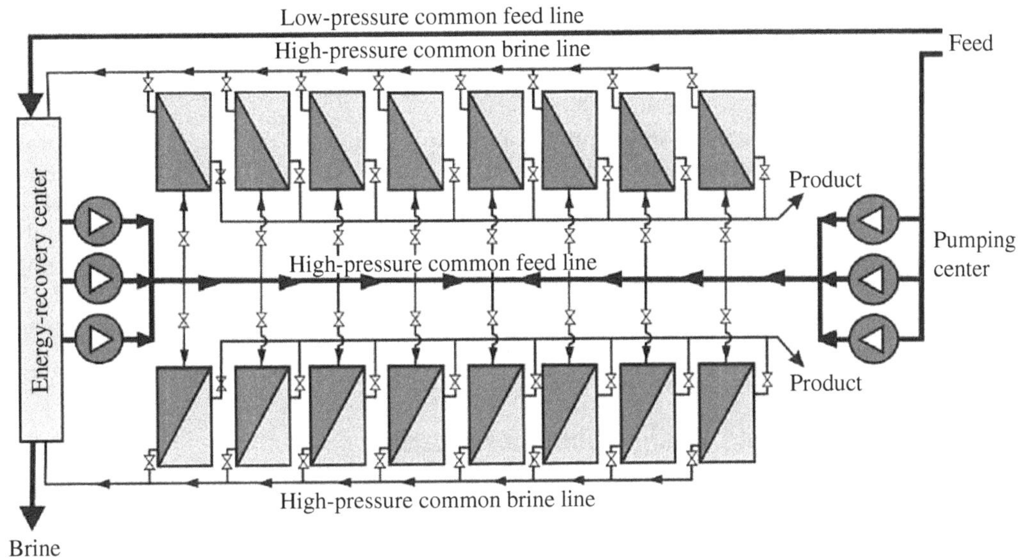

FIGURE 14.40 Three-center SWRO system configuration.

While in a conventional RO train design the membrane vessels are typically grouped in 100 to 200 units per train and in 2 to 20 RO trains, the membrane center configuration contains two to four times larger number of RO vessel groups (banks) and a smaller number of membrane vessels per bank. Under this configuration the individual vessel banks are directly connected to the high-pressure pump feed lines and can be taken off service one at a time for membrane replacement and cleaning.

Although the feed water distribution piping for such membrane center configuration is more elaborate and costly than the use of individual RO trains that contain two to three times more vessels per train, what is lost in capital expenditure is gained in overall system performance reliability and availability. A reliability analysis completed for a 95,000 m^3/day (25 mgd) SWRO plant (Liberman, 2002) indicates that the optimum number of vessels per bank for this scenario is 54 and number of RO banks per plant is 20. A typical RO-train-based configuration would include two to four times more (108 to 216) vessels per RO train and two to four times less (five to 10) RO trains. According to this analysis, the use of the three center configuration instead of the conventional RO-train-based approach allows us to improve RO system availability from 92 to 96 percent (avg. 95 percent) to 98 percent, which is a significant benefit in terms of additional amount of water delivered to the customers and improvement in water supply reliability.

The centralized energy recovery system included in the three-center configuration (Fig. 14.40) uses high-efficiency pressure-exchanger-based energy-recovery technology. The proposed configuration allows us to improve the overall energy recovery efficiency of the RO system and to reduce system power, equipment, and construction costs. While typically the energy recovery of the conventional Pelton wheel systems drop significantly when the reduction of the overall SWRO plant recovery, the recovery efficiency of the pressure-exchanger systems improve with lowering the plant recovery rate. This allows operating the SWRO plant cost-effectively while delivering variable product water flow. For example, if the SWRO plant output has to be reduced by 40 percent to accommodate low diurnal demand, a SWRO system with RO-train-based configuration has to shut down 40 percent of its trains, and, if this low demand persists, it has to flush these trains to prepare them for the next start-up.

An RO system with a three-center configuration would only need to lower its overall recovery to achieve the same reduction of the diurnal demand. Although temporary operation at lower recovery would result in elevated costs for pumping and pretreatment of larger volumes of source water, these extra operation expenses are typically compensated for by the improved energy-recovery efficiency that results from operating the SWRO system at a lower water-recovery ratio. In addition to providing flexibility in operating cost-effectively the desalination plant at varying production flows, the three-center design yields low-fouling RO plant operations of a high availability factor.

14.12 Planning and Design Considerations

The RO system design aims at achieving flexible and cost-effective plant operation in the entire range of the intake water quality conditions and meeting product water quality and quantity specifications.

14.12.1 General Design Methodology

Key steps associated with the design of RO desalination system include: (1) source water quality characterization, (2) determination of target product water quality, quantity and operations regime, (3) selection of RO system configuration, and (4) selection of key RO system-performance parameters.

Step 1. Source Water Quality Characterization The first step of the RO system design is to establish the water quality that will be introduced into the RO membranes. Chapter 2 provides information of what source water quality data need to be collected for the prudent design of RO systems. The mineral content of the source water will define whether the RO system will be nanofiltration, brackish, or seawater type. The foulants contained in the source water will indicate how conservative the key design criteria such as permeate flux, RO system recovery, and number of RO system passes may be needed to accommodate this source water quality.

Besides determining the raw source water quality, the project designer will also need to determine how this water quality will be changed by the plant pretreatment processes in terms of content of salts, foulants, oxidants, and temperature. Chapters 9 through 13 of this book provide the background information for determining the impact of plant pretreatment on source water quality and projecting of the quality of the pretreated water entering the RO system.

Step 2. Determination of Target Product Water Quality, Quantity, and Plant Operations Regime The next step in the design process is to determine the product water quality specifications of the desalination plant and its target annual production capacity, installed capacity, and availability factor. Chapter 4 provides guidance regarding the selection of target product water quality and quantity. It should be pointed out that each desalination plant has two important fresh water production capacity parameters: annual average production capacity and installed capacity. Annual average production capacity is calculated by dividing the total annual volume of product water that the plant will need to be designed to generate by the number of days per year the plant is planned to be operational. This capacity accounts for the fact that the plant will be shut down for certain period of time (usually between 10 and 20 days) each year for preventive or emergency maintenance.

The plant-installed capacity is the actual daily maximum volume of water the plant can produce with all duty units in service during any time of its useful life. Usually, the plant-installed capacity is designed to be 10 to 20 percent larger than the plant design average annual capacity in order to accommodate routine operational conditions, during which one or more components would be down. For example, RO train will be down for membrane cleaning, and/or replacement and/or intake conduit will be removed out of service for inspection and cleaning. The safety factor built into the plant-installed capacity is a function of the target plant capacity availability factor (i.e., percentage of time per year when the plant will generate product water volume at or above its average annual production capacity).

For example, for a plant with an annual design production capacity of 40,000 m^3/day (10.6 mgd), a design availability factor of 96 percent means that 96 percent of the time (i.e., 350 out of 365 days) the desalination plant will have to be designed to produce 40,000 m^3/day (10.6 mgd) or more. Most desalination plants are designed with a target availability factor of 90 to 95 percent. The higher the design availability factor, the more conservative the RO system design should be, and the higher the design safety factor/installed redundant production capacity.

This step of the design process will also have to define how the plant would be operated—continuously (24 hours/day and 365 days per year) or intermittently. If the plant is expected to be operated intermittently and/or with varying diurnal and/or monthly flows, it will be important to understand/decide if the plant would need to be defined to operate for only a portion each day (say, eight hours per day) or on a 24-hour schedule. If the plant capacity has to be turned down for prolonged periods of time (weeks or months), the project designer will have to determine what is the design plant turndown ratio (i.e., what is the smallest percentage of the total installed plant capacity at which the RO system has to be designed to perform reliably and cost-effectively).

Step 3. Selection of RO System Configuration and Membrane Element Type An RO system type—NF, BWRO, or SWRO—will be selected based on the source water quality salinity and other mineral/contaminant content and on target product water quality. As indicated in the previous sections of this chapter the NF system would be selected if source water salinity is low, and the product water quality requirements mainly target removal of hardness, organics, nitrates, color, or other specific contaminants from the source water.

The BWRO system will be selected if the source water salinity is between 500 and 10,000 mg/L. As indicated previously within this salinity range, the BWRO system could be designed as a low-salinity BWRO system with a 5 to 50 percent source water bypass if the feed water salinity is below 2000 to 2500 mg/L, and, depending on the actual product water and source water quality and target overall system recovery, the BWRO system could be configured as a single-stage or two-stage RO system.

A general rule of thumb is that for six, seven, and eight elements per vessel in BWRO systems, the maximum recovery that could be achieved per stage is 55, 65, and 75 percent, respectively. For example, if the source water salinity is relatively low, and we could bypass 35 percent of the flow, then we could select a single-stage BWRO system with seven or eight elements. If higher recovery is needed because, for example, the available concentrate disposal system can only receive 25 percent of the feed flow (i.e., BWRO plant recovery has to be increased from 65 to 75 percent), then the RO system would need to be designed with two stages.

Similar logic applies for the selection of the configuration of SWRO reverse osmosis systems; however, there the main configuration factor is the target product water quality and the number of passes needed to achieve this water quality. Selected design RO system recovery is mainly driven by the source water salinity (the higher the salinity the lower the target recovery), the fouling propensity of the water (more fouling waters favor lower plant recovery and two-pass, two stage systems) and the type of the selected energy recovery system and unit power costs (isobaric type of energy recovery devices and high energy costs favor designing at lower RO system recovery).

Selection of the number of plant RO trains is mainly driven by the plant design capacity turndown ratio and the target plant capacity availability factor. For example, if a given RO plant will need to be designed for a minimum turndown capacity ratio of 10 percent, this requirement would necessitate the size of the individual RO trains to be designed to produce 10 percent of the total plant permeate flow, so if the plant production capacity is turned down to 10 percent, then only RO train will be kept in operation. In this case, a 40,000 m^3/day (10.6 mgd) plant would be designed with 10 RO trains, each with production capacity of 4000 m^3/day (1.1 mgd)—i.e., 10 percent of the flow. If the plant has relatively low target availability factor—say, 90 percent—then the plant train is usually oversized with only 8 to 10 percent, especially if each RO train produces only 10 percent of the total flow—i.e., the individual RO trains will be sized for target installed capacity of 1.1×4000 m^3/day = 4400 m^3/day (1.16 mgd).

If the plant reliability factor is closer to its current industry-wide target of 95 to 96 percent, usually the plant installed capacity is 15 to 20 percent of the annual production flow target—i.e., in this case, the individual RO trains will be designed for 4600 to 4800 m^3/day (1.22 to 1.30 mgd).

Step 4. Selection of Key RO System Performance Parameters Once the RO system source and product quality parameters are known, the type of pretreatment is selected; target plant installed capacity, size, and number of individual trains; and target recovery are established, RO system performance projection computer software available from major RO membrane manufacturers could be used in order to project and optimize the type and number of RO system membrane elements per vessel and vessels per train, as well as key RO feed pump design criteria and permeate and concentrate water quality. Such RO performance projection software can be downloaded from a membrane supplier's website. Popular software packages commonly used for this purpose are IMSDesign from Hydranautics, ROSA from Dow Filtmtec, TorayDS from Toray Membranes, and ROPRO from Koch Membrane Systems. Similar software packages are also available from Trisep, Woongjin, and GE Water and Process Technologies. Membrane manufacturers update their performance projection software frequently, and the use of the latest software version is advisable.

The type of input data and user interface is virtually the same for all equipment manufacturer software. The use of the software packages requires input of source water quality data, type of intake, and type of source water pretreatment. The software also allows us to input target permeate flow, recovery rate, membrane model, membrane array, salt passage increase factor, flux decline coefficient, and membrane age. Most models allow to analyze the performance of multipass and multistage RO systems

Component/Parameter	Specifications/Design Criteria
• Feed water • Design flow rate, m^3/day (mgd) • TDS • Temperature	51,000 m^3/day (13.5 mgd) 3600 mg/L 20 to 25°C (68 to 77°F)
RO membranes	
Unit product water capacity	9500 m^3/day (2.5 mgd)
Number of RO skids	4
Number of stages	2
Number of pressure vessels in first & second stage	40/20
Number of RO elements per vessel	7
Total number of RO elements	420
RO membrane element size	8 × 40 in
Type of BWRO elements	Standard BWRO spiral wound
Design system recovery	75%
Average membrane flux	25.5 Lmh/15 gfd
Energy Recovery Device	
Type	Turbocharger
Number of units	4 (One per RO Skid)
First-stage concentrate flow	295 m^3/h (1300 gpm)
Second-stage concentrate flow	170 m^3/h (750 gpm)
Second-stage concentrate pressure	21 bars (300 lb/in^2)
Concentrate system back pressure	1.8 bars (25 lb/in^2)
Interstage booster pressure	4.6 bars (65 lb/in^2)
RO system energy use, kWh/m^3 (kWh/1000 gal)	1.2 kWh/m^3 (4.5 kWh/1000 gal)
Materials of construction	
Frame	Epoxy-coated aluminum
Pressure vessels	FRP
Feed water piping	Duplex stainless steel
Concentrate piping	Duplex stainless steel–high pressure/ PVC schedule 80, low pressure
Permeate piping	PVC schedule 80
Energy recovery devices	Duplex stainless steel

Table 14.21 Example of Key Design Criteria for 40,000 m^3/day (10.6 mgd) BWRO Desalination System

and introduction of RO system features such as permeate blending, permeate throttling, concentrate recirculation, and use of RO systems with interstage booster pumps. All RO system performance projection software produces the values of key RO system performance parameters, including flux of individual RO elements within the vessels, recommended RO pump feed pressure and flow, and projected permeate and concentrate water quality and flows. This software also projects energy use of the high-pressure RO feed pumps and allows us to account for the efficiency and backpressure of the selected ERD. The software also generates membrane-scaling related information such as Langelier and Stiff & Davis Saturation Indexes, ionic strength, osmotic pressure and percentage saturation of calcium, strontium, and barium sulfates and silica saturation.

During the design process, the membrane supplier software could be used iteratively to analyze the feasibility of various RO system configurations and membrane elements. Typically, the design engineer runs software from several membrane manufacturers and compares the RO system performance projections to select the most suitable RO system configuration and membranes for the site-specific conditions of a given project.

14.12.2 Design Example: BWRO System

Table 14.21 provides a summary of the design criteria for a 40,000 m^3/day (10.6 mgd) BWRO plant planned to treat brackish well water with TDS concentration of 3600 mg/L; temperature of 20 to 25°C (68 to 77°F); total hardness of 900 mg/L; TOC of 1.5 mg/L; silica of 12 mg/L; calcium and magnesium of 240 and 550 mg/L, respectively; sodium of 980 mg/L; chloride of 1700 mg/L; nitrate, ammonium and bromide of 0 mg/L; barium of 0.01 mg/L, strontium of 12 mg/L and sulfate of 460 mg/L. The source water is anaerobic and contains total sulfide concentration of 5 mg/L. The target product water quality TDS concentration is 250 mg/L. The BWRO plant will need to produce finished water suitable for potable use.

Since the source water salinity is relatively high and is made up mainly of sodium and chlorides, this water will need to be treated with BWRO membrane system. This system will be designed to treat the entire source water flow without bypass. Because of the high cost of the only available concentrate discharge alternatives—deep injection wells—the system would be designed for relatively high recovery of 75 to 80 percent. Achieving such recovery will necessitate the use of a two-stage BWRO system.

14.12.3 Design Example: SWRO System

Design criteria for 40,000 m^3/day (10.6 mgd) SWRO system using source water, which is collected via open intake, are summarized in Table 14.22. The source seawater has TDS concentration of 42,000 mg/L; temperature of 15 to 38°C (59 to 100°F); total hardness of 160 mg/L; TOC of 2.5 mg/L; silica of 2 mg/L; calcium and magnesium of 500 and 1550 mg/L, respectively; sodium of 13,400 mg/L; chloride of 22,825 mg/L; nitrate, ammonium, and bromide of 0 mg/L; barium of 0.01 mg/L, strontium of 2 mg/L, and sulfate of 3100 mg/L. Boron and bromide in the source seawater are 5.2 and 80 mg/L, respectively. The source water is aerobic (ORP = 300 mV) and contains dissolved oxygen concentration of 6.5 mg/L. The target product water quality TDS concentration is 250 mg/L, and boron is less than 0.75 mg/L. The SWRO plant is designed to produce drinking water.

Component/Parameter	Specifications/Design Criteria
Feed Water • Design flow rate, m³/day (mgd) • TDS • Temperature	119,100 m³/day (31.5 mgd) 42,000 mg/L 15 to 38°C (59 to 100°F)
RO membrane skids	
Unit product water capacity	9200 m³/day (2.43 mgd)
Number of RO skids	5
Number of passes	2 (first-pass SWRO and full second-pass BWRO)
First (SWRO) pass	
Number of stages	1
First-pass recovery rate	43%
Number of pressure vessels per skid	107
Number of RO elements per vessel	7
Total number of SWRO elements	749
RO membrane element size	8 × 40 in
Type of SWRO Elements	SWC4B max (see Table 14.8)
Flow produced by the first pass	10,240 m³/day (2.7 mgd)
Average membrane flux	14 Lmh/8.2 gfd
High-pressure RO feed pump • Number of pumps • Unit pump capacity • Operating feed pressure • Maximum delivery pressure • Pump efficiency	 5 (one per RO skid) 430 m³/h (1900 gpm) 60.4 bars (875.8 lb/in²) 75 bars (1087 lb/in²) 85%
Second (BWRO) pass	
Number of stages	2
Second-pass recovery rate	90%
Number of pressure vessels in stage 1	26
Number of pressure vessels in stage 2	13
Number of RO elements per vessel	7
Total number of BWRO elements per skid	273
RO membrane element size	8 × 40 in
Type of SWRO Elements	ESPAB max

TABLE 14.22 Example of Key Design Criteria for 40,000 m³/day (10.6 mgd) SWRO Desalination System

Component/Parameter	Specifications/Design Criteria
Flow produced by the second pass	9200 m^3/day (2.43 mgd)
Average membrane flux	34.4 Lmh/20.2 gfd
Second-pass feed pump • Number of pumps • Unit pump capacity • Operating feed pressure • Maximum delivery pressure • Pump efficiency	5 (one per RO skid) 430 m^3/h (1900 gpm) 18 bars (261 lb/in^2) 22 bars (319 lb/in^2) 82%
Energy recovery device	
Type	Pressure exchanger: ERI PX 260
Number of pressure exchangers per RO train	12
Concentrate/permeate flow per 12 PX units	640 m^3/h (2800 gpm) 53.3 m^3/h (560 gpm)/PX unit
RO feed pressure	60.4 bars (875.8 lb/in^2)
Energy-recovery efficiency	95%
RO system energy use, kWh/m^3 (kWh/1000 gal)	3.0 kWh/m^3 (11.3 kWh/1000 gal)
Materials of construction	
Frame	Epoxy-coated steel
Pressure vessels	FRP
Feed water piping	Super duplex stainless steel
Concentrate piping	Super duplex stainless steel–high pressure/ PVC schedule 80–low pressure
Permeate piping	PVC schedule 80
Energy recovery devices	Super duplex stainless steel

Table 14.22 Example of Key Design Criteria for 40,000 m^3/day (10.6 mgd) SWRO Desalination System (*Continued*)

Because of the high salinity and temperature of the source seawater and the relatively low target product water TDS, boron and bromide concentration of 250 mg/L, 0.75 mg and 0.5 mg/L, respectively, the SWRO system will be designed with two passes—first-pass SWRO and second-pass BWRO with two stages. The second pass will have two stages in 2:1 array. The total plant recovery is 40.5 percent. Assuming that the plant has to be designed for very high level of availability (i.e., 96 percent or more), the actual installed plant capacity is selected to be 15 percent higher than the average annual production capacity—i.e., the installed RO system capacity is 1.15 × 40,000 m^3/day = 46,000 m^3/day (12.15 mgd).

Item	Construction Cost ($/Item or as Indicated)
8-in brackish RO membrane elements	\$250–\$350/element
8-in SWRO membrane elements	\$400–\$600/element
16-in SWRO membrane elements	\$2800–\$3300/element
Brackish RO pressure vessels for 8-in elements	\$1000–\$1300/vessel
SWRO pressure vessels for 8-in elements	\$1300–\$1800/vessel
SWRO pressure vessels for 16-in elements	\$3600–\$5000/vessel
RO Train piping	\$250,000–\$750,000/ RO train
RO Train support frame	\$150,000–\$550,000/ RO train
RO Train instrumentation and controls	\$30,000–\$150,000/RO train
High-pressure pumps	\$150,000 –\$2,400,000/RO train

Note: All costs in year 2012 \$US.

Table 14.23 Construction Costs of Key Membrane RO System Components

The SWRO system is equipped with ERI ERDs. The SWRO system consists of five RO trains, each with production capacity of 9200 m^3/day (2.43 mgd). The RO trains have conventional design configuration with individual high-pressure RO feed pump and one ERI unit for each train. The plant is equipped with a full second-pass.

14.13 RO System Desalination Costs

The construction costs of some of the key membrane SWRO system elements are summarized in Table 14.23. Approximately 10 to 20 percent has to be added to those costs shown in Table 14.23 for shipping, handling, installation oversight, and insurance. The cost of the membrane RO modules (trains) is proportional to the design capacity and flux of the SWRO system. Typically, one RO module contains 50 to 200 membrane vessels.

While there is limited economy of scale in terms of the costs of the RO modules, the costs of the other RO system components (high-pressure pumps, energy-recovery devices, stainless-steel piping and valves, and membrane cleaning system) can benefit significantly from the use of large-size units. For example, one membrane cleaning system can be used for several RO membrane trains, the high-pressure feed pump efficiency and cost improves with their size, and the economy of scale between two sizes of stainless-steel pipe is usually 10 to 15 percent. Therefore, as the RO membrane module size increases, the relative cost of the RO system per unit volume of produced permeate decreases.

There are two limitations of the maximum size of the membrane train (module): system reliability and the available off-the-shelf equipment and RO membranes that can be used to build a very large module. The main limiting factor with using large-size modules is the loss of production capacity when the RO module is shut down for membrane cleaning, replacement, or equipment and piping repairs. The larger the individual module, the larger the potential loss of capacity and therefore the lower the availability factor of the SWRO plant. Since the availability factor is directly related to the cost of water, a SWRO system with lower availability factor yields higher cost of water.

Another factor that limits the benefits of constructing very large RO trains is the need to use custom-made rather than off-the-self equipment (mainly high-pressure pumps, motors, and energy-recovery devices). Although the manufacturing of this equipment is possible, the one-time custom design and production of such equipment is significantly more costly than the use of the standard "off-the-shelf" equipment with well-known production costs, performance parameters, and proven track records. Therefore, in such applications, often the gain of the economy of scale due to the use of large custom-made equipment and trains is negated by the additional expenditures for equipment production and risks associated with the lack of long-term track record of equipment performance.

The introduction of large-diameter (16 to 19 in) RO membranes and pressure vessels in 2006 has increased the envelope of the maximum-size RO trains that can be used for seawater desalination. Potential cost benefits and challenges associated with the use of large-diameter membranes are discussed in further detail in the following section.

14.14 Desalination Systems with Large-Diameter RO Membranes

One of the key innovations in the field of membrane desalination since 2004 is the development and commercialization of large-size (16, 18, and 19 in) reverse osmosis membrane elements, which are aimed to respond to the recent desalination industry trend toward construction of large- and mega-desalination projects (i.e., projects with fresh water production capacity of 75,000 m^3/day (20 mgd) and 380,000 m^3/day (100 mgd) or more, respectively. Since 2009, such projects constituted approximately 40 percent of the total new commissioned desalination capacity worldwide.

Large-diameter RO elements have four to ten times higher membrane area and unit production then conventional size 8-in membranes, which allows us to reduce significantly the number of RO system components (membranes, vessels, piping, fittings, instrumentation, RO trains, O-rings, brine seals, and pumps) and to decrease the total footprint of the RO system. Other potential benefits include reduced maintenance and improved reliability because of the fewer element connections (O-rings and brine seals).

14.14.1 Large RO Membranes: Commercial Products

At present, a number of membrane manufacturers offer large-diameter RO membrane products. Table 14.24 summarizes key performance parameters of commercially available large-diameter membrane elements for brackish water desalination and water reclamation. Table 14.25 provides similar information for SWRO membranes.

Membrane Element Model	Nominal Diameter × Length (in)	Nominal Surface Area (Ft^2)	Feed Spacer Thickness (mm)	Permeate Flow at Standard Test Conditions (gpd)	Nominal Salt Rejection at Standard Test Conditions(%)	Weight, Dry (lb)	Weight, Wet (lb)
DOW/Filmtec							
BW30-1725	16 × 40	1725	0.71	45,000	99.5	115	150
BW30-1570/34*i*	16 × 40	1570	0.86	41,000	99.5	115	150
Hydranautics (Nitto Denko)							
ESPA2 1640HYD	16 × 40	1600	0.76	38,000	99.6	114	139
Toray							
TMG40-160	16 × 40	1600	0.79	40,700	99.5	145	167
TM740-160	16 × 40	1600	0.79	40,700	99.7	145	167
TML40-160	16 x 40	1600	0.79	40,700	99.7	145	167
Woongjin Chemical (formerly Saehan)							
CSM RE16040-BE	16 × 40	1600	0.71	41,000	99.7	123	136
CSM RE16040-FE[n]	16 × 40	1600	0.71	41,000	99.7	123	136
CSM RE16040-FE	16 × 40	1600	0.71	38,400	99.7	123	136
CSM RE16040-BLR	16 × 40	1600	0.71	36,000	99.6	123	136
Koch Membrane Systems (TFC—MegaMagnum[(1)] and MegaMagnum Plus[(2)] Membranes)							
18061-HR-3050[(1)]	18 × 61	3050	0.70	86,200	99.55	200	250
18061-ULP-3050[(1)]	18 × 61	3050	0.70	66,200	98.65	200	250
18061-S-3130[(1)]	18 × 61	3130	0.70	69,200	90.00	200	250
19061-HR-3525[(2)]	19 × 61	3525	0.70	98,900	99.55	300	360

Table 14.24 Large-Diameter BWRO and Softening Membrane Elements

Membrane Element Model	Nominal Diameter × Length (in)	Nominal Surface Area (ft^2)	Feed Spacer Thickness (mm)	Permeate Flow at Standard Test Conditions (gpd)	Nominal Salt Rejection at Standard Test Conditions (%)	Weight, Dry (lb)	Weight, Wet (lb)
DOW/Filmtec							
SW30HRLE-1725	16 × 40	1725	0.71	32,000	99.75/ 91% boron	125	150
Hydranautics (Nitto Denko)							
SWC4 1640	16 × 40	1600	0.76	26,000	99.80/ 93% boron	114	139
SWC5 1640	16 × 40	1600	0.76	34,000	99.80/ 92% boron	114	139
Toray							
TM840C-160	16 × 40	1600	0.71	26,700	99.75/ 93% boron	145	167
TM840E-160	16 × 40	1600	0.71	30,000	99.75/ 91% boron	145	167
Woongjin Chemical (formerly Saehan)							
CSM RE16040-SHN	16 × 40	1600	0.71	24,600	99.75	132	145
CSM RE16040-SHF	16 × 40	1600	0.71	36,000	99.70	132	145
CSM RE16040-SR	16 × 40	1600	0.71	24,400	99.60	132	145
Koch Membrane Systems (TFC–MegaMagnum(1) and MegaMagnum Plus(2) Membranes)							
18061-SW-3050(1)	18 × 61	3050	0.70	53,000	99.75	200	250
18061-HF-3050(1)	18 × 61	3050	0.70	69,500	99.70	200	250
19061-SW-3525(2)	19 × 61	3525	0.70	60,800	99.75	300	360

Table 14.25 Large-Diameter SWRO Membrane Elements

Tables 14.24 and 14.25 incorporate products from five key manufacturers of large-diameter membranes: Dow/Filmtec, Hydranautics, Toray, Woongjin Chemical (formerly Saehan), and Koch Membrane Systems (KMS). The first three membrane manufacturers have participated in a consortium that, in 2003/2004, has developed a "standard" large element of 16-in (400 mm) diameter and 40-in (1016 mm) length under the guidance of the US Bureau of Reclamation (USBR, 2004). While the 16-in (400 mm) elements of the manufacturers listed in Tables 14.24 and 14.25 have few differences, they are commoditized in size and diameter and could be used interchangeably in the same membrane vessel.

Woongjin Chemical did not participate in the USBR-led consortium but has adopted the 16-in (400 mm) "standard" for its large-size RO membranes. Tri-Sep Corporation has participated in the consortium but has not developed a large-size RO membrane elements as of yet.

Koch Membrane Systems have independently developed large RO elements of 18-in (460 mm) diameter and 61-in (1549 mm) length ("MegaMagnum"), which are not compatible in size and length with the other large-size membrane elements available on the market today. In November 2009, Koch introduced 19-×-61-in (480 mm × 1549 mm) brackish water and seawater RO elements, which have enhanced production capacity ("MegaMagnum Plus" models 19061-HR-3525 and 19061-SW-3525, respectively).

14.14.2 Key Features of Large RO Elements and Systems

Diameter, Length, Membrane Area, and Productivity

Sixteen-inch RO Elements. Dow/Filmtec, Hydranautics, Toray and Woongjin have adopted the "standard" 16 × 40-in (400 mm × 1016 mm) membrane size developed by the USBR-led consortium of membrane manufacturers (USBR, 2004). While the consortium considered the feasibility of 20-in (508 mm) or larger vessels, the USBR study concluded that the preferred diameter large-membrane element is 16 in (400 mm) based on the fact that cost savings decrease and manufacturing risks increase asymptotically for membrane systems with larger diameter elements.

The consortium has selected 40-in (1016 mm) length of the 16-in (400 mm) elements for several reasons: (1) this size element has exactly four-times-higher surface area as compared with 8-in (200 mm) element, (2) the length of the 16-in (400 mm) pressure vessels for five and seven elements will be the same as that of existing 8-in pressure vessels, which would facilitate the retrofit of 8-in (200 mm) RO installations with larger vessels within the same RO building. GrahamTek, a Singapore-based company, has developed two enhancements to 16-in (400 mm) RO systems: (1) a patented flow distributor located on the inlet and outlet ends of the vessels to achieve a more uniform distribution of the source seawater flow within the membrane element feed spacer, (2) electromagnetic field (EMF) inducing coils embedded in the pressure vessels to enhance membrane flux and suppress membrane scale formation and biofouling.

Each integrated flow distributor has a 45° angle and evenly distributed holes to control the angle of entry and flow velocity in the membrane spacers. The electromagnetic field created in the vessels generates net movement in the direction of the concentrate stream through the membrane surface, thereby increasing permeability as well as inhibiting scale formation.

GrahamTek claims that two technological enhancements have the following benefits: (1) use of lower quality water for desalination, (2) operation at up to two-times-higher flux, which allows us to reduce the total number of elements needed to produce target permeate flow, (3) lower membrane scaling rate due to diminished concentration polarization on the membrane surface as a result of the more uniform spacer flow distribution, the scrubbing effect of micro-bubbles created by the flow distributors and the electromagnetic field, and (4) reduced scale formation and biofouling.

Eighteen- and Nineteen-Inch RO Elements

In 2003, KMS have developed series of large-diameter RO membrane elements with 18-in diameter and 61-in length ("MegaMagnum"). These elements have over seven times larger membrane area and fresh water production capacity as compared with the traditional 8-in elements 283 m^2 versus 37.1 m^2 (3050 ft^2 versus 400 sq ft). Up to five 18-in (460 mm) membrane elements can be installed into one large pressure vessel. As a result, one five-element vessel with 18-in (460 mm) MegaMagnum RO membranes can produce approximately five times more permeate flow than one seven-element vessel with 8-in membranes.

The 19-in (480 mm) KMS MegaMagnum Plus RO elements, introduced in November 2009, have a membrane area of 327.5 m^2 (3525 ft^2), which is 8.8 times higher than that of a standard 8-in element. Productivity of one five-element vessel with MegaMagnum Plus elements is over six times higher than that of seven-element vessel with 8-in (200 mm) membranes. The 18-in (460 mm) KSM MegaMagnum elements have approximately 30 percent greater filtration area than the 16-in (400 mm) RO membranes provided by other vendors. Similarly, the 19-in (480 mm) MegaMagnum Plus elements have over 50 percent higher filtration area than the 16-in (400 mm) RO elements.

Membrane Materials and Performance

All membrane manufacturers use the same membrane flat sheet (leaf) materials for their large size RO elements and their 8-in (200 mm) elements. They also employ the same feed/brine spacer configuration and thickness. As a result, large-size membrane elements are produced with the same performance characteristics (rejection, standard production capacity; permeability; feed spacer size, etc.) as their 8-in (200 mm) equivalents.

Recent side-by-side studies of 8-in (200 mm) and large-diameter membrane elements for water reclamation and seawater desalination applications (Hallan et. al, 2007; Ng et al, 2008; Johnson et al, 2009; Bergman and Lozier, 2010) indicate that the latest-generation large-size elements perform equally well in terms of salt rejection, permeability, flux, and fouling rate. The same filament winding process for 8-in (200 mm) and large-size elements produces the outer shell of the membrane elements. However, the outer wrap of the large-size elements is stiffer and thicker in order to obtain a stiffer shell laminate.

Seal Carrier

This membrane element component (also called antitelescoping device) is positioned at both ends of the fiberglass-wrapped spiral-wound element, and its main function is to support the downstream side of the membrane leafs and to prevent them from telescoping due to pressure differential across the element.

The seal carriers of the large-size elements are several times thicker than those of their 8-in counterparts because they are exposed to significantly higher loads. Hydranautics seal carriers for 16-in (400 mm) elements incorporate vents, which facilitate removal of air from the annular gap between the outside of the element and the pressure vessel wall and thereby prevent over-pressurization and damage of the RO elements.

Element Interconnection

In all large-size elements, the permeate seals at each membrane-to-membrane connection are reduced from two to one. A single O-ring between each two elements is used instead. Taking into consideration the reduction in total number of membrane elements and the fact that only one instead of two O-rings are used to connect the elements, the total number of O-rings relative to the standard 8-in (200 mm) elements is reduced sevenfold. This reduction would have a beneficial effect on the potential O-ring leakage, which is one of the most frequent causes for RO membrane-system performance integrity loss.

The Toray 16-in (400 mm) elements have an "axial labyrinth" seal between the membranes (patent-pending), which allows us to avoid the installation of radial seals on each element, reduces friction during loading/unloading, facilitates air displacement, and controls bypass flow.

The 16-in (400 mm) Dow/Filmtec RO elements are available with interlocking devices similar to their 8-in (200 mm) equivalents (see Chap. 3). However, the permeate coupler is eliminated in favor of a permeate seal locked on the end cap. This configuration simplifies membrane loading and eliminates the difficulties associated with the routine probing of the elements. It also reduces the pressure drop created by the coupler internal to the product water tube. The membrane elements are coupled via interlocking tabs located on the complementary upstream and downstream end caps. A pair of modules is locked by rotating the newly loaded element approximately 30°. Aligned markings on the end-cap perimeter allow us to verify the membrane locking visually.

In the 18-in (460 mm) MegaMagnum elements, coupling between elements is accomplished by an external sleeve design. The coupling unit is locked within a cavity that is an integral part of the two element seal plates. The coupling is external to the core tube, which allows a large cross-section O-ring to be used for connecting two elements. The elements are joined together at the outer surface of the seal plates by several fastener keys.

Brine Seal

Typical 8-in (200 mm) elements use a radially loaded cup seal between the inside wall of the pressure vessel and the element. Such configuration would create excessively high friction for large-elements and would make membrane element loading more difficult. Therefore all membrane suppliers use a brine seal configuration where the seal is moved to the face of the seal plate. With this configuration, the flow path within the large RO elements is identical to 8-in (200 mm) membranes but without the significant drag force against the pressure tube walls.

Membrane Element Costs

At present, the costs of large-size RO membrane elements per unit filtration area are higher than those of 8-in (200 mm) elements. Membrane materials used for 8-in and larger-diameter elements are identical, and both standard- and large-size elements are typically designed at similar flux and produce approximately the same volume of permeate per square foot of membrane area. However, the size of the core tube and membrane element wrapping are larger, and the production costs of rolling larger size elements are higher. As a result, the price of large-size membrane element per unit production capacity is higher.

For example, based on recent bids for large-size projects, the unit cost of an 8-in SWRO element with standard permeate flow production capacity of 26.5 m^3/day (7000 gpd) is $400 to $550/8-in element (i.e., $57.1/1000 gpd to $78.6/1000 gpd—avg. $67.9/1000 gpd). The typical price of 16-in (400 mm) SWRO element, with permeate production capacity of 113.5 m^3/day (30,000 gpd) at standard test conditions, is $2200 to $2500/16-in element ($73.3/1000 gpd to $83.3/1000 gpd—avg. $78.3/1000 gpd). Thus on average large SWRO membrane elements are expected to cost 15 percent more than 8-in elements for the same size plant. Similar unit cost difference between 8-in (200 mm) and 16-in (400 mm) elements is expected for brackish and seawater applications as well. This difference is reflective of the higher production costs of large-size membrane elements.

Membrane Vessels

Overview Membrane vessels for large-size elements are currently offered by four manufacturers: Beakaert (Protec), Codeline, Harbin ROPV, and BEL. The forces on the vessels and end caps are proportional to the square of the vessel diameter. As a result, the vessel wall and end-cap thickness and weight are four to six times higher than those of 8-in (200 mm) vessels. For example, the estimated weight of an end-cap assembly designed for SWRO vessels meeting ASTM code requirements is 68 kg (145 lb) and cannot be handled manually. For comparison, the end-cap assembly for an 8-in (200 mm) SWRO vessel is 11 kg (25 lb) and can be installed single-handedly by one operator.

Because of the higher end-cap weight loads, vessel manufacturers have adopted the use of the configuration and design of the existing 8-in (200 mm) seawater end caps and shimming for the large-diameter pressure vessels offered for brackish water desalination and water reuse applications. However, the design and production technology of large-diameter end caps for seawater applications are still in its early stages of development, and the membrane vessel suppliers do not have industry standards or in-house experience with manufacturing of such end caps and shimming. Therefore the traditional membrane vessel manufacturers have to order custom-made end-caps for seawater vessels they deliver. The specialty manufacturers of such end caps have several-times-higher profit margin expectations than the profit margins the vessel suppliers can sustain from the sale of large vessels.

A very important cost-benefit consideration for all large-size pressure vessels today is that they are only offered in end-port and side-port configurations (i.e., no multiple-port configuration large-diameter pressure vessels are currently available on the market). Taking into consideration that the use of 8-in (200 mm) multiple-port configuration provides significant savings of high-quality stainless-steel piping as compared with end- and side-port configurations, this disadvantage of large-diameter systems diminishes their overall cost benefits.

Membrane Vessel Costs

As indicated previously, the vessels for large-size RO membranes have significantly thicker and heavier walls and end caps than those for 8-in (200 mm) elements. As a result, the vessel costs for these elements are higher than 8-in (200 mm) vessels. For example, the cost of a seven-element 8-in (200 mm) pressure vessel for large SWRO project is typically in a range of $1400/vessel to $1800/vessel (avg. $1600/vessel). A four-element, 16-in (400 mm) pressure vessel costs $7000 to $9000/vessel (avg. $8000/vessel). Taking into consideration that one seven-element, 8-in (200 mm) SWRO vessel has a standard production capacity of 185.5 m^3/day (49,000 gpd) and a four-element 16-in (400 mm) vessel would produce 454 m^3/day (120,000 gpd), the average vessel cost per unit production capacity is for an 8-in (200 mm) vessel is $32.7/1000 gpd, while this for a 16-in vessel is $66.7/1000 gpd. This analysis indicates that the use of 16- versus 8-in (400 mm versus 200 mm) elements will cost on average two times more for SWRO plants of the same size. The vessel cost difference for large brackish water and water reclamation projects is expected to be in a range of 50 to 80 percent higher.

RO Train Number, Size, and Configuration

It should be pointed out that economies of scale from the use of large RO elements can only be obtained when the number of the RO trains of a plant with large elements is smaller than the number of RO trains using 8-in elements. For example, for the 190,000 m^3/day (50 mgd) brackish water, water reclamation, and SWRO plants, the USBR study team (USBR, 2004) have selected size of the individual large element trains of 47,000 m^3/day (12.5 mgd); 38,000 m^3/day (10 mgd), and 32,000 m^3/day (8.33 mgd), respectively, and have compared it against a 16,000 m^3/day (4.17 mgd) 8-in train [i.e., in all cases the 16-in RO trains were at least two times smaller number than the 8-in (200 mm) RO trains]. This allowed the USBR team to conclude that, for this size plant the use of large RO elements will have a clear life-cycle cost advantage as compared with an 8-in (200 mm) element-based system.

Cost-benefit analysis for RO systems with an identical number of RO trains would have shown an unfavorable outcome. For example, a RO system of a 190,000 m^3/day (50 mgd) SWRO plant with ten 19,000 m^3/day (5-mgd) 8-in (200 mm) element trains, would likely cost 10 to 30 percent more than the same size plant with ten 19,000 m^3/day (5-mgd) 16-in (400 mm) RO element trains despite the fact that fewer elements and vessels are used. In order for the 16-in (400 mm) RO system to become more competitive than the 8-in RO system, the large-element RO system would have to have at least two times fewer trains than the 8-in (200 mm) system. The main reason for this disparity is the fact that the costs for large RO elements per unit production capacity are 10 to 20 percent higher than the costs of 8-in (200 mm) elements, and the costs for membrane vessels are approximately two times higher. The main cost savings that can offset these significantly higher membrane and vessel costs can mainly come from the shorter-length stainless-steel interconnecting piping and fewer fittings (valves, elbows, etc.) and instrumentation, and lower number of racks resulting from the use of fewer RO trains.

In the case of the 190,000 m^3/day (50 mgd) plant, in order for a large RO element system to be more competitive than an 8-in system of the same capacity (i.e. 50 mgd), it has to have five RO trains or less (i.e., each train would have to have capacity of at least 38,000 m^3/day (10 mgd). The problem with such large RO train capacity for this size plant is that when one train is taken out of service for cleaning, the plant would lose 20 percent of its production capacity. In order for the plant to maintain its overall production capacity during RO train shutdown, the RO system design flux has to be selected conservatively (i.e., plant design flux is such that the other RO trans can be operated at

20 percent higher than design flux), and the train transfer pumps, high-pressure pumps, and energy-recovery devices have to be oversized. These additional costs however, will greatly negate the benefits of the use of larger trains.

When applied to BWRO and water reclamation projects, the general analysis of the relationship between costs, size of membranes, and size and number of trains is likely to yield a lower-capacity threshold, above which large-size RO systems will have clear cost advantage over 8-in-based RO plants. This threshold is likely to be in the 76,000 to 115,000 m^3/day (20 to 30 mgd) capacity range.

It should be pointed out that the cost-benefit analysis and threshold of beneficial use of large-diameter of over 8-in elements would be project specific, and constraints such as space availability and land costs may become an important factor that would make large-element RO systems more attractive for plants of capacity lower than 76,000 m^3/day (20 mgd). However, under the present economics of large-element vessels, for plants smaller than 76,000 m^3/day (20 mgd) it is likely that the use of 8-in elements would be more cost beneficial, and economy of scale derived from the use of fewer trains and interconnecting piping and fittings can be obtained by the use of large-size individual 8-in trains rather than by using large-size RO elements.

The benefits associated with economies of scale that eight-element systems can yield are limited to plants of approximately 189,000 m^3/day (50 mgd), and therefore large-element RO systems would have a clear cost advantage for larger plants only. A typical 8-in RO train of a large desalination plant contains 100 to 200 vessels and 700 to 1600 membrane elements. Even feed flow distribution beyond 200 vessels per train is very difficult to achieve and not practiced. The complexity of fabricating, transporting, and installing large-size RO racks also becomes more complex with increase in train size. At present, these hydraulic, construction, and physical constraints limit the maximum production capacity of individual RO trains with 8-in elements to 21,000 to 25,000 m^3/day (5.5 to 6.5 mgd)/train.

The large number of connections, elements, pressure vessels, and seals significantly reduces the cost benefits derived from the economy of scale for large- and mega-desalination projects. Full-scale experience to date indicates that very little economy of scale could be achieved when constructing 8-in (200 mm) RO desalination plants with production capacity larger than 200,000 m^3/day (53 mgd).

From a practical point of view, large-size RO membrane trains can be constructed with capacity of 50,000 to 100,000 m^3/day (13 to 26 mgd) per train. While trains larger than 100,000 m^3/day (26 mgd) are possible to build, their use would face the same economy of scale limitations as the 8-in (200 mm) RO trains have at present but at a higher threshold. In summary, the use of large-size elements would move the economy of scale threshold of 200,000 m^3/day (53 mgd) associated with 8-in (200 mm) elements to up to 500,000 to 1,000,000 m^3/day (190 to 380 mgd), if larger elements are used.

Depending on the RO membrane supplier, one large membrane vessel houses four to seven membrane elements. Table 14.26 presents a typical one-vessel configuration of large RO elements offered by key membrane suppliers. Analysis of this table indicates that a single vessel can produce between 550 to 1900 m^3/day (0.15 and 0.50 mgd) of permeate, depending on the membrane supplier and configuration.

Usually the smallest-size trains used in full-scale plants to date have two to three vessels only. This size train, however, is too small to be more cost-competitive as compared with the same size RO plant with 8-in (200 mm) elements. In fact, because of the significantly higher costs associated with the large-size membrane elements and vessels, use of large membrane elements for such small plants would be disadvantageous.

Membrane Manufacturer/ Membrane Element Size	Typical Number of Elements per Vessel	Product Water Capacity per Vessel (mgd)	
		BWRO & Water Reuse	SWRO
Dow/Filmtec 16 × 40 in	7	0.28–0.30	0.22
Hydranautics 16 x 40 in	4	0.12–0.15	0.10–0.14
Toray 16 × 40 in	7	0.28	0.19–0.21
Woongjin Chemical 16 × 40 in	4	0.15	0.10–0.15
KMS–MegaMagnum 18 × 61 in	5	0.33–0.43	0.26–0.35
KMS–MegaMagnum Plus, 19 × 61 in	5	0.40–0.50	0.30–0.40

Table 14.26 Typical Production Capacity of One Large RO Vessel

Horizontal versus Vertical Membrane Configuration

All large-diameter full-scale RO projects in operation to date are configured with membrane vessels installed horizontally on membrane racks. The membrane rack structure for RO systems of such configuration is relatively large, heavy, and costly. The recently announced SWRO project in Sorek, Israel, is configured with RO racks that have vertically installed vessels. This vertical configuration allows us to minimize the size of the otherwise heavy RO train support structure and to further reduce RO system costs.

Individual Trains versus Three-Center Design

To date, all existing large-diameter plants have been designed with individual train configuration where each RO train is serviced by a separate high-pressure feed pump and energy-recovery system. However, over the past 10 years, a three-center design concept is being more widely accepted and implemented by the desalination industry. Under this configuration, the RO membrane vessels, the high-pressure pumps, and the energy-recovery equipment are no longer separated in individual RO trains, but are rather combined in three functional centers: a high-pressure RO feed pumping center, a membrane center, and an energy-recovery center.

The three functional centers are interconnected via service piping. The SWRO feed pumping center includes only a few large-capacity high-pressure pumps that deliver seawater to the RO membrane center. One of the benefits of the three-center design is that it allows membrane cleaning to be completed in small sections of the RO system and to take less than 10 percent of membrane elements out of service for cleaning at any given time.

Because of the previously discussed issues associated with significant plant capacity reduction due to cleaning for RO systems with large elements, the three-center design would be a more advantageous RO system configuration for large-element RO systems. In fact, the large-element USBR study (USBR, 2004) has clearly recognized this benefit and has endorsed its consideration for such RO systems. All cost-benefit analyses completed by the USBR study team are based on the three-center design of the RO system.

The three-center design configuration is planned for use in the 420,000 m^3/day (110-mgd) Sorek SWRO plant, which has adopted the use of 16-in (400 mm) elements. When constructed, this plant is projected to produce the lowest-cost desalinated water delivered over the past decade—\$0.53/m^3 (\$2.02/1000 gal). For comparison, the cost of water production of the 95,000 m^3/day (25-mgd) Tampa Bay SWRO plant is \$0.87/m^3 (\$3.30/1000 gal).

Element Loading and Unloading

As compared with 8-in (200 mm) elements, the large RO elements cannot be loaded manually because of their heavy weight 52 to 66 kg (115 to 145 lb) dry and 62 to 76 kg (136 to 167 lb wet) for 16-in (400 mm) elements and 91 kg dry/113 kg wet (200 lb dry/250 lb wet) for 18-in (460 mm) elements. End caps for large element vessels are also several times heavier than those for eight-element vessels and require special handling equipment.

Membrane manufacturers differ by the methods and equipment they have developed for large-size membrane lifting, staging, and loading. Some manufacturers (Hydranautics and Woongjin) currently do not offer membrane loading equipment, and their membranes are lifted and loaded using a crane assembly designed to facilitate their handling. Currently, these membrane manufacturers are working on the development of proprietary loading/unloading devices, which are expected to be introduced on the market in the near future.

Toray has developed a patented wheel-mounted material lift to load the elements. This lift is fully motorized and can be operated by a single operator. KSM also offers an automatic loading device to accommodate the loading of its 18-(460 mm) and 19-in (480 mm) elements. Dow/Filmtec also has a patented loading system, which includes a lightweight cradle that attaches to anchors placed in the end face of the pressure vessel. At this time they do not provide an automatic loading device, but offer design drawings and specifications for customers that may be interested in a motorized membrane loading tool. KMS large elements are designed to be connected in series, and therefore they can be loaded and unloaded from one side of the pressure vessels only. The large elements of all other membrane manufacturers are designed for loading from both ends of the vessel.

Full-scale experience shows that the two-person crew necessary to load a seven-element/8-in vessel is approximately the same as the loading or unloading time for large-size five-element vessel—approximately 15 to 20 minutes.

While large-size RO membrane elements are mainly considered a cost-competitive alternative to 8-in membranes for large projects, to date they have found applications only in relatively small-and medium-size desalination and water reclamation plants. Most of the full-scale projects with large RO membrane elements in operation to date are water reclamation plants. The highest capacity water reclamation plant that employs large-size elements is the 66,000 m^3/day (17.5-mgd) Bundamba Advanced Water Treatment Plant in Queensland, Australia. This plant was built in two stages: 1A and 1B. Both stages use 18-in (460 mm) MegaMagnum BWRO elements. Stage 1A has capacity of 30,000 m^3/day (8.0 mgd) and includes four RO trains arranged in 7:4:2 arrays. Each train has the capacity of approximately 7570 m^3/day (2.0 mgd) and contains 65 membrane elements. Stage 1B includes the addition of five more trains of the same size with a total capacity of 36,000 m^3/day (9.5 mgd). The second largest project is the 56,000 m^3/day

(14.8 mgd) Bedok NEWater reclamation plant in Singapore. This facility has a total of seven identical RO trains of which six trains are equipped with Hydranautics ESPA2 1640 membranes, and one RO train is furnished by Woongjin Chemical RE106040-BRL membranes. Each RO train contains 12 pressure vessels in the first stage and six vessels in the second stage. The size of the individual RO trains is approximately 4000 m^3/day (1.1 mgd). The plant has been in operation since 2008 and has worked well.

The largest-capacity seawater desalination plant with 16-in membrane elements in operation at present (the 10,000 m^3/day (2.6-mgd) PowerSeraya SWRO facility in Singapore) has two independent 5000 m^3/day (1.3-mgd) RO trains, each of which is configured as a two-pass, two-stage system with a 2:1 staging array. Each RO train has 24 first-stage vessels and 12 second-stage vessels. This plant has two passes and uses Hydranautics SWC3-1640 membranes in the first pass and ESPA2-1640 membranes in the second pass.

Most other existing large RO membrane installations for water reclamation are relatively small and typically have trains with 2:1 array configuration (i.e., concentrate from the first two bank vessels is blended and fed into the third vessel of the same bank). The smallest RO trains have three-vessel configurations, and larger RO trains are often configured in multiple three-bank (2:1) arrays. Existing water reclamation plants and brackish water desalination plants with large elements are designed for membrane flux in a range of 20 to 25 Lmh (12 to 15 gfd). Seawater desalination plants usually have design membrane flux of 14 to 16 Lmh (8.5 to 9.0 gfd).

Advantages of Large RO Element Systems

Potential Cost Saving A detailed cost-benefit analysis comparing 8- and 16-in (200 mm and 400 mm) RO elements for brackish and seawater applications was completed by the US Bureau of Reclamation in 2004 as a part of a membrane industry consortium study aimed at the development of standard-size large RO membrane elements (USBR, 2004) The results from this study indicate that the potential construction cost savings associated with element diameter increase from 8- to 16-in (200 mm to 400 mm) size for a plant capacity range of 47,000 to 570,000 m^3/day (12.5 to 150 mgd) are: (1) 18.5 to 27 percent for brackish water and water reuse projects, and (2) 7 to 17 percent for seawater projects.

According to the USBR study, the main construction cost savings are associated with the reduction of the physical size and number of the RO trains. For a hypothetical 189,000 m^3/day (50-mgd) brackish water desalination project, the estimated RO train-related construction cost savings associated with the use of 16- versus 8-in (400 mm versus 200 mm) elements is 50 percent ($0.22 gpd versus $0.33/gpd in $2004).

The USBR study concluded that the O&M costs of desalination and water reclamation plants using 8- and 16-in (200 mm and 400 mm) element systems are comparable. This study stipulated that use of larger elements, however, could yield potential O&M benefits due to the overall reduction of repair and maintenance activities associated with the fewer elements, pressure vessels, and RO skid components (valves, instrumentation, etc.).

The 2004 USBR Study has also concluded that the overall life-cycle cost savings for using 16- versus 8-in (400 mm versus 200 mm) elements for the desalination plant capacity range of 47,000 to 570,000 m^3/day (12.5 to 150 mgd) are projected to be: (1) 8 to 11 percent for brackish groundwater projects, (2) 5 to 8 percent for brackish surface water and water reclamation projects, and (3) 4 to 6 percent for seawater projects.

It should be pointed out that the cost analyses included in the USBR study were completed assuming the use of front- or side-port 8- and 16-in (200 mm and 400 mm) vessels. Taking into consideration that multiple-port entry vessels are currently available only for 8-in (200 mm) elements and not available for large-diameter vessels, and that a multiple-port configuration could yield up to 50 percent reduction of stainless-steel piping costs, the potential construction cost savings associated with the use of large RO vessels will likely be diminished with 10 to 20 percent and the life-cycle cost benefits would be reduced with 5 to 10 percent.

RO Building Area Reduction

The results of the USBR large-element study (USBR, 2004) indicate that for a typical 189,000 m^3/day (50-mgd) plant using 16-in (400 mm) elements, the total RO system building footprint savings are projected to be as follows: (1) 9 to 22 percent for BWRO systems using groundwater source, (2) 4 to 13 percent for BWRO systems using surface source water and for water reclamation plants, (3) 0 to 6 percent for SWRO plants. The use of 20-in elements did not provide significant additional RO building area savings over those yielded by the 16-in RO membrane systems.

Based on cost-benefit analysis completed during the design phase of the 10,000 m^3/day (2.6 mgd) PowerSeraya SWRO plant, it was estimated that the use of 16-in (400 mm) instead of 8-in (200 mm) elements would result in 30 percent RO system footprint reduction (Bergman and Lozier, 2010). The overall reduction of the RO building footprint, however, is smaller due the additional space needed for the large-size RO membrane cleaning system and additional circulation area needed around the RO trains for membrane handling.

Based on the experience from existing smaller-size plants, the total RO building footprint savings are likely to be within 5 to 15 percent of the footprint of the same size RO system using 8-in (200 mm) elements. The space savings are expected to be more significant for RO plants with production capacity larger than 189,000 m^3/day (50 mgd).

Potential Challenges Associated with Use of Large RO Element Systems

While the use of large-diameter RO membranes has a number of advantages (mainly construction cost, building area, and maintenance savings), it may also have potential challenges for the site-specific conditions of a given project.

Reduced Production Capacity Due to Membrane Cleaning and Equipment Downtime

As indicated previously, because of their large unit production capacity, RO trains using large RO elements have the disadvantage to cause reduction of a larger portion of the plant production capacity when RO trains are down for service. This issue holds true especially for plants with production capacity smaller than 189,000 m^3/day (50 mgd) and individual RO trains that have capacity larger than 10 percent of the total plant production flow. In order to address this issue, large RO plants have to be designed with a minimum of five to ten RO trains or to be configured in a three-center design configuration. However, the decrease of individual RO train capacity to address the overall plant production reliability concerns for small plants compromises the cost benefits associated with the use of large-diameter trains.

Increased Annual RO Membrane Replacement Costs

As indicated previously, the unit costs of large-diameter RO membranes per 1000 gallons of produced permeate are typically 10 to 20 percent higher than the 8-in element unit costs for the same size plant. As a result, the annual RO membrane replacement costs for large-size membrane plants are expected to be 10 to 20 percent higher as well.

Elevated Operations Risks

As indicated previously, large-size elements and end caps cannot be handled manually and require more operator attention and reliance on auxiliary equipment to load and unload. Accidental drop of large elements on equipment and operations staff could cause significantly higher damage and serious injury than that of standard 8-in (200 mm) elements.

Because of the lack of industry standard for design and installation of end-cap assemblies for large SWRO vessels at this time, there is a higher risk and potential for damage and staff injury associated with the accidental release of the end caps due to faulty design, production, or installation.

Potential Additional Structural Costs

While the total footprint of the RO building housing large-diameter membrane system is projected to be smaller than that of an 8-in RO system, the construction load over the same RO membrane train footprint would increase several times because the RO elements and end caps are significantly heavier per unit production capacity. As a result, depending on the bearing capacity of the soils under the RO building, savings from reduced RO building footprint may be lost due to additional expenditures associated with the need to build heavier building foundations and to construct the RO building foundations on piles. For example, cost-benefit analysis for use of 16- versus 8-in (400 mm versus 200 mm) RO system completed during the initial planning phases of the Gold Coast SWRO Project in Queensland, Australia, have concluded that the net savings associated with the construction of the RO building will be minimal because the desalination plant is located on inactive landfill site in which soils have very low load-bearing capacity.

14.15 References

Bartels G., K., Lai, and M. Wilf, "New Generation Low Fouling Nanofiltration Membranes," Proceedings of April 2007 Conference of the European Desalination Society, Halkidiki, Greece, 2007.

Bates, W. T., C. Bartels, and R. Franks, "Improvements in Spiral Wound RO and NF Membrane Element Construction for High Fouling Feed Water Applications," Annual Conference, American Membrane Technology Association, 2007.

Bergman, R. A., and J. C. Lozier, "Large-Diameter Membrane Elements and Their Increasing Global Use," *IDA Journal*, First Quarter, 16–23, 2010.

Bursilovsky, M., and M. Faigon, "The Impact of Varying the Number of Elements in PV in SWRO Plants—Actual and Future Considerations," *Desalination*, 184, 233–240, 2005.

Dreizin, Y., A. Tenne, and D. Hoffman, "Integrating Large Scale Seawater Desalination Plants within Israel's Water Supply System," *Desalination*, 220, 132–149, 2008.

Gorenflo, A., M. Brusilovsky, M. Faigon, and B. Liberman, "High pH Operation in Seawater Reverse Osmosis Permeate: First Results from the World's Largest SWRO Plant in Ashkelon," *Desalination*, 203, 82–90, 2007.

Greenlee L. F., D. F. Lawler, B. D. Freeman, B. Marrot, and P. Moulin, "Reverse Osmosis Desalination: Water Sources, Technology, and Today's Challenges," *Water Research*, 43, 2317–2348, 2009.

Hallan, M. J., J. E. Johnson, M. S. Koreltz, and M. H. Peery, "Design, Development and Evaluation of Sixteen Inch Diameter RO Modules," Proceedings of IDA World Congress, Gran Canaria, Spain, October 21–26, 2007. Ref: IDAWC/MP07-011.

Henthorne, L., F. Wang, U. Sanjeev, and K. Deshmukh, "Demonstration Testing of Innovative 16-in SWRO in Arabian Gulf and Singapore Waters," Proceedings of IDA World Congress, Gran Canaria, Spain, October 21–26, 2007. Ref: IDAWC/MP07-188.

Johnson, J. E., P. Shen, M. Brusilovsky, and R. S. Ong, "Sixteen-in Reverse Osmosis Module Performance in Seawater and Water Reuse," Proceedings of IDA World Congress–Atlantis, The Palm-Dubai, UAE, November 7–12, 2009. Ref: IDAWC/MP09-212.

Li, N. N., A. G. Fane, W. S. Winston Ho, and T. Matsuura, eds,. *Advanced Membrane Technology and Applications*, John Wiley & Sons, Inc., Hoboken, New Jersey, 2008.

Liberman, B. *The Importance of Energy Recovery Devices in Reverse Osmosis Desalination*, "The Future of Desalination in Texas—Volume 2," Technical Papers, Case Studies Desalination Technology Resources (report # 363), Texas Water Development Board, December, 2002.

Liberman, B., and M. Wilf, "Evolution of Configuration of RO Seawater Desalination Elements," Proceedings of IDA World Congress–Singapore, 2005. Ref: SP05-059.

Mickols, W. E., M. Busch, Y. Maeda, and J. Tonner, "A Novel Design Approach for Seawater Plants," IDA World Congress, Singapore, 2005.

Ng, H. Y., K. G. Tay, S. C. Chua, and H. Shea, "Novel 16-in Spiral-Wound RO Systems for Water Reclamation—A Quantum Leap in Water Reclamation Technology," *Desalination*, 225, 275–287, 2008.

Sachaf, N., and T. Haarburger, "Comparison of Multiple Port and End Port Connections for Pressure Vessels," *Desalination*, 221, 202–206, 2008.

Subramani, A., M. Badruzzaman, J. Oppenheimer, and J. G. Jacangelo, "Energy Minimization Strategies and Renewable Energy Utilization for Desalination: A Review," *Water Research*, 45, 1907–1920, 2011.

"USBR Industry Consortium Analysis of Large Reverse Osmosis/Nanofiltration Element Diameters," Report No. 114, 2004.

Wang, F., L. Henthorne, H. C. Yon, Y. P. Seng, and T. S. Chye, EPC, "Commissioning, Operation: the World's First Full-Scale Large-Diameter SWRO Plant," Proceedings of IDA World Congress, Gran Canaria, Spain, October 21–26, 2007. Ref: IDAWC/MP07-187.

Yun, T., I., C. J. Gabelich, M. R. Cox, A. A. Mofidi, and R. Lesan, "Reducing Costs for Large-scale Desalting Plants Using Large-Diameter Reverse Osmosis Membranes," *Desalination*, 189, 141–154, 2006.

Wilf, M., L. Awerbuch, C. Bartels, M. Mickley, G. Pearce, and Voutchkov, *The Guidebook To Membrane Desalination Technology: Reverse Osmosis, Nanofiltration And Hybrid Systems Process Design, Applications And Economics*. Rehovot: Balaban Publishers, 2007.

Verhuelsdonk M., T. Attenborough, O. Lex, and T. Altmann, "Modeling the Impact of Using Multiple-Port RO Pressure Vessels iIn Seawater Reverse Osmosis Desalination Plants Using Special Simulation Software," *Desalination and Water Treatment*, 5, 192–197, 2009.

CHAPTER 15

Post-Treatment of Desalinated Water

15.1 Introduction

Product water from desalination plants is characteristically low in mineral content, hardness, alkalinity, and pH. Therefore, desalinated water must be conditioned (post-treated) prior to final distribution and use. Post-treatment of fresh water produced by desalination has two key components: (1) mineral addition in order to protect public health and to safeguard integrity of the water distribution system, and (2) disinfection.

In addition, the product water quality of desalination plants is often also determined by other beneficial uses such as agricultural irrigation, industrial use, and water reuse. In this case, post-treatment typically involves enhanced removal of some minerals such as boron, sodium, and chlorides, and/or supplemental addition of other minerals such calcium and magnesium.

Usually, the actual application dosage of post-treatment chemicals to desalinated water is selected based on the minimum quantities needed to achieve all purposes for which the conditioning chemical is added. If the use of the same chemical is not found to be cost-effective to achieve both public health and other water quality goals, then a combination of chemicals that yield the lowest overall cost of water production is typically used to meet all post-treatment targets.

Typically, post-treatment of product water includes one or more of the following processes: (1) remineralization for corrosion protection, (2) disinfection for biological stability and public health protection, and (3) water quality polishing for enhanced removal of specific water constituents [e.g., boron, silica, gases that cause taste and odor, N-nitrosodimethylamine (NDMA), etc.].

15.1.1 Remineralization for Corrosion Protection

Corrosion-Related Issues of Desalinated Water

The lack of carbonate alkalinity as well as the low content of calcium and magnesium (i.e., very low hardness) causes desalinated water to be unstable and prone to wide variations in pH due to its low buffering capacity and its inability to form protective calcium carbonate films on pipe walls, which makes it corrosive. When blended with other water sources in the distribution system, desalinated water deficient in carbonate alkalinity may cause previously deposited calcium carbonate films to dissolve. Monovalent ions such as chlorides as well as gases such as hydrogen sulfide, oxygen, ammonia, and carbon dioxide may pass through the reverse osmosis membranes to a higher degree than other ions or molecules and also contribute to the corrosion potential of the desalinated water.

Product water remineralization for corrosion protection has two aspects: (1) protection of water distribution systems and household piping and fixtures from damage caused by corrosion, and (2) maintaining the desalinated water quality in terms of aesthetic appearance (color and taste). Over time, practically all water distribution systems using metal pipes build up sediments (deposits), which consist of source water impurities as well as precipitated iron and manganese created as a result of long-term pipeline corrosion.

When distribution system water quality changes significantly (such as could be the case of introducing desalinated water of different mineral content, pH, alkalinity, and hardness), pipeline deposits may be released into the distributed water thereby causing episodes of "red water" or "black water," which makes the finished water aesthetically displeasing and in some cases unsafe to drink. Remineralizing desalinated water to match the water quality of the other water sources, which are delivered to the same distribution system is therefore of critical importance for maintaining the high quality of this water all the way to the consumer's tap.

Corrosion can affect many aspects of a drinking water supply, including pumping costs, public acceptance of treated water, disinfection efficacy, and public health due to exposure to heavy metals (e.g., lead, copper, and cadmium). In conventional water treatment, corrosion is defined as degradation of pipe materials due to a reaction with water. This reaction can be physical, chemical, and electrically or biologically induced. Similarly, in a desalinated water supply systems, chemical reactions can cause degradation of metallic pipes that comes in direct contact with water. Corrosion of metallic surfaces is affected by many water quality parameters, including pH, alkalinity, calcium, hardness, chlorides, silica, phosphate, and temperature. The significance of each of these parameters is summarized in Table 15.1.

Water Corrosion Potential Indexes

Water stability and corrosion potential may be characterized by parameters (corrosion indices) indicating the potential of the desalinated water to precipitate calcium carbonate ($CaCO_3$) and by parameters that address corrosivity caused by specific compounds in the desalinated water. The mostly used calcium carbonate-based corrosion (stability) indices relevant to water distribution pipes are the Langelier Saturation Index (LSI) and the calcium carbonate precipitation potential (CCPP). Elaboration on the mathematical development of the two indices is presented elsewhere (Lahav et al., 2012).

Langelier Saturation Index

The Langelier Saturation Index (LSI) is a provides qualitative assessment of water's potential to precipitate calcium carbonate. This index is based on the difference between the pH of the unconditioned desalinated water and the pH of the desalinated water at the point of its saturation with calcium carbonate. A negative LSI indicates that the water is undersaturated (i.e., calcium carbonate will dissolve), whereas a positive LSI may indicate that the water is oversaturated (i.e., calcium carbonate will precipitate). However, this index does not actually account for the amount of carbonate in water. Therefore, while calculations may indicate a positive LSI, it is possible that very little calcium carbonate may actually precipitate.

Variations of the LSI calculation exist. The original LSI was developed to predict the potential precipitation of calcium carbonate in a specific fresh surface water source and

Water Quality Parameter	Significance
pH	Low pH (typically below 7.0) may increase corrosion rates High pH (typically in a range of 7.8 to 8.3) may reduce corrosion rates
Alkalinity	Provides water stability and prevents variations in pH May contribute to the deposition of protective films Highly alkaline water may cause corrosion in lead and copper pipes
Calcium	May deposit as a calcium carbonate film on pipe walls to provide a physical barrier between metallic pipe and water Excess concentrations of calcium may decrease the water transmission capacity of pipes
Hardness	Hard water is generally less corrosive than soft water if calcium and carbonate alkalinity concentrations are high and pH conditions are conducive to calcium carbonate deposition
Chlorides	High concentrations may increase corrosion rates of iron, lead, and galvanized steel pipes
Silica	Can form a protective film when present in dissolved form
Phosphate	Can form protective film
Temperature	Can have an impact on solubility of protective films and corrosion rate

Table 15.1 Factors Affecting Corrosion of Desalinated Water

it has limited suitability for assessment of the corrosion potential of desalinated water. The LSI overestimates saturated conditions at high pH when additional carbonate may not actually be available for precipitation, which may be interpreted as highly scaling conditions when water is undersaturated.

Calcium Carbonate Precipitation Potential

As opposed to the other carbonate-based corrosion indices, the calcium carbonate precipitation potential (CCPP) is an index that quantifies the amount of calcium carbonate that may dissolve or precipitate. This index provides the most accurate indication of water's potential to deposit calcium carbonate and to form protective coating on pipe walls thereby minimizing corrosion. The CCPP index is the most useful tool for developing corrosion control and post-treatment strategies for desalinated seawater. Positive value of the CCPP indicates the concentration of calcium carbonate that exceeds the saturated condition, whereas negative value denotes the amount of calcium carbonate that must be added and dissolved to reach a saturated condition.

Based on a number of references (AWWA, 2007; Cotruvo et al., 2010; WHO, 2008) the recommended range for the CCPP in desalinated water is between 4 and 10 mg/L as calcium carbonate. In South Africa, where soft waters are very common, lower CCPP range of 2 to 5 mg/L as $CaCO_3$ (Loewenthal et al., 1988), and even 1 to 2 mg/L as $CaCO_3$ are applied (de Souza et al., 2002). The low end of the CCPP value is determined by the threshold concentration at which $CaCO_3$ precipitation is initiated on the surface of the conveyance piping. Increasing the CCPP level enhances $CaCO_3$ precipitation of the pipe's inner service but typically requires the water pH to be elevated beyond 7.8. However, overly high pH interferes with the chlorine disinfection process and is not desirable. Therefore, the maximum CCPP level is typically limited by pH increase to up to 8.5. Furthermore, although a higher CCPP value increases the driving force for $CaCO_3$ precipitation on the pipe's internal surface and thus increases the potential for the formation of a denser, more effective passivation layer, an upper CCPP limit should also be set in order to prevent the buildup of excessive $CaCO_3$ scales on pipes, pumps, etc.

Higher CCPP values can be acceptable when combined with the use of calcium complexing agents, such as polyphosphate-based scale inhibitors. However, when the CCPP is too high and scale inhibitors are not used, the water transmission capacity of a pipeline may be reduced due to excessive calcium carbonate precipitation.

The CCPP has not been widely used by water treatment engineers mainly because of the analytical complexity of its manual calculation. Spreadsheet models and commercial software can be used to calculate the CCPP for a wide range of water qualities (i.e., AWWA, 2011).

Larson Ratio

Research completed by Larson in 1970 has demonstrated that chloride and sulfate ions can increase corrosion rates and iron concentrations in water when it is conveyed in iron and alloy pipes. It is important to understand the effect of chloride in particular when designing desalination systems that treat high-chloride brackish waters and seawater.

Material of pipes used to convey water to the post-treatment system and permeate piping must be compatible with relatively high chloride concentrations. PVC material is typically selected for low-pressure piping, whereas high molybdenum content (6 to 8 percent or higher) stainless-steel alloys are used for high-pressure piping. However, corrosion in distribution systems cannot be prevented by the use of suitable pipe material material alone. Post-treatment is still required, as distribution piping may include metals that are prone to chloride attack. The Larson Ratio (LR) is defined as follows:

$$LR = [(Cl^-) + (SO_4^{2-})]/[(HCO_3^-)], \tag{15.1}$$

where all parameters in ( ) are expressed in milliequivalents per liter.

Chloride (Cl-) concentration in desalinated water is often higher than that in other water sources, but the sulfate (SO_4^{2-}) concentration is typically one order of magnitude lower, thereby reducing the Larson Ratio. If the need arises, the LR value of desalinated water can be reduced by increasing the bicarbonate (HCO_3^-) concentration in the post-treatment step.

A number of sources recommend LR values lower than 5 to minimize steel pipe corrosion (Delion et al., 2004; Loewenthal et al., 2004; WHO; 2004). However, due to the fact that all factors affecting passivation/corrosion in waters which contain sulfates and chlorides are not yet fully understood, the LR threshold of 5 should be considered only an indicative threshold.

Alkalinity

The commonly used term "alkalinity" is defined as the proton-accepting capacity of the water with respect to H_2CO_3* as reference species. The mathematical expression of this value is given in Eq. 15.2. Water alkalinity is typically determined by titration with strong acid to a target pH value of 4.5 (i.e., close to the H_2CO_3 equivalence point) or by applying the Gran titration technique, which circumvents the need to know the exact location of the equivalence point.

$$\text{Alkalinity } (H_2CO_3^*) = 2[CO_3^{2-}] + [HCO_3^-] + [OH^-] - [H^+] \tag{15.2}$$

For a given pH level, the higher the alkalinity of given water the higher the ability of this water to withstand a change in pH due to a release of H^+ or OH^- ions to the water –i.e., the higher is its buffering capacity. Buffering capacity of water is defined as the concentration of strong acid (or base) needed to be added in order to change the water pH with one unit. Minimization of pH variability in the distribution system as a result of its increased alkalinity promotes the formation of a denser scale structure on the pipe's wall and decreases the rate of iron release from the pipe.

In addition, a higher alkalinity at a given pH translates into a higher inorganic carbon concentration and thus a higher concentration of CO_3^{2-}. This is a potential advantage because various models for corrosion and red water prevention point out that the precipitation of siderite ($FeCO_3$) is imperative for the development of an effective passivation layer. From this perspective, a higher alkalinity value also appears to be advantageous.

Another benefit of high alkalinity waters stems from the fact that they can reach the target CCPP value for a given Ca^{2+} concentration, at relatively lower pH which, is advantageous with respect to disinfection. The results from a comprehensive pilot study conducted on the matter of red water prevention indicate that maintaining alkalinity concentration above 80 mg/L as $CaCO_3$ is the most important individual parameter for preventing the release of metal ions to water (Imran et al., 2005).

Dissolved Calcium

Maximum and minimum levels of dissolved calcium (Ca^{2+}) in water are not necessarily related to its chemical stability. A minimal dissolved calcium level of 50 to 60 mg/L as $CaCO_3$ is desirable for health reasons (Berghult et al., 1999; Kozisek, 2003). The practical maximum content of dissolved calcium (recommended level of 120 mg/L) in the water is determined by economic factors attributed to the need to supply water that is not excessively hard. A dissolved calcium concentration (for a given alkalinity and CCPP values) closer to the maximum level of 120 mg/L, allows maintaining a relatively lower pH value in the product water, which is advantageous from a biostability (chlorine disinfection) standpoint.

pH

Desalinated water usually has lower pH than the saline source from which it has originated because the desalination process removes almost completely the carbonate (CO_3^{2-}) and bicarbonate (HCO_3^-) ions, while allowing dissolved carbon dioxide (CO_2) to partially pass to the permeate side of the RO membranes, thereby lowering pH. It should be pointed out that seawater and brackish water usually differ significantly in terms of their natural dissolved carbon dioxide content. Brackish water typically contains high levels of CO_2, while the CO_2 content of seawater is very low. As a result, the pH of desalinated brackish water is usually lower than that of desalinated seawater. In both cases, however, the pH of the desalinated water is lower than the minimum pH threshold needed to provide adequate corrosion protection. This minimum product water pH threshold is a function of the target product water alkalinity, the dissolved Ca^{2+} concentration, and target CCPP level. The actual minimum pH value depends also on the ionic strength and temperature. The maximum pH level of the finished water, has to be controlled because of its negative impact on disinfection efficiency. In addition, pH (along with alkalinity) also determines water's buffering capacity. Within the typical pH range of remineralized desalinated water of 7.5 to 8.4, the higher the pH the lower the buffering capacity (for a given alkalinity concentration).

Ionic Strength and Temperature

The ionic strength and temperature of the product water have an impact on the pH and the value needed to reach target CCPP at constant alkalinity and dissolved calcium concentrations. For a given level of dissolved calcium and alkalinity concentrations, the rate of $CaCO_3$ precipitation increases with the increase of temperature and decrease of ionic strength.

15.1.2 Mineral Supplementation

Both thermal and reverse osmosis desalination processes, widely used at present to produce fresh water from saline water, remove over 90 percent of most of minerals from the source water. As a result, unless desalinated water is supplemented with minerals by post-treatment, it often contains significantly smaller amounts of micronutrients than traditional water sources (rivers, lakes, ground water). Maintaining/supplementing certain minimum levels of nutrients such as calcium, magnesium, zinc, copper, chromium, manganese, and potassium is important for human health and agricultural and horticultural uses of the desalinated water (Birnhack et al., 2011).

15.1.3 Disinfection

At present, chlorine in various forms (e.g., sodium hypochlorite, calcium hypochlorite, and chlorine gas) is the typical conditioning chemical used for disinfection of desalinated water because of its recognized pathogen inactivation efficiency and the reasonably low levels of disinfection byproducts (DBPs) it generates in desalinated water. However, other disinfectants, such as chlorine dioxide and chloramines as well as ultraviolet (UV) light irradiation, could also be used for desalinated water disinfection (AWWA, 2011).

Desalinated water typically has a lower organic content than most fresh surface water sources. Since a large portion of the DBPs in the finished drinking water originate from organics, desalinated water usually has significantly lower DBP formation potential than traditional fresh water supplies. It should be pointed out, however, that in brackish reverse osmosis (BWRO) desalination plants where blending of source water and permeate is commonly practiced, DBP formation may still be an issue if the brackish source water used for blending contains a large quantity of organics. In addition, depending on the type of the desalination system used for fresh water production, desalinated water may contain higher levels of bromides than other water sources and may form more brominated DBPs (Augus et al., 2009).

While reverse osmosis (RO) membranes reject most of the organics contained in the source water, they are not as effective in rejecting DBPs, which are already formed when chlorine is used for source water pretreatment. While seawater reverse osmosis (SWRO) membranes would remove 80 to 90 percent of the DBPs in the source water, BWRO membranes have a lower DBP rejection (50 to 80 percent), and therefore the desalinated water quality may need to be polished by enhanced post-treatment.

15.1.4 Water Quality Polishing

Water quality polishing is used for enhanced treatment of specific compounds (e.g., boron, silica, NDMA) when these compounds have to be removed from the water to meet water quality targets for drinking or industrial uses. Depending on the compounds targeted for removal, the treatment technologies may include ion exchange, granular activated carbon filtration, additional multistage/multipass membrane RO treatment or a combination of treatment processes, which could include advanced oxidation. Li et al. (2008), Parekh (1988), and Wilf et al. (2007) provide additional information on desalinated water polishing. Advanced treatment of desalinated water is most often associated with industrial applications of desalinated water, which are not included in the scope of this book.

Brackish water from aquifers often contains odorous gases such as sodium bisulfide, which also have to be removed in order to produce water quality acceptable for human consumption in terms of taste and odor. Existing RO membrane and electrodialysis reversal (EDR) systems do not remove dissolved gases well. Therefore, gases that present public health or taste and odor challenges have to be removed by additional treatment of the desalinated water, most often by air stripping. In a typical stripping process, the desalinated water falls down through a tower packed with contact media while air is conveyed in an opposite direction (upward) through the tower, thereby stripping the dissolved gasses from the water. In order to achieve complete removal of hydrogen sulfide, the pH of the desalinated water has to be reduced to 5 or less at which pH level practically 100 percent of the hydrogen sulfide is in gaseous state. Air stripping not only removes odorous gases such as hydrogen sulfide but also aerates the finished drinking water, thereby improving its taste. Technologies for removal of odorous gases are outside of the scope of this book and are discussed in greater detail elsewhere (Watson, et al., 2003).

15.1.5 General Product Water Quality Guidelines

Based on worldwide experience and taking into account practical (economic) considerations, the following set of post-treatment water quality criteria could be considered for desalinated water that is intended to have multiple uses:

1. Alkalinity > 80 mg/L as $CaCO_3$
2. $80 < Ca^{2+} < 120$ mg/L as $CaCO_3$
3. $3 < CCPP < 10$ mg/L as $CaCO_3$
4. $7.5 < pH < 8.5$
5. Larson Ratio < 5 (not obligatory)

The key considerations behind the choice of these specific criteria are as follows:

- From process engineering perspective, the alkalinity and Ca^{2+} concentrations should be comparable.
- The maximum Ca^{2+} concentration could be set at 120 mg/L as $CaCO_3$ based on economic considerations (water not becoming excessively hard).
- Adequate positive CCPP in the range listed above and slightly positive Langlier Saturation Index by adjustment of alkalinity and pH to promote formation of protective pipe layer.
- Target alkalinity of 80 to 100 mg/L as $CaCO_3$ to provide adequate buffering capacity.
- To address possible nitrification problems in wastewater treatment plants receiving wastewater originating from desalinated water, an alkalinity value of > 100 mg/L as $CaCO_3$ is recommended.
- The minimum Ca^{2+} concentration is recommended to be 100 mg/L as $CaCO_3$. However, a lower level of 80 mg/L as $CaCO_3$ may be acceptable to minimize water production costs.

- pH is recommended to be maintained at a value lower than 8.5 at all times and preferably in a range of 7.5 to 8.3 to allow for efficient chlorine disinfection and to minimize household plumbing corrosion.
- The Larson Ratio is recommended (but not obligatory) to be lower than 5. This implies that the use of H_2SO_4 as an acidifying chemical is less desirable than the use of CO_2. However, water devoid of sulfates is less favorable from an agricultural irrigation standpoint.

15.1.6 Overview of Typical Post-Treatment System

Figure 15.1 depicts a schematic of a typical desalination plant post-treatment system. This system includes RO permeate conditioning with lime and carbon dioxide followed

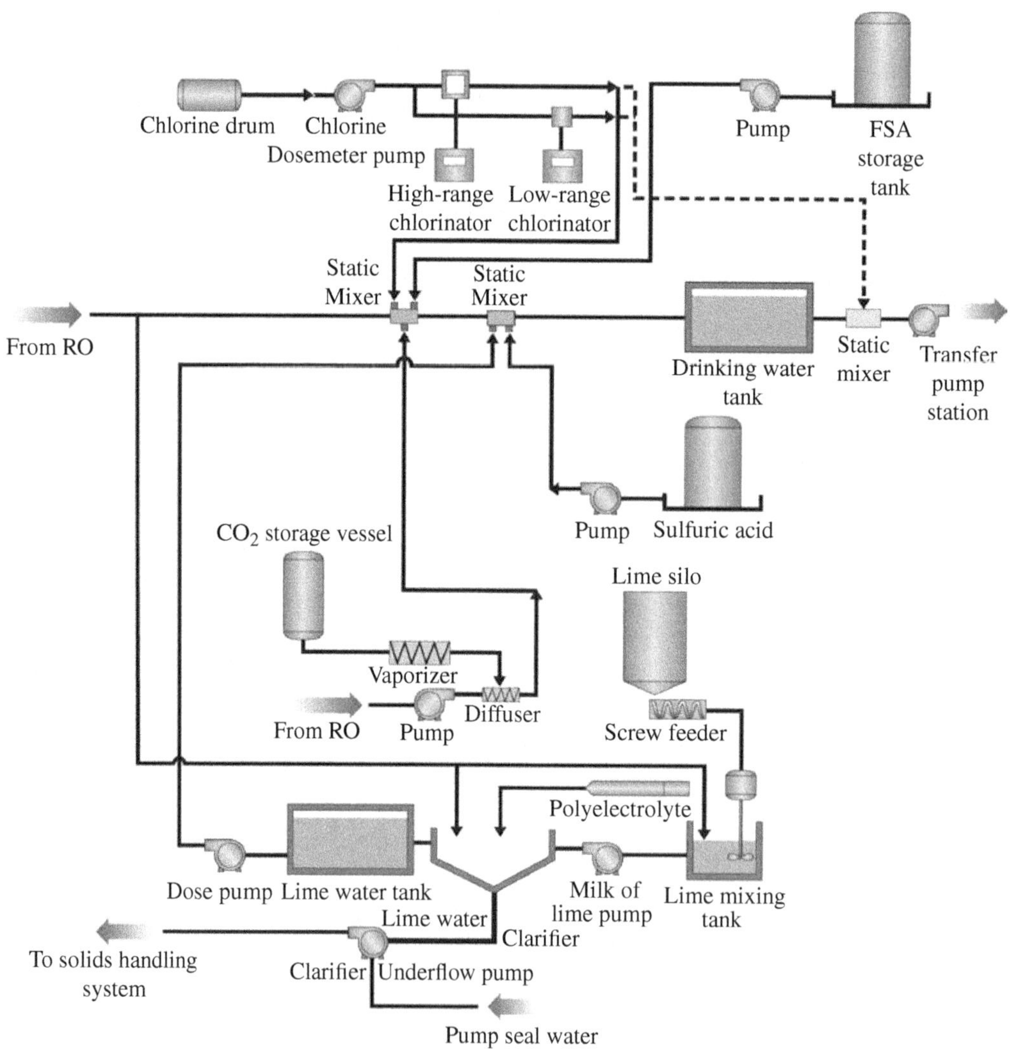

FIGURE 15.1 General schematic of desalination plant post-treatment system.

by addition of chlorine (stored as chlorine gas) and fluorosilicic acid (FSA) for product water fluoridation (if required based on local regulations). The pH of the final product water is adjusted by addition of sulfuric acid. Carbon dioxide, chlorine, and fluororsilicic acid are added before the addition of lime, which reduces the pH of RO permeate and facilitates a faster dissolution of lime in the RO permeate.

Addition of lime not only provides alkalinity and hardness to the RO permeate but also increases pH of the product water. Since lime dosage is primarily driven by the target calcium hardness of the product water, if such target is relatively high (i.e., 100 to 150 mg/L), the mix of lime and RO permeate may have a pH beyond the typical target range of 7.6 to 8.3. Addition of sulfuric acid to the lime-conditioned RO permeate allows to bring the pH of the finished water down into the target range. In addition, pH reduction helps to combat residual turbidity caused by small amounts of lime that is not completely dissolved in the RO permeate.

15.2 Remineralization Systems

Remineralization of desalinated water is typically completed by three main groups of technologies: (1) processes that are based on direct addition of chemicals containing calcium (i.e., calcium hydroxide/lime) and magnesium (i.e., magnesium chloride and sulfate), (2) processes where remineralization is accomplished by mixing of desalinated water with a portion of the source water used for desalination, or with other fresh water sources with high calcium and magnesium content, and (3) processes where magnesium and/or calcium are added by dissolving naturally occurring minerals such as limestone (calcium carbonate/calcite) and dolomite (calcium and magnesium carbonate).

15.2.1 Remineralization by Chemical Addition

Calcium Addition

Most seawater and brackish water desalination plants typically add calcium to the desalinated water in the form of lime or calcite.

Over 90 percent of the existing seawater desalination plants worldwide use a sequential feed of calcium hydroxide (hydrated lime) and carbon dioxide to supply hardness and alkalinity to the product water needed to protect distribution system and household plumbing from corrosion. The remineralization using lime and carbon dioxide follows the chemical reaction presented below:

$$2CO_2 + Ca\,(OH)_2 \rightarrow Ca\,(HCO_3)_2 \qquad (15.3)$$

Based on the Eq. (15.3), 0.74 mg/L of hydrated lime (0.56 mg/L of quicklime) and 0.88 mg/L of carbon dioxide would need to be added in order to increase desalinated water hardness and alkalinity by 1.0 mg/L (as $CaCO_3$) each. Therefore, for a target recommended dosage of alkalinity and hardness in the product water of 100 mg/L, the water produced by the desalination system will need to be treated with 74 mg/L of lime and 88 mg/L of carbon dioxide.

The lime product used for remineralization is usually delivered and stored at the desalination plant site in silos as either powdered hydrated lime or as pebble-lime (CaO), which is than slaked to generate hydrated lime [$Ca(OH)_2$]. In smaller desalination plants, powdered lime is often stored in 25-kg bags. Hydrated lime is fed into lime saturators in the form of lime slurry (milk of lime). This lime slurry is blended with the fresh water produced by the desalination process and is thoroughly mixed in lime saturator tanks to create saturated limewater, which is then injected into the unconditioned desalinated water (Fig. 15.2).

Figure 15.3 depicts the lime storage silos and saturator of the 272,000 m^3/day (72 mgd) groundwater replenishment reverse osmosis plant located in Orange County, California. The lime saturator is shown on the left, while the lime silos can be seen on the right.

Carbon dioxide is typically delivered to the desalination plant site in a liquefied form and is stored under pressure in metal storage tanks (Fig. 15.4). In thermal desalination plants, however, carbon dioxide released from the source water can be recycled and reused for the remineralization process described above instead of adding commercial carbon dioxide product. A solution of carbon dioxide and water (carbonic acid) is injected into the desalinated water downstream of the point of introduction of the saturated limewater.

The majority of desalination plants are designed to produce finished water of total (calcium and magnesium) hardness of 80 to 120 mg/L as $CaCO_3$. Since desalinated water usually contains less than 2 mg/L of magnesium, over 90 percent of the total hardness of drinking water conditioned with lime is calcium hardness.

When adding lime to desalinated water, it is important to keep in mind that the solubility of calcium carbonate is dependent upon pH, temperature, and ionic strength. Lime may not dissolve easily and may add residual turbidity of 0.05 to 0.5 NTU (or higher) to the finished water, which is a disadvantage of this type of remineralization process.

Lime-based remineralization typically requires the addition of acid (e.g., carbonic or sulfuric acid) to enhance lime solubility and to produce finished water with desired hardness and calcium carbonate precipitation potential adequate to provide corrosion protection. If the desalinated water is too warm [i.e., its temperature is 25°C (77°F) or higher],

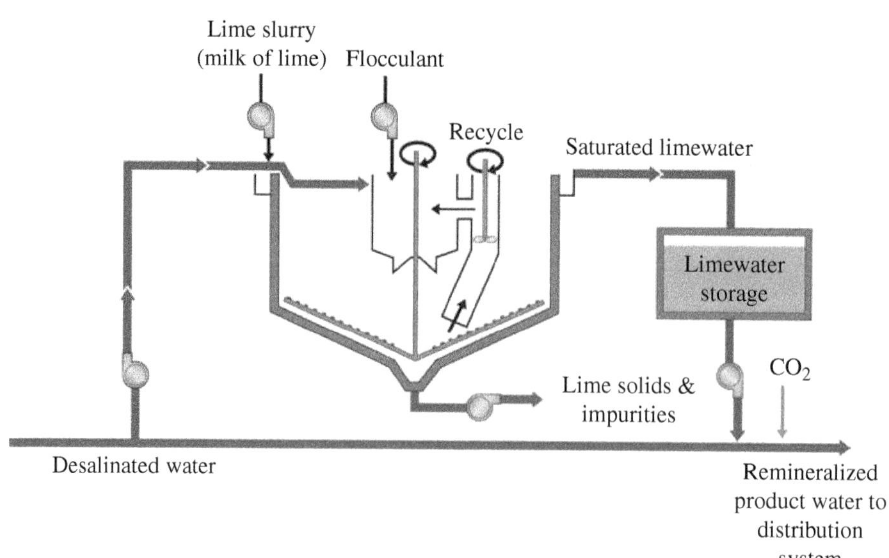

Figure 15.2 Schematic of typical lime/carbon dioxide addition system.

Figure 15.3 Lime storage silos and saturator.

Figure 15.4 Carbon dioxide storage tanks.

the rate of lime solubility slows down, and therefore acid addition for pH reduction to enhance lime solubility would be beneficial.

There are a few methods to enhance solubility of slaked lime in water with temperatures higher than 25°C (77°F). One method is to introduce multiple points for carbonic acid and/or sulfuric acid injection into the desalinated water and to use a separate lime contact chamber that creates highly turbulent conditions and provides contact times of 5 to 10 minutes or more. Another approach used to enhance solubility of lime in relatively warm plant desalinated water is turbulent mixing of the lime suspension with the plant desalinated water in the product water storage tank using large recirculation pumps. This approach, however, is cost-effective only if the unit power cost is relatively low (i.e., $0.03 to 0.05/kWh).

Magnesium Addition

While it is acceptable for total hardness to be added only as calcium hardness for the purpose of protecting the water distribution system against corrosion, such water provides somewhat limited human health protection and is of lesser agricultural value. Therefore some countries, such as Cyprus and Israel, are currently practicing or planning for addition of magnesium to desalinated water. At present, in desalination plants magnesium is added as a commercially available food-grade product of magnesium sulfate or magnesium chloride.

15.2.2 Remineralization by Mixing of Desalinated and Source Waters

Minerals, including calcium and magnesium, could be added to desalinated water by blending it with seawater or brackish source water. This practice is frequently used for both brackish water reverse osmosis plants and thermal desalination plants and is acceptable only when the source water is of high quality and is pretreated appropriately—and when the blend meets all applicable water quality standards.

When desalinated water and pretreated source seawater are blended, the amount of source seawater is typically limited to 1 percent or less due to taste and other water quality considerations. For example, blending ratio of 1-to-99 of Pacific Ocean water of salinity of 33,500 mg/L, and desalinated water produced by a single pass seawater reverse osmosis system of salinity of 220 to 350 mg/L would yield finished water with calcium content of 4 to 5 mg/L and magnesium of 19 to 20 mg/L. However, this water will have relatively high TDS content (550 to 680 mg/L) as well as high sodium (180 to 230 mg/L) and chloride (310 to 370 mg/L) concentrations. A high level of bromide in seawater (typically in a range of 60 to 90 mg/L) may also lead to excessive formation of disinfection byproducts in the finished water, which are considered carcinogenic and are therefore regulated by the US Environmental Protection Agency, the WHO and other regulatory agencies worldwide. Because of these implications, blending of permeate produced by SWRO plants with source seawater is typically not practiced. However, blending of low-salinity brackish water and desalinated water is a common practice in many parts of the world—including in the United States, the Middle East, and Europe.

The saline source water used for blending must be treated prior to its mixing with the desalinated water. The type and complexity of source water treatment depend on its quality. At a minimum, the saline source water used for desalinated water remineralization has to be filtered through cartridge filters. Enhanced source water treatment such as granular activated carbon filtration is recommended for source water exposed to potential contamination from excessive algal growth, surface runoff, or other human-made sources of elevated organics or turbidity in the water. Pretreatment chemicals (such as acid) may need to be added, depending on where the blending source water is split from the feed water piping.

15.2.3 Remineralization by Dissolving Minerals in Desalinated Water

Calcium Addition: Limestone (Calcite) Contactors

Limestone is a natural mineral made of calcite (calcium carbonate). Processing water through calcite media dissolves this calcium source and, in reaction with carbon dioxide, adds calcium hardness and bicarbonate alkalinity to the product water. Remineralization using calcite and carbon dioxide follows the following chemical reaction:

$$CO_2 + CaCO_3 + H_2O \rightarrow Ca(HCO_3)_2 \tag{15.4}$$

Based on the Eq. 15.4, 1.00 mg/L of calcite and of 0.44 mg/L of carbon dioxide would need to be added in order to increase the desalinated water hardness and alkalinity by 1.0 mg/L (as $CaCO_3$) each. Therefore, for a target recommended dosage of alkalinity and hardness in the product water of 100 mg/L, the water produced by the desalination system will need to be treated with 100 mg/L of calcite and 44 mg/L of carbon dioxide.

Comparison of Eqs. 15.3 and 15.4 indicates that remineralization using calcite requires two times less carbon dioxide. While only 0.74 mg/L of lime versus 1.0 mg/L of calcite is required to add 1.0 mg/L of alkalinity and hardness, the cost of lime usually is over two times higher than that of calcite. In addition, since lime is typically produced from high-temperature treatment of calcite, the carbon footprint of producing 1 kg of lime is several times higher than that associated with the production of calcite suitable for drinking water applications. Therefore, the use of calcite is usually more cost effective and environmentally palatable than the application of lime for remineralization.

An alternative to the direct use of carbon dioxide is the addition of sulfuric acid, which indirectly creates the carbon dioxide needed for formation of bicarbonate alkalinity following the chemical reactions described below:

$$CaCO_3 + H_2SO_4 \rightarrow CaSO_4 + H_2O + CO_2 \tag{15.5}$$

$$CaCO_3 + CO_2 + H_2O \rightarrow Ca\ (HCO_3)_2 \tag{15.6}$$

In summary:

$$2\ CaCO_3 + H_2\ SO_4 \rightarrow Ca\ (HCO_3)_2 + CaSO_4 \tag{15.7}$$

As seen from the comparison of the process Eqs. 15.3 and 15.7, the use of sulfuric acid instead of carbon dioxide would require addition of two times more calcite to provide the same amount of alkalinity to the finished water, because one half of the added calcite is used in the formation of calcium sulfate. However, sulfuric acid is often preferred because it reduces desalinated water pH more easily to practically any level needed for desalinated water enrichment with calcium, and because the rate of solubility of calcite with sulfuric acid is much higher than that associated with using carbon dioxide. This allows treating only a fraction of the desalinated water flow through the calcite filters and reducing the size of these filters. However, if the unit chemical costs for sulfuric acid are significantly higher than those of carbon dioxide, than the trade-off between higher capital costs for larger calcite contactors and lower O&M expenditures may warrant the use of carbon dioxide instead of sulfuric acid.

Another advantage of using sulfuric acid for enhanced calcite solubility is that this process yields a dissolved calcium to alkalinity concentration ratio of 2:1 (or higher) measured in equivalent units, while the use of carbon dioxide would yield to only 1:1 calcium to alkalinity ratio. Using sulfuric acid for calcite solubility increases the sulfate (SO_4^{2-}) concentration of this water. This may be advantageous if the water is intended for agricultural irrigation because typically desalinated water has significantly lower sulfate content than most surface water sources.

A schematic of typical calcite contactor system for remineralization is presented in Fig. 15.5. Most existing desalination plants with low pH target for finished water (i.e., pH of 7.5 or less) are usually designed to process 100 percent of the RO desalinated water through the calcite filters, especially when the cost of sulfuric acid or carbon dioxide used for desalinated water acidification is relatively high. However, for desalination plants with higher target pH range of the finished water (8.0 to 8.5) it usually is more cost effective to treat a portion (20 to 50 percent) of the entire desalination flow through the calcite contactors and to blend this highly saturated stream with the remaining desalinated water in order to achieve target hardness, alkalinity, and pH in the finished product water.

In a typical calcite contactor remineralization system, the pH of the desalinated water is first reduced down to 4.5 or less by the addition of carbon dioxide or sulfuric acid; this water is then conveyed through a filter bed composed of calcite granules (referred to also as beads, grains, or pebbles) at a contact time of 10 to 30 minutes, thereby achieving target hardness and alkalinity levels in the finished water. The calcium concentration of the remineralized water can be reliably controlled by adjusting the acid dosage of the desalinated water fed to the calcite contactor.

Calcite contactors are proven technology and have found implementation at the 330,000 m^3/day (86 mgd) SWRO desalination plant in Ashkelon, Israel (Fig. 15.6), which, at present, is one of the largest operational membrane desalination facility in the world.

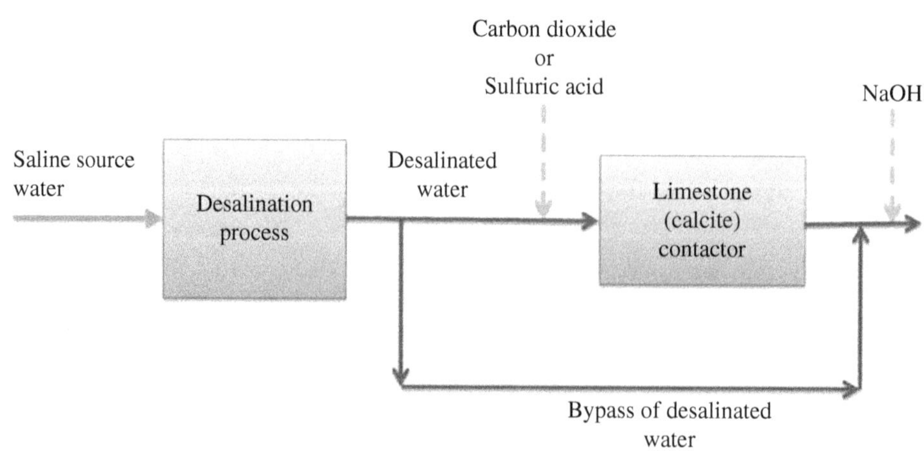

Figure 15.5 Schematic of typical limestone (calcite) contactor remineralization system.

Figure 15.6 Calcite contactors of Ashkelon desalination plant, Israel.

This facility only processes approximately 25 percent of the RO permeate through the limestone contactor, saturates the content of calcium carbonate content of the processed RO permeate to 400 to 450 mg/L by pH reduction of the permeate down to 3 to 4, and then blends this saturated RO permeate with the rest (75 percent) of the RO permeate flow to achieve a target final hardness in a range of 80 to 100 mg/L.

Calcite contactors are also used at the 60,000 m^3/day (16 mgd) Larnaka SWRO plant in Cyprus and the 200,000 m^3/day (53 mgd) Barcelona SWRO plant. Calcite for water remineralization has been successfully applied at a number of other desalination plants worldwide.

The largest thermal desalination plant using calcite contactors for remineralization of desalinated water is located in Bahrain [the 340,000 m^3/day (90 mgd) Hidd phases 1 and 3 plant]. This plant employs pressure-driven calcite contactors divided in three parallel treatment trains with 14 contactors per train (42 contactors in total). This plant uses CO_2 to lower the pH of the distillate prior to filtration through the calcite vessels. Sodium hydroxide is added as a final step of the remineralization process to adjust the pH of the finished water to the target level. An interesting challenge associated with the operation of this remineralization system was the relatively high content of organic and particulate residues in the natural limestone used for remineralization, which resulted in intermittent episodes of increased TOC and turbidity levels of the finished water. Such performance challenges were resolved by using higher-quality limestone and modifying the flushing procedures for the calcite contactors.

As compared with the lime-based remineralization systems, calcite contactor systems are usually less costly in terms of capital and chemical expenditures, require less carbon dioxide, and typically produce lower turbidity finished water. However, in many locations worldwide, high-quality food-grade calcite is not as readily available as lime, which is one of the main reasons why this technology has not been used as frequently as lime/carbon dioxide conditioning to date. Therefore, most of the existing large seawater desalination plants in Australia, the United States, the Middle East, Spain, and North Africa have adopted conditioning of desalinated water using a combination of lime and carbon dioxide.

Calcium and Magnesium Addition: Dolomite Contactors

Dolomite is a natural mineral that contains calcium and magnesium carbonate. Passing desalinated water through a dolomite contactor, similar in configuration and design to calcite contactors, adds both minerals to the finished product water. In nature, the dolomite rock mineral is often nonhomogenous and is interbedded with limestone. Therefore, the exact quality of the product water is more difficult to predict than when using pure limestone.

Solubility of dolomite to enrich desalinated water with an adequate amount of calcium and magnesium requires pH reduction of this water to less than 4.5 by adding sulfuric acid or carbon dioxide. Because magnesium does not dissolve well in water with pH higher than 5.5, the pH adjustment to a target level of 8 to 8.5 needed for corrosion protection often results in relatively low level of carbonate alkalinity and of magnesium in the water as compared with calcium hardness. For example, in order to achieve a target magnesium level of 10 mg/L, the calcium hardness of this water has to be several times higher than necessary. Therefore, if the water quality regulations applicable for a given desalination project have a maximum limit for total hardness in the water, , the use of dolomite contactor may not be viable.

Water remineralization with dolomite is less feasible than using limestone for several reasons: (1) usually dolomite is more costly and less readily available than limestone, (2) solubility rate of dolomite is over three times slower than that of limestone, which necessitates the dolomite contactors to be significantly larger in size and therefore more costly to construct and install, and (3) dolomite is naturally more nonhomogenous than limestone, which requires dolomite-based remineralization systems to be designed with higher contingency, a factor that additionally increases the application costs of this technology.

Because of the process constraints listed above and the limited availability of food-grade dolomite, this mineral has not been used for remineralization of desalinated water to date. It should be pointed out however, that dolomite has found full-scale applications for treatment of soft, fresh surface water.

Another reason for the limited application of dolomite remineralization contactor system is that using this technology for combined calcium and magnesium addition will be more costly than traditional lime/carbon dioxide/magnesium sulfate process described previously.

15.3 Design Considerations for Lime Feed Systems

Lime feed systems consist of lime storage silos; slurry preparation equipment; and lime saturators. The main purpose of these systems is to generate homogenous lime solution which is introduced and mixed into the RO permeate to increase its alkalinity and hardness to target product water levels.

15.3.1 Key Design Criteria

Lime Silo and Slurry System

As indicated previously, typically lime is delivered to the desalination plant site as bulk powdered hydrated lime, which is stored on site in cylindrical metal silos with a 60° cone bottom (lime bins). These silos are often the tallest structures on the desalination plant site and usually have a diameter between 2.0 and 6.0 m (6.6 and 20.0 ft) and a height of 5.0 to 15.0 m (16.0 to 50 ft).

The silos are designed to store lime for 15 to 60 days and are reloaded by delivery trucks suited with pneumatic conveyance equipment. They are equipped with air vent filters and with bin vibrators to promote flow of lime from the silo into a transition bin hopper. The feed of lime is controlled by a rotary discharge (airlock) valve, which is motorized and closed or opened automatically based on the level of lime in the hopper.

A volumetric feeder draws from the transition bin hopper and delivers lime proportionally to the pH/flow into a slurry (milk of lime) mixing tank where the dry powdered lime is mixed with desalinated water or process water to generate lime slurry. This lime slurry is then conveyed to the feed line of the lime saturators using slurry pumps/eductors. In order to prevent the scaling and accumulation of the slurry in the conveyance pipelines, this slurry is transferred to the lime saturator at very high velocity.

Lime silos are typically welded/bolted stainless-steel structures equipped with level sensors, a fill line with long-radius elbows, and a truck hose loading fitting with dust cap. The minimum lime silo capacity is 50 m^3 (13,210 gallons). The maximum lime silo capacity is often determined by the height of the maximum size of the silos. Key design criteria for a typical lime silo and slurry system are presented in Table 15.2.

Lime Saturator System

Lime slurry is pumped from the slurry tank to the lime saturation system by a progressive cavity type of pumps. The lime saturation system is also fed with permeate, which is used to dissolve and dilute the lime slurry from 10 down to 0.1 percent. In addition, polymer is frequently added at a dosage of 0.5 to 1.0 mg/L to the feed water in order to reduce the turbidity of the limewater. The lime saturator system consists of saturator tanks, recirculation pumps or mixers, saturated limewater tanks, and limewater dosing pumps.

Lime saturators are thickener-clarifier tanks and their main purpose is to produce homogenous and fully dissolved lime solution (saturated limewater) and to remove the solid lime impurities inherently contained in the commercial lime product. Typically, these facilities are constructed as circular metal tanks that consist of a central feed well (reaction zone) to dissolve and mix the lime slurry with the desalinated water, and a settling zone where the undissolved lime material and impurities are separated from the saturated lime water and removed by a bottom scraper and hopper (see Fig. 15.2).

Lime saturators are equipped with mechanical propeller-operated (turbine) mixers or pumps with variable speed drives, which recirculate lime slurry from the tank bottom to the reaction zone in order to accelerate the lime solubility process. Lime slurry is fed continuously into the reaction zone and mixed to enhance and accelerate the solubility process. Key design criteria of a typical lime saturator system are summarized in Table 15.3.

Component/Parameter	Specifications/Design Criteria
Lime dosage (as 100% pure product)	0.74 mg/L per 1.0 mg/L of target alkalinity and hardness concentrations (as $CaCO_3$)
Lime consumption (kg/day as 100% pure product) per 1000 m^3/day of desalinated water for addition of alkalinity and hardness in a range of 80 to 120 m/L (as $CaCO_3$)	59.2–88.8
Lime purity, %	85–94
Silo vessel • Diameter, m • Height, m • Storage time, days • Structure–material type	 2–6 5–15 15–60 Coated carbon or stainless steel
Silo vent filter • Material • Type	 Stainless steel cartridge-type with polyester felt cartridges
Lime flow facilitating equipment	Bin vibrator fluidized air pads
Lime slurry • Concentration, % • Slurry tank retention time, h • Pumps	 10 3–5 Progressive cavity, equipped with flushing system

Note: 1 m = 3.3 ft

TABLE 15.2 Lime Silo and Slurry System

The maximum lime solubility in water is approximately 1500 mg/L and saturated limewater is designed to be at concentration of 900 to 1400 mg/L (typically 1000 mg/L). The saturated limewater is collected in overflow launders at the top of the saturator tanks from where it flows by gravity to a limewater tank or tanks. The volume of the limewater tank should be selected such that it provides storage for 18 to 24 hours of the limewater needed to condition the entire daily volume of desalinated water. This volume allows ample time to periodically take the lime saturator out of service for cleaning. Limewater is pumped to the dozing point/s along the length of the product water pipeline using centrifugal pumps equipped with variable frequency drives.

Lime solids are collected at the bottom of the settling zone. A portion of these solids is recycled to the feed well to enhance solubility and particle flocculation and settling. Polymer is often added to accelerate the flocculation process and to minimize the carry-over of fine solids into the saturated limewater because these solids contribute turbidity to the final product water. Polymer addition also assists in increasing the solids concentration of the lime bed, which, in turn, reduces the volume of the sludge generated by the lime saturator system.

Component/Parameter	Specifications/ Design Criteria
Lime slurry feed concentration, %	10
Lime slurry feed volume	Estimated for the target lime dosage at 10% concentration
Saturator solids loading rate, kg lime/m^2·h	1.5–4.0
Saturator hydraulic loading rate, m^3/m^2·h	0.8–1.8
Saturator retention time, min	100–160
Saturator diameter, m	2–20
Saturator depth	
• Depth of water, m	3.0–4.5
• Total depth, m	3.5–5.5
Retention time in reaction zone, min	8–15
Reaction zone diameter, % of total lime saturator diameter	25–35
Reaction zone height, % of depth of water	75–85
Limewater storage tank	
• Total retention time, h	18–24
• Limewater concentration, mg/L	900–1400

Note: 1 m = 3.3 ft ; 1 kg/m^2.h = 4.92 lb/ft^2.day; 1 m^3/m^2. h = 0.41 gpm/ft^2

TABLE 15.3 Lime Saturator System

15.3.2 Design and Operational Issues

Lime Purity

Depending on its quality and origin, commercially available lime products usually have a purity of 85 to 94 percent. The actual lime purity is an important factor to consider when assessing the finished desalination plant product water quality.

For example, if a lime product of 90 percent purity is used, the addition of 74 mg/L of 100 percent pure lime needed to produce water of 100 mg/L of alkalinity and hardness (as $CaCO_3$) would also result in the introduction of 8.2 mg/L of impurities, typically in the form of suspended solids. Unless these impurities are completely removed by the lime saturator tanks, they will find their way into the plant finished product water and will increase its turbidity/suspended solid content.

Use of lime of highest commercially available purity (typically 94 percent) is recommended, despite of its relatively higher costs because of the numerous negative implications associated with removing and disposing of the solids associated with the lime impurities. These solids tend to accumulate in the lime feed system pipelines and equipment and must be removed periodically because they could result in equipment malfunction over time. In addition, the costs of treatment and disposal of solids generated by lime impurities could be significant.

For example, a 40,000 m^3/day (10.6 mgd) desalination plant, which is using lime of commercial grade purity of 85 percent and is adding 74 mg/L of lime as a pure product (87.1 mg/L of 85 percent purity commercial product), will generate 191 dry tons/year of impurities, which would need to be separated and disposed of. The use of high (94 percent) purity lime instead will reduce this solids content to only 69 tons/year.

Maintenance

Dosing lines for saturated limewater are exposed to potential blockages from lime solids buildup and therefore are recommended to be equipped with a flushing system and to be flushed every one to two weeks with carbon dioxide saturated water. Dosing valves are recommended to be exercised weekly in order to prevent buildup.

Saturators have to be monitored routinely and desludged as needed. The turbidity of the limewater from well-operated saturators typically does not exceed 30 to 35 NTU. The sludge solids concentration is usually between 10 and 15 percent.

Elevated limewater turbidity is an indication for potential buildup of sludge in the tanks and/or lime bed turnover, which may be triggered by colder temperatures. Such turbidity increase may also be caused by buildup of lime in the mixer lift piping over time. Therefore, the mixer system would need to be cleaned once every 6 to 12 months.

Saturator launders also accumulate lime over time and need periodic cleaning. In order to accommodate their routine maintenance, lime saturators have to be designed with an access platform for launder cleaning.

15.4 Design Considerations for Carbon Dioxide Feed Systems

Carbon dioxide is used in addition of alkalinity to desalinated water in combination with both lime and limestone and for pH adjustment to enhance the limestone solubility process. As indicated previously, 0.88 mg/L of carbon dioxide would need to be added to the desalinated water to increase its alkalinity by 1 mg/L (as $CaCO_3$).

15.4.1 Key Design Criteria

Carbon dioxide is delivered to and stored at the desalination plant in compressed liquid form [20 bars (290 lb/in^2) of pressure at 20°C (68°F)]. The liquid chemical is converted into gaseous form in evaporators at 3 bars (43.5 lb/in^2) at 12°C (54°F), dissolved in carrier water of 4 to 5 bars (58 to 73 lb/in^2), and the carbonic acid solution is injected into the desalinated water.

A typical carbon dioxide feed system consists of liquid carbon dioxide storage tanks equipped with refrigeration units, vaporizers, vapor heaters, carbon dioxide dosing system with pressurized gas solubility panel, and an in-line gas feed diffuser. Table 15.4 summarizes key design criteria for a typical carbon dioxide feed system.

15.4.2 Design and Operational Issues

For operational flexibility, a carbon dioxide system is recommended to be designed with two points of delivery: one upstream of the lime feed system for pH adjustment and one after the point of addition of lime. For limestone or dolomite contactors, carbon dioxide is often added to reduce the pH of the desalinated water and thereby to accelerate their solubility.

Component/Parameter	Specifications/Design Criteria
Carbon dioxide dosage (as 100% pure product)	For lime/CO_2 systems–88 mg/L For calcite/CO_2 systems–44 mg/L per 1.0 mg/L of target alkalinity and hardness concentrations (as $CaCO_3$)
CO_2 consumption (kg/day as 100% pure product) per 1000 m^3/day of desalinated water for addition of alkalinity and hardness in a range of 80 to 120 m/L (as $CaCO_3$)	70.4–105.6
CO_2 purity, %	100
Liquid CO_2 storage tank (refrigerated)	
• Storage pressure/temperature, bars/°C	20 bars/–20°C
• Diameter, m	0.8–3.0
• Length/height, m	5–15
• Storage time, days	15–60 days
• Structure material type	Carbon steel
Vaporizer	
• Type	Electric
• Capacity, horsepower/kg CO_2.day	0.03–0.05
• Delivery pressure/temperature, bars/°C	3 bars/12°C
Vapor Heater	
• Type	Electric
• Capacity, horsepower/kg CO_2.day	0.01–0.02

Note: 1 m = 3.3 ft; 1 bar = 14.5 psi; 1 kg = 2.2 lb

TABLE 15.4 Carbon Dioxide Storage and Feed System

The dosage of carbon dioxide fed to the desalinated water is adjusted proportionally to the desalinated water flow rate by flow control valves installed on the gas feed lines from the vaporizer to the point of chlorine dioxide gas injection in the desalinated water conveyance pipe.

15.5 Design Considerations for Limestone (Calcite) Contactors

Limestone (calcite) contactors are well suited for remineralization of desalinated water produced by all types of desalination processes (thermal, RO, and EDR desalination). Typically they are contactors designed to operate in an upflow mode and are configured similar to gravity- or pressure-driven granular media filters. Because of the low solubility of calcite at the near-neutral pH of the desalinated water, the addition of adequate calcium concentration requires a pH reduction of permeate to usually less than 4.5 (corresponding to CCPP of at least 200 mg/L as $CaCO_3$) before the water enters the contact tanks.

15.5.1 Calcite Contactor Configuration

A typical calcite contactor system consists of calcite bed with a supporting layer and a feed water distribution system, feed (inlet) line, backwash system, feed (inlet), and outlet and overflow lines (Fig. 15.7). The desalinated water is introduced through the distribution system located at the bottom of the contactor and saturated with calcium carbonate as it flows upward through the calcite bed.

In order to enhance the solubility process, the pH of the feed water is reduced typically below 4.5, and, as the water travels upward, its pH, alkalinity, and hardness increase. The pH, alkalinity, and hardness of the conditioned water depend on a number of factors, including the initial pH, CO_2 content, alkalinity, TDS, water temperature, upflow velocity, and the contact time the reactor.

Distribution System

The feed water distribution system is located at the bottom of the tank and has the same configuration and design as those of an upflow granular media filter. Some designs have infiltration platforms with nozzles distributed uniformly on its surface for even flow distribution.

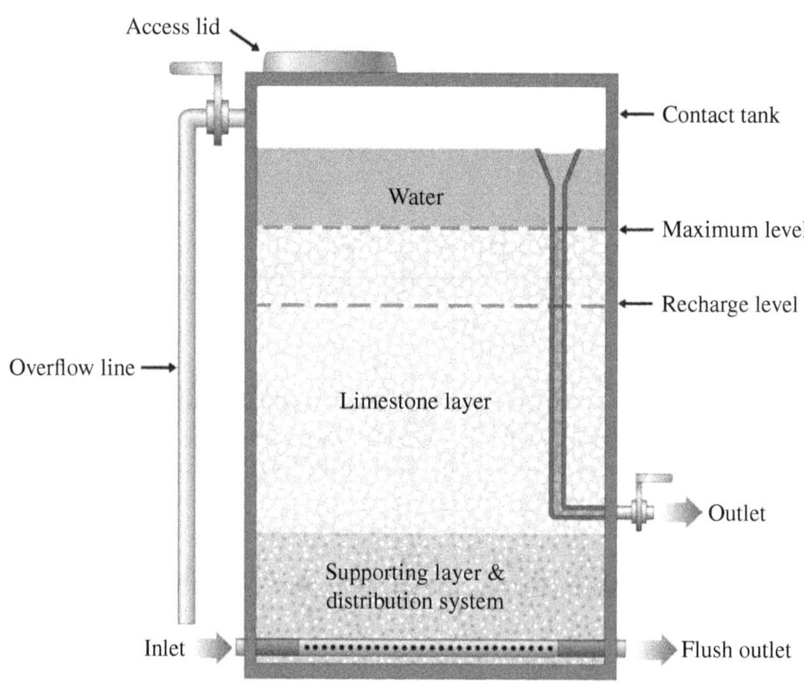

Figure 15.7 Calcite contactor schematic.

Supporting Layer

The supporting layer is usually 0.15 to 0.35 m (0.5 to 1.1 ft) deep and consists of gravel, granite aggregate, or other inert materials. The size of the support media should be larger than the size of the limestone.

Limestone (Calcite) Bed

The calcite bed consists of either crushed limestone pebbles or limestone granules (grains) with a diameter of 1.5 to 2.5 mm (0.06 to 0.1 in). Table 15.5 provides recommendations for the quality of limestone to be used in calcite contactors.

The depth of the contact zone of the calcite bed is usually between 1.5 and 3.0 m (5 to 10 ft) and is a function the source water temperature, pH, the target mineral content of the conditioned water, and the calcite grain size. The contact zone portion of the limestone bed is the volume of calcite between the recharge level and the surface of the support layer (Fig. 15.7).

Since limestone is consumed during the remineralization process, an additional working volume of calcite has to be added on the top of the contact zone. This calcite bed working (reload) zone is denoted in Fig. 15.7 as the zone between the maximum and recharge levels of calcite. The volume of this zone is determined based on the daily consumption of calcite and the time between two calcite bed-reloads (typically two to eight weeks).

Calcite Media Consumption and Working (Reload) Volume

As indicated previously, 1 mg/L of calcite (as $CaCO_3$) is needed to add 1 mg/L of hardness to the desalinated water. If the amount of the desalinated water contains a measurable amount of calcite, the additional calcite that has to be added could be reduced. For example, permeate produced by a single-pass RO system from Pacific Ocean seawater usually contains a minimum of 0.6 mg/L of calcium ion. Taking into consideration that 1 mg/L of calcium ion equates to 2.493 mg $CaCO_3$/Ca, the actual available content of $CaCO_3$ in the desalinated water would be only 0.6 mg/L × 2.943 = 1.8 mg/L as $CaCO_3$.

If the target hardness of the desalinated water is 100 mg/L (as $CaCO_3$), then the actual maximum amount that would be consumed daily is 100 mg/L − 1.8 mg/L = 98.2 mg/L of calcite as 100 percent pure product. At calcite purity of 99 percent, the quantity of the

Parameter	Value
Grain size, mm	1.5–2.5
Purity, %	99 or more
Iron oxide content, %	Less than 0.1
SiO_2, %	0.3
Al_2O_3, %	0.1
MgO, %	0.1
Hardness, Mohs	3
Specific weight, tons/m^3	1.5

Table 15.5 Recommended Calcite Grain Specifications

commercially available chemical used will be 98.2/0.99 = 99.2 mg/L (g/m^3). The volume of the calcite media, which has to be reloaded, is calculated based on the quantity of daily calcite consumption, the specific weight of the product (Table 15.5), and time between two media reloads.

Conditioned Water Zone

This is the zone of conditioned water between the limestone layer and the outlet overflow of the filter. Usually this zone is 0.6 to 1.0 m (2 to 3 ft) high.

Limestone Feed Zone

The limestone bed has to be replenished periodically because it is dissolved during the remineralization process. Limestone media is typically refilled from the top of the filters (Fig. 15.8). Refilling is completed manually if limestone is delivered in 55 lb (25 kg) packages (typical for small plants) or the bed is reloaded using a gentry crane installed above the filter cells if the limestone is packaged in 1.1-ton bags (common for large plants).

Empty Bed Contact Time

The hydraulic residence time of the desalinated water in the limestone bed is a key design parameter for this system. Empty bed contact time (EBCT) is calculated by dividing the volume of the limestone layer (bed) by the feed flow rate. EBCT is typically in a range of 10 to 30 minutes, depending on the temperature, CO_2, and HCO_3^- content of the desalinated water.

The needed EBCT increases with decrease in desalinated water temperature. In order to provide adequate product water quality at all times, this contact time should be sized for the daily minimum temperature of the desalinated water and for daily maximum product water flow.

Figure 15.8 Refilling of calcite media.

Surface Loading Rate

The typical average design surface loading rate is in a range of 4 to 8 $m^3/m^2 \cdot h$ (1.6 to 3.2 gpm/ft^2), and the maximum rate is 10 $m^3/m^2 \cdot h$ (4.1 gpm/ft^2) or higher. Surface loading rates typically result in elevated turbidity of the remineralized product water.

Calcite Bed Backwash System

It is recommended to backwash the limestone contactors weekly in order to maintain consistent product water quality. Backwash is completed through the distribution system and is recommended to include the following sequence: (1) air purge for three to five minutes at a rate of 60 $m^3/m^2 \cdot h$ (24 gpm/ft^2), (2) air and water backwash for 5 to 10 minutes at air rate of 60 $m^3/m^2 \cdot h$ (24 gpm/ft^2), and water rate of 10 to 12 $m^3/m^2 \cdot h$ (4 to 5 gpm/ft^2), and (3) water backwash only for 10 to 15 minutes at rate of 15 to 25 $m^3/m^2 \cdot h$ (6 to 10 gpm/ft^2).

In addition, it is recommended to down-flush the filter bed at least once per month to remove fine dust and waste material introduced during filter cell loading. Calcite filter cells should be provided with flush outlet for this purpose (see Fig. 15.7).

15.5.2 Design and Operational Issues

Calcite Grain Size Impact on Calcite Bed Depth

It is important to note that calcite grain size has an impact on the needed depth of the contact zone of the calcite bed, especially if limestone grains are larger than 2 mm (0.08 in). The required depth of the contact zone increases proportionally with the increase of the calcite grain size.

For example, if calcite grains of 3 mm (0.12 in) are used instead of 2.5 mm (0.1 in), the depth of the contact zone of the calcite bed would need to be increased approximately 30 percent. Similarly, the required height of the contact zone of the bed could be reduced by approximately 25 percent if calcite grain size is reduced from 2.5 down to 2.0 mm (0.1 down to 0.8 in). Further grain reduction to 1.5 mm (0.06 in) or less, however, has lesser impact on the needed depth of the calcite bed. Therefore the optimum size of the calcite grains is between 1.5 and 2.0 mm (0.06 and 0.08 in).

Calcite Quality Impact on Performance

Practical experience indicates that calcite contactor turbidity is closely related to the type of calcite used and the surface loading rate of the calcite filters. If well washed calcite is used (i.e., calcite that contains less than 1 percent fines), the product water turbidity is not affected until the surface loading rate of the filters exceeds 14 $m^3/m^2 \cdot h$ (6 gpm/ft^2). For calcite with higher content of fine particles (2 percent or more), the maximum surface loading rate of the calcite contactor cells under which the finished water quality is not affected is 11 to 12 $m^3/m^2 \cdot h$ (4.5 to 5.0 gpm/ft^2).

Refilling of Calcite Cells

Contactor cells should be filled with calcite only when water is present in the cells because, otherwise, the calcite grains may be compacted and form clusters, which will hinder the solubility process.

Calcite Media Replacement

Over time calcite media in the contact bed deteriorates and accumulates surface scale, and therefore it would have to be replaced completely once every 5 to 10 years.

Operations Monitoring

At a minimum it is recommended to monitor the pH of the remineralized water frequently in order to ensure that the design equilibrium pH is maintained at all times. A steady decrease of pH usually is an indication for the need to reload calcite. Another important monitoring parameter is the calcite bed media headloss. Over the course of the remineralization process, the filter headloss increases steadily and, once it reaches its maximum design level, the calcite bed would need to be backwashed.

15.5.3 Calcite Beds with Continuous Feed

Description

DrinTec Solutions (Spain) has developed and recently implemented an innovative upflow calcite reactor with continuous feed of limestone into the calcite bed for remineralization of permeate produced of the SWRO system of the 200,000 m^3/day (53 mgd) Barcelona desalination plant in Spain. A cross-section of such a calcite bed cell is shown in Fig. 15.9.

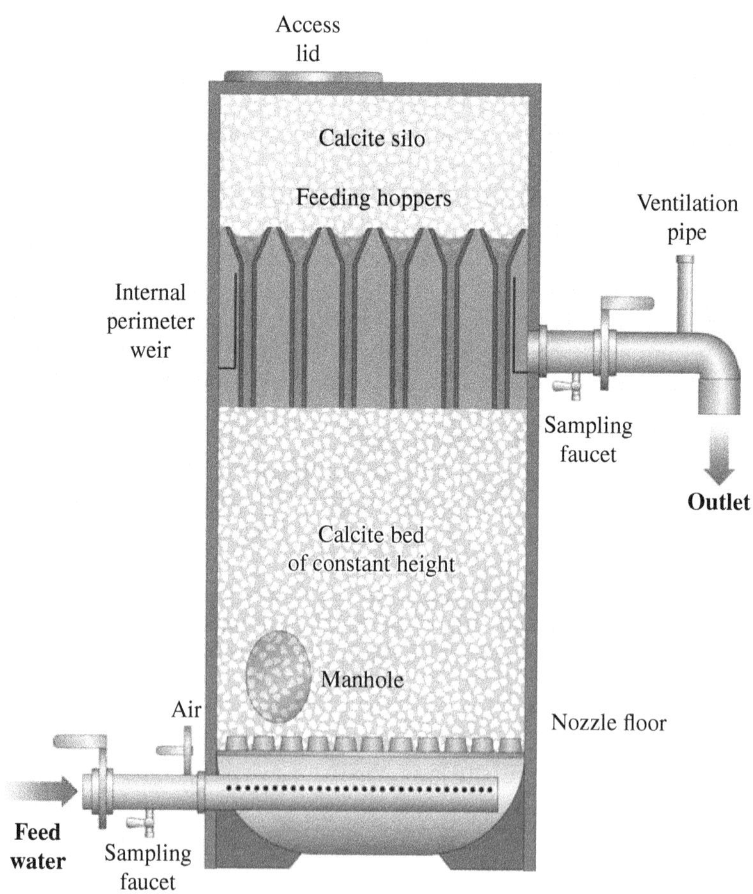

FIGURE 15.9 Cross-section of DrinTec calcite contactor cell.

The general configuration of the calcite bed in Fig. 15.9 is similar to that of typical calcite contactors (see Fig. 15.7). The desalinated water is fed at the bottom of the calcite bed. However, as compared with conventional calcite contactors, which are reloaded intermittently, the DrinTec calcite bed is reloaded continuously underwater by a series of small guiding funnels installed on the bottom of the feeding hoppers and located above the bed (Fig. 15.10).

Calcite grains are fed to the bed through the funnels from calcite storage silos, which are an integral part of the filter cell structure. The silos typically contain several weeks of calcite supply. The calcite cells for large plants such as the Barcelona desalination facility are concrete structures. Enclosed fiberglass-reinforced plastic (FRP) cells are used for smaller facilities. This novel technology can operate for gravity and pressurized calcite contactors.

The calcite grains introduced through the funnels are delivered to the bed at a low velocity (one 2-mm calcite grain per minute), which eliminates the turbulence and turbidity spike of the remineralized water, which is typically created during the reloading of conventional calcite beds. This feature allows us to increase the design hydraulic surface loading rate (upflow velocity) of the calcite bed without the penalty of elevated turbidity, which is a key limitation in conventional limestone systems.

Desalinated water saturated with carbon dioxide enters the calcite cells through a perforated floor equipped with FRP flow distribution nozzles. As the desalination water moves upward through the calcite bed, the carbon dioxide contained in this water dissolves the calcite-forming calcium bicarbonate. As a result the bicarbonate alkalinity of the desalinated water increases until it reaches equilibrium. After the water passes through the calcite bed, it enters a quiescent zone from where it is conveyed out of the cells. The alkalinity content of the finished water quality is controlled by the dosage of carbon dioxide (feed water pH). Similar to most conventional calcite contactors, the DrinTec units are also backwashed periodically with air and water (once every two to

Figure 15.10 Calcite hopper and feed funnels.

three weeks). However, the DrinTec system has a provision for turbid water recirculation to the entrance of the calcite facility in order to prevent significant turbidity increase of the finished water during backwash events. Table 15.6 shows the key design criteria of the DrinTec calcite contactors for the Alicante II and Barcelona seawater desalination plants in Spain.

Comparison of DrinTec to Traditional Calcite Contactor Technologies

The low and uniform calcite loading rate of the cells allows us to increase the surface loading rate of the filters from a typical range for conventional designs of 4 to 8 $m^3/m^2 \cdot h$ (1.6 to 3.3 gpm/ft^2) to as high as 18 $m^3/m^2 \cdot h$ (7.4 gpm/ft^2) while still maintaining the turbidity of the finished water below 1 NTU. The typical design loading rate of these units is 8 to 12 $m^3/m^2 \cdot h$ (3 to 5 gpm/ft^2).

Filter bed headloss at surface loading rate of 12 $m^3/m^2 \cdot h$ (5 gpm/ft^2) is 23.4 cm per meter (2.8 in/ft) of calcite bed depth. This headloss is proportional to the loading rate and drops to 20.0 cm/per one meter (2.4 in/ft) of bed depth at surface loading rate of 10 $m^3/m^2 \cdot h$ (4 gpm/ft^2). The higher design surface loading rate the DrinTec calcite contactors allows us to reduce their size and associated capital costs.

The continuous loading of the calcite reactors also allows us to achieve more uniform and predictable product water quality with minimal operator attention. This feature also results in reduced filter backwash and flushing frequencies—and in simplified and fully automated filter maintenance.

	Value	
Parameter	**Alicante II Plant**	**Barcelona Plant**
Plant production capacity, m^3/day/mgd	67,200/18	200,000/53
Calcite consumption/alkalinity addition, mg/L as $CaCO_3$	59	56
Carbon dioxide dosage, mg/L	26	25
pH of remineralized water	8.0–8.2	8.0–8.2
Total surface of calcite contactors, m^2	320	648.4
Number of cells	32	32
Cell dimensions (width × length), m	2.0 × 5.0	3.02 × 7.02
Design flow rate per cell, m^3/day	2100	6250
Surface loading rate, $m^3/m^2 \cdot h$	8.75	9.64
Bed height, m	1.6	2.5
EBCT, minutes	11.0	12.2

Note: 1.0 m = 3.3 ft; 1 mgd = 3785 m^3/day; 1 gpm/ft^2 = 2.445 $m^3/m^2 \cdot h$

Table 15.6 Design Criteria for DrinTec Calcite Contactors of the Alicante II and Barcelona Seawater Desalination Plants

15.6 Impact of Remineralization on Product Water Quality

Increase in Alkalinity and Hardness

The impact of lime, carbon dioxide, and calcite on alkalinity and hardness of the finished fresh water produced by the desalination plant is dependent on the pH, temperature, and calcium and carbon dioxide content of the RO permeate. Because of the complex relationship between these parameters, typically the product water alkalinity and hardness are calculated as a function of these parameters using product water quality modeling software. The Tetra Tech (RTW) model software for water process and corrosion chemistry offered by the American Water Works Association (AWWA, 2011) is one of the most popular software packages used by practitioners worldwide.

Increase in Total Dissolved Solids

Besides alkalinity and/or hardness, remineralization chemicals also add TDS content to the desalination plant product water. Table 15.7 indicates the amount of TDS of turbidity per mg/L of various chemicals used for addition of alkalinity.

Increase in Turbidity

Product water turbidity is mainly a function of the content and quality in the minerals used for RO permeate conditioning (i.e., lime, calcite, dolomite). The solubility of these minerals depends of temperature, mixing conditions, and the type of conditioning system. In general, limestone contactors produce water of lower turbidity than lime conditioning systems because of the longer contact time and more favorable chemistry.

Turbidity of the lime conditioned permeate could be controlled by addition of acid to the conditioned permeate and/or by filtration of this water through sand media filters. In limestone contactor systems, product water turbidity is typically controlled by the filtration rate through the contact vessels.

15.7 Design Examples

This section presents two examples for conditioning systems of permeate produced by the RO system of a hypothetical 40,000 m^3/day (10.6 mgd) desalination plant with permeate temperature of 18 to 30°C. Both systems are designed to produce drinking water with total alkalinity and hardness of 80 mg/L and pH in a range of 7.6 to 8.3 with a target pH of 8 at 20°C as well as LSI of up to + 0.3.

Conditioning Chemical	TDS Increase (mg/mg)
Lime [$Ca(OH_2)$]	1.84
Sodium hydroxide [NaOH]	1.91
Calcite [$CaCO_3$]	3.27
Sodium carbonate [N_2CO_3]	3.40

Table 15.7 TDS Increase in Product Water Caused by Conditioning Chemicals

15.7.1 Example of Lime-Carbon Dioxide Conditioning System

Lime Feed System

As per Table 15.2, the amount of lime needed to produce permeate of alkalinity of 80 mg/L is 0.74 × 80 mg/L = 59.2 mg/L as $CaCO_3$. Assuming that commercial product of 94 percent purity is used, than the concentration of the applied lime product is 59.2 mg/L / 0.94 = 63 mg/L.

As indicated previously, the key components of the lime feed facility are the lime silo sand slurry system and lime saturator system. The main design criteria for these systems are shown in Tables 15.8 and 15.9, respectively.

Carbon Dioxide Feed System in Combination with Lime Addition

Based on information provided in Table 15.10, the dosage of CO_2, which will need to be added to obtain product water of alkalinity of 80 mg/L, is 0.88 × 80 mg/L = 70.4 mg/L as 100 percent purity product. The key design criteria for this system are presented below.

Carbon dioxide is planned to be added at two feed locations along the RO permeate pipeline—one before and one after the point of addition of lime.

Component/Parameter	Specifications/ Design Criteria
Lime dosage, mg/L (as 100/94% product)	59.2/63
Lime consumption (kg/day)	2520
Lime purity, %	94
Silo vessel • Number • Capacity, tons • Diameter, m • Height, m • Storage time, days	 2 65 4 5 50
Silo vent filter size, m²	25
Lime slurry tank volume	5 m³ (1320 gal)
Lime slurry • Concentration, % • Slurry tank retention time, h • Pump capacity (1 duty + 1 standby)	 10 4.8 25.1 m³/day (6630 gpd)

Note: 1 m = 3.3 ft

TABLE 15.8 Key Design Criteria of Example Lime Silo and Slurry System [40,000 m³/day (10.6 mgd) Desalination Plant]

Component/Parameter	Specifications/Design Criteria
Lime slurry feed concentration, %	10
Lime slurry feed rate	2520 kg/day (5550 lb/day)
Lime slurry flow rate	25.2 m^3/day (6630 gpd)
Permeate flow rate	2520 m^3/day (663,000 gpd)
Total flow rate into saturator	2545.2 m^3/day (669,630 gpd)
Number of saturator clarifiers	2
Saturator diameter	6.4 m/21 ft
Saturator depth • Depth of water • Total depth	 4.0 m/13.0 ft 4.4 m/14.5 ft
Saturator unit volume	140 m^3/37,108 gal
Saturator retention time	158 min
Saturator solids loading rate	1.63 kg lime/m^2·h
Saturator hydraulic loading rate	1.65 m^3/m^2·h
Retention time in reaction zone, min	11 min
Reaction zone diameter and height	2.1 m/3.0 m (7 ft/10 ft)
Limewater storage tank • Number of units • Volume	 2 5.7m^3/ 1500 gallons
Limewater feed pumps	2 duty + 1 standby 53 m^3/h/230 gpm @ 7.5 hp

Note: 1 m = 3.3 ft

Table 15.9 Key Design Criteria of Example Lime Saturator System [40,000 m^3/day (10.6 mgd) Desalination Plant]

Component/Parameter	Specifications/Design Criteria
Carbon dioxide dosage (as 100% pure product)	For lime/CO_2 systems–70.4 mg/L
CO_2 consumption	2816 kg/day (6200 lb/day)
CO_2 purity, %	100
Feed rate	118 kg/h (260 lb/h)
Carrier water flow and pressure	47.7 m^3/h (210 gpm) @ 5 bars/71 psig
Liquid carbon dioxide storage tank • Number • Capacity	 1 50 tons @ 21 bars/300 psig
Vaporizer • Number • Capacity	 1 155 kg @ 21 bars (340 lbs @ 300 lb/ft^2)
Other CO_2 system components • Vapor heater • Feed panels and diffusers	 1 2 (one for each feed point)

Table 15.10 Key Design Criteria of Example Carbon Dioxide System [40,000 m^3/day (10.6 mgd) Desalination Plant]

15.7.2 Example of Calcite-Carbon Dioxide Conditioning System

Calcite Contactors

As previously discussed, the amount of calcite needed to produce permeate of alkalinity of 80 mg/L is 1.0 × 80 mg/L = 80 mg/L as $CaCO_3$. Assuming that commercial product of 99.1 percent purity is used, the concentration of the applied concentration of the actual lime product is 80 mg/L / 0.99 = 81 mg/L. Table 15.11 presents key criteria for a calcite contactor system, which processes 100 percent of the RO permeate flow and is designed as a conventional gravity limestone contactor.

Carbon Dioxide Feed System for Calcite Contactor Conditioning

One of the advantages for using calcite instead of lime for the addition of hardness is that it also adds bicarbonate alkalinity to the water. This allows us to reduce the amount of CO_2 in a half (i.e., only 0.44 mg of CO_2 needs to be added in order to gain 1 mg/L of alkalinity). Therefore, in this case, the amount of CO_2 that needs to be added to reach product water of alkalinity of 80 mg/L is 0.44 × 80 mg/L = 35.2 mg/L as 100 percent purity product. The key design criteria for this system are presented below. Because the dosage is two times smaller, the size of the CO_2 feed system will be exactly two times smaller than that shown in Table 15.9.

15.8 Remineralization Costs

15.8.1 Overview of Construction Costs

The capital costs of lime/carbon dioxide systems vary between \$50 to \$150 per m^3/day (\$0.2 and \$0.6 mm/mgd) of finished desalinated water. For comparison, the capital cost

Component/Parameter	Specifications/Design Criteria
Calcite dosage, mg/L (as 100/99% product)	80/81
Calcite consumption	3240 kg/day (7140 lb/day)
Calcite contactors • Number • Length • Width • Surface area per unit • Total surface area	 32 5.0 m (16.5 ft) 2.0 m (6.6 ft) 10 m^2 (108 ft^2) 320 m^2/(3450 ft^2)
Surface loading rate	5.2 $m^3/m^2 \cdot h$ (2.1 gpm/ft^2)
Contact zone (bed) height	2.0 m (6.6 ft)
Depth of working volume	0.5 m (1.6 ft)
Depth of conditioned water zone	0.8 m (2.6 ft)
EBCT, minutes	23

Note: 1.0 m = 3.3 ft; 1 mgd = 3785 m^3/day; 1 gpm/ft^2 = 2.445 $m^3/m^2 \cdot h$

Table 15.11 Key Design Criteria of Example Limestone (Calcite) Contactor System [40,000 m^3/day (10.6 mgd) Desalination Plant]

of calcite remineralization systems producing the same finished water quality is between \$25 and \$75/m^3/day (\$0.1 and \$0.3 mm/mgd) of produced finished water. The capital costs of most other remineralization systems are typically in a range of \$80 to \$180 per m^3/day (\$0.3 mm to \$0.7 mm/mgd). Figure 15.11 presents construction costs for lime–carbon dioxide and calcite–carbon dioxide post-treatment systems.

The cost curves presented in Fig. 15.11 are derived based on comparative review and statistical analysis of a number of full-scale projects and should be used for initial cost estimates only. Actual project costs may vary in arrange of +/– 30 percent of these cost curves depending on the site-specific project conditions, size, and location.

15.8.2 Example Cost Estimate of Lime-Carbon Dioxide Conditioning System

Breakdown of the capital and O&M costs of a typical lime/carbon dioxide remineralization system for the hypothetical 40,000 m^3/day (10.6 mgd) discussed in the previous section is shown in Table 15.12.

As seen in Table 15.12, the total construction cost of the lime/CO_2 remineralization system is \$2.965 million, which corresponds to \$74/m^3/day (\$0.28 mm/mgd). These capital costs are amortized using capital recovery factor (CRF) estimated for an amortization rate of 5 percent over a period of 20 years (CRF = 12.462) to determine the capital cost portion of the water production cost associated with this remineralization system—\$0.016/m^3 (\$0.062/1000 gallons). The O&M costs for remineralization are \$0.044/m^3 (\$0.166/1000 gallons). The total cost of drinking water production associated with remineralization is \$0.060/m^3 (\$0.228/1000 gallons).

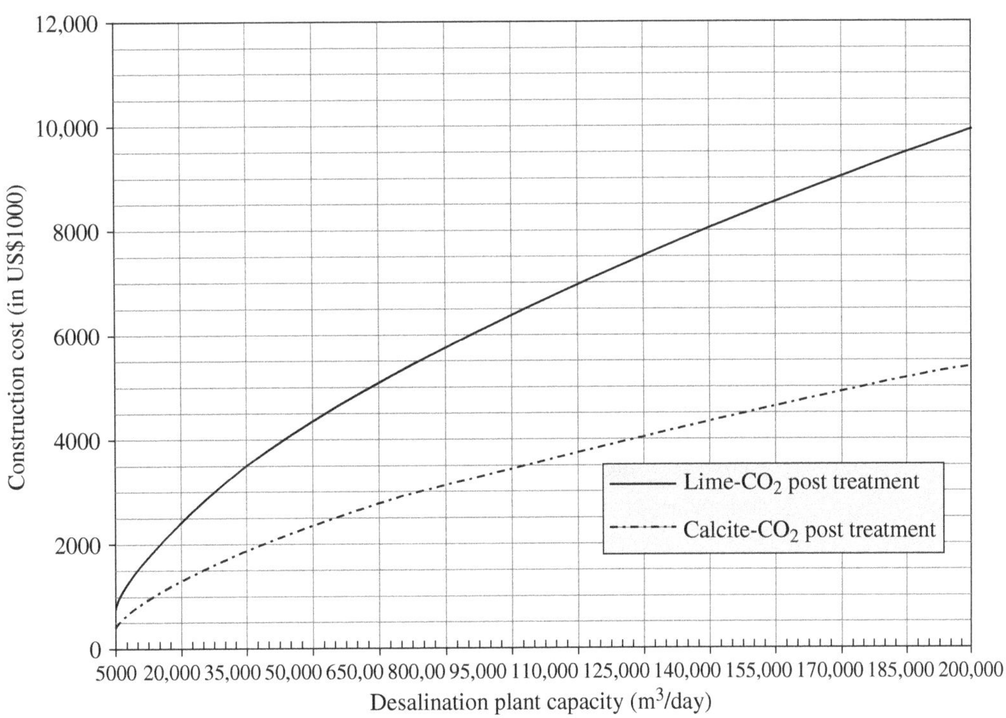

FIGURE 15.11 Capital Costs for remineralization.

Capital Costs	Lime/CO_2 System (in 1000 US$/yr)
Lime silos and slurry preparation system	480
Lime slurry tanks	120
Lime saturators	650
Limewater storage tank	65
Lime feed system	55
Carbon dioxide storage system	400
Carbon dioxide evaporators	60
Carbon dioxide feed system	50
Lime clarifier sludge handling system	60
Other auxiliary and service facilities	200
Land costs	90
Engineering and construction management	340
Start-up and commissioning	75
Other costs	320
Total capital costs	2965
Amortized capital Costs, US$/m^3 (US$/1000 gallons)	*0.016/0.062*
Operation and maintenance costs	**Lime/CO_2 system (1000 US$/yr)**
Labor	40
Lime	200
Carbon dioxide	80
Polymer for lime clarification	50
Polymer for lime sludge dewatering	25
Lime sludge disposal	35
Maintenance and spare parts	90
Power use	50
Other O&M costs	75
Total annual O&M costs, 1000 US$/yr	645
Annual O&M costs, US$/m^3 (US$/1000 gallons)	*0.044/0.166*
Total cost of water remineralization, US$/m^3 *(US$/1000 gallons)*	0.06/ 0.228

Table 15.12 Capital and O&M Costs of Lime–Carbon Dioxide Remineralization System for 40,000 m^3/day (10.6 mgd) RO Plant

15.8.3 Example Cost Estimate of Calcite–Carbon Dioxide Conditioning System

This section provides a cost estimate for the example 40,000 m^3/day (10.6 mgd) calcite/carbon dioxide system (design described in the previous section). The capital costs presented in Table 15.12 include the expenditures for all key components of the calcite/CO_2 post-treatment system along with associated interconnecting piping, fittings, monitoring, instrumentation, and control systems and equipment, electrical system, and other service and auxiliary facilities needed for the normal operation of the systems.

Please note that the costs for such system also include the expenditures for construction of sodium hydroxide system needed to adjust the pH of the remineralized water up to the target range of pH of 8.0 to 8.3. Usually, calcite contactors produce water with pH lower than this target range.

The cost estimate presented in Table 15.2 is prepared for remineralization system, which is designed to process the entire volume of the RO system permeate. As indicated previously, permeate remineralization costs could be reduced by conditioning only 15 to 30 percent of the permeate flow through the limestone contactors and then blending of this permeate with the rest of the RO system flow.

Comparison of Tables 15.12 and 15.13 clearly indicates that the use of calcite/carbon dioxide systems is significantly less costly both in terms of capital and O&M expenditures (\$0.060 versus \$0.035/m^3).

15.8.4 Costs of Remineralization Chemicals

Remineralization costs are sensitive to the unit costs of chemicals added for conditioning of the desalinated water, which, in turn, can vary widely from one location to another. Therefore the cost information provided herein will need to be considered as a guideline rather than as a design or budgeting tool.

As indicated previously, cost of chemicals is one of the key expenditures for remineralization of desalinated water. Typically, most desalination projects target the addition of 60 to 120 mg/L of total alkalinity to the desalinated water. Desalinated water alkalinity can be increased using a number of different commercially available chemicals. However, these chemicals add different amounts of alkalinity for the same amount of delivered chemicals, and their unit costs differ as well.

Table 15.14 presents a summary of the key chemicals used for alkalinity addition and their typical unit prices. Analysis of this table indicates that calcite is the most cost-effective compound for adding alkalinity to desalinated water because it has the lowest costs per 1 mg/L of $CaCO_3$ added. Use of calcite has the advantage of adding alkalinity and total hardness to the finished water.

The combination of hydrated lime and carbon dioxide is the most widely used remineralization method today, although the total costs for this combination of conditioning chemicals is typically one and a half to two times higher than that for calcite and carbon dioxide or sulfuric acid. Use of hydrated lime instead of quick lime is usually two to three times more costly for the same amount of alkalinity and total hardness increase of the desalinated water. Therefore quicklime-based remineralization of desalinated water has found limited use.

Carbon dioxide, soda ash, sodium hydroxide, and sodium bicarbonate are the most costly applied chemicals for delivery of target alkalinity to the desalinated water. Because, in general, the unit costs of soda ash and sodium hydroxide are comparable,

Capital Costs	Calcite/CO_2 System (in 1000 US$/yr)
Calcite contactors	800
Sodium hydroxide feed system	110
Carbon dioxide storage system	220
Carbon dioxide evaporators	40
Carbon dioxide feed system	30
Other auxiliary and service facilities	100
Land costs	120
Engineering and construction management	150
Start-up and commissioning	50
Other costs	170
Total capital costs	1790
Amortized capital costs, US$/m³ (US$1000 gallons)	*0.008/0.031*
Operation and Maintenance Costs	**Calcite/CO_2 System (in 1000 US$/yr)**
Labor	35
Carbon dioxide	40
Limestone (calcite)	70
Sodium hydroxide	110
Maintenance and spare parts	50
Power use	30
Other O&M costs	60
Total annual O&M costs, US$/yr (1000 US$/yr)	395
Annual O&M costs, US$/m³ (US$/1000 gallons)	*0.027/0.100*
Total cost of water remineralization, US$/m³ (US$/1000 gallons)	0.035/ 0.131

TABLE 15.13 Capital and O&M Costs of Calcite/CO_2 Remineralization System for 40,000 m³/day (10.6 mgd) Desalination Plant

and because sodium hydroxide is easier to handle, it is more commonly used than soda ash for final pH adjustment of the finished desalinated water.

15.9 Disinfection Systems

The two types of chlorine-based disinfectants most commonly used in municipal water distribution systems worldwide are: (1) chlorine gas or its derivatives (HOCl and OCl^-), and (2) chloramines.

Chemical	Alkalinity Addition (as $CaCO_3$) per mg/L of Chemical	Unit Chemical Costs (US$/ton)	Unit Costs in US$/ton per 1 mg/L of Added Alkalinity as $CaCO_3$
Calcite	1.00	30–40	30–40
Carbon dioxide	1.14	70–90	61–78
Sulfuric acid	1.02	50–80	49–78
Quicklime	1.78	120–150	67–84
Hydrated lime	1.35	260–280	193–207
Soda ash	0.94	540–580	574–617
Sodium hydroxide	1.25	700–750	560–600
Sodium bicarbonate	0.60	900–950	1500–1583

TABLE 15.14 Cost of Common Chemicals Used for Increase of Water Alkalinity

15.9.1 Chlorination

At present, chlorination with chlorine gas and sodium hypochlorite is the most widely used disinfection method for disinfection of desalinated water. The desalination process usually eliminates over 99.99 percent of the pathogens and other undesirable microorganisms contained in the source water. The typical target chlorine dosage that provides adequate disinfection depends on two key factors: desalinated water temperature and contact time. Usually, the chlorine dosage used for disinfection is 0.5 to 2.5 mg/L.

Although popular worldwide, the use of chlorine gas is associated with potential safety considerations associated with accidental gas releases. Therefore chlorine gas disinfection facilities have to be equipped with gas detection, containment, and treatment facilities that provide adequate protection of public health.

A 10 to 15 percent solution of sodium hypochlorite is safer to use, handle, and store than chlorine gas, and therefore sodium hypochlorite has found wider application than chlorine gas in desalination plants worldwide. Sodium hypochlorite used for disinfection either can be delivered to the desalination plant as a commercial product or can be generated on-site using commercially available sodium chloride (salt).

The main advantage of on-site generation of sodium hypochlorite at the desalination plant, especially for large plants, is that it minimizes space requirements for storing large quantities of sodium hypochlorite solution, thus reducing hypochlorite decay and formation of chlorate during long storage. Usually, sodium hypochlorite solutions decay rapidly over time and lose 10 to 20 percent of their strength over a period of 10 to 15 days. Such decay is accompanied with an increase in chlorate content in the sodium hypochlorite solution, especially in warm climates. The rate of solution strength decay depends mainly on the initial concentration of the sodium hypochlorite, the ambient temperature, and the exposure to sunlight.

On-site sodium hypochlorite generation using commercially available high-grade sodium chloride of low bromide content instead of seawater is recommended. Although the use of seawater as a source of chloride for the sodium hypochlorite generation process is less costly and simpler, it results in the generation of higher concentrations of DBPs and bromate because of the naturally high level of bromide in the ambient seawater. When blended with desalination plant desalinated water, the sodium hypochlorite generated from seawater increases the concentration of DBPs and bromate in the produced fresh water. Therefore, electrolysis of seawater to produce hypochlorite for final disinfection of desalinated water should be avoided.

Chlorine and sodium hypochlorite have more effective bactericidal effect at lower pH values. When Cl_2 is added to water, it rapidly and completely hydrolyses to HOCl, which is in equilibrium with OCl^- as a function of pH. Since HOCl is much more effective as a bactericide than OCl^-, and since this compound exists at a higher concentration at lower pH values, some experts recommend to maintain the pH of the desalinated water at or below 8.0 (Barbeau, 2004).

The effect of pH on chlorine disinfection is acknowledged by most disinfection guidelines: the United States Environmental Protection Agency (US EPA) guidelines recommend raising CT_{99} (i.e., the product of chlorine concentration times the retention time required to get 99 percent pathogen removal) by approximately three times when pH is increased from 6 to 9 (US EPA, 2004). WHO guidelines are more conservative: they recommend a 40-time increase in CT_{99} value when pH increases from 7.0 to 8.5 and also state that efficient chlorine disinfection is attained at pH values < 8.0 (WHO, 2004).

15.9.2 Chloramination

Chloramination is widely used principally as a secondary disinfectant because of its lower biocide potency but higher stability. This disinfection method includes the sequential addition of chlorine and ammonia to the product water to form chloramines. Chloramines have a significantly slower rate of decay than free chlorine and therefore, are often favored especially for product water delivered to large distribution systems with high temperatures and long retention times, which have high potential for chlorine residual loss. Chloramination typically results in the creation of lower concentration of DBPs than with free chlorine disinfection. It may contribute to nitrite or NDMA production under some conditions.

Since desalinated seawater has very low content of organics, the use of chloramines for seawater desalination is not as advantageous as it may be for disinfection of drinking water produced from brackish or freshwater sources and therefore, it is not widely practiced. However, chloramination of desalinated water may be necessary if this water is planned to be blended with other water sources disinfected with chloramines.

Chloramines are used for prevention of microorganism regrowth in the product water distribution system and their disinfection efficiency is also pH sensitive, although at lesser extend. WHO guidelines distinguish between chloramine disinfection of bacteria, which is pH sensitive, and thus CT_{99} is higher at higher pH values, and disinfection for virus and protozoa inactivation, which is not sensitive to pH for a range of $6.0 < pH < 9.0$ (WHO, 2004).

An important factor to take under consideration when desalinated water is chloraminated, or it is blended with other chloraminated water sources in the distribution system, is the buffering capacity of this desalinated water (Lahav et al., 2012). Buffering capacity is the ability of water to "absorb" the impact of the introduction of acids and bases into the finished water. Typically, the buffering capacity of desalinated water is directly related to its content of bicarbonate alkalinity in this water. If, for example, the desalinated water has high pH and alkalinity, but a large portion of this alkalinity is of nonbicarbonate origin (i.e., addition of NaOH), then the addition of desalinated water could trigger corrosion in a distribution system exposed to nitrification or containing other water sources of significantly lower pH.

This concept can be illustrated in the example of two desalinated waters that have the same water quality in terms of pH and hardness but have different alkalinity (50 versus 80 mg/L) and therefore different buffering capacity (Lahav et al., 2012). If such waters are chloraminated and contain an excessive content of ammonia, the elevated ammonia content can trigger nitrification and associated pH reduction in the distribution system. The oxidation of ammonia during the nitrification process is described by the following equation:

$$NH_4^+ + 1.5O_2 \rightarrow NO_2^- + 2H+ + H_2O \tag{15.8}$$

Based on Eq. 15.8, the oxidation of 0.2 mg/L of ammonium decreases alkalinity (and increases acidity) by 1.43 mg/L as $CaCO_3$. Table 15.15 illustrates the impact of nitrification on the water quality the low- and high-alkalinity waters.

The results shown in Table 15.15 demonstrate the importance of buffering capacity/bicarbonate alkalinity of the desalinated water. The CCPP of the water with the lower alkalinity/buffering capacity is reduced to −1.9 mg/L as $CaCO_3$ when pH drops from 8.15 to 7.78 as a result of the nitrification process, whereas the pH of the higher-alkalinity (80 mg/L) water drops to only 7.9 and CCPP remains positive (+ 0.7 mg/L as $CaCO_3$) despite the nitrification impact. As a result, the lower alkalinity water may trigger corrosion in the distribution system, while the

Water Quality Type	Parameter	Units	Before Nitrification	After Nitrification
Lower alkalinity (50 mg/L) /lower buffering capacity water	Alkalinity	mg/L as $CaCO_3$	50	48.6
	(Ca^{2+})	mg/L as $CaCO_3$	110	110
	pH	–	8.15	7.78
	CCPP	mg/L as $CaCO_3$	0.8	−1.9
	Buffering capacity	mM/ΔpH unit	0.13	–
Higher alkalinity/ higher buffering capacity water	Alkalinity	mg/L as $CaCO_3$	85	83.6
	(Ca^{2+})	mg/L as $CaCO_3$	100	100
	pH	–	8.15	7.90
	CCPP	mg/L as $CaCO_3$	3.5	0.7
	Buffering capacity	mM/ΔpH unit	1.8	–

Table 15.15 Effect of Nitrification on Water Quality with Low and High Alkalinity [Temperature = 22°C (72°F) and TDS = 192 mg/L]

higher alkalinity water will provide adequate protection against this potential challenge. It should be pointed out that this example is for a temperature of 22°C (72°F), and the nitrification rate could be accelerated at higher temperatures, thereby even rendering the higher-alkalinity water not adequately protective for the corrosion of the distribution system. The example presented in Table 15.15 is to illustrate the importance of alkalinity in the desalinated water when it is introduced into distribution systems that may experience nitrification.

When different water sources are blended in the distribution system, the chemical stability of the blend is determined by the individual buffering capacity of the blended waters. Chemical stability, as determined by the CCPP value, depends on the relationship between alkalinity, dissolved calcium, and pH. The alkalinity and dissolved calcium concentrations (and also the acidity value) are conservative parameters in the sense that their concentrations in the blend can be determined by a simple weighted average.

In contrast, pH and CCPP are not conservative parameters. The pH value of the blend can be determined from the relationship between the alkalinity and acidity values in the blend. CCPP, in turn, is determined by the alkalinity, dissolved calcium and pH values of the blend. Because of the nonlinear relationship between these parameters, when desalinated water with a relatively low buffering capacity and a relatively high pH is blended with ground water that has high alkalinity (and high buffering capacity) and a relatively low pH (around pH 7.0), certain blends might result in a negative CCPP value even if the original waters had a positive CCPP.

In contrast, combining the same ground water with desalinated water with higher buffering capacity would result in a blend that in all scenarios is positive with respect to CCPP, and the water will provide adequate corrosion protection. A simple rule of thumb to follow in such cases is to match as closely as possible the pH and the alkalinity of the desalinated water with the pH and alkalinity of the water in the distribution system. If the existing distribution system is already experiencing nitrification, the desalinated water is recommended to have higher bicarbonate alkalinity than the alkalinity of the water in the distribution system.

15.9.3 Chlorine Dioxide Disinfection

Chlorine dioxide is widely used in pre-oxidation and post-disinfection of drinking water as an alternative to chlorine gas or sodium hypochlorite. Chlorine dioxide does not form significant quantities of total trihalomethanes (TTHMs) and adsorbable organic halogens in comparison with other chlorinated disinfectants. Bromate is not formed even if the desalinated water is blended with other sources containing bromide ion.

The main chlorine dioxide byproducts are chlorite and chlorate, as well as some organic oxidation products. To minimize chlorate formation, it is necessary to improve the on-site generation of chlorine dioxide by using properly designed generators capable of producing pure chlorine dioxide solutions and reaching high conversion of the reagents (Gordon, 2001).

Chlorite is the main chlorine dioxide residue/byproduct. When chlorine dioxide undergoes chemical reduction in water treatment processes, about 60 percent is converted to chlorite ion. However, in desalinated water, the dosage of chlorine dioxide necessary to maintain a residual in the distribution network is quite low (< 0.4 mg/L, which is about one fourth of the required chlorine dosage) (Belluati et al. 2007), and therefore the chlorite residue is expected to be much lower than the current WHO limit (0.7 mg/L) (WHO, 2008).

15.9.4 Ozonation

Ozonation is a widely accepted practice for the disinfection of product water from freshwater sources. However, ozonation of RO-desalinated water is associated with the potential formation of excessive amounts of assimilable organic carbon (AOC) and bromate as a result of the relatively high content of bromide in desalinated water compared with that in drinking water from other surface water sources.

However, bromate formation could be addressed by reducing the bromide level in the desalinated water to less than 0.3 mg/L before ozonation. While such reduction is technically feasible, it would be costly, especially for desalinated water produced by membrane separation because it would require the treatment of the source seawater through a two-pass RO system.

15.9.5 Ultraviolet Light Disinfection

(Ultraviolet) irradiation of desalination plant desalinated water is a viable disinfection alternative that is particularly useful for Cryptosporidium oocyst inactivation. The disinfection of desalinated water will typically require lower UV dosages than those used for UV disinfection of other surface water sources because of the lower turbidity and lower content of pathogens in the desalinated water. Another advantage of UV disinfection is that it does not add any chemicals to the product water and, therefore the desalinated water has a low content of DBPs. However, it also does not leave a disinfectant residual to control bacterial regrowth in the distribution system.

15.9.6 Disinfection Byproducts Formation and Control

Blending of low-DBP desalinated seawater with surface water with high DBP content can reduce the overall DBP levels of the drinking water. Desalinated water from brackish water sources, however, could contain a relatively higher concentration of organics and organics-related DBPs than desalinated seawater, and therefore blending of such water may not have the same beneficial impact on the DBP levels in the distribution system.

The compatibility of the various water sources in terms of DBP formation potential must be taken into consideration prior to their blending. Specific issues that must be investigated include: (1) bromide and TOC concentrations in the various waters, and their effect on DBP formation and concentration in the blend, (2) type of disinfection method used for the various water sources and the effect on DBP formation and chlorine residual stability, and (3) temperature of the blended waters (desalinated waters of high temperature may result in accelerated nitrification and corrosion in the distribution system). The negative impact of the elevated temperature of the desalinated water may, however, have a net positive effect on nitrification if this water is of very low organic content.

Conveyance of desalinated water in long pipelines may pose chlorination and corrosion problems, especially in warm climates. Loss of disinfectant residual is one of the main challenges in such systems, mainly because of residual disinfectant decay caused by high-temperature water. This may be addressed by super-chlorination (i.e., adding chlorine at dosages resulting in breakpoint chlorination, i.e., 3.5 to 4 mg/L); reinjection of chlorine along the length of the pipeline, which could be activated when the chlorine residual drops below 0.5 mg/L, or by the use of chloramine instead of chlorine for disinfection because of its lower chemical reactivity. Chloramines have significantly slower decay rates than free chlorine.

Loss of calcium alkalinity may occur over the length of the pipeline and, therefore may result in corrosion problems. There are two main alternatives to address this challenge: (1) reinject calcium conditioning chemicals or corrosion inhibitors along the pipeline route at locations where the water LSI is reduced to a negative level, or (2) use nonmetal pipeline materials such as high-density polyethylene (HDPE), which are not sensitive to low levels of calcium alkalinity in the water.

15.9.7 Chlorine Residual Stability and Product Water Quality

Disinfection of desalinated seawater with chlorine results in a stable and long-lasting chlorine residual that provides adequate disinfection in the distribution system in many conditions. However, ammonia addition to desalinated water in order to form chloramines may cause an accelerated decay of the total chlorine residual if the bromide concentration of desalinated water is above 0.4 mg/L (McGuire Environmental & Poseidon Resources, 2004). Applying a combination of chlorine and ammonia to desalinated water with bromide levels above 0.4 mg/L may yield an unstable residual that decays at an accelerated rate because of the rapid conversion of chloramines to bromamines, which are much more chemically reactive.

If chlorinated desalinated water is blended with chloraminated drinking water produced from a fresh surface water source, mixing of the two types of water may also result in accelerated decay of chlorine residual of the blend of two waters for the same reason (i.e., formation of unstable bromamines, which decay to chlorine and bromide over a period of several hours).

The destabilizing effect of bromide on the chloramination process can be mitigated by super-chlorination or by producing desalinated water with a bromide level below 0.4 mg/L. The former would increase byproducts, including organobromines. The latter, however, will increase the cost of desalination, and, if reverse osmosis is used for membrane separation, it will require the installation of full second-pass RO.

15.9.8 Design of Disinfection Systems

Design of disinfection systems for desalination plant permeate is similar to that used for conventional water treatment plants. At present, sodium hypochlorite is the most commonly used disinfectant in desalination plants worldwide. Unless the desalinated water is blended with other water sources, typically disinfection is completed by chlorination rather than chloramination because most desalinated waters do not contain organics, and their chlorine residual is fairly stable. Chloramination is typically practiced with the desalinated water mixed with other chloraminated waters in the distribution system.

Component/Parameter	Specifications/Design Criteria
Chlorine dose	0.5–2.5 mg/L
Bulk storage tanks • Minimum number • Material of the storage • Tanks • Storage capacity	 2 High-density cross-linked polyethylene (HDXLPE) 15–30 days
Sodium hypochlorite • Concentration, %	 5–12
Day tank • Number • Material	 1 HDXLPE
Hypochlorite solution transfer pumps • Minimum number (duty/standby) Type • Control	 1/1 Centrifugal constant speed
Metering pumps • Minimum number (duty/standby) • Type • Control	 1/1 Diaphragm constant speed
Piping	Schedule 80 PVC

Table 15.16 Key Design Criteria for Sodium Hypochlorite Disinfection System

Key Design Criteria and Considerations

Table 15.16 provides an overview of key design criteria and considerations for sodium hypochlorite disinfection system for desalinated water.

Design Example

Table 15.17 provides a summary of key design parameters of a sodium hypochlorite feed system for a hypothetical 40,000 m^3/day (10.6 mgd) desalination system. This system is designed with one bulk storage tank and one-day tanks equipped as a minimum with an ultrasonic level transmitter, float switches, pressure gauge, pressure/vacuum relief system, visual level indicator, and high-level warning instrumentation. The system metering pumps are equipped with pressure gauges, magnetic flowmeter, and speed controls.

Since RO permeate does not have chlorine, and the sodium hypochlorite system has to be designed for the maximum dosage (4 mg/L = 4 g/m^3), then the total amount of chlorine needed for disinfection per day is 4 g/m^3 × 40,000 m^3/day/1,000 g/kg = 160 kg/day. Since a 12 percent solution of sodium hypochlorite contains 0.12 kg of chlorine per one liter of solution, then the total maximum daily volume of sodium hypochlorite is 160 kg/0.12 kg/L = 1334 L/day = 1.334 m^3/day. For 30-day storage of the solution, the tank would need to be designed for 1.334 m^3/day × 30 days = 40 m^3.

Component/Parameter	Specifications/Design Criteria
Chlorine dose	
• Minimum	0.5 mg/L
• Maximum	4.0 mg/L
Sodium hypochlorite	
• Concentration, %	12
• Grams of Cl per liter of solution	120
Bulk storage tanks	
• Number	2
• Capacity	20 m^3/5280 gallons
• Diameter	3 m/10 ft
• Storage time	30 days
Day tank	
• Number	1
• Capacity	1.4 m^3/264 gallons
• Diameter	0.6 m/1.6 ft
Hypochlorite solution transfer pumps	
• Number (duty/standby)	2 + 1
• Capacity	10 gpm
• Control	Constant speed
Metering pumps	
• Minimum number (duty/standby)	2 + 1
• Type	Diaphragm
• Control	variable stroke

Table 15.17 Example of Sodium Hypochlorite Disinfection System for 40,000 m^3/day (10.6 mgd) Desalination Plant

15.9.9 Disinfection System Costs

Graphs for the year 2012 construction costs of chlorine gas and sodium hypochlorite feed systems for disinfection of desalination plant permeate are presented in Fig. 15.12. The analysis of this figure indicates that the capital expenditures for chlorine gas feed systems are typically several times higher than those for sodium hypochlorite. Such cost difference could be even more significant if the chlorine gas disinfection system is equipped with containment and scrubber facilities for treatment of potential gas leaks.

On the other hand, the cost of commercially available chlorine gas is typically three to six times lower than that of sodium hypochlorite, and, as a result, the overall life-cycle cost for disinfection with chlorine gas could be lower. In addition, if the desalination plant is planned to be operated intermittently, the use of chlorine gas may be preferable because sodium hypochlorite solution loses strength over time.

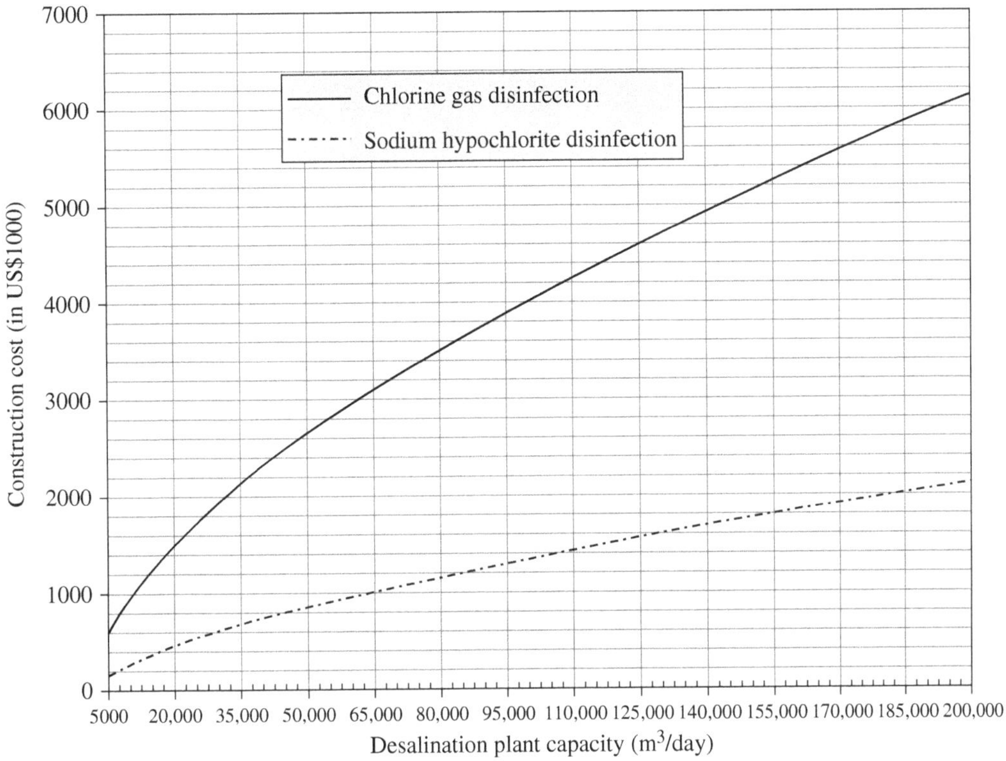

FIGURE 15.12 Construction costs of disinfection systems.

15.10 References

AWWA *Manual of Water Supply Practices M46*, 2nd ed., "Reverse Osmosis and Nanofiltration," American Water Works Association, Denver, Colorado, 2007.

AWWA "Tetra Tech (RTW) Model for Water Process & Corrosion Chemistry," 2011.

Augus E., N. Voutchkov, and D. L. Siedlak "Disinfection Byproducts and Their Potential Impact on the Quality of Water Produced by Desalination Systems: A Literature Review," *Desalination*, 237: 214–237, 2009.

Barbeau, B., D. Huffman, C. Mysore, R. Desjardins, and M. Prevost "Examination of Discrete and Confounding Effects of Water Quality Parameters During the Inactivation of Ms2 Phages and Bacillus Subtilis Spores With Free Chlorine," *Journal of Environmental Engineering and Science*, 3: 255–268, 2004.

Berghult, B., T. Hedberg, and A. E. Broo, Drinking Water Distribution: Corrosion control in Swedish Municipalities," *Journal of Water Supply Research and Technology–Aqua*, 48: 44–52, 1999.

Belluati, M., E. Danesi, G. Petrucci et al., "Chlorine Dioxide Disinfection Technology to Avoid Bromate Formation in Desalinated Seawater in Potable Waterworks," *Desalination* 203(1–3): 312–318, 2007.

Birnhack, L., N. Voutchkov, and O. Lahav, "Fundamental Chemistry and Engineering Aspects of Post-Treatment Processes for Desalinated Water: A Review," *Desalination*, 273: 6–22, 2011.

Cantor, A. F. D. Denig-Chakroff, R. R Vela, M. G. Oleinik, and D.L. Lynch, "*Use of polyphosphate in corrosion control*," *AWWA*, 92: 95–102, 2002.

Catling, L. A., I. Abubaker, I. R. Lake, L. Swift, and P. R. Hunter, "A Systematic Review of Analytical Observational Studies Investigating the Association Between Cardiovascular Disease and Drinking Water Hardness," *Journal of Water & Health*, 6: 433–442, 2008.

Cotruvo, J., N. Voutchkov, J. Fawell, P. Payment, D. Cunliffe, and S. Lattemann, *Desalination Technology, Health and Environmental Impacts*, CRC Press and IWA Publishing, Boca Raton, London, New York, 2010.

Delion, N., G. Mauguin, and P. Corsin, "Importance And Impact Of Post Treatments On Design And Operation Of SWRO Plants," *Desalination*, 165: 323–334, 2004.

de Souza, P. F., G. J. du Plessis, and G. S. Mackintosh, "Water SA–Special Edition," WISA Proceedings, 2002.

Imran, S. A., J. D. Dietz, G. Mutoti, S. J. Taylor, A. A. Randall, and C. D. Cooper, "Optimizing Source Water Blends for Corrosion and Residual Control in Distribution Systems,"*Journal AWWA*, 97: 93–100, 2005.

Kozisek, F., "Health Significance of Drinking Water Calcium and Magnesium," http://www.szu.cz/chzp/voda/pdf/hardness.pdf, 2003.

Lahav, O., N. Voutchkov, and L. Birnhack, *Post-Treatment of Desalinated Water*, Balaban Desalination Publications, Hopkinton, Massachusetts, USA, 2012.

McGuire Environmental & Poseidon Resources, *Disinfection Byproduct Formation in Simulated Distribution System: Blending Desalinated Seawater from The Poseidon Resources Inc. Pilot Facility with Local Drinking Water Sources*, Final Report, March, 2004.

Parekh, S., *Reverse Osmosis Technology Application for High Purity Water Production*, Marcel Dekker, Inc., New York, 1988.

Wilf, M., L. Awerbuch, C. Bartels, M. Mickley, G. Pearce, and Voutchkov, *The Guidebook To Membrane Desalination Technology: Reverse Osmosis, Nanofiltration and Hybrid Systems Process Design, Applications and Economics* Balaban Desalination Publications, Rehovot, Israel, 2007.

US Environmental Protection Agency (USEPA) (2004) Edition of the Drinking Water Standards and Health Advisories, Office of Water, Washington, DC, 2004: http://www.epa.gov/waterscience/criteria/drinking/standards/dwstandards.pdf.

WHO (2008) *Guidelines for Drinking-Water Quality*. 3rd ed., Geneva: World Health Organization, 2008: http://www.who.int/water_sanitation_health/dwq/gdwq3rev/en/index.html.

CHAPTER 16

Desalination Plant Discharge Management

16.1 Introduction

Desalination plants generate discharge that contains the plant's source water treatment byproducts, including concentrate, spent pretreatment filter backwash water, and membrane cleaning solutions. Concentrate is the desalination process byproduct of the largest volume and the greatest management challenges. According to a 2009 report on concentrate treatment prepared by the United States Bureau of Reclamation (Mickley, 2009), the five most commonly used concentrate management alternatives in the United States are: (1) surface water discharge, (2) sewer disposal, (3) deep-well injection, (4) land application, and (5) evaporation ponds (Fig. 16.1). The desalination concentrate management practices shown in this figure have similar frequency of application worldwide.

As indicated in Figure 16.1, surface water discharge is the most common method for disposal of desalination plant waste streams because it is applicable for practically all sizes of desalination projects. Sewer (wastewater collection system) disposal is the most widely applied method for disposal of discharges from small desalination plants. Deep well injection has found application as one of the most suitable methods for disposal of concentrate from medium- and large-size inland brackish water desalination plants. Land application and evaporation ponds are concentrate management alternatives typically applied for small- and medium-size plants in areas where climate and soil conditions provide for high evaporation rates and year-round growth and harvesting of halophytic vegetation.

None of the discharge management methods shown in Fig. 16.1 can be applied universally to every size and type of desalination project at every plant site. Therefore selecting the most suitable and cost effective method or combination of methods for management of plant discharge is one of the greatest implementation challenges for brackish and seawater desalination projects. This chapter provides an overview of the most commonly used alternatives for desalination plant discharge management and outlines their areas of application and key advantages and disadvantages.

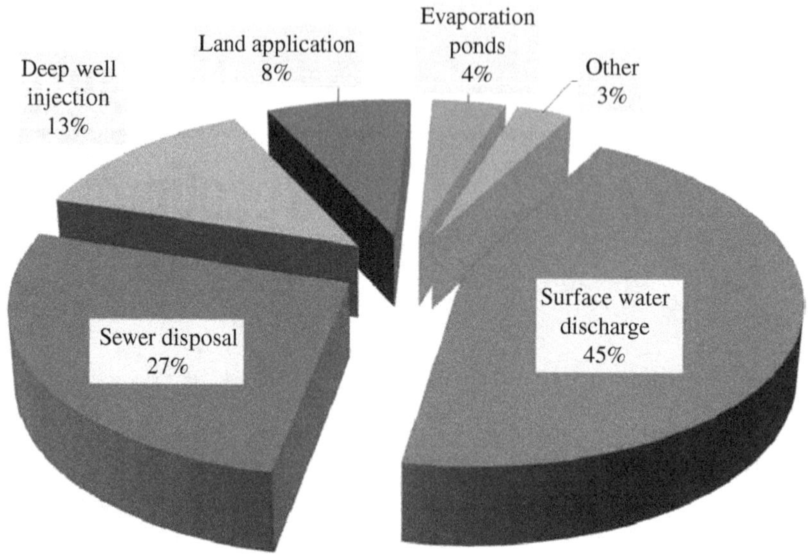

FIGURE 16.1 Current concentrate management practices.

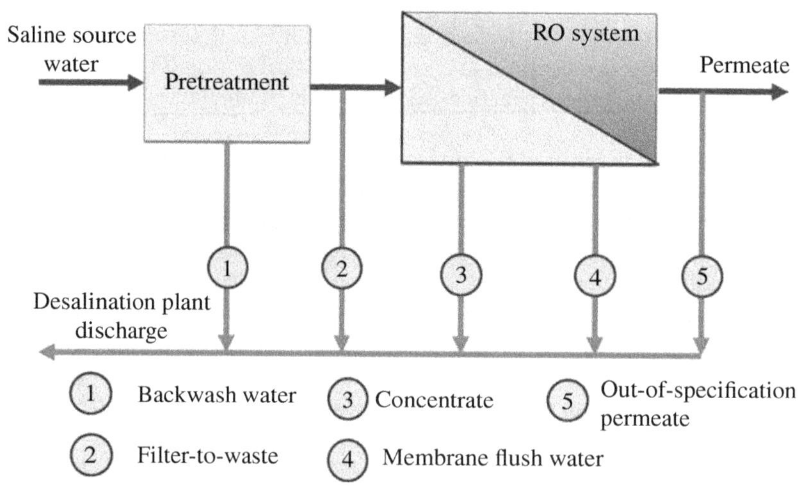

FIGURE 16.2 Desalination process side streams.

16.2 Desalination Plant Discharge Characterization

16.2.1 Desalination Process Side Streams

Typically, desalination plants generate the following three key side streams: (1) concentrate from membrane salt separation, (2) backwash water from the plant system, and (3) membrane flush water from the periodic chemically enhanced cleaning [clean-in-place (CIP)] of the RO and pretreatment membranes [if membrane pretreatment is used). (See Fig. 16.2.)

The other two side streams shown in this figure (filter-to-waste and out-of-specification permeate) are of intermittent nature and of several orders of magnitude smaller volume and content of contaminants as compared with the three main side streams.

During normal desalination plant operation, concentrate is produced continuously while spent filter backwash water is generated after every backwash cycle, which can be as long as 24 to 48 hours for conventional granular media filters to 30 to 60 minutes for membrane pretreatment systems. Spent RO and pretreatment membrane cleaning side streams are generated intermittently (typically every one to six months).

16.2.2 Concentrate

Concentrate is generated as a by-product of the separation of the minerals from the source water used for desalination. This liquid stream contains in concentrated form most of the source water's dissolved solids as well as some pretreatment additives (i.e., residual amounts of coagulants, flocculants, and antiscalants), and other chemicals, microbial contaminants, and particulates rejected by the RO membranes.

Quantity

Concentrate quantity is a function of the desalination plant size and recovery. Desalination plant size is defined as the fresh water production capacity of the plant. Typically, plant recovery is expressed as the percentage of the total volume of saline source water, which is converted into fresh water by the desalination plant. Brackish water RO plants usually operate at recoveries of 70 to 90 percent. The recovery rate of seawater RO plants is lower (typically 40 to 55 percent). Operation at higher plant recovery results in the generation of smaller concentrate volume of higher salinity and vice versa.

The daily volume of concentrate produced by the desalination plant can be calculated by the following formula:

$$Q_c = Q_p \times \frac{1-R}{R} \tag{16.1}$$

where Q_p and Q_c are the volumes of the plant fresh water production flow and concentrate respectively, and R is the plant recovery in decimal.

Applying Formula (16.1), a seawater desalination plant producing 40,000 m^3/day (10.6 mgd) of fresh water and operating at 45 percent recovery will generate concentrate of the following volume:

$$Q_{c\ \text{seawater plant}} = 40{,}000\ \text{m}^3/\text{day} \times \frac{(1-0.45)}{0.45} = 48{,}889\ \text{m}^3/\text{day (12.9 mgd)}$$

A brackish water desalination plant of the same production capacity and 80 percent recovery will generate approximately five times smaller volume of concentrate:

$$Q_{c\ \text{brackish water plant}} = 40{,}000\ \text{m}^3/\text{day} \times (1 - 0.80/0.80) = 10{,}000\ \text{m}^3/\text{day (2.6 mgd)}$$

Quality

Concentrate water quality depends on the water quality of the saline source water, the salt rejection characteristics of the desalination membranes, and the desalination plant recovery. Higher source water salinity, RO membrane salt rejection, and desalination plant recovery yield higher concentrate salinity.

The salinity of a brackish desalination plant concentrate is typically 4 to 10 times higher than that of the source water. Seawater concentrate usually has 1.5 to 2.0 times higher mineral content than the source water. Concentrate TDS (TDS_c) can be calculated as a function of source water and product water TDS concentrations (TDS_s and TDS_p), and plant recovery (R) as follows:

$$TDS_c = TDS_s \times \frac{1}{1-R} - \frac{R \times TDS_p}{100 \times (1-Y)} \tag{16.2}$$

The ion concentration factor based on 100 percent rejection can be calculated from the following equation:

$$CF = \frac{1}{1-R} \tag{16.3}$$

where CF = concentration factor, dimensionless; R = RO system recovery, expressed as a decimal.

For example, assuming a recovery of 80 percent, the concentration factor is CF = 1/(1 – 0.80) = 5. For more accurate calculation, if the membrane salt passage (SP) is known, the concentration factor can be calculated using the formula below:

$$\text{CF} = \frac{1-(R \times \text{SP})}{(1-R)} \tag{16.4}$$

where SP, the salt passage = 1 – % salt rejection = permeate TDS (TDS_p)/feed TDS (TDS_s), expressed as a decimal.

For example, for an RO system with recovery of 80 percent, and overall 90 percent salt rejection (10 percent salt passage), the CF is

$$\text{CF} = \frac{1-(0.80 \times 0.10)}{(1-0.8)} = 4.6$$

Since RO membranes reject some chemicals better than others, variable concentration factors may apply for specific chemicals. Exactly how the salinity concentration factor impacts the disposal of concentrate depends mainly on the means of disposal. In some cases, volume minimization (high brine concentration factor) is preferred, whereas in cases where the concentrate is to be discharged to waterways, achieving lower TDS concentration is usually more important than low volume.

The salinity concentration factor is primarily limited by the increasing osmotic pressure of the generated concentrate. For reverse osmosis seawater desalination systems, this limit is approximately 65,000 to 80,000 mg/L. The combined effect of membrane rejection and source water concentration typically renders the optimum recovery from a single-pass SWRO system as low as 40 to 45 percent. Therefore concentration factors for single-pass seawater desalination processes are often in a range of 1.5 to 1.8.

For comparison, the considerably lower salinity concentrations of brackish groundwater and municipal wastewater allow us to achieve much greater recoveries. Brackish ground water RO plants typically operate at recoveries of 70 to 90 percent, corresponding to a concentration factor of 4.0 to 10.

The following rules can be used to predict concentrate quality based on source water characteristics: (1) RO membranes reject heavy metals in a similar ratio as calcium and magnesium, (2) most organics are rejected in excess of 95 percent (except for organics of low molecular weight), (3) concentrate from brackish groundwater desalination plants would likely be anaerobic and would contain hydrogen sulphide, and (4) the pH of concentrate is generally higher than the pH of source water because concentrate has higher alkalinity.

If pretreatment is used, the RO membrane feed water would have lower levels of certain constituents (i.e., dissolved metals, microorganisms, and particles) than source water. However, source water pretreatment may result in a slight increase in the content of inorganic ions such as sulphate, chloride, and iron in the RO system feed water, if coagulants are used. Concentrate may also contain residual organics from source water conditioning with polymers and antiscalants.

Concentrate has low turbidity (usually < 2 NTU) and low total suspended solids (TSS) and biochemical oxygen demand (BOD) levels (typically < 5 mg/L) because most of the particulates contained in the source water are removed by the desalination plant pretreatment system. However, if plant pretreatment side streams are discharged along with the concentrate, the blend may contain elevated turbidity, TSS, and occasionally BOD. Acids and scale inhibitors added to the desalination plant source water are rejected by the seawater reverse osmosis membranes in the concentrate and also have an impact on its overall mineral content and quality. Scale inhibitor levels in the concentrate are typically < 20 mg/L.

16.2.3 Backwash Water

Spent filter backwash water is generated during the periodic backwashing of the desalination plant pretreatment filtration system.

Quantity

Backwash quantity mainly depends on the type of pretreatment (granular media or membrane filtration) and the solids content in the source water. At present, the two most widely used types of pretreatment technologies are granular media filtration and membrane microfiltration (MF) or ultrafiltration (UF).

Usually, granular media filtration pretreatment systems use 3 to 6 percent of the intake source water for backwash. For comparison, the backwash water generated by membrane pretreatment systems is 5 to 10 percent of the total volume of the intake source water.

The daily volume of backwash water can be calculated as a function of the production capacity of the desalination plant, the plant recovery, and the percentage of backwash water as a function of the intake water flow as follows:

$$Q_{bw} = Q_{\mathrm{p}} \times \frac{\mathrm{BW}}{R} \tag{16.5}$$

where Q_{bw} and Q_p are the daily flows of backwash water and plant product water capacity, respectively, expressed in m^3/day, R is the desalination plant recovery in percentage of intake water, and BW is the volume of backwash water expressed as a percentage of the volume of the plant intake.

For example, for a desalination plant with fresh water production capacity of 40,000 m^3/day (10.6 mgd), plant recovery of 45 percent, and daily backwash water volume of 5 percent of the plant intake daily flow, the daily volume of backwash will be:

$$Q_{bw} = 40{,}000\ \text{m}^3/\text{day} \times (5\%/45\%) = 4444\ \text{m}^3/\text{day}\ (1.2\ \text{mgd})$$

Backwash volume increases with the increase of source water turbidity because the filter cycles between two backwashes shorten, and the filters need to be backwashed more frequently.

Quality

The main constituents of the spent filter backwash water are the source water solids removed by the pretreatment system and the spent coagulant (if coagulant is used for source water conditioning prior to filtration). When ferric salts (ferric chloride and ferric sulphate) are used for source water coagulation, backwash water contains a mix of coagulated solid and colloidal particles and ferric hydroxide. The concentration of total suspended solids in the spent backwash water can be calculated as a function of the TSS concentration of the saline source water and the dosage of the applied iron coagulant using the following formula:

$$TSS_{bw} = \frac{(TSS_s + 0.8 \times DOSE_{FE}) \times Q_S}{Q_{bw}} \tag{16.6}$$

where TSS_{bw} and TSS_s are the total suspended solids concentrations of backwash water and source water, respectively, in mg/L, $DOSE_{Fe}$ is the dose of ferric salt expressed as iron concentration, in mg/L, and Q_{bw} and Q_s are the daily flows of desalination plant backwash water and intake source water, respectively, in m^3/day.

Using Formula (16.6), the TSS concentration of the backwash water generated by the pretreatment system of the 40,000 m^3/day (10.6 mgd) desalination plant described in the previous example with TSS concentration of the source water of 2.5 mg/L, which is treated with ferric chloride coagulant at a dosage of 5.0 mg/l (as iron) before pretreatment, will be:

$$TSS_{bw} = \frac{(2.5\ \text{mg}/\text{L} + 0.8 \times 5.0\ \text{mg}/\text{L}) \times 40{,}000\ \text{m}^3/\text{day}}{4444\ \text{m}^3/\text{day}} = 58.9\ \text{mg}/\text{L}$$

The example above indicates that backwash water could contain a significant amount of solids, and its concentration could exceed the 30 mg/L TSS discharge standard commonly applied for surface water discharges.

However, if mixed with the desalination plant concentrate, the TSS backwash concentration could be reduced below the regulatory threshold. As indicated in a previous example, a 40,000 m^3/day (10.6 mgd) seawater desalination plant operating at 45 percent recovery will have a daily concentrate discharge volume of 48,889 m^3/day (12.9 mgd). Since concentrate can be assumed to be void of suspended solids (TSS = 0 mg/L), the concentration of the blend of 48,889 m^3/day of concentrate and 4444 m^3/day of backwash water of 58.5 mg/L of TSS, the TSS of the blend will be:

$$TSS_{blend} = \frac{(58.5\ \text{mg/L} \times 4444\ \text{m}^3/\text{day}) + (0\ \text{mg/L} \times 48{,}889\ \text{m}^3/\text{day})}{(4444\ \text{m}^3/\text{day} + 48{,}889\ \text{m}^3/\text{day})} = 4.9\ \text{mg/L}$$

This calculation indicates that blending of the concentrate and the backwash water will be beneficial. However, the continuously low solids content can only be achieved if the backwash water and the concentrate are mixed and equalized before their discharge. In practical terms, backwash water from washing of the media of the individual pretreatment filter cells is generated periodically, and solids load discharge is not evenly distributed unless it is equalized. As a result, if the spent filter backwash is released as it is generated, even if blended with the concentrate, it will cause discharge TSS spikes of undesirable magnitude. Therefore backwash water is typically stored in equalization tanks and released from these tanks at a near constant rate.

Since ferric hydroxide (also commonly known as rust) which is generated as a result of the addition of iron-based coagulant in the water is red in color, the backwash water is discolored, if ferric salt is used for coagulation. Therefore a direct discharge of the backwash water into a surface water body would also cause discoloration of the entire plant discharge. While iron contained in the backwash water is typically not harmful to the marine environment, concentrate discoloration usually is not acceptable from an aesthetic point of view. To address this challenge, regulatory requirements often necessitate backwash treatment in an on-site solids handling system to remove iron hydroxide. Such treatment is of critical importance if the desalination plant discharge will be disposed by well injection or by shallow or onshore surface water discharge.

While surface water discharge after blending with the concentrate is the most commonly practiced method for spent filter backwash disposal at present, alternatively, for small plants, this side stream could be discharged to the sanitary sewer for further treatment at the local wastewater treatment plant (WWTP). Usually, coagulant contained in the backwash water has a positive impact on the WWTP primary clarification process. However, backwash salinity may inhibit the secondary biological treatment of the wastewater because of its high salinity. Therefore discharging spent filter backwash water to the sanitary sewer requires careful consideration of the impact of this discharge on the operation of the receiving WWTP.

Besides iron, other conditioning chemicals that may be contained in the spent filter backwash water include flocculants, chlorine compounds, acids, and biocides. These conditioning chemicals are usually in quantities that do not have a significant impact on the overall desalination plant discharge water quality after the dilution of the spent backwash water with desalination plant concentrate. Since such concentrate dilution is of critical importance for the mitigation of the environmental impact of the spent filter backwash water, most desalination plants have an interim retention (buffer) tank for blending of the desalination plant waste streams prior to their discharge.

16.2.4 Membrane Flush Water

Membranes used for source water pretreatment and RO separation have to be cleaned periodically with chemicals in order to remove foulants accumulated on the membrane surface during routine plant operations. Because such cleaning is completed without removing the membranes from the membrane vessels, such cleaning process is referred to as clean-in-place (CIP). The type and combination of CIP chemicals are selected based on the predominant type of fouling occurring on the membrane surface (particulate, colloidal, organic, and microbiological). Because often more than one type of fouling occurs on the membrane surface, a combination of CIP chemicals may be needed to recover membrane performance. The configuration and function of a typical CIP system as well as typical membrane cleaning procedures are discussed in greater detail in Chap. 14.

Quantity

The total quantity of the spent CIP chemicals depends on the size of the desalination plant, the number and type of the plant pretreatment and RO membranes, the quantity and type of the membrane foulants, the fouling potential of the source water, and the type of fouling that accumulates on the membrane surface.

Typically, the cleaning solution volumes generated during a CIP of RO membranes are 1.0 to 1.8 L/m^2 (0.025 – 0.045 gal/ft^2) of membrane surface. This volume does not include the flush water volumes. Typical cleaning solution volume is estimated by adding the total volume of the RO system and interconnecting pipe volume. The volume of the RO system is calculated as follows:

$$V_{\text{RO system}} = N_t \times N_{vpt} \times N_{epv} \times A_{ro} \times U_{cl} \tag{16.7}$$

where $V_{\text{RO system}}$ is the volume of the RO system, N_t is the number of RO trains, N_{vpt} is the number of vessels per train, N_{epv} is the number of elements per vessel, A_{ro} is the total membrane surface area of one RO element (m^2), and U_{cl} is the unit cleaning volume (L/m^2).

For example, 40,000 m^3/day (10.6 mgd) RO system with six RO trains with 72 RO vessels per train and seven 8-in elements per vessel, and RO elements with a surface area of 37.2 m^2 (400 ft^2), as well as 1500 m (4900 ft) of average distribution system diameter of 200-mm (8-in) pipe, the total volume of cleaning solution for all RO trains is approximately 215,840 L/57,020 gal (168,740 L for RO system and 47,100 L in piping) per cleaning chemical and per cleaning event. This volume is calculated assuming cleaning volume of 1.5 L/m^2 of membrane area. The volume is estimated as follows:

$V_{\text{RO system}}$ = 6 RO trains × 72 vessels per train × 7 RO elements per vessel × 37.2 m^2 per RO element × 1.5 L/m^2 of cleaning solution = 168,740 L (44,580 gal).

The volume of the cleaning solution for the 1500 m of 200-mm (0.2 m) diameter pipe is (3.14 × 0.2 × 0.2/4) × 1500 m = 47.1 m^3 = 47,100 L (12,440 gal).

This volume is specific for each chemical solution and RO system configuration. RO system cleaning is often completed in multiple steps, so the total annual volume is the sum of the volumes used in each step. Depending on the foulants, a low-pH solution is usually followed by one with a high pH. The trains are also cleaned in steps.

A commonly applied approach for large RO trains (100 vessels or more) is to first clean the modules in one half of the vessels in the first stage, then the other half of the first stage, and finally all modules in the second stage.

Membrane cleaning is followed by draining of the spent cleaning chemicals and flushing of the RO membranes. Therefore the waste streams generated during the RO train cleaning are (1) concentrated waste cleaning solution, (2) first flush, (3) spent flush water permeate from consecutive flushes, and (4) flush water concentrate.

Concentrated waste cleaning solution contains the actual spent membrane cleaning chemicals. The quality and quantity of this stream is described in detail above. Flush water residual cleaning solution (first flush) is the first batch of clean product water used to flush the membranes after the recirculation of cleaning solution is discontinued. This first flush contains diluted residual cleaning solution. Flush water permeate is the spent cleaning water used for several consecutive membrane flushes after the first flush. This flush water is of low salinity and contains only trace amounts of cleaning solution. Flush water concentrate is the flush water removed from the concentrate lines of the membrane system during the flushing process. This water contains very little cleaning chemicals and is of slightly higher salinity concentration than flushing permeate.

The total volume of flushing water for cleaning of the RO system would depend on the size of the RO system and individual trains and the number of different cleaning chemicals applied per cleaning—as a rule of thumb flushing water volume would be 5 to 10 times larger than the volume of the cleaning chemicals.

It should be pointed out that the total annual volume of the membrane flush water is usually less than 0.1 percent of the total volume of the total discharge flow, and therefore its impact on the discharge water quality is insignificant. In many cases, however, this side stream is discharged to the wastewater collection system.

Assuming that three different chemicals are used for cleaning, and the volume of flush water is seven times the volume of the cleaning chemicals, for the example above, the total volume of membrane flush water generated for one cleaning of the entire RO system of the example 40,000 m^3/day plant will be three chemicals $\times$ 215,840 L $\times$ (1 + 7) (for chemicals and flushing water) = 1,726,720 L/cleaning = 456,200 gal/RO system cleaning. This averages approximately 76,000 gal per RO train for this example.

If all RO membrane trains are cleaned four times per year, then the total volume of the membrane flush water for the entire year is 1,726,720 L/1000 $\times$ 4 times = 6910 m^3/yr (1,824,800 gal/yr). Taking into consideration that the desalination plant will produce 40,000 m^3/day $\times$ 365 days = 14,600,000 m^3 (3,860 million gallons) of fresh water per year, for this example, the total annual volume of the membrane flush water is only 0.05 percent of the plant annual production flow and less than 0.04 percent of the total plant discharge flow.

Quality

The water quality of the spent membrane cleaning solution (CIP residuals) reflects the chemical characteristics of both the spent cleaning solution and the material removed from the membrane system during CIP. Reactions with foulants will tend to raise the pH of acid solutions and lower that of basic ones.

Chapter 14 provides discussion of typical cleaning formulations developed to remove various types of foulants. Some of the cleaning solutions, such as citric acid, may have relatively high BOD concentration (2000 to 3000 mg/L) and therefore may contribute to the increase in the BOD level of the desalination plant discharge. Others, such as phosphoric and nitric acid, can add undesirable nutrients to the discharge.

It should be pointed out, however, that when blended with the desalination plant concentrate, which has very low nutrient and BOD content and several order of magnitude larger volume, these streams would not result in a measurable impact on the surrounding environment. If increase in nutrient and/or BOD load in the discharge from spent CIP chemicals is limited due to site-specific regulatory requirements, these waste streams are typically directed to sanitary sewer for further treatment in a local WWTP.

16.3 Surface Water Discharge of Concentrate

Surface water discharge involves disposal of concentrate from the desalination plant to an open water body such as a bay, tidal lake, brackish canal, or ocean. The three most widely used alternatives for disposal of concentrate from desalination plants to surface water bodies at present are: (1) direct surface discharge through a new near-shore outfall structure or an off-shore outfall, (2) discharge through existing wastewater treatment plant outfall, and (3) co-disposal with cooling water of existing power plant (co-location). Each of these concentrate management alternatives has limitations and potential environmental impacts on an aquatic environment (Hoepner and Windelberg, 1996; Hoepner, 1999; Rhodes, 2006). This chapter focuses on design features and considerations for surface water discharges.

16.3.1 New Surface Water Discharge

Description

Discharge of concentrate and other desalination plant waste streams through a new surface water discharge system (near-shore discharge structure or offshore outfall) is widely used for desalination projects of all sizes. Such discharges are more common for seawater rather than brackish water desalination plants.

Over 90 percent of the large seawater desalination plants worldwide dispose their concentrate through a new outfall specifically designed and built for that purpose. Examples of large SWRO desalination plants with ocean outfalls for concentrate discharge are the 462,000 m^3/day (122 mgd) plant in Hadera, Israel (Fig. 16.3), the 136,000 m^3/day (36 mgd) Tuas seawater desalination plant in Singapore, the 64,000 m^3/day (17 mgd) Larnaka desalination facility in Cyprus, and the majority of large SWRO plants in Spain, Australia, and the Middle East.

The main purpose of outfalls is to discharge the plant concentrate to a surface water body in an environmentally safe manner, which in practical terms means to minimize the size of the zone of the discharge in which the salinity is elevated outside of the typical TDS range of tolerance of the aquatic organisms inhabiting the discharge area.

The two key options available to accelerate concentrate mixing with the water of the receiving water body is to either rely on the naturally occurring mixing capacity of the

Figure 16.3 Near-shore discharge of Hadera SWRO Plant, Israel. (*Source: IDE.*)

tidal (surf) zone or to discharge the concentrate beyond the tidal zone and to install diffusers at the end of the discharge outfall in order to improve mixing. Although open-ocean near-shore tidal zones usually carry a significant amount of turbulent energy and provide much better mixing than the end-of-pipe-type diffuser outfall system, such zones have limited capacity to transport and dissipate the saline discharge load into the surface water body.

If the mass of the saline discharge exceeds the threshold of the tidal zone's salinity load transport capacity, the excess salinity would begin to accumulate in the tidal zone and could ultimately result in a long-term salinity increment in this zone beyond the level of tolerance of the aquatic life in the area of the discharge. Therefore the tidal zone is usually a suitable location for salinity discharge only when it has adequate capacity to receive, mix, and transport this discharge into the surface water body (ocean, river, bay, etc.). The site-specific salinity threshold mixing/transport capacity of the tidal zone in the area of the desalination plant discharge can be determined using hydrodynamic modeling (Bleninger and Jirka, 2010).

Examples of large desalination plant discharges in the tidal zone are those of the 330,000 m^3/day (86 mgd) Ashkelon seawater desalination plant, the 462,000 m^3/day (122 mgd) Hadera SWRO Plant in Israel (Fig. 16.3), and the 170,000 m^3/day (45 mgd) Fujairah SWRO plant in the United Arab Emirates (UAE).

For small desalination plants (i.e., plants with production capacity of 1000 m^3/day or less), the outfall is typically constructed as an open-ended (sometimes perforated) pipe that extends several hundred meters into the tidal (high mixing intensity) zone of the receiving water body. This type of discharge usually relies on the mixing turbulence of the tidal zone (for ocean discharges) to dissipate the concentrate and to reduce the discharge salinity to ambient conditions.

Most of the ocean outfalls for large seawater desalination plants usually extend beyond the tidal zone. Large ocean outfalls are equipped with diffusers in order to provide the mixing necessary to prevent the heavy saline discharge plume from accumulating at the ocean bottom in the immediate vicinity of the discharge.

The length, size, and configuration of the outfall and diffuser structures for large desalination plants are typically determined based on hydrodynamic or physical modeling of the discharge diffuser structure for the site-specific conditions of the outfall location (Purnama and Al-Barwani, 2004; Purnama et al., 2007; Bleninger and Jirka, 2010).

16.3.2 Potential Environmental Impacts

Overview of Environmental Issues and Considerations

The main challenges associated with selecting the most appropriate location for a desalination plant's outfall discharge are finding an area devoid of endangered species and stressed aquatic habitats, identifying a location with strong underwater currents that allows quick and effective dissipation of the concentrate discharge, avoiding areas with frequent naval vessel traffic that could damage the outfall facility and change mixing patterns, and identifying a discharge location in relatively shallow waters, which at the same time is close to the shoreline in order to minimize outfall construction expenditures. Key environmental issues and considerations associated with concentrate disposal to surface waters include salinity tolerance of aquatic species inhabiting the discharge area, concentration of some source water constituents to harmful levels, and discharge discoloration and low oxygen content.

The key issues that should be addressed during the feasibility evaluation of disposal of desalination plant concentrate to a surface water body include: (1) assessment of discharge dispersion and recirculation of the discharge plume to the plant intake, (2) evaluation of the potential for whole effluent toxicity of the discharge, (3) determination of whether the discharge water quality meets the numeric and qualitative effluent water quality standards applicable to the point of discharge and established by regulatory agencies, and (4) determination of the aquatic organism salinity tolerance threshold for the site-specific conditions of the discharge location and outfall configuration in order to design the outfall for dilution that meets this threshold within a minimal distance from the point of discharge. An overview of key environmental challenges and solutions associated with the surface water disposal of desalination plant discharges is presented in Chap. 5.

16.3.3 Concentrate Treatment Prior to Surface Water Discharge

Potential BWRO Concentrate Treatment Requirements

A comprehensive study completed by the American Water Works Research Foundation (Mickley, 2000) indicates that discharge of BWRO plant concentrate to surface waters may exhibit toxicity and therefore, in some cases it may not a viable method for concentrate disposal. This research also concludes that brackish concentrate toxicity is not caused by the membrane treatment process itself but results from the nature of the groundwater/brackish water source and its major ion makeup. In comparison to brackish brine, high-salinity concentrate generated during the desalination of seawater by reverse osmosis membranes does not exhibit ion-imbalance related toxicity.

A method which can be used to predict the potential toxicity of groundwater brackish concentrate discharge is based on comparative evaluation of concentrate salinity relative to seawater and is referred to as a percent difference from balance (PDFB) approach (Mickley, 2001). Based on this approach from a prospective of toxicity having an impact on the aquatic environment, seawater is considered balanced water. The PDFB approach compares the ion composition of seawater diluted or concentrated to the brackish water concentrate TDS level to the ion composition of the concentrate from the brackish water desalination plant. The PDFB of seawater of TDS concentration of 33,000 mg/L is assumed equal to zero (i.e., typical seawater would not exhibit ion-imbalance-triggered toxicity).

If the concentrations of major ions (i.e., chloride, sodium, sulfate, magnesium, calcium, and bicarbonate) in the brackish water concentrate or blend of wastewater and seawater concentrate, deviate from the concentrations of the same ions in the reference seawater (i.e., seawater diluted of concentrated to the same TDS concentration) than the probability of toxicity triggered by ion imbalance of the concentrate increases.

The site-specific threshold level of PDFB of the discharge depends on the aquatic organisms living in the outfall discharge. Therefore, the most reliable and practical approach to assess the potential impact of desalination plant concentrate discharge generated by a specific project is to complete a pilot test, which allows to produce concentrate representative of the full-scale plant operations and then to complete chronic and acute WET testing at various dilutions of this concentrate and the receiving water. The level of dilution at which toxicity is not observed would be considered an acceptable discharge TDS level. Once this level is known, the next step is to determine the dilution ratio that needs to be achieved in the zone of initial dilution of the discharge, which is a critical parameter for sizing of discharge configuration.

For example, if chronic WET testing shows that under-average annual salinity and flow concentrate would need to be diluted three times with receiving water, in order to eliminate its toxicity impact, then the outfall structure would need to be designed in such a manner that three-time dilution is achieved within a 300-m (1000-ft) radius of the discharge.

Acute WET testing allows us to assess the environmental impact of the discharge at maximum salinity. While with chronic toxicity time is not a factor (i.e., chronic toxicity limit has to be met at all times), the acute toxicity limitation is time-related because usually extreme salinity conditions occur over relatively short periods of time. For example, if the average annual salinity of a desalination plant discharge is 8000 mg/L and the maximum daily salinity is 12,000 mg/L and this extreme salinity would not occur for more than three days, then an acute WET test should be completed for salinity of 12,000 mg/L and for a duration of three days. The dilution at which acute toxicity does not occur in three days of exposure of the test organisms to 12,000 mg/L of concentrate discharge will be considered the threshold acute toxicity dilution ratio. For example, if such ratio is equal to 5, then the acute toxicity dilution threshold (i.e., dilution ratio of 5) will be the design dilution ratio for which the desalination plant discharge configuration would have to be designed to accommodate both acute and chronic toxicity dilution thresholds.

Potential SWRO Concentrate Treatment Requirements

Usually, concentrate from seawater desalination plants has ion composition similar to than of the ambient seawater and its direct ocean discharge does not pose ion-imbalance-driven toxicity challenges. Therefore, seawater concentrate typically can be discharged to the ocean without additional treatment, especially if the source seawater is collected by an open ocean intake. In this case, concentrate is either discharged using a diffuser system or is blended with source seawater down to a salinity level that is safe for direct discharge (usually 40,000 mg/L or less) without the need for complex diffusion structure. While blending concentrate with ambient seawater is relatively simple to implement, it may result in an elevated impingement and entrainment of marine organisms and energy use to collect source water needed for concentrate dilution.

Typically, seawater concentrate from open ocean intakes typically does not require treatment prior to discharge; however, if subsurface (well) intake is used to collect source seawater, the plant concentrate may be discolored due to an elevated concentration of iron, have a low oxygen concentration, or contain other contaminants that may trigger the need for additional source water or concentrate treatment.

Often source seawater collected from alluvial coastal aquifers by beach wells may contain high levels of iron and manganese in reduced form. In many applications, such source seawater is processed through the desalination plant pretreatment and RO facilities without exposure to air/oxygen, which keeps iron and manganese in a dissolved reduced form in which they are colorless. Because RO membranes easily remove dissolved iron and manganese, after membrane separation these contaminants are retained in the concentrate. However, if this concentrate is exposed to air, iron will convert from reduced form (typically ferric sulfide) to oxidized form (ferric hydroxide). Since ferric hydroxide is red in color, it would discolor the concentrate, which degrades the visual appearance of the discharge area. Therefore iron contained in the source water in reduced form would need to be oxidized and removed in the pretreatment system in order to address the elevated iron content, or concentrate would need to be treated by sedimentation to remove ferric hydroxide.

If a large desalination plant delivers low-DO concentrate to the surface water body, this discharge could cause oxygen depletion and stress to aquatic life. Therefore this desalination plant concentrate has to be re-aerated before surface water discharge.

Potential sources of pollution of source water supply aquifers or surface water bodies are existing landfills, septic tank leachate fields, industrial and military installations, cemeteries, etc. Intakes and, therefore, discharges from desalination plants using such water sources would contain elevated content of these contaminants. The compounds of concern could be treated by a number of available technologies, including activated carbon filtration, UV irradiation, hydrogen peroxide oxidation, ozonation, etc. However, because these treatment systems will need to be constructed in addition to the RO system, this supplemental concentrate treatment may increase the overall desalinated water production cost measurably.

16.3.4 Design Guidelines for Surface Water Discharges

Outfall Pipeline

The concentrate disposal site should be located as near to the desalination plant as practically possible. Concentrate discharge pipes should be made of corrosion- and crush-resistant material. At present, plastic HDPE, GRP, and PP pipe materials are most commonly used for outfalls of small-, medium-, and large-size desalination plants. The most widely used low-cost construction of outfalls includes the use of plastic pipes installed directly on the ocean floor and secured to the bottom with concrete blocks (Fig. 16.4).

Figure 16.4 Outfall discharge pipeline supported on concrete blocks.

Table 16.1 lists the type and maximum size of plastic pipes most commonly used in outfall construction. Over the past 10 years, HDPE, PP, and GRP piping has replaced traditional materials for construction of ocean outfall piping systems (concrete, steel, and cast iron). Key advantages of plastic pipe materials include higher corrosion resistance, chemical inertness, lighter weight, resistance to galvanic attack, and lower unit cost. In many cases, HDPE and GRP outfalls are less expensive than traditional piping materials, such as concrete or steel pipe.

Usually, the use GRP pipe is more cost advantageous then HDPE pipe. However, GRP pipe is positively buoyant in water, cracks more easily, and is not resistant to negative pressures. If the outfall pipe is located in a beach area and exposed to accelerated erosion or wave action, GRP pipe has to be buried and installed in trenches on special bedding, which often makes it more costly than the installation of HDPE, concrete, or steel pipe on the ocean bottom.

While plastic pipes are typically preferred to concrete or steel pipe, often for large- and mega size desalination plants with outfall located in areas with strong underwater currents and environmentally sensitive marine habitats, and sites with active beach erosion or intense ship traffic, the construction of large-diameter reinforced concrete tunnels under the ocean bottom is the preferred outfall discharge alternative. Usually such concrete tunnel discharges are several times more expensive than construction of discharge, which consists of multiple plastic outfall pipes. Therefore construction of discharge tunnels is recommended to be avoided if site-specific conditions are favorable for pipeline construction.

Typically, outfall pipelines are designed to maintain velocity of 1 m/s (3 ft/s) or more in order to prevent formation of deposits and scale on the inner surface of the pipes. The maximum velocity/minimum pipe size is determined based on the total available discharge head and the goal to avoid pumping of concentrate into the discharge line, if possible.

Discharge outfall pipe is usually sized to convey the entire maximum design intake volume because during start-up and commissioning, as well as after routine plant shutdowns, the pretreated seawater is often discharged back to the ocean until it reaches quality suitable to be directed to the RO membranes. Similarly, in cases of source water quality contamination of magnitude that makes this water unsuitable for processing through the desalination plant (i.e., large oil spill or other temporary source water contamination), the water that has entered the desalination plant intake can be returned directly back to the ocean without contaminating the desalination plant treatment facilities and equipment.

In many cases, the discharge outfall pipe is designed to handle only concentrate, spent filter backwash water, and CIP solutions, thereby reducing discharge facility size and costs. While this design approach decreases the plant capital expenditures, it also reduces a plant's operational flexibility, especially if the plant is planned to be operated intermittently.

Plastic Material	Typical Maximum Acceptable Diameter, in/mm
High-density polyethylene (HDPE)	78/2000
Glass reinforced plastic (GRP)	156/4000
Polypropylene (PP)	24/600

TABLE 16.1 Plastic Piping Materials Used for Outfalls

Concentrate Conveyance

Concentrate exits the RO system at pressures ranging from ambient atmospheric pressure to 2.5 bars (36 lb/in^2) depending on the type of energy recovery device. In most cases, the available concentrate head is sufficient to overcome frictional losses within the pipe, allowing transport of the concentrate flow to the disposal site without the need for additional pumping. When pumping is necessary, energy use and maintenance associated with the concentrate pump station and conveyance system become important cost factors. The need for surge control should also be considered in the design.

Outfall Diffuser Design

Outfall pipes typically terminate with a multiport diffuser, a perforated discharge section, or a simple open end. A multiport diffuser is designed so that the end of the transport pipe is capped, and the last sections of the pipe contain lines of small ports (openings or diffuser nozzles) around the circumference of the pipe. The purpose of the diffuser is to provide a greater initial dilution of the concentrate as it enters the surface water.

Most small outfalls, as well as larger ones built before 1980, have simple open ends with perforations along the last 10 to 30 percent of the pipeline length. In recent years, multiport diffusers have become the accepted design norm for larger-diameter outfalls. Simple open-end outfalls are recommended when the initial dilution that is achieved naturally at the point of exit is adequate to meet applicable discharge water quality standards. If dilution requirements are not met at the point of exit, installation of diffusers at the end of the discharge pipeline/s becomes necessary.

The most commonly used concentrate discharges have a series of diffusers, which are designed to direct the desalination plant concentrate toward the surface of the ocean and to release it with energy that is adequate to facilitate concentrate plume dissipation within a predetermined distance from the point of discharge referenced as zone of initial dilution (ZID).

The key parameters for desalination discharge outfalls, which need to be determined during design, include:

- Diameter and length of concentrate discharge pipe
- Configuration of diffuser system
  - Number of diffuser ports
  - Distance between ports
  - Port diameter
  - Port angle from pipeline
- Pipe and diffuser port material
- Distance of diffuser system from shore
- Diffuser depth
- Diffuser exit velocity

The optimum configuration and design of the outfall diffuser system listed above can be determined using hydrodynamic models such as CORMIX, EPA PLUMES (Visual Plumes), and others.

Some key guidelines for a diffuser system design are as follows:

- Design diffuser exit velocity to be in a range of 2 to 4 m/s (7 to 13 ft/s). This velocity is determined by the force needed to eject the desalination concentrate near the surface of the ocean, which allows us to maximize the concentrate mixing/contact time with the ambient water column and to engage the largest possible volume of ambient water in the mixing process.
- Place the diffuser system perpendicular to the prevailing ocean current.
- Select distance between ports should be such that their individual discharge plumes do not overlap.
- Install diffusers at an angle of 45 to 60° from the horizontal pipe (Fig. 16.5 for configuration of 50° diffusers of the Victorian SWRO plant).

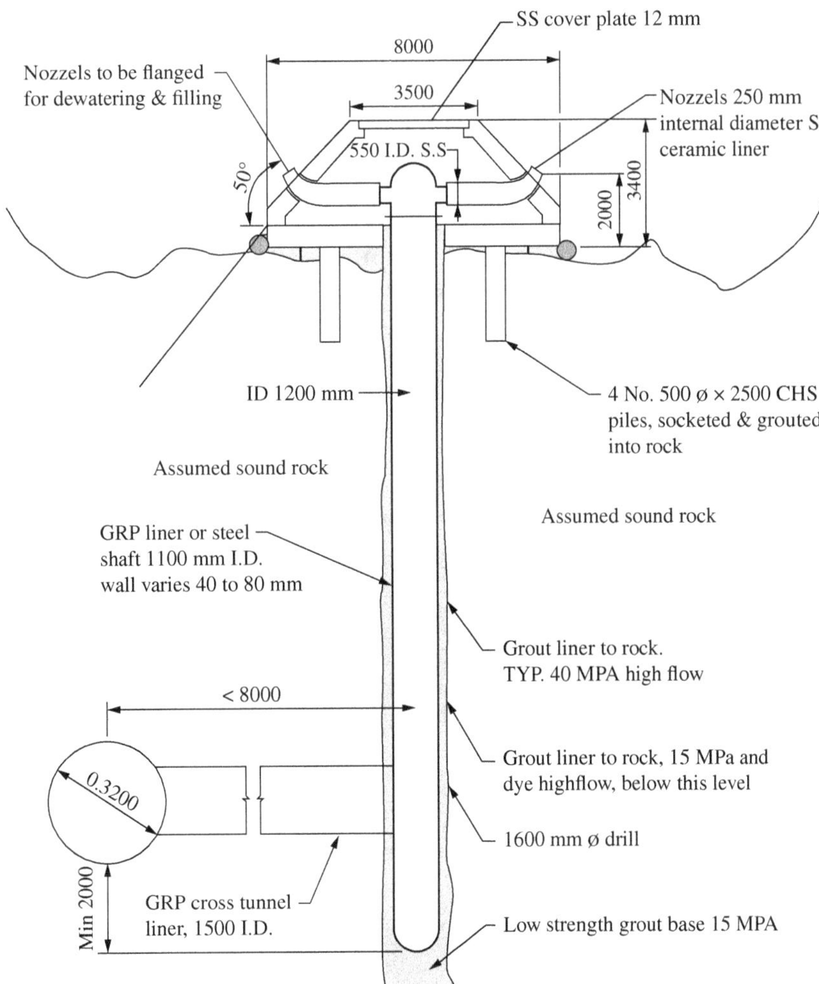

Figure 16.5 Outfall diffusers of Victorian SWRO plant (*Source: Melbourne Water*).

- Increase gradually the size of the ports toward the end of the pipe to maintain sufficient flow in each diffuser.
- Maintain the total cross-sectional area of the diffuser ports below 70 percent of the cross-section of the outfall pipe.
- Install diffuser ports with diameter of 75 mm (3 in) or larger in order to prevent their blockage.

Since dilution is accelerated with the decrease in the density difference between the concentrate and ambient water, concentrate pre-dilution could simplify or completely eliminate the need for diffuser structure.

Various diffuser configurations have found applications in seawater desalination projects. For example, the Sydney Water SWRO project uses four outlet structures ejecting seawater approximately 60° upward (Fig. 16.6). The outlet structures are located at depth of 20 to 30 m (66 to 98 ft).

Other diffuser structures, such as the Perth and Gold Coast SWRO plants in Australia have risers with ports directing concentrate discharge upward (Fig. 16.7).

16.3.5 Costs for New Surface Water Discharge

The costs for construction of new surface water concentrate discharge are influenced by a number of site-specific factors such as: (1) concentrate discharge flow rate, (2) type of surface discharge—near-shore or offshore, (3) materials of construction of the discharge structure and outfall pipeline/tunnel, (4) the complexity of the discharge diffuser system, (5) the costs to convey the concentrate from the desalination plant to the surface water discharge structure/outfall, (6) concentrate treatment costs (if needed), and (7) costs associated with the environmental monitoring of the concentrate discharge. In addition to these cost-related factors, installation of outfall pipeline above or below ground also has a measurable impact on the overall outfall cost.

FIGURE 16.6 Sydney water SWRO diffuser structure.

Unusual ground conditions can significantly increase the cost of pipeline system installation. Underwater trenching is usually three to five times more expensive than trench excavation on dry land. Therefore, instead of installing the outfall in a trench, it is often laid down on the ocean bottom and secured by concrete blocks located at every 5 to 10 m (17 to 33 ft) along the entire outfall length (see Fig. 16.7).

The costs for concentrate conveyance are typically proportional to the concentrate flow rate and the distance between the desalination plant and the discharge outfall. The outfall construction costs are site specific and, in addition to the outfall size and diffuser system configuration (which is driven by the concentrate volume, salinity, and the hydrodynamic conditions in the area of the discharge), these costs are dependent on the outfall length and material, which, in turn, are function of the site-specific surface water body hydrodynamics conditions.

An order of magnitude construction cost for near-shore ocean discharges as a function of the concentrate flow rate is presented on Fig. 16.8. Figure 16.9 depicts the unit construction cost of HDPE pipeline outfalls and of concrete tunnel outfalls expressed in US$/linear meter of outfall length.

Both cost estimates do not incorporate the expenditures for concentrate conveyance from the desalination plant to the outfall structure and for treatment of concentrate, if such treatment is needed. Construction costs associated with elaborate offshore monitoring of concentrate discharge (i.e., stationary or floating water quality monitoring equipment) are not included as well.

The costs associated with environmental monitoring in the case of surface water discharge may be significant, especially if the discharge is in the vicinity of an impaired water body, environmentally sensitive area or area of limited natural flushing.

Cost Example

An order of magnitude cost estimate of near-shore and offshore concentrate discharge systems from a hypothetical reference brackish desalination plant is presented below. The BWRO plant has fresh water production capacity of 40,000 m^3/day (10.6 mgd); it is

FIGURE 16.7 Gold Coast SWRO plant diffuser.

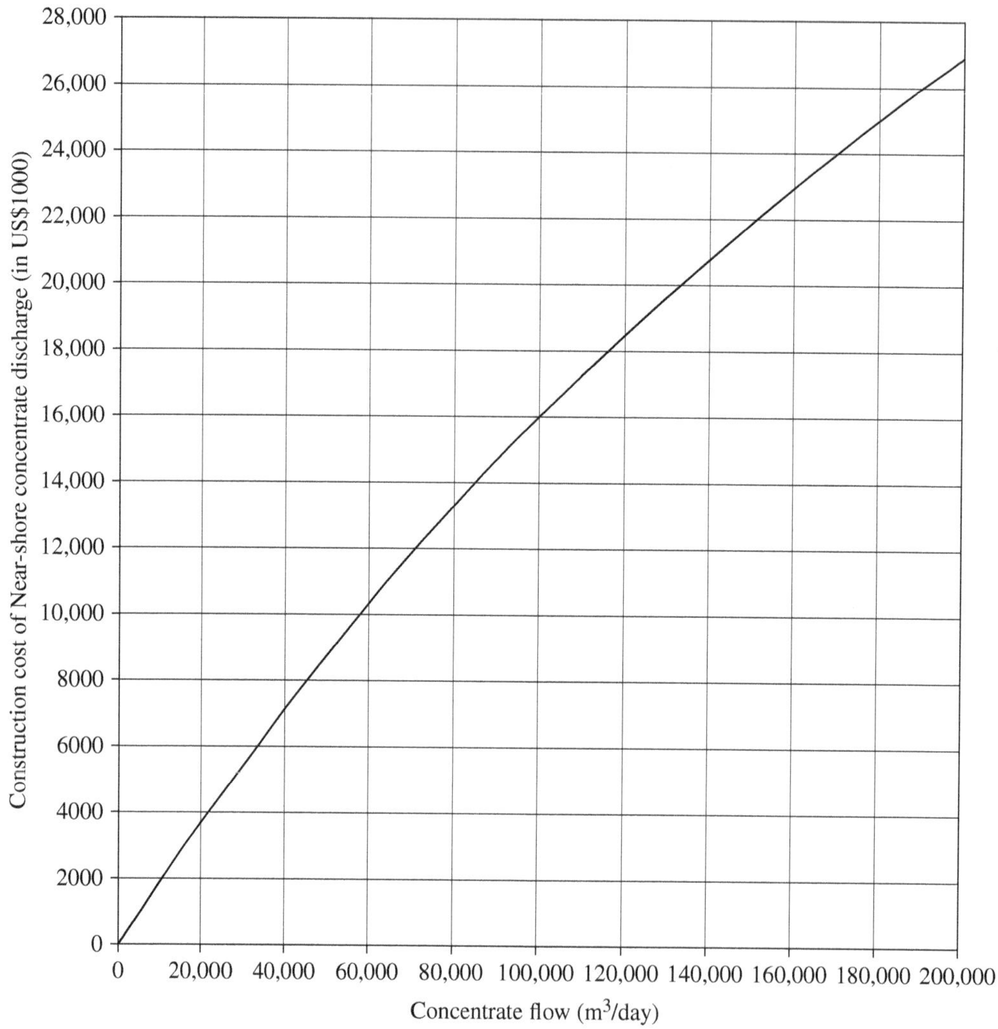

FIGURE 16.8 Construction cost of near-shore discharge.

designed for 80 percent recovery and collects brackish source water from a surface water body. The concentrate discharge projected to be generated by this plant is 10,000 m^3/day (2.6 mgd).

If this concentrate is planned to be disposed via near-shore discharge, using Fig. 16.8, the construction cost for such discharge is estimated at \$2.0 million. If local conditions are not suitable for construction of near-shore discharge, the outfall could be constructed using HDPE pipe or built as a concrete outfall tunnel. Assuming outfall length of 1000 m (3280 ft), at unit cost of HDPE outfall of \$3100/m (see Fig. 16.9), the total outfall construction cost is estimated at \$3.1 million. The construction of directionally drilled concrete tunnel instead of HDPE outfall pipeline would be more expensive, and, at a unit tunnel cost of \$10,200/m (Fig. 16.9), the construction of the 1000-m (3280 ft) long tunnel would cost \$10.2 million.

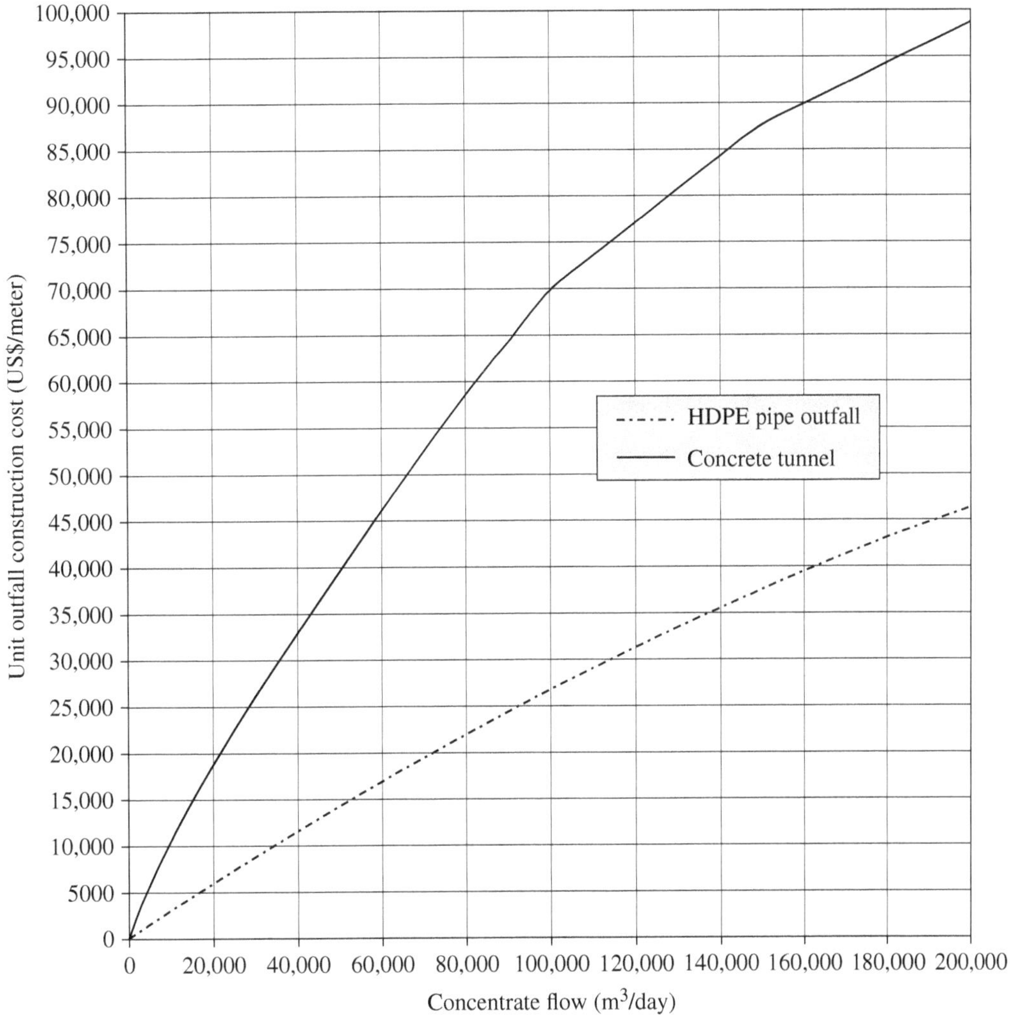

FIGURE 16.9 Unit outfall costs.

This example illustrates the fact that near-shore discharges are usually the least costly surface water discharge scenario. Construction of HDPE outfall for the same size discharge in this case is over 30 percent more expensive. Building an underground tunnel is the most costly scenario.

The costs presented above do not include the expenditures associated with acquisition of land needed for facility construction, concentrate treatment (if needed), and the costs for conveyance of concentrate form the desalination plant site to the surface discharge location.

16.3.6 Case Studies of New Surface Water Discharges

This section presents worldwide case studies of surface water discharges from seawater desalination plants.

Gold Coast Desalination Plant, Australia

This 136,000 m^3/day (36 mgd) desalination plant is located in southeast Queensland, Australia, in an area that is a renowned tourist destination (Fig. 16.10).

The desalination plant has been in operation since November of 2008 and employs an open-intake, pretreatment system and reverse osmosis desalination system. The Gold Coast plant is a stand-alone facility, which discharges concentrate of 67,000 mg/L through a multiple diffuser system. The zone of initial dilution of this plant is 120 × 320 m (400 × 1050 ft).

The Gold Coast plant discharge diffusers are located at the ocean bottom and discharge concentrate upward into the water column to a height of approximately 10 m (33 ft) (Fig. 16.11). This discharge is conveyed another 5 m (16 ft) upward toward the ocean surface.

According to a recent publication presented at the 2009 World Congress of the International Desalination Association (Cannesson, 2009), the aquatic habitat in the area of

Figure 16.10 Gold Coast seawater desalination plant. (*Source: Water Corporation.*)

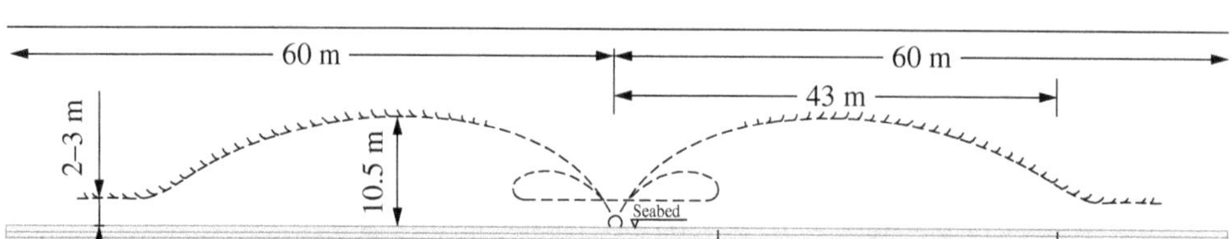

Figure 16.11 Discharge of Gold Coast seawater desalination plant.

the Gold Coast desalination plant discharge is sandy bottom inhabited primarily by widely scattered tube anemones, sipunculid worms, sea stars, and burrowing sponges.

For 18 months prior to the beginning of the desalination plant operations, the project team had completed baseline monitoring to document the original existing environmental conditions, flora, and fauna in the area of the discharge. Figure 16.12 shows plant intake and outfall configurations as well as the location of the reference sites used for comparison.

Since the plant began operations in November 2008, the project team has completed marine monitoring at four sites around the discharge diffuser area at the edge of the mixing zone and at two reference locations 500 m (1640 ft) away from the edge of the mixing zone in order to determine environmental impacts and verify salinity projections.

The water quality and benthic in-fauna abundance and diversity results after the start of the Gold Coast plant operations were compared with the baseline monitoring results as well as with the results of the monitoring sites. The results of pre- and post-plant commissioning clearly indicate that the desalination plant operations did not have a measurable impact on the marine habitat in the area of the discharge—the aquatic fauna has practically remained the same in terms of both abundance and diversity. The Gold Coast plant has been in operation for over one year, and monitoring to date has confirmed that the plant's discharge is environmentally safe.

Perth Seawater Desalination Plant, Australia

As reported at the November 2009 World Congress of the International Desalination Association (Christie and Bonnelye, 2009), the 143,000 m^3/day (38 mgd) Perth seawater desalination plant, has been in continuous operation since November 2006. This plant supplies over 17 percent of the drinking water for the city of Perth, Australia, which has over 1.6 million inhabitants (Fig. 16.13).

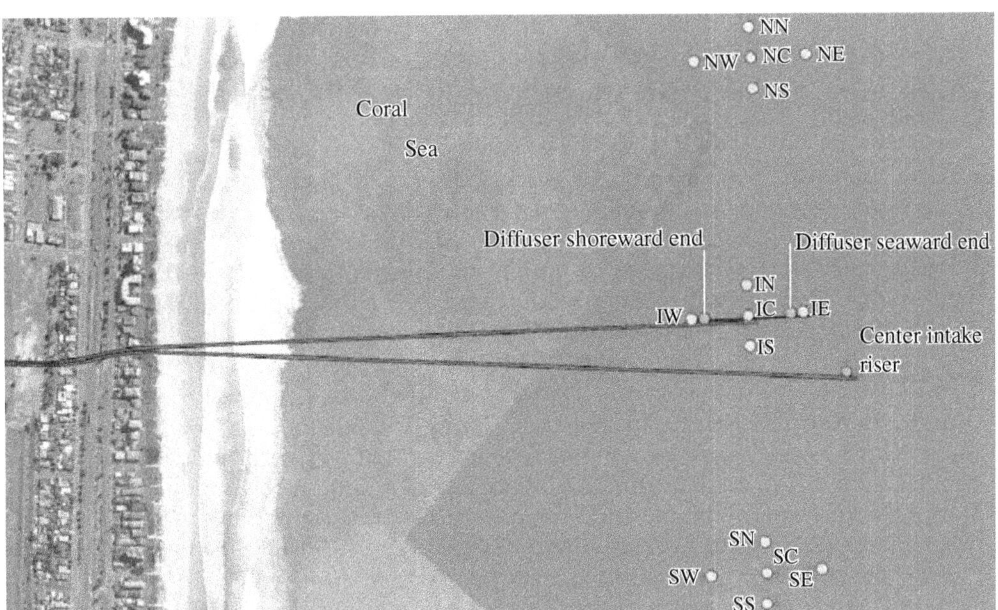

FIGURE 16.12 Gold Coast plant discharge and monitoring sites. (*Source: Water Corporation.*)

Figure 16.13 Perth seawater desalination plant. (*Source: Water Corporation.*)

The treatment facilities of the Perth seawater desalination plant are typical for state-of-the-art desalination plants worldwide. This plant has a velocity-cap-type open-intake structure extending 200 m (660 ft) from the shore. Source seawater is pretreated with single-stage granular media filters, 5-micron cartridge filters, and a two-pass reverse osmosis membrane system with pressure exchangers for energy recovery.

Perth SWRO plant discharge is located in Cockburn Sound, which is a shallow and enclosed water body with limited water circulation. Cockburn Sound frequently experiences naturally occurring low oxygen levels. Since the Perth SWRO plant discharge area has limited natural mixing, the desalination plant project team has constructed a diffuser-based outfall, which is located approximately 500 m (1640 ft) offshore and has 40 ports along the final 200 m (660 ft) at about 0.5 m (1.6 ft) from the seabed surface at a 60° angle.

The diffuser ports are spaced at 5-m (16 ft) intervals with 220-mm (9 in) nominal port diameter at a depth of 10 m (33 ft) (Fig. 16.14). Diffuser length is 160 m (520 ft). The outfall is a single GRP pipeline with a 1600-mm (63 in) diameter.

This diffuser design was adopted with the expectation that the plume would rise to a height of 8.5 m (28 ft) before beginning to sink due to its elevated density. It was designed to achieve a plume thickness at the edge of the mixing zone of 2.5 m (8 ft) and, in the absence of ambient cross-flow, to extend to approximately 50 m (160 ft) laterally from the diffuser to the edge of the mixing zone (Figs. 16.14 and 16.15).

Extensive real-time monitoring was undertaken in Cockburn Sound since the plant began operation in November 2006 to ensure that the marine habitat and fauna are

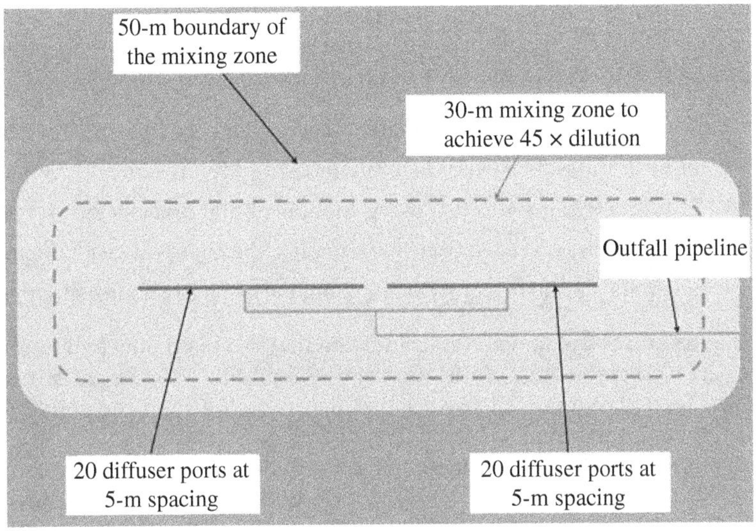

Figure 16.14 Perth SWRO plant discharge configuration. (*Source: Water Corporation.*)

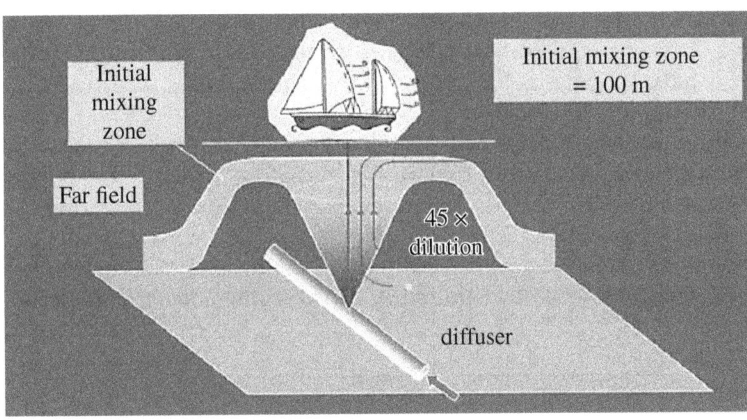

Figure 16.15 Perth desalination plant mixing zone.

protected. This monitoring includes continuous measurement of dissolved oxygen levels via sensors located on the sandy bed of the sound.

Visual confirmation of the plume dispersion was achieved by the use of 52 L (14 gal) of Rhodamine dye added to the plant discharge. The dye was reported to have billowed to within approximately 3 m (10 ft) of the water surface before falling to the seabed and spilling along a shallow sill of the sound toward the ocean. The experiment showed that the dye had dispersed beyond what could be visually detected within a distance of approximately 1.5 km (0.9 miles), well within the protected deeper region of Cockburn Sound, which is located approximately 5 km (3 miles) from the diffusers. The environmentally benign dye experiment was first commissioned in December 2006 and repeated in April 2007 when discharge conditions were calm.

In addition to the dye study, the project team has completed a series of toxicity tests with a number of species in larval phase to determine the minimum dilution ratio needed to be achieved at the edge of the zone of initial dilution:

- 72-hour macro-algal germination assay using the brown kelp *Ecklonia radiate*
- 48-hour mussel larval development using *Mytilis edulis*
- 72-hour algal growth test using the unicellular algae *Isochrysis galbana*
- 28 day copepod reproduction test using the copepod *Gladioferens imparipes*
- Seven-day larval fish growth test using the marine fish pink snapper *Pagrus auratus*

The results of the toxicity tests indicate that the plant concentrate dilution needed to be achieved at the edge of the zone of initial dilution in order to protect the sensitive species listed above is 9.2:1 to 15.1:1, which is well within the actual design diffuser system mixing ratio of 45:1.

In addition to the toxicity testing, the Perth desalination project team has also completed two environmental surveys of the desalination plant discharge area in terms of macrofaunal community and sediment (benthic) habitat (Okel et al., 2007; Oceanica Consulting, 2009). The March 2006 baseline survey covered 77 sites to determine the spatial pattern of the benthic macrofaunal communities, while the repeat survey in 2008 covered 41 sites originally sampled in 2006 and five new reference sites. Some of the benthic community survey locations were in the immediate vicinity of the discharge diffusers, while others were in various locations throughout the bay. The two surveys have shown no changes in benthic communities that can be attributed to the desalination plant discharge.

Water quality sampling completed in the discharge area has shown no observable effect of ocean water quality, except that the salinity at the ocean bottom increased up to 1000 mg/L, a salinity level is well within the naturally occurring salinity variation (Christie and Bonnelye, 2009).

Figure 16.16 depicts the conductivity of the Perth SWRO plant discharge over the period of January 2007 to September 2009. Taking into consideration that the ratio

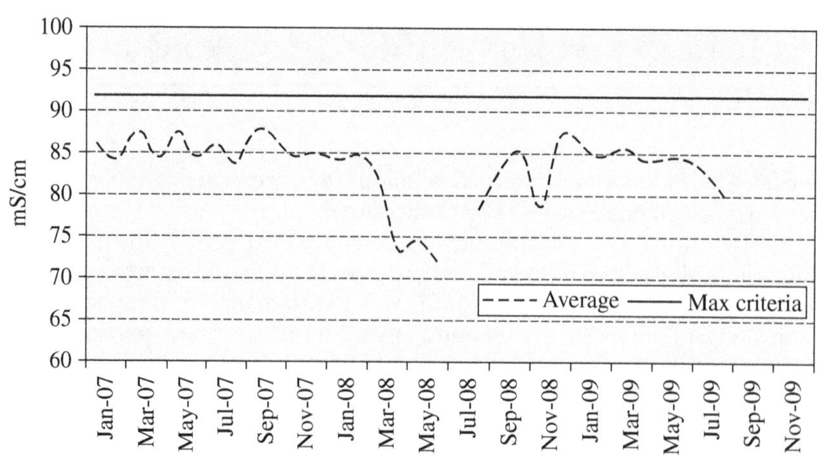

FIGURE 16.16 Perth desalination plant, discharge conductivity. (*Source: Christie and Bonnelye, 2009.*)

between salinity and conductivity is 0.78, the plant discharge salinity varied between 64,500 mg/L (88 mS/cm) and 56,200 mg/L (72 mS/cm).

Dissolved oxygen concentration of the discharge for the same period was between 7.6 and 11.0 mg/L and was always higher than the minimum regulatory level of 5.0 mg/L. Similarly, concentrate pH was between 7.2 and 7.6, which was well within 10 percent of the ambient ocean water pH.

Discharge turbidity for the same period (January 2007 to September 2009) was always less than 3 NTU (Fig. 16.17). It should be pointed out that the spent filter backwash water from the plant's pretreatment system is treated on site in lamella settlers, and the supernatant from this treatment process is discharged with the desalination plant concentrate. The solids generated as a result of the backwash treatment process are dewatered using belt filter press and disposed to a landfill.

In summary, all studies and continuous environmental monitoring completed at the Perth seawater desalination plant to date indicate that the desalination plant operations do not have a significant environmental impact on the surrounding marine environment.

Overview of Discharge Impacts of Desalination Plants in Spain

An independent overview of the discharges of three desalination plants in Spain [22,000 m^3/day (6 mgd) Javea SWRO Plant; 68,000 m^3/day (18 mgd) Alicante 1 SWRO Plant; and 68,000 m^3/day (18 mgd) San Pedro Del Pinatar] completed by the University of Alicante, Spain (Torquemada, 2009), provides additional insights related to environmental impacts of desalination plant discharges. The three plants are located within 80 km (50 miles) from each other, and the salinity of their discharges is 68,000 to 70,000 mg/L.

The Alicante 1 plant is located in a turbulent and very well tidally mixed area. This feature of the desalination plant discharge allows the Alicante plant to operate without measurable environmental impacts even at a relatively low mixing ratio of 1.5 to 5 between concentrate and ambient seawater at the edge of the zone of initial dilution.

The discharge of the Javea SWRO plant is in an open canal, which then carries the concentrate into the ocean. The concentrate from this plant is diluted in the channel

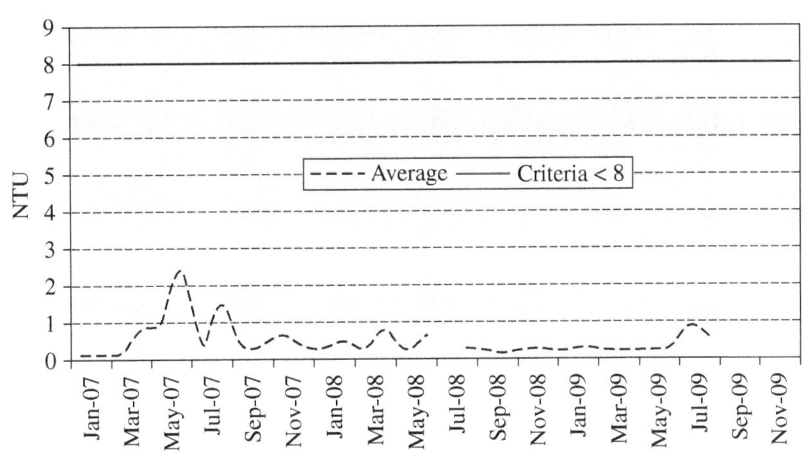

FIGURE 16.17 Perth desalination plant, discharge turbidity. (*Source: Christie and Bonnelye, 2009.*)

from 69,000 down to 44,000 mg/L in a 4:1 mixing ratio. This salinity level was found not to have a negative impact on the marine habitat in the discharge area. In the case of the Alicante SWRO plant, the discharge is located directly on the shoreline to take advantage of the turbulent tidal mixing that naturally occurs in the discharge area. The discharge of the San Pedro del Pinatar Plant is through a diffuser located 5 km away from the shore at a 38-m depth.

All three desalination plants have been in operation for over three years. The water quality and environmental monitoring of the three discharges indicates that the size and time for dispersion of the salinity plume varied seasonally. These variations, however, did not affect the benthic organisms inhabiting the seafloor. The desalination discharge of the Javea plant has high oxygen levels that diminish the naturally occurring apoxia in the area of the discharge. The independent overview emphasizes the fact that well-designed desalination discharge can result in minimal environmental impacts and, in some cases, can be beneficial to the environment due to its high oxygen content.

Maspalomas II Desalination Plant, Canary Islands, Spain

This desalination plant is located in Gran Canarias and has two concentrate outfalls, which extend 300 m (1000 ft) away from the shore (Talavera and Ruiz, 2001). The outlet of the discharge outfalls does not have diffusers (Fig. 16.18), and the mixing between the concentrate and ambient seawater is mainly driven by the velocity of the discharge and the fact that the discharge is located in an area with naturally occurring underwater currents of high intensity. The depth of the discharge is 7.5 to 8.0 m (25 to 26 ft).

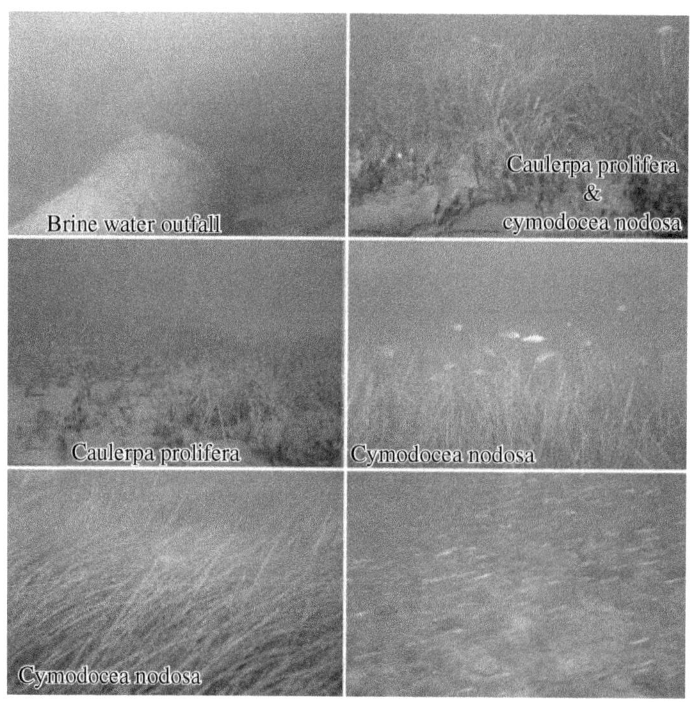

Figure 16.18 Discharge of Maspalomas desalination plant. (*Source: Talavera and Ruiz, 2001.*)

The Maspalomas discharge conditions are challenging: (1) very high salinity of the concentrate (90,000 mg/L), and (2) seagrass habitat for fish and other marine organisms. Due to the naturally occurring near-shore mixing, the salinity of the discharge is dissipated down to 38,000 mg/L (38 PSU) within 20 m (66 ft) from the discharge point as shown in Fig. 16.19. The salinity noted in this figure is presented in psu (practical salinity units), which have the same value as ppt (parts per thousand) of salinity concentration. Concentration of 1 psu equals 1000 mg/L of total dissolved solids.

The zone of initial dilution of the Maspalomas II desalination plant is a sandy bed with practically no flora. However, this zone is surrounded by seagrass beds, which, based on environmental study of the discharge area, are not significantly affected by the desalination plant discharge.

Antigua Desalination Plant Discharge Study

In 1998, the Southwest Municipal Water District of Florida and the University of South Florida completed a study entitled, "Effects of the Disposal of Seawater Desalination Discharges on Near Shore Benthic Communities" (Hammond et al., 1998). The purpose of this study was to identify the environmental impact of discharge of existing desalination plants on the benthic, plant, and animal communities that inhabit the discharge area. The selected test site was located at a 7000 m^3/day (2 mgd) seawater desalination plant in Antigua, the Caribbean. The discharge salinity of this plant is 57,000 mg/L.

The desalination plant outfall extends approximately 100 m (330 ft) from the shore and does not have diffusers—the concentrate exits the open pipe directly and is mixed by the kinetic energy of the discharge and the ocean tidal movement. The salinity within 1.0 m (3.3 ft) from the point of discharge was in a range 45,000 to 50,000 mg/L.

The research team has developed six transects extending radially from the point of discharge and has completed two monitoring studies of the condition of the marine organisms encountered along the six transects—including seagrass, macroalgae, benthic microalgae, benthic foraminifera, and macrofauna—within a six-month period. The results of these studies indicate the desalination plant discharge did not have a detectible effect on the density, biomass, and production of seagrass. In addition, the discharge did not have a statistically significant impact on the biomass and the numerical abundance of the benthic microalgal community, benthic foraminifera, and macrofauna (polychaetes, oligochaetes, bivalves, gastropods, pelagic fish, anemones, worms, sea stars, and other species inhabiting the discharge).

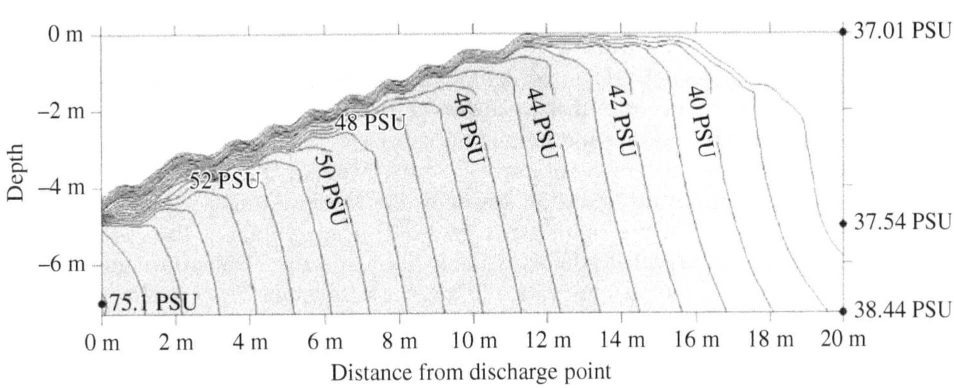

FIGURE 16.19 Discharge of Maspalomas desalination plant (*Source: Talavera and Ruiz, 2001.*)

16.3.7 Co-Disposal with Wastewater Effluent

Description

The key feature of this surface water discharge alternative is the benefit of accelerated mixing that stems from blending the heavier than ocean water concentrate with the lighter wastewater effluent. Depending on the volume of the concentrate and on how well the two waste streams are mixed prior to the point of discharge, the blending may allow us to reduce the size of the wastewater discharge plume and to dilute some of its constituents. Co-discharge with the lighter-than-seawater wastewater effluent would also accelerate the dissipation of the saline plume by floating this plume upward and expanding the volume of the ocean water with which it mixes.

Use of existing wastewater treatment plant outfalls for concentrate discharge has the key advantages of avoiding costs and environmental impacts associated with the construction of new outfall for the seawater desalination plant. Mixing of the negatively buoyant wastewater discharge with the heavier than ocean water concentrate promotes the accelerated dissipation of the wastewater plume, which tends to float to the ocean surface, and the concentrate, which tends to sink toward the ocean bottom. In addition, concentrate often contains metals, organics, and pathogens, which are of order-of-magnitude lower levels than those in the wastewater discharge, which helps reduce the overall waste discharge load of the mix.

Potential Environmental Impacts

Brackish water and seawater concentrate may trigger ion-imbalance-based toxicity when blended with wastewater and discharged to a surface water body with significantly different ion composition of the receiving water. This impact is site-specific and will need to be investigated on a case-by-case basis.

Bioassay tests completed on blends of desalination plant concentrate and wastewater effluent from the El Estero wastewater treatment in Santa Barbara, California, indicate that this blend can exhibit toxicity on fertilized sea urchin (*Strongylocentrotus purpuratus*) eggs. Parallel tests on desalination plant concentrate diluted to similar TDS concentration with seawater rather than wastewater effluent did not show such toxicity effects on sea urchins. Long-term exposure of red sea urchins to the blend of concentrate from the Carlsbad seawater desalination demonstration plant and ambient seawater discharged by the adjacent Encina power plant confirm the fact that sea urchins can survive elevated salinity conditions when the discharge is void of wastewater.

The most likely factor causing the toxicity effect on the sensitive marine species is the difference in ratios between major ions (calcium, magnesium, sodium, chloride, and sulfate) and TDS that occur in the wastewater effluent-concentrate blend as compared with the blend of concentrate and ambient ocean water. Such difference may trigger an effluent toxicity effect as a result of ion imbalance (Mickley, 2000).

The SWRO membranes reject all key seawater mineral ions at approximately the same level. As a result, the ratios between the concentrations of the individual key mineral ions that contribute to the seawater salinity and the TDS of the concentrate are approximately the same as these ratios in ambient seawater. Therefore marine organisms are not exposed to conditions of ion-ratio imbalance, if this concentrate is directly disposed to the ocean.

An additional environmental concern of combining wastewater and desalination plant discharges is that the high salinity may cause wastewater contaminants and other constituents to aggregate in particles of different sizes than they would otherwise. This could result in an enhanced sedimentation or some of the metals and solids contained in the wastewater treatment plant effluent and could potentially have an impact on benthic organisms and phytoplankton in the vicinity of the existing discharge.

Feasibility Considerations

Although the use of existing wastewater treatment plant outfalls may seem attractive for its simplicity and low construction costs, this disposal method has to be evaluated for its site-specific challenges. Due to potential toxicity effects of the concentrate-wastewater effluent blend, the direct discharge of the seawater concentrate through existing wastewater discharge outfalls may be limited to relatively small concentrate discharge flows.

For this concentrate disposal option to be feasible, there has to be an existing wastewater treatment plant in the vicinity of the desalination plant, and this plant has to have available extra outfall discharge capacity. In addition, the fees associated with the use of the wastewater treatment plant outfall have to be reasonable, and the wastewater treatment plant owner, who would allow the use of their outfall for concentrate discharge, has to accept the potential liabilities associated with the environmental impacts of the blended discharge. The WWTP owner would also have to be agreeable to any potential modifications of the existing outfall and the downtime associated with the implementation of these modifications. Usually, this beneficial combination of conditions is not easy to find, especially for discharging large seawater concentrate volumes.

Other feasibility considerations related to the use of existing wastewater treatment plant outfall for desalination plant concentrate discharge are: (1) the potential need for modification of the outfall diffuser system of the existing seawater desalination plant due to altered buoyancy of the concentrate-wastewater mix, and (2) the compatibility of the diurnal fluctuation of the secondary effluent flow with the diurnal fluctuation of the concentrate discharge flow.

The change of the buoyancy of the mixed wastewater effluent-concentrate plume, and the ability of the existing wastewater outfall diffuser system to provide proper mixing, is a key factor associated with co-discharge feasibility. Since the heavier concentrate discharge will reduce the buoyancy of the wastewater effluent, the initial momentum and mixing energy that are delivered by the existing effluent diffuser structure will be altered. Depending on the volumes of the concentrate discharge and the wastewater discharge, the existing wastewater outfall may need to be modified (i.e., by closing diffuser nozzles or by changing diffuser configuration and direction of the nozzles) in order to accommodate the wastewater concentrate discharge. Therefore the impact of the concentrate discharge on the ability of the existing wastewater outfall to provide adequate dispersal of the mixed concentrate-wastewater plume should be evaluated by hydrodynamic modeling for the size-specific conditions of a given project.

Often seawater desalination plants are operated at a constant production rate and, as a result, they generate concentrate discharge with little or no diurnal flow variation. On the other hand, wastewater treatment plant effluent availability for dilution of the desalination plant concentrate typically follows a distinctive diurnal variation pattern.

Adequate protection of marine life requires a certain minimum concentrate dilution ratio in the ZID to be maintained at all times. However, during periods of low wastewater effluent flows (i.e., at night), the amount of concentrate disposed by the desalination plant (and, therefore, the plant production capacity) may be limited by the lack of secondary effluent for blending.

In order to address this concern, the desalination plant operational regime and capacity may need to be altered in order to match the wastewater effluent availability patterns, or diurnal concentrate storage facility may need to be constructed at the desalination plant.

Cost Factors and Analysis

Construction costs associated with desalination plant concentrate co-discharge through an existing WWTP outfall are site-specific because they depend on the capacity of the WWTP outfall, the distance between the desalination plant and the WWTP, the complexity of the structure needed to connect the concentrate discharge line to the outfall, the expenditures associated with modifying the WWTP outfall diffusers to accommodate the buoyancy change of the blended discharge, and the need to construct concentrate retention tank in order to accommodate differences in diurnal discharge flow variations and minimum ratio.

Case Studies of Concentrate Co-disposal with Wastewater Effluent

Co-discharge of brackish RO plant concentrate through existing wastewater treatment plant outfall has found successful implementation in the United States. One example is the Santa Ana River Interceptor (SARI) located in Southern California, which collects and conveys over 65,000 m^3/day (17 mgd) of BWRO concentrate from six inland desalination plants and several wastewater and power plant discharges to Orange County Sanitation District (OCSD) wastewater treatment plant outfall for co-discharge. The SARI line currently exports over 130,000 tons of salt per year from the Santa Ana River Watershed and has a nominal hydraulic capacity of 114,000 m^3/day (30 mgd). This regional desalination concentrate management line consists of approximately 150 km (90 miles) of 40 to 2100 mm (16 to 84 in) gravity pipeline. The SARI line content is discharged into the existing 3000-mm (120 in) ocean outfall of OCSD's WWTP in the city of Huntington Beach. The outfall has a maximum capacity of 1,817,000 m^3/day (480 mgd), of which 480,000 m^3/day (130 mgd) is currently utilized.

Discharge of seawater desalination plant concentrate through an existing wastewater treatment plant outfall has found limited application to date. The largest plant in operation at present, which practices co-discharge of desalination plant concentrate and wastewater effluent, is the 200,000 m^3/day (53 mgd) Barcelona SWRO facility in Spain (Compte et al., 2009) (Fig. 16.20).

16.3.8 Co-disposal with Power Plant Cooling Water

Description

At present, co-disposal of desalination plant power plant cooling discharges is mainly practiced for seawater desalination plants co-sited with large coastal power plants with open intakes. Figure 6.21 shows a typical configuration where a desalination plant is co-located with a power plant, and the discharge of this plant is used as a source of saline water for the desalination plant and as dilution water for concentrate mixing and co-disposal. A similar configuration could also be used for inland power generation plants with once-through cooling and brackish desalination plants (Fig. 6.22).

As shown in Figs. 16.21 and 16.22, under typical operational conditions saline water enters the power plant intake facilities and after screening is pumped through

Figure 16.20 Barcelona desalination plant. (*Source: Degremont*)

the power plant condensers to cool them and thereby to remove the waste heat generated during the electricity generation process (Voutchkov, 2004). Typically the cooling water discharged from the condensers is 5 to 10°C warmer than the source ocean water, which could be beneficial for the desalination process because warmer saline water has lower viscosity and, therefore, lower osmotic pressure/energy for salt separation.

Co-location of SWRO desalination plants with existing once-through cooling coastal power plants yields four key benefits: (1) the construction of a separate desalination plant outfall structure is avoided, thereby reducing the overall cost of desalinated water, (2) the salinity of the desalination plant discharge is reduced as a result of the mixing and dilution of the membrane concentrate with the power plant discharge, which has ambient seawater salinity, (3) because a portion of the discharge water is converted into potable water, the power plant thermal discharge load is decreased, which, in turn, lessens the negative effect of the power plant thermal plume on the aquatic environment, and (4) the blending of the desalination plant and the power plant discharges results in accelerated dissipation of both the salinity and the thermal discharges.

As a result of the co-location the desalination plant unit, power costs could be further decreased by avoiding the use of the power grid and the associated fees for power transmission to the desalination plant. Typically, the electricity tariff (unit power cost) structure includes two components: power production and power grid transmission. Often, the power transmission grid portion of the tariff is 30 to 50 percent of the total unit power cost. By connecting the desalination plant directly to the power plant electricity generation equipment, the grid transmission portion of the power fees could be

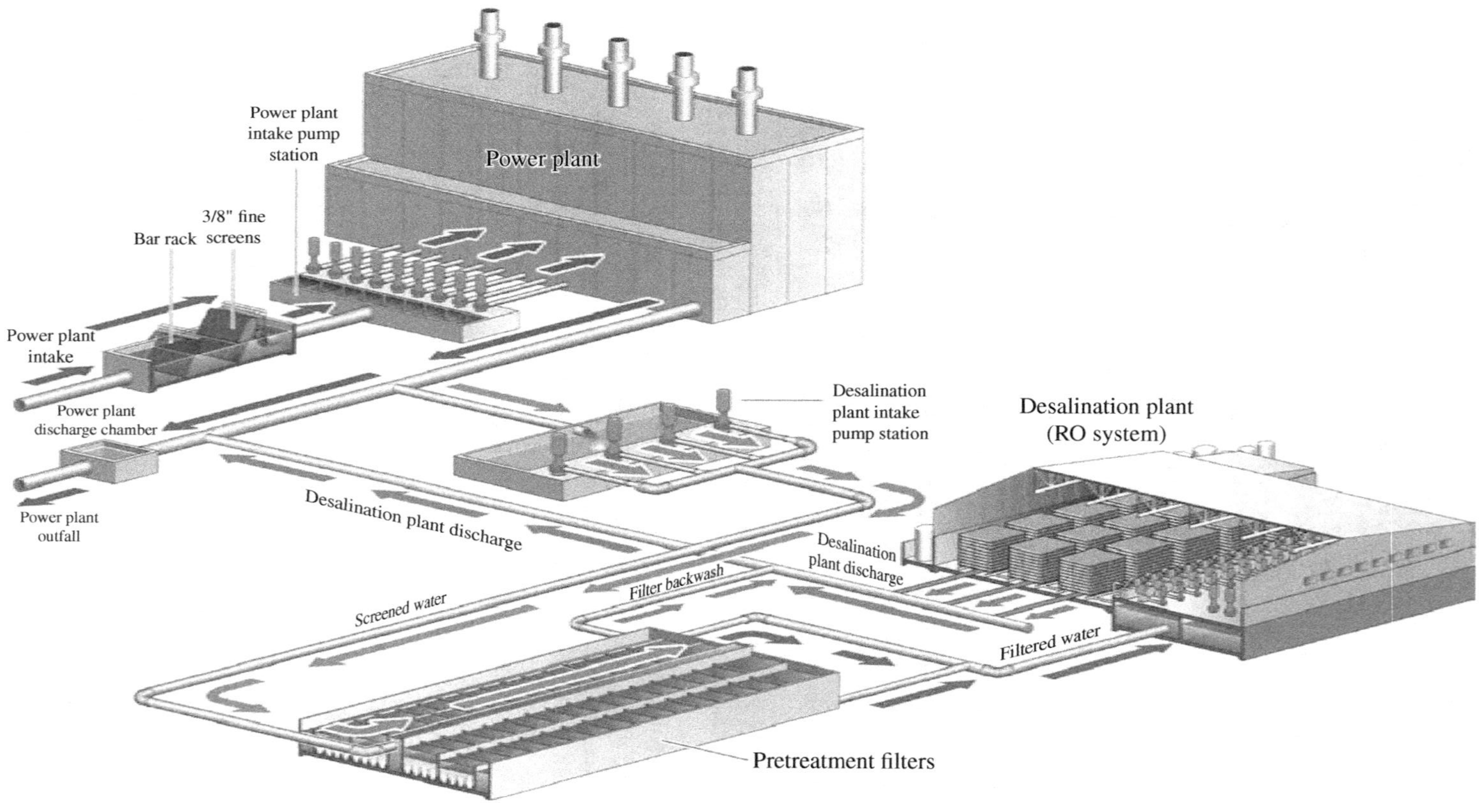

Figure 16.21 Configuration of co-located coastal desalination plant.

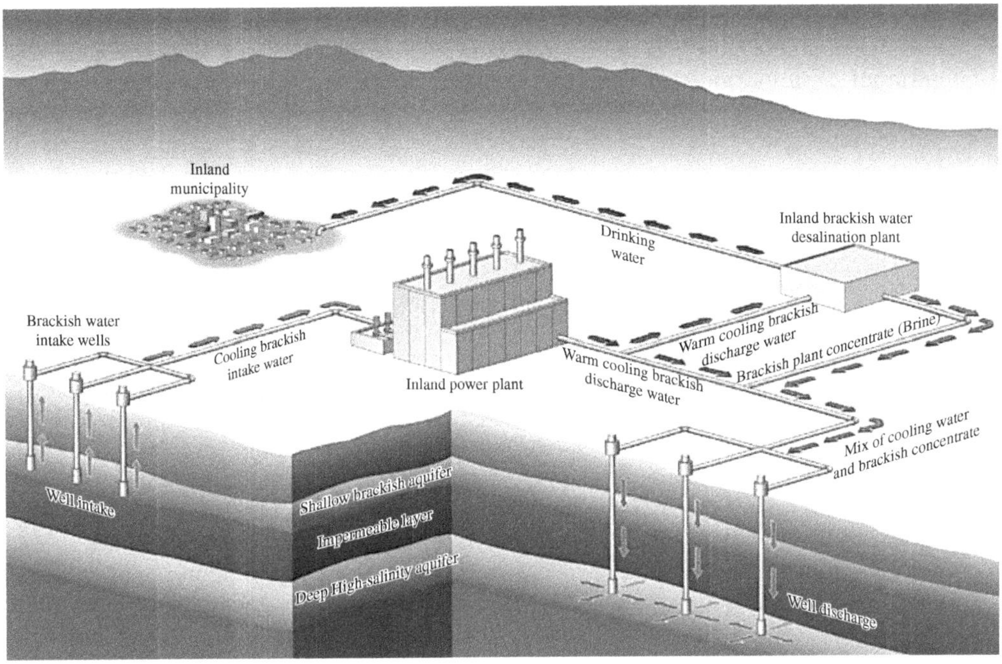

FIGURE 16.22 Configuration of co-located inland desalination plant.

substantially reduced or completely avoided, thereby further reducing the overall seawater desalination cost.

Co-location of power and desalination plants may also have advantages for the power plant host. In addition to the benefit of gaining a new customer and generating revenue by leasing power plant property to locate the desalination plant, the power plant host also gets a customer of favorable power use profile—a steady and continuous power demand and a high-power load factor. This continuous high-quality power demand allows the power plant host to operate the plant's electricity generation units at optimal regime which, in turn, reduces the overall costs of power generation.

Under a typical co-location configuration, the desalination plant uses the power plant discharge water as a source for the desalination process as well as dilution water for the desalination plant concentrate. An example of co-location configuration where the power plant discharge is used only for dilution of the concentrate is the 120,000 m^3/day (32 mgd) Carboneras desalination plant in Spain. This plant's concentrate is discharged to the cooling water canal of a nearby coastal power generation plant and thereby diluted to an environmentally safe level before its return to the sea. The Carboneras seawater desalination plant has a separate open-intake independent from the intake and discharge of the power plant.

Sharing intake infrastructure has environmental benefits because it avoids the need for new intake and outfall construction in the ocean and the seashore area near the desalination plant. The construction of a separate new open-intake structure and pipeline for the desalination plant could cause a measurable disturbance of the benthic marine organisms on the ocean floor.

Another clear environmental benefit of the co-location of power generation stations and desalination plants is the overall reduction of entrainment, impingement, and entrapment of marine organisms as compared with the construction of two separate open-intake structures—one for the power plant and one for the desalination plant. This benefit stems from the fact that total biomass of the impacted marine organisms is typically proportional to the volume of the intake seawater. By using the same intake seawater twice (once for cooling and the second time for desalination), the net intake inflow of seawater and marine organisms is minimized.

The length and configuration of the desalination plant concentrate discharge outfall are closely related to the discharge salinity. Usually, the lower the discharge salinity, the shorter the outfall and the less sophisticated the discharge diffuser configuration needed to achieve environmentally safe concentrate discharge. Blending the desalination plant concentrate with the lower-salinity power plant cooling water often allows reducing the overall salinity of the ocean discharge within the range of natural variability of the seawater at the end of the discharge pipe, thereby completely alleviating the need for complex and costly discharge diffuser structures.

The power plant thermal discharge is lighter than the ambient ocean water because of its elevated temperature, and therefore it tends to float on the ocean surface. The heavier saline discharge from the desalination plant draws the lighter cooling water downward and thereby engages the entire depth of the ocean water column into the heat and salinity dissipation process. As a result the time for dissipation of both discharges shortens significantly, and the area of their impact is reduced.

Seawater density is a function of both temperature and salinity. While seawater density increases with salinity, it decreases with the increase in temperature. A close to ideal condition for co-location of desalination and power plants is a configuration where the increase in density of the blend of desalination plant concentrate and power plant cooling water as compared with the salinity of the ambient water is compensated by the decrease in density of this blend due to higher than ambient temperature.

For example, in the case of the Carlsbad desalination project illustrated in Fig. 5.10, the average annual ambient seawater temperature in the open ocean near the power plant is 18°C, and the seawater salinity is 33,500 mg/L. The seawater density at this temperature and salinity is 1024.12 kg/m^3. The desalination plant concentrate salinity is 67,000 mg/L. If this concentrate is not blended with the warmer and lighter cooling water from the power plant and instead is discharged directly into the ocean at 18°C (64°F), the density of the concentrate would be 1050.03 kg/m^3. Because the concentrate has significantly higher density than the ambient ocean water, after discharge it will quickly sink to the ocean floor and expose the bottom marine habitat to significantly higher salinity, which may have a detrimental effect on the aquatic life.

In the case of the co-located discharge, the concentration of the desalination plant concentrate will be reduced from 67,000 down to 36,200 mg/L as a result of the blending with the cooling water, which has ambient salinity. In addition, the blend would typically have a temperature that is 8°C higher than the ambient seawater temperature (i.e., 26 versus 18°C/79 versus 64°F). As a result of the co-location and mixing of the two discharges, rather than sinking down toward the ocean floor, the concentrate will actually float and quickly mix and dissipate within the water column as it moves upward toward the ocean surface.

For comparison, the discharge of concentrate through diffusers has to be released at a high velocity [2 to 4 m/s (7 to 13 ft/s)] in order to achieve adequate mixing, which, in turn, requires significant energy expenditure associated with pumping concentrate discharge (Fig. 16.23).

Potential Environmental Impacts

The potential environmental impacts associated with co-located desalination facilities are similar to these of open ocean outfalls. Depending on the site-specific mixing conditions, for power plant outfalls equipped with diffusers, the plant outfall diffuser structure may need to be modified in order to accommodate the heavier concentrate discharge.

The environmental impacts of desalination plant operations may increase if the power plant operation is discontinued because the desalination plant cannot benefit from the mixing effect of its concentrate and warm and buoyant power plant cooling water. As a result, more source seawater may need to be collected in order to provide pre-dilution

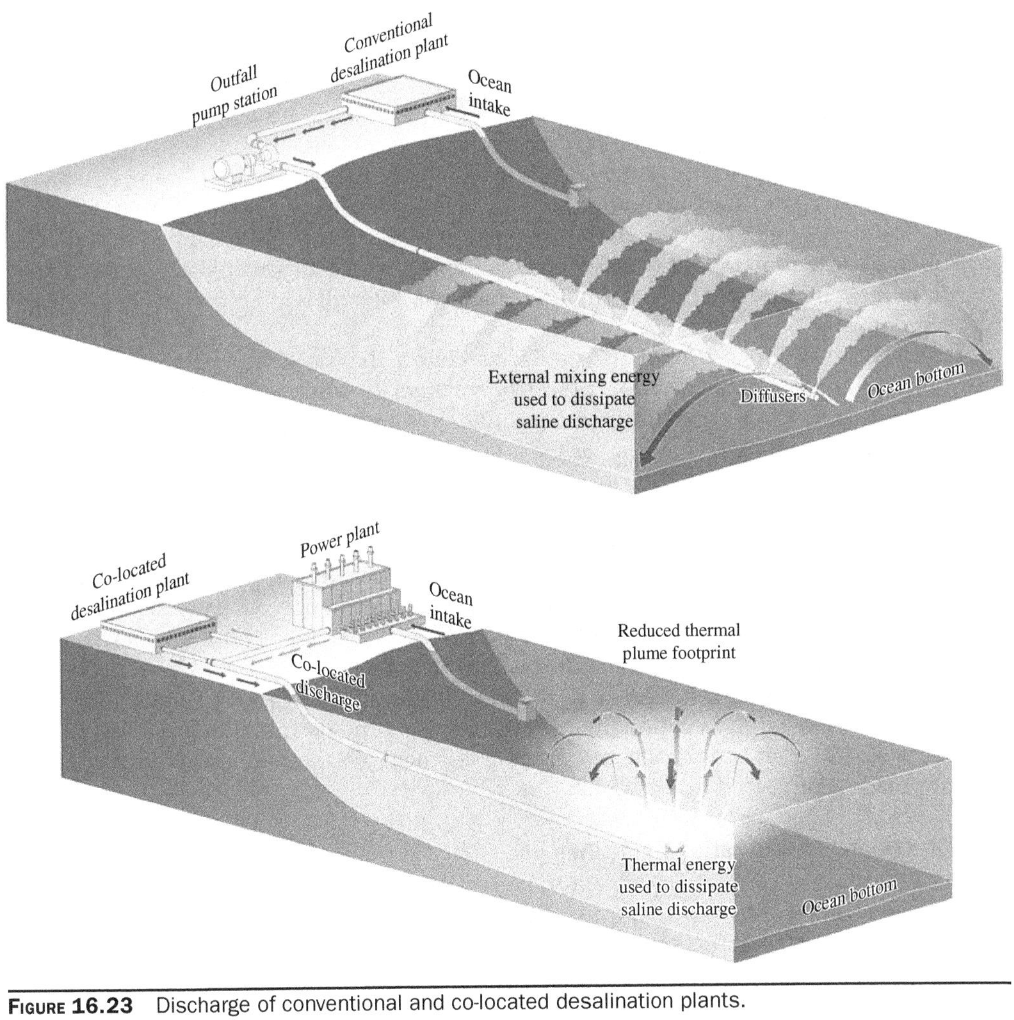

FIGURE 16.23 Discharge of conventional and co-located desalination plants.

of the concentrate to environmentally safe salinity level prior to its discharge. Collection of dilution water may result in an additional impingement and entrainment of marine organisms.

Source Water Treatment Requirements

When assessing the feasibility of co-location of desalination and power plants, particular consideration should be given to the effect of power plant cooling water quality on desalination plant operations. For example, if the power plant cooling water contains levels of copper, nickel, or iron significantly higher than those in ambient source water, this cooling water may be not be suitable for desalination because the above-mentioned metals may cause irreversible fouling and damage of the membrane elements.

Another potential challenge with co-location could be the method of disposal of power plant intake screenings. In most power plants, the debris removed out by the intake screens is disposed offsite to a landfill. However, this disposal practice may change during the course of power plant and desalination plant operations.

For example, in the case of the Tampa Bay seawater desalination plant, during the final phase of desalination plant construction, the power plant host (Tampa Electric Power Company, TECO) decided to change its intake screenings disposal practices and to discharge the screenings just upstream of the point of connection of the desalination plant intake rather than to continue disposing them off-site.

This change in power plant operations had a dramatic effect on Tampa Bay water desalination plant start-up and operations and especially on the pretreatment system performance. Since the desalination plant was pilot tested and designed around the original method of power plant operations, under which all screenings were removed from the cooling water, the desalination plant was not built with its own separate intake screening facilities.

The presence of power plant waste screenings in the desalination plant intake water had a detrimental effect of the pretreatment filter operations because the screening debris frequently clogged the filter distribution piping, airlifts, and sand media. This, in turn, was one of the key causes for the low quality of the filter effluent and the related short useful life of the plant's cartridge filters.

Although this problem had a significant effect on the desalination plant operations, it also had relatively straightforward solutions—either installing separate fine screening facilities at the desalination plant or moving the point of the power plant screening debris discharge downstream of the location of the desalination plant intake. The project owner of the Tampa Bay SWRO plant installed separate screens for the water entering the desalination plant in order to address this challenge.

These observations indicate that the successful implementation of co-location project requires close and continuous coordination with power plant operations. A summary of key issues and considerations for assessment of the feasibility of the co-location approach is presented in Table 16.2.

Design and Configuration Guidelines

In order for the co-location concept to be cost-effective and possible to implement, the minimum power plant cooling water discharge flow has to be at least several times larger than the desalination plant production capacity. In addition, the power plant outfall configuration and hydraulics have to be such that entrainment and recirculation of concentrate into the desalination plant intake is avoided under all power plant operational

Advantages	Disadvantages and Feasibility Considerations
• Capital cost savings by avoiding construction of new intake discharge outfall. • Decrease of the required RO system feed pressure and power cost savings as a result of using warmer water. • Reduction of marine organism impingement and entrainment because the desalination plant does not collect additional seawater from the ocean. • Reduction of impact on marine environment as a result of faster dissipation of thermal plume and concentrate. • Reduction of the power plant thermal discharge to the ocean because a portion of this discharge is converted to potable water. • Use of already disturbed land at the power plant minimizes environmental impact.	• Use of warmer seawater may accelerate membrane biofouling. • RO membranes may be exposed to iron, copper, or nickel fouling from power plant condensers. • Source seawater has to be cooled if its temperature increases above 40°C (104°F) in order to protect RO membrane integrity. • Permeate water quality diminishes slightly with the increase of source water temperature. • Use of warmer water would result in lower boron rejection. • RO plant source water screening may be required if the power plant disposes of its screenings through its outfall and the point of disposal is upstream of the desalination plant intake. • Desalination plant operations may need to be discontinued during periods of heat treatment of the power plant facilities.

Table 16.2 Issues and Considerations of Desalination Plant Co-location

conditions and ocean tide elevations, including high tide levels in combination with low power plant discharge flows.

It is preferable that the distance of the power plant outfall from the point of connection of the desalination plant discharge to the point of entrance of the discharge outfall into the ocean be long enough in order to achieve complete mixing of the concentrate and the cooling water. Complete mixing of the two streams upstream from the point of discharge minimizes the negative effect of the streams on the environment. The minimum distance required for complete mixing depends on numerous factors, including location and angle of entrance of the concentrate pipe discharge into the power plant outfall; size of the concentrate discharge pipe and the power plant outfall; flow rates, temperatures, and salinities of the cooling water and concentrate discharge streams.

Because of the complexity of the outfall mixing phenomenon, use of computational fluid dynamics (CFD) models is recommended to identify the optimum location and entrance of the concentrate discharge pipe into the power plant outfall/canal. Figures 16.24 and 16.25 present the results of CFD modeling analysis of two alternative entrance configurations of the 760-mm (30 in) concentrate discharge pipe of the Tampa Bay seawater desalination plant into the 2743-mm (108 in) outfall of the Tampa Electric Power Plant.

Figure 16.24 depicts the level of concentrate discharge mixing with the power plant cooling water when the concentrate line enters the power plant outfall under a 45° angle protruding 0.75 m (2.5 ft) into the outfall (Case 1). This concentrate discharge entrance configuration was found to be optimal and was actually implemented for this project because it allows complete mixing to be achieved at minimum distance

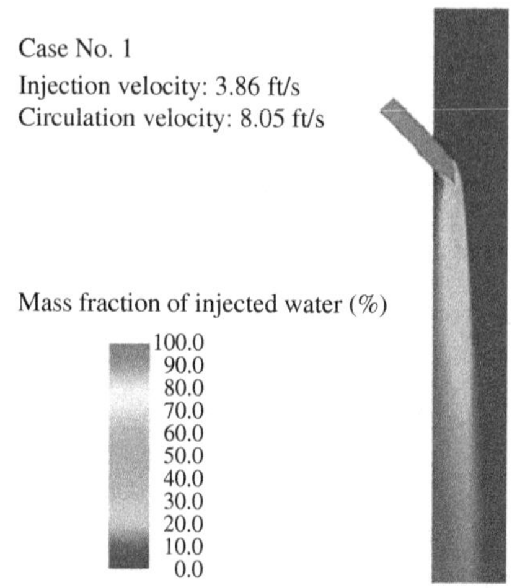

FIGURE 16.24 Tampa Bay SWRO discharge entrance configuration: Case 1.

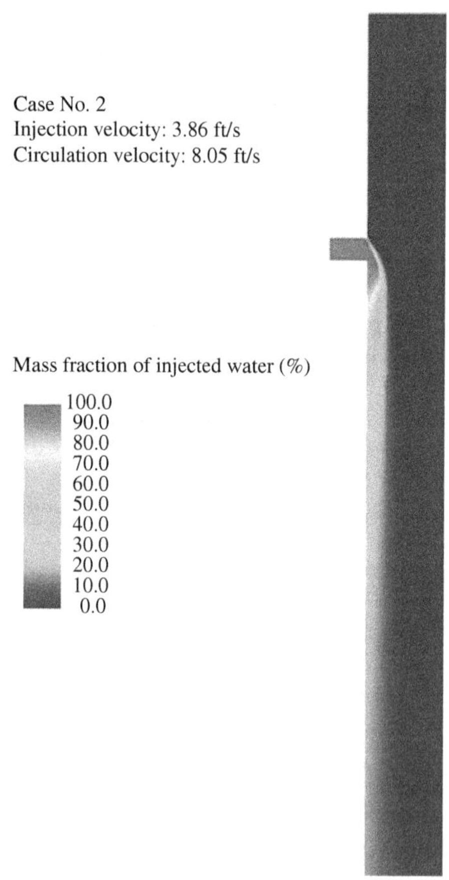

FIGURE 16.25 Tampa Bay SWRO discharge entrance configuration: Case 2.

[less than 25 m (82 ft)] from the point of concentrate discharge into the power plant outfalls. In this case, this distance was a physical limitation that had to be accommodated in the design of the concentrate discharge pipe.

Figure 16.25 illustrates the less efficient mixing achieved when the same size concentrate pipe enters into the power plant outfall without protrusion and at a 90° angle (Case 2), which is the lowest cost and the easiest to construct configuration.

Comparison of the two figures clearly indicates the benefits of angled concentrate entrance and the projection of this entrance into the power plant outfall for this project.

Cost Factors and Analysis

One of the key additional benefits of co-location is the overall reduction of the desalination plant power demand and associated costs of water production as a result of the use of warmer source water. The source water of the RO plant is typically 5 to 10°C warmer than the temperature of the ambient ocean water. This is a significant benefit, especially for desalination plants with cold source seawater, because the feed pressure required for RO membrane separation decreases with 6 to 8 percent for every 10°C of source water temperature increase. Since the power costs are approximately 30 to 40 percent of the total costs for production of desalinated water, the use of warmer source water could have a measurable beneficial effect on the overall water production expenditures.

It should be pointed out that the concentrate mixing benefit of co-location would no longer be available if and when the co-located power plant discontinues using once-through cooling. If such modification occurs, the concentrate mixing effect of the warm water would need to be compensated by either diluting the concentrate by additional intake seawater or by modifying the existing power plant outfall structure and installing an appropriate diffuser structure.

Opponents of seawater desalination plants often present the argument that if a coastal power generation plant discontinues its once-through cooling practices, co-location with a desalination plant would no longer be viable. This argument, however, is unfounded in reality because even if the host power plant abandons once-through cooling, the desalination project will still retain the main cost-benefits of co-location and avoidance of the need to construct a new intake and outfall. The capital cost savings from the use of the existing power plant intake and outfall facilities are typically 5 to 30 percent of the total plant construction costs.

Case Study of Co-Located Desalination Plant

Tampa Bay Seawater Desalination Plant Co-location with a power station in a large scale was first used for the Tampa Bay seawater desalination project, and since then has been considered for numerous plants in the United States and worldwide. The intake and discharge of the Tampa Bay seawater desalination plant are connected directly to the cooling water discharge outfalls of the Tampa Electric's Big Bend Power Station (Fig. 16.26).

The TECO power generation station discharges an average of 5.3 million cubic meters (1.4 billion gallons) of cooling water per day, of which the desalination plant takes an average of 166,500 m^3/day (44 mgd) to produce 95,000 m^3/day (25 mgd) of fresh drinking water. The desalination plant concentrate is discharged to the same TECO cooling water outfalls downstream from the point of seawater desalination plant intake connection.

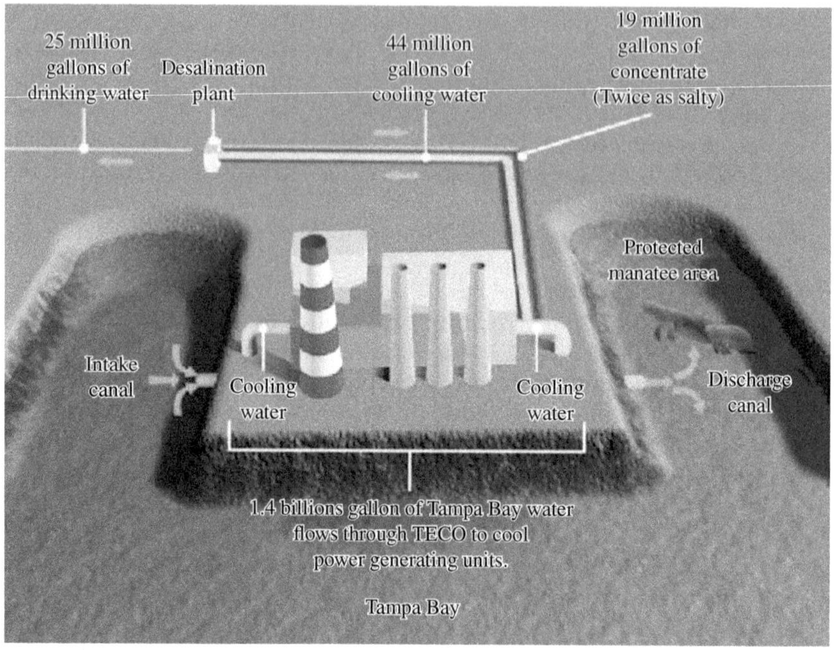

FIGURE 16.26 Tampa Bay SWRO plant co-location schematic.

The source seawater is treated through fine screens, coagulation, and flocculation chambers, sand media, and diatomaceous filters in series, and an SWRO system with a partial second-pass. The spent filter backwash water from the desalination plant is processed through lamella settlers and dewatered using a belt filter press. Treated backwash water and concentrate are blended and disposed through the power plant outfalls.

Environmental monitoring of the desalination plant discharge has been ongoing since the plant first began operation in 2002 (McConnell, et al., 2009). The desalination plant discharges 72,000 m^3/day (19 mgd) of concentrate of salinity of 54,000 to 62,000 mg/L, which is blended with the reminder of the power plant cooling water prior to its disposal to Tampa Bay. Because of the large dilution volume of the cooling water, the blend of concentrate and cooling water has a salinity that is well within 2000 mg/L of the ambient bay water salinity.

Tampa Bay water implements an environmental monitoring program in the area of the desalination plant discharge independently from the desalination plant operator, in fulfillment of plant discharge permit requirements. Overall objectives for the monitoring program are to detect and evaluate effects of discharge through comparison to a control area and time periods defined by facility operation (pre-operational, operational, and off-line periods).

The plant discharge permit requires additional supplemental sampling to be performed as part of Tampa Bay water's hydrobiological monitoring program. Water quality and benthic invertebrate monitoring includes fixed and random sites and is focused on areas most likely to be affected by the discharge. A control area considered representative of ambient background bay water quality conditions has been used for comparison. For fish and seagrass, data collected by other government agencies monitoring in the vicinity of the desalination facility have been used to evaluate potential changes.

Monitoring of the desalination facility began in April 2002. The desalination facility first began operation in 2003, and, since that time, it had operated at varying production levels until being taken off-line for remediation in May 2005. The facility came back on-line in March 2007. Evaluation of monitoring data from 2002 to 2008 shows that, even during periods of maximum water production, changes in salinity in the vicinity of the discharge were within or below the maximum thresholds (less than 2000 mg/L increase over background) predicted by the hydrodynamic model developed during the design and permitting phases of the plant. Review of monitoring data to date indicates that the plant operation does not have any adverse impact on Tampa Bay's water quality and abundance and diversity of the biological resources near the facility discharge.

While benthic assemblages varied spatially in terms of dominant taxa, diversity, and community structure, the salinity did not vary among monitoring strata, and the observed spatial heterogeneity of marine life distribution has been found to be caused by variables not related to the discharge from the desalination facility (i.e., temperature and substrate). Patterns in fish community diversity in the vicinity of the facility were similar to those occurring elsewhere in Tampa Bay, and no differences between operational and nonoperational periods have been observed.

16.4 Discharge to Sanitary Sewer

16.4.1 Description

Discharge to the nearby wastewater collection system is one of the most widely used methods for disposal of concentrate from small brackish and seawater desalination plants worldwide (Mickley, 2006). This indirect wastewater plant outfall discharge method, however, is only suitable for small volumes of concentrate into large-capacity wastewater treatment facilities, mainly because of the potential negative impact of the concentrate's high TDS content on the operations of the receiving wastewater treatment plant. Discharging concentrate to the sanitary sewer in most countries is regulated by the requirements applicable to industrial discharges of the utility/municipality, which is responsible for wastewater collection system management.

16.4.2 Potential Environmental Impacts

Desalination plant discharge to a sanitary sewer could potentially have environmental impacts similar to those of co-discharge of concentrate and WWTP effluent (see Sec. 16.3.2).

16.4.3 Effect on Sanitary Sewer Operations

Usually, concentrate water quality is compliant with typical requirements for discharging wastewater to a sanitary sewer. Therefore the application of this concentrate disposal method is not anticipated to have significant impacts on the sanitary sewer system.

16.4.4 Impact on Wastewater Treatment Plant Operations

Feasibility of this concentrate disposal method is limited by the hydraulic capacity of the wastewater collection system and by the treatment capacity of the wastewater treatment plant receiving the discharge. A detailed analysis of the potential impacts of concentrate discharges on WWTP treatment processes is provided elsewhere (Rimer et al., 2008).

Typically, a wastewater treatment plant's biological treatment process is inhibited by high salinity when the plant influent TDS concentration exceeds 3000 mg/L. Therefore, before directing desalination plant concentrate to the sanitary sewer, the increase in the wastewater treatment plant's influent salinity must be assessed, and its impact on the plant's biological treatment system should be investigated.

Taking into consideration that wastewater treatment plant influent TDS may be up to 1000 mg/L in many facilities located along the ocean coast, and that the seawater desalination plant concentrate TDS level would be above 65,000 mg/L, the capacity of the wastewater treatment plant has to be at least 30 to 35 times higher than the daily volume of concentrate discharge in order to maintain the wastewater plant influent TDS concentration below 3000 mg/L. This means that, for example, a 40,000 m^3/day (10.6 mgd) wastewater treatment plant would likely not be able to accept more than 1000 m^3/day (0.3 mgd) of concentrate.

16.4.5 Effect on Water Reuse

If the effluent from the wastewater treatment plant is used for water reuse, the amount of concentrate that can be accepted by the wastewater treatment plant is limited not only by the concentrate salinity but also by the content of sodium, chlorides, and boron in the blend. All of these compounds could have a profound negative impact on the reclaimed water quality, especially if the effluent is used for irrigation. Treatment processes of a typical municipal wastewater treatment plant—such as sedimentation, activated sludge treatment, and sand filtration—do not remove a measurable amount of these concentrate constituents.

A number of crops and plants cannot tolerate irrigation water that contains over 1000 mg/L of TDS. However, TDS is not the only water quality parameter of concern when the desalinated water is used for irrigation. High levels of chloride and sodium may also have significant negative impacts on the irrigated plants. Most plants cannot tolerate chloride levels above 250 mg/L.

Typical wastewater plant effluent has chloride levels of 150 mg/L or less, while seawater treatment plant concentrate could have chloride concentration in excess of 40,000 mg/L. For example, using the chloride levels indicated above, a 40,000 m^3/day (10.6 mgd) wastewater treatment plant cannot accept more than 80 m^3/day (0.02 mgd) of seawater desalination concentrate if the plant effluent would be used for irrigation. This limitation would be even more stringent if the wastewater effluent is used for irrigation of salinity-sensitive ornamental plants, which often have tolerance threshold levels for sodium of $\leq$ 80 mg/L and chloride of $\leq$ 120 mg/L.

16.4.6 Design and Configuration Guidelines

Conveyance pipeline for this concentrate disposal alternative is designed similar to any other wastewater discharge pipeline. The pipeline material is usually HDPE, GRP, or PVC. Because desalination plant concentrate is typically safe to dispose to the wastewater collection system, usually no specific concentrate pretreatment is needed.

If spent RO membrane cleaning solutions are planned to be discharged to the sanitary sewer, their pH would have to be adjusted to a range of 6 to 9 in order to protect the integrity of the wastewater collection system.

If the concentrate discharge volume is such that after blending with the WWTP effluent, the salinity of the blend exceeds 3000 mg/L during the periods of daily low wastewater flows (off-peak hours), then this concentrate will need to be stored and equalized in order to prevent the excessive increase in the WWTP influent salinity. As indicated previously, influent salinity over 3000 mg/L could inhibit the biological activated sludge wastewater treatment process.

16.4.7 Costs for Sanitary Sewer Discharge

Discharge to a sanitary sewer is usually the lowest-cost concentrate disposal method, especially when a wastewater collection system is available in the vicinity of the desalination plant site, and the wastewater treatment plant receiving concentrate discharge has adequate capacity to handle it.

Sanitary sewer discharge conditions and therefore costs are site-specific, and the key cost factors for this disposal method are the expenditures for discharge conveyance (pump station and pipeline) and the costs and fees for connecting to the sanitary sewer and for treatment/disposal of concentrate.

While concentrate conveyance costs are mainly driven by its volume, wastewater collection system connection fees can vary significantly for a given location from none to several orders of magnitude larger than the conveyance costs. The wastewater collection system connection fees usually are related to the available capacity of the sewer facilities and the effect of desalination plant discharge on the operational costs of the wastewater treatment plant, which would provide ultimate treatment and disposal of this discharge.

16.5 Deep Well Injection

16.5.1 Description

This disposal method involves injection of desalination plant concentrate into an acceptable, confined deep underground aquifer adequately separated from freshwater or brackish water aquifers above it. The depth of such wells usually varies between 500 and 1500 m (1600 and 4900 ft). A variation of this disposal alternative is the injection of concentrate into existing oil and gas fields to aid field recovery. Deep well injection is frequently used for disposal of concentrate from all sizes of brackish water desalination plants.

Shallow exfiltration beach well systems could be used for seawater concentrate disposal. Compared to deep well injection, beach well disposal consists of concentrate discharge into a relatively shallow unconfined coastal aquifer that ultimately conveys this discharge into the open ocean through the bottom sediments. Discharge beach wells are mainly used for small- and medium-size seawater desalination plants.

Concentrate disposal wells typically consist of three or more concentric layers of pipe: surface casing, long string casing, and injection tubing. Figure 16.27 illustrates the key components of a typical concentrate deep injection well (Carollo, 2009). A deep injection well consists of a wellhead (equipped with pump, if needed) and a lined well shaft protected by multiple layers of casing and grouting.

Well Shaft

The type of materials selected for well shaft construction should be compatible with desalination plant discharge water quality. Materials often used for the inner liner of a well shaft include fiberglass, plastic, stainless steel, and extra-thick steel pipe.

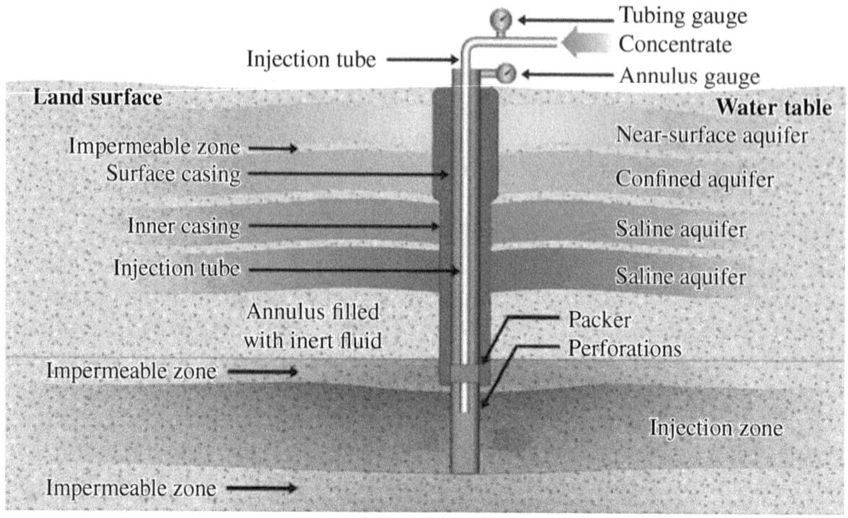

Figure 16.27 Schematic of deep injection well.

Injection wells are generally constructed by the same process used to construct extraction wells. Cable tools and rotary drilling have been used successfully to construct deep wells. Completion of the well involves testing the casing and cement grouting to make sure they do not leak and can sustain design pressures.

Casing

Deep injection wells are multi-cased, with the innermost casing set at the top of the injection zone. Three to four casings are typically used. The depth of each casing depends on the geological environment surrounding the well. The main purpose of multi-staged casings is to protect the upper freshwater zones from deeper, brackish zones and to reduce the possibility of fluid exchange between the different aquifers.

The well casing prevents the borehole from caving in and houses the tubing. Typically, casing is constructed of corrosion-resistant material such as steel or fiberglass-reinforced plastic (FRP). Surface casing is the outermost of the protective layers; it extends from the surface to below the lowermost underground source of drinking water (USDW). The long-string casing extends from the surface to or through the injection zone. This casing terminates in the injection zone with a screened, perforated, or an open-hole completion, where injected concentrate exits the tubing and enters the receiving formation.

The well casing design and materials may vary based on the physical and chemical nature of the concentrate and naturally occurring saline water in the subsurface soil formation. Concentrate must be compatible with the well materials that come into contact with it. Cement made of latex, mineral blends, or epoxy is used to seal and support the casing.

Grouting

Cement grouting surrounding each casing protects it from external corrosion, increases its strength, and prevents the injected waste from traveling to areas other than the designated injection zone. The type of cement and width of each cement layer surrounding the well casing are typically regulated by the government agencies issuing permits (licenses) for well construction and operation.

Injection Zone

The characteristics of the receiving formation (injection zone) determine the appropriate well assembly—a perforated or screen assembly is appropriate for unconsolidated formations such as sand and gravel, while an open-hole completion is used in wells that inject into consolidated sandstone or limestone.

The innermost layer of the well, the injection tubing, conducts concentrate from the surface to the injection zone. Because it is in continuous contact with concentrate, this tubing is constructed of corrosion-resistant material (e.g., fiberglass-reinforced plastic, coated or lined alloy steel, or zirconium, tantalum, or titanium).

The annular space between the tubing and the long string casing, which is sealed at the bottom by a packer and at the top by the wellhead, isolates the casing from the injected concentrate and creates a fluid-tight seal. The packer is a mechanical device set above the injection zone that seals the outside of the tubing to the inside of the long string casing. The packer may be a simple mechanically set rubber device or a complex concentric seal assembly. Constant pressure is maintained in the annular space. This pressure is continuously monitored to verify the well's mechanical integrity and proper operational conditions.

Pumping

Concentrate discharge pressure is usually adequate to convey concentrate to and down into the injection well. If the concentrate head is insufficient, additional pumping will be required. The material of the injection well pump should be compatible with the physical and chemical properties of the injected concentrate. Past experiences with injection systems indicate that improperly selected materials cause many difficulties, resulting in corrosion of the injection pumps.

Storage

Temporary storage of concentrate or an alternative method of disposal is needed to allow for maintenance and repairs of the injection well system. Additionally, the well system may be shut down if monitoring systems and monitoring wells indicate leakage. The type of storage facility or standby disposal method is highly dependent on the location of the well and the conditions surrounding the well site. If the injection well system is located near the coast, a discharge canal or pipeline could be used to temporarily discharge the concentrate flow to a saline water body.

Deep well concentrate injection systems also include a set of monitoring wells to confirm that concentrate is not migrating into the adjacent aquifers.

16.5.2 Potential Environmental Impacts

Based on over a 20-year track record with operation of deep injection wells for concentrate disposal in the United States, this disposal method is fairly reliable and has a low probability of negative environmental impacts (Mickley, 2006). However, there are at least five possible circumstances under which concentrate could migrate upward and could potentially contaminate shallow aquifers above it: (1) failure of injection well casing due to corrosion or excessive feed pressure could result in upward migration of the concentrate through the well bore, (2) concentrate could propagate vertically outside of the well casing from the injection zone into the USDW aquifer, (3) concentrate could migrate vertically from the injection zone through the overlaying confining bed if this bed has high permeability, solution channels, joints, faults, or fractures, and (4) concentrate could reach the USDW aquifer via nearby wells, which are inappropriately cemented or plugged or have an inadequate casing.

Detailed geophysical logs of the well site are typically performed after construction. Before well start-up, television surveys of the well shaft and radioactive tracer surveys are usually completed to validate the integrity of the well. Accidental contamination of surface and subsurface fresh waters is avoided by continuous monitoring of concentrate flow and wellhead pressure.

Readings of increasing pressure during steady operation could indicate possible clogging, whereas a sudden decrease in pressure is indicative of leaks within the casing, grout, or seal. Monthly testing of monitoring wells is also typically performed to ensure that the well system is not leaking into underground soils or water sources. During operation of the injection wells, plugging, contamination, and wide variations in concentrate flow rates and pressures should be monitored and avoided. Plugging can be caused by various occurrences, such as bacterial growth, suspended solids precipitation, or entrained air.

16.5.3 Criteria and Methods for Feasibility Assessment

Deep well injection systems for concentrate disposal are only viable for confined aquifers of large storage capacity, which have good soil transmissivity. They are not feasible for areas of elevated seismic activity or sites near geologic faults that can result in a direct hydraulic connection between the discharge aquifer and a water supply aquifer.

Concentrate disposal using deep injection wells may result in contamination of groundwater with concentrated pollutants, if the discharge aquifer is not adequately separated from the water supply aquifer in the area of discharge. If the injection wells develop leakages and are exposed to heavy scaling, their discharge capacity may decrease over time.

Deep well injection of concentrate in the United States is regulated by an underground injection control program. Regulatory considerations include the receiving aquifer's transmissivity and TDS, the presence of a structurally isolating and confining layer between the receiving aquifer, and any overlying USDW considered to be used as water bearing formation that has < 10,000 mg/L TDS.

16.5.4 Design and Configuration Guidelines

Site Selection

Site selection is the first step of designing a deep injection well system for concentrate disposal. Pertinent regulatory requirements in the United States require injection wells to be sited in such fashion that they deliver concentrate into a formation that is beneath the lowermost aquifer used for a drinking water supply, which is located within 400 m (quarter mile) of the well site.

The location of a deep injection well is determined by the proximity of an acceptable injection zone. In order to avoid eventual plugging of the well, the water quality of the underground injection zone must be compatible with the water quality of the membrane concentrate, and the injection zone receiving concentrate must have salinity over 10,000 mg/L.

The injection zone should be characterized by high permeability and high transmissivity, which allows large volumes of concentrate to be injected without significant pressure buildup. The injection zone should also be located away from abandoned wells, faults, or other hydrogeological short-circuits.

The first step in the site selection for a discharge deep injection well system is the evaluation of the condition, type, and transmissivity of the geological formations and the salinity of the deep groundwater aquifers in the vicinity of the desalination plant site. The next step is to determine the location and depth of the shallow aquifers in the vicinity of the target well intake site as well as the current uses of these aquifers such as water supply, aquifer recharge, wastewater disposal, etc.

Deep injection wells should be located in geologically stable areas without fractures or faults in the confining rock layer(s) through which injected concentrate could propagate to drinking water sources. The target well discharge area should be investigated for other wells or artificial pathways between the injection zone and USDWs through which concentrate can travel. USEPA regulations prevent deep injection wells from being sited in areas where earthquakes could occur and can compromise the ability of the injection zone and confining zone to contain the injected concentrate.

Selection of Geological Formation

Deep injection wells should deliver concentrate into geological formations with proper rock types and configuration to ensure they can safely receive it. Pre-siting geological tests should be completed in order to confirm that the injection zone is of sufficient thickness and storage capacity and has adequate porosity and permeability so that the concentrate injected through the well can enter the receiving rock formation without an excessive buildup of pressure and displacement of injected concentrate outside of the intended zone. Typically, highly porous rock formations such as sandstone are suitable for concentrate injection zones because they can retain large volumes of liquid.

The injection zone should be confined by one or more layers of relatively impermeable rock (i.e., dolomite) that can hold the concentrate in place and not allow it to move vertically toward a USDW; this rock layer defines the confining zone. Confining zones are typically composed of shale or clay, which are "plastic" (i.e., they are less likely to be fractured). Brittle rock-type formations (i.e., sandstone) are not suitable to serve as confining layers.

Deep wells typically inject concentrate into geologic formations that are located thousands of meters below the land surface. In the United States, the most suitable deep aquifers for concentrate discharge are located in the Gulf Coast, Texas, Great Lakes, and Florida. Florida has a distinctive underground environment that favors the use of deep injection wells.

It is important to check the groundwater quality of the aquifer selected for deep well injection. The salinity of this injection zone aquifer should be higher or equal to that of the concentrate being injected into it in order not to degrade aquifer quality. Usually deep aquifers have very high salinity, and meeting this requirement is not a challenge.

Sizing of Injection Wells

Well sizing involves the determination of well depth, diameter, and number of wells.

Well Depth

This well parameter is determined by the depth of the injection zone to which the desalination plant concentrate is delivered. The injection zone depth, in turn, is established based on the available deep aquifers in the vicinity of the desalination plant site, which are suitable for concentrate discharge. In most developed countries, the deep confined aquifers are well-studied, and information of their capacity and the size and location of other existing installations discharging to the same aquifers is usually readily available from the state and/or local regulatory agencies responsible for groundwater resource management.

If such information is not readily available, the design engineer will need to complete an on-site hydrogeological investigation in order to determine the depth and capacity of the deep confined aquifer/s to which desalination plant discharge could be discharged. Injection well depth could vary from several hundred to several thousand meters.

Well Diameter and Number

Well diameter and number are established based on the maximum and average volumes of concentrate planned to be discharged. The number of wells is typically determined as a function of the desalination plant annual operation pattern and the RO system configuration. Typically, the total number of duty discharge wells is designed to match the number of RO trains of the desalination plant, if possible. In addition, a number of standby wells of discharge capacity of 20 to 30 percent are constructed to accommodate periodic well maintenance and inspection along with a potential decrease in well capacity over time.

Well diameter is typically determined based on a maximum well tubing velocity of 3 m/s (10 ft/s), the total maximum concentrate discharge flow rate, and the total number of duty wells. The maximum design well velocity of 3 m/s (10 ft/s) is established based on good engineering practices and the regulatory requirements of some states (i.e., Florida). The average well tube velocity should be in a range of 1.5 to 3 m/s (5 to 10 ft/s). Figure 16.28 presents injection well discharge capacity as a function of well diameter and tubular velocity. This graph can be used for determining the size of individual injection wells.

For example, a brackish desalination plant with fresh water production capacity of 40,000 m^3/day (10.6 mgd) designed at 80 percent recovery will generate 10,000 m^3/day (2.6 mgd) of concentrate. It is assumed that the BWRO plant has four 10,000 m^3/day (2.6 mgd) RO trains and should be designed to operate at a minimum capacity of 25 percent of the total plant production capacity with one RO train in service. The desalination plant concentrate is planned to be disposed using deep injection wells, which will deliver the concentrate to a confined aquifer at a depth of 800 m (2600 ft).

The example below presents an estimate of the number and size of concentrate injection wells needed for this project. The number of wells will be selected to be the same as the number of RO trains (i.e., four) so at minimum production capacity the plant will have one RO train and one discharge well in operation.

As a result, the unit capacity that a single duty well has to be designed to discharge is 10,000 m^3/day/4 = 2500 m^3/day (0.66 mgd) per well. Using Fig. 16.28, for this size injection well and well velocity of 2.5 m/s (8.2 ft/s), the well diameter is selected to be 150 mm (6 in). At average velocity of 2.5 m/s (8.2 ft/s), this well can discharge 3900 m^3/day (1.03 mgd), which is well above its average design capacity of 2500 m^3/day (0.66 mgd). At maximum discharge velocity of 3.0 m/s (10 ft/s), the well can safely dispose up to 5000 m^3/day (1.3 mgd) of concentrate. Because of the large installed maximum discharge capacity of the four duty 150-mm (6-in) diameter wells, the desalination plant

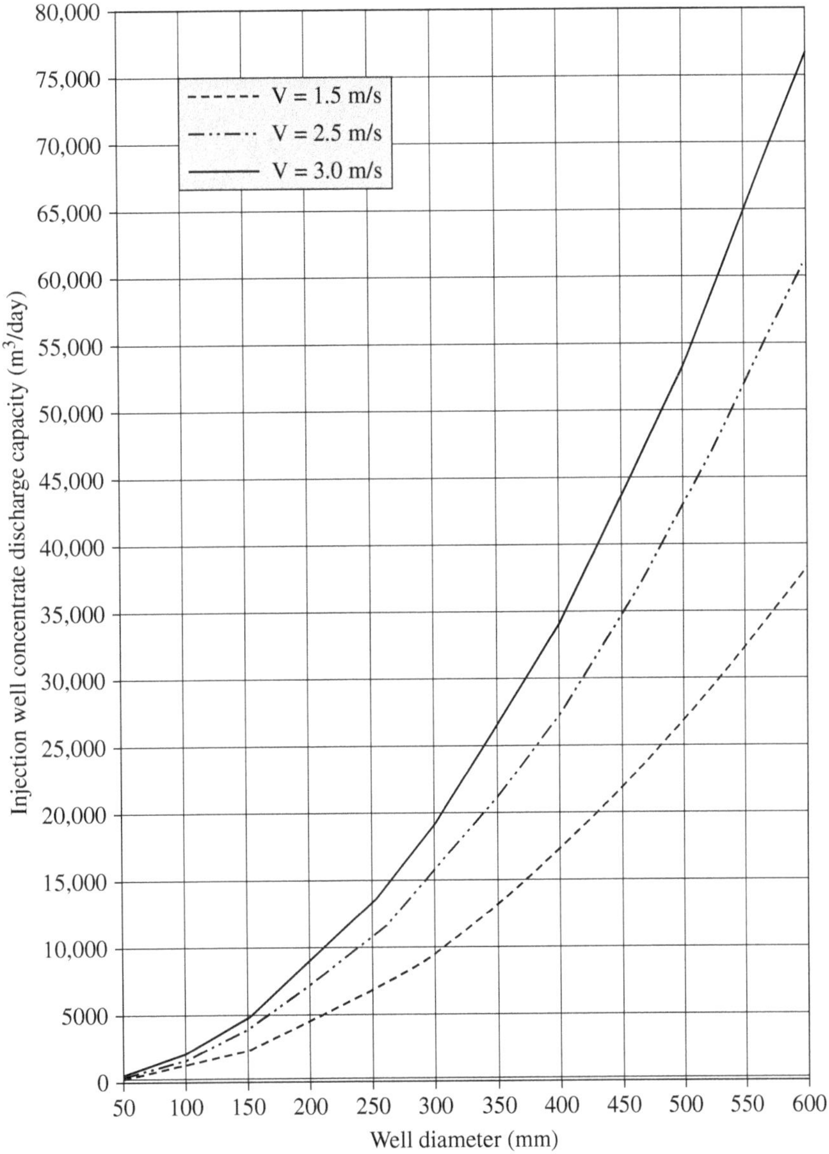

FIGURE 16.28 Injection well discharge capacity as function of diameter.

does not require additional standby wells. Depending on the discharge water quality, however, it may be prudent to install one standby well, especially if the scaling potential of the source brackish water and the water quality of the receiving aquifer are not well-known.

In summary, for this example, a conservatively designed injection well system for concentrate disposal of the reference 40,000 m^3/day (10.6 mgd) desalination plant will have four duty and one standby wells of 150 mm (6 in) diameter and 800 m (2600 ft) depth, each.

16.5.5 Injection Well Costs

The key factors that influence deep injection well costs are the well depth and the diameter of well tubing and casing rings. Table 16.3 presents construction costs for deep injection wells with diameters ranging between 100 and 500 mm (4 and 20 in) as a function of concentrate discharge flow (in m^3/day) and well depth (in meters).

Cost Example

For the example of the reference 40,000 m^3/day (10.6 mgd) brackish water desalination plant described in the previous section, which has four duty and one standby 150-mm (6 in) diameter wells of 500 m (1600 ft) depth, the construction costs of such a deep well injection system calculated using Table 16.3 can be determined as follows: (4 + 1 wells) × [190 × (10,000 m^3/day/4 duty wells) + (550 × 800 m depth) + 120,000] = $5,175,000.

Costs in Table 16.3 and the cost estimate presented above are reflective of a typical deep well injection system, which does not require concentrate pretreatment, pumping, or storage prior to well disposal. In actual projects, however, such facilities may need to be provided. In addition, operation of the well injection system will require the installation of groundwater monitoring wells, which are also not included in the costs above.

Concentrate Pretreatment

Concentrate pretreatment prior to disposal is typically needed when the receiving formation may be plugged by the concentrate discharge as a result of chemical incompatibility. Typical pretreatment includes removal of total suspended solids from the desalination plant discharge, which may be accomplished using cartridge or bag filters or more sophisticated solids removal system, such as contact clarifiers or lamella settlers.

Another type of pretreatment that may be needed is the reduction of the concentrate pH in order to prevent scale formation along the well walls and in the injection zone. Typically, scaling compounds, which may create disposal challenges, are, sulfates of calcium, barium and strontium, calcium fluoride, as well as salts of iron, manganese, and aluminum.

Depending on the actual pretreatment needed, the additional costs for concentrate pretreatment may vary between $20 and $50/$m^3$/day of plant production capacity. For example, a pH adjustment system for the 40,000 m^3/day (10.6 mgd) reference desalination plant would cost $950,000 in 2012 $.

Well Diameter (mm)	Typical Well Concentrate Discharge Capacity (m^3/day)	Construction Costs in 2012 $ as a Function of Concentrate Flow, Q (m^3/day) and Well Depth, H (m)
100	1000–2000	165 Q + 310 H + 100,000
150	2000–4500	190 Q + 550 H + 120,000
200	4500–6500	180 Q + 1250 H + 160,000
250	6500–10,000	170 Q + 1700 H + 230,000
300	10,000–15,000	165 Q + 2000 H + 290,000
400	15,000–30,000	160 Q + 2800 H + 330,000
500	30,000–50,000	150 Q + 4500 H + 370,000

Table 16.3 Construction Costs of Concentrate Injection Wells

Concentrate Pumping

In most cases concentrate disposal in deep wells does not require pumping. Many deep discharge wells operate at pressures of less than 1 bar (14.5 lb/in^2). However, depending on the geologic conditions and depth of the injection zone, often the well-feed pressure needed is in a range of two to four bars (29 to 58 lb/in^2). The additional well pump costs are usually in a range of \$2000 to \$2500/hp. For the reference example seawater desalination plant, an injection well pump station discharging 10,000 m^3/day (2.6 mgd) of concentrate at three bars (43 lb/in^2) will cost \$140,000.

Environmental Monitoring Well System

In order to ascertain the proper performance of the deep well injection system for concentrate disposal, this system will need to incorporate one or more monitoring wells. Typically a deep and shallow monitoring well or dual zone well are installed in the vicinity of the concentrate discharge system. The construction costs of the monitoring wells are function of their depth and range between \$600 and \$800/m of depth for wells less than 1000 m deep (3,280 ft). For deeper monitoring wells, the costs are between \$400 and \$600/m. The monitoring well costs for the reference 40,000 m^3/day (10.6 mgd) project will include the expenditures for construction two wells: a deep monitoring well at an injection zone level of 500 m of \$380,000 and a shallow monitoring well for the groundwater supply aquifer (at depth of 100 m) = \$80,000, for a total cost of \$460,000 (US\$0.46 mm).

When the costs of the well construction are added to the additional concentrate disposal costs, the total cost of the deep well discharge system is injection wells @ \$5.175 mm + pretreatment @ \$0.950 mm + pumps @ \$0.140 mm + monitoring wells @ \$0.460 mm = \$6.725 mm. Taking into consideration that the total construction cost of the 40,000 m^3/day (10.6 mgd) BWRO plant will be in a range of \$55 to \$65 mm, the use of deep injection wells will encompass 10 to 12 percent of this cost.

16.6 Land Application

16.6.1 Description

Land application is a concentrate disposal alternative that involves: (1) spray irrigation of concentrate on salt-tolerant plants, or (2) infiltration of concentrate through earthen rapid infiltration basins (RIBs). Land application is typically used for small volumes of brackish water concentrate only, and its full-scale application is limited by climate conditions, seasonal application, and by availability of suitable land and groundwater conditions.

Irrigation

Irrigation with concentrate involves its application to vegetative surface and collection of the residual drainage water by a runoff control system, if needed. The concentrate stream is applied to a vegetative area by a distribution system. There are three broad categories of concentrate distribution systems: (1) sprinkler or spray systems, (2) surface systems, and (3) drip irrigation systems. Sprinkler systems are most commonly used for concentrate disposal. These systems have spray nozzles that move across the land (Fig. 16.29).

FIGURE 16.29 Spray irrigation system.

Spray irrigation systems cannot be used on variable soils, shallow soil profiles, rolling terrain, erosion-prone soils, and areas where high water tables exist. Disadvantages of sprinkler systems include their higher initial capital cost, higher energy costs, mechanical failures, wind drift problems, and excessive evaporation losses. Also, crops irrigated with sprinklers are subject to injury not only from the salts in the soil but also from the salts directly adsorbed on the wetted leaf surfaces. In general, plants with waxy leafs are less susceptible to injury from contact with concentrate than others.

Slowly rotating sprinklers that allow drying between spray cycles should be avoided since this irrigation pattern increases the wetting-drying frequency. Sprinkling should be completed at night or in the early morning when evaporation rate is lower. Surface systems use narrow-graded [less than 5 m (16 ft)] and wide-graded [30 m (100 ft) or greater] borders or furrows for irrigation water distribution. In general, surface irrigation systems are more suitable for irrigation with higher salinity concentrate than sprinklers. Drip systems have the greatest advantages when saline water is sprayed, but they have found limited application because the system emitters clog easily. Drip irrigation avoids wetting the leaves with saline water.

The volume of runoff generated by an irrigation process depends on the type of irrigation system used. Spray distribution systems do not generally cause surface runoff, whereas surface systems produce some runoff. Ditches or drainage canals can be constructed to retain runoff, or tailwater return systems can be used instead. A tailwater return system consists of pump or reservoir, pump station, and return pipeline. The pumps servicing these systems are typically sized for 25 percent of the distribution system.

Sprinkler System

The predominant type of sprinkler systems is solid set (Fig. 16.30). A solid-set system consists of main distribution and lateral pipes that cover the irrigation field with the sprinklers spaced along each lateral. Pumping is usually needed to deliver the concentrate to the lateral pipelines and sprinkler heads.

FIGURE 16.30 Solid-set sprinkler irrigation systems.

Solid-set systems remain in one position during concentrate application. The major advantages of these sprinkler irrigation systems are their low labor requirements and maintenance costs. The main disadvantage is their high installation cost. Figure 16.31 presents a typical configuration of a solid-set sprinkler system (USEPA, 1984).

Concentrate Storage

Storage facilities are typically needed to retain concentrate during heavy rainfalls or periods when concentrate cannot be applied on the vegetation. Storage is usually provided in earthen-lined holding tanks or in steel structures with protective coating. Sometimes temporary concentrate storage is provided in percolation ponds or earthen storage lagoons, which allow us to achieve initial reduction of concentrate volume.

Subsurface Drainage

Irrigation systems located in areas with a high water table [i.e., water levels located 3 m (10 ft) or less from the ground surface] are often designed with subsurface drainage to provide a root zone area conducive to good vegetative growth. The proximity of the irrigation site to canals, rivers, and other bodies of water should be considered when the irrigation site is chosen because seepage from other water bodies can contribute to subsurface drainage problems.

Subsurface drainage systems consist of a network of buried perforated drainage pipes that are designed to collect concentrate that has not been retained in the irrigated upper soil layer and vegetation. The collected concentrate is conveyed to a basin and is either reused for irrigation or discharged into a surface water body.

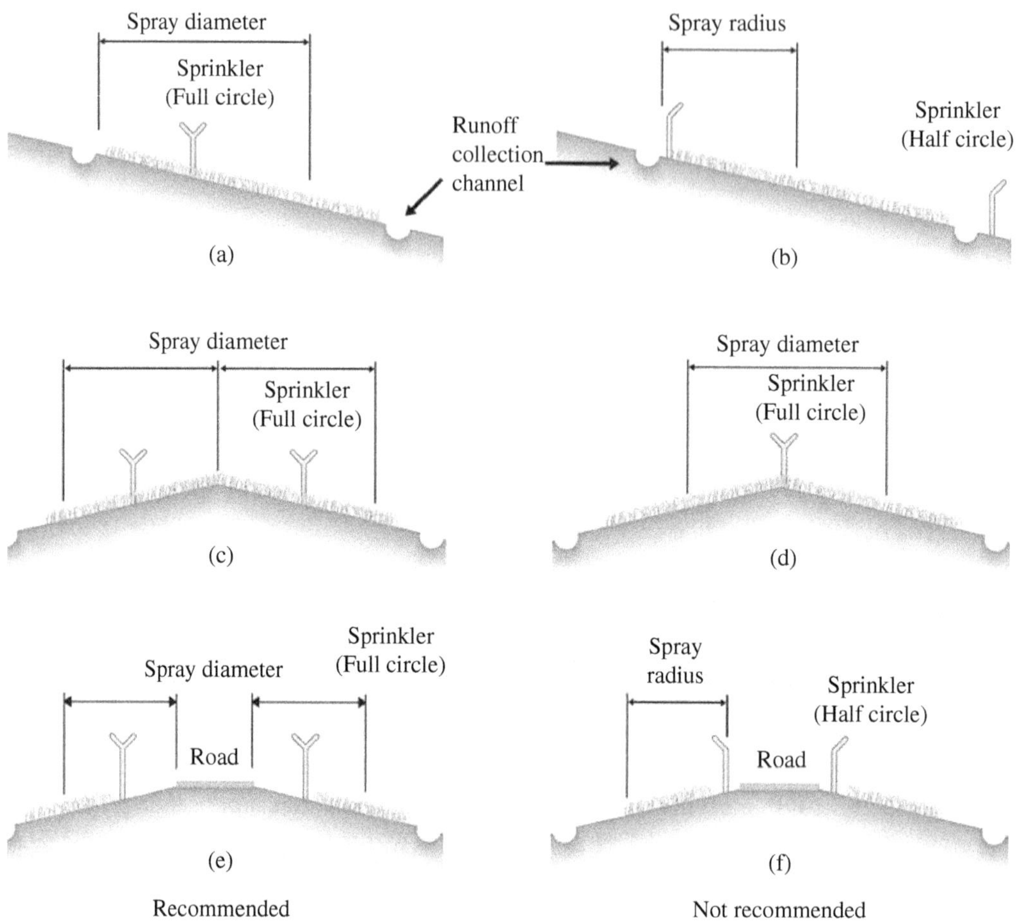

FIGURE 16.31 Sprinkler system configuration. (*Source: US EPA, 2004.*)

Rapid Infiltration Basins

Rapid infiltration basins typically are a series of earthen basins with highly permeable soil bottoms, which allow for high-rate percolation and infiltration of the concentrate into the ground (Fig. 16.32).

Concentrate is delivered to the individual infiltration basins via conveyance pipeline; it enters the basins, quickly infiltrates through the porous surface soil, and then rapidly percolates into the underlying soils. In addition to the basins, the RIB system includes dikes, access ramps, inlet structures, outlet structures, flow control devices, and depth measurement devices.

Uniform application of the desalination plant discharge on the basin surface is necessary to avoid erosion. A simple splash block at the point of discharge may be used for small basins, whereas larger basins typically have a concentrate distribution system (Mickley, 2006).

FIGURE 16.32 Rapid infiltration basins.

16.6.2 Potential Environmental Impacts

Irrigation

Irrigation may have a negative impact on the groundwater aquifer beneath the irrigated area. Shallow groundwater aquifers are typically less saline than desalination plant concentrate, and therefore in most cases surface runoff and ground percolation of concentrate may increase aquifer salinity. However, this would not be a limiting factor in the case of shallow saline coastal aquifers (i.e., if the irrigation site is located close to the ocean shore) or deep confined aquifers isolated from direct or indirect interaction with the concentrate.

Rapid Infiltration

Disposal of concentrate via infiltration is controlled by groundwater discharge permit issued under the groundwater regulations of the pertinent government body with jurisdiction over groundwater resources management. Regulations governing groundwater quality and protection of drinking water aquifers should be investigated as early as possible to confirm the acceptability of this alternative. Significant concerns may arise if concentrate contains arsenic, nitrates, or other contaminants regulated in drinking water. If allowed, concentrate may be diluted to meet groundwater standards.

Rapid infiltration systems must also demonstrate that percolating concentrate meets water quality standards as well as primary and secondary drinking water standards. Monitoring wells are required to assess the environmental impact of RIB systems on groundwater quality.

16.6.3 Criteria and Methods for Feasibility Assessment

Key feasibility factors associated with the use of land application for concentrate disposal include climate, availability and cost of land, percolation rate, irrigation needs, water-quality of the underlying groundwater aquifers, salinity tolerance of the irrigated vegetation, and the ability of the land application system operation to comply with pertinent regulatory requirements and groundwater quality standards.

Successful multiyear use of such concentrate disposal by land application is contingent upon availability of a site with relatively low ground water level and a warm, dry climate as well as a large amount of low-cost land in the vicinity of the desalination project generating concentrate for disposal. Year-round conditions for land application of concentrate usually exist in inland desert-like environments. In colder climate conditions and for specific vegetation, storage facilities may be needed to retain concentrate during the period when it cannot be land applied (typically two to six months). Alternatively, a backup concentrate disposal option should be considered for periods of the year when land application is not possible.

Concentrate salinity is a key limiting factor for the feasibility of land application. As concentrate salinity increases, the feasibility of this scenario decreases. In many cases, concentrate has to be diluted prior to application in order to meet applicable groundwater quality constraints and/or vegetation salinity tolerance limits. Often treated wastewater effluent or low-salinity water extracted from shallow aquifers near the land application site is used to dilute the concentrate prior to land application.

Soil type is of critical importance for the feasibility of land application. Typically, loamy and sandy soils are suitable for this concentrate disposal method. Neutral and alkaline soils are preferable because they would minimize trace metal leaching. Sites with a groundwater level lower than 2 m (7 ft) are preferred. If site groundwater level is less than 3 m (10 ft) from the surface, installation of drainage system would be needed. Typically, sites with slopes of up to 20 percent are suitable for land application. Sites with higher slopes would need to be levelled.

Irrigation

Concentrate salinity and levels of other contaminants determine if irrigation is a viable option. An assessment of the compatibility with target vegetation should be conducted, including review of the acceptable maximum sodium adsorption ratio (SAR), trace metals uptake, and other vegetative and percolation factors. When salinity level of the concentrate is higher than 1000 mg/L, special salt-tolerant species (halophytes) could be considered for irrigation.

Spray irrigation may be a viable land application alternative when the desalination plant, which is the source of concentrate, is located in the vicinity of an agricultural area where salt-tolerant crops are grown year-round. While in most cases, concentrate cannot be applied directly for the irrigation of lawns, golf courses, and public parks due to its high salinity/sodium content, after blending and dilution with reclaimed water or other low-salinity water source down to less than 1000 mg/L of TDS, such application may become feasible.

Sodium Adsorption Ratio

A parameter referred to as a sodium adsorption ratio (SAR) is typically used to determine the maximum level of sodium in the concentrate that could be safely applied to the soil without an adverse long-term effect on soil structure and permeability. SAR is defined by the following formula:

$$SAR = \frac{Na}{\sqrt{\frac{(Ca + Mg)}{2}}} \tag{16.8}$$

where Na = sodium concentration in milliequivalent per liter (meq/L), Ca = calcium concentration (meq/L), and Mg = magnesium concentration in (meq/L).

Usually, SAR higher than 9 may have an adverse impact on soils and is not recommended. In the case of land application to low-salinity tolerance crops and plants, SAR often will have to be mentioned below 6.

TDS

Salinity decreases the water intake of plants by lowering the osmotic potential of the soil. The presence of salts in the soil reduces the rate at which water moves into the soil and also diminishes soil aeration. As a result, increase in salinity of the irrigation water results in decrease in the plant productivity.

Practically all plants can tolerate TDS lower than 500 mg/L. Some salinity-sensitive species (i.e., beans, strawberries, almonds, carrots, onions, avocado, and most golf-course grasses) are affected by a TDS concentrate higher than 1000 mg/L. Some crops (i.e., sugar beet, sugar cane, dates, cotton, and barley) are tolerant to salinities of 2000 mg/L or more.

Typically, only high-salinity tolerance plants (halophytes) can be irrigated with concentrate of salinity higher than 2000 mg/L. These plants can not only tolerate high salinity levels but can also extract salt from the water and store it in the plant tissue. Since most desalination plant concentrates have salinities higher than 2000 mg/L, spray irrigation typically can be applied only in limited number of occasions.

Trace Metals

In addition to the effects of total salinity on vegetative growth and soil, individual ions can cause a reduction in plant growth as well. Toxicity caused by a specific ion occurs when that ion is taken up and accumulated by the plants. The recommended long- and short-term use limits of key trace metals in the concentrate applied for irrigation are shown in the Table 16.4 (adapted from USEPA, 2004).

Salt is continually added to the soil with each irrigation water application, a practice that would eventually harm vegetation. The rate of saline accumulation depends on the quantity of salt applied and the rate at which it is removed from the soil by leaching. Adequate subsurface drainage is also necessary to avoid shallow water tables, which become an additional source of salts.

pH

The pH of the concentrate typically has an indirect effect on the soils mainly by leaching trace metals at low pH. The minimum pH threshold of the concentrate is recommended at 6.

Other Considerations

Other conditions must also be met before irrigation with concentrate can be considered a practical disposal option. First, there must be a need for irrigation water of salinity tolerant vegetation in the vicinity of the desalination plant. Second, a backup disposal

Constituent	Long-Term Use (mg/L)	Short-Term Use (mg/L)	Notes
Aluminum	5.0	2.0	Can cause nonproductivity in acid soils
Arsenic	0.1	2.0	Toxicity threshold varies, Sudan grass limit = 12 mg/L
Beryllium	0.1	0.5	Toxicity threshold varies–kale limit = 5 mg/L
Boron	0.75	2.0	Most grasses tolerant at 2 to 10 mg/L
Cadmium	0.01	0.05	Toxic to beans and beets at 0.1 mg/L
Cobalt	0.05	5.00	Toxicity inactivated in neutral and alkaline soils
Copper	0.2	5.0	Toxic to a number of plants at 0.1 to 1.0 mg/L
Iron	5.0	20.0	Could contribute to soil acidification and loss of phosphorus
Lead and manganese	5.0	10.0	Can inhibit plant growth
Nickel	0.2	2.0	Reduced toxicity in neutral and alkaline soils
Selenium	0.02	0.02	Toxic to many plants at relatively low concentrations.
Vanadium	0.1	1.0	Toxic to many plants at relatively low concentrations
Zinc	2.0	10.0	Reduced toxicity in soils with pH above 6

Table 16.4 Recommended Limits for Trace Metal Constituents

method or storage must be available during periods of heavy rainfall. Third, nearby surface waters have to be protected from the runoff generated from the irrigation site. The soil must also be able to support a vegetative surface. The need to prepare irrigation land by clearing or grubbing adds to overall disposal site costs and should be considered in selecting potential irrigation sites.

Rapid Infiltration

The feasibility and size of rapid infiltration systems for concentrate disposal is typically determined based on land availability, soil and groundwater conditions, existing land uses, site flooding potential, and proximity to the desalination plant that is the source of concentrate. Examples of suitable soils for construction of rapid infiltration basins are sands, sandy loams, and other coarse-textured soils (Mickley, 2009).

The topography and surface conditions of the ground, the characteristics of the surface material, and the rate of precipitation are the key factors that determine the practical applicability of this disposal method. Percolation is defined as the horizontal and vertical flow of water within the soil. It is mainly controlled by soil characteristics. Soil terrain topography is also important because extensive cut-and-fill requirements can dramatically increase construction costs. Before a rapid infiltration basin system can be constructed, extensive geological surveys are needed to determine site soil permeability and hydrogeological conditions.

16.6.4 Design and Configuration Guidelines

Sizing of Irrigation Systems

Selection of Vegetation Type As indicated previously, most crops and plants typically tolerate salinity levels lower than 1000 mg/L. This low threshold makes concentrate disposal by irrigation of non-salt-tolerant plants impractical.

A class of salt-tolerant species (halophytes), however, could grow sustainably at higher levels of salinity. Halophytes are salt-tolerant plants that grow in the world's salt marshes and deserts. Table 16.5 contains maximum TDS thresholds of various salt-tolerant crops (Svensson, 2005). The salinity thresholds presented in this table should be considered as guidelines only. Actual crops tolerance would also vary, dependent on the site-specific climate and soil conditions.

As indicated in Table 16.5, rye and rapeseed could be successfully cultivated using concentrate of 6000 to 7000 mg/L TDS. Date palms, which are commonly cultivated in the Middle East and other arid parts of the world, could tolerate salinity of up to 2550 mg/L. While the plants presented in Table 16.5 have salinity tolerance in a

Crops	TDS Threshold (mg/L)	TDS at Which Yield Declines With 25% (mg/L)
Rye	7300	8800
Rapeseed	7000	8250
Guar	5600	6550
Kenaf	5200	6600
Barley(1)	5100	8300
Guayule	5000	6500
Cotton	4900	8000
Sugar beet(2)	4500	7200
Sorghum	4350	5350
Triticale	3900	10,300
Date palm	2550	7000

Notes: (1)Sensitive during seeding stage (max salinity 2600 mg/L);
(2)less tolerant during germination (max salinity 2000 mg/L)

Table 16.5 Guideline for Salinity Tolerance of Common Crops

Plant species	Productivity grams DW/ m²·year
Atriplex lentiformis	1794
Batis maritima	1738
Atriplex canescens	1723
Salicornia europea	1539
Attriplex barclayana	863
Attriplex nummularia	801

Table 16.6 Annual Productivity of Halophytes Irrigated with 40,000 mg/L Seawater

range of 2500 to 7300 mg/L, other halophytes can tolerate salinities of up to 40,000 mg/L (O'Leary et al., 1985).

Table 16.6 presents the sustainable yield of high-productivity halophytes irrigated with water with salinity of 40,000 mg/L. These halophytes have a productivity of 1539 to 1794 g of dry weight (DW) per square meter per year.

The most productive halophytes shown in Table 16.6 yield an equivalent of 8 to 17 tons of dry matter per hectare (ha), which compares favorably to common forage crops such as alfalfa grown on fresh water, which has a yield of 5 to 20 tons of dry matter per hectare annually.

Bluegreen saltbush (*Attriplex nummularia*) (Fig. 16.33) is considered one of the most promising halophyte species for use as forage crop on marginal lands (O'Leary et al., 1985). Prior experience shows that this species can be irrigated practically indefinitely at loading rates of 2.1 to 2.8 m (6.9 to 9.2 ft) of water per year and salinity of up to 41,000 mg/L under low leaching fraction (< 10 %) without reaching the threshold salinity for yield reduction.

Because of the salinization of the root zone, use of concentrate for irrigation will require periodic application of fresh water for irrigation. Usually, the higher the salinity of the irrigation water the greater the need for periodic fresh water irrigation and drainage. A successful solution to the high-salinity challenges associated with golf course irrigation with brackish water concentrate is its dilution with reclaimed water in a 30/70 percent ratio (Messner et al., 1999).

Irrigation Area

Spray irrigation of concentrate is more land-intensive than rapid infiltration basin disposal. If an existing land area requiring irrigation, such as a golf course, is not available, then land areas surrounding the plant must be purchased or leased for concentrate disposal.

The total active irrigation area needed for concentrate disposal is a function of the water application rate, which, in turns, depends on the type of the irrigated vegetation and soils. Typical application rates are 0.5 to 5.0 m/yr (1.6 to 16.4 ft/yr).

Spray irrigation systems also require land for service roads, buffer zones, storage lagoons, and equipment storage, in addition to the land area needed for the irrigation field. These additional land requirements should be taken into account in estimates of the total land requirements and costs for the land application system.

Figure 16.33 Bluegreen saltbush (*Attriplex nummularia*).

For example, if a spray irrigation system is designed to grow *Attriplex nummularia* for use as forage crop, at a year-round application rate of 2.0 m/yr (6.6 ft/yr) and an annual flow of 365 days for the previously referenced 40,000 m^3/day (10.6 mgd) hypothetical BWRO desalination project, which generates 10,000 m^3/day (2.6 mgd) of concentrate, the minimum amount of land that will be needed for the irrigation field of this project is (365 days × 10,000 m^3/day)/(2 m/yr) = 1,825,000 m^2 (182 ha/451 ac). The actual amount of land needed for this concentrate disposal system will be at least 30 percent larger due to the additional area needed for access roads and other support facilities [i.e., the total amount of land needed would be at least 240 ha (593 ac)].

Concentrate Storage

Temporary storage facilities will need to be provided for periods when concentrate cannot be applied to the irrigated vegetation (rainfall events or times when irrigation is not applied). Usually, concrete storage tanks or lined ponds are used for concentrate storage. Concentrate retention volume is typically driven by the amount of rainfall expected for a particular site, but for operational flexibility a minimum two to five days of storage are provided under any climate conditions. Figure 16.34 provides a general guidance for the recommended size of the concentrate storage tanks depending on their location in the United States (USEPA, 1984).

Additional detailed recommendations for the design of spray irrigation systems for concentrate disposal are presented elsewhere (Mickley, 2006).

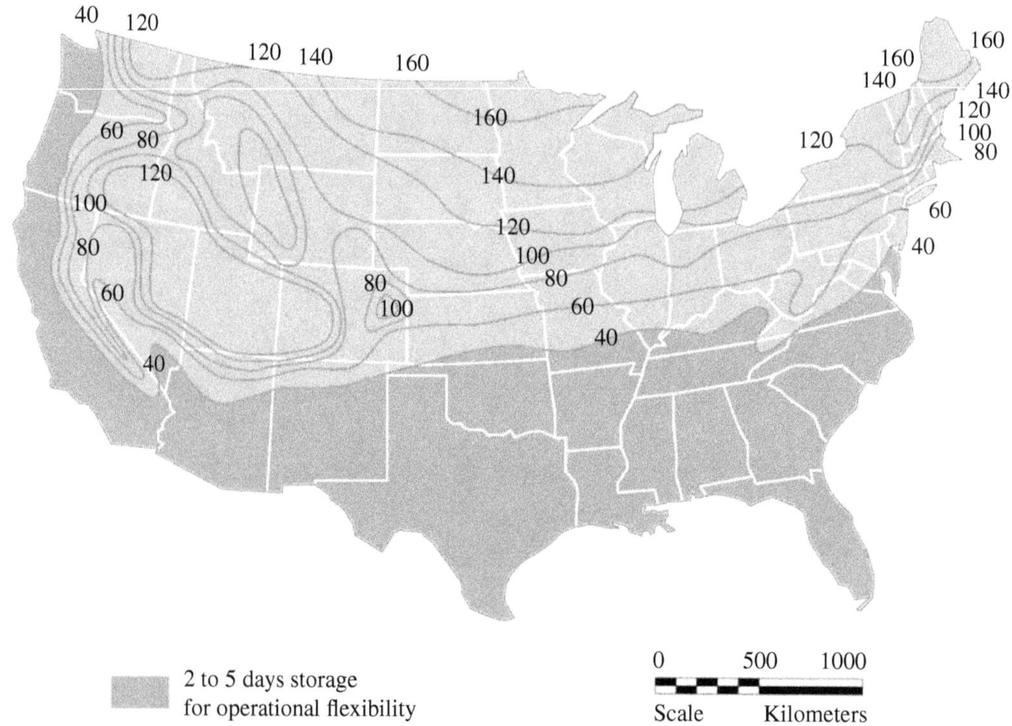

FIGURE 16.34 Recommended concentrate storage time.

Sizing of Rapid Infiltration Systems

Site Selection Selecting a suitable site for the RIB system is of critical importance for the successful use of this concentrate disposal method. Potential RIB sites should be characterized in terms of topography, soil classification to 3.0 m below the bottom of the RIB, lithology of the vadose zone, aquifer quality and gradient, existing vegetation, and distance to nearest seeps and surface waters.

The site hydrogeological conditions will need to be investigated based on information from several boreholes extending to depth of the groundwater surface or maximum of 50 m (160 ft). Infiltration and permeability of the site soils will need to be tested at the bottom of the basin, 2 and 4 m (6.6 to 13.1 ft) below the bottom of the RIBs in order to identify the most suitable depth of concentrate delivery.

RIB Area The total area of a rapid infiltration basin system is determined by the amount of land needed for transmission pipe easement, infiltration basins, access roads, pumping, buffer zones, a maintenance building, and future expansion. The active concentrate application surface area of the RIBs is calculated based on a hydraulic surface loading rate, which, in turn, depends on the effective conductivity of the soils.

Figure 16.35 provides guidance for determining the hydraulic surface loading rate of RIBs (USEPA, 1984). The hydraulic conductivity (K_v) is defined as the amount of water that can move through a unit cross-section in the soil, at unit gradient, and under saturated conditions. For example, a soil with vertical conductivity of 5 cm/h (2 in/h) could transmit 438 m^3/yr (0.12 million gallons/yr) of water through every square meter of horizontal area:

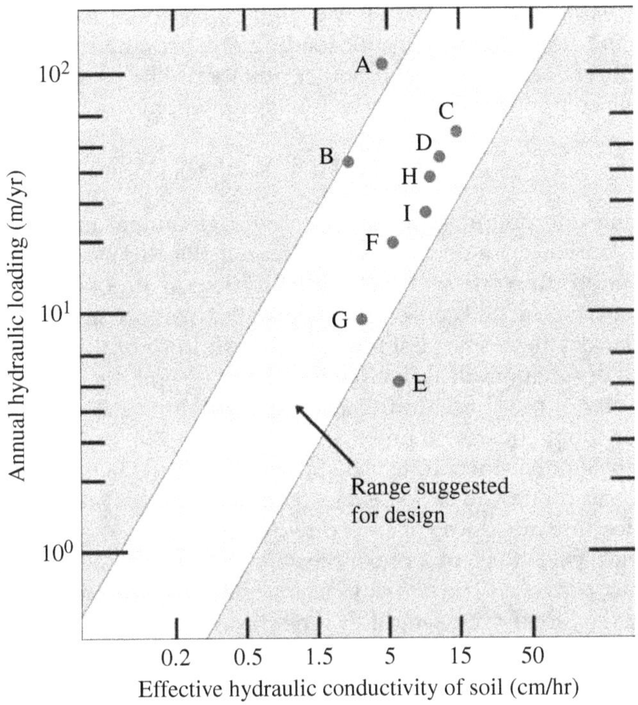

Locations: A - Pilot Basins, Phoenix, AZ
B - Lake George, NY
C - Ft. Devens, MA
D - Boulder, CO
E - Hollister, CA
F - Corvallis, MT
G - East Glacier, MT
H - Jackson, WY
I - Eagle, ID

FIGURE 16.35 RIB loading rate as a function of soil conductivity.

For $K_v = 5$ cm/h (2 in/h), the hydraulic surface loading rate (HLR) is calculated as follows:

$$\text{HLR}_{cv} = \frac{(5\ \text{cm/h})\ (24\ \text{h/day})\ (365\ \text{days/yr}\ (1\ \text{m}^2)}{(100\ \text{cm/m})} = 438\ \text{m}^3 (0.12\ \text{mg})$$

This rate can also be expressed as depth of water on a unit surface area:

$$\text{HLR}_{cv} = \frac{438\ \text{m}^3/\text{yr}}{1\ \text{m}^3} = 438\ \text{m/yr}\ (1437\ \text{ft/yr})$$

The value of the hydraulic conductivity of clean water (K_v) can be determined based on a basin flooding test, which involves construction of a small pilot scale test cell on the selected site or sites and applying clear water for a minimum of two to four months. Typically, the design hydraulic conductivity is selected at 5 to 15 percent of the conductivity measured during the clean water test.

As a result, the actual annual design concentrate loading can be determined based on the clear water hydraulic loading rate presented above and the adjustment factor that reflects the impact of water quality on the long-term application rates.

Annual concentrate loading rate for adjustment factor assumed at 9 percent for is

$$LR_{conc} = (9\%)\ (HLR_{cv}) = (0.09)(438) = 39 \text{ m/yr (127 ft/yr)}$$

The loading rate presented above is an average annual rate assuming that RIBs are operated continuously throughout the year. If the RIB system will be operated for a shorter period, this rate will have to be decreased proportionally to the number of months for which actual concentrate application will occur.

Regular drying is essential for the successful use of the RIB systems. The ratio of loading to drying periods is usually less than one and depends on the season and the water quality of the concentrate (i.e., if the concentrate also contains spent backwash water from the pretreatment process).

Usually, if only concentrate is disposed, the ratio is between 0.5 and 1.0, while, if a blend of concentrate and backwash water is applied, this ratio will be 0.2 or less. In any case, the loading rate does not exceed two days.

A typical cycle for a mix of concentrate and pretreatment backwash water during the summer period will be two days of concentrate application followed by seven days of drying (i.e., total cycle length of nine days). In the winter, the typical application schedule is two days of concentrate feed followed by 12 days of infiltration (i.e., total cycle length of 14 days).

The calculation of the RIB area is illustrated for a desalination plant with a total concentrate discharge flow of 10,000 m^3/day (2.6 mgd); soil hydraulic conductivity, K_v of 5 cm/h (2 in/yr); and summer and winter application cycle lengths of 9 and 14 days, respectively.

For these assumptions the number of summer application cycles (i.e., April through October, 214 days) is 214/9 = 24 cycles. The number of application cycles in the winter (November through March, 151 days) is 151/14 = 11 cycles. As a result, the total annual number of concentrate application cycles is 24 + 11 = 35.

For concentrate hydraulic loading rate of 39 m/yr, the RIB surface loading rate per cycle is (39 m/yr)/(35 cycles/yr) = 1.1 m/cycle. As a result, the application rate (R) during a typical two-day wet period is:

$$R = \frac{1.1 \text{ m/cycle}}{2 \text{ days/cycle}} = 0.56 \text{ m/day (1.84 ft/day)}$$

Since in this example concentrate is generated year-round, and no seasonal storage is assumed to be provided, the area needed for concentrate application during the winter months will govern the total size of the RIBs.

The RIB areas needed for concentrate disposal during the summer and winter period are as follows:

$$\text{Summer area} = \frac{(10{,}000 \text{ m}^3\text{day})\ (214 \text{ days})}{(1.1 \text{ m/cycle})\ (24 \text{ cycles})\ (10{,}000 \text{ m}^3\text{/ha})} = 8.1 \text{ ha (20.0 ac)}$$

$$\text{Winter area} = \frac{(10{,}000 \text{ m}^3\text{/day})\ (151\text{days})}{1.1 \text{ m/cycle (11 cycles) } (10{,}000 \text{ m}^3\text{/ha})} = 12.5 \text{ ha (30.9 ac)}$$

Because the area needed for concentrate application in the winter is larger (12.5 versus 8.1 ha), this is the total active filtration area needed for the RIB system. The RIB system will need to include a circulation area for servicing the ponds and for the concentrate distribution system. Therefore the total RIB system area typically is increased with 40 to 60 percent to accommodate these service needs [i.e., the total amount of land needed for the example RIB system is 1.6 × 12.5 ha = 20 ha (49.4 ac)]. This area compares favorably to the area needed for the example spray irrigation project for disposal of the same amount of concentrate (240 ha/593 ac).

The design loading rate must be based on the least permeable soil layer in the soil profile and on expected worst-case weather conditions. Concentrate discharge into the RIBs should be intermittent to maintain the design loading rate and soil capacity.

Other Key RIB Design Criteria

The RIB system should be designed to comply with the following key design requirements:

- Minimum number of RIBs = 3
- Minimum basin depth = 1.5 m (5 ft)
- Minimum distance from RIBs to site boundary = 150 m (490 ft)
- Minimum basin bottom permeability at 30 cm = 1.4 cm/s (0.55 in/s)
- Maximum depth of ground water below basin bottom = 3 m (10 ft)
- Minimum depth of impermeable layer below basin bottom = 10 m (33 ft)
- Minimum distance from water supply wells = 300 m (1000 ft)
- Minimum number of monitoring wells = 3 (one up-gradient and one down-gradient of the RIB)

Typically, the number of RIBs varies between 3 and 17, and individual basins can range from 0.2 to 2.0 ha (0.5 to 5.0 ac) for small- and medium-size applications and 2.0 to 5.0 ha (5.0 to 12.3 ac) for large projects. RIBs are recommended to be located perpendicular to the groundwater flow direction in order to reduce groundwater mounding.

Dikes

The dikes for the RIBs have to be at least 0.5 meter (1.6 ft) deeper than the maximum design water depth. Most dikes are 1.0 to 1.5 meter (3.3 to 4.9 ft) deep, and in some cases they are shallower. Higher dikes are not beneficial and contribute to operation problems due to the extra runoff and potential for erosion of soil fines. The dikes should be compacted to prevent seepage through them. The top of the dikes is usually deigned for a vehicular access and should be at least 6.0 m (20 ft) wide. Use of a silt fence or similar porous barrier at the toe of the dikes is recommended to protect against washout of soil fines. Additional design details for RIB systems are provided elsewhere (USEPA, 1984; Mickley, 2006).

Concentrate Storage

Similar to irrigation systems, a minimum two to five days of operational storage is recommended to be provided for RIB facilities. Depending on the local climate conditions, larger storage volume may be needed (see Fig. 16.34).

16.6.5 Land Application Costs

Spray Irrigation System Costs

Key cost parameters associated with spray irrigation systems for concentrate disposal include concentrate flow rate, hydraulic loading rate, distance to the irrigation site, irrigation land cost, size of storage tank, and the need/size of the subsurface drainage.

Land application by spray irrigation on crops is typically cost effective only if concentrate is blended with a fresh water source to reduce its salinity to a level acceptable for irrigation, and its feasibility depends on the type of the crops/vegetation and on the soil uptake rates.

The construction cost of land irrigation systems as a function of the concentrate flow rate and the annual hydraulic loading rate (LR_{conc}) is presented in Fig. 16.36.

The costs depicted in Fig. 16.36 do not include expenditures associated with the purchase and clearing of the land for the irrigation system, for construction of temporary concentrate storage facilities (tanks or lagoons), or for concentrate conveyance from the desalination plant to the irrigation site because these costs are site-specific.

Cost Example

An order of magnitude cost estimate of a spray irrigation system for a hypothetical reference brackish desalination plant with fresh water production capacity of 40,000 m^3/day (10.6 mgd) designed at 80 percent recovery is provided below. Such a desalination plant will generate 10,000 (2.6 mgd) m^3/day of concentrate. If this concentrate is planned to be disposed using a spray irrigation system applied to the halophyte *Attriplex nummularia* at a loading rate of 2.0 m (6.6 ft) of water per year, then from Fig. 16.36, the construction cost for the spray irrigation system is estimated at $3.30 million.

Assuming that the concentrate can be applied year-round, and that an earthen storage lagoon with a retention time of two days (20,000 m^3/5.2 mg) is built for this project, at a unit storage lagoon cost of $0.05 mm/1000 m^3, the additional costs associated with concentrate storage = $1.0 mm (20,000 m^3 × $0.05 mm/1000 m^3 = $1.0 mm).

As indicated previously, a year-round application rate of 2.0 m/yr the total amount of land needed for this project would be 240 ha (593 ac). Assuming a unit cost of land of $5000/ac, the additional construction cost for land acquisition is $2.97 mm.

In addition to the costs listed above, the environmentally safe operation of the spray irrigation system will require installation of a groundwater monitoring well system. Such a system would have a unit cost of $3000 to 5000/ha, or for this example $1.0 mm (240 ha × $4160/ha = $1.0 mm).

In summary, the total construction costs for the spray irrigation system for disposal of 10,000 m^3/day of concentrate for the reference 40,000 m^3/day BWRO project is $3.30 mm (spray irrigation system) + $1.00 mm (storage) + $2.97 mm (land) + $1.00 mm (monitoring system) = $8.27 mm.

This construction cost is higher than the cost for the same size project using deep injection wells for concentrate disposal [$6.725 mm (see Sec. 16.5.4)]. While the costs for such systems are significant, if specifically developed for concentrate disposal, some of the capital expenditures listed above will not be incurred if the concentrate is applied on existing vegetation (i.e., forage crops, golf course grass), which uses fresh water for irrigation.

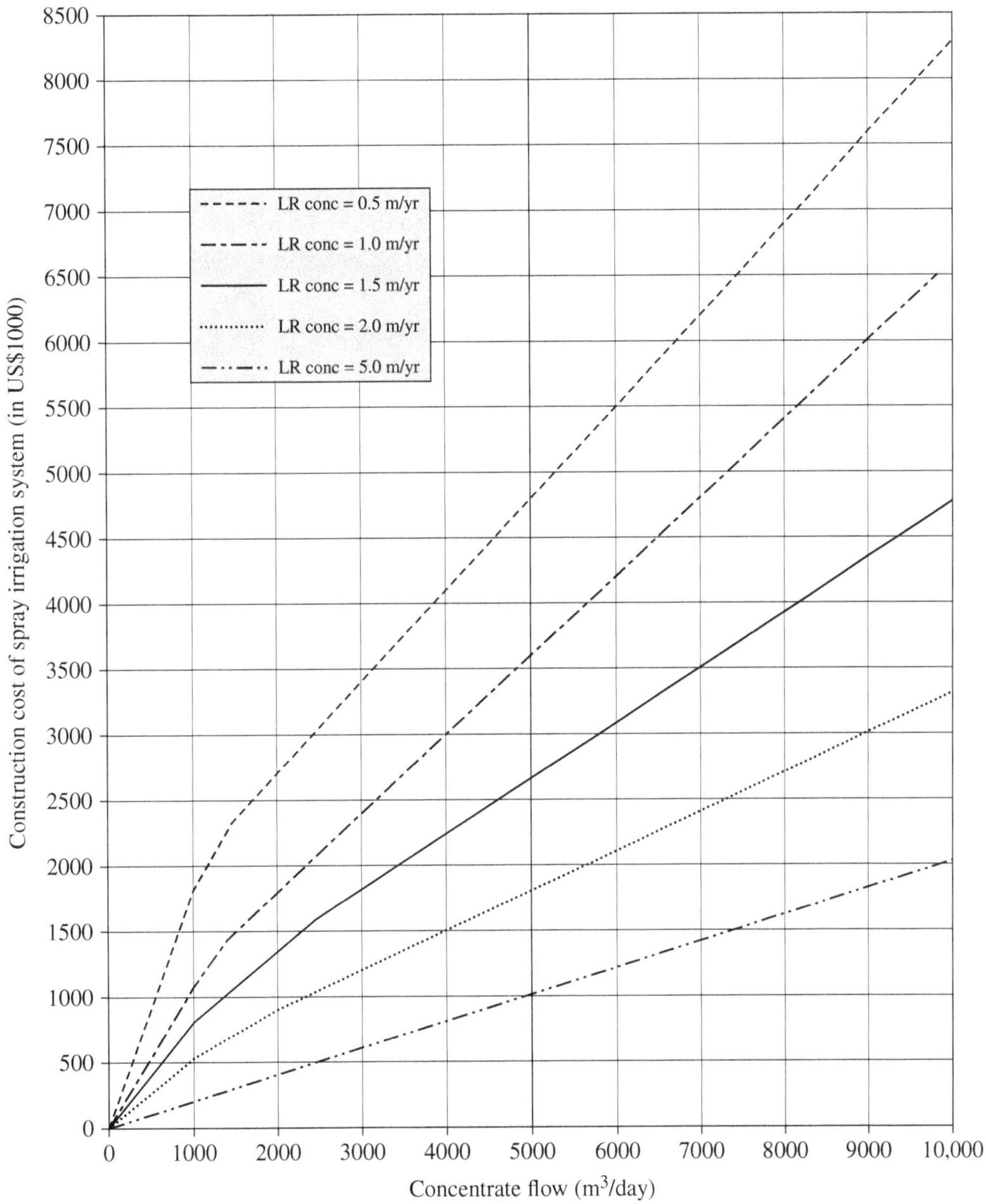

FIGURE 16.36 Construction costs of spray irrigation system.

It should be pointed out that spray irrigation is not only costly in terms of construction costs, but also involves significant operations and maintenance expenditures. Operation and maintenance of a concentrate spray irrigation system is more labor-intensive than the other concentrate disposal methods. Labor requirements include sprinkler system repair and vegetative surface maintenance. Energy costs for pump operations also add to the system's total operational costs.

One of the large O&M expenditures is the harvesting and disposal of the halophyte vegetation, unless this vegetation can be used as a forage crop. For example, the halophyte bluegreen saltbush [*Attriplex nummularia* (see Fig. 16.34)] used in the cost example above produces 801 g DW/m^2·yr of vegetation mass.

For the specific conditions of the cost example, 1,825,000 m^2 of irrigated land will produce (0.801 kg/m^2·yr × 1,825,000 m^2)/1000 kg per ton = 1460 dry tons of vegetation per year, which equates to approximately 15,000 tons of green vegetation. This vegetation will need to be harvested, stored, and delivered for use as a forage crop or ultimately disposed to a landfill. The costs for these O&M related activities would exceed \$0.5 mm/yr.

Rapid Infiltration System Costs

The construction costs for RIB systems as a function of the concentrate flow and land application rates are shown on Fig. 16.37. Since RIB systems are constructed on soils with significantly higher infiltration rates, the hydraulic loading rates of these systems are usually higher than those of spray irrigation systems.

For the RIB example presented in Sec. 16.6.4 (concentrate discharge flow of 10,000 m^3/day and loading rate of 39 m/yr), the construction cost of such RIB system is estimated of \$4.90 mm (Fig. 16.37). This cost does not include storage tank expenditures as well as costs for land acquisition, delivery of the concentrate to the RIB site, and groundwater monitoring.

Using the same assumptions as in the cost example for spray irrigation systems (i.e., two-day storage tank, unit storage tank cost of \$0.05 mm/1000 m^3, and unit land cost of \$5000/ac), the additional costs are: (1) storage tank costs of \$1.00 mm (10,000 m^3/day × 2 days × \$0.05 mm/1000 m^3 = \$1.00 mm); and (2) land acquisition costs of \$ (49.4 ac × \$5000/ac = \$0.25 mm). The cost of the groundwater monitoring system is \$0.02 mm (20 ha × \$5000/ha = \$0.10 mm).

In summary, the total construction cost for the implementation of this project (excluding concentrate delivery to the site) is = \$4.90 mm (RIB system) + \$1.0 mm (storage tank) + \$0.25 mm (land acquisition) + \$0.10 (monitoring) = \$6.25 mm. This cost compares favorably with the cost for construction of a spray irrigation system for disposal of the same quantity of concentrate (\$8.27 mm) and is comparable to the costs of a deep well injection system (\$6.725 mm).

16.7 Evaporation Ponds

16.7.1 Description

Evaporation ponds are shallow, lined earthen basins in which concentrate evaporates naturally as a result of solar irradiation. As fresh water evaporates from the ponds, the minerals in the concentrate are precipitated in salt crystals, which are harvested periodically and disposed off-site.

Evaporation ponds could be classified in two main groups: (1) conventional evaporation ponds, and (2) salinity gradient solar ponds. The fundamental difference of the two types of ponds is that while conventional evaporation ponds are primarily designed for concentrate disposal, the main function of solar ponds is to generate electricity from solar energy.

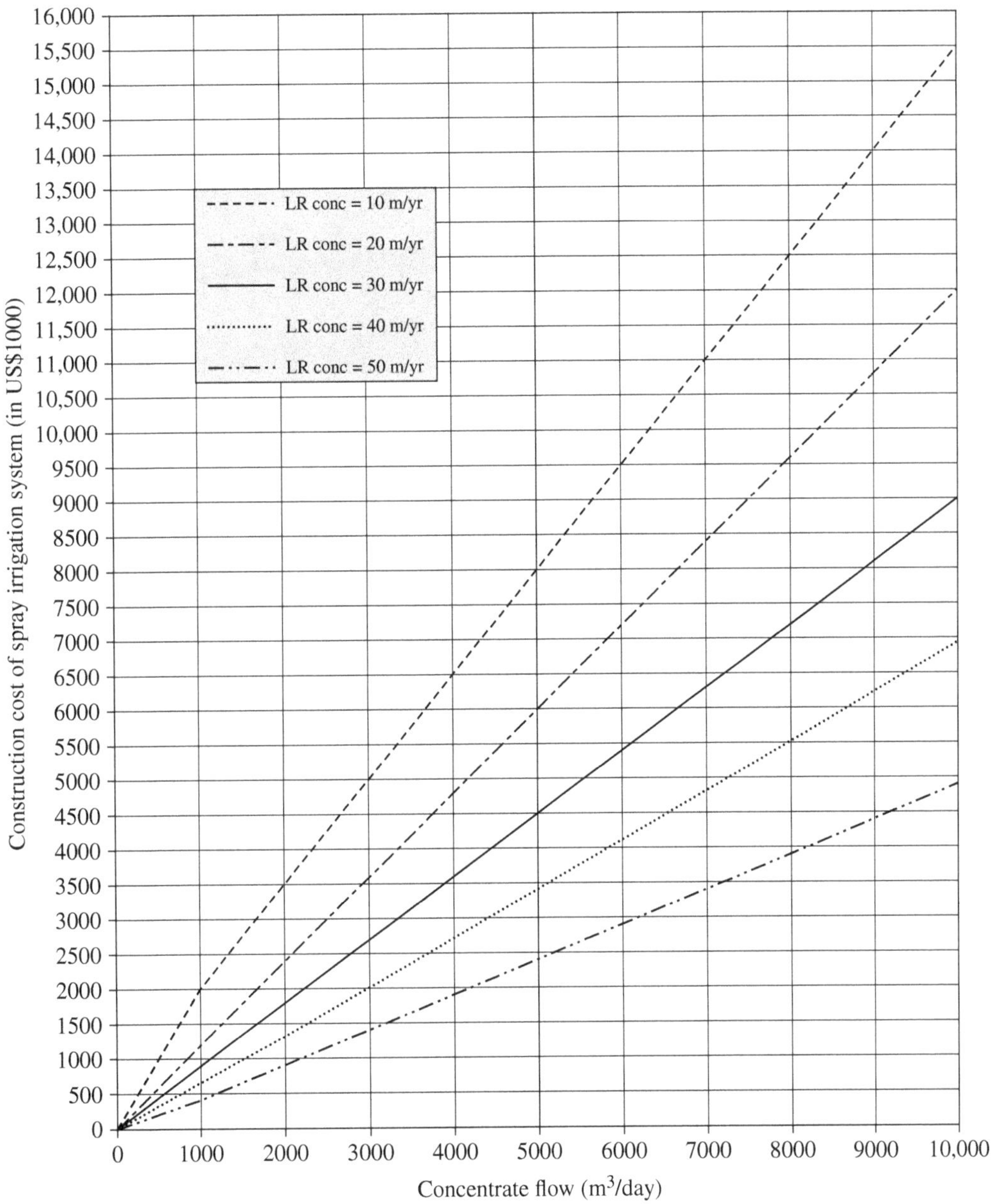

FIGURE 16.37 Construction costs for RIB system.

Conventional Evaporation Ponds

Conventional evaporation ponds consist of a series of lined or unlined earthen or concrete structures designed to maximize water evaporation (see Figs. 16.38 and 16.39). These ponds have found applications mainly for seawater concentrate disposal. They operate using natural solar evaporation of concentrate periodically fed to man-made, lined

FIGURE 16.38 Conventional evaporation ponds.

earthen basins. Holding ponds are needed for concentrate storage, while the evaporation pond reaches the high salinity needed for normal pond operations. The ponds should be fenced to prevent entrance and potential harm of people and animals in the area.

Evaporation Pond Performance Enhancements

Several approaches have been studied to date to enhance evaporation rates from concentrate disposal ponds including spray evaporation (Fig. 16.40), pond aeration (Fig. 16.41), and addition of dye to elevate pond water temperature.

Spray Evaporation

This evaporation rate can be enhanced with the use of mechanical spray evaporators, which disperse concentrate over the pond surface in the form of fine mist. While the use of this technology would allow increasing pond evaporation rate with over 20 percent,

FIGURE 16.39 Salt crystals removed from the bottom of evaporation pond.

FIGURE 16.40 Spray evaporation pond.

it also has high energy requirements and may overall not be cost competitive for locations where access to electricity is not readily available or costly.

Pond Aeration

Pond aeration accelerates evaporation by increasing the contact surface between the air and the concentrate. Both floating and submerged aerators can be used for this application.

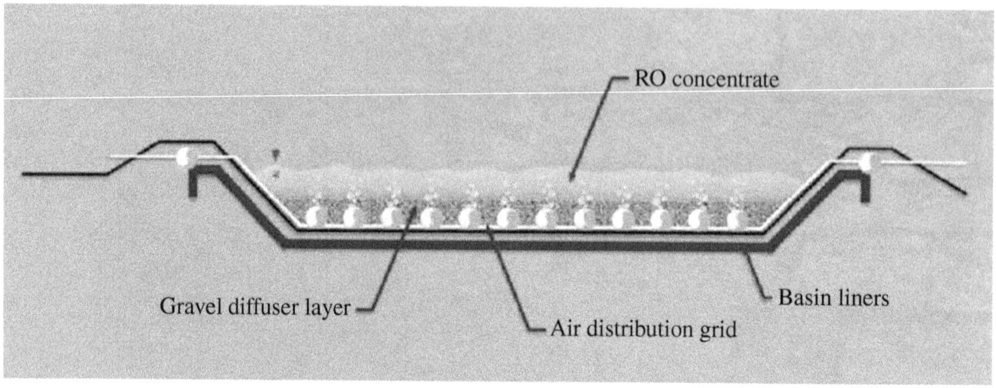

FIGURE 16.41 Evaporation pond with submerged aeration system.

FIGURE 16.42 Solar bee pond aeration system.

Figure 16.41 presents a schematic of a pond with a submerged coarse bubble aeration system.

The solar-powered pond mixing equipment shown on Fig. 16.42 is a water circulation system, which draws concentrate from the bottom of the pond and spreads it across its surface. Testing of this system in Salton Sea, California (TDS concentration of 45,000 mg/L), has shown that the overall evaporation rate could be increased in average by 30 percent (Carollo, 2009).

Use of Dye for Enhanced Evaporation

Adding Naphthol green dye in the top 0.2 m (0.7 ft) at a concentration of 2 mg/L has been found to increase the evaporation rate of a 500 m^2 (0.12 ac) pond by 13 percent (Ahmed et al., 2000). This evaporation rate enhancement measure, however, could be costly for large ponds.

Solar Ponds

Solar ponds are deep, lined earthen lagoons containing high-salinity water and are designed and operated to collect solar energy and convert it into electricity (Fig. 16.43).

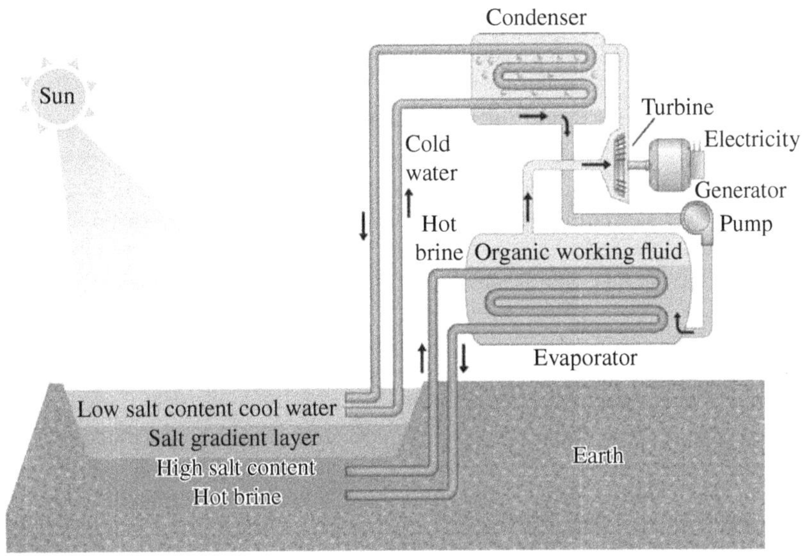

FIGURE 16.43 Solar pond schematic.

While conventional evaporation ponds are configured to maximize heat convection and evaporation, solar ponds are deeper lagoons designed to retain heat and therefore have a lower evaporation rate. Therefore solar ponds are often considered a system for beneficial use of concentrate (i.e., generation of electricity) rather than an efficient concentrate disposal method.

Three layers of different salinity water naturally form in solar ponds (from top to bottom): surface zone, gradient zone, and lower zone. The surface zone is also referred to as an upper convective zone and is comprised of cool water of low salt content (Walton and Swift, 2001). This zone is typically 0.3 to 0.5 m (1.0 to 1.6 ft) deep.

The lower (salt-gradient) layer is a homogeneous, concentrated salt solution that is typically salinity and temperature stratified. The temperature and salinity of the concentrate in this layer increase from top to bottom. The thickness of this layer is typically 0.5 to 1.5 m (1.6 to 4.9 ft).

The bottom (high-salt content) layer contains concentrate with salinity near saturation level (TDS of 250,000 to 260,000 mg/L). If the salinity gradient in the salt-gradient layer is large enough, there is no convection in the gradient zone even when heat is absorbed in the lower zone and on the bottom, because the hotter, saltier water at the bottom of the gradient remains denser than the colder, less salty water above it.

Because water is transparent to visible light but opaque to infrared radiation, the energy in the form of sunlight that reaches the lower zone and is absorbed there can escape only via conduction. Because the heat conductivity of concentrate is low, the salt-gradient layer above the lower level works as an insulation, which retains the heat accumulated in the bottom layer. As the temperature of this layer reaches 85°C (185°F), the hot concentrate can be used to generate thermal electricity (Fig. 16.43).

Solar ponds have been successfully tested in El Paso, Texas, and Victoria, Australia. A 10,000 m^2 (2.5 ac) solar pond in Australia was reported to produce electricity of 200,000 kWh/yr (Arakel et al., 2001). Another 5000 m^2 (1.25 ac) solar pond system in

Australia has been documented to produce electricity of 130,000 kWh/yr at power generation cost of \$0.12/kWh (Highnett et al., 2002).

16.7.2 Potential Environmental Impacts

Groundwater quality regulations in the United States require evaporation ponds to be constructed with impervious lining for protection of underlying aquifers. Typically, concentrate is not contaminated with hazardous materials, and a single layer liner is adequate for groundwater protection. However, if concentrate is contaminated (i.e., contains high levels of trace metals), then a double-lined pond may need to be constructed.

If the ponds are not lined or point liner is damaged, a portion of the concentrate may percolate to the water aquifer beneath the pond and deteriorate its water quality. Therefore evaporation pond systems, especially these using geo-membrane liners, should be equipped with underground leak-detection systems that are installed beneath the liner. Alternatively, pond leakage can be monitored via a groundwater monitoring well system with at least three monitoring wells: one installed up-gradient to the groundwater flow, one down-gradient, and one in the middle of the pond system. Monitoring must be conducted monthly.

16.7.3 Criteria and Methods for Feasibility Assessment

Solar evaporation is a feasible concentrate disposal alternative only in relatively warm, dry climates with high evaporation rates; low precipitation rates and humidity; flat terrain; and low land cost. Typically, evaporation ponds are not feasible for regions with an annual evaporation rate lower than 1.0 m/year (3.3 ft/yr) and annual rainfall rate higher than 0.3 m/yr (1.0 ft/yr). Factors affecting evaporation rate are humidity, temperature, solar irradiation intensity, wind, rainfall, and concentrate salinity.

Humidity has a significant impact on pond evaporation rate—the higher the humidity the lower the evaporation rate. Usually when the average annual humidity of a given location exceeds 60 percent the use of evaporation ponds is not likely to be a viable concentrate disposal option.

Evaporation ponds are climate dependent. The higher the temperature and solar irradiation intensity the more viable this option is. Dry equatorial and subequatorial regions of the world would be suitable for such concentrate disposal alternative. Figure 16.44 presents a map of the solar irradiation intensity in the United States. Arizona, Nevada, and New Mexico as well as parts of Texas and Southern California are conducive to the use of solar evaporation ponds. Wind speed and duration have a significant impact on evaporation rate—windier locations are more suitable for installation of evaporation ponds. However, wind often carries solids (sand and dust) that could fill the ponds during sandstorms. Significant rainfall reduces evaporation rates. In high-rainfall portions of the world, the actual annual rainfall rate should be subtracted from the annual evaporation rate, when determining the actual design pond evaporation rate. The difference between the standard annual evaporation rate and rainfall is referred to as evapotranspiration potential. For example, in southern Florida the standard evaporation rate is between 1.0 and 2.0 m/yr (3.3 and 6.6 ft/yr). However, when corrected for the rainfall impact, the actual pond design evaporation rate has to be is reduced down to 0.6 m/yr (2.0 ft/yr) (Carollo, 2009). This rate corresponds to a land requirement of over 70 ha/1000 m^3/day (655 ac/mgd) of concentrate.

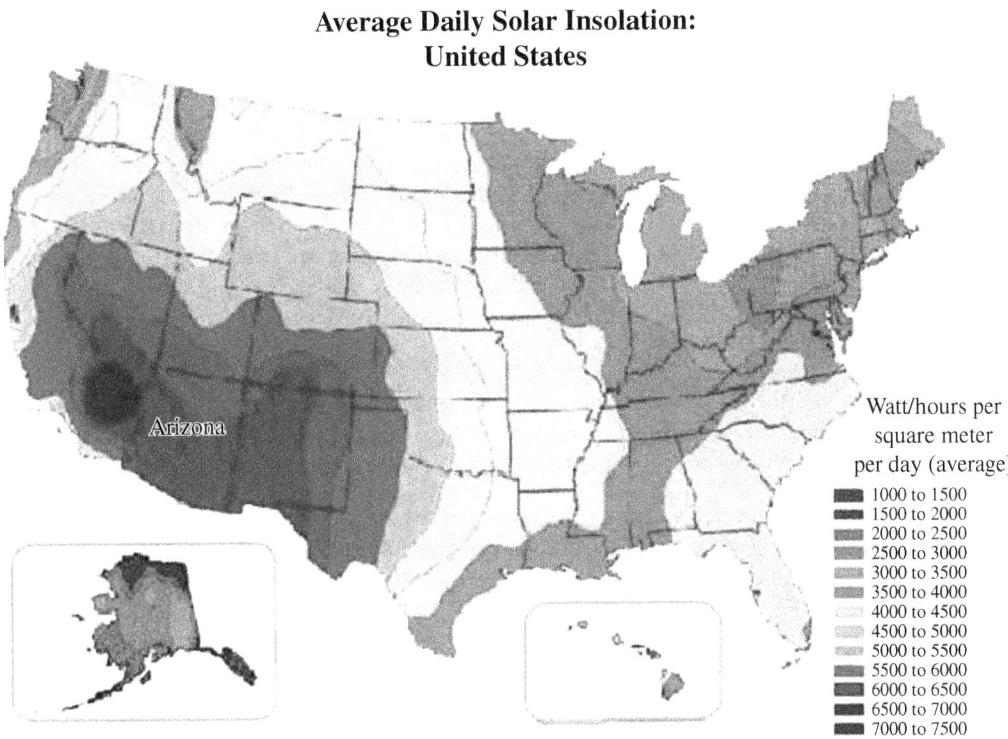

FIGURE 16.44 Map of Solar Irradiation Intensity in the United States.

Evaporation rate decreases as solids and salinity levels in the ponds increase. However, typically it is less costly to evaporate higher salinity concentrate of smaller volume than lower salinity concentrate of larger volume (i.e., minimization of concentrate volume is beneficial if this concentrate will be disposed using evaporation ponds).

16.7.4 Design and Configuration Guidelines

The disposal capacity of conventional evaporation ponds is a function of concentrate flow, evaporation rate at the location of the ponds, and average annual rainfall. Evaporation ponds are typically sized to ensure the containment of the maximum operating volume of concentrate and an inflow from a 100-year, 24-hour storm event, under which conditions the ponds should have a minimum of 0.5 m (1.6 ft) freeboard.

The basic design recommendations for conventional evaporation ponds are as follows: minimum of two ponds, dikes constructed of impervious material and compacted to at least 90 percent of its maximum dry proctor density, minimum depth = 2.5 m (8.2 ft), minimum freeboard at average annual flow = 1.0 meter (3.3 ft), and removal of salt deposits every two years.

Most ponds are designed in a square or rectangular shape. Usually it is more beneficial to construct a larger number of smaller size ponds than to have one or two large evaporation ponds because the smaller-size pond configuration allows minimizing wind-triggered wave damage on the pond dikes.

Sizing of Conventional Evaporation Ponds

Pond Depth As indicated previously, shallower ponds result in increased evaporation rate. However, the lower the pond depth the larger evaporation area is needed, which, in turn, translates to higher concentrate disposal costs. Taking into consideration that increase of pond depth from 0.1 to 2.5 m (0.3 to 8.2 ft) would result in only 4 percent reduction in evaporation rate (Mickley, 2006), deeper ponds are overall more cost effective.

Optimum pond depth in terms of evaporation rate is approximately 0.5 m (1.6 ft), but often deeper (2.5 to 5.0 m/8 to 16 ft) ponds are used in order to reduce their construction costs and to accommodate salt accumulation at the bottom of the ponds, as well as to provide for accumulation of water from precipitation and for contingency water storage.

Pond Dikes Earthen dikes surround the perimeter of the evaporation ponds. The dikes are typically compacted earthen structures with a slope of 2:1 to 4:1 and a 4- to 6-m (13 to 20 ft) wide road on the top. Dike height usually varies between 1.5 and 4.0 m (4.9 to 13.1 ft).

Pond Liner Typically, concentrate evaporation ponds are lined with clay, clay/bentonite mix, or plastic (PVC, HDPE, and Hypalon) liners. Liners should be designed to cover pond bottom, dikes, and 2 to 3 m (7 to 10 ft) of additional area between the dike walls and the road (Fig. 16.45).

FIGURE 16.45 Geo-membrane evaporation pond liner.

Evaporation pond liners should be designed to have low hydraulic conductivity [$< 10^{-7}$ cm/sec ($< 4.0 \times 10^{-8}$ in/sec)] and seepage rate [< 5 mm/day (0.2 in/day)] and at least a 20-year durability on exposure to high-salinity concentrations and ultraviolet (UV) light. Suitable liners are *in-situ* clay with a thickness of 1.0 m or more, compacted clay with a thickness of 0.5 m (1.6 ft) or more, soil and bentonite mix with a minimum thickness of 0.10 m (4 in), or a geo-membrane liner with thickness of 30 mil or more (see Fig. 16.45). If a HDPE liner is used, the minimum liner thickness should be at least 1.5 mm (60 mil).

The clay (*in-situ* or compacted) used as a pond liner must comply with the following requirements: (1) more than 30 percent of the material passing the #200 sieve (0.074 mm); (2) liquid limit of 30 percent or more; and (3) plasticity index higher or equal to 15 percent. In addition, the clay liner must be applied in at least four successive layers ("lifts") of not more than 20 cm (8 in) in thickness (uncompacted) each, which should be compacted to 95 percent of its standard Proctor maximum dry density to meet the maximum hydraulic conductivity requirement of 10^{-7} cm/sec (4.0×10^{-8} in/sec) at minimum compacted thickness of 15 cm (6 in). Most pond liners have a pH tolerance range of 6 to 9, and if the pH of the concentrate is outside of this range, it has to be properly adjusted before its application.

Pond Area The evaporation pond surface area is primarily the function of the evaporation rate, which, in turn, is determined by local climate conditions. Standard evaporation rate is typically presented in m/yr and is measured for fresh water (1 m/yr = 27.4 m^3/day·ha = 2900 gal/day.ac).

Figure 16.46 shows typical evaporation rates for locations in the United States most favorable for the construction of evaporation ponds: southern Arizona, western Texas, and Southern California. As shown in this figure, the average annual evaporation rates

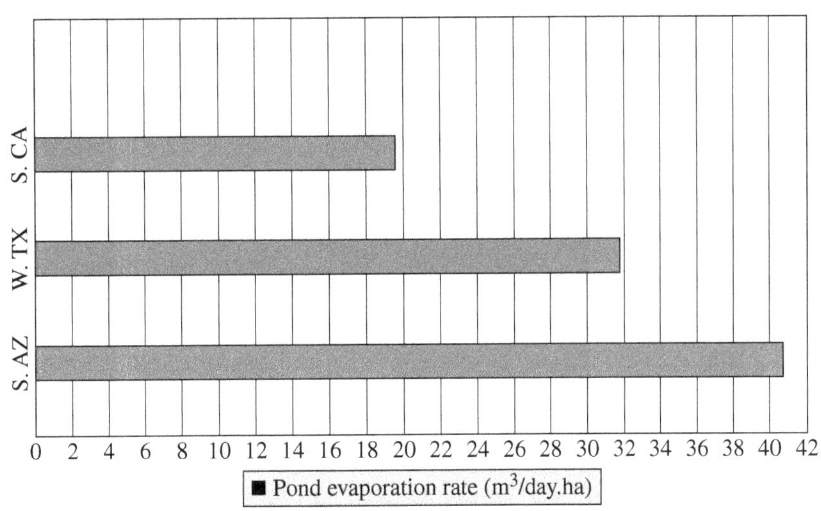

FIGURE 16.46 Evaporation rates in the three arid US regions.

vary between 0.7 and 1.5 m/yr (19.8 and 40.7 m^3/day·ha or 2000 to 4350 gal/day·ac. For comparison, the evaporation rate is Aswan, Egypt, is 5.0 m/yr or 14,500 gal/day. ac (Swanson, 2005).

Evaporation rate also depends on the pond depth: shallow ponds enhance evaporation rate but may dry more frequently and expose their liners to the environment. Clay liners exposed to frequent wetting and drying cycles are more likely to desiccate and crack.

Standard evaporation rates readily available in the technical literature are determined for fresh water. Since evaporation rate decreases with salinity concentration, such "fresh water" evaporation rates should be reduced when applied to concentrate. The reduction ratio/actual evaporation rate will be site-specific, and therefore it is recommended to be determined through pilot testing.

If no specific data are available, concentrate evaporation rate can be assumed to be 70 percent of the fresh water evaporation rate for a given location (Mickley, 2006). While 70 percent is a conservative estimate, for low-salinity brackish SWRO plants this ratio could be significantly lower (80 to 90 percent).

Often an additional 20 to 25 percent of contingency is added to the capacity of the ponds to accommodate rain events and varying concentrate production and water quality over the useful life of the project. In summary, the total pond surface area as a function of the fresh water evaporation rate is expressed as follows:

$$A_p = \frac{(Q_C \times S_F)}{(C_F \times SER)} \tag{16.8}$$

where A_p = active evaporative pond area (ha), Q_c is concentrate flow rate in m^3/day, SER = standard evaporation rate for fresh water, m^3/day·ha, C_F = contingency factor, and S_F = factor for conversion of fresh water evaporation rate to concentrate evaporation rate. As indicated previously, unless the specific C_F is determined based on pilot testing, a conservative value of C_F = 0.70 (i.e., 70 percent) should be used for pond sizing. The S_F typically has a value of 1.2 to 1.3.

For the reference example, for the 40,000 m^3/day 10.6-mgd BWRO plant, which generates 10,000 m^3/day (2.6 mgd) of concentrate and is located in southern Arizona, which has a local fresh water evaporation rate of 40.70 m^3/day·ha (1.5 m/yr) (see Fig. 16.46), the active evaporation pond area needed is:

$$A_{\text{ep active}} = \frac{(10{,}000\,\text{m}^3/\text{day})\,(1.2)}{(0.7)\,(40.70\,\text{m}^3/\text{day}\cdot\text{ha})} = 421\ \text{ha}(1040\ \text{ac})$$

Taking into consideration that evaporation pond systems also have dikes, the total pond area, including the dikes, could be determined using the following formula (Mickley, 2006):

$$A_{\text{ep total}} = A_{\text{ep active}}\left\{1 + \left[\frac{0.325(H_{dike})}{\sqrt{A_{ep\ active}}}\right]\right\} \tag{16.9}$$

Where H_{dike} is the height of the dikes in meters and $A_{ep\ total}$ and $A_{ep\ active}$ are in hectares. For the example above and dike height of 2.5 m (8.2 ft):

$$A_{ep\ \ total} = (421\ \text{ha})\left\{1 + \frac{0.325(2.5\ \text{m})}{\sqrt{421}}\right\} = 438\ \text{ha}\ (1082\ \text{ac})$$

This area is over 1.8 times higher than the area needed for spray irrigation (240 ha) and over 20 times higher than the land needed for RIBs (20 ha) for disposal of the same volume of concentrate disposal. While the actual differences between these concentrate disposal methods would vary from one project location to another, this example underlines the fact that evaporation ponds are the most land-intensive concentrate disposal alternatives and are practically feasible for small desalination plants in arid equatorial regions of the world.

16.7.5 Evaporation Pond Costs

The evaporation rate (local climate), concentrate volume, land and earthwork costs, liner costs, and the salinity of the concentrate mainly drive the costs of evaporation pond systems. The main cost variable is the evaporative area.

Figure 16.47 depicts the construction cost of evaporation pond system as a function of the evaporation rate and the concentrate flow. This figure is generated assuming the use of a geo-composite liner, which typically contributes 25 to 30 percent of the total construction cost. The costs for a state-of-the-art geo-composite liner are \$42,000/ac. This cost compares favorably to the cost of using a 1-m thick clay liner (\$48,000/ac). If a 0.6-m (2 ft) clay liner is used instead of 1–m (3.3 ft) liner, the unit liner cost is reduced to \$32,000/ac.

Cost Example

For the example of an evaporation pond system for disposal of 10,000 m^3/day (10.6 mgd) of brackish water concentrate and evaporation rate of 1.5 m/yr (i.e., southern Arizona location), using Fig. 16.47, the cost of the evaporation pond system is estimated at \$30.0 mm. This cost does not include the expenditures for land acquisition, delivery of the concentrate to the pond site, and the cost for the installation of a leak-detection system.

The cost of a leak-detection system, which is usually required to comply with groundwater protection-related regulatory requirements, is estimated at \$8500/ac (Nicot et al., 2005).

Using the unit land cost assumption of \$5000/ac and leak-detection system cost of \$8500/ac, for the example referenced above, the land acquisition cost is \$5.4 mm (1082 ac × \$5000/ac = \$5.4 mm), and the cost for installation of a leak-detection system is \$9.2 mm (1082 ac × \$8500/ac = \$9.2 mm).

In summary, the total construction cost for the implementation of this project (excluding concentrate delivery to the site) = \$30.0 mm (evaporation pond system) + \$5.4 mm (land acquisition) + \$9.2 mm (monitoring) = \$44.6 mm. This cost is over five times higher than that for the construction of a spray irrigation system (\$8.25 mm) or RIB system (\$6.25 mm) for disposal of the same quantity of concentrate.

The significant cost differences are due to the fact that, typically, evaporation rates are significantly lower than soil uptake rates, and therefore disposal of the same volume of concentrate using evaporation ponds requires more land. In addition, cost of pond lining, which is required for construction of evaporation ponds but not needed for land application, also contributes to the significant cost differences of the two types of concentrate disposal systems.

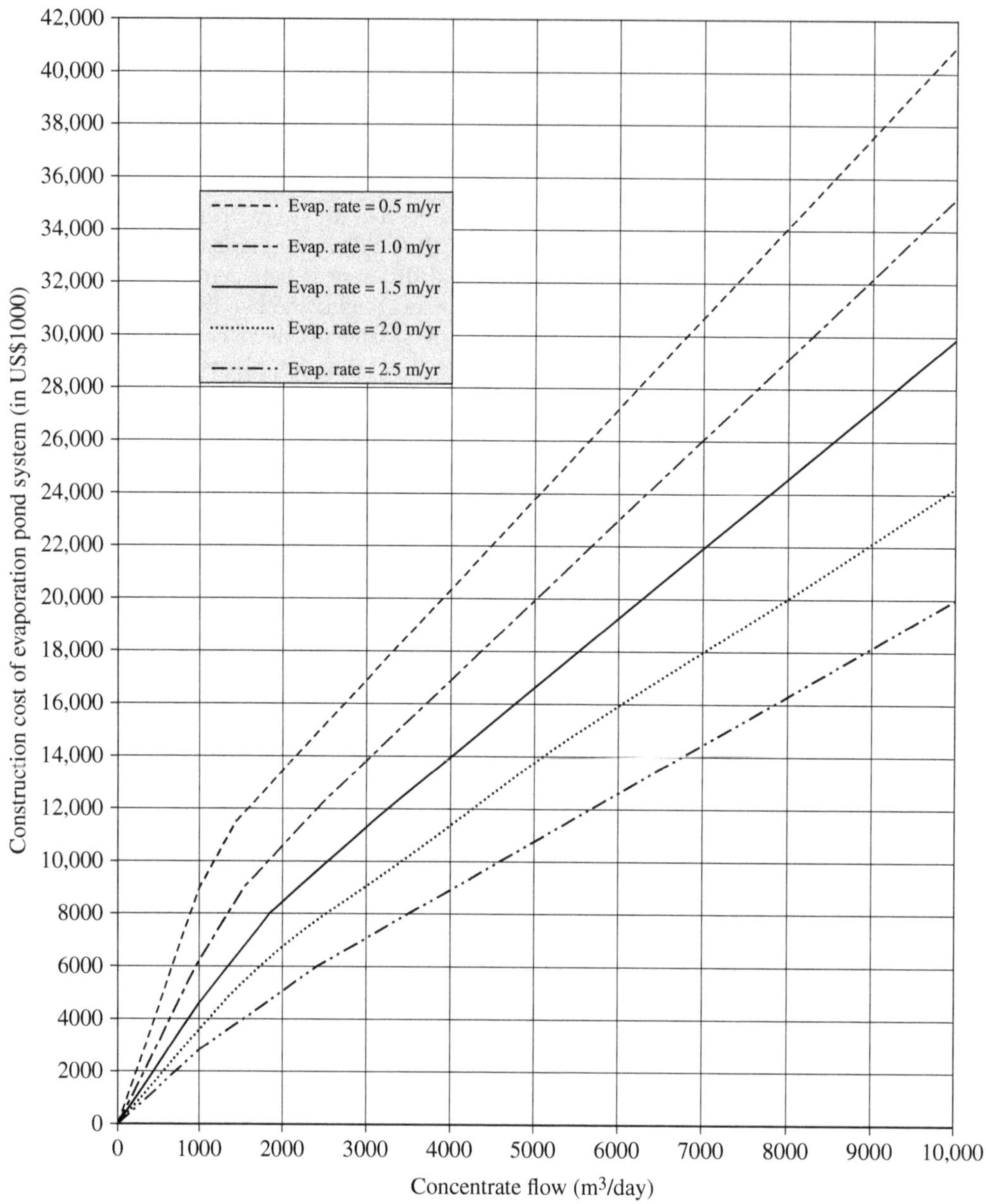

FIGURE 16.47 Construction cost for evaporation pond system.

16.8 Zero Liquid Discharge Concentrate Disposal Systems

16.8.1 Description

Zero liquid discharge (ZLD) technologies, such as brine concentrators and crystallizers, convert concentrate by thermal evaporation into highly purified water and solid dry product suitable for landfill disposal or for recovery of useful salts.

These systems typically consist of concentrate conveyance pipelines to and from the equipment, concentrator and or crystallizer towers, heat exchangers, de-aerators, seed slurry storage and delivery system, and vapor compressors and recirculation pumps. If a crystallizer system is included, this also has concentrate slurry dewatering equipment.

While evaporator/crystallizer systems are the most commonly used zero liquid discharge technologies, other such technologies, high recovery processes, and their combination into cost-competitive concentrate management systems are discussed in detail elsewhere (Mickley, 2008).

Brine Concentrators (Evaporators)

Brine concentrators are single-effect thermal evaporator systems that convert concentrate from liquid phase into dense slurry by boiling it in a tall packed tower. In these systems, a compressor pressurizes the vapor produced from boiling of concentrate, which is then recirculated for more vapor production (Fig. 16.48).

Before entering the evaporator, the feed concentrate passes through a heat exchanger with the distillate and through a de-aerator that removes noncondensable gases such as carbon dioxide and oxygen from concentrate. A scale inhibitor is typically added to the feed concentrate to prevent scaling of the evaporator and heat exchanger chambers. The conditioned concentrate is pumped to the top of the evaporator and distributed over the surface of the evaporation chamber heat transfer tubes (heating element) in the form of thin film, which travels by gravity to the bottom of the evaporator chamber. A portion of this traveling thin film of concentrate is evaporated, and the rest is collected at the bottom of the evaporator to be recirculated for another evaporation cycle.

The vapor generated in the evaporator is evacuated from the evaporator chamber and is pressurized by a vapor compressor. This vapor condenses on the outside surface of the heat transfer tubes, and the condensate (distillate) formed on the tube surface is collected and evacuated from the evaporator. Usually distillate TDS concentration is lower than 10 mg/L.

With every cycle, a small portion of the concentrated saline stream is removed from the reactor in the form of slurry. The high-salinity slurry generated in the brine evaporator could be either solidified in an evaporation pond or crystalized by mechanical drying equipment (crystallizer or drier) and disposed of to a landfill in a solid form as dry residual.

Usually, existing concentrator technology can evaporate 90 to 98 percent of the concentrate (i.e., can reduce concentrate volume 10 to 50 times). As a result, TDS content of the high-salinity concentrate produced by these systems can reach 20,000 to 100,000 mg/L. The maximum salinity that can be achieved by evaporators is limited by formation of precipitates of various mineral salts such as glauberite [$Na_2Ca(SO_4)_2$], sodium chloride (NaCl), and sodium sulphate (Na_2SO_4).

The concentrated stream can be further dewatered and disposed to a landfill as a solid waste. Ultimately, the concentrated salt product could be designated for commercial

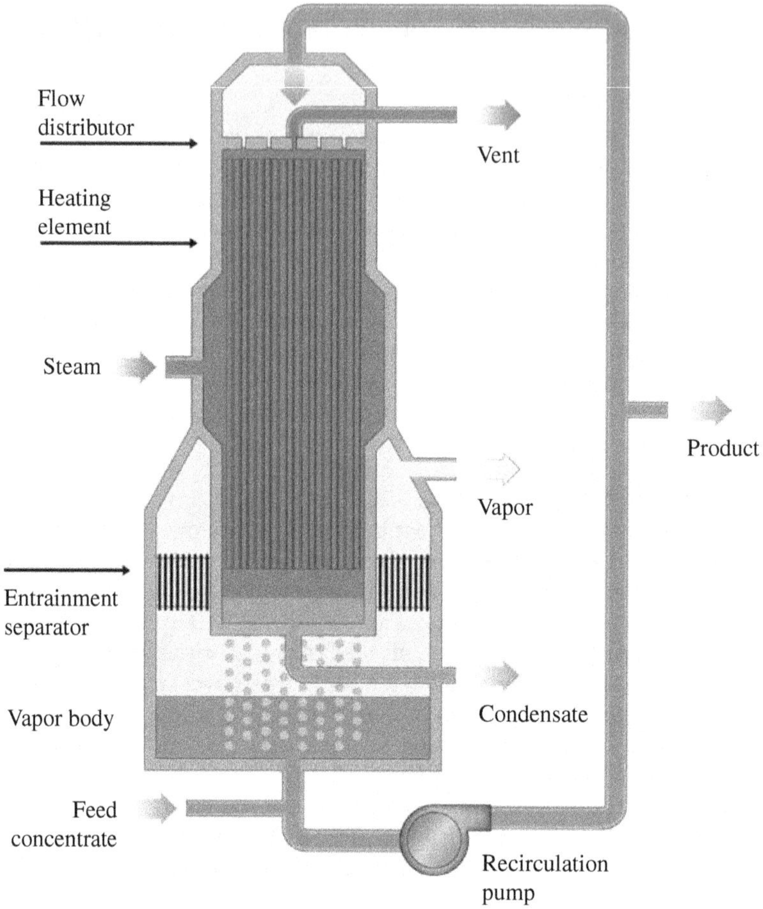

FIGURE 16.48 Schematic of brine concentrator.

applications. Vapor compression driven concentrators are energy efficient; they use approximately 10 times less energy than single-effect steam-driven evaporators. Typically energy for the concentrators is supplied by a mechanical vacuum compression system.

It is important to note that both capital and O&M costs associated with the use of brine concentrators are closely dependent on the volume of the concentrate processed by these systems. Since usually the expenditures for pre-concentration of concentrate before thermal evaporation are lower than the costs for providing additional evaporator capacity, concentrate volume minimization processes are commonly used in such systems.

Crystallizers

Crystallizers precipitate highly soluble salts from concentrate such as sodium carbonate, sodium sulphate, and sodium chloride into solid residuals. This technology applies vacuum compression and produces salt crystals and distilled water by forced circulation of slurry or dense concentrate in tall cylindrical reactors (crystallization vessels).

The crystallization vessels are vertical units operated using steam supplied by a boiler or heat provided by vacuum compressors for evaporation (Fig. 16.49).

Concentrate or concentrate slurry from an evaporator or a pretreatment evaporation pond system is fed to the crystallizer vessel, passed through shell-and-tube heat exchanger, and heated by vapor introduced by the vacuum compressor. The heated concentrate then enters the crystallizer where it is rotated in a vortex. Concentrate crystals are formed in the vessel, and the crystalline mineral mass is fed to a centrifuge or a filter press to be dewatered to a solid state. The mineral cake removed from the concentrate contains 85 percent solids and is the only waste stream produced by the crystallizer.

The low-salinity water separated from the concentrate is collected as distillate at the condenser. The filtrate from the filter press or concentrate from the dewatering centrifuge is typically blended with the RO feed or permeate. The recovery of salts and reuse of the liquid separated from the concentrate is practically 100 percent.

Usually, for small volumes of concentrate (10 to 50 m^3/day—2600 to 13,200 gal/day) steam-driven evaporators are used. Steam can either be produced using a standard boiler system, or it could be supplied by a nearby industry that generates steam for its main production processes. Typically, steam for larger crystallizer systems is produced by electrically driven vacuum compressors or supplied by industrial installations in the vicinity of the desalination plant.

Similar to brine concentrators, crystallizers are made of corrosion-resistant materials and therefore are very costly. They also consume more energy per unit of processed concentrate than concentrators because removing water from the concentrate by evaporation becomes more difficult with the increase in concentrate salinity.

Commercially available crystallizers typically have a unit processing capacity in a range of 50 to 500 m^3/day (13,200 and 132,000 gal/day) and are available from the same manufacturers that offer brine concentrator systems and driers.

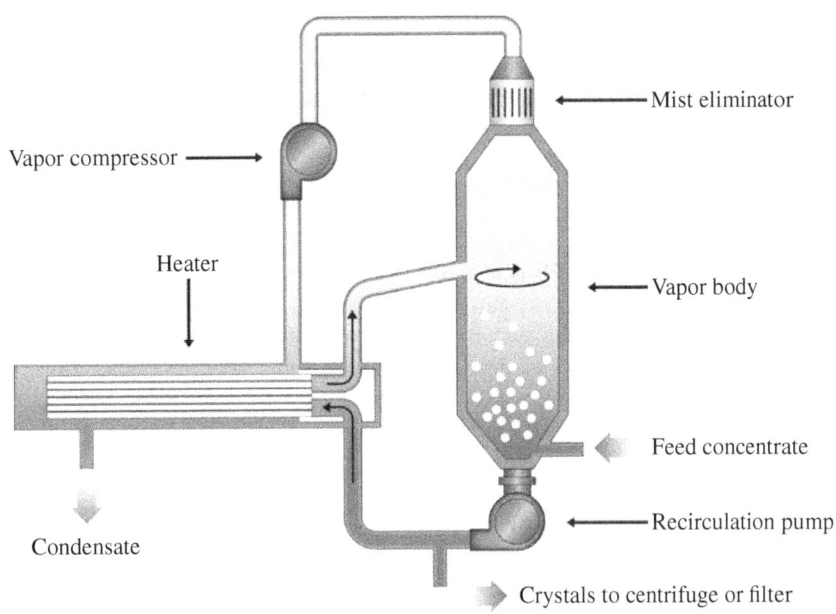

FIGURE 16.49 Schematic of concentrate crystallizer.

Evaporator-Crystallizer System

Often brine concentrator and crystallizer systems are combined into one evaporator-crystallizer system. In this system, the brine slurry produced by the evaporator is fed into the crystallizer (Fig. 16.50). Figure 16.51 depicts an evaporator-crystallizer system designed to process 1000 m^3/day (0.26 mgd) of concentrate.

16.8.2 Potential Environmental Impacts

The evaporator-crystallizer system for zero liquid discharge management of concentrate is the highest energy use and carbon footprint type of all concentrate management alternatives and often exceeds the total power demand for production of desalinated water by the plant generating concentrate.

The brine concentrator increases the content of all constituents in the source saline water by 10 to 100 percent, which, depending on the quality of the source water, may make the crystalline product from concentrate treatment a hazardous waste.

16.8.3 Criteria and Methods for Feasibility Assessment

Usually, ZLD systems are used when other options for concentrate management are not feasible mainly because of their high construction and O&M costs. Since concentrate is corrosive, all equipment used in this type of system is built from corrosion-resistant materials such as titanium, molybdenum, and super-duplex stainless steel. This makes zero liquid discharge systems quite costly.

The generation of steam for the concentrate evaporation process could also add significant expense to the ZLD system operation. Therefore most exiting evaporator-crystallizer systems are operated using waste steam from nearby power plant or industrial facility that generates steam as a site product (i.e., oil refineries).

While zero liquid discharge has received significant attention over the past 10 years, its cost challenges have not been successfully solved to date. Often, the total cost for construction of the zero discharge concentrate processing system is comparable to or higher than the cost of the actual desalination facility. Therefore, zero discharge technologies are not likely to find large-scale application in the near future, except for specific conditions where there are no other feasible alternatives.

16.8.4 Design and Configuration Guidelines

Evaporator and crystallizer systems for concentrate management apply proprietary technologies, and therefore the system manufacturer should be contacted to determine the design parameters and configuration of such systems. At present the three leading suppliers of concentrator and crystallizer systems are GE-Ionics-RCC, HPD, and Aquatech. Typically, the size of brine concentrators and crystallizers is a function of the concentrate feed rate.

Most commercially available systems have unit concentrate processing capacity in a range of 500 to 4000 m^3/day (0.1 to1.1 mgd) . In 2008, there were over 80 ZLD plants in the United States.

Evaporator-crystallizer systems are usually much more space efficient than most of the other concentrate management methods. Typically the land needed for the construction of such systems is 10 to 20 percent of the total footprint of the desalination plant.

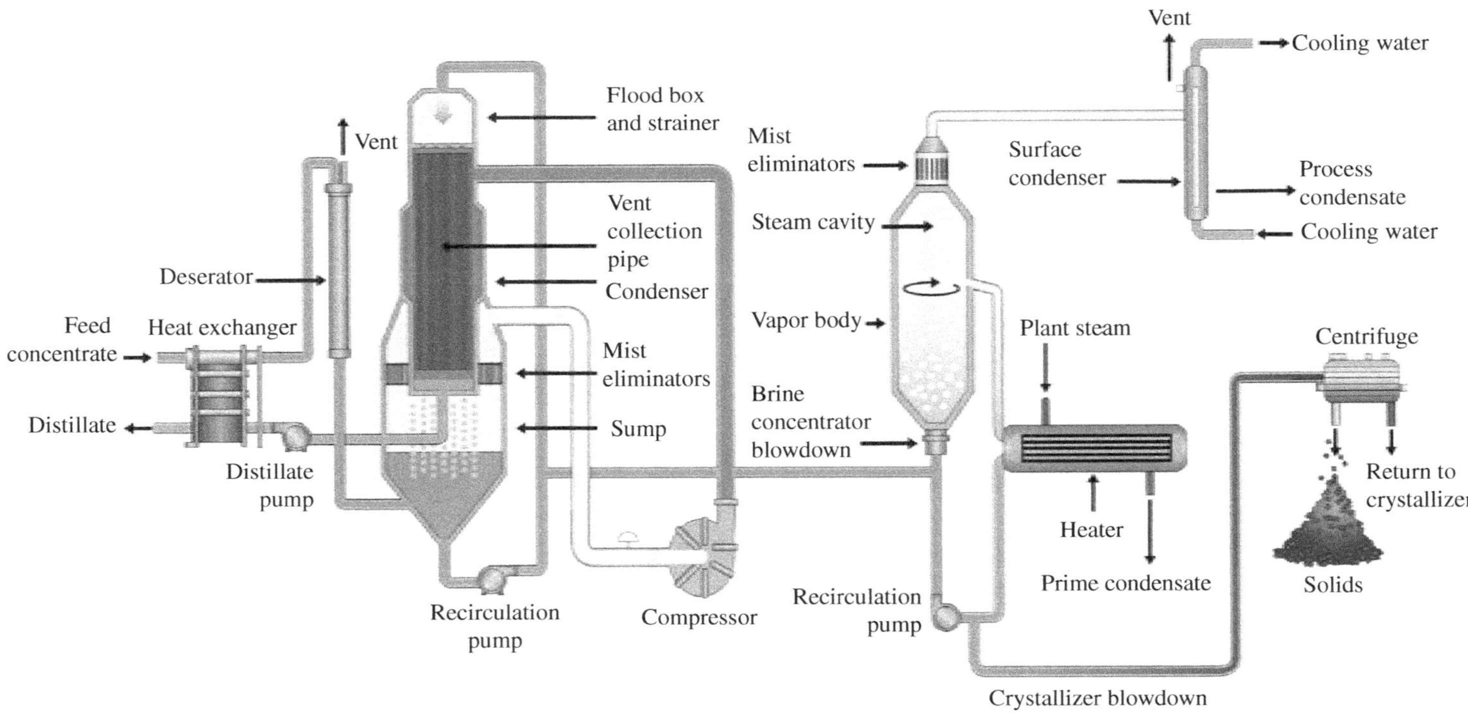

FIGURE 16.50 General schematic of evaporator crystallizer system.

FIGURE 16.51 1000 m^3/day (0.26 mgd) evaporator-crystallizer.

16.8.5 Zero Liquid Discharge Costs

Typically, zero liquid discharge is the least cost-effective concentrate management method because it requires the use of costly mechanical equipment and a large amount of energy for evaporation, crystallization, and dewatering of the salts in the concentrate.

The construction cost of zero liquid discharge by evaporator-crystallizer system is mainly a function of the flow rate of the processed concentrate (Fig. 16-52). This figure should be used for an order of magnitude cost estimate only because project-specific factors, such as concentrate salinity and availability of waste steam, could have a measurable impact on these costs.

The energy use of evaporator and crystallizer systems is usually an order of magnitude higher than that of any of the other concentrate management alternatives and often exceeds the energy used for fresh water production by RO separation. The energy demand of brine concentrators is typically in a range of 15 to 25 kWh/m^3 (57 to 95 kWh/1000 gal) of processed concentrate. Crystallizers use 50 to 70 kWh/m^3 (190 to 260 kWh/1000 gal) of feed concentrate.

Cost Example

The construction cost of an evaporator-crystallizer system for disposal of 10,000 m^3/day (2.6 mgd) of brackish water concentrate can be determined using Fig. 16.52 ($58.08 mm). This cost does not include the expenditures for land acquisition and for delivery of the brine to the concentrator.

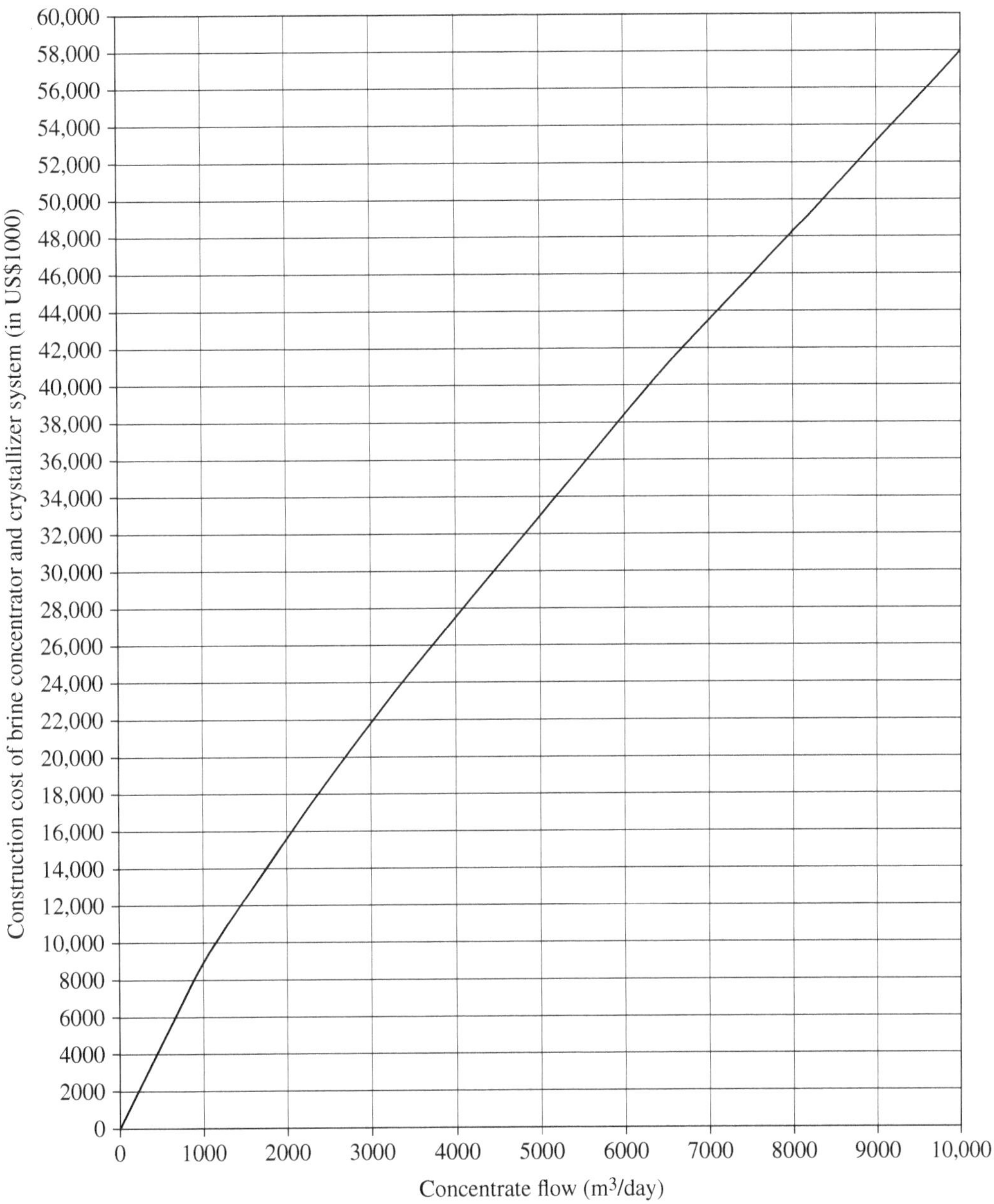

FIGURE 16.52 Construction cost of evaporator-crystallizer system.

Land requirement for construction of the example ZLD system of 3.6 acres and assuming unit land cost of \$5000/ac, for the example referenced above, the land acquisition cost is \$0.02 mm (3.6 ac × \$5000/ac = \$0.02 mm).

In summary, the total construction cost for the implementation of this project (excluding concentrate delivery to the site) = \$58.08 mm (brine concentrator/crystallizer system) + \$0.02 mm (land acquisition) = \$58.1 mm. The total expenditure for construction of brine concentrator/crystallizer system is higher than the construction cost for implementation of any of the other alternatives presented previously. This cost would be comparable to the cost for construction of 40,000 m^3/day (10.6 mgd) BWRO plant, which would generate this concentrate (\$50 to \$60 mm).

Because of the high capital and O&M costs, zero discharge technologies are not practical unless no other concentrate management alternatives are available. Usually, zero discharge concentrate management systems are justifiable for inland brackish water desalination plants where site-specific constraints limit the use of natural evaporation, deep well injection, or evaporation ponds. The most commonly applied approach to reduce the overall costs of the ZLD system overall is to minimize the volume of concentrate delivered to the ZLD system by concentrate pretreatment in a high recovery desalination system.

Case Study: Tracy, California

The largest evaporator system for treatment of concentrate from an RO brackish water desalination plant providing drinking water supply in the United States was installed in 2007 for the Deuel Vocational Institute (DVI) in Tracy, California. This system uses evaporation technology to treat 1400 m^3/day (0.4 mgd) of brine from the groundwater BWRO system supplying water to DVI. The evaporator reduces concentrate volume by 97 percent and recycles high-quality drinking water back to the facility. The remaining 3 percent of concentrate slurry is disposed to a small, on-site evaporation pond for drying and ultimate disposal to a landfill to achieve zero liquid discharge. A mechanical vapor recompression system drives the falling film evaporator to concentrate the brine. The unit power use of this system is 21 kWh/m^3 (79 kWh/1,000 gal) of concentrate.

16.9 Beneficial Use of Concentrate

16.9.1 Technology Overview

Concentrate from desalination plants contains large quantities of minerals that may have commercial value when extracted. The most valuable minerals are magnesium, calcium and sodium chlorides, and bromine. Magnesium compounds in seawater have agricultural, nutritional, chemical, construction, and industrial applications. Calcium sulphate (gypsum) could be used as a construction material for wallboard, plaster, building cement, and road building and repair. Sodium chloride can be applied for production of chlorine and caustic soda, highway de-icing, and food products.

Technologies for beneficial recovery of minerals from concentrate can be used for management of concentrate from inland brackish water desalination plants and coastal seawater desalination plants. These technologies have the potential to decrease the volume and cost of transporting concentrate as well. Specific emerging technologies for beneficial reuse of desalination plant concentrate are discussed below.

Salt Solidification and Recovery

Unlike brine concentration, salt solidification and recovery focuses on the selective recovery of beneficial salts of high purity from concentrate. A general schematic of salt solidification and recovery is presented in Fig. 16.53. The existing salt recovery technologies extract salts by fractional crystallization or precipitation. Crystallization of a given salt can be achieved by concentrate evaporation or temperature control.

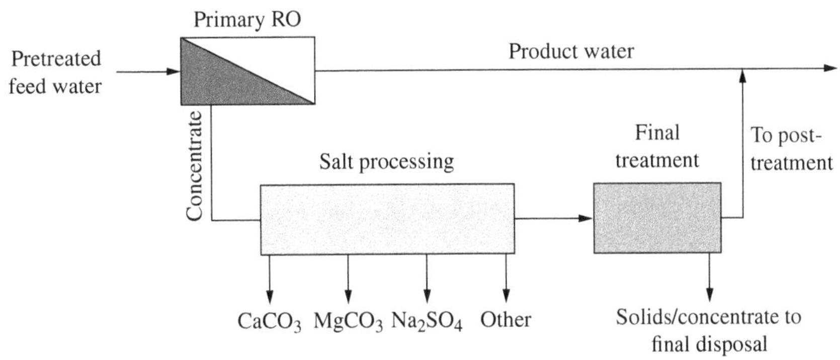

FIGURE 16.53 Schematic of salt solidification and recovery system.

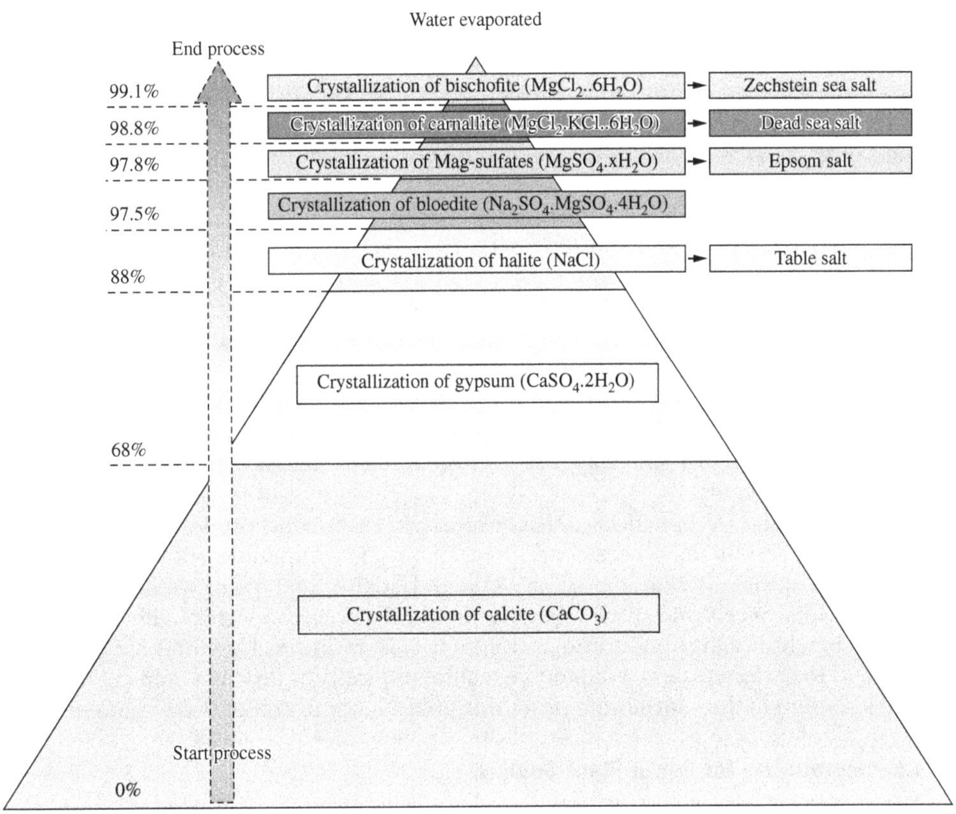

FIGURE 16.54 Sequence of precipitation of minerals from concentrate.

Figure 16.54 depicts the evaporation precipitation order of key minerals contained in seawater concentrate. As seen from this table, the two salts that are easiest to extract by evaporation are calcium carbonate (calcite) and calcium sulphate (gypsum). Calcite can be used to produce lime and plastics, while gypsum has found application in the production of various construction materials such as drywall.

Fractional precipitation is attained by adding chemical agents to selectively remove a target mineral from the concentrate solution. For example, there are a number of commercially available technologies for the extraction of magnesium and calcium salts from concentrate and for production of structural materials from these salts (Mickley, 2009).

A chemical precipitation technology that has been demonstrated in the United States and Australia to successfully recover salts such as magnesium carbonate, calcium carbonate, and gypsum from concentrate is the SALPROC process developed and patented by GEO-Processors (Svensson, 2005; Carollo, 2009). This process involves multiple evaporation and cooling steps, supplemented by mineral and chemical processing.

Disposal to Saltwater Wetlands

When trace element contamination is not a concern, concentrate and saline drainage water could be used to support wetland habitats. In general, wetland application is feasible when the TDS of the concentrate is lower than 2500 mg/L. Sometimes, saline water with TDS concentration of up to 5000 mg/L could be applied over short periods of time.

Brackish RO plant concentrate of 4500 mg/L salinity has been successfully used for small coastal marsh restoration project in Oxnard, California (Bays et al., 2005). The halophytes, which have shown the highest adaptability to concentrate irrigation, were the salt grass yerba mansa (*Anemopsis californica*) and bulrush (*Scirpus americanus and Scirpus californicus*).

Standard concentrate application practice to saltwater wetlands is autumn flooding to a depth of 0.2 to 0.5 m (0.7 to 1.6 ft), which depth could be maintained until January/February. In late winter the wetland ponds are drained to release accumulated salts. Then the wetland ponds are refilled with new water as deeply as possible. After 14 days the water is drained again. This cycle is repeated two to three times in the winter. The ponds are drawn down to mudflat state in March or April to facilitate germination of desirable salt-tolerant marsh plants such as alkali bulrush, salt grass, and tulles. Since periodic wetland drainage is critical to sustain its habitat, there must be an environmentally safe way (ocean, large river, or salt lake) of disposal of the wetland drainage water.

This method for beneficial reuse of brackish water concentrate is site-specific and suitable for conditions where concentrate quality is compatible with the native flora and fauna of the saltwater marsh or wetland. Usually, the type of wetlands that could be used for concentrate discharge is hydraulically interconnected with the ocean or with a brackish water body, and therefore this is an indirect method for concentrate disposal to surface waters. Wetland vegetation typically assimilates some of the nitrate and selenium in the concentrate providing effective reduction of these contaminants.

Concentrate Use for Power Plant Cooling

Use of concentrate for cooling of the condensers of power generation plants is typically practiced for small facilities (100 MW or less) with limited cooling needs and cooling towers that are made of materials suitable to withstand the highly corrosive concentrate. A key concern is the high scaling potential of the concentrate. In addition, only a small portion (2 to 10 percent) of the concentrate is actually converted into vapor and disposed to the air. The remaining concentrate would ultimately need to be discharged, or the system will need to be sized with a power plant cooling system that allows disposal of the target volume of concentrate. This option also involves large storage and pumping energy costs to circulate the concentrate through the cooling tower system.

Other Beneficial Uses

Small volumes of concentrate have been used occasionally for dust suppression, road-bed stabilization, soil remediation, and de-icing (Mickley, 2010). These applications are, however, of limited duration and cannot be relied upon as the main concentrate disposal alternative.

In addition, a number of states (Texas, Utah, Arizona, Kanas, New Mexico, and Utah) have inactive salt mines, which, unless refilled, could collapse and potentially cause structural damage of buildings and other facilities in the vicinity of their location. Such salt mines could be filled up with solidified concentrate to provide structural integrity of the mine caverns. This and the other applications listed above are site-specific and can only be used as supplemental concentrate disposal alternatives.

16.9.2 Feasibility of Beneficial Reuse

The key challenges of current technologies for beneficial reuse of concentrate are the large capital costs, energy, and chemical expenditures needed to reduce concentrate volume and to extract valuable minerals from it. Concentrate could be used for production of construction materials and other products of commercial significance and practical use (i.e., gypsum, metals, sodium chloride, etc.).

However, at this time, the industrial production of these materials by traditional technologies is significantly less costly. Therefore, while environmentally attractive, the large-scale beneficial reuse of minerals produced from desalination plant concentrate is highly unlikely to gain significant grounds in the near future. As the costs of construction materials and other products that can be generated from concentrate increase in the long-term (next 5 to 10 years), and more cost-competitive technologies for their production are developed, the beneficial reuse of concentrate may become viable.

16.10 Regional Concentrate Management

16.10.1 Types of Regional Concentrate Management Systems

Regional concentrate management includes two alternative approaches that can be used separately or co-implemented at the same facility: (1) regional collection and centralized disposal of concentrate at one location applying one or more of the methods described in the previous sections, and (2) use of concentrate from brackish water desalination plants as source water to a receiving regional seawater desalination plant.

The first approach takes advantage of site-specific beneficial conditions for disposal that may not be available at other locations and of the economies of scale of constructing larger concentrate disposal facilities. An example of such system is the Santa Ana Regional Interceptor in Southern California, which conveys concentrate from inland brackish desalters and disposes it through an existing WWTP ocean outfall (Corollo, 2008). It is important to note that regional disposal of concentrate would be more likely to be viable if the available concentrate disposal alternatives yield economy of scale cost benefits.

The two concentrate disposal options that can result in the highest economies of scale are surface water discharge (i.e., discharge through an existing outfall) and deep injection wells. The alternatives with the lowest economies of scale are land application and pond evaporation.

The second approach to regional brine management is based on the fact that concentrate from brackish water plants is of significantly lower salinity than seawater, and, when blended with the ocean water fed to a seawater desalination plant, will reduce the overall plant salinity.

16.10.2 Use of Brackish Water Concentrate in SWRO Plants

Disposal of brine from brackish seawater desalination plants is usually one of the key limiting factors associated with the wider implementation of inland brackish water desalination. Currently, in many locations brackish water concentrate from inland desalters is disposed most often by either deep well injection into high-salinity aquifers or it is conveyed using a regional interceptor pipeline to a wastewater treatment plant and discharged to the ocean using the treatment plant's ocean outfall.

The first disposal method (i.e., disposal to deep saline aquifer) is often limited by the capacity of this aquifer and is dependent on the availability of such aquifer in the vicinity of the ocean desalter. The second approach (i.e., disposal through the outfall of existing WWTP) is also relatively costly, and, more importantly, it occupies outfall capacity, and thereby it indirectly limits the treatment capacity of the host WWTP. Both alternatives treat brine from inland desalters as waste and involve significant expenditures for the disposal of this brine.

An innovative alternative approach for integrated regional concentrate management is to convey brine generated from one or more inland desalters to a coastal seawater desalination plant, blend this brine with seawater collected from the ocean, and then desalinate this blend in the seawater desalination plant (Fig. 16.55).

The key components of such regional concentrate management system include: (1) inland brackish water desalination plants, (2) regional brine interceptor/collectors, and (3) centralized coastal seawater desalination plants. The purpose of a regional brine collector is to convey the concentrate from the inland desalters to the regional seawater desalination plant, where this concentrate is used as supplemental feed water to the source seawater used for desalination.

Although Fig. 16.55 presents a combination of seawater desalination plant co-located with a coastal power generation plant, this approach could be used for SWRO plants with conventional intakes and outfalls as well. Use of concentrate from brackish water desalination plants as feed water to a seawater desalination plant is mutually beneficial for both plants.

Usually, inland brackish water desalination plant capacity is limited by lack of suitable discharge locations for the plant concentrate. If the seawater desalination plant can accept the brackish water desalination plant concentrate and process it, the brackish water desalination plant capacity could be increased beyond the threshold driven by brine discharge limitations, and the desalination plant source salinity could be reduced at the same time. This regional concentrate management approach has a number of benefits.

Brine from inland desalters using brackish ground water sources typically does not contain pathogens (bacteria, Giardia, cryptosporidium, etc.), and therefore it could be a safe and suitable source of water for seawater desalination. As a result, rather than being disposed as a waste product to the ocean or to deep aquifers, brackish water concentrate could be reused for drinking water production.

Brine from inland desalters usually has an order of magnitude lower total dissolved solids concentration than seawater (i.e., 2000 to 5000 mg/L versus 33,500 to 35,000 mg/L). As a result, mixing of brine and seawater will reduce the overall salinity of the source water fed to the seawater desalination plant, and therefore it will decrease the total amount of energy needed to desalinate seawater.

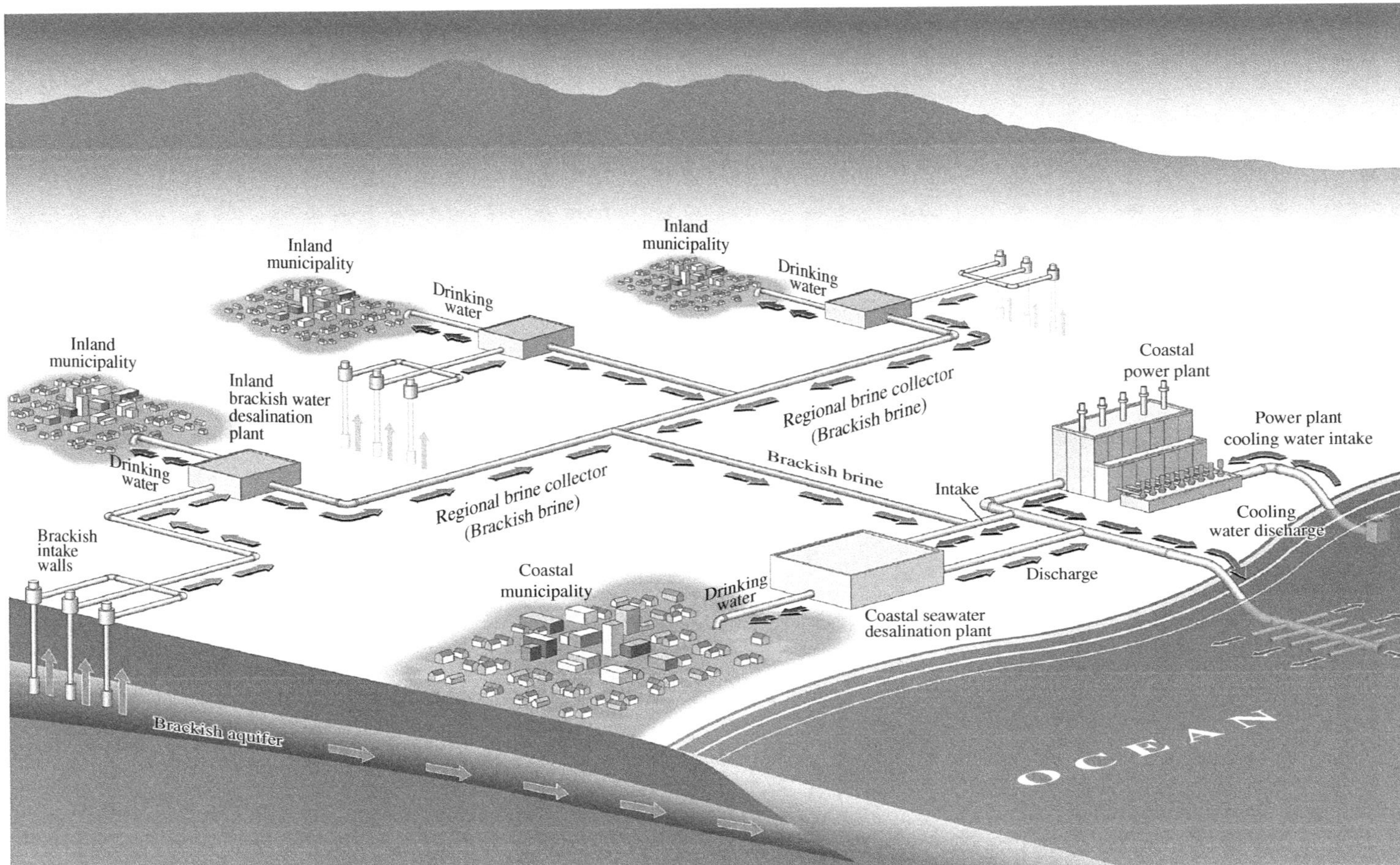

FIGURE 16.55 Integrated inland desalter brine disposal and seawater desalination.

Typically, brine from inland desalters contains antiscalants, which will allow us to reduce or to completely eliminate the expenditures for addition of such chemicals at the seawater desalination plant and will increase seawater recovery. This benefit, in turn, would allow increasing the overall recovery of the desalination plant. Increased recovery means producing more fresh water from the same amount of feed water, which, in turns, yields lower unit production costs.

By replacing some of the source seawater with brine from inland desalters, the total amount of new seawater that needs to be collected for the desalination plant operations will be reduced proportionally, which would lower the overall impingement and entrainment of marine organisms associated with collection of ocean water for seawater desalination.

Because the brackish water desalter brine will be put to beneficial use, rather than being a disposal burden, as it is today, it will become a valuable resource, which will reduce the operational costs of the brackish water desalters and, at the same time, will enhance the affordability of seawater desalination.

Diverting brine from exiting WWTP ocean outfalls will enhance the available outfall capacity and thereby could decrease wastewater treatment and disposal costs, especially if the WWTP capacity is limited by outfall discharge capacity availability.

Operating SWRO plants at higher recovery as a result of integrated brine management would result in reduction of the overall discharge volume and salinity of the SWRO plants, which, in turn, could yield potential environmental benefits in the zone of plant discharge.

One of the key constraints of the practical use of this regional concentrate management approach is the fact that brackish water desalination plant concentrate may exhibit whole effluent toxicity due to ion imbalance of the brackish concentrate, which may have an impact on the ability of the seawater desalination plant to discharge its concentrate.

The main factor that governs the brackish brine toxicity is the ratio of the concentration of one or more key ions (calcium, magnesium, fluoride, strontium, sodium, chloride, potassium, sulfates, and bicarbonates) in the brackish brine and the total dissolved solids concentration of the brine (ion/TDS ratio) (Mickley, 2000). If the ion/TDS ratio for one or more of these key ions contained in the brackish brine is above a certain threshold value, the brine exhibits toxicity. If the ion/TDS ratio is lowered below a certain level by either removing the ion from the brine solution by precipitation or absorption or increasing the brine salinity, the brackish brine becomes nontoxic.

For example, if a standard whole effluent toxicity test organisms (mycid shrimp) is exposed to brackish brine that contains calcium ion of 500 mg/L and has a TDS concentration of 10,000 mg/L (i.e., an ion/TDS ratio of (500 mg/L)/(10,000 mg/L) = 0.05), the brine causes mortality of 100 percent of the test organisms (Mickley, 2000).

However, when the brine TDS concentration is increased to 20,000 mg/L at the same calcium ion concentration (500 mg/L), the testing organisms survive (i.e., the increase in brine TDS concentration renders the same brackish brine nontoxic by decreasing the ion/TDS ratio below the threshold value for calcium ion of 0.05). Using this principle, brackish brine can be detoxified cost-effectively by mixing it with seawater or higher-salinity concentrate generated during seawater desalination with reverse osmosis membranes in a certain mixing ratio. This mixing ratio depends on a number of factors, including the TDS concentrations of the brackish brine and the seawater, and the concentration of the major ions in the brackish water brine. The maximum ratio of brackish water concentrate to seawater that does not render the regional seawater desalination plant concentrate toxic can be established by pilot testing.

16.11 Nonconcentrate Side Stream Management

As previously discussed, desalination plants generate three key types of side streams: concentrate, backwash water, and membrane flush water. The previous sections of this book focus on concentrate management. This section addresses the treatment and disposal of the other two nonconcentrate residual streams.

16.11.1 Backwash Water

Spent filter backwash water (backwash water) is a waste stream produced by the desalination plant's pretreatment filtration system. Depending on the type of pretreatment system used (granular or membrane filters), the backwash water will vary in quantity and quality. In general, the membrane pretreatment systems produce 1.5 to 2 times larger volume of spent filter backwash water than the granular media filters. However, compared with MF or UF membrane pretreatment filters, granular media filters typically require larger dosages of coagulant for pretreatment, and therefore they contain larger amount of solids. Depending on the source water quality and the pretreatment technology, membrane pretreatment may operate successfully without the addition of coagulant. Spent pretreatment filter backwash water may also include filter aid and coagulants.

Spent filter backwash water is typically handled either by direct discharge to a surface water body after blending with concentrate or by on-site treatment. Discharge to a surface water body along with plant concentrate without treatment is the most widely practiced disposal method. Typically this also is the lowest-cost disposal method because it does not involve any treatment prior to disposal except for dechlorination and pH neutralization (if needed). This disposal method is usually suitable for deep discharge into large water bodies with good flushing, such as open oceans or large rivers.

On-site treatment prior to surface water discharge or recycled upstream of the filtration system is currently becoming a widely practiced backwash management alternative. The filter backwash water must be treated at the membrane treatment plant when its direct discharge does not meet surface body water quality requirements or at deep injection wells if it is not suitable for a direct disposal.

At present, the most widely used backwash treatment process is gravity settling in conventional or lamella plate sedimentation tanks followed by solids thickening and dewatering on by belt filter presses or centrifuges (Fig. 16.56).

Spent wash water from membrane pretreatment systems is usually treated in separate MF or UF membrane modules or lamella settlers. Filter backwash sedimentation tanks are often designed for a retention time of three to four hours and allow removal of more than 90 percent of the backwash solids.

The settled filter backwash water can be either disposed with the desalination plant concentrate or recycled at the head of the pretreatment filtration system for reuse. It may be more cost-effective to recycle and reuse the settled filter backwash water rather than to dispose of it with the concentrate.

However, blending and disposal with the concentrate may be more beneficial, if the concentrate water quality is inferior, and it cannot be disposed of to a surface water

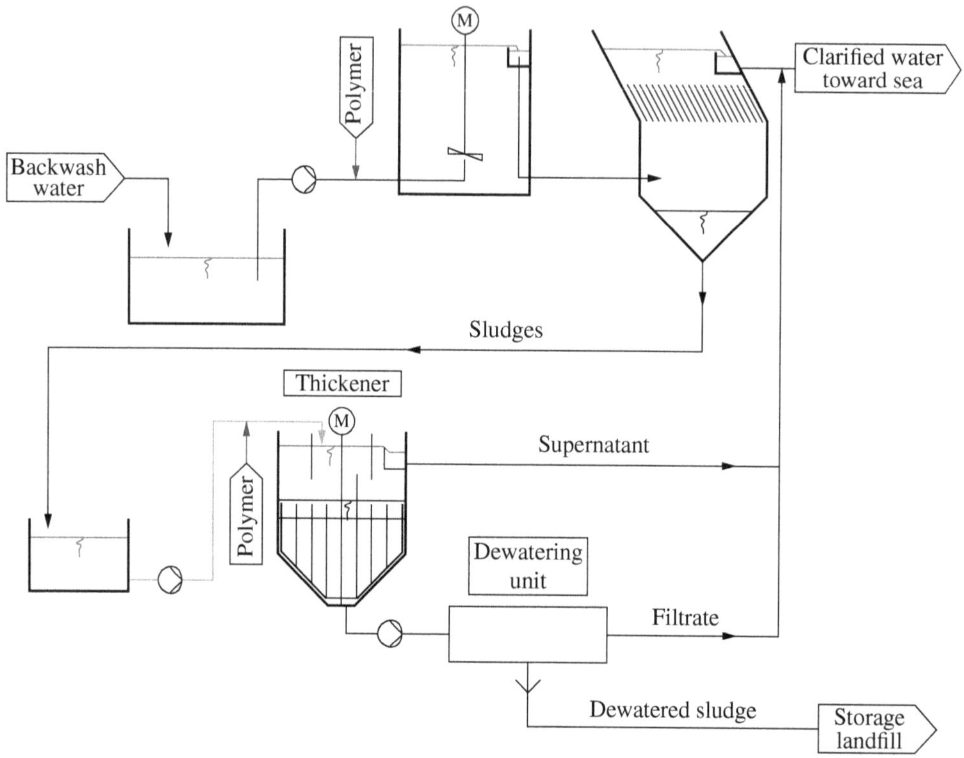

FIGURE 16.56 Schematic of typical backwash treatment system.

body without prior dilution with a stream of lesser salinity. The solid residuals (sludge) retained in the sedimentation basin are often discharged to the sanitary sewer in liquid form (typically practiced at small- to medium-size plants) or dewatered on-site in a designated solids handling facility.

16.11.2 Membrane Flush Water

The accumulation of silt or scale on the membranes causes fouling, which reduces membrane performance. Desalination system membranes must be cleaned periodically to remove foulants and extend the membrane's useful life. Typical cleaning frequency of the membranes is two to four times per year. Membrane trains are usually cleaned sequentially. A chemical cleaning solution is circulated through the membrane train for a preset time. After the cleaning solution circulation is completed, the spent cleaning solution is evacuated from the train to a storage tank, and the membranes are flushed with permeate (flush water). The flush water is used to remove all the residual cleaning solution from the RO train in order to prepare the train for normal operation. The flush water is stored separately from the rest of the plant permeate in a flush tank.

All the membrane cleaning streams listed above are typically conveyed to one wash water tank often named "scavenger tank" for waste cleaning solution retention and treatment. This tank must be able to retain the waste cleaning solution from the simultaneous

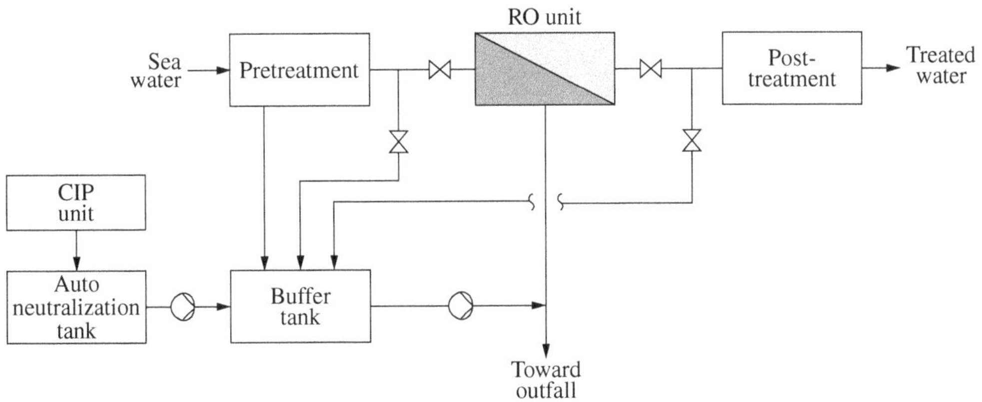

FIGURE 16.57 Schematic of waste stream management system.

cleaning of a minimum of two membrane trains. The scavenger tank should be equipped with mixing and pH neutralization systems. The mixing system should be installed at the bottom of the tanks to provide complete mixing of all four cleaning solution streams listed above. After mixing with flush water, the concentration of the cleaning solution chemicals will be reduced significantly. The used cleaning solution should be neutralized to a pH compatible with the pH requirements for discharge to the wastewater collection system. At many plants, only the most concentrated first flush is discharged to the wastewater collection system. The rest of the flush water usually has only trace levels of contaminants and is most often suitable for a surface water discharge (i.e., discharge to the ocean or other nearby water body). Often desalination plants are provided with a buffer tank that receives and blends all plant waste streams prior to discharge (Fig. 16.57).

The buffer tank is sometimes equipped with pH adjustment system to control discharge pH and an aeration system to mix tank content and to boost the oxygen of the discharge. Such configuration is most common for ocean water discharges.

16.12 Comparison of Concentrate Management Alternatives

16.12.1 Selection of Concentrate Management Approach

Key advantages and disadvantages of the most commonly used concentrate management alternatives presented in the previous sections of this book are summarized in Table 16.8.

A general decision tree for selection of desalination plant discharge management alternatives is presented in Fig. 16.58 (AWWA, 2007).

Key criteria for selection of the most viable alternative or combination of alternatives for concentrate management are costs, environmental impacts, regulatory acceptance, ease of implementation, site footprint, reliability and operational constraints, and energy use.

While concentrate water quality is of key importance in the selection process, the criterion of highest significance, which is the most widely applied for selection of the most viable concentrate management alternative, is the life-cycle project cost.

Concentrate Management Alternative	Key Advantages	Key Disadvantages and Challenges
Surface water discharge	• Can be used for all size plants • Cost effective for medium and large projects	• Concentrate may have an impact on marine habitat • Complex and costly to permit
Sanitary sewer discharge	• Low construction and operation costs • Easiest to implement • Low energy use	• Applicability limited to small-size plants • Potential negative impact on WWTP operations
Deep well injection	• Suitable for inland desalination plants • Moderate costs • Low energy use	• Only feasible if deep confined saline aquifers are available • Potential for groundwater contamination
Land application	• Relatively easy to implement and operate • Beneficial use of concentrate	• Seasonal and climate dependent • Limited to small plants potential for groundwater contamination
Evaporation ponds	• Easy to implement and operate • Inland and coastal use	• High footprint and costs • Limited to small plants
Zero liquid discharge	• No liquid waste • Minimum land needed	• High energy use and costs • Complex operation

Table 16.8 Comparison of Concentrate Management Alternatives

16.12.2 Costs

A number of site-specific factors have an impact on the costs for the concentrate disposal methods listed above, and therefore a general cost estimate analysis is difficult to complete. Table 12 presents the construction costs for an example concentrate disposal system for 40,000 m^3/day (10.6 mgd) brackish and seawater desalination plants, respectively. The BWRO plant is assumed to operate at 80 percent recovery and produce 10,000 m^3/day (2.6 mgd) of concentrate, while the SWRO plant has 45 percent recovery and generates 48,900 m^3/day (12.9 mgd) of concentrate. Review of this table indicates that the sanitary sewer and surface water discharge are the two most cost-effective methods for concentrate disposal, which explains their popularity.

Depending on the site-specific conditions, deep well injection, surface water discharge, and spray irrigation could be competitive concentrate disposal alternatives. A zero liquid discharge system typically has the highest construction and operation costs. However, under specific circumstances (such as cold climate, low evaporation and soil uptake rates, high land costs, and low power costs), the zero liquid discharge systems could be cost competitive to evaporation pond and spray irrigation disposal alternatives.

Since life-cycle costs are often the prime criterion for selection of concentrate management alternative, it is not surprising that concentrate disposal to the sanitary sewer, which usually is the lowest-cost concentrate management method is also the most widely used alternative. Historically over 75 percent of the concentrate from brackish

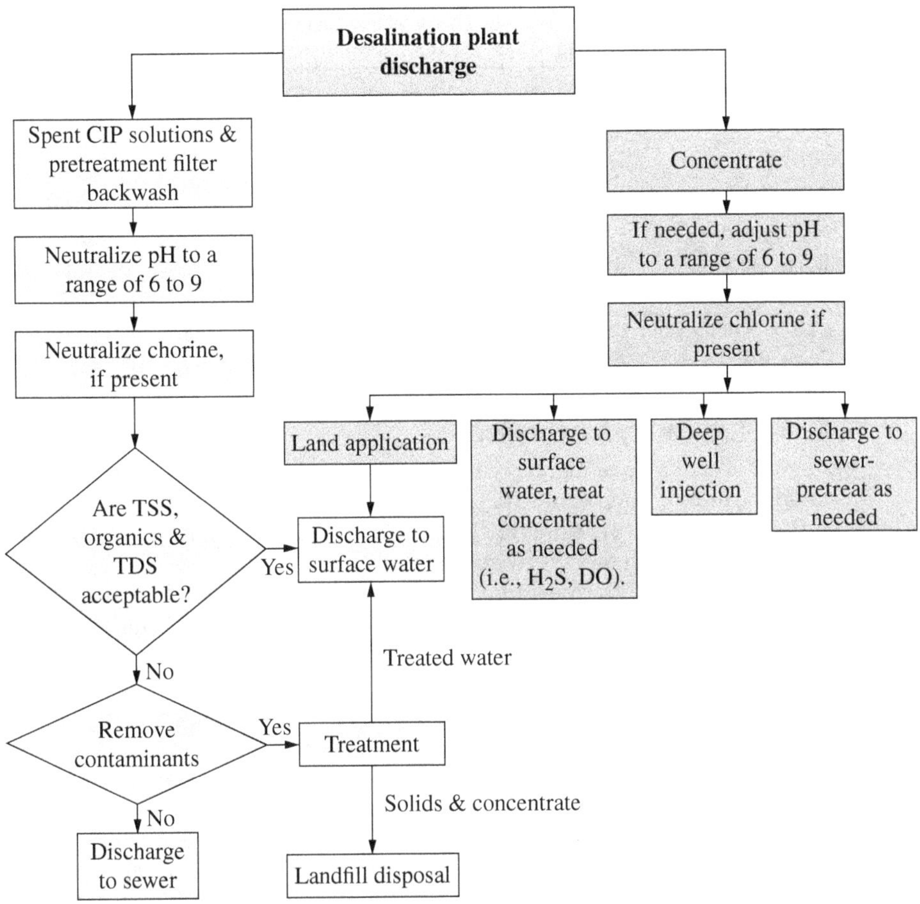

FIGURE 16.58 Decision tree for desalination plant discharge management. (*Source: AWWA, 2007.*)

desalination plants in use is disposed to the sewer (Mickley, 2008). This alternative is followed by surface water discharge, which has two key advantages over most of the other concentrate management methods: it is suitable for all sizes desalination plants, and it is not climate dependent.

16.12.3 Environmental Impacts

The environmental impacts of the various concentrate management alternatives are site-specific. Evaporation-crystallization typically has the lowest environmental impacts in terms of waste stream volume and quality. However, this alternative typically has over a 10 times higher carbon footprint than any other concentrate management scenario.

When suitable, deep high-salinity aquifer is available near the location of the desalination plant, well injection is an environmentally attractive alternative with a reasonably low carbon footprint.

Concentrate Disposal Method	Brackish Water Desalination Plant (US$ mm)	Seawater Desalination Plant (US$ mm)
Surface water discharge	2.0–10.0	6.5–30.0
Sanitary sewer discharge	0.5–2.0	1.5–6.0
Deep well injection	4.0–8.0	15.0–25.0
Evaporation ponds	30.0–50.0	140.0–180.0
Spray irrigation	8.0–10.0	30.0–40.0
Zero liquid discharge	50.0–70.0	160.0–200.0

Table 16.9 Construction Costs for Key Concentrate Disposal Methods of Hypothetical 40,000 m^3/day (10.6 mgd) Desalination Plant

16.12.4 Regulatory Acceptance

Regulatory acceptance of a given concentrate disposal alternative could be evaluated based on the number of permits (licenses) needed to construct and operate the concentrate disposal system, the time needed to obtain the regulatory permits/licenses, the complexity and length of the environmental studies required by the governing agencies in order to issue the permits/licenses, the environmental monitoring requirements associated with the operation of the concentrate disposal system, and the overall environmental and construction permitting costs.

Concentrate discharge to the sewer or to surface waters (sea, ocean, or river) are usually the most well understood disposal alternatives by environmental regulators worldwide because they are the most common. While discharge to a sanitary sewer is usually the easiest to receive approvals for, this concentrate management alternative is viable only for small desalination plants.

Construction of lined evaporation ponds with an appropriate leakage monitoring system typically has wider regulatory acceptance than land application (RIB disposal and spray irrigation) because it is more protective of local groundwater resources.

16.12.5 Ease of Implementation

This criterion plays an important role in the selection of the most viable concentrate management alternative when time is of the essence for the implementation of the desalination project that will be the source of concentrate. The length of construction of some concentrate disposal systems, such as long ocean outfalls with complex diffuser structures, is often comparable to the time needed to build the desalination plant and involves prolonged environmental studies and regulatory review. Therefore in such projects the selection of discharge management alternative is often driven by its ease of implementation and environmental review.

Similarly, the construction of RIBs for concentrate disposal and deep injection wells involve detailed and often six-month to one-year-long studies of site suitability and constraints. Discharge to a sanitary sewer is usually the easiest way to implement a concentrate management alternative.

16.12.6 Site Footprint

The total area needed to implement alternative concentrate disposal methods varies significantly and could be an important constraint and factor for selecting the most viable concentrate disposal alternative or combination of alternatives. The smallest site footprint concentrate disposal alternative is usually concentrate discharge to the sanitary sewer. Construction of evaporation ponds usually is the alternative with the largest site requirements.

16.12.7 Reliability and Operational Constraints

This selection criterion refers to the mechanical reliability of the equipment and performance reliability of the treatment technologies incorporated into a given disposal method as well as its dependency on natural changes in the surrounding environment such as temperature, wind speed, humidity, precipitation, solar irradiation intensity, strength and direction of underwater currents, natural changes in source water salinity, and exposure to storms, tornadoes, hurricanes, earthquakes, floods, and other natural disasters.

For example, deep injection wells are not suitable for discharge of concentrate in seismic zones and require the availability of deep and high-saline-confined aquifers to be feasible. Similarly, shallow beach wells for concentrate disposal are not suitable for seashore locations exposed to significant beach erosion.

Some of concentrate management alternatives (i.e., evaporation ponds, land application) may be seasonal in nature, and, in this case, a backup alternative is needed to improve their reliability. If deep well injection is used, the injection wells will need to be inspected and maintained periodically, which requires either a backup disposal alternative or installation of backup wells to sustain continuous operation.

16.12.8 Energy Use

The energy use and carbon footprint associated with the construction and operation of concentrate disposal alternatives varies significantly. Usually energy use for concentrate disposal is 5 to 20 percent of that for seawater desalination. However, achieving zero liquid discharge with an evaporator-crystallizer concentrate treatment system usually involves the use of electricity, which could be larger than the electricity needed for RO desalination.

16.13 References

Ahmed, M., W. H. Shays, D. Hoey, A. Mahendran, R. Morris, and J. Al-Handaly, "Use of Evaporation Ponds for Brine Disposal in Desalination Plants," *Desalination* 130: 155–168, 2000.

American Water Works Association, *Reverse Osmosis and Nanofiltration, Manual of Water Supply Practices*, M46, 2007.

Arakel, A. M., A. Hoey, and D. M. Coleman, "Integrated Power, Water and Salt Generation: a Discussion Paper," *Desalination*, 134: 37–45, 2001.

Bays, J., K. Ortega, and P. Frank, "Oxnard Membrane Concentrate Pilot Wetland Study: Phase 3—Preliminary Results," Proceedings of Annual AWWA Membrane Conference, 2005.

California State Water Board, Procedures Manual for Conducting Toxicity Tests Developed by the Marine Bioassay Project, 96-1WQ, 1996.

Cannesson, N., "Community, Environmental and Marine Impact Minimization at the Gold Coast Desalination Plant," International Desalination Association, Biennial Conference, IDAWC/DB09-242, Dubai, UAE, November 7–12, 2009.

Carollo Engineers, Water Desalination Management and Piloting, South Florida Water Management District, Sunrise, Florida, 2009.

Chapman, G.A., D. L. Denton, and J. M. Lazorchak, "Short-term Methods for Estimating the Chronic Toxicity of Effluents and Receiving Waters to West Coast Marine and Estuarine Organisms," 1995. USEPA Report No EPA/600/R-95/136.

Christie, S., and V. Bonnelye, "Perth, Australia: Two-year Feed Back on Operation and Environmental Impact, IDA International Desalination Association, Biennial Conference," Dubai WC/DB09-278, Dubai, UAE, November 7–12, 2009.

Cotruvo, J., N. Voutchkov, J. Fawell, P. Payment, D. Cunliffe, and S. Lattemann, *Desalination Technology–Health and Environmental Impacts*, CRC Press, Boca Raton, Florida, 2010.

Einav, R., K. Harussi, and D. Perry, "The Footprint of the Desalination Processes on the Environment," *Desalination*, 152: 141, 2002.

Einav, R., and F. Lokiec, "Environmental Aspects of a Desalination Plant in Ashkelon," *Desalination*, 156: 79–85, 2003.

Graham J., "Marine Biological Considerations Related to the Reverse Osmosis Desalination Project at the Encina Power Plant," Carlsbad, CA, Environmental Impact Report for Carlsbad Seawater Desalination Plant, City of Carlsbad, 2004.

GWA, Government of Western Australia, Department of Environment and Conservation, License for Prescribed Premises, Perth Seawater Desalination Plant, License Number 8108/1; Perth, WA, 63, 2007.

Jenkins, S. A., and J. Wasyl, "Hydrodynamic Modeling of Dispersion and Dilution of Concentrated Seawater Produced by the Ocean Desalination Project at the Encina Power Plant," Carlsbad, California, Environmental Impact Report for Carlsbad Seawater Desalination Plant, City of Carlsbad, 2001.

Jordhal, J., "Beneficial and Nontraditional Use of Concentrate," Report # WRF-02-006b, WateReuse Research Foundation, Alexandria, Virginia, 2006.

Highnett, C., R. White, and H. Rechten, "Solar ponds – Technology and Economics, 2002": www.senrac.sa.gov.au/Presentations%20Conference%202002/Highnett%20 SolarPond%2 0Paper.pdf 2004-07-06

Hammond, M., N. Blake, P. Hallock-Muller, M. Luther, D. Tomasko, and G. Vargo, "Effects of Disposal of Seawater Desalination Discharges on Near Shore Benthic Communities," Report of Southwest Florida Water Management District and University of South Florida, 1998.

Hoepner, T., "A Procedure for Environmental Impact Assessments (EIA) For Seawater Desalination Plants," *Desalination*, 124: 1, 1999.

Hoepner, T., and J. Windelberg, "Elements of Environmental Impact Studies On Coastal Desalination Plants," *Desalination*, 108: 11, 1996.

Le Page, S., "Salinity Tolerance Investigations: A Supplemental Report for the Carlsbad, CA Desalination Project," M-Rep Consulting, Environmental Impact Report for Carlsbad Seawater Desalination Plant, City of Carlsbad, 2004.

Mackey, E. D., and T. Seacord, "Regional Solutions for Concentrate Management," Report # WRF-02-006d, WateReuse Research Foundation, Alexandria, Virginia, 2008.

Mauguin, G. and P. Corsin, "Concentrate and Other Waste Disposals from SWRO Plants: Characterization and Reduction of Their Environmental Impact," *Desalination*, 182: 355-364, 2005.

McConnell, R., "Tampa Bay Seawater Desalination Facility–Environmental Impact Monitoring," Proceedings of 2009 Annual WateReuse Conference, Seattle, September 2009.

Messmer, S., G. Hart, J. Netzel, and J. Dietrich, "Membrane Concentrate Reuse by Controlled Blending," *Florida Water Resources Journal*, January 1999.
Mickley, M. C., "Major Ion Toxicity in Membrane Concentrate," AWWA Research Foundation, Denver, Colorado, USA, 2000.
Mickley, M. C., "Membrane Concentrate Disposal: Practices and Regulation, Desalination and Water Purification Research and Development Program," Report N. 123 (Second Edition), U.S. Department of Interior, Bureau of Reclamation, 2006.
Mickley, M. C., "Survey of High-Recovery and Zero Liquid Discharge Technologies for Water Utilities," WateReuse Research Foundation, Report WRF-02-006a, 2008.
Mickley, M. C. "Treatment of Concentrate," U.S. Department of the Interior, Bureau of Reclamation, Technical Service Center, Water Treatment and Research Group, Denver, Colorado, 2009.
Mickley, M. C., "Brackish Water Concentrate Management," State-of-the-Science White Paper Prepared for NMSU and CHIWAWA, 2010.
Nicot, J-P., B. Gross, S. Walden, and R. Baier, "Self-Cleaning Evaporation Ponds for Desalination Facilities in Texas," Texas Water Development Board, Austin, Texas, 2007.
Oceanica Consulting, Perth Metropolitan Desalination Plant—Cockburn Sound Benthic Macrofouna Community and Sediment Habitat, Repeat Macrobenthic Survey, 2009.
Okel, P. N., J. P. Antenucci, and J. Imberger, "Field Investigation of the Impact of the Perth Seawater Desalination Plant Discharge on Cockburn Sound During Summer," Center for Water Research, University of Western Australia, Perth, 2007.
O'Leary J. W., E. P. Glenn, and M. C. Watson, "Agricultural Production of Halophytes Irrigated with Seawater," *Plant and Soil* 89: 311–321, 1985.
Purnama, A., H. H. Al-Barwani, and M. Al-Lawatia, "Modeling Dispersion of Brine Waste Discharges From a Coastal Desalination Plant," *Desalination*, 155: 41, 2003.
Purnama, A. and H. H. Al-Barwani, " Some Criteria to Minimize the Impact of Brine Discharge into the Sea," *Desalination*, 171: 167, 2004.
Rhodes, M., "Marine Management Is High Priority," *The International Desalination and Water Reuse Quarterly*, 16: 30, 2006.
Rimmer, A. E., E. A. Kobylinski, G. L. Hunter, F. A. DiGiano, and B. Bierick, "The Impacts of Membrane Process Residuals on Wastewater Treatment," Guidance Manual, WateReuse Foundation, Alexandria, Virginia, US, 2008.
Sadhwani, J. J., J. M. Veza, and C. Santana, "Case Studies on Environmental Impact of Seawater Desalination," *Desalination*, 185: 1, 2005.
Svensson, M., *Desalination and the Environment: Options and Considerations for Brine Disposal in Inland and Coastal Locations*, SLU, Dept. of Biometry and Engineering, Sweden, 2005.
Talavera, J. L., J. J. Q. Ruiz, "Identification of the Mixing Process in Brine Discharges Carried Out in Barranco del Toro Beach, South Gran Canarias (Canary Island)," *Desalination*, 139: 277–286, 2001.
Torquemada, F. Y., "Dispersion of Brine Discharge from Seawater Reverse Osmosis Desalination Plants," *Journal of Desalination and Water Treatment*, 5: 137–145, 2009.
USEPA, "Process Design Manual for Land Treatment of Municipal Wastewater, Supplement on Rapid Infiltration and Overland Flow," EPA 625/1-81-013a, Cincinnati, Ohio, 1984.
USBR, "Zero Discharge Waste Brine Management for Desalination Plants, Desalination Research and Development Program Report No. 89," University of El Paso, Texas, US Department of the Interior, Bureau of Reclamation Technical Service Center, 2002.

USEPA, "Guidelines for Water Reuse," EPA 625/R-04/108, U.S. Agency for International Development, Washington, DC, 2004.

Vila, J., J. Compte, T. Cazurra, N. Ontanon, M. Sola, and F. Urrutia, "Environmental Impact Reduction in Barcelona's Desalination and Brine Disposal," In Proceedings of World Congress in Desalination and Reuse, International Desalination Association, IDAWC/DB09-309, Dubai, UAE, November 7–12, 2009.

Voutchkov, N., "Seawater Desalination Costs Cut through Power Plant Co-location," *Filtration and Separation Magazine*, 41(7): 24, 2004.

Walton L. H., J. C. Swift, "Desalination Coupled with Salinity-Gradient Solar Ponds," *Desalination* 136: 13–23, 2001.

Weber, C. I., W. B. Horning, D. J. Klemm, T. W. Nieheisel, P. A. Lewis, E. L. Robinson, J. Menkedick, and F. Kessler, "Short-term Methods for Estimating the Chronic Toxicity of Effluents and Receiving Waters to Marine and Estuarine Organisms," EPA/600/4-87/028, National Information Service, Springfield, Virginia, 1998.

CHAPTER 17

Desalination Project Cost Estimates

17.1 Introduction

Desalination project cost estimates typically include three key components: (1) capital costs, (2) operation and maintenance costs, and (3) cost of water production. Capital costs are reflective of all expenditures directly related to the construction of the project ("direct" or "hard" capital costs) and to the project planning, engineering, environmental review, and funding ("indirect" capital costs).

Operation and maintenance (O&M) costs are all expenditures associated with RO plant operations (power, chemicals, labor, and replacement of consumables, such as membranes and cartridge filters); with maintenance of plant equipment, buildings, grounds and utilities; and with compliance with all plant operation and environmental monitoring requirements and other pertinent administrative and management costs. The O&M costs associated with a given project are typically expressed as the all-inclusive operational expenditures for a period of one year (i.e., US$/yr) or as operational costs for the production of unit volume of desalinated water (i.e., US$/m^3 or US$/1000 gal).

The cost of water is an economic parameter that incorporates all project capital and annual O&M expenditures associated with water production and is typically presented as monetary units per unit volume of desalinated water (i.e., US$/m^3). The total cost of fresh water production (cost of water) is calculated by dividing the sum of the amortized (annualized) capital costs (i.e., US$/yr) and the annual O&M costs (i.e., US$/yr) by the total annual desalination plant fresh water production volume (m^3/yr or MG/yr).

This section provides an overview of the key cost components for seawater and brackish water desalination projects and discusses factors and considerations for their determination. Desalination costs vary in a wide range and are driven by many site-specific factors, which make them difficult to simply extrapolate from one project to another. The cost information provided in this chapter is based on actual data from full-scale desalination projects, which are adjusted for comparative analysis.

Source water salinity is one of the most important factors in determining desalination project design and costs (AWWA, 2007). As discussed in Chap. 3, desalination plants can be divided into three broad categories: low- and high-salinity brackish water desalination plants and seawater desalination plants.

Low-salinity brackish water desalination plants often have a relatively simple single-stage RO system configuration and are typically designed to treat water of

TDS concentration between 500 and 2500 mg/L. For such plants, it is common that 5 to 30 percent of the source water flow is bypassed and blended with permeate produced by the RO system. Therefore such facilities are relatively less costly to build and operate. Depending on the target water quality and method of concentrate disposal, low-salinity BWRO plants may employ more than one RO stage in order to reduce concentrate volume and costs. Most of the BWRO plants in Florida and Texas are low-salinity groundwater desalination plants. It should be pointed out that the low-salinity surface water BWRO plants usually produce desalinated water at 10 to 20 percent higher cost usually because of the more costly and complex pretreatment.

High-salinity BRWO plants are configured to process brackish source waters with TDS content in a range of 2500 to 10,000 mg/L, usually treat the entire source water flow and, at a minimum, incorporate a two-stage RO system. Usually, fresh water production costs of high-salinity desalination plants are 15 to 35 percent higher than those of low-salinity desalination projects. The main cost differences originate from the higher-energy use associated with the elevated source water salinity, the more complex RO system configuration, and the lower fresh water recovery at which such plants typically operate.

Seawater desalination projects are designed to process source water of salinity between 15,000 and 46,000 mg/L. Such plants are typically designed as multipass, multistage RO systems, which operate at significantly lower recoveries and have higher energy use than brackish water desalination plants. In addition, SWRO membrane elements and vessels are more costly because they are designed to withstand higher pressures. As a result, the costs for desalinating seawater are usually measurably higher than those for producing the same quality of fresh water by brackish water desalination.

Depending on the target product water quality and site-specific conditions such as energy costs and concentration of other source water constituents besides sodium and chloride, saline waters of TDS concentration between 10,000 and 15,000 mg/L could be processed by seawater and brackish water desalination systems.

17.2 Overview of Water Production Costs

17.2.1 Cost of Water Produced by BWRO Desalination Plants

Table 17.1 provides a summary of the costs of fresh water production of low- and high-salinity brackish water desalination plants. The cost summary presented in this table is

	Cost of Water (US$/m³)	
Classification	**Low-Salinity BWRO Plants**	**High-Salinity BWRO Plants**
Low-end bracket	0.2–0.4	0.3–0.6
Medium range	0.5–0.8	0.7–1.0
High-end bracket	1.0–1.5	1.3–1.8
Average	0.7	0.9

Note: $1.0/m^3 = $3.785/1000 gal

Table 17.1 Water Production Costs of Medium- and Large-Size BWRO Desalination Plants

based on comparative analysis of over 40 brackish water desalination plants worldwide. The actual costs of the individual projects used to generate Table 17.1 were adjusted for time scale, scope, and location to provide a common base for comparison.

Review of Table 17.1 indicates that at present (i.e., year 2012) the industry-wide average cost for production of fresh water by low- and high-salinity BWRO plants costs is $0.7/m^3 ($2.6/1000 gal) and $0.9/m^3 ($3.4/1000 gal), respectively. As anticipated, use of low-salinity brackish water yields lower fresh water production costs. However, it is interesting to note that the cost difference is not proportional to salinity. Often, low-salinity sources may contain additional contaminants such as silica, cyanide, iron, manganese, or large quantity of organics and dissolved gases that have a profound impact on plant construction costs because their removal usually requires additional treatment steps and expenditures. In addition, typically both low- and high-salinity brackish water plants use the same type of RO membrane elements, vessels, and pumps, which have the same unit costs per processed capacity (i.e., their costs are mainly determined by plant production flow and recovery and are not as significantly impacted by source water salinity as they are by production flow).

Figures 17.1 and 17.2 present breakdown of the water production costs of low- and high-salinity BWRO plants by main components: direct (construction) and indirect capital costs, power, and other O&M costs.

For low-salinity desalination plants, construction costs (i.e., direct capital costs) are typically the largest component of the water production costs. The wide range of these costs is mainly attributed to the economy of scale and differences in intake and concentrate disposal cost components.

Comparison of Figs. 17.1 and 17.2 indicates that in high-salinity BWRO plants power expenditures are a slightly larger portion of the total water production costs (typically because of the higher source water salinity). However, the energy cost component is not incrementally proportional to the salinity because high-salinity BWRO plants often apply energy-recovery devices, which typically are not cost attractive for low-salinity BWRO plants and also operate at lower recoveries, which results in elevated construction costs and lower energy use.

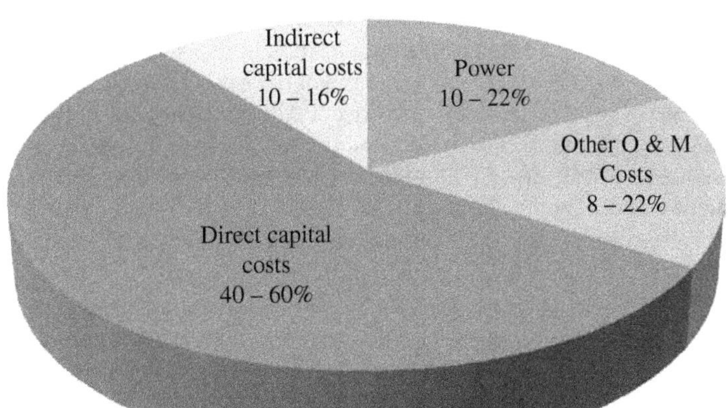

FIGURE 17.1 Typical cost of water breakdown for low-salinity BWRO plants.

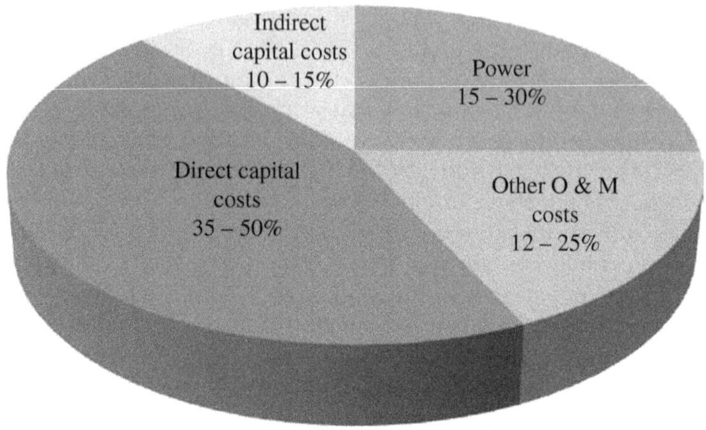

Figure 17.2 Typical cost of water breakdown for high-salinity BWRO plants.

17.2.2 Cost of Water Produced by SWRO Desalination Plants

Table 17.1 presents the range of water production costs of medium- and large-size seawater desalination projects. Information for this table is compiled based on comparative review of over 50 desalination projects in the United States, Australia, Europe, the Middle East, the Caribbean, and other parts of the world. As seen in this table, at present (in year 2012 US$) the average industry-wide cost of production of desalinated water by reverse osmosis is approximately $1.1/m^3$ ($4.2/1000 gal).

Comparative analysis of Tables 17.1 and 17.2 indicates that on average seawater desalination production costs are 1.2 to 1.6 times higher than those for high- and low-salinity brackish water desalination, respectively. When comparing some individual projects, however, this difference could be significantly higher. For example, the cost of water production of a low-salinity BWRO project in the low-end cost bracket [i.e., $0.2 to $0.4 m^3/day ($0.8 to $1.6/1000 gal)] could be over 10 times higher than the water production cost of SWRO desalination project in the high-end cost bracket [$1.6 to $3.0/$m^3$ ($6.1 to $11.3/1000 gal]. While factually accurate, such comparisons, however, are misleading if they are taken out of context of the site-specific project conditions, which may differ significantly from one project to another.

Figure 17.3 depicts a typical breakdown of the fresh water production costs of medium- and large-size seawater desalination projects. Although the ratio between the key cost components varies from project to project, the largest pieces of the cost pie are usually the plant construction expenditures (i.e., the direct capital costs), power, and the other O&M costs (i.e., maintenance, chemicals, membranes, etc.). The indirect capital costs, which mainly include expenditures for project engineering, development, and finance, are also a measurable portion (typically 10 to 20 percent) of the water production costs.

Comparison of Figs. 17.1, 17.2, and 17.3 indicates that capital costs for BWRO facilities are usually a higher portion of the total water production expenditures than those for SWRO plants (45 to 76 percent versus 40 to 60 percent). In BWRO projects, energy contributes 10 to 30 percent of the total costs, as compared with SWRO projects where the energy contribution is usually in a range of 20 to 35 percent, and, in extreme conditions for remote plant locations with high unit energy costs, energy expenditures for SWRO desalination could exceed 50 percent of the total costs of water production.

Classification	Cost of Water (US$/m³)
Low-end bracket	0.5–0.8
Medium range	0.9–1.5
High-end bracket	1.6–3.0
Average	1.1

Note: $1.0/m³ = $3.785/1000 gal

TABLE 17.2 Water Production Costs of Medium and Large-Size SWRO Desalination Plants

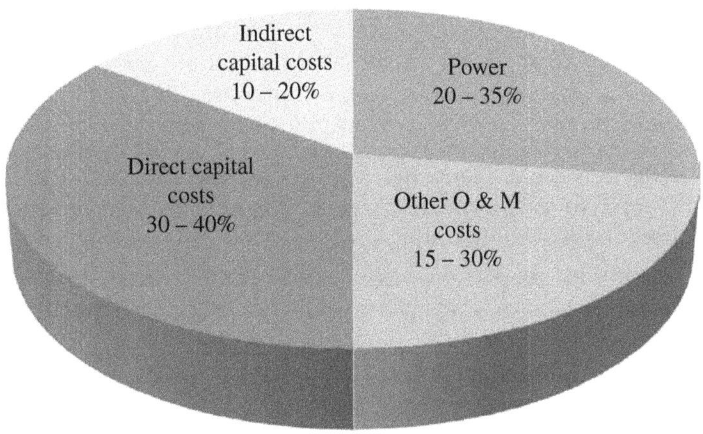

FIGURE 17.3 Seawater reverse osmosis plant, cost of water breakdown.

17.3 Capital Cost Estimates

Capital costs include all expenditures associated with desalination project implementation: from the time of conceptual development through design, permitting, financing, construction, commissioning, and acceptance testing for continuous operation. Construction costs encompass all direct expenditures needed to build plant source water intake and concentrate discharge systems and all project-related structures; procure and install all facility equipment, install and connect plan piping and service utilities; and deliver desalinated water to final user/s. Because of their direct association with the construction of physical facilities, construction costs are also referred to as "direct" or "hard" capital costs.

The remaining capital costs are often referred to as "indirect" or "soft" costs. These costs are associated with all engineering, administrative, permitting, and funding efforts necessary to bring the project to fruition as well as expenditures needed to secure contractors for design, construction, and operation of the desalination project.

Total project capital costs are typically presented in monetary units (i.e., US$) and are estimated either for the year when project construction is initiated or are referenced to the middle of the construction period. Depending on the type, length, and term of project funding, capital costs are often converted into monetary units per year and referred to as amortized or annualized costs (US$/yr). In addition, total capital and construction costs are sometimes presented as expenditures per unit of desalination project fresh water production capacity (i.e., US$/$m^3$.day or US$/1000 gal).

17.3.1 Capital Cost Breakdown for BWRO Desalination Projects

Capital cost breakdown for low- and high-salinity brackish water desalination projects are presented in Table 17.3.

17.3.2 Capital Costs Breakdown for SWRO Desalination Projects

A typical breakdown of the project capital costs for low- and high-complexity desalination projects is presented in Table 17.4. Project complexity is determined based on project size (i.e., fresh water production capacity), source water quality and its variability, type of plant intake, method of disposal of concentrate and other plant waste streams, complexity of permitting regulations governing project implementation, and financing sources and structure. Usually low-complexity projects are:

- Relatively small plants [i.e., projects of production capacity of 20,000 m^3/day (5 mgd) or less] located in non-environmentally sensitive areas with a project-friendly local community.
- Plants with good source water quality: turbidity (measured in NTU) and SDI of less than 1, trace levels of organics and bacterial contamination, and very low content of fouling and scaling constituents.
- Plants with subsurface or open intakes that collect seawater without significant interference from contaminated surface fresh water sources, groundwater aquifers, or waste discharges.
- Plants with simple, low-cost concentrate disposal methods, such as direct sewer or near-shore ocean discharge with suitable environmental conditions that do not require waste stream treatment prior to discharge and construction of complex discharge structures such as long outfalls equipped with diffusers.
- Regulatory environment where the key regulating agencies involved in the project permitting process have experience with similar size desalination projects and adequate expertise to complete project environmental review in an expeditious and timely manner.
- Projects that have simple and well-developed financing and tariff structure where project costs, revenues, risks, and rewards are well balanced and where the cost of desalinated water is competitive to that of other available water sources.

The cost breakdown brackets presented in Table 17.4 are based on data from actual seawater desalination projects.

If the site-specific conditions of a given individual project differ significantly from those encountered in desalination projects completed over the past 10 years, the actual cost breakdown for this project may be outside of the cost brackets presented in Table 17.4.

Tables 17.3 and 17.4 should be used for preparation of conceptual and preliminary cost estimates only and are intended to reflect current "market" conditions for the individual cost items. With the advancement of membrane technology and maturing of the markets involved in funding and serving the development, construction, and operation

Cost Item	Percentage of Total Capital Cost (%)	
	Low-Salinity BWRO Project	High-Salinity BWRO Project
Site preparation, roads, and parking	0.5–1.5	0.5–1.0
Intake	16.0–24.0	10.0–14.0
Pretreatment	0.7–1.0	0.5–1.5
RO system equipment	30.0–35.0	33.0–40.0
Post-treatment	0.5–2.0	1.0–1.5
Concentrate disposal	0.2–1.0	0.2–2.7
Waste and solids handling	0.1–0.5	0.3– 0.8
Electrical and instrumentation systems	5.0–8.5	6.5–10.5
Auxiliary equipment and utilities	1.5–2.5	1.0–2.0
Buildings	4.5–5.5	3.0–4.0
Start-up, commissioning, and acceptance testing	1.0–1.5	1.0–2.0
Subtotal direct (construction) costs (% of total capital costs)	*60.0–83.0*	*57.0–80.0*
Project engineering services		
Preliminary engineering	0.5–1.5	0.5–1.5
Pilot testing	0.0–1.0	0.0–1.0
Detailed design	4.0–6.0	4.5–7.5
Construction management and oversight	1.5–2.5	2.0–3.0
Subtotal engineering services	*6.0–11.0*	*7.0–13.0*
Project development		
Administration, contracting, and management	1.0–2.0	1.0–2.5
Environmental permitting (licensing)	0.5–5.5	1.0–6.0
Legal services	0.5–3.5	0.5–3.5
Subtotal project development	*2.0–11.0*	*2.5–12.0*
Project financing costs		
Interest during construction	1.5–3.0	1.5–3.0
Debt service reserve	2.0–6.0	2.5–6.0
Other financing costs	0.5–1.0	0.5–1.0
Subtotal project financing	*4.0–10.0*	*4.5–10.0*
Contingency	*5.0–8.0*	*6.0–8.0*
Subtotal indirect capital costs (% of total capital costs)	*17.0–40.0*	*20.0–43.0*
Total capital costs	100	100

Table 17.3 Typical Direct Capital (Construction) Cost Breakdown for BWRO Projects

Cost Item	Percentage of Total Capital Cost (%)	
	Low-complexity Project	High-complexity Project
Site preparation, roads, and parking	1.5–2.0	0.5–1.0
Intake	4.5–6.0	3.0–5.0
Pretreatment	8.5–9.5	6.0–8.0
RO system equipment	38.0–44.0	30.5–36.0
Post-treatment	1.5–2.5	1.0–2.0
Concentrate disposal	3.0–4.0	1.5–3.0
Waste and solids handling	2.0–2.5	1.0–1.5
Electrical and instrumentation systems	2.5–3.5	1.5–2.5
Auxiliary equipment & utilities	2.5–3.0	1.0–2.0
Buildings	4.5–5.5	3.0–5.0
Start-up, commissioning, and acceptance testing	1.5–2.5	1.0–2.0
Subtotal direct (construction) costs (% of total capital costs)	*70.0–85.0*	*50.0–68.0*
Project engineering services		
Preliminary engineering	0.5–1.0	0.5–1.5
Pilot testing	0.0–0.5	1.0–1.5
Detailed design	3.5–4.5	5.0–6.0
Construction management and oversight	1.0–2.0	2.5–3.5
Subtotal engineering services	*5.0–8.0*	*9.0–12.5*
Project development		
Administration, contracting, and management	1.0–1.5	2.0–3.0
Environmental permitting (licensing)	0.5–3.5	4.5–5.0
Legal services	0.5–1.0	1.5–2.0
Subtotal project development	*2.0–6.0*	*8.0–10.0*
Project financing costs		
Interest during construction	0.5–2.5	1.0–4.5
Debt service reserve	2.0–5.5	4.5–8.5
Other financing costs	0.5–1.0	3.5–4.5
Subtotal project financing	*3.0–9.0*	*9.0–17.5*
Contingency	*5.0–7.0*	*6.0–10.0*
Subtotal indirect capital costs (% of total capital costs)	*15.0–30.0*	*32.0–50.0*
Total capital costs	100	100

TABLE 17.4 Typical Direct Capital (Construction) Cost Breakdown for SWRO Projects

of desalination projects, the ratios between the individual cost items is expected to change over time. Gradual changes are expected every two to five years. A more dramatic change is likely within a 10-year time frame.

17.3.3 Direct Capital (Construction) Costs

Plant Site-Related Construction Expenditures

Site-related construction costs include expenditures for land acquisition and for site preparation for construction (clearing, grubbing, filling, grading, and fencing) as well as costs for construction of access roads to the desalination plant and to all buildings, facilities, and equipment within the desalination plant. The cost of land, the expenditures for site clearing, soil contamination mitigation, and dewatering as well as the cost and length of access roads are site-specific and could vary significantly from one location to another. In general these costs are in a range of \$15 to \$200/m^3·day (\$0.06 to \$0.8 mm/mgd) of plant production capacity.

The land requirements for a typical desalination plant are summarized in Table 4.1. This table can be used for initial planning of brackish and seawater desalination projects. However, it should be pointed out that, in general, brackish desalination plants may require 10 to 15 percent less land than seawater desalination plants of the same fresh water production capacity, mainly because of their simplified pretreatment facilities and higher recovery. On the other hand, many BWRO plants have post-treatment facilities for hydrogen sulfite gas removal, which typically are not needed for SWRO desalination.

Intake Construction Costs

The intake construction costs include expenditures for the plant saline intake structure and pipeline, intake pump station, and screening facilities. These costs vary depending on the type of source saline water intake: subsurface (vertical wells, horizontal directionally drilled (HDD) wells, Raney-type wells, and infiltration galleries), surface (open) intake, or co-located intake using existing power plant outfall or intake. Capital costs for construction of various types of intakes for brackish and seawater desalination plants are presented in Chap. 6.

The least-costly type of brackish desalination plant intake is a shallow vertical groundwater well that collects water from low-salinity aquifers. Similarly, vertical beach wells are also a cost-competitive type of intake facility for source seawater supply as compared with open ocean intakes. However, their use is typically limited to small- and medium-size SWRO plants (AWWA, 2011).

Typically, the lowest-cost intake for seawater desalination plants is that co-located with the discharge of an existing coastal power plant that uses seawater for cooling. The co-located desalination plant taps into the power plant discharge to collect source seawater. This co-location approach allows avoiding construction of new desalination plant intake structure, pipeline, and screens, which reduces approximately 60 to 80 percent of the total intake construction expenditures.

The use of HDD wells for large-size facilities may prove beneficial for conditions where the HDD wells can collect high-quality saline water at a steady rate. The main challenge with HDD wells is maintaining their capacity and water quality over time. Cost comparisons of alternative source water intakes are provided in other sources (Watson et al., 2003; WRF, 2011). Typically, intake construction costs are between \$100 and \$800/m^3·day (\$0.4 and \$3.0 mm/mgd).

Pretreatment Construction Costs

Pretreatment construction costs include expenditures for removal of all contaminants in the saline water that may have an impact on the normal operation of the membrane separation process and cause accelerated membrane fouling and/or premature replacement. The magnitude of these costs depends mostly on the content of solids (turbidity/total suspended solids), biodegradable organics, and nonorganic membrane fouling compounds in the source water, and the selected type of pretreatment technologies and equipment needed for their effective removal.

The pretreatment process may involve physical removal of contaminants by coarse and fine screening and microscreening, grit separation, sedimentation, dissolved air flotation, granular media or membrane filtration as well as chemical conditioning of the source water to prevent nonorganic scale formation (addition of antiscalants), membrane biofouling (biocides and UV irradiation), enhanced boron removal (by pH adjustment), and for improved solids removal by source water conditioning with coagulants and flocculants.

Many brackish and seawater desalination plants with subsurface intakes have a simplified and inexpensive pretreatment system, which mainly includes cartridge filtration and chemical conditioning with scale inhibitors. Therefore, the pretreatment facility construction costs of such plants are minimal (typically less than 2 percent of the total plant construction costs). However, desalination plants with surface water intakes usually have to process lower-quality source water and employ elaborate pretreatment to reduce various foulants contained in the source water.

Because of the significant differences in source water quality and variety of available technologies and equipment for solids and organics removal and chemical conditioning, the costs associated with source water pretreatment may vary in a wide range. Typically, the pretreatment costs are between \$150 and \$450/m^3·day (\$0.6 and \$1.7 mm/mgd).

Chapters 11 and 12 provide construction cost information for conventional and membrane pretreatment, respectively. Detailed discussion of the cost differences between conventional granular media pretreatment systems and membrane pretreatment is presented in Chap. 13.

RO System Equipment Costs

This cost item includes the expenditures associated with the procurement, purchase, installation, and construction of the following facilities and equipment: cartridge filters, high-pressure pumps and motors to feed the RO system, energy-recovery system, RO pressure membrane vessels and racks, RO membrane elements, membrane cleaning system, membrane flush system, and interconnecting piping.

The RO system is the most complex portion of the desalination plant and usually contributes 30 to 50 percent of the total plant construction costs. The design and construction costs of this system are mainly influenced by the source water salinity and temperature and by the target product water quality the system is designed to produce. The construction cost of the RO system is predominantly determined by the cost of the membrane vessels and racks, high-pressure pumps and piping, cost of the energy recovery system, and the price of the RO membrane elements. Typically, the construction costs for low- and high-salinity BWRO desalination systems vary between \$275 and \$550/m^3day (\$1.0 and 2.1 mm/mgd) and \$325 and \$675/m^3day (\$1.2 and \$2.5 mm/mgd), respectively. For comparison, SWRO system equipment costs vary between \$450 and \$1150/m^3·day (\$1.7 and \$4.3 mm/mgd). More details related to RO system equipment costs are included in Chap. 14.

Post-Treatment Costs

Post-treatment costs incorporate expenditures for construction of a chemical conditioning system for permeate stabilization, disinfection system, and facilities for product water quality polishing. The post-treatment costs are mainly driven by the target product water quality and the final use of the desalinated water. Typically, the costs for construction of post-treatment facilities for BWRO and SWRO plant permeate stabilization and disinfection range between \$100 and \$300/m³·day (\$0.4 and \$1.1 mm/mgd) and between \$80 and \$275/m³·day (\$0.3 and \$1.0 mm/mgd), respectively. The higher costs for post-treatment of permeate from brackish water RO systems are related to the fact that such waters often have to be treated for removal of hydrogen sulfide and sometimes consume more chlorine because they have lower oxygen content.

If desalinated water has to be polished to consistently maintain boron levels below 0.4 mg/L, or other specific constituents have to be removed to produce high purity water (i.e., silica, dissolved gases), then these costs may increase beyond the range indicated above. Construction costs of alternative post-treatment systems are discussed in further detail in Chap. 15.

Plant Discharge Costs

Concentrate disposal costs encompass expenditures for the conveyance and disposal of the concentrate and other waste streams generated at the desalination plant. These costs can vary significantly depending on the concentrate disposal method. Construction costs associated with various concentrate disposal methods are discussed in detail in Chap. 16.

Table 17.5 presents a typical construction cost range for concentrate disposal alternatives most commonly used in seawater desalination plants. Review of this table indicates that the sanitary sewer and surface water discharge are the two most cost-effective methods for concentrate disposal. Depending on the site-specific conditions, deep well

Concentrate Disposal Method	Disposal Construction Cost (US\$/m³·day)
New surface water discharge (new outfall with diffusers)	50–750
Collocation of desalination plant and power plant discharge	10–30
Co-disposal with wastewater treatment plant discharge	30–150
Sanitary sewer discharge	5–150
Deep/beach well injection	200–625
Evaporation ponds	300–4500
Spray irrigation	200–1000
Zero liquid discharge	1500–5000

Note: \$1/m³·day = \$3785/mgd

Table 17.5 Construction Costs for Common Concentrate Disposal Alternatives

injection, evaporation ponds, and spray irrigation could be competitive concentrate disposal alternatives.

A zero liquid discharge system typically has the highest construction and operation costs. However, under specific circumstances (such as cold climate, low evaporation, soil uptake rates, high land costs, and low power costs), the zero liquid discharge systems could be cost competitive than evaporation pond and spray irrigation disposal alternatives. A large number of site-specific factors influence the costs for surface water discharge and are difficult to generalize. The key factors that determine the costs of concentrate discharge to surface water are: (1) the conveyance costs to transport the concentrate from the desalination plant to the surface water discharge outfall, (2) the costs for outfall construction and operation, and (3) the costs associated with the monitoring of the environmental effects of the concentrate discharge on the surface waters.

The costs for concentrate conveyance are typically closely related to concentrate volume and the distance between the desalination plant and the discharge outfall. The outfall construction costs are site specific and in addition to the outfall size and diffuser system configuration (which is driven by the concentrate volume and salinity), these costs are dependent on the outfall configuration, length and material, which, in turn, are determined by the site-specific surface water body hydrodynamics and environmental conditions.

Sanitary sewer discharge cost is site-specific, and the key cost components for this disposal method are the cost of conveyance (pump station and pipeline) and the costs and fees for connecting to the sanitary sewer and for treatment/disposal of the concentrate. In some instances, when the wastewater treatment plant (WWTP) discharge volume is comparable to the volume of the desalination plant discharge, and the outfall has limited mixing capabilities, the use of this disposal method may require equalization of the concentrate discharge flow to match the diurnal variability of the WWTP effluent flow.

Key factors that influence deep well injection construction costs are well depth and diameter of well tubing and casting rings. Well diameter seems to have a limited influence on the costs. Several other key cost factors are: (1) the need for concentrate pretreatment prior to disposal; (2) the concentrate feed pump size and pressure, which vary depending on the type of the desalination plant energy recovery system, the geological conditions, and the depth of the injection zone, (3) the environmental monitoring well system size and configuration, and (4) the complexity of site preparation, mobilization, and demobilization.

Evaporation rate (local climate), concentrate volume, land and earthwork costs, liner costs, and the salinity of the concentrate mainly drive the costs of concentrate evaporation pond systems. The main cost variable is the evaporation area. Typically, evaporation rates are lower than soil uptake rates, and therefore disposal of the same volume of concentrate using evaporation ponds requires more land than disposal by spray irrigation.

Spray irrigation is usually is cost effective only if the concentrate is blended with a fresh water source to reduce its salinity to a level acceptable for crops/vegetation irrigation, and its feasibility depends on the type of the crops/vegetation and on the soil uptake rates. The key cost factors of this disposal method are the costs of land, the storage and distribution system costs, and the irrigation system installation costs. Most of these costs are driven by concentrate volume and salinity.

Achieving zero liquid discharge is usually the most costly method for concentrate disposal because it requires the use of elaborate mechanical equipment for evaporation and crystallization and concentration (dewatering) of the salts in the concentrate. Although this method has found practical application in industrial water reuse facilities, it has not yet been used for disposal of concentrate from large seawater desalination plants.

Waste and Solids Handling Costs

These costs include expenditures for construction of facilities for collection, conveyance, and disposal of solid waste (spent membranes, cartridge filters, and waste solids) from the plant site as well as for the construction of a solids handling system for treatment and disposal of residuals generated during the pretreatment process (screenings, sludge settled in the sedimentation tanks, solids from the spent filter backwash water). In addition, these expenditures also encompass costs for equipment and storage tanks for collection, conveyance, and treatment (if necessary) of the waste membrane cleaning chemicals and flush water to their final disposal site (typically the sewer system or the desalination plant discharge after pretreatment by neutralization).

Usually, the system for collection and disposal of waste membrane cleaning chemicals consists of storage tank and pumps and piping used to convey the spent cleaning chemicals to the storage tank and from the tank to the nearby sewer system. The cost of this system is in a range of \$15 to \$75/m^3·day (\$0.06 to \$0.28 mm/mgd).

Reverse osmosis plants with open intakes, would generate a large quantity of solids, which are removed from the source water by the plant pretreatment system. If regulatory constraints limit the disposal of these solids back to the surface water body of their origin, the filter backwash solids will have to be settled, dewatered, and disposed to a landfill. The expenditures for construction of the solids handling facility for backwash water residuals are typically in a range of \$20 to \$180/m^3·day (\$0.08 to \$0.68 mm/mgd).

Costs of Electrical and Instrumentation Systems

These costs encompass expenditures for the desalination plant's electrical supply system (electrical substation, equipment, and conduits connecting the desalination plant to the electrical grid or to a power generation facility), the equipment transformers and motor control centers, and all electrical conduits and equipment connecting the plant electrical system to the individual electrically driven equipment. The electrical system construction costs incorporate the expenditures for emergency power generation equipment as well. These costs also include funds for plant instrumentation and controls. The plant electrical and instrumentation costs are usually in a range between \$100 and \$250/m^3·day (\$0.4 and \$1.0 mm/mgd).

Costs of Auxiliary and Service Equipment and Utilities

The facilities in this category are the plant chemical storage and feed systems, process air and water supply facilities, the plant fire protection system, sanitary wastewater collection system, storm water management system, and all utilities needed for the normal plant operation (potable and utility water, telephone, gas, etc.). These costs also incorporate the expenditures for an initial set of spare parts for the desalination plant facilities. The expenditures for construction of auxiliary and service equipment and utilities are usually between \$30 and \$150/m^3·day (\$0.1 and \$0.6 mm/mgd).

Building Costs

Typically, the desalination plant has one or more buildings that house plant administration and management, laboratory, operator locker and shower facilities, maintenance shop, equipment and chemical storage area, and the key equipment of the RO system (high-pressure pumps, membrane vessels and racks, energy-recovery system, etc.). Depending on the type, complexity, and size of the desalination plant as well as its location, appearance, and ambient environment, the construction costs for the desalination plant buildings may vary from \$10 to \$20/m² (100 to 200/ft²) of the building footprint and range between \$50 and \$100/m³·day (\$0.2 and \$0.4 mm/mgd).

Start-up, Commissioning, and Acceptance Testing Costs

These costs include all expenditures for labor, consumables (electricity, chemicals, etc.), and equipment used during the plant commissioning, start-up, and acceptance testing process. These expenditures typically also incorporate the costs for construction-related permitting and insurance, for preparation of plant operation and maintenance manuals, for initial training of the permanent desalination plant O&M staff, and for equipment and other items that are required for normal plant operations (tools for the workshop, service vehicles for plant operations staff and management, furnishings and equipment for the plant laboratory and administration building, etc.).

These costs also incorporate all expenditures associated with the use of outside services, such as lab analysis of all source and product water quality parameters that cannot be completed in-house. Depending on the complexity of the project, these costs can vary in a wide range—\$40 to \$80/m³·day (\$0.15 to \$0.30 mm/mgd).

17.3.4 Indirect Capital Costs

Costs for Project Engineering Services

Preliminary Engineering Preliminary engineering costs encompass all expenditures associated with initial assessment of project feasibility, definition of project scope and size, as well as studies required to determine the project location, the type of project intake and discharge, and the configuration of key project facilities and equipment (i.e., intake, pretreatment, RO separation, concentrate disposal, permeate post-treatment, and product water conveyance and delivery). The preliminary engineering costs are dependent on the project size and complexity. These costs range from \$30 to \$100/m³·day (\$0.1 to \$0.4 mm/mgd) of project's product water capacity.

Pilot Testing Pilot testing is highly recommended for medium and large desalination projects [i.e., projects of production capacity of 40,000 m³/day (10.6 mgd)]. Although pilot testing costs are relatively high—\$10/m³·day to \$50/m³·day (\$0.04 to 0.20 mm/mgd)—they usually are a good investment toward the successful implementation of large desalination projects. In addition to the costs for constructing a pilot plant, additional operational costs of \$15,000 to \$30,000 per month for pilot operations and maintenance have to be budgeted.

Detailed Design Development of detailed project drawings and specifications typically costs \$75/m³·day to \$175/m³·day (\$0.3 to 0.7 mm/mgd). Detailed project design also includes preparation of as-built drawings and specifications that document the actual project implementation and deviations from the original design during construction.

Construction Management and Oversight Construction management and oversight include all engineering activities associated with project construction as well as the management of the construction contractors and suppliers involved in project implementation. The construction management and oversight costs range between \$40 and \$80/m^3.day (\$0.15 and \$0.30 mm/mgd).

Project Development Costs Project development costs comprise of all desalination plant owner indirect expenditures associated with project implementation prior to project construction and initiation of full-time operation—from its inception and conceptual development to initial planning, administrative review and budgeting, environmental permitting, procurement of contractors for project construction and implementation, project funding, and staffing of desalination plant operations. These costs could vary significantly from project to the project depending on its size, complexity, permitting environment, public acceptance, land and right of way ownership, funding sources, and other site-specific factors.

Project Administration, Contracting, and Management Project administration, contracting, and management are owner responsibilities, which usually involve in-house expenditures for owner staff and overhead associated with project implementation as well as costs for contracting of outside engineering consultants and other advisors to provide specialized support services to project owner as needed. Expenditures associated with these efforts depend on the owner's in-house capabilities and experience with the implementation of desalination projects and may vary between \$25 and \$50/m^3.day (\$0.1 and \$0.2 mm/mgd).

Environmental Permitting Expenditures associated with environmental permitting (licensing) include two key components: (1) costs for preparation of environmental studies and engineering impact assessment needed to obtain environmental permits (licenses) and, (2) fees associated with environmental permit filing and processing. Environmental permitting efforts and associated costs depend on the size and complexity of the desalination project, on the methods planned to be used for disposal of the desalination plant concentrate, and on the site-specific environmental conditions of the area of plant intake and discharge.

Extensive waste discharge modeling studies are often necessary to ascertain the environmental viability of the construction and operation of large outfalls and deep injection wells. Usually, the completion of this type of study for large desalination projects is a multiyear effort and involves significant cost expenditures, expert reviews, and a multistep evaluation process.

Environmental permitting costs and efforts also depend on the experience of the regulatory agencies with permitting similar desalination projects and the advancement of the regulatory law addressing intake impingement and entrainment and concentrate discharge permitting and monitoring. Because of the significant differences in desalination plant discharge permitting experience in various countries, the cost of environmental permitting may vary in a wide range from project to project and from country to country. Overall, the costs associated with project permitting may vary from \$20 and \$200/m^3.day (\$0.08 to \$0.8 mm/mgd).

Legal Services Costs for legal services include expenditures associated with legal review and processing of environmental permits/licenses and with the preparation and negotiation of contracts for water supply, engineering, construction, and O&M services. In addition, these expenditures encompass costs for review and processing of contractual agreements for acquisition of land for the desalination plant site, obtaining easements and rights of way for source water, product water pipelines, and electrical supply lines to and from the site; for negotiation of power supply contract/s; and for preparation of other contracts for services, equipment, and goods needed for construction and operation of the desalination plant. The cost of legal services is directly related to the complexity of the project and usually varies between \$20 and \$150/m^3·day (\$0.08 and \$0.60 mm/mgd).

Project Financing Costs Project financing costs include expenditures for obtaining of all funds and insurance needed for project implementation, from its inception and development through construction, start-up, and commissioning.

Interest During Construction Debt/bond obligations are typically repaid using revenue from the sale of the desalinated water to the consumers of this water. However, during the period of time when the project is under construction no revenue is available to repay debt obligations. Therefore, the owner of the project often borrows additional funds to pay the interest on the money used for construction.

Typically, interest during construction is calculated by multiplying the construction cost of the project by the annual interest rate of the loan and by 50 percent of the length of the construction period in years. This estimate assumes that 50 percent of the loan on average will be outstanding. Depending on the type of financing used for funding of the desalination project, interest during construction is usually between 0.5 and 4.5 percent of the total capital costs—\$20 to \$180/m^3·day (\$0.08 and \$0.68 mm/mgd).

Debt Service Reserve The debt service reserve is intended to protect project lenders against the inability of the owner to repay debt because the revenue generated by the project is insufficient. Depending on the type of financing, the complexity of the project, and the revenues of the water sales as compared with the debt obligations, the debt service reserve is typically set as one of the following three values: (1) maximum annual debt service, (2) 125 percent of the average debt service, or (3) 10 percent of the principal. The debt service reserve typically ranges between 2.0 and 8.5 percent to the project capital costs [\$80 and \$340/m^3·day (\$0.3 and \$1.3 mm/mgd)].

Other Financing Costs Other project financing costs comprise of expenditures associated with the funding of other reserve funds in addition to the debt service reserve fund if needed to satisfy lender requirements; of administrative and legal costs related to issuing project bonds or arranging project loans and administering payments; and of costs associated with arranging project equity, if equity contributions are used for project financing.

Other financing costs also include expenditures associated with purchasing insurance and obtaining performance and payment bonds to protect the owner and contractors against construction failures and problems, and for payment of various taxes associated with project implementation as well as for encompassing shipping costs for delivering plant components to the site. These costs range between 0.5 and 4.5 percent of the total capital costs [\$20 to \$180/m^3·day (\$0.08 to \$0.68 mm/mgd)].

Contingency

Contingency provisions in the project cost estimate reflect the fact that even when a detailed cost estimate is completed, a number of unknown factors may influence the actual expenditures associated with project implementation. The size of contingency funds included in a given cost estimate depends on the level of accuracy of this estimate as well as on project complexity, size, funding structure, contractor experience with similar projects and other project-related risks (see Chap. 4). A detailed cost estimate usually carries a contingency factor of 5 to 10 percent depending on the complexity and size of the project. Higher contingency levels are used in lower accuracy cost estimates.

17.4 Operation and Maintenance (O&M) Cost Estimates

Desalination plant operation and maintenance (O&M) costs incorporate all expenditures associated with facility operations and maintenance. Usually O&M costs are assessed over a period of one year and are often referred to as annual O&M costs (US$/year).

The key O&M cost components are energy (power), maintenance, chemicals, labor, RO membrane and cartridge filter replacement, and concentrate disposal. In total these costs typically encompass over 80 percent of the annual O&M expenditures. Typically, O&M costs are divided in two categories depending on their relation to the actual plant production of desalinated water: variable (function of the quantity of the produced flow) and fixed (costs not related to the actual plant fresh water production flow).

17.4.1 O&M Cost Breakdown for BWRO Desalination Plants

A typical O&M cost breakdown for low- and high-salinity brackish water desalination plants is presented in Table 17.6.

Cost Item	Percentage of Total O&M Cost (%)	
	Low-Salinity Project	High-Salinity Project
Variable O&M Costs		
Power	28.0–52.0	35.0–55.0
Chemicals	8.0–10.0	10.0–12.0
Replacement of membranes and cartridge filters	3.0–4.5	4.0–6.0
Waste stream disposal	1.5–5.5	2.5–6.0
Subtotal, variable O&M costs	*40.5–72.0*	*51.5–79.0*
Fixed O&M costs		
Labor	12.5–18.5	7.0–13.5
Maintenance	9.0–20.0	7.0–15.0
Environmental and performance monitoring	1.0–6.5	2.0–5.0
Indirect O&M costs	5.5–14.5	5.0–16.0
Subtotal, fixed O&M costs	*28.0–59.5*	*21.0–49.5*
Total O&M costs	100	100

Table 17.6 Typical Annual O&M Cost Breakdown for BWRO Desalination Plants

Review of Table 17.6 indicates that high-salinity projects have higher variable costs and lower fixed O&M costs than low-salinity projects. This difference is mainly due to the elevated salinity of the source water and associated higher energy use.

17.4.2 O&M Cost Breakdown for SWRO Desalination Plants

Table 17.7 presents key annual O&M cost components for low- and high-complexity SWRO desalination projects.

Comparison of the data presented in Tables 17.6 and 17.7 reveals that for both brackish and seawater desalination plants, energy costs are the largest component of the plant's annual O&M expenditures. The relative percentage of power, chemical and membrane replacement O&M costs increases, and percentage of maintenance and labor costs decreases with the increase in source water salinity.

17.4.3 Power Costs

Annual desalination plant power costs are dependent on two key parameters: (1) the power tariff (and associated unit cost of power, usually expressed in monetary units per kWh), and (2) the amount of power used to produce desalinated water, typically presented in kWh per m^3 or 1000 gal of fresh product water.

Power Costs and Desalination Plant Energy Use

As discussed in Chap. 4, power costs are directly related to the source water salinity and temperature, and the associated osmotic pressure that has to be overcome in order to produce fresh water. Table 4.4 provides an overview of unit energy use for low- and

Cost Item	Percentage of Total O&M Cost (%)	
	Low-Complexity Project	High-Complexity Project
Variable O&M costs		
Power	45.0–61.0	35.0–58.0
Chemicals	3.0–6.5	5.5–9.0
Replacement of membranes and cartridge filters	5.0–9.0	6.5–11.0
Waste stream disposal	2.5–5.5	3.5–7.0
Subtotal, variable O&M costs	*55.5–82.0*	*50.5–85.0*
Fixed O&M costs		
Labor	5.0–9.5	4.0–11.0
Maintenance	6.5–12.5	3.0–13.0
Environmental and performance monitoring	0.5–4.0	1.0–5.0
Indirect O&M costs	7.5–18.5	7.0–20.5
Subtotal, fixed O&M costs	*19.5–44.5*	*15.0–49.5*
Total O&M costs	100	

TABLE 17.7 Annual O&M Cost Breakdown

high-salinity BWRO desalination plants and for SWRO facilities. As shown in this table, the desalination industry-wide averages for these energy uses are 0.8 kWh/m^3 (3.0 kWh/1000 gal), 1.4 kWh/m^3 (5.3 kWh/1000 gal), and 3.1 kWh/m^3 (11.7 kWh/1000 gal), respectively.

The actual energy use for a given desalination project may vary in a fairly wide range and mainly depends on the source water salinity and temperature. In general, source seawater of lower salinity and higher temperature yields lower power use for production of the same volume of fresh water mainly due to the reduction of RO feed water osmotic pressure.

Often for brackish desalination plants, high content of scaling compounds, hydrogen sulphide, iron, and manganese may also have a measurable impact on the overall plant energy use. In many low-salinity brackish SWRO and membrane softening projects, the distance of the desalination plant from the source water well field and associated pumping power costs may be comparable with the energy expenditures for the RO salt separation process.

Another key factor associated with overall energy use is the efficiency of the applied energy recovery system. Many existing low-salinity brackish desalination plants do not have an energy-recovery system because of the relatively long payback period associated with the installation of such systems. However, the majority of high-salinity brackish water desalination plants and practically all SWRO facilities are equipped with energy-recovery systems. Such systems allow to reuse a large portion of the energy applied for desalination by recovering it from the plant concentrate and applying it for production of new desalinated water. The efficiency of energy transfer from concentrate to source water varies with the type of energy-recovery technology (pressure exchanger, Pelton wheel, turbocharger, or reverse running pump) and with the overall water recovery and configuration of the RO system. A more detailed discussion of alternative energy recovery systems is presented in Chap. 14.

Power Costs and Electricity Tariff

When electricity is purchased from an independent power generation supplier, the unit cost-of-power tariff is typically outside the control of the desalination plant owner. In this case, the desalination plant could be designed to take advantage of the cost reduction associated with the off-peak power rate, which usually is lower than this rate during the peak hours of power consumption.

Usually, the peak power rate timeframe coincides with the periods of peak of water demand, during which the desalination plant often has to operate at maximum rather than minimum capacity. Therefore provision of an adequate amount of product water storage would be essential to take advantage of the benefits of maximum off-peak power tariff operation of the desalination plant. Construction of additional plant product water storage capacity to accommodate off-peak power tariff benefits would increase plant construction costs, and therefore its viability has to be accessed on a life-cycle cost basis.

Some power generation utilities provide additional power tariff incentives if the desalination plant owner is willing to significantly curtail or completely discontinue plant operations during periods of the year when the power generation utility can sell this power at high prices to other users. Power curtailment conditions, if offered by the electrical company, would vary from one power supplier to another, but in general would involve a requirement for reduction of over 90 percent of the desalination plant power use for a period of 6 to 12 hours for at least two times per month. In order to accommodate such curtailment schedule, the desalination plant design, operations, and water supply delivery commitments have to have built-in flexibility and extra product water storage capacity, which usually come with an increased capital expense.

Another potential alternative for reduction of the unit power rate is to co-locate the desalination plant with an existing power plant and to connect the desalination plant's electrical system directly to the power plant generation units, thereby completely avoiding the use of the power grid for electrical supply. Often the power tariff consists of two components: a power generation and a power grid distribution charge component. Depending on the regulations governing power generation, supply, and distribution, the direct connection to the power plant's generation units may allow us to avoid the payment of the power grid component of the tariff. Since this component may be as large as half of the power rate, the co-location approach could allow a substantial reduction in the unit power costs and therefore of the total costs for desalination.

Power grid associated charges could be eliminated, and unit power costs could be reduced by self-generating electricity at the desalination plant site. This approach is usually viable for very large plants (for example, the Ashkelon seawater desalination facility) because the generation of small quantities of electricity is typically not as cost effective as power generation on a large commercial scale by an experienced power generation company.

Power self-generation may be cost-effective for small-size desalination plants in cases when there is no easy access to a nearby electrical power grid and/or when the commercially available power rate is high and self-generation of electricity is cost-competitive. Another important issue associated with power self-generation is the risk the desalination plant owner and investors take with the increase in the unit cost of fuel (usually natural gas) used for power generation over time, and the sustained availability of a particular type of fuel over the useful life of the desalination project. Taking these risks is usually prudent only if they can be shared with the water consumer, mitigated by the government, or taken by a major supplier of this fuel product via long-term fuel-supply contract that expands over the useful life of the project.

Cost of power is a variable annual expenditure. For low- and high-energy BWRO desalination plants, it is typically in a range of \$0.04 to \$0.08/m^3 (\$0.15 to \$0.30/1000 gal) and \$0.07 to \$0.13/m^3 (\$0.26 to \$0.50/1000 gal), respectively. For SWRO desalination plants, the cost of power varies between \$0.15 and \$0.25/m^3 (\$0.57 and \$0.95/1000 gal). The cost variation may be wider for site-specific conditions where power supply is difficult or power self-generation is applied. Alternative approaches for reducing desalination plant power expenditures are further discussed in Chap. 14.

17.4.4 Chemical Costs

Chemical costs are highly variable from one location to another and are mainly dependent on the source water quality, the selected pretreatment processes, and the target product water quality. Table 17.8 presents unit costs for various chemicals frequently used in seawater desalination plants. The actual chemical cost values for a given project have to be established based on quotes from local suppliers of the site-specific chemicals.

Cost of chemicals is a variable expenditure and typically is in a range of \$0.025/m^3 to \$0.075/m^3 (\$0.1 to \$0.3/1000 gal) of product water.

Chemical	Unit Cost (US$/kg)
Chlorine gas	0.6–1.1
Sodium hypochlorite	2.2–3.2
Ferric sulfate and ferric chloride	0.4–1.2
Sulfuric acid (93% H_2SO_4)	0.05–0.10
Citric acid	1.6–2.6
Biocide	2.8–5.2
Sodium hydroxide (50% NaOH)	0.8–1.0
Sodium bisulfite	0.4–0.6
Antiscalant (scale-inhibitor)	1.8–4.4
Ammonium hydroxide	0.6–1.2
Hydrated lime	0.30–0.35
Calcite	0.05–0.08
Carbon dioxide	0.08–0.10
Sodium tri-polyphosphate (corrosion inhibitor)	1.8–3.4
Other cleaning chemicals (US$/$m^3$ of permeate)	0.005–0.008

Note: $1/$m^3$ = $3.785/1000 gal

Table 17.8 Unit Costs of Commonly Used Chemicals

17.4.5 Labor Costs

Plant operation labor costs are closely related to plant size, complexity and number of treatment processes and equipment, and to the overall level of plant automation. Typically, desalination plants are highly automated and reliable facilities, which use a limited amount of specialized staff for overall plant performance monitoring and control, equipment maintenance, preparation of chemical batches for various treatment processes, and collection and analysis of water quality samples.

Usually, every desalination plant is staffed with a plant manager, shift supervisors, operators, one or more mechanics and electricians, and laboratory and administrative employees. Often several smaller facilities [i.e., package desalination plants with production capacity of 500 m^3/day (130,000 gal/day) or less] are supervised by one regional plant manager and serviced by a central laboratory and instrumentation and control group.

Table 17.9 summarizes the typical plant staffing requirements for desalination plants as a function of the level of plant automation, treatment process complexity, and labor skills. As seen in this table, the number of plant staff varies with plant capacity and is strongly influenced by economy of scale. The staffing information presented in Table 17.9 is applicable to brackish and seawater desalination plants.

Typically, desalination plant staff is organized in one to three shifts, and in some smaller and fully automated plants, plant operations are unmanned at night. Large plants are typically staffed 24 hours per day, 365 days per year, with at least two

Plant Capacity (m^3/day)	Plant Automation and Labor Skill Level	
	High	Low
1000	2–3	4–6
5000	4–6	8–10
10,000	7–10	12–15
20,000	9–12	16–18
40,000	12–16	18–20
100,000	14–18	20–24
200,000	18–28	30–40
300,000	35–50	60–80

Table 17.9 Desalination Plant Staffing Requirements

operators on duty at all times. The labor costs are fixed for a given plant and are typically in a range of \$0.015/m^3 to \$0.040/m^3 (\$0.06 to \$0.15/1000 gal) of treated water.

17.4.6 Maintenance Costs

This cost item includes all expenditures associated with routine plant operations and preventive and emergency maintenance of plant equipment, structures, buildings, and piping. Typically, the useful life of most of the key desalination plant equipment is between 25 and 50 years. Therefore the average annual maintenance expenditure is approximately 2 percent (100 percent/50 years) to 4 percent (100 percent/25 years) of the cost of the installed equipment. Usually, annual costs for maintaining structures and piping are 1 to 2 percent of their construction costs. Maintenance costs vary from year to year because the key high-cost desalination equipment such as high-pressure pumps, energy-recovery system, and other large-capacity pumps undergo routine equipment refurbishment every 5 to 10 years in order to maintain their high efficiency and consistent performance.

Since most of the plant equipment is maintained routinely on a preset schedule independent of the actual water production, some (typically 40 to 60 percent) of the routine equipment maintenance costs are often considered fixed O&M costs. The remaining portion of the maintenance costs is accounted as a variable component and is related to the actual equipment run time. Plant total maintenance costs are typically in a range of \$0.035 to \$0.075/m^3 (\$0.13 to \$0.28/1000 gal) of desalinated water.

17.4.7 Membrane and Cartridge Filter Replacement Costs

This O&M cost component incorporates expenditures for replacement of pretreatment membranes (if membrane pretreatment is used), RO membranes, and cartridge filters. Annual membrane and cartridge filter replacement costs are proportional to the replacement frequency of these consumables, which, in turn, depend on the source water quality and plant design.

The typical useful life of ultra- and microfiltration pretreatment membranes is five to eight years. Therefore their annual replacement costs range between 12 and 20 percent of the initial installed membrane cost.

The useful life of RO membranes is typically between five and seven years, and, in the case of high-quality source seawater, it may extend to up to 10 years. As a result, the typical annual average RO membrane replacement rate is 14.3 to 20 percent of their initial installed costs.

Both pretreatment and RO membranes are replaced when the membrane media fouls irreversibly to levels that require excessive power use for their operation and/or reduce membrane productivity or produced water quality below a certain acceptable threshold. Membranes are also replaced when they lose their integrity, and their performance declines irreversibly. The unit costs for 8-in BWRO and SWRO elements vary between $250 and $350/element and $400 to $600/element.

Cartridge filters for SWRO plants have a typical minimum useful life of six to eight weeks. However, in many applications where the source water is of high quality, cartridge filter replacement is less frequent (once every 6 to 12 months). Depending on the cartridge filter size, the unit cartridge filter cost is between $8 and $30 per filter. The total membrane and cartridge filter replacement costs are typically in a range of $0.020 to $0.070/m^3 ($0.08 to $0.26/1000 gal).

17.4.8 Plant Waste Stream Disposal Costs

The main waste stream of every membrane desalination plant is the RO system concentrate. Depending on the concentrate disposal practices, the total waste stream disposal costs are typically in a range of $0.015/m^3 to $0.035/m^3 ($0.06 to $0.13/1000 gal).

In most applications, the ocean discharge of concentrate from desalination plants with an open intake is acceptable without any additional treatment and at minimal or no costs. For surface water discharges from desalination plants with well intakes, in which water contains high levels of iron and manganese and low content of oxygen, disposal costs are dependent on the need to aerate the concentrate before its discharge to a surface water body or to otherwise treat it if the concentrate exhibits toxicity or has other measurable environmental impacts.

Concentrate disposal costs include expenditures associated with operation and maintenance of the selected disposal method (concentrate injection wells, evaporation ponds, or mechanical evaporation equipment) if such disposal methods are used. These costs may vary widely depending on the disposal method and project size.

For sanitary sewer disposal of plant discharge, the volume of concentrate mainly drives the conveyance costs. Often, however, O&M costs would also include sewer/outfall capacity use fees. The sewer capacity use fees usually are related to the available capacity of the sewer facilities and the effect of the concentrate discharge on the operational costs of the wastewater treatment plant, which would provide ultimate treatment and disposal of the concentrate.

Desalination plants generate a number of other waste streams in addition to the plant concentrate. The main waste streams are the pretreatment waste filter backwash and the spent RO membrane cleaning solution. In the case of membrane pretreatment, desalination plants would also generate two additional waste streams: chemically enhanced backwash (CEB) and spent pretreatment membrane cleaning solution.

The cost of waste disposal depends on the method of waste disposal planned to be used at the desalination plant and the size of the waste streams. In many applications worldwide, all waste streams are discharged to a nearby surface water body (i.e., the ocean). Therefore, under the best-case scenario, no additional costs for waste disposal are incurred.

Frequently, the waste filter backwash along with the plant concentrate are the only two desalination plant process streams allowed to be discharged to the surface water body from which the saline source water originated, and the rest of the waste streams have to be conveyed to the sanitary sewer for disposal and further treatment. In this case, the expense of waste stream disposal is usually the sewer discharge fee established by the local wastewater collection and treatment agency. This cost may be between $0.005/m^3 and $0.015/m^3 ($0.02 and $0.06/1000 gallons).

In some large seawater desalination plants, spent filter backwash water has to be treated (typically by sedimentation) before discharge to the ocean. The residuals (sludge) generated during the filter backwash treatment are usually dewatered to solids content of 20 percent or higher via mechanical dewatering equipment (belt filter presses, centrifuges, or plate-and-frame presses) and disposed to a sanitary landfill. Depending on the capacity and distance of the available landfills in the area of the desalination plant, the residual disposal costs usually vary between $16 to $80/wet ton of sludge, which would correspond to an additional cost of water production between $0.005 and $0.025/m^3 ($0.02 and $0.10/1000 gal).

17.4.9 Environmental and Performance Monitoring Costs

Every desalination plant has discharge water quality monitoring requirements. These requirements may be applicable to the entire discharge and/or to the individual plant waste streams. In addition, in many environmentally sensitive areas the monitoring requirements encompass not only the discharge but the receiving water body (ocean, groundwater aquifer, or estuary) as well.

Depending on the complexity and frequency of the environmental monitoring required for permit compliance, the discharge monitoring costs could be substantial and should be taken under consideration in determining the overall plant O&M costs. Plant discharge monitoring costs may vary between $0.005/m^3 and $0.020/m^3 ($0.02 and $0.08/1000 gal).

Plant performance monitoring costs are expenses needed to measure and analyze key process performance parameters (i.e., SDI, temperature, pH, salinity of plant feed water, etc.). These O&M costs depend on the level of automation and plant complexity. Product water monitoring costs are expenditures associated with sample collection, laboratory analysis, and data management and reporting, which are required to be completed in order to comply with all applicable regulatory requirements associated with the product water supply. Typically, plant performance and product water quality monitoring costs are between $0.005/m^3 and $0.015/m^3 ($0.02 and $0.06/1000 gal).

17.4.10 Indirect O&M Costs

Indirect O&M costs include annual expenditures for staff training, professional development, and certification; expenditures for consumables and maintenance of plant service vehicles; administrative and utility/service (water, sewer, telephone, etc.) expenses; taxes associated with plant operations; operations insurance; contingency and other O&M reserve funds. These costs also incorporate the fees for plant operation, if a private company operates the desalination plant. Typically plant indirect O&M costs vary in a wide range—$0.025 to $0.075/m^3 ($0.1 to $0.3/1000 gal).

17.5 Water Production Cost Estimate

The cost of water production encompasses all expenditures associated with project implementation (including funding) and operation and maintenance, and consists of fixed and variable components. The fixed water costs are expenditures for plant construction and for repayment of the capital investment in the plant (i.e., capital cost recovery) and also include the portion of the annual O&M expenditures that are independent of the actual volume of water produced by the desalination plant (labor, maintenance, environmental, performance monitoring, and indirect O&M costs). The variable component of the cost of water incorporates O&M expenditures that are directly related and usually proportional to the actual volume of produced desalinated water (power, chemicals, replacement of membranes and cartridge filters, and waste stream disposal).

When the desalination plant is delivered under a BOOT contract between a public agency and a private contractor, the water tariff structure is typically reflective of the cost of water structure described above. The tariff usually includes capacity payment component, which compensates the private contractor for the fixed cost associated with water production, and a commodity (output) tariff payment component, which provides compensation for contractor's variable O&M expenditures. Example capital, O&M, and cost of water estimates for a fictional 40,000 m³/day (10.6 mgd) seawater desalination project are provided in Sec. 17.6.

17.5.1 Fixed Cost Components

Capital Cost Recovery

The capital cost for construction of a given desalination plant is usually amortized over the term of repayment of the capital used to build the plant (typically a period of 5 to 20 years). To determine the amortized value of the capital costs, these costs are divided by a capital recovery factor (CRF) and by the plant's design capacity availability factor. The CRF is a function of the interest rate of the capital and the number of years over which the investment is recovered (i.e., the plant capital expenditures are repaid). The CRF can be calculated using the following relationship:

$$\mathrm{CRF} = \{(1+i)^{n}-1\}/\{i\,(1+i)^{n}\} \tag{17.1}$$

where n = period of repayment of capital expenditures, and i = interest rate of the amortized investment.

For example, the CRF for a 40,000 m³/day (10.6 mgd) seawater desalination project that has total capital costs of \$78 million, repayment period of 20 years, and amortization rate of 5.7 percent, is 11.752. Therefore the project's annual amortized (annualized) capital cost is \$74 mm/11.752 = \$6,296,800/yr. The capital cost recovery portion of the cost of water for this example project is \$6,296,800/yr/(40,000 m³/day × 365 days) = \$0.43/m³ (1.63/1000 gal). This cost estimate assumes the plant has 100 percent availability. If, for example, the plant has a design capacity availability factor of 96 percent, then the capital cost recovery charge will be increased accordingly to \$0.43/m³/96% = \$0.45/m³ (\$1.70/1000 gal).

In many projects, the capital investment is a combination of equity and debt, which has different interest rates of return on investment. In addition, these interest rates may vary over the repayment period. As a result the calculation of the capital cost recovery for such project may not be as straightforward as shown above and typically requires the development of a financial model that reflects all specific features and terms of the various investments used for the project. Development of financial model for a given desalination project is usually the responsibility of the project developer/owner. If the project developer does not have adequate in-house capabilities to develop a financial model level of sophistication needed to obtain competitive financing, typically the developer/owner retains a specialized company to provide the necessary expertise.

Other Fixed Costs

As indicated previously, the other fixed components of the cost of water, besides capital cost recovery, include labor costs, maintenance costs, plant environmental and performance monitoring costs, and indirect O&M costs. These costs are typically calculated by dividing the annual fixed O&M expenditures by the design average annual production capacity of the desalination plant and by the plant design capacity availability factor.

For example, if for the 40,000 m^3/day (10.6 mgd) plant referenced above, the annual labor costs at 100 percent availability are determined to be \$420,000/yr, the maintenance costs are \$700,000/yr, the plant environmental and performance monitoring costs are \$120,000/yr, and the indirect O&M costs are \$700,000/yr, then the other fixed water costs are estimated at (\$420,000 + \$700,000 + \$120,000 + \$700,000)/(40,000 m^3/day × 365 days) = \$0.133/m^3 (\$0.50/1000 gal). As a result, the total fixed water costs for this example at 100 percent desalination plant availability are \$0.43/m^3 + \$0.133/m^3 = \$0.563/m^3 (\$2.13/1000 gal). For 96 percent availability, these costs will be \$0.563/m^3/ 0.96 = \$0.586/m^3 (\$2.22/1000 gal).

The fixed cost of water component is independent of the actual amount of water that is produced by the desalination plant. Therefore, this cost component has to be minimized as much as possible. A high level of automation typically reduces labor costs. Selecting high-quality materials, equipment and piping and implementing proactive and systematic preventive maintenance program minimizes maintenance costs. Plant environmental and monitoring costs are maintained at low levels by using environmentally safe, low-cost concentrate disposal methods and by automation of most plant performance monitoring functions.

Using highly qualified operations staff or subcontracting plant operations to a private company specialized in seawater desalination plant operation typically reduces indirect O&M costs. Since the reduction of the other fixed costs requires higher capital expenditures, and therefore increases the capital recovery costs, the total fixed costs have to be optimized to find the prudent balance between these two key fixed cost components.

17.5.2 Variable Cost Components

Variable portion of the cost of water typically includes the following O&M expenditures: power, chemicals, replacement of membranes and cartridges, and waste stream disposal.

Power expenditure is the largest variable cost component and usually accounts for 15 to 35 percent of the total cost of water. Depending on the power tariff structure, the fixed portion of the power costs, such as the electrical grid connection charges, may sometimes be accounted for as a portion of the fixed water cost component. On the other hand, some of the maintenance costs, which traditionally are considered fixed costs, may be accounted for in the cost of water production as variable costs. This holds especially true for equipment, which has routine maintenance/replacement schedule that is based on the actual number of operating hours.

As indicated previously, chemical costs are related not only to the desalination plant source water and production flows, but to the source water quality as well. Usually, treatment of source water of good quality (low SDI, turbidity, and organic content) requires a lower amount of pretreatment chemicals and less frequent membrane cleaning, which, in turn, reduce plant chemical costs.

The difference between the chemical pretreatment and membrane cleaning costs for good and worst-than-average source water quality could be significant—often two to four times lower. This difference, however, has to be put in perspective. Since the chemical costs are usually less than 10 percent of the total water production costs, a twofold chemical cost reduction due to improved source water quality may not amount to a very large reduction of the overall fresh water production cost.

Where source water quality makes a measurable cost difference, however, is the extent of RO membrane fouling and the associated increase in membrane cleaning frequency, and the plant RO train downtime. If the source water quality is poor, and it requires frequent membrane cleaning and replacement due to fouling, the excessive membrane maintenance needs typically result in plant production interruptions and ultimately in a reduced overall plant capacity availability factor. In addition, accelerated membrane fouling increases the average plant power use.

The O&M costs associated with waste stream disposal usually are relatively small. However, in some cases operation of the concentrate disposal facilities could constitute a significant portion of the plant water production costs, and malfunctioning of these facilities could significantly reduce the plant capacity availability factor (i.e., could increase downtime). Therefore the use of simple and environmentally sound methods of concentrate disposal such as co-discharge with power plant cooling water, sanitary sewer discharge, or direct open discharge to surface waters when viable are recommended over deep well injection, evaporation pond disposal, or zero liquid discharge.

The variable water costs are typically calculated by dividing the total annual variable O&M costs by the actual average annual production capacity of the desalination plant. For the purposes of budgetary cost estimates and determination of the water tariff of new desalination projects, the variable water costs are calculated by dividing the projected annual variable O&M expenditures by the design average annual plant water production flow and availability factor.

For example, if the hypothetical 40,000 m^3/day (10.6 mgd) seawater desalination plant has actual annual plant power costs of \$3,370,000/yr, chemical costs of \$440,000/yr, annual costs for membrane and cartridge filter replacement of \$780,000/yr, and waste stream disposal costs of \$330,000/yr, then the variable water costs are estimated at \$4,920,000/yr / (40,000 m^3/day × 365 days × 100%) = \$0.337/m^3(\$1.28/1000 gal).

In summary, the total (fixed and variable) cost for production of desalinated water for this example at 100 percent plant availability will be \$0.563/m^3 + \$0.337/m^3 = \$0.90/m^3 (\$3.41/1000 gal). For the same hypothetical plant designed for 96 percent availability, however, this cost water production will be \$0.563/0.96 + \$0.337/m^3 = \$0.923/m^3 (\$3.49/1000 gal).

17.6 Example Cost Estimate

This section presents a budgetary cost estimate for a 40,000 m^3/day (10.6 mgd) seawater desalination project. All costs included in this example are in year 2012 US\$ and are based on actual data from similar size projects supplemented with cost information from budgetary vendor quotes and cost estimates for all key equipment, piping, materials, and buildings.

This cost estimate is provided for illustrative purposes only. As indicated in the previous chapters of this book, many factors and site-specific differences for a given project may cause other projects of similar capacity, source, and product water quality to yield costs significantly different from those presented in this illustrative example.

17.6.1 Project Description

Plant Capacity and Availability

The example project is a seawater desalination plant with an average annual plant production capacity of 40,000 m^3/day (10.6 mgd), and maximum installed production capacity of 48,000 m^3/day (12.7 mgd), when operated at 50 percent recovery. The plant is designed to have an availability factor of 96 percent [i.e., to produce 40,000 m^3/day or more for 96 percent of the time (350 days per year) and operate in a recovery range of 45 to 50 percent]. A plant's minimum daily production capacity is 32,000 m^3/day (8.5 mgd).

Plant Location, Intake, and Discharge

The plant is located on a 20,000-m^2 (5-acre) site in a commercially zoned area and is approximately 800 m from the shore, and the plant site is an abandoned commercial property, which has an elevation of 10 m (33 ft) above the mean ocean tide level. The plant has open ocean intake that extends 200 m (670 ft) beyond the shoreline. The plant discharge is a 950-m (3120-ft) pipeline, of which 150 m (490 ft) extend in the ocean. The last 50 m of the outfall are equipped with diffusers for concentrate dissipation.

Intake Water Quality

Key plant intake water quality parameters are summarized in Table 17.10. The source seawater is typical Pacific Ocean water which, because of its depth, is not influenced significantly by algal blooms, hydrocarbon contamination, and other potential sources of pollution that may exist in other circumstances. Review of Table 17.10 indicates that

Parameter	Design Minimum Value	Design Maximum Value	Design Average Value
Intake flow, m^3/day	68,000	114,000	84,000
Salinity (TDS), mg/L	32,500	34,500	33,500
Chloride, mg/L	16,900	20,800	18,000
Bromide, mg/L	52	79	73
Boron, mg/L	3.6	5.0	4.5
Temperature, °C	10	26	18
Turbidity, NTU	0.2	24	2
Total suspended solids, mg/L	0.5	30	4
pH	7.3	8.1	7.8

Note: All design characteristics are daily average values.

Table 17.10 Key Intake Seawater Design Characteristics

the source seawater for the hypothetical desalination project has relatively low-fouling potential and consistent water quality.

Product Water Quality

The desalination plant will supply product water of water quality that is in compliance with the key parameters specified in Table 17.11. Product water quality in this table is typical for drinking water applications in the Unites States. The level of boron in Table 17.11 is driven by drinking water quality standards in California. Boron level requirements sometimes are more stringent if the water will be used for agricultural and horticultural irrigation.

The boron level requirement listed in Table 17.11 would allow the desalination plant to be designed with a single-pass RO membrane system. During the summer the source seawater pH will need to be elevated to approximately 8.8 in order to achieve the desired boron water quality goal. Such pH adjustment will be accomplished using sodium hydroxide at a dosage of 15 mg/L.

Allowed bromide levels in Table 19 are acceptable for desalinated, which will be disinfected by chlorination. If the desalinated water is disinfected using chloramines, this water will need to be chlorinated at relatively high dosages (2 to 4 mg/L) in the summer when the bromide level in the SWRO permeate exceeds 0.5 mg/L in order to prevent the negative impact of the relatively high bromide levels on the stability of the chloramine residual.

Key Plant Treatment Facilities

The seawater desalination plant will have the following key treatment facilities:

- Intake pipeline: high-density polyethylene pipe
- Bark racks, 100-mm openings
- Intake screens, 10-mm openings
- Intake pump station equipped with vertical turbine pumps
- Pretreatment facility combining coagulation and flocculation chambers and dual-media (sand and anthracite) gravity filters

Quality Parameter	Analytical Method	Sampling: Sample Period	Sampling: Sample Method	Units	Concentration limits: Central Tendency	Concentration limits: Extreme
Total dissolved solids	2540 C	One year	Weekly grab	mg/L	350	400
Chloride	4110 B	One year	Weekly grab	mg/L	180	210
Bromide	4110 B	One year	Weekly grab	mg/L	0.5	0.8
Boron	3120 B	One year	Weekly grab	mg/L	No limit	1
Turbidity	2130 B	One month	Continuous	NTU	0.3	0.5

TABLE 17.11 Key Product Water Quality Specifications

- Four duty and one standby RO trains of 8000 m^3/day (2.11 mgd) production capacity, each designed to operate at recovery range of to 50 percent. Each RO train includes a filter effluent transfer pump, cartridge filter, high-pressure pump coupled with Pelton wheel energy-recovery turbine, and an RO rack with membrane vessels and associated piping and equipment.
- Post-treatment system with limestone filters
- Chemical feed and storage systems
- Solids handling system, which consists of clarifiers for settling of the spent pretreatment filter backwash and belt filter presses for dewatering of clarifier residuals
- Administration and RO system building
- Electrical substation
- Auxiliary facilities.

Plant construction is planned to be completed in 22 months. The project will be implemented under a BOOT method of delivery. The debt financing for the project will be secured using commercial construction loan of 5.2 percent interest rate and 20-year term. The project will be financed with 10 and 90 percent debt. Project equity return on investment is 10 percent. The overall interest rate of the amortized investment is 5.7 percent (CRF = 11.752).

This is a high-complexity project, which will require a two-year permitting process, a detailed hydrodynamic modeling of the plant discharge area and extensive source water sample collection and analysis. The project is likely to face legal challenges from local environmental groups.

Plant Operations

The desalination plant will be highly automated and will be operated by a staff comprising of 15 employees. The unit cost of power is \$0.06/kWh, while the total power plant energy demand is 3.85 kWh/m^3 (14.57 kWh/1000 gal). Dewatered sludge from the spent filter backwash will be disposed to a sanitary landfill in the vicinity of the plant. The spent cleaning solution from the reverse osmosis membrane cleaning will be discharged to the sanitary sewer.

Plant effluent discharge water quality will be measured at the point of exit from the desalination plant, and the effect of this discharge on the marine environment will be monitored by collection and analysis of water quality samples at 10 monitoring stations located in the vicinity of the plant discharge.

17.6.2 Capital Costs

The capital costs for construction, start-up, and commissioning of the 40,000 m^3/day (10.6 mgd) seawater desalination plant are presented in Table 17.12. The total capital costs for this desalination plant are estimated at \$74 million (\$1850/m^3·day or \$7 mm/mgd).

17.6.3 Operation and Maintenance Costs

A breakdown of the annual O&M costs for the 40,000 m^3/day (10.6 mgd) project is presented in Table 17.13. These costs total \$6.86 million per year (\$0.47/m^3–\$1.78/1000 gal).

Cost Item	Capital Cost	
	US$	% of Total
Direct Capital (Construction) Costs		
Site preparation, roads, and parking	730,000	1.0
Intake	3,480,000	4.7
Pretreatment	5,850,000	7.9
RO system equipment	25,600,000	34.6
Post-treatment	1,460,000	2.0
Concentrate disposal	1,830,000	2.5
Waste and solids handling	1,100,000	1.5
Electrical and instrumentation systems	1,650,000	2.2
Auxiliary and service equipment and utilities	1,560,000	2.2
Buildings	3,240,000	4.4
Start-up, commissioning, and acceptance testing	1,460,000	2.0
Subtotal, direct (construction) costs (% of total capital costs)	*$47,960,000*	*65.00*
Project engineering services		
Preliminary engineering	780,000	1.0
Pilot testing	720,000	1.0
Detailed design	3,650,000	4.9
Construction management and oversight	2,200,000	3.0
Subtotal, engineering services	*7,350,000*	*9.9*
Project development		
Administration, contracting, and management	1,500,000	2.0
Environmental permitting	2,100,000	2.8
Legal services	490,000	0.5
Subtotal, project development	*4,090,000*	*5.3*
Project financing costs		
Interest during construction	2,200,000	3.0
Debt service reserve fund	3,900,000	5.3
Other financing costs	1,100,000	1.5
Subtotal, project financing	*7,200,000*	*9.8*
Contingency	*7,400,000*	*10.0*
Subtotal indirect capital costs (% of total capital costs)	*$26,040,000*	*35.0*
Total capital costs	$74,00,000	100

TABLE 17.12 Project Capital Cost Breakdown

Cost Item	Annual O&M Costs US$/year	US$/m³	% of Total
Variable O&M costs			
Energy	3,370,000	0.231	49.1
Chemicals	440,000	0.030	6.4
Replacement of membranes and cartridge filters	780,000	0.053	11.4
Waste stream disposal	330,000	0.023	4.8
Subtotal, variable O&M costs	*4,920,000*	*0.337*	*71.7*
Fixed O&M costs			
Labor cost	420,000	0.029	6.1
Maintenance	700,000	0.048	10.2
Environmental and performance monitoring	120,000	0.008	1.8
Indirect O&M costs	700,000	0.048	10.2
Subtotal, fixed O&M costs	*1,940,000*	*0.133*	*28.3*
Total O&M Costs	$6,860,000/yr	0.470/m³	100

Note: $1/m³ = $3.785/1000 gal

TABLE 17.13 Project Annual O&M Cost Breakdown

17.6.4 Water Production Cost

A summary of the fixed and variable components of the water cost for the 40,000 m³/day (10.6 mgd) project is presented in Table 17.14. The cost of water production for this project is estimated at $0.9/m³ ($3.41/1000 gal).

The total cost of water in the example above is estimated assuming that the desalination plant produces water 100 percent of the time. The actual desalination plants, however, usually have less than 100 percent availability. Therefore the cost of water has to be adjusted to account for the fact that for a portion of the time the plant will not deliver desalinated water to the final users (i.e., will not generate revenue from water sales) while incurring expenses associated with fixed plant costs.

Because the variable cost component is proportional to flow, the plant availability factor would not have an effect on the variable water cost component expressed as unit cost. However, the plant fixed unit cost will increase because the same amount of fixed expenses would need to be recovered at reduced water sales.

For the example above, let's assume that the plant availability factor is 96 rather than 100 percent. In this case the actual annual average volume of desalinated water produced by the 40,000 m³/day desalination plant would be 40,000 m³/day × 96% = 38,400 m³/day. As a result, the fixed component of the desalination cost would increase from $0.563/m³ to $0.586/m³ ($0.563/m³/96% = $0.586/m³). Therefore the total cost of water will increase from $0.90/m³ to $0.586/m³ + $0.337/m³ = $0.923/m³ ($3.49/1000 gal). This 2.5 percent

Cost of Water Item	Cost of Water	
	US$/$m^3$	% of Total
Fixed Costs		
Capital cost recovery	0.430	47.8
Labor costs	0.029	3.2
Maintenance	0.048	5.3
Environmental and performance monitoring	0.008	0.1
Indirect O&M costs	0.048	5.3
Subtotal, fixed costs	*0.563*	*61.7*
Variable costs		
Energy	0.231	25.7
Chemicals	0.030	3.3
Replacement of RO membranes and cartridge filters	0.053	5.9
Waste stream disposal	0.023	2.4
Subtotal, variable costs	*0.337*	*37.3*
Total cost of water	0.90/m^3	100

Note: \$1/m^3 = \$3.785/1000 gal

Table 17.14 Project Cost of Water Production

increase of the desalination plant water production cost would allow to account for the design plant availability factor and to recover plant fixed expenses during times the plant is not delivering desalinated water to the final user.

17.7 References

American Water Works Association (AWWA), *Reverse Osmosis and Nanofiltration, Manual of Water Supply Practices M46, Second Addition*, Denver, Colorado, 2007.

American Water Works Association (AWWA), *Desalination of Seawater, Manual of Water Supply Practices M 61*, Denver, Colorado, 2011.

Watson, I., O. Morin, and L. Henthorne, *Desalting Handbook for Planners*, 3rd ed., Report No. 72, USBR, July, 2003.

Wright, R., and T. Missimer, "Alternative Intake Systems for Seawater Membrane Water Treatment Plants," In Proceedings of International Desalination Association Conference, Madrid, 1997.

WRF Assessing Seawater Intake Systems for Desalination Plants, Water Research Foundation, WateReuse Research Foundation and US Bureau of Reclamation, 2011. http://ntlm.waterrf.org/ProjectsReports/PublicReportLibrary/4080.pdf.

Glossary

Acre-foot The volume of water that would cover a 1-ac area 1 ft deep; equivalent to approximately 1233.6 m^3 or 325,800 gal.

Aerobic Containing oxygen.

Ambient Surrounding or in the background.

Ambient seawater Seawater in the open ocean used for desalination.

Anaerobic Not containing oxygen.

Anthropogenic Caused by human activity.

Antiscalant A chemical added to the saline source water fed to an RO system that inhibits or prevents precipitation of minerals on the RO membrane surface.

Applied research Systematic study to gain knowledge or understanding necessary to determine the means by which a recognized and specific need may be met.

Aquifer An underground formation that is saturated with water.

Asymmetric Description of a membrane that has increased porosity from the surface to the base.

Bactericide A chemical capable of destroying bacteria.

Biocide A chemical used to inactivate microbiological organisms (e.g., chlorine).

Biofouling Membrane fouling caused by the excessive growth and accumulation of microorganisms and their secretions on the membrane surface.

Biomass Living organic matter.

Brackish water Water with a total dissolved solids concentration of 1000 to 15,000 mg/L.

Brine (concentrate) The concentrated stream separated from the saline source water during the desalination process.

Brine seal A rubber or plastic device of special design that is installed on the outer side of a membrane near the feed side to seal the space between the membrane elements and the pressure vessel and to prevent the bypassing of feed water or concentrate around the elements.

CCPP Calcium carbonate precipitation potential.

Co-location Location of a desalination plant with an existing power generation station; the desalination plant intake and outfall are connected to the cooling water discharge of the power generation station.

Coagulation A pretreatment process used in some desalination plants—a substance (e.g., ferric chloride) is added to a solution to cause suspended particles to agglomerate and form larger particles, which are easier to remove from the source water than small particles.

Colloid A suspended solid with a diameter less than 1 μm that cannot be removed by sedimentation alone.

Concentrate See *brine (concentrate).*

Concentrate management The handling and disposal or reuse of concentrated seawater generated by a desalination system during the salt separation process.

Concentration polarization The phenomenon in which solutes form a more concentrated layer next to the membrane surface and, restrict flow through the membrane.

Contaminant An undesirable substance contained in the source water, permeate, or concentrate.

Desalination (desalting) A process that removes dissolved solids (salts) from saline source water.

Diffuser The offshore end portion of the outfall, which consists of discharge ports configured to maximize the mixing of the desalination plant discharge with the ambient receiving waters.

Double-pass RO system An RO system that consists of two sets of RO trains configured in series, in which permeate from the first set of RO trains is processed through the second set of RO trains.

Ecosystem A community of living organisms interacting with each other and with nonliving components of the environment they inhabit.

Effective size The media grain diameter for which 10 percent of the media (by weight) is smaller than the diameter, as determined by sieve analysis.

Feed water The influent water that is fed into a treatment process or system.

Filtrate The purified water that is produced by a membrane pretreatment system.

Flux The rate of water flow across a unit of membrane surface area, expressed in liters per hour per square meter ($L/m^2 \cdot h$ or lmh) or gallons per square foot per day (gfd).

Fouling The gradual accumulation of contaminants on and/or within the RO or pretreatment membrane surface that inhibits the passage of water, thus decreasing membrane permeability and productivity.

Hardness The concentration of calcium and magnesium salts in water.

Inert Description of matter that does not dissolve in water nor react chemically with water or other substances.

Injection zone A geological formation receiving desalination plant discharge via a deep injection well.

Inorganic Description of all matter that does not originate from living organisms (animals, plants, bacteria, etc.); commonly also referred to as *mineral.*

Intake The facility through which source water is collected to produce freshwater in a desalination plant.

Ion An atom or group of atoms or molecules that has a positive charge (cation) or negative charge (anion) as a result of having lost or gained electrons.

Ion strength A measure of the overall electrolytic potential of a solution.

Langelier Saturation Index (LSI) A parameter indicating the tendency of a water solution to precipitate or dissolve calcium carbonate.

Mass transfer coefficient A coefficient quantifying material passage through a membrane.

Membrane A thin film of polymer material permeable to water and capable of separating contaminants from the source seawater as a function of their chemical and physical properties when a driving force is applied; microfiltration and ultrafiltration membranes have measurable porous structures that physically remove particles and microorganisms larger than the size of the pores; ultrafiltration membranes also remove molecules larger than a specified molecular weight; reverse osmosis membranes remove both soluble and particulate matter from the source water.

Membrane element An individual membrane unit of standard size and performance.

Membrane system A complete system of membrane elements, pumps, piping, and other equipment that can treat feed water and produce filtrate (UF and MF systems) or permeate (RO systems).

Microfiltration Filtration through membranes with a pore size between 0.1 and 0.5 μm.

Mitigation Prevention of significant environmental impact and/or repair of such impact on an aquatic habitat exposed to desalination plant discharge; often involves restoration of an existing habitat or creation of a new habitat similar to the one that is impacted on the same or a different location.

Near-shore discharge Disposal of a desalination plant's waste streams through structures (channel, pipe, weir, etc.) located on the shore or within several hundred meters of the shore in the tidal zone.

Offshore discharge Disposal of a desalination plant's waste streams via a long outfall structure extending beyond the tidal zone.

Open intake An intake collecting source water directly from the water column of a surface water body.

Organic Description of matter that includes both natural and man-made molecules containing carbon and hydrogen; all organisms living in water are made up of organic molecules.

Osmosis The naturally occurring transport of water or other solvents through a semipermeable membrane from a less concentrated solution to a more concentrated solution.

Osmotic pressure The pressure applied on the surface of a semipermeable membrane as a result of the naturally occurring transport of water from the side of the membrane with lower salinity to the side of the membrane with higher salinity.

Percent recovery The ratio of desalinated low-salinity water (filtrate or permeate) flow to feed water flow of a filtration system; in RO systems this is the ratio between permeate and feed water, whereas in UF and MF systems it is the ratio of filtered water to feed water.

Permeate The purified water of low mineral content produced during the reverse osmosis separation process; the portion of the feed seawater that passes through the RO membranes.

pH The negative logarithm of the concentration of hydrogen ions; a pH lower than 7 is acidic, whereas a pH higher than 7 is alkaline.

PREN Pitting resistance equivalent number.

Pressure filtration Filtration aided by the imposition of a pressure differential across an enclosed filter vessel.

Pressure vessel A housing containing membranes in a preset configuration that operates under pressure; for RO systems, pressure vessels are plastic or metal tube-shaped devices that house six to eight RO elements.

Pretreatment A process that includes one or more source water treatment technologies (e.g., screening, coagulation, sedimentation, filtration, chemical addition, etc.) that aim to remove foulants from the source water prior to RO separation in order to protect the membranes and improve the desalination plant's performance.

Product water The low-salinity (fresh) water, usually with a TDS concentration of 500 mg/L or less, produced by a desalination plant and suitable for distribution system delivery; in order for the desalination plant's permeate to be converted to product water, it has to be disinfected and conditioned for requirements related to corrosion and predetermined water quality.

Reverse osmosis The pressure-driven movement of water through a semipermeable membrane from the side of the membrane with the more concentrated solution to the side with the less concentrated solution.

Salinity The concentration of total dissolved solids in water.

Salinity tolerance threshold The maximum TDS concentration of the desalination plant discharge at which the concentrate will not exhibit harmful effects on the aquatic environment; the salinity tolerance threshold is usually established based on the most salinity-sensitive species living in the discharge area.

Salt passage The ratio of the concentration of salt (ions) in the permeate and the concentration of the same salt (ions) in the saline source water; typically, salt passage is expressed as a percentage of the feed water concentration of the salt.

Salt rejection The ratio of salt (ions) removed (rejected) by the RO membrane to the salt (ions) of the source water; salt rejection is equal to 100 percent minus the salt passage.

SCADA supervisory control and data acquisition.

Scale Mineral deposits formed on the surface of a membrane and/or membrane matrix as a result of concentration (saturation) of the minerals to a level at which they form insoluble amorphous or crystalline solids.

Scale inhibitor See *antiscalant*.

Scaling The process of scale formation on the surface or in the matrix of an RO membrane.

Semipermeable membrane A membrane that has a structure that allows small molecules, such as water, to pass while rejecting a large portion of the salts contained in the saline source water.

Silt density index (SDI) A dimensionless parameter widely used to quantify the potential of saline water to cause particulate and colloidal fouling of RO membranes.

Specific permeability (flux) The capacity of a membrane material to transit flow, also named *specific flux*; expressed as the membrane flux normalized for temperature and pressure, in liters per square meter per hour per bar (lmh/bar) or gallons per square foot per day or flow per pound per of square inch of pressure (gfd/psi).

Spiral-wound element An RO or NF membrane element that consists of membrane leaves wound around a central permeate collection tube and including feed and permeate spacers, antitelescoping devices, and a brine seal.

Stage A set of pressure vessels installed and operated in parallel.

Subsurface intake An intake located below the ground surface collecting source water from a groundwater aquifer; examples of subsurface intakes are vertical, horizontal, and slant wells and infiltration galleries.

Suspended solids Particulate solids suspended in the water.

Total dissolved solids (salinity) A measure of the total mass of all dissolved solids contained in the water.

Total suspended solids The concentration of filterable particles in water (retained on a 0.45 μm filter), reported by mass per volume.

Train A membrane system that consists of a rack housing a number of pressure vessels that have a common feed and permeate and concentrate piping and control equipment, and can be operated independently; an RO system or MF or UF membrane system consists of multiple trains operating in parallel.

Turbidity A measure of the concentration of suspended solids in water determined by the amount of light scattered by these solids.

Ultrafiltration Filtration through membranes with a pore size between 0.01 and 0.05 μm.

Uniformity coefficient The ratio of the 60th percentile media grain diameter to the effective size of the filer media.

Vacuum filtration Filtration through an MF or UF membrane created in an enclosed filter vessel by applying a vacuum.

Viscosity The tendency of fluid to resist flow (movement) as a result of molecular attraction (cohesion).

Zero liquid discharge A concentrate management alternative in which the concentrate is converted from the liquid phase to the solid phase (salt residual) by evaporation, freezing, or other means, allowing crystallization of the salts contained in the concentrate.

Zone of initial dilution The area around the discharge of a desalination plant at whose boundary the concentration of the TDS of the mix between concentrate and ambient water reaches 10 percent of the TDS level of the ambient water.

Index

www.ingramcontent.com/pod-product-compliance
Lightning Source LLC
LaVergne TN
LVHW082051011026
849162LV00019B/344